Boris M. Smirnov

Plasma Processes and Plasma Kinetics

1807–2007 Knowledge for Generations

Each generation has its unique needs and aspirations. When Charles Wiley first opened his small printing shop in lower Manhattan in 1807, it was a generation of boundless potential searching for an identity. And we were there, helping to define a new American literary tradition. Over half a century later, in the midst of the Second Industrial Revolution, it was a generation focused on building the future. Once again, we were there, supplying the critical scientific, technical, and engineering knowledge that helped frame the world. Throughout the 20th Century, and into the new millennium, nations began to reach out beyond their own borders and a new international community was born. Wiley was there, expanding its operations around the world to enable a global exchange of ideas, opinions, and know-how.

For 200 years, Wiley has been an integral part of each generation's journey, enabling the flow of information and understanding necessary to meet their needs and fulfill their aspirations. Today, bold new technologies are changing the way we live and learn. Wiley will be there, providing you the must-have knowledge you need to imagine new worlds, new possibilities, and new opportunities.

Generations come and go, but you can always count on Wiley to provide you the knowledge you need, when and where you need it!

William J. Pesce
President and Chief Executive Officer

Peter Booth Wiley
Chairman of the Board

Boris M. Smirnov

Plasma Processes and Plasma Kinetics

586 Worked Out Problems for
Science and Technology

WILEY-VCH Verlag GmbH & Co. KGaA

The Author

Prof. Boris M. Smirnov
Institute for High Temperatures
Russian Academy of Sciences
Moscow, Russia
bsmirnov@orc.ru

Library of Congress Card No: applied for

British Library Cataloging-in-Publication Data
A catalogue record for this book is available from the British Library

Bibliographic information published by the Deutsche Nationalbibliothek
The Deutsche Nationalbibliothek lists this publication in the Deutsche Nationalbibliografie; detailed bibliographic data are available in the Internet at http://dnb.d-nb.de

Typesetting Hilmar Schlegel, Berlin
Printing betz-druck GmbH, Darmstadt
Bookbinding Litges & Dopf Buchbinderei GmbH, Heppenheim

Printed in the Federal Republic of Germany
Printed on acid-free paper

ISBN: 978-3-527-40681-4

Contents

Plasma Processes and Plasma Kinetics. Boris M. Smirnov

ISBN: 978-3-527-40681-4

Preface

Various aspects of a low-temperature plasma are represented in the form of problems. Kinetics of this plasma is determined by elementary processes involving electrons, positive and negative ions, excited atoms, atoms, and molecules in the ground states. Along with collision processes involving these atomic particles, radiative processes are of importance for a low-temperature plasma and excited gas, both elementary radiative processes and transport of radiation through a gas that includes reabsorption processes. The collective processes, oscillation plasma properties, and nonlinear plasma processes are represented in the corresponding problems. Transport of particles in a plasma is of importance for a nonequilibrium plasma. A cluster plasma, an aerosol plasma, and other plasma forms with a dispersive phase are considered in the book. Because all these processes and phenomena are given in a specific form for each plasma, we consider separately two plasma types, a plasma of the atmosphere together with atmospheric phenomena due to this plasma and some types of gas-discharge plasma. Appendices contain information which is useful for the analysis of specific plasma types, and this information is represented in the convenient form for the user. The book is intended for students and professionals in the field of plasma physics, plasma chemistry, and plasma applications.

Boris M. Smirnov

Plasma Processes and Plasma Kinetics. Boris M. Smirnov

ISBN: 978-3-527-40681-4

1
Distributions and Equilibria for Particle Ensembles

1.1
Distributions of Identical Atomic Particles

▶ **Problem 1.1** An ensemble of n weakly interacting identical particles is located in a close space and does not interact with a surrounding environment. As a result of interactions, particles can change their state. Find the probability of a certain distribution of particles by states.

Distributing identical particles over states, if in each state many particles are found, we account for particles that can change their states, but an average number of particles in a given state almost conserve, and the better this is fulfilled, the more the number of particles found in this state on average. Let us distribute n identical particles over k states assuming that the probability for a test particle to be found in a given state, as well as the average number of particles in this state, is proportional to the number of versions which lead to this Gibbs principle.

Let us denote by $P(n_1, n_2, \ldots, n_i, \ldots)$ the number of ways to place n_1 particles in the first group of states, n_2 particles in the second group of states, n_i particles in the i-th group of states, etc. For determining this probability we use the character of distributions for the location of a particle in a certain group of states does not influence the character of distributions for other particles (*Boltzmann statistics*). Under these conditions, the probability of locating n_1 particles in the first state, n_2 particles in the second state, etc. is given by

$$P(n_1, n_2, \ldots, n_i, \ldots) = p(n_i)p(n_2) \cdots p(n_i) \cdots , \tag{1.1}$$

where $p(n_i)$ is the number of ways to distribute n_i particles in the i-th group of states. Evidently, the number of ways to place n_1 particles from the total number n of particles in the first group of states is

$$C_n^{n_1} = \frac{n!}{(n - n_1)!n_1!} \, .$$

Correspondingly, the number of ways to place n_2 particles from the remaining $n - n_1$ particles in the second group of states is $C_{n-n_1}^{n_2}$. Continuing this operation,

Plasma Processes and Plasma Kinetics. Boris M. Smirnov

ISBN: 978-3-527-40681-4

we determine the probability of the indicated distribution of particles in states,

$$P(n_1, n_2, \ldots, n_i, \ldots) = \text{const}\frac{n!}{\prod_i n_i!}, \tag{1.2}$$

where "const" is a normalization constant. The basis of this formula is the assumption that particles are free, so that the distribution of one particle does not influence the distribution of others.

▶ **Problem 1.2** Derive the Boltzmann distribution for an ensemble of weakly interacting particles.

The distribution under consideration relates to almost free classical particles when the number of states in a given state group is large compared to the number of particles which are found in states of a given group. Then the location of some particles in states of a given group does not influence the possibility of finding test particles in these states. Next, this distribution corresponds to conservation of the total number of particles in all the states,

$$n = \sum_i n_i, \tag{1.3}$$

and the total energy E for all the particles,

$$E = \sum_i \varepsilon_i n_i, \tag{1.4}$$

because particle's energy does not change with an environment. Here ε_i is the energy of a particle located in the i-th group of states.

For determining an average $\overline{n}_i$ or the most probable number of particles for a given i-th group of states, we account for any probable distribution that the variation of the number of particles in these states from the average value $\delta n_i = n_i - \overline{n}_i$ is relatively small. Next, according to formulas (1.3) and (1.4), these variations satisfy the relation

$$\sum_i \delta n_i = 0 \tag{1.5}$$

and

$$\sum_i \varepsilon_i \delta n_i = 0. \tag{1.6}$$

In addition, on the basis of the relation

$$\ln n! = \ln \prod_{m=1}^{n} m =\approx \int_0^n \ln x dx \,,$$

we have $d \ln n!/dn = \ln n$. Using this relation with the expansion of formula (1.2) over a small parameter δn_i, we obtain

$$\ln P(n_1, n_2, \ldots, n_i, \ldots) = \ln P(\overline{n}_1, \overline{n}_2, \ldots, \overline{n}_i, \ldots) - \sum_i \ln \overline{n}_i \delta n_i - \sum_i \frac{\delta n_i^2}{2\overline{n}_i}. \tag{1.7}$$

Since the real distribution is near the maximum, the linear terms with respect to δn_i disappear, which corresponds to the relation

$$\sum_i \ln \overline{n}_i \delta n_i = 0. \tag{1.8}$$

In order to find the average number of particles in a given group of states, we use the formal operation of multiplying equation (1.5) by a constant $-\ln C$, equation (1.6) by the parameter $-1/T$, and adding these equations to (1.8). The resultant equation has the form

$$\sum_i \left(\ln \overline{n}_i - \ln C + \frac{\varepsilon_i}{T}\right) \delta n_i = 0. \tag{1.9}$$

Since variations δn_i are random, the expression in the parentheses is zero, which gives

$$\overline{n}_i = C \exp\left(-\frac{\varepsilon_i}{T}\right). \tag{1.10}$$

As a result, we obtain the Boltzmann formula. During its deduction we introduce two characteristic parameters, C and T. The first one is the normalization constant that follows from the relation

$$C \sum_i \exp(-\varepsilon_i/T) = n\,.$$

The energetic parameter T is the temperature of the system.

► **Problem 1.3** Represent the Boltzmann distribution by taking into account the statistical weight of a particle.

Dividing the states of a particle ensemble in groups, we as above assume equal number of states for each group. If these are different, we introduce the statistical weight of the particle state g_I as the number of states per particle. For example, if this particle is a diatomic molecule, and we characterize its state by the rotational momentum J, the number of its projections $g_i = 2J + 1$ onto the molecular axis is the statistical weight of a given state. As is seen, the statistical weight is the number of states per particle of this particle ensemble.

By taking into account the statistical weight for a given state of a particle, the Boltzmann distribution (1.10) is transformed to the form

$$\overline{n}_i = C g_i \exp\left(-\frac{\varepsilon_i}{T}\right). \tag{1.11}$$

From this we have the following relation between the number densities of particles in two states when an ensemble consists of the infinite number of particles,

$$N_i = N_0 \frac{g_i}{g_0} \exp\left(-\frac{\varepsilon_i}{T}\right), \tag{1.12}$$

where N_0, N_i are the number densities of particles in these states, and g_0, g_i are the statistical weights of these states.

► **Problem 1.4** Find the distribution of molecules over vibrational states considering vibrations of the molecule like harmonic oscillators.

In this approximation, the energy of excitation ε_v of the v-th vibrational state is

$$\varepsilon_v = \hbar\omega\, v\,,$$

where $\hbar\omega$ is the difference of energies for neighboring vibrational states. On the basis of the Boltzmann formula (1.12) we obtain the number density of molecules located in the v-th vibrational state,

$$N_v = N_0 \exp\left(-\frac{\hbar\omega v}{T_v}\right)\,,$$

where N_0 is the number density of molecules in the ground vibrational state, and T_v is the vibrational temperature. The total number density of molecules is

$$N = \sum_{v=0}^{\infty} N_v = N_0 \sum_{v}^{\infty} \exp\left(-\frac{\hbar\omega v}{T_v}\right) = \frac{N_0}{1-\exp\left(-\frac{\hbar\omega}{T_v}\right)}\,, \tag{1.13}$$

which allows us to express the number density of molecules in a given vibrational state, N_v, through the total number density N of molecules,

$$N_v = N \exp(-\frac{\hbar\omega v}{T_v})\left[1-\exp(-\frac{\hbar\omega}{T_v})\right]\,. \tag{1.14}$$

From this one can find the average vibrational excitation energy ε_{vib}

$$\varepsilon_{\text{vib}} = \hbar\omega\overline{v} = \frac{1}{N}\sum_{v=0}^{\infty} v\, N_v = \frac{\hbar\omega}{\exp(\frac{\hbar\omega}{T_v})-1}\,. \tag{1.15}$$

► **Problem 1.5** Find the distribution of diatomic molecules over rotational states.

The energy of excitation of a rotational state with a rotational momentum J is

$$\varepsilon_J = BJ(J+1)\,,$$

where B is the rotational constant. The statistical weight of the state with a momentum J, which is the number of momentum projections onto a given axis, equals $g_J = 2J+1$. On the basis of this and the Boltzmann formula (1.12), we obtain the number density of molecules with a given rotational momentum and vibrational state,

$$N_{vJ} = N_v\,(2J+1)\frac{B}{T}\exp\left[-\frac{BJ(J+1)}{T}\right]\,, \tag{1.16}$$

where we assume $B \ll T$, as it is usually, and N_v is the total number density of molecules in a given vibrational state. From this one can also find the average rotational energy of molecules,

$$\varepsilon_{\text{rot}} = \overline{BJ(J+1)} = T\,. \tag{1.17}$$

In this analysis we account for that typical vibrational energies exceed significantly typical rotational energies, which allows us to separate vibrational and rotational degrees of freedom. We also note that we assume a diatomic molecule to be consisting of other isotopes. Otherwise, because of the molecule symmetry, only certain values of the rotation momentum can be realized.

► **Problem 1.6** Determine the statistical weight for a free particle.

Let us place an ensemble of free particles in a rectangular box with the edge size L, so that particles are reflected from the box's walls and cannot penetrate outside wall boundaries. Each particle is free and moves freely inside the box. Hence the wave function of a particle moving inside the box in the axis x direction can be composed of two waves, $\exp(ip_x x/\hbar)$ and $\exp(-ip_x x/\hbar)$, propagated in opposite directions, where p_x is the particle momentum. Placing the origin at lower-left corner of the box cube and requiring the wave function to be zero at cube facets, we obtain from the first boundary condition $\psi(0) = 0$ for the particle wave function ψ,

$$\psi = \sin \frac{p_x x}{\hbar} .$$

The second boundary condition $\psi(L) = 0$ leads to quantization of the particle momentum

$$\frac{p_x L}{\hbar} = \pi k ,$$

where k is an integer. This relation gives the prohibited values of the particle momentum if it is moving in a rectangular box of size L.

This gives the number of states for a particle with a momentum in the range from p_x to $p_x + dp_x$, which is equal to $dg = L dp_x/(2\pi\hbar)$, where we take into account two directions of the particle momentum. Introducing a coordinate range to be dx, we find the number of states for a free particle to be

$$dg = \frac{dp_x dx}{2\pi\hbar} . \tag{1.18}$$

Generalization of this formula to the three-dimensional case leads to the following number of states for a free particle:

$$dg = \frac{dp_x dx}{2\pi\hbar}\frac{dp_y dy}{2\pi\hbar}\frac{dp_z dz}{2\pi\hbar} = \frac{d\mathbf{p} d\mathbf{r}}{(2\pi\hbar)^3} , \tag{1.19}$$

where the quantity $d\mathbf{p}d\mathbf{r}$ is an element of the phase space, and the notation is used for three-dimensional elements of space $d\mathbf{r} = dxdydz$ and momentum $d\mathbf{p} = dp_x dp_y dp_z$. Formula (1.19) gives the statistical weight of the continuous spectrum—the number of states per element of the phase space.

► **Problem 1.7** Find the velocity distribution function for free particles—the Maxwell distribution.

We now use the Boltzmann formula (1.11) for the distribution of kinetic energies of free particles. In the one-dimensional case the kinetic energy of a particle whose velocity is v_x equals $\varepsilon_i = mv_x^2/2$, and the statistical weight of states when the particle velocity ranges from v_x to $v_x + dv_x$ is proportional to dv_x. Then formula (1.11) gives for the number of particles whose velocities are found in the range v_x to $v_x + dv_x$

$$f(v_x)dv_x = C\exp\left(-\frac{mv_x^2}{2T}\right)dv_x, \tag{1.20}$$

where C is the normalization factor. This is the Maxwell distribution for the one-dimensional case.

Transferring to the three-dimensional case by taking into account the independence of different directions of motion, we obtain

$$f(\boldsymbol{v})d\boldsymbol{v} = C\exp\left(-\frac{mv^2}{2T}\right)d\boldsymbol{v}. \tag{1.21}$$

Here the vector $\boldsymbol{v}$ has components v_x, v_y, v_z, and $d\boldsymbol{v} = dv_x dv_y dv_z$. The kinetic energy of a particle, $mv^2/2$, is the sum of the particle kinetic energies for all the directions of motion. Thus, the independence of different directions of particle motion results in the isotropy of the distribution function.

Let us rewrite the Maxwell distribution for the number density of particles of a given velocity, normalizing this distribution to the total number density N of particles. Then formula (1.21) gives

$$f(v) = N\left(\frac{m}{2\pi T}\right)^{3/2}\exp\left(-\frac{mv^2}{2T}\right). \tag{1.22}$$

It is convenient to rewrite the Maxwell distribution function (1.22) through one-dimensional distribution functions

$$f(v) = N\varphi(v_x)\,\varphi(v_y)\,\varphi(v_z), \tag{1.23}$$

where the functions $\varphi(v_i)$ are normalized to 1 and has the form

$$\varphi(v_x) = \sqrt{\frac{m}{2\pi T}}\exp\left(-\frac{mv_x^2}{2T}\right)\quad \int_{-\infty}^{\infty}\varphi(v_x)dv_x = 1. \tag{1.24}$$

Figure 1.1 gives this dependence.

▶ **Problem 1.8** On the basis of the Maxwell distribution connect an average kinetic energy of a free particle with the temperature.

We introduce above the energetic parameter of the Boltzmann distribution function, T, which is the temperature of a given ensemble of particles and the temperature is expressed in energetic units. Below we find the average kinetic energy for an ensemble of free particles, and then the temperature of the particle ensemble will be expressed in terms of the average particle energy.

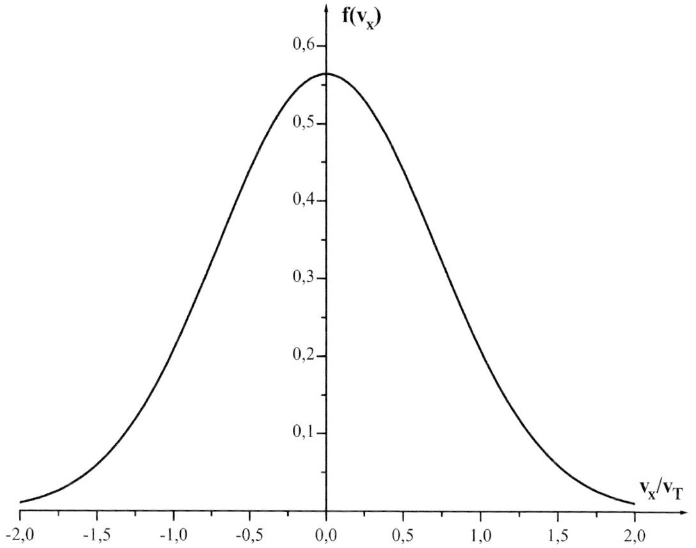

Fig. 1.1 The Maxwell distribution function $f(v_x)$ as a function of the reduced velocity $x = v_x\sqrt{m/(2T)}$, where v_x is the particle velocity in a given direction, m is the particle mass, T is the temperature.

Indeed, the average kinetic energy of free particles in a direction x is according to its definition

$$\frac{\overline{mv_x^2}}{2} = \frac{\int\limits_{-\infty}^{\infty} \frac{mv_x^2}{2} \exp\left(-\frac{mv_x^2}{2T}\right) dv_x}{\int\limits_{-\infty}^{\infty} \exp\left(-\frac{mv_x^2}{2T}\right) dv_x} = -\frac{d \ln \int\limits_{-\infty}^{\infty} \exp\left(-\frac{mv_x^2}{2T}\right) dv_x}{d(-1/T)}$$
$$= -\frac{d \ln\,(aT^{1/2})}{d(-1/T)} = \frac{T}{2}, \tag{1.25}$$

where the bar means an average over particle velocities, and the constant a does not depend on the temperature. Thus, the particle kinetic energy per unit degree of freedom is equal to $T/2$.

Transferring to the three-dimensional case, we take into account the isotropy of particle motion, and the total kinetic energy of a particle is given by

$$\frac{\overline{mv^2}}{2} = \frac{\overline{mv_x^2}}{2} + \frac{\overline{mv_y^2}}{2} + \frac{\overline{mv_z^2}}{2} = \frac{3\,\overline{mv_x^2}}{2} = \frac{3T}{2}. \tag{1.26}$$

Thus, the average particle kinetic energy in the three-dimensional space is $\overline{mv^2}/2 = 3T/2$. Formulas (1.25) and (1.26) may be used as the temperature definition.

▶ **Problem 1.9** Find the energy distribution function for free particles.

Our task is to rewrite the distribution function (1.22) in the energy space $\varepsilon = mv^2/2$, where we denote it by $f(\varepsilon)$. This distribution function is normalized according to the condition

$$\int_0^\infty f(\varepsilon)\varepsilon^{1/2} d\varepsilon = N, \tag{1.27}$$

where N is the number density of particles. In these terms the distribution function (1.22) takes the form

$$f(\varepsilon) = \frac{2\,N}{\sqrt{\pi}T^{3/2}} \exp\left(-\frac{\varepsilon}{T}\right) \tag{1.28}$$

and is represented in Fig. 1.2.

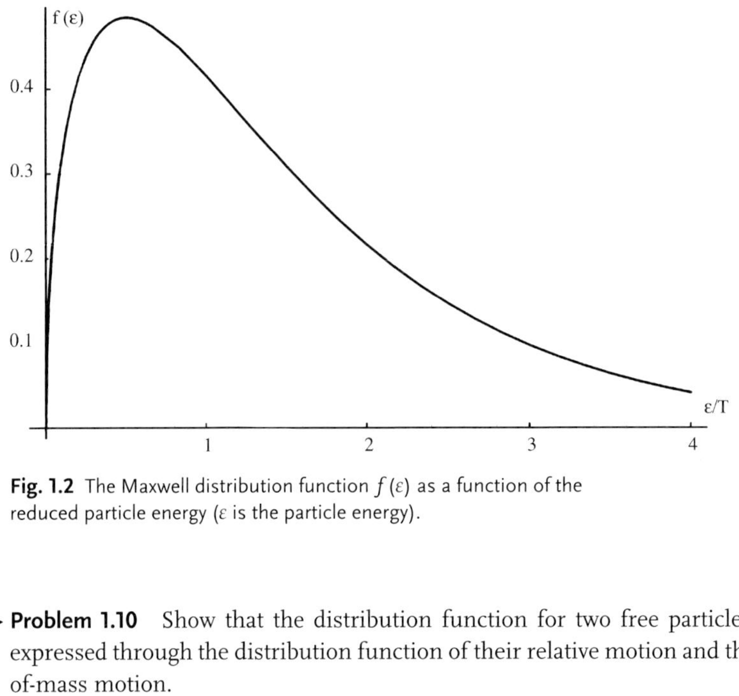

Fig. 1.2 The Maxwell distribution function $f(\varepsilon)$ as a function of the reduced particle energy (ε is the particle energy).

▶ **Problem 1.10** Show that the distribution function for two free particles can be expressed through the distribution function of their relative motion and the center-of-mass motion.

If the particles under consideration belong to two different groups, we have the product of their distribution functions

$$f(v_1)f(v_2)dv_1dv_2\ ,$$

where $v_1,\ v_2$ are the velocities of the corresponding particles, $f(v_1)f(v_2)$ are their distribution functions, and $dv_1,\ dv_2$ are the elements in the velocity space. Let us introduce the relative velocity $\mathbf{g}$ of the particles and the velocity $\mathbf{V}$ of their center of mass according to the relations

$$\mathbf{g} = v_1 - v_2\ ,\ \mathbf{V} = \frac{m_1v_1 + m_2v_2}{m_1 + m_2}\ ,$$

where m_1, m_2 are the masses of these particles. One can see from these formulas that

$$dv_1 dv_2 = dg dV .$$

Next, the total kinetic energy of the particles is

$$\frac{m_1}{2}v_1^2 + \frac{m_2}{2}v_2^2 = \frac{\mu g^2}{2} + \frac{M}{2}V^2 ,$$

where the reduced mass of the two particles is $\mu = m_1 m_2/(m_1 + m_2)$, and their total mass is $M = m_1 + m_2$. From this we obtain for the Maxwell distribution function $f(v)$, which is given by formula (1.21),

$$f(v_1)f(v_2)dv_1 dv_2 = f(g)f(V)dg dV, \tag{1.29}$$

and the relative particle motion is characterized by the reduced mass, whereas the motion of the center of mass is connected with the total particle mass.

1.2
Statistics of Bose–Einstein and Fermi–Dirac

► **Problem 1.11** Find the distribution function over states for an ensemble of particles in the case of the Bose–Einstein statistics if any number of particles can be located in one state.

The Bose–Einstein statistics relates to an ensemble of identical particles with a whole spin and permits us to find in the same state two and more particles. We take for this case the probability (1.11), w_i, of the location of a particle in a given state i reducing this formula to one state of this group

$$w_i = \exp\left(\frac{\mu - \varepsilon_i}{T}\right) . \tag{1.30}$$

We introduce here the chemical potential μ, which is determined by the normalization of the distribution function and therefore is expressed through the normalization constant of formula (1.11) as $C = \exp(\mu/T)$.

From this we find, for example, that the probability of the location of m particles in a given state is w_i^m, and therefore the average number of particles in this state is given by

$$\overline{n_i} = \sum_{m=1}^{\infty} m w_i^m = \frac{w_i}{1 - w_i} = \frac{1}{\exp\left(\frac{\varepsilon_i - \mu}{T}\right) - 1} . \tag{1.31}$$

One can derive the Bose–Einstein distribution function (1.31) in the other manner by placing particles over states, as was used for deduction of the Boltzmann distribution (1.11). Indeed, let us find the probability of placing n_i particles in g_i states when the position of one particle does not depend on the positions of others. To this end we take n_i particles and g_i states as elements of the same set and construct sequences from these elements such that the first place is occupied by a state, and other elements are arranged in a random order. Then we assume the number of particles which are found after the corresponding state and before the next ones which belong to this state, and this is the method of placing particles over states. Then the number of ways to obtain different distributions of particles over states is equal to $(g_i + n_i - 1)!$, and among them some are identical which can be obtained by permutation of states or particles. Hence, the total number of ways to distribute particles over states for the Bose–Einstein statistics is

$$p(n_i) = \frac{(n_i + g_i - 1)!}{n_i!(g_i - 1)!} . \tag{1.32}$$

The optimal number of particles $\overline{n_i}$ in a given state corresponds to the maximum of the function $p(n_i)$ or $\ln p(n_i)$. Hence from the condition that the derivative $\frac{d \ln p(n_i)}{dn_i}$ is zero at $n_i = \overline{n_i}$ in the limit $g_i \gg 1, \; n_i \gg 1$ we obtain formula (1.31) for the average number of particles in one state $\overline{n_i}/g_i$.

► **Problem 1.12** Find the distribution function over states for an ensemble of particles in the case of the Fermi–Dirac statistics if only one particle can be located in one state.

The Fermi–Dirac statistics relates to particles with half-integer spin and does not permit two particles to be located in the same state. In order to find the distribution function of particles in this case, we place n_i particles over g_i states with the same energy ε_i $(n_i \ll g_i)$. It can be done by $p(n_i)$ ways, and the number of such ways is

$$p(n_i) = C_{g_i}^{n_i} = \frac{g_i!}{n_i!(g_i - n_i)!} , \qquad n_i \leq g_i . \tag{1.33}$$

The optimal number of distributions follows from the condition

$$\frac{d \ln p(n_i)}{dn_i} = \ln \frac{n_i}{g_i - n_i} = 0 , \tag{1.34}$$

where we consider the limiting case $g_i \gg 1, n_i \gg 1$. Introducing the optimal number of particles in one state

$$\overline{n_i} = \frac{n_i}{g_i} ,$$

we obtain the average number of particles in one state for the Fermi–Dirac distribution

$$\overline{n_i} = \frac{1}{\exp\left(\frac{\varepsilon_i - \mu}{T}\right) + 1} . \tag{1.35}$$

▶ **Problem 1.13** Find the condition for the transition from the Bose–Einstein and Fermi–Dirac distributions to the Boltzmann distribution.

In the case of the Boltzmann distribution the probability of finding particles' location in each state is small. This holds true if

$$\varepsilon - \mu \gg T\,. \tag{1.36}$$

Using this criterion in formulas (1.32) and (1.35) for the Bose–Einstein and Fermi-Dirac distributions, we transfer them to the Boltzmann distribution

$$\overline{n_i} = \exp\left(\frac{\mu - \varepsilon_i}{T}\right). \tag{1.37}$$

This coincides with the Boltzmann distribution (1.10).

▶ **Problem 1.14** Obtain the electron distribution over momenta in a dense electron gas at low temperature (a degenerate electron gas).

The distribution for a dense cold electron gas is governed by the Pauli principle according to which two electrons cannot be located in one state. We determine below the distribution of electrons over momenta p at zero temperature when formula (1.35) in a space of electron momenta takes the form

$$f(p) = f_o\,\eta(p - p_F)\,.$$

This formula means that all electron states are occupied until $p \le p_F$, where p_F is the Fermi momentum. Correspondingly, the maximum electron energy, Fermi energy ε_F, is equal to

$$\varepsilon_F = \frac{p_F^2}{2m_e}\,.$$

This electron distribution corresponds to the location of electrons inside a ball that is restricted by the Fermi sphere. One can connect the parameters p_F and ε_F of this distribution with the electron density N_e. Indeed, the number of electrons in an element of the phase space is given by

$$n = 2\int\limits_{p \le p_F} \frac{d\boldsymbol{p}d\boldsymbol{r}}{(2\pi\hbar)^3}\,,$$

where the factor 2 accounts for two directions of the electron spin, and $d\boldsymbol{p}$ and $d\boldsymbol{r}$ are elements of the electron momentum and volume. Taking the electron number density as $N_e = n/\int d\boldsymbol{r}$, we obtain the relation between the electron number density and the maximum electron momentum and maximum electron energy,

$$p_F = (3\pi^2\hbar^3 N_e)^{1/3}\,, \qquad \varepsilon_F = \frac{p_F^2}{2m_e} = \frac{(3\pi^2 N_e)^{2/3}\hbar^2}{2m_e}\,. \tag{1.38}$$

Note that the chemical potential of electrons in the Fermi–Dirac formula (1.35) for this distribution is

$$\mu = \varepsilon_F\,.$$

▶ **Problem 1.15** Determine the total energy per unit volume of a degenerate electron gas at low temperatures.

At zero temperature the total energy per unit volume of a degenerate electron gas is equal to

$$E_o = \int_0^{\varepsilon_F} \frac{\varepsilon \cdot 2d\mathbf{p}}{(2\pi\hbar)^3} = \frac{2\sqrt{2}}{5\pi^2}\frac{m_e^{3/2}\varepsilon_F^{5/2}}{\hbar^3}. \tag{1.39}$$

At low temperatures the distribution of a degenerate electron gas is determined by the Fermi–Dirac formula (1.35) and is characterized by a small parameter

$$\eta = \frac{T}{\varepsilon_F}. \tag{1.40}$$

We find below the next term of formula (1.39) for the expansion of the total electron energy over the small parameter above.

We consider a general formula for the electron energy per unit volume that at low temperatures has the form

$$E = \int_0^{\infty} \frac{\varepsilon \cdot 2d\mathbf{p}}{(2\pi\hbar)^3} \frac{1}{\exp\left(\frac{\varepsilon-\mu}{T}\right)+1}, \tag{1.41}$$

where we use formula (1.35) for the electron distribution with the chemical potential $\mu = \varepsilon_F$, which corresponds to zero temperature, and the energy of an individual electron is $\varepsilon = p^2/(2m_e)$. Note that under the condition $T \ll \varepsilon_F = \mu$ the integral

$$E - E_o = \frac{m_e^{3/2}\sqrt{2}}{\pi^2\hbar^3}\left(\int_0^{\infty} \varepsilon^{3/2} d\varepsilon \frac{1}{\exp\left(\frac{\varepsilon-\mu}{T}\right)+1} - \int_0^{\varepsilon_F} \varepsilon^{3/2} d\varepsilon\right) \tag{1.42}$$

converges near $\varepsilon = \varepsilon_F$. Introducing a new variable $x = (\varepsilon - \mu)/T$, we transform this expression to the form

$$E - E_o = \frac{m_e^{3/2}\sqrt{2}T^{5/2}}{\pi^2\hbar^3}\left(\int_0^{\infty}\left(x+\frac{\mu}{T}\right)^{3/2} dx \frac{1}{1+\exp x} - \int_{-\mu/T}^{0}\left(x+\frac{\mu}{T}\right)^{3/2} dx \frac{\exp x}{1+\exp x}\right).$$

Changing the variable in the second integral $x \to -x$ and the lower limit of integration $-\mu/T$ by $-\infty$, we obtain the expansion over a small parameter T/ε_F,

$$E - E_o = \frac{m_e^{3/2}\sqrt{2}T^{5/2}}{\pi^2\hbar^3}\int_0^{\infty}\frac{dx}{1+\exp x}\left[\left(\frac{\mu}{T}+x\right)^{3/2} - \left(\frac{\mu}{T}-x\right)^{3/2}\right]$$

$$= \frac{3m_e^{3/2}\sqrt{2}T^2\sqrt{\mu}}{\pi^2\hbar^3}\int_0^{\infty}\frac{x dx}{1+\exp x} = \frac{m_e^{3/2}T^2\sqrt{\mu}}{\sqrt{2}\hbar^3},$$

and the expansion of the total electron energy over a small parameter $\eta = \frac{T}{\varepsilon_F}$ takes the form

$$E = E_o \left(1 + \frac{5\pi^2}{4} \frac{T^2}{\varepsilon_F^2}\right). \quad (1.43)$$

In particular, this gives the heat capacity of a degenerate electron gas per unit volume,

$$C = \frac{dE}{dT} = \frac{5\pi^2}{2} \frac{T}{\varepsilon_F^2} E_o = \frac{m_e^{3/2} T \sqrt{2\varepsilon_F}}{\hbar^3}. \quad (1.44)$$

1.3 Distribution of Particle Density in External Fields

▶ **Problem 1.16** Derive the barometric formula for the distribution of particles in the gravitation field of the Earth.

Let us use the Boltzmann formula (1.12) and use the particle potential energy U in an external field as the particle energy ε_i in this formula. For particles located in the gravitational field of the Earth we have $U = mgh$, where m is the particle mass, g is the free fall acceleration and h is the altitude above the Earth surface. Hence the Boltzmann formula (1.12) takes the form

$$N(h) = N(0) \exp\left(-\frac{mgh}{T}\right), \quad (1.45)$$

where $N(z)$ is the molecule number density at an altitude z. This is the barometric formula.

From this formula it follows that a typical altitude where the number density of particles varies noticeably is $\sim (mg)^{-1}$. In particular, for air molecule we have $mg = 0.11\,\text{km}^{-1}$, which tells us that a significant drop of the atmospheric pressure proceeds at altitudes of several kilometers.

▶ **Problem 1.17** Find the relation between the drift velocity of particles in a gas in a weak external field and the diffusion coefficient of particles in a gas.

If a weak external field acts on particles located in a gas, it causes a flux $\boldsymbol{j}$ of these particles that is proportional to the number density N of these particles and a force $\boldsymbol{F}$ that acts on an individual particle. So, we have

$$\boldsymbol{j} = N\boldsymbol{w} = Nb\boldsymbol{F}, \quad (1.46)$$

where $\boldsymbol{w}$ is the drift velocity of a particle, which is the definition of the mobility b for a neutral particle. For a charged particle in a gas the following mobility definition is used:

$$\boldsymbol{w} = K\boldsymbol{E}, \quad (1.47)$$

where $\mathbf{E}$ is the electric field strength and K is the mobility of a charged particle in a gas.

If the distribution of admixture particles in a gas is nonuniform, the diffusion flux $\mathbf{j}_{\text{dif}}$ arises,

$$\mathbf{j}_{\text{dif}} = -D\nabla N \,, \tag{1.48}$$

which tends to remove the gradient. Here D is the diffusion coefficient for test particles in a gas. When a gas is located in an external field that acts on its particles, a nonuniform distribution of particles occurs. But since it is a stationary distribution, the flux due to the external field is compensated by the diffusion flux, and we have

$$\mathbf{j} = Nb\mathbf{F} - D\nabla N = 0 \,.$$

Since the test particles are found in thermodynamic equilibrium with the gas, the distribution for the number density of the test particles is given by the Boltzmann formula $N = N_0 \exp(-U/T)$, where U is the potential due to the external field, and T is the temperature of the gas. Using it in the above equation and accounting for the force acting on the test particle is $\mathbf{F} = -\nabla U$, we find from the last equation

$$b = \frac{D}{T} \,. \tag{1.49}$$

This expression is known as the Einstein relation. It is valid for small fields that do not disturb the thermodynamic equilibrium between the test and gaseous particles. For the mobility of a charged particle K in a gas this gives

$$K = \frac{eD}{T} \,. \tag{1.50}$$

▶ **Problem 1.18** Find the character of distributions of positively charged particles located in a weakly ionized gas.

The electric potential of a particle of charge e in vacuum is

$$\varphi = \frac{e}{r} \,. \tag{1.51}$$

However, if this particle is surrounded by charge particles of a quasineutral plasma, this field is screened since negatively charged particles are attracted to a test particle and positively charged particles are repulsed from it. In order to ascertain the result of this interaction, we analyze the Poisson equation for the field of a positively charged test particle that has the form

$$\Delta\varphi = 4\pi e(N_e - N_i) \,,$$

where we consider a plasma to be consisted of electrons and ions of a charge e, and their number densities are denoted as N_e and N_i, respectively. The distributions of electrons and ions near a test particle are determined by the Boltzmann formula (1.12) and are given by

$$N_- = N_0 \exp\left(\frac{e\varphi}{T}\right), \quad N_+ = N_0 \exp\left(-\frac{e\varphi}{T}\right), \tag{1.52}$$

where N_0 is the average number density of electrons and ions in the plasma (its average charge equals zero), and T is the temperature of electrons and ions. Substituting this in the Poisson equation, reduce it to the form

$$\Delta\varphi = 8\pi e N_0 \sinh\left(\frac{e\varphi}{T}\right). \tag{1.53}$$

This equation is valid at distances from a test particle where electrons and ions are located, i. e., at distances larger than $N_0^{-1/3}$, while at small distances from a test particle, where the pressure of electrons and ions is negligible on average, the right-hand side of the Poisson equation is zero, and the electric potential of a test particle is given by formula (1.51).

Because of the problem symmetry, the particle electric potential is spherically symmetric on average, which gives at large distances where $e\varphi \ll T$

$$\frac{1}{r}\frac{d^2}{dr^2}(r\varphi) = \frac{8\pi N_0 e^2}{T}\varphi. \tag{1.54}$$

We consider the case when the solution of this equation is substituted into (1.51) at small distances r from a test particle. Then this solution has the form

$$\varphi = \frac{e}{r}\exp\left(-\frac{r}{r_D}\right), \quad r_D = \sqrt{\frac{T}{8\pi N_0 e^2}}. \tag{1.55}$$

The value r_D is the Debye–Hückel radius that characterizes the character of screening of electric fields in the plasma. The solution obtained is valid if the Debye–Hückel radius for this plasma is large compared to an average distance between the charged particles of the plasma, $N_0^{-1/3}$, which corresponds to the criterion

$$\frac{e^2 N_0^{1/3}}{T} \ll 1. \tag{1.56}$$

If this criterion holds true, many electrons and ions take part in the shielding process that corresponds to the physical nature of this phenomenon. A plasma that satisfies to the criterion (1.56) is the ideal plasma. In this plasma a typical energy of the interaction of charged particles or the interaction energy of two charged particles at an average distance between charged particles, $e^2 N_0^{1/3}$, is small compared to a typical thermal energy of particles ($\sim T$). This criterion tells that the main part of time charged particles of the plasma is free. The same criterion is fulfilled for neutral atomic particles of a gas.

▶ **Problem 1.19** Obtain the expression for the Debye–Hückel radius r_D if the electron T_e and ion T_i temperatures are different.

In this case formula (1.52) for number densities of electrons and ions has the form

$$N_- = N_0 \exp\left(\frac{e\varphi}{T_e}\right), \qquad N_+ = N_0 \exp\left(-\frac{e\varphi}{T_i}\right).$$

Repeating the steps of the previous problem, we finally obtain formula (1.55) for the potential of a positively charged test particle, but the expression for the Debye–Hückel radius now takes the form

$$r_D = \sqrt{\frac{\left(\frac{1}{T_e} + \frac{1}{T_i}\right)^{-1}}{4\pi N_0 e^2}}. \tag{1.57}$$

▶ **Problem 1.20** A dense plasma propagates in a buffer gas and conserves its quasineutrality, i. e., electrons as more mobile particles come off the ions and create in this way a field that breaks electrons and accelerates ions. As a result, electrons and ions propagate in a buffer gas together and a plasma almost conserves its quasineutrality. Find the diffusion coefficient for this plasma in a buffer gas.

One more general plasma property relates to the character of its propagation in a neutral gas. Electrons as light particles move faster in a gas than ions, which violates the plasma quasineutrality, and electric fields are generated. Only these fields make the plasma almost quasineutral. As a result, the plasma propagates in a buffer gas as a whole, and we consider below such a process.

In the regime under consideration, when electrons and ions propagate in a buffer gas and the mean free paths of electrons and ions in the gas are relatively small, we have the following expressions for the electron flux $\boldsymbol{j}_e$ and the ion flux $\boldsymbol{j}_i$,

$$\boldsymbol{j}_e = -D_e \nabla N_e - K_e \boldsymbol{E} N_e \; ; \; \boldsymbol{j}_i = -D_i \nabla N_i + K_i \boldsymbol{E} N_i \, .$$

Here N_e, N_i are the number densities of electrons and ions, respectively, D_e, D_i are their diffusion coefficients, and K_e, K_i are their mobilities. Because the electric field acts on electrons and ions in opposite directions, the field enters into the flux expressions with different signs. The electric field strength $\boldsymbol{E}$ satisfies the Poisson equation

$$\operatorname{div} \boldsymbol{E} = 4\pi e(N_i - N_e) \, .$$

In the case of propagation of a dense plasma, the plasma converges to quasineutrality during evolution, i. e., the charge difference is small $\Delta N = |N_i - N_e| \ll N_e$, which gives $N_e \approx N_i \approx N$. Hence the plasma motion is self-consistent, and we have $\boldsymbol{j}_e \approx \boldsymbol{j}_i$. Next, because of the higher mobility of electrons, the electron flux is zero $\boldsymbol{j}_e = 0$ on the scale of electron quantities. This means $D_e \nabla N_e \gg j_i$ and $eEK_e N_e \gg j_i$, so that in terms of the magnitudes of electron parameters we have $\boldsymbol{j}_e = 0$. This gives for the electric field strength that arises due to plasma motion as a single whole,

$$\boldsymbol{E} = -\frac{D_e}{eK_e}\frac{\nabla N}{N} .$$

From this we have for the flux of charged particles

$$\boldsymbol{j} = \boldsymbol{j}_i = -\left(D_i + D_e\frac{K_i}{K_e}\right)\nabla N = -D_a \nabla N .$$

In this way we define D_a, the coefficient of ambipolar diffusion. Thus the plasma evolution has a diffusive character with a self-consistent diffusion coefficient. In particular, when electrons and ions are found in thermodynamic equilibrium that allows us to define the electron T_e and ion T_i temperatures for each subsystem on the basis of the Einstein relation (1.50), we have for the diffusion coefficient of a collective plasma motion

$$D_a = D_i\left(1 + \frac{T_e}{T_i}\right) . \tag{1.58}$$

One can see that in the regime under consideration a plasma propagates in a buffer gas with the speed of the ions rather than that of the electrons.

In this regime of plasma motion, the plasma remains almost quasineutral, so the relation $\Delta N = |N_i - N_e| \ll N$ holds true. Then the Poisson equation gives $\Delta N \sim E/(4\pi e L)$, where L is a typical plasma dimension. The above equation for the electric field strength that follows from $j_e = 0$ together with the Einstein relation (1.50) gives $E \sim T/(e^2 L)$, where for simplicity we assume the electron and ion temperatures to be equal. From this we have $\Delta N \sim N r_D^2/L^2$. Thus the criterion of the regime of ambipolar diffusion corresponds to the plasma criterion $L \gg r_D$.

▶ **Problem 1.21** Find the distribution function over the potential energies for ions in a quasineutral ideal plasma.

In considering a quasineutral ideal plasma, we assume that neutral atomic particles (atoms or molecules) whose number density exceeds that of electrons and ions provide the stability of this plasma. But because they do not influence the electric properties of such a plasma, they will not be considered below.

An electric potential of this plasma is determined by charged particles, and according to formula (1.55) the average potential energy for a test ion with other electrons and ions is equal to

$$\overline{U} = e\overline{\varphi} = \int_0^\infty \frac{e^2}{r}\exp\left(-\frac{r}{r_D}\right)\left[N_o \exp\left(-\frac{e\varphi}{T}\right) - N_o \exp\left(\frac{e\varphi}{T}\right)\right] 4\pi r^2 dr$$
$$= \frac{e^2}{4 r_D} , \qquad e\varphi \ll T . \tag{1.59}$$

As is seen, for an ideal plasma that submits to the criterion (1.56) the average potential energy of ions or electrons is small compared to its thermal kinetic energy

$$\overline{U} \ll T .$$

In addition, this average potential energy is identical for positively charged ions and electrons.

In the same manner we have for the mean square of the potential energy for an ion or electron

$$\overline{U^2} = \int_0^\infty \frac{e^4}{r^2} \exp\left(-\frac{2r}{r_D}\right) \left[N_o \exp\left(-\frac{e\varphi}{T}\right) + N_o \exp\left(\frac{e\varphi}{T}\right)\right] 4\pi r^2 dr$$

$$= 4\pi N_o e^4 r_D = \frac{T}{2} \frac{e^2}{r_D}, \qquad e\varphi \ll T . \tag{1.60}$$

One can see that this value is small compared to the square of the thermal energy,

$$\frac{\overline{U^2}}{T^2} \sim \frac{e^2}{r_D T} \ll 1 ,$$

for an ideal plasma, but is large compared to the square of the average potential energy,

$$\frac{\overline{U^2}}{\left(\overline{U}\right)^2} = 16\pi N_o r_D^3 \gg 1 .$$

The last relation allows us to neglect the divergence of the average ion potential energy from zero, and the distribution function over ion potential energies has the form of the Gauss distribution

$$f(U)dU = \frac{1}{\sqrt{2\pi \Delta U^2}} \exp\left[-\frac{U^2}{2\Delta U^2}\right] ,$$

where $\Delta U = \sqrt{\overline{U^2}/2}$ is the fluctuation of the ion potential energy.

Substituting in this formula the average square of the ion potential energy (1.60), we find the distribution over ion potential energies,

$$f(U)dU = \frac{1}{\sqrt{2\pi \Delta U^2}} \exp\left(-\frac{U^2}{2\Delta U^2}\right) = \frac{dU}{2\pi e^2 \sqrt{2 N_o r_D}} \exp\left(-\frac{U^2}{8\pi N_o e^4}\right) . \tag{1.61}$$

This value $f(U)dU$ is the probability that the potential energy for a test ion and an electron located inside a plasma ranges from U up to $U + dU$.

1.4 Laws of Black Body Radiation

▶ **Problem 1.22** Determine the distribution of thermal photons over frequencies.

Let us consider thermal or black body radiation that is characterized by a temperature T. Such a radiation is located inside a vessel with the wall temperature T, and an equilibrium for photons of different frequencies there results from absorption and radiation of the walls. The number of photons is not fixed, and the relative probability that n photons of an energy $\hbar\omega$ are found in a given state according to the Boltzmann formula (1.11) is equal to $\exp(-n\hbar\omega/T)$. This gives the average number of photons $\overline{n_\omega}$ in this state,

$$\overline{n_\omega} = \frac{\sum_n n \exp(-\frac{\hbar\omega n}{T})}{\sum_n \exp(-\frac{\hbar\omega n}{T})} = \frac{1}{\exp(\hbar\omega/T) - 1}. \tag{1.62}$$

This is the Planck formula. As is seen, it corresponds to the Bose–Einstein distribution (1.31) with zero chemical potential. Thus, black body radiation as an ensemble of photons is characterized by zero chemical potential.

▶ **Problem 1.23** Determine the spectral density of black body radiation, i. e., the energy of this radiation per unit time, unit volume, and unit frequency.

On the basis of the definition of the spectral density of radiation, the radiation energy per unit time and unit volume with a range of frequencies from ω up to $\omega + d\omega$ is $\Omega U_\omega d\omega$. On the other hand, on the basis of the statistical weight of continuous spectrum (1.19), one can represent this value as $2\hbar\omega n_\omega \Omega d\boldsymbol{k}/(2\pi)^3$, where $\boldsymbol{k}$ is the photon wave number, $d\boldsymbol{k}/(2\pi)^3$ is the number of states per unit volume and for a given element of the wave numbers, the factor 2 accounts for two polarizations of an electromagnetic wave, and n_ω is the number of photons located in one state. We take into account that the electromagnetic wave is the transversal one, i. e., its electric field strength $\boldsymbol{E}$ is directed perpendicular to the propagation direction that is determined by the vector $\boldsymbol{k}$. On the basis of the dispersion relation for photons $\omega = ck$ between the photon frequency ω and its wave number k (c is the light velocity), we obtain the Planck radiation formula

$$U_\omega = \frac{\hbar\omega^3}{\pi^2 c^3} n_\omega. \tag{1.63}$$

On the basis of formula (1.62) one can rewrite formula (1.63) in the form

$$U_\omega = \frac{\hbar\omega^3}{\pi^2 c^3 \left[\exp(\hbar\omega/T) - 1\right]}. \tag{1.64}$$

Let us consider the limiting case of this formula. In the case of small frequencies $\hbar\omega \ll T$ this formula is converted into the Rayleigh–Jeans formula

$$U_\omega = \frac{\omega^2 T}{\pi^2 c^3}, \quad \hbar\omega \ll T. \tag{1.65}$$

Because this formula corresponds to the classical limit, it does not contain the Planck constant.

In the other limiting case of high frequencies $\hbar\omega \gg T$, the above formula is transformed to the Wien formula

$$U_\omega = \frac{\hbar\omega^3}{\pi^2 c^3} \exp\left(-\frac{\hbar\omega}{T}\right). \tag{1.66}$$

► **Problem 1.24** Evaluate the energy flux for radiation emitted from a black body surface (the Stephan–Boltzmann law).

A black body surface emits a number of photons which propagate inside a volume occupied by thermal radiation. The energy flux per unit frequency range ω is equal to $d\omega$ to $cU_\omega d\omega$, and the energy flux per unit elementary solid angle $d\Theta = d\varphi d\cos\theta$ is

$$\frac{d\Theta}{4\pi} c \int_0^\infty U_\omega d\omega \, .$$

Because of the symmetry, the total flux is directed perpendicular to the surface, and projecting the flux in each solid angle in this direction, we obtain the total energy flux per unit frequency range of black body radiation from a surface,

$$j_\omega d\omega = \int_0^{\pi/2} \frac{c}{4\pi} U_\omega \, d\omega \, 2\pi \cos\theta d\cos\theta = \frac{c}{4} U_\omega d\omega \, ,$$

where θ is the angle between the normal to the surface and the direction of motion of an emitting photon. Integrating over frequencies, we obtain from this on the basis of the Planck formula (1.63)

$$J = \int_0^\infty j_\omega d\omega = \sigma \, T^4. \tag{1.67}$$

This is the Stephan–Boltzmann law, and its component, Stephan–Boltzmann constant σ, is equal to

$$\sigma = \frac{1}{4\pi^2 c^2 \hbar^3} \int_0^\infty (e^x - 1)^{-1} x^3 dx = \frac{\pi^2}{(60 c^2 \hbar^3)} = 5.67 \times 10^{-12} \frac{\text{W}}{\text{cm}^2\,\text{K}^4} \, . \tag{1.68}$$

In deriving this formula, we use the relation

$$\int_0^\infty \frac{x^3 dx}{e^x - 1} = \frac{\pi^2}{15} \, .$$

► **Problem 1.25** Find the dependence of the Stephan–Boltzmann constant on the parameters from the dimensionality consideration.

The emitting energy flux depends on three-dimensional parameters—the radiation temperature T, the Planck constant $\hbar$, and the light velocity c. From these parameters one can compose only one combination with the energy flux dimensionality, and this gives

$$J \sim \frac{T^4}{\hbar^3 c^2} \, .$$

As is seen, this gives the temperature dependence as in the Stephan–Boltzmann law (1.67) and the dependence on parameters for the Stephan–Boltzmann constant as in formula (1.68).

1.5
Ionization and Dissociation Equilibrium

► **Problem 1.26** Find the equilibrium number density of electrons and ions in an equilibrium of a weakly ionized gas with the number density of atoms N_a and temperature T.

We considered above the equilibrium between discrete states (1.12) and between states of continuous spectrum (1.23) for particle ensembles. We now analyze an equilibrium between bound and free states of atomic particles.

We consider a quasineutral plasma located in a finite volume Ω and denote the number of electrons, ions, and atoms in this volume as n_e, n_i, n_a ($n_e = n_i$), respectively. These atomic particles are characterized by the statistical weights $g_e = 2$ for electrons that account for two spin directions, g_i for ions and g_a for atoms, and these statistical weights relate to electron states of these atomic particles. According to the Boltzmann formula (1.12) and expression (1.19) for the statistical weight of continuous spectrum states we have the ratio between free and bound states of electrons,

$$\frac{n_i}{n_a} = \frac{g_e g_i}{g_a} \int \exp\left(-\frac{J + p^2/(2m_e)}{T}\right) \frac{d\mathbf{p}\, d\mathbf{r}}{(2\pi\hbar)^3} .$$

Here p is the electron momentum, J is the atom ionization potential, and $J + p^2/(2m_e)$ is the energy of transition from the ground atom state to the state of a free electron of a given momentum. We assume the atoms to be in the ground state only.

Integrating this expression over the electron momenta, we obtain

$$\frac{n_i}{n_a} = \frac{g_e g_i}{g_a} \left(\frac{m_e T}{2\pi\hbar^2}\right)^{3/2} \exp\left(-\frac{J}{T}\right) \int d\mathbf{r} .$$

During integration over the volume, we take into account the symmetry of an electron gas, so that the exchange of two electrons by their states does not change the state of the electron system. Therefore, $\int d\mathbf{r} = \Omega/n_e$, and introducing the number densities of electrons $N_e = n_e/\Omega$, ions $N_i = n_i/\Omega$, and atoms $N_a = n_a/\Omega$, we reduce the obtained expression to the form of the Saha distribution,

$$\frac{N_e N_i}{N_a} = \frac{g_e g_i}{g_a} \left(\frac{m_e T}{2\pi\hbar^2}\right)^{3/2} \exp\left(-\frac{J}{T}\right) . \tag{1.69}$$

One can introduce from this the equilibrium constant for the ionization equilibrium that has the form

$$K(T) = \frac{g_e g_i}{g_a} \left(\frac{m_e T}{2\pi\hbar^2}\right)^{3/2} \exp\left(-\frac{J}{T}\right) . \tag{1.70}$$

▶ **Problem 1.27** Show that states of continuous spectrum are characterized by a large statistical weight compared to discrete atom states.

Rewriting the Saha formula (1.69) in the form of the Boltzmann formula (1.12)

$$\frac{N_i}{N_a} = \frac{g_c}{g_a} \exp\left(-\frac{J}{T}\right) ,$$

we define the statistical weight g_c of continuous spectrum. Comparing this formula with the Saha formula (1.69), we find for the statistical weight of the continuous spectrum,

$$g_c = \frac{g_e g_i}{N_e} \left(\frac{m_e T}{2\pi\hbar^2}\right)^{3/2} . \tag{1.71}$$

One can see that for an ideal plasma (1.56) this value is enough large. Hence, the presence of free electrons in an equilibrium gas becomes noticeable at low temperatures when the probability of atom excitation is small. This allowed one to account for the ground atom state during derivation of the Saha formula.

▶ **Problem 1.28** Analyze the ionization equilibrium for a large metallic particle of the spherical shape.

Introducing the work function W of a metallic surface as the electron binding energy, we reduce the problem of ionization equilibrium near the metallic surface to the above problem of ionization equilibrium for atomic particles. Considering the ionization equilibrium for a spherical metallic particle of charge $Z \gg 1$ and radius r_0, we account for the electron binding energy that includes a work for electron removal to infinity and is

$$W_Z = W + \frac{Ze^2}{r_0} .$$

Correspondingly, the Saha formula (1.69) takes the form

$$\frac{P_Z\, N_e}{P_{Z+1}} = 2\left(\frac{m_e T}{2\pi\hbar^2}\right)^{3/2} \exp\left(-\frac{W_Z}{T}\right) = 2\left(\frac{m_e T}{2\pi\hbar^2}\right)^{3/2} \exp\left(-\frac{W}{T} - \frac{Ze^2}{r_0 T}\right) , \tag{1.72}$$

where P_Z is the probability for the particle to have a charge Z, N_e is the electron number density, the factor 2 accounts for two spin projections for an electron.

In reality, this formula gives the distribution of metallic particles over their charges when a typical charge is large. Let us rewrite this formula in the form

$$P_Z = P_{Z-1} A \exp\left(-\frac{Ze^2}{r_0 T}\right) = P_0\, A^Z \exp\left(-\frac{Z^2 e^2}{2 r_0 T}\right) , \tag{1.73}$$

where

$$A = \frac{2}{N_e}\left(\frac{m_e T}{2\pi\hbar^2}\right)^{3/2} \exp\left(-\frac{W}{T}\right) .$$

For large particle sizes this formula gives a sharp maximum for the particle distribution function over charges. Indeed, the average particle charge $\overline{Z}$ coincides with the maximum of P_Z, and the latter according to formula (1.70) satisfies the relation

$$\overline{Z} = \frac{r_0 T}{e^2} \ln A \,, \tag{1.74}$$

and $\overline{Z} \gg 1$. Expanding formula (1.73) near the maximum P_Z, we obtain

$$P_Z = P_{\overline{Z}} \exp\left[-\frac{(Z-\overline{Z})^2}{2\Delta Z^2}\right], \quad \Delta Z^2 = \frac{r_0 T}{e^2} \,. \tag{1.75}$$

We have for the relative width of the particle distribution function over charges

$$\frac{\Delta Z}{\overline{Z}} = \frac{e}{\sqrt{r_0 T \ln A}} \,,$$

and for large particle sizes this ratio is small. As a result, the particle distribution function over charges takes the form of the Gauss distribution.

▶ **Problem 1.29** Evaluate the flux of emitting electrons from a hot metallic surface considering an electron ensemble inside the metal and near its surface as almost degenerate gas.

In considering an electron ensemble near the metal surface as almost degenerate gas, we have for the electron distribution function over momenta according to formula (1.37),

$$f(\boldsymbol{p})d\boldsymbol{p} = \frac{2d\boldsymbol{p}}{(2\pi\hbar)^3} \exp\left(-\frac{\varepsilon-\mu}{T}\right) \,.$$

This formula gives the number electrons with momenta ranged from $\boldsymbol{p}$ up to $\boldsymbol{p}+d\boldsymbol{p}$, and the chemical potential for this distribution is equal to the Fermi energy $\mu = \varepsilon_F$. Evidently, an electron can leave the metal surface if its kinetic energy exceeds the value $\varepsilon_F + W$. Since the electron flux outside the metallic surface is $\int v_x f(\boldsymbol{p})d\boldsymbol{p}$, where v_x is the electron velocity in the perpendicular direction to the surface and the integral is taken for electron energies $m_e v_x^2/2 \geq \varepsilon_F + W$, the total electron flux is equal to

$$j = 2\pi m_e T \int\limits_{\varepsilon=\varepsilon_F+W}^{\infty} v_x \frac{m_e dv_x}{4\pi^3\hbar^3} \exp\left(-\frac{m_e v_x^2}{2T} + \frac{\mu}{T}\right) = \frac{m_e T^2}{2\pi^2\hbar^3} \exp\left(-\frac{W}{T}\right) \,,$$

and during evaluation of the integral we use the condition $\varepsilon - \mu \gg T$. Correspondingly, the current density of emitting electrons is

$$i = ej = \frac{e m_e T^2}{2\pi^2\hbar^3} \exp\left(-\frac{W}{T}\right) \,. \tag{1.76}$$

This is the Richardson–Dushman formula that gives the density current for thermoemission of a metal surface.

▶ **Problem 1.30** Evaluate the flux of emitting electrons from a hot metallic surface from the analysis of ionization equilibrium for a large metallic particle.

Let us transfer a spherical metal particle to a flat surface by turning a particle radius to infinity. Then the electric potential of the particle is small compared to a typical thermal energy, the parameter $Ze^2/(r_0 T)$ in formula (1.74) becomes small, and $A = 1$. This gives the equilibrium electron density near the surface,

$$N_e = 2\left(\frac{m_e T}{2\pi\hbar^2}\right)^{3/2} \exp\left(-\frac{W}{T}\right) . \quad (1.77)$$

We assume that free electrons near a metallic surface can penetrate inside the metal, and because of equilibrium, this current density is equal to the current density of emitting electrons. From the equality of these current densities we obtain the emitting current density

$$i = e\sqrt{\frac{T}{2\pi m_e}}\, N_e = \frac{e m_e T^2}{2\pi^2\hbar^3} \exp\left(-\frac{W}{T}\right) . \quad (1.78)$$

We derive again the Richardson–Dushman formula for thermoemitting electron current density from other considerations. Usually it is represented in the form

$$i = A_R T^2 \exp\left(-\frac{W}{T}\right) , \quad A_R = \frac{e m_e}{2\pi^2\hbar^3}$$

and the Richardson constant A_R is equal to $120\,\mathrm{A/(cm^2\,K^2)}$ according to this formula.

▶ **Problem 1.31** Analyze the dissociation equilibrium for a gas consisting of diatomic molecules.

With the transition of a discrete state to a continuous spectrum state, the dissociation equilibrium becomes analogous to the ionization one. Therefore, one can use the Saha formula (1.69) for the dissociation equilibrium with changing atomic parameters by parameters of the molecule. As a result, we get the relation between the number density of molecules N_m and atoms N_a on which the molecule is disintegrated,

$$\frac{N_a^2}{N_m(v=0, J=0)} = \frac{g_a^2}{g_m}\left(\frac{\mu T}{2\pi\hbar^2}\right)^{3/2} \exp\left(-\frac{D}{T}\right) .$$

Here g_a, g_m are the statistical weights of an atom and a molecule with respect to their electron state, μ is the reduced mass of atoms, and D is the dissociation energy of the molecule. We assume the vibrational and translational temperatures to be equal, and atoms are different isotopes, which allows us to escape the prohibition of some rotational states.

In contrast to the ionization equilibrium, molecules can be found in excited vibration-rotational states, so we change the number density of molecules in the ground state that is used in the above formula, by the total number density N of molecules. As a result, we get

$$\frac{N_a^2}{N_m} = K_{dis}(T) = \frac{g_a^2}{g_m}\left(\frac{\mu T}{2\pi\hbar^2}\right)^{3/2}\frac{B}{T}\left[1-\exp\left(-\frac{\hbar\omega}{T}\right)\right]\exp\left(-\frac{D}{T}\right), \quad (1.79)$$

where K_{dis} is the equilibrium constant for the dissociation process.

▶ **Problem 1.32** A cluster is located in a buffer gas with an admixture of a vapor consisting of cluster atoms. Find the dependence of the equilibrium number density of atoms on a cluster size.

Let us introduce the equilibrium number density of atoms over a flat surface $N_{sat}(T)$ at a given surface temperature T and the binding energy ε_o of the atom with the surface. Considering a cluster consisting of n atoms with the binding energy ε_n of surface atoms, we obtain the equilibrium number density of atoms,

$$N = N_{sat}(T)\exp\left(-\frac{\varepsilon_n - \varepsilon_o}{T}\right). \quad (1.80)$$

At this number density the rate of atom evaporation from the cluster surface and the rate of atom attachment to the cluster surface are equal. The difference of the atom binding energies for the flat surface and cluster surface is determined mostly by the surface tension.

1.6 Ionization Equilibrium for Clusters

▶ **Problem 1.33** Find the ratio between numbers of metallic clusters of neighboring charges in a hot vapor.

In considering the equilibrium between charged clusters

$$M_n^{+Z+1} + e \leftrightarrow M_n^{+Z}. \quad (1.81)$$

Introducing the probability $P_Z(n)$ that a cluster consisting of n atoms to have a charge Z, we have on the basis of the Saha formula (1.69)

$$\frac{P_Z(n)N_e}{P_{Z+1}(n)} = 2\left(\frac{m_e T_e}{2\pi\hbar^2}\right)^{3/2}\exp\left[-\frac{I_Z(n)}{T_e}\right]. \quad (1.82)$$

Here T_e is the electron temperature, and we assume the temperatures of free and bound electrons to be identical, m_e is the electron mass, N_e is the electron number density, and $I_Z(n)$ is the ionization potential of the cluster of a charge Z consisting of n atoms.

Note that because the third particle in the recombination process (1.81) is a bound electron, the number density of internal electrons of a metallic clusters greatly exceeds that of plasma electrons. Moreover, internal electrons of a metallic plasma can be responsible for release of initially bound electrons. Thus, if the temperatures of internal and plasma electrons are identical, the Saha formula is valid, but the electron release for the equilibrium (1.81) can be determined by internal electrons. Next, because the work function of metals (the binding energy of electrons with a metallic surface) is lower than the ionization potential of corresponding atoms, the ionization potential of metal clusters lies between these values. Hence, ionization of metallic clusters occurs at relatively low electron temperatures of the plasma.

▶ **Problem 1.34** Find the charge distribution of large metallic clusters in a hot vapor in the limit $e^2/r \ll T_e$.

The equilibrium (1.81) between free electrons and bound electrons of a metallic cluster relates to a large mean free path of electrons in a gas compared to the cluster radius. Therefore, the interaction energy of a removed electron and a charged cluster during electron removal from a charged cluster must be included in the cluster ionization potential. Hence, the ionization potential $I_Z(n)$ of a large cluster of n atoms differs from that of a neutral cluster $I_0(n)$ by the energy that is consumed as a result of the removal of electrons from the cluster surface to infinity, i. e.,

$$I_Z(n) - I_0(n) = \frac{Ze^2}{r} ,$$

where $I_0(n)$ is the ionization potential of a neutral cluster.

Substituting this into formula (1.82), one can represent this formula in the form of the Gauss formula if $Z \gg 1$,

$$\frac{P_Z}{P_{Z-1}} = A \exp\left(-\frac{Ze^2}{rT_e}\right) , \quad A = \frac{2}{N_e}\left(\frac{m_e T_e}{2\pi\hbar^2}\right)^{3/2} \exp\left(-\frac{I_0}{T_e}\right) .$$

Representing this formula in the form

$$\frac{P_Z}{P_0} = A^Z \exp\left(-\frac{Z^2 e^2}{2rT_e}\right) ,$$

we expand this expression near the maximum of the function $\ln P_Z(n)$, reducing it to the Gauss formula

$$P_Z(n) = P_{\overline{Z}}(n) \exp\left[-\frac{(Z-\overline{Z})^2}{2\Delta^2}\right] , \tag{1.83}$$

where the mean particle charge is given by

$$\overline{Z} = \frac{rT_e}{e^2}\left\{\ln\left[\frac{2}{N_e}\left(\frac{m_e T_e}{2\pi\hbar^2}\right)^{3/2}\right] - \frac{I_0(n)}{T_e}\right\} , \quad \Delta^2 = \frac{rT_e}{e^2} . \tag{1.84}$$

This formula holds true if $\Delta \gg 1$, that is valid for large clusters.

▶ **Problem 1.35** Express the ionization potential of a charged cluster through the atom ionization potential I and the work function of a corresponding metal W, and on this basis determine the cluster mean charge.

Defining the average cluster charge Z on the basis of the relation $P_Z(n) = P_{Z+1}(n)$, we have on the basis of the above formulas

$$Z = \frac{T_e r}{e^2}\left\{\ln\left[\frac{2}{N_e}\left(\frac{m_e T_e}{2\pi\hbar^2}\right)^{3/2}\right] - \frac{I_0(n)}{T_e}\right\} .$$

Taking the cluster ionization potential such that $I_0(\infty) = W$, the metal work function, and $I_0(1) = I$, the atom ionization potential, one can represent the ionization potential of a large neutral cluster in the form $I_0(n) = W + \text{const}/n^{1/3}$, where $\text{const} = I - W$. This gives

$$Z = \frac{T_e r}{e^2}\left\{\ln\left[\frac{2}{N_e}\left(\frac{m_e T_e}{2\pi\hbar^2}\right)^{3/2}\right] - \frac{W}{T_e} - \frac{I - W}{T_e n^{1/3}}\right\} .$$

One can see that the basic dependence of the cluster charge on its size is $Z \sim n^{1/3}$, and the proportionality coefficient depends on the cluster temperature.

▶ **Problem 1.36** Find the conditions when a large metal cluster located in a plasma is neutral.

Let us take the ionization potential of a neutral cluster as $I_0(n) = W + (I - W)/n^{1/3}$ and the electron affinity to a neutral cluster as $EA(n) = W + (EA - W)/n^{1/3}$, where I is the atom ionization potential, EA is the atom electron affinity, and W is the metal work function. Comparing formulas for positively and negatively charged clusters according to the Saha formulas (1.69) with these ionization potential and electron affinity, we find the following ratio of the number densities of positive N_+ and negative N_- clusters of a charge e:

$$\frac{N_+}{N_-} = \zeta^2 \exp\left(-\frac{\Delta}{T_e n^{1/3}}\right), \quad \zeta = \frac{2}{N_e}\left(\frac{m_e T_e}{2\pi\hbar^2}\right)^{3/2} \exp\left(-\frac{W}{T_e}\right) .$$

Here $\Delta = I_0(1) + EA - 2W$, N_e is the electron number density, and T_e is the electron temperature. If a cluster is large and the parameter $\Delta/(T_e n^{1/3})$ is small, the electron temperature T_*, at which the average cluster charge is zero, is given by the relation $\zeta(T_*) = 0$. At higher temperatures the cluster is charged positively, while at lower temperatures it is charged negatively.

Let us demonstrate it on a copper example, when $I = 7.73$ eV, $EA = 1.23$ eV and $W = 4.4$ eV. This gives $\Delta = 0.16$ eV, and the ratio $\Delta/(T_e n^{1/3})$ is small for large clusters. The neutralization temperature is $T_* = 2380$ K for $N_e = 1 \times 10^{11}\ \text{cm}^{-3}$, $T_* = 2640$ K for $N_e = 1 \times 10^{12}\ \text{cm}^{-3}$, and $T_* = 2970$ K for $N_e = 1 \times 10^{13}\ \text{cm}^{-3}$. The neutralization temperature T_* increases with an increase in the number density of plasma electrons.

▶ **Problem 1.37** Find the maximum negative charge of an isolated dielectric particle if the charge state is stable.

Charging of a dielectric particle has a different nature than that for metal clusters, where electrons form a degenerated gas. The active knots or centers on the surface of a dielectric particle are traps for electrons, and negative or positive ions. The process of electron and ion attachment to the surface of a dielectric particle proceeds according to the scheme

$$\begin{aligned} e + A_n^{-Z} &\to \left(A_n^{-(Z+1)}\right)^{**} , \\ \left(A_n^{-(Z+1)}\right)^{**} + A &\to A_n^{-(Z+1)} + A , \\ B^+ + A_n^{-(Z+1)} &\to B + A_n^{-Z} , \end{aligned} \tag{1.85}$$

and an autodetachment state $(A_n^{-(Z+1)})^{**}$ is quenched by collisions with surrounding atoms. Because the rate constant of pair attachment of an electron to a dielectric particle greatly exceeds the ionization rate constant of the particle by electron impact, these particles are charged negatively.

In contrast to metallic particles, the binding energies for active centers do not depend on the particle size, because the action of each center is concentrated in a small region of space. Evidently, the number of such centers is proportional to the area of the particle's surface, and for particles of micron size this value is large compared to that occupied by the charges. Hence, here we consider the regime of charging of a small dielectric particle far from the saturation of active centers. Then positive and negative charges can exist simultaneously on the particle's surface. They spread over the surface and can recombine there.

Usually, the binding energy of electrons in negative active centers is in the range $EA = 2–4$ eV, and the ionization potential for positive active centers is about $J_o \approx 10$ eV. Hence, attachment of electrons is more profitable for electrons of a glow discharge, and a small dielectric particle has a negative charge in a glow gas discharge. Due to the particle charge Z, it has the electric potential $\varphi = Ze/r$, where r is the particle radius. In the case $e\varphi < EA$ the electron state is stable, while in the case $e\varphi > EA$ an electron tunnel transition is possible, which leads to the decay of the electron state. Thus, the stable charge state of a dielectric particle is possible if new electrons can attach to the particle. An isolated charged particle emits electrons until it reaches the limiting charge

$$Z_* = r \cdot EA/e^2 . \tag{1.86}$$

We consider it as the maximum charge of a dielectric particle. As a demonstration, we consider an example of a dielectric particle of radius $r = 2\,\mu\text{m}$ and $EA = 2$ eV, which leads to its maximum charge $Z_* = 4 \times 10^3$ of the particle, and its electric potential is equal to 2 V.

▶ **Problem 1.38** Estimate a typical time of the electron transition for a dielectric particle whose negative charge exceeds the critical one (1.86).

If the negative charge of a micron dielectric particle exceeds the critical one (1.86), it is energetically profitable for some surface electrons to liberate and to be removed from the particle. Simultaneously, at the critical charge of a micron dielectric particle active centers of the particle are populated weakly. Indeed, in the example of the previous problem, the distance between nearest surface charges is approximately 0.4 μm, which exceeds by one to two orders of magnitude a typical distance between neighboring active centers. Hence, the critical particle charge (1.86) may be exceeded, and we estimate below the lifetime of such autodetachment states as a result of tunnel electron transitions. If this time exceeds the time of electron capture, such states may be realized.

Let us estimate a typical time of the electron tunnel transition through the potential barrier that has the following exponential dependence:

$$\frac{1}{\tau} \sim \exp(-2S), \quad S = \int_r^{R_c} dR \sqrt{\frac{2m_e}{\hbar^2}\left[EA - U(r) + U(R)\right]} .$$

Here r is the particle radius, EA is the electron binding energy (the electron affinity of an active center), $U(R) = Ze^2/R$ is the interaction potential of an electron with the Coulomb field of the particle if its distance from the particle's center is R, and R_c is the turning point, that is

$$R_c = \frac{r}{1 - EA/\varepsilon_o} ,$$

where $\varepsilon_o = Ze^2/r$. Thus we have

$$S = \frac{\pi}{2} r \sqrt{\frac{2m_e}{\hbar^2} \frac{\varepsilon_o}{1 - EA/\varepsilon_o}} .$$

Since $Z \sim Z_*$, according to formula (1.86) the value ε_o is of the order of a typical atomic value, which gives $S \sim r/a_o$, where a_o is the Bohr radius. Being guided by dielectric particles of micron sizes, we obtain a very high lifetime of surface negative ions with respect to their barrier decay with electron release. Hence, in reality the limit value Z_* (1.86) of the particle charge may be exceeded.

▶ **Problem 1.39** Estimate a charge of a dielectric particle that results from the equilibrium of the particle charge with a surrounding electron gas.

This equilibrium under consideration proceeds according to the scheme

$$A + e \longleftrightarrow A^- , \tag{1.87}$$

where A is an active center of the surface, and A^- is its negative ion. Denoting the total number of active centers on the particle's surface by p and the electron binding energy by EA, we have from this equilibrium on the basis of the Saha formula (1.69)

$$\frac{(p - Z)N_e}{Z} = g\left(\frac{m_e T_e}{2\pi\hbar^2}\right)^{3/2} \exp\left(-\frac{EA}{T_e}\right) .$$

Here N_e is the number density of free electrons, Z is the particle charge or the number of surface negative ions, and $g \sim 1$ is the combination of statistical weights of an electron, active center, and negative ion; for simplicity we take below $g = 1$. From this we find the particle charge in the limit $Z \ll p$,

$$Z = N_e p \left(\frac{2\pi\hbar^2}{m_e T_e} \right)^{3/2} \exp\left(\frac{EA}{T_e} \right) .$$

Since the number of active centers on the surface is proportional to its area, we have

$$Z \sim n^{2/3} .$$

Correspondingly, because the particle radius is $r \sim n^{1/3}$, the electric potential of a large dielectric particle is

$$\varphi \sim \frac{Ze}{r} \sim n^{1/3} .$$

From this it follows that the interaction of surface electrons with the electric potential of a negatively charged dielectric particle may be responsible for the negative charge of the large particle.

▶ **Problem 1.40** Analyze detachment of surface negative ions of a dielectric particle in collisions with electrons and positive ions of a surrounding plasma.

We now compare the rates of detachment of surface negative ions of a negatively charged dielectric particle in collisions with electrons and ions of a surrounding plasma. According to scheme (1.87), a surface negative ion decays in collisions with electrons of a surrounding plasma, and the rate of this process is estimated as

$$\nu_e \sim Z v_e N_e \sigma_0 \exp\left(-\frac{EA}{T_e} \right) ,$$

where $v_e \sim \sqrt{T_e/m_e}$ is a typical electron velocity, T_e is the electron temperature, and m_e is the electron mass; σ_0 is of the order of the cross section of the negative ion, and EA is the electron binding energy in the surface negative ion.

Along with this process, detachment of surface negative ions of the dielectric particle may result from charge exchange of plasma positive ions and surface negative ions, and also from capture of plasma positive ions by other active centers. In the latter case, surface positive and negative ions can recombine after ion motion on the particle surface. The rate of the charge exchange process is estimated as

$$\nu_i \sim Z v_i \sigma_{ex} N_i ,$$

where $v_i \sim \sqrt{T_i/m_i}$ is a typical ion velocity, T_i is the ion temperature, and m_i is the ion mass; σ_{ex} is the cross section of the charge exchange process for an incident positive ion and a surface negative ion, and N_i is the number density of plasma positive ions. Taking $\sigma_{ex} \sim \sigma_0$ and accounting for the plasma quasineutrality $N_i = N_e$, we obtain for $T_e > T_i$ that the criterion $\nu_e \gg \nu_i$ that the charge equilibrium corresponds to scheme (1.87),

$$T_e \ll \frac{2EA}{\ln((T_e m_i)/(T_i m_e))} ,$$

and the equilibrium (1.87) can be realized at high electron temperatures.

If the dielectric particle has an active centers which can capture plasma positive ions, the rate of the latter process is estimated as

$$\nu_i \sim \pi r^2 v_i N_i ,$$

if the number of occupied active centers for positive ions is small compared with their total number. Because the cluster radius $r \sim n^{1/3}$, a typical particle charge is $Z \sim r \sim n^{1/3}$. Hence, if active centers for positive ions exist on the dielectric particle surface, the process of capture of positive ions with their subsequent recombination with surface negative ions is responsible for the decay of surface negative ions. Thus, the charge equilibrium for a dielectric particle may be determined by different processes.

2
Elementary Processes in Plasma

2.1
Elementary Act of Particle Collision

▶ **Problem 2.1** Define the characteristic of the elementary inelastic process for the transition of a particle A between the states i and f as a result of collision with a particle B.

Let us consider a motionless particle A located in a gas of particles B, and interaction with these particles cause the transition between the indicated states. Evidently, the probability of surviving of a particle A in a state i up to time t is described by the equation

$$\frac{dP}{dt} = -\nu_{\text{if}} P, \tag{2.1}$$

where ν_{if} is the rate (the probability per unit time) of a transition.

The rate ν_{if} of transition is proportional to the flux j of particles B, which collide with a test particle A. One can define the ratio of these values,

$$\sigma_{\text{if}} = \frac{\nu_{\text{if}}}{j}, \tag{2.2}$$

the cross section of an inelastic transition, as the parameter of elementary act of collision. It is of importance that this parameter relates to the elementary act of pair collision rather than to the parameters of a gas of particles B.

One can introduce the characteristic of the elementary collision as the ratio of the transition rate ν_{if} to the number density $[B]$ of particles B, and this is the rate constant k_{if} of this process,

$$k_{\text{if}} = \frac{\nu_{\text{if}}}{[B]}. \tag{2.3}$$

If all the particles are moving with an identical velocity v, the particle flux is $j = v[B]$. Hence

$$k_{\text{if}}(v) = v \cdot \sigma_{\text{if}}(v) .$$

Plasma Processes and Plasma Kinetics. Boris M. Smirnov

ISBN: 978-3-527-40681-4

The cross section $\sigma_{\rm if}$ can depend on the relative velocity of the particles, as well as the rate constant of transition $k_{\rm if}$. Both characteristics of elementary collision, $\sigma_{\rm if}$ and $k_{\rm if}$, may be used equivalently.

► **Problem 2.2** A test particle A is located in a gas consisting of particles B and can change its state as a result of collisions. Write the balance equation for a particle A in a given state.

We take into consideration the processes

$$A_{\rm i} + B \rightleftarrows A_{\rm f} + B ,$$

where the subscript shows a particle state. This leads to the balance equation for the number density of particles in a given state,

$$\frac{dN_{\rm i}}{dt} = \sum_f \left[\overline{k_{\rm fi}(v)}N_{\rm f}[B] - \overline{k_{\rm if}(v)}N_{\rm i}[B]\right] , \tag{2.4}$$

where an overline means averaging over relative velocities of colliding particles A and B. If we take into consideration other processes, new terms may be added to balance equation in the same manner.

► **Problem 2.3** Derive the expression for the differential cross section of elastic scattering of classical particles through the impact parameter of collision.

Considering the two-particle scattering process as the motion of a single particle in the center-of-mass frame, we characterize scattering of this particle by a scattering angle ϑ, and the scattering has a cylindrical symmetry for the isotropic interaction potential $U(R)$ of particles. Let us find the differential scattering cross section, which is the number of scattering events per unit time and unit solid angle divided by the flux of incident particles. Evidently, for a monotonic interaction potential we obtain the monotonic dependence of the scattering angle on the impact parameter of collision of classical particles. Due to the cylindrical symmetry of the scattering process, the elementary solid angle is $d\Theta = 2\pi d\cos\vartheta$, and particles are scattered into this element of solid angle from the range of impact parameters between ρ and $\rho + d\rho$. The particle flux is Nv, where N is the number density of incident particles and v is the relative velocity of collision. The number of particles scattered per unit time into a given solid angle is $2\pi\rho d\rho Nv$, so the differential cross section is (see Fig. 2.1)

$$d\sigma = 2\pi\rho d\rho . \tag{2.5}$$

► **Problem 2.4** A beam of electrons with identical velocities v moves perpendicular to a gas boundary and penetrates in a gas as a result of elastic scattering of electrons in atoms. Analyze the character of penetration of the electron beam in a gas.

Since the electron mass is small compared the atom mass, one can assume atoms to be motionless and consider electron scattering in the laboratory frame of axes

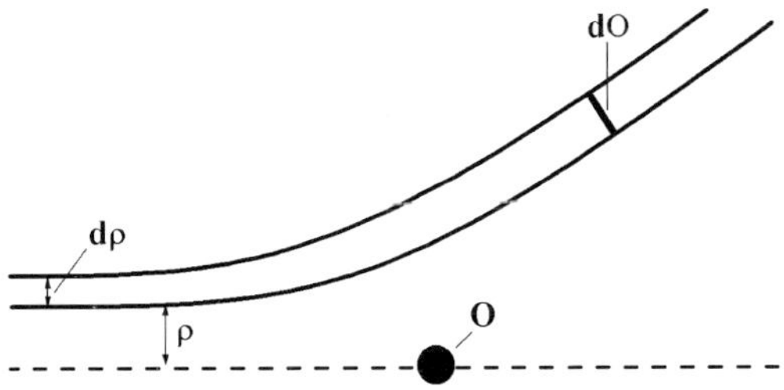

Fig. 2.1 The definition of the differential cross section σ as the ratio of number of scattering acts in a given solid angle to the flux of incident particles. O is the scattering center.

(see Fig. 2.2). If an electron is scattered by an angle ϑ, a variation of its momentum along the beam is $m_e v(1 - \cos\vartheta)$. Correspondingly, the change of the electron flux as the electron penetrates inside the gas is

$$\frac{dj}{dx} = -\frac{1}{v}\frac{dj}{dt} = -j\, N_a \int (1 - \cos\vartheta) d\sigma = -j\, N_a\, \sigma^* ,$$

where x is the direction along the flux, and

$$\sigma^* = \int (1 - \cos\vartheta) d\sigma \tag{2.6}$$

is the diffusion or transport cross section of electron–atom scattering. Rewriting this relation in the form

$$\frac{dj}{dx} = -\frac{j}{\lambda} ,$$

we introduce in this way the mean free path of an electron in a gas as

$$\lambda = (N_a\, \sigma^*)^{-1} . \tag{2.7}$$

As is seen, the mean free path for an electron is expressed through the diffusion cross section of elastic scattering as a cross section of scattering by a large angle.

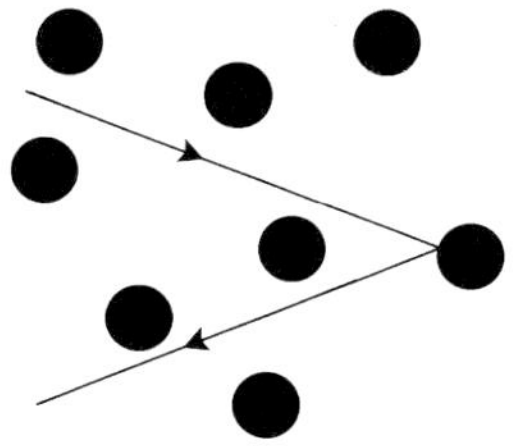

Fig. 2.2 Penetration of a flux of atomic particles in a gas. A sphere is drawn around each atom that corresponds to the collision cross section.

▶ **Problem 2.5** Obtain the condition of gaseousness for an ensemble of weakly interacting particles.

The criterion that an ensemble of atoms is a gas requires the mean free path of particles $\lambda = (N\sigma)^{-1}$ to be large compared to an average distance between particles $N^{-1/3}$, where N is the number density of particles and σ is the cross section of scattering by a large angle in collision of particles. From this we obtain the gaseousness criterion in the form

$$N\sigma^{3/2} \ll 1. \tag{2.8}$$

To derive this criterion, one can require the mean free path of particles $\lambda = (N\sigma)^{-1}$ to be large compared to the interaction radius $\sqrt{\sigma}$. This leads to the gaseousness criterion (2.8).

One can represent the gaseous state condition requiring that the interaction potential of a test particle with its neighbors be small compared to its mean kinetic energy. This criterion has the form

$$U(N^{-1/3}) \ll \varepsilon\,.$$

If this criterion is fulfilled, neighboring particles interact weakly. Comparing this formula with (2.8), we have $U(N^{-1/3}) \ll U(\rho_0)$. Assuming the interaction potential of two particles as a function of distances between the particles to be monotonic, we obtain from this $N^{-1/3} \gg \rho_0$. As a result, this criterion leads to the criterion (2.8).

▶ **Problem 2.6** Obtain the gaseousness condition for a plasma – an ensemble of weakly interacting charged particles.

Let us apply the gaseousness criterion for a plasma, i. e., an ensemble of charged particles. Because of the Coulomb interaction between charged particles $\mid U(R) \mid = e^2/R$, the criterion (2.8) gives

$$\sigma \sim e^4/T^2$$

for a typical large-angle scattering cross section, where the temperature T, which is expressed in energy units, is a typical kinetic energy of the particles. From this we have the criterion of the plasma gaseousness, that is this plasma is ideal. Specifically, T is their temperature. The criterion (2.8) that a particle system is a true gas is transformed in the plasma case to

$$N_e\, e^6/T^3 \ll 1, \tag{2.9}$$

where N_e is a typical number density of charged particles. This criterion of the plasma gaseousness coincides with the plasma ideality criterion (1.56).

2.2 Elastic Collision of Two Particles

▶ **Problem 2.7** Show that in elastic collision of two classical particles their scattering is characterized by the interaction in the center-of-mass frame of reference.

As a result of elastic collision, internal states of particles are not changed. We characterize evolution of colliding particles by their coordinates $\boldsymbol{R}_1$, $\boldsymbol{R}_2$ whose variation in time is described by Newton's equations

$$M_1 \frac{d^2 \boldsymbol{R}_1}{dt^2} = -\frac{\partial U}{\partial \boldsymbol{R}_1}, \quad M_2 \frac{d^2 \boldsymbol{R}_2}{dt^2} = -\frac{\partial U}{\partial \boldsymbol{R}_2}.$$

Here M_1 and M_2 are masses of the colliding particles, and the interaction potential U between the particles depends on the relative distance $\boldsymbol{R} = \boldsymbol{R}_1 - \boldsymbol{R}_2$ between them. In addition, we introduce the coordinate of the center of mass, $\boldsymbol{R}_c = (M_1 \boldsymbol{R}_1 + M_2 \boldsymbol{R}_2)/(M_1 + M_2)$. In new variables, the Newton equations take the form

$$(M_1 + M_2) \frac{d^2 \boldsymbol{R}_c}{dt^2} = 0, \quad \mu \frac{d^2 \boldsymbol{R}}{dt^2} = -\frac{\partial U(R)}{\partial \boldsymbol{R}},$$

where $\mu = M_1 M_2/(M_1 + M_2)$ is the reduced mass of the particles.

As follows from these equations, the center of mass travels with a constant velocity, and the relative motion of the particles determines their scattering. As a result, the problem of collisions of two particles reduces to the problem of the motion of one particle with a reduced mass in the center-of-mass frame of axes. Although the analysis is made for classical particles, this conclusion holds true in the quantum case also because the variables are separated for the Schrödinger equation in the same manner.

▶ **Problem 2.8** Show that the problem of elastic collision of two particles in the absence of an external field is reduced to the problem of scattering of one reduced particle in a force center in the quantum case.

Taking the interaction potential of the two particles as $U(\boldsymbol{R}_1 - \boldsymbol{R}_2)$, we have the following Schrödinger equation for colliding particles:

$$\left[-\frac{\hbar^2}{2m_1}\Delta_1 - \frac{\hbar^2}{2m_2}\Delta_2 + U(\boldsymbol{R}_1 - \boldsymbol{R}_2)\right] \Psi(\boldsymbol{R}_1, \boldsymbol{R}_2) = E\Psi(\boldsymbol{R}_1, \boldsymbol{R}_2).$$

The variables in this equation are separated if we introduce the same combinations of distances as we use in the classical case. Namely, we use the coordinate of the center of mass, $\boldsymbol{R}_c = (m_1 \boldsymbol{R}_1 + m_2 \boldsymbol{R}_2)/(m_1 + m_2)$ and the relative distance between the particles, $\boldsymbol{R} = \boldsymbol{R}_1 - \boldsymbol{R}_2$. In these variables the sum of Laplacians is equal to

$$\frac{\hbar^2}{2m_1}\Delta_1 + \frac{\hbar^2}{2m_2}\Delta_2 = \frac{\hbar^2}{2(m_1 + m_2)}\Delta_{R_c} + \frac{\hbar^2}{2\mu}\Delta_R,$$

and we obtain two independent Schrödinger equations. The first equation corresponds to motion of the center of mass for colliding particles with a constant velocity, and the second equation for the relative distance takes into account the interaction of particles, and only this equation is responsible for particle scattering. This equation has the form

$$-\frac{\hbar^2}{2\mu}\Delta_{\boldsymbol{R}}\psi(\boldsymbol{R}) + U(\boldsymbol{R})\psi(\boldsymbol{R}) = \varepsilon\psi(\boldsymbol{R}) ,$$

and $\varepsilon = \hbar^2 q^2/(2\mu)$ is the particle energy in the center-of-mass frame of reference. We represent this equation in the form

$$(\Delta_{\boldsymbol{R}} + q^2)\psi(\boldsymbol{R}) = \frac{2\mu}{\hbar^2}U(\boldsymbol{R})\psi(\boldsymbol{R})$$

and will consider the right-hand side of this equation as a nonlinearity. Based on the Green function of the uniform equation that is given by

$$G(\boldsymbol{R}, \boldsymbol{R}') = -\frac{\exp\left(iq\left|\boldsymbol{R} - \boldsymbol{R}'\right|\right)}{4\pi\left|\boldsymbol{R} - \boldsymbol{R}'\right|} ,$$

one can formally write the solution of this equation as

$$\psi(\boldsymbol{R}) = C\left[e^{i\boldsymbol{q}\boldsymbol{R}} - \frac{\mu}{2\pi\hbar^2}\int\frac{\exp\left(iq\left|\boldsymbol{R} - \boldsymbol{R}'\right|\right)}{\left|\boldsymbol{R} - \boldsymbol{R}'\right|}U(\boldsymbol{R}')\psi(\boldsymbol{R}')d\boldsymbol{R}'\right] .$$

As a matter of fact, it is an integral form of the Schrödinger equation. But this representation of the wave function is convenient, because it allows us to transfer to the limit of large R. In the limit $R \to \infty$ we have $\left|\boldsymbol{R} - \boldsymbol{R}'\right| = R - \boldsymbol{R}'\boldsymbol{n}$, where $\boldsymbol{n}$ is the unit vector directed along $\boldsymbol{R}$. The expansion of the above equation in the limit of large R can be represented as

$$\psi(\boldsymbol{R}) = C\left[e^{i\boldsymbol{q}\boldsymbol{R}} + f(\vartheta)\frac{e^{iqR}}{R}\right] , \tag{2.10}$$

and this is a general expression for the wave function far from the scattered center. Here the first term is an incident wave, the second term is a scattered wave, the value $f(\vartheta)$ is the scattering amplitude, which is responsible for particle scattering, and ϑ is the scattering angle between the vectors $\boldsymbol{R}$ and $\boldsymbol{q}$. This wave function is normalized such that it tends to $e^{i\boldsymbol{q}\boldsymbol{R}}$ when $R \to \infty$. As follows from the above expression for the wave function, the scattering amplitude is given by

$$f(\vartheta) = -\frac{\mu}{2\pi\hbar^2}\int\exp(-iq\boldsymbol{n}\boldsymbol{R}')U(\boldsymbol{R}')\psi(\boldsymbol{R}')d\boldsymbol{R}'. \tag{2.11}$$

This expression does not allow us to find the scattering amplitude for a given interaction potential since it includes the accurate wave function $\psi(\boldsymbol{R})$ of particles. Nevertheless, this form of the scattering amplitude is convenient for approximations.

► **Problem 2.9** Express the differential cross section of elastic scattering through the scattering amplitude.

We obtain this expression from the analysis of formula (2.11). The first term of this formula is a plane wave, i. e., the particle wave function without scattering, and the second term is responsible for scattering. Let us obtain from this the flux of incident particles and the rate of scattering particles. The flux of incident particles is

$$j = \frac{\hbar}{2mi}\left(\psi^* \nabla \psi - \psi \nabla \psi^*\right) = |C|^2 \boldsymbol{v} \,,$$

where $\boldsymbol{v} = \hbar \boldsymbol{q}/m$ is the particle velocity. In the same manner we find the flux of scattering particles, which is determined by the second term. Multiplying it by $R^2 d\Theta$, we find the scattering rate in the solid angle $d\Theta$, and dividing by the flux of incident particles, we find the cross section of scattering $d\sigma$ in an element $d\Theta$ of the solid angle, which is given by

$$d\sigma = \left|f(\vartheta)\right|^2 d\Theta. \tag{2.12}$$

Thus, the differential cross section of elastic scattering is connected with the scattering amplitude in a simple way.

► **Problem 2.10** Find the amplitude of elastic scattering of particles within the framework of the perturbation theory.

Using the perturbation theory, we insert in formula (2.11) the wave function of particles in the absence of interaction, when the wave function has the form of a plane wave $\psi(\boldsymbol{R}) = e^{i\boldsymbol{q}\boldsymbol{R}}$. This gives for the scattering amplitude

$$f(\vartheta) = -\frac{\mu}{2\pi\hbar^2}\int e^{-i\boldsymbol{K}\boldsymbol{R}}\,U(\boldsymbol{R}')d\boldsymbol{R}'. \tag{2.13}$$

Here $\boldsymbol{K} = q\boldsymbol{n} - \boldsymbol{q}$ is the variation of the wave vector for the relative motion of particles due to collision, and formula (2.13) is named the Born approximation. The relation between the wave vector variation $\boldsymbol{K}$ and the scattering angle ϑ is $K = 2q\sin(\vartheta/2)$.

The Born approximation is valid if the interaction potential is small compared to the kinetic energy for distances between colliding particles, $R \sim 1/q$, where scattering proceeds. This gives the following criterion for the Born approximation:

$$U\left(\frac{1}{q}\right) \ll \varepsilon \,. \tag{2.14}$$

2.3 Elastic Scattering of Classical Particles

► **Problem 2.11** Represent the relation between the impact parameter of collision and the distance of the closest approach for elastic scattering of particles when their interaction potential is isotropic.

Thus scattering of colliding particles can be represented as a motion of a single particle of the reduced mass μ of colliding particles in the center-of-mass frame of reference, where this particle is moving in the central interaction potential. Figure 2.3 gives the trajectory of classical particles in the center-of-mass system when the central interaction potential is isotropic, i. e., it depends only on the scalar distance between the particles $| \mathbf{R}_1 - \mathbf{R}_2 |$. The basic parameters of the collision are the impact parameter ρ and the distance of the closest approach r_0. We determine below the relation between these parameters.

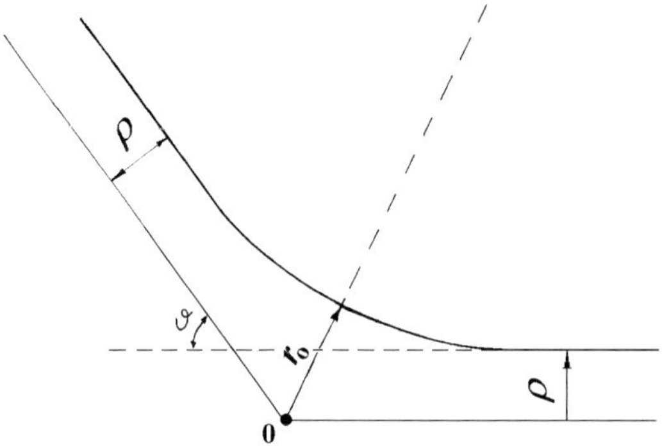

Fig. 2.3 Trajectory for relative motion of particles in the center-of-mass frame of reference and collision parameters: 0 is a position of a scattering center, ρ is the impact parameter of collision, r_0 is the distance of closest approach, ϑ is the scattering angle.

Let us use conservation of the momentum of a reduced particle moving in the central field if $U = U(R)$. Its momentum at large distances between the particles is $\mu\rho v$ and coincides with the momentum $\mu v_\tau r_0$ at the distance of the closest approach, where $v =| \mathbf{v}_1 - \mathbf{v}_2 |$ is the relative velocity of the particles, and v_τ is the tangential component of the velocity at the distance of the closest approach. Since the normal component of the velocity is zero at the distance of the closest approach, we have from the energy conservation that $\mu v_\tau^2/2 = \mu v^2/2 - U(r_0)$. From this we obtain the expression

$$1 - \frac{\rho^2}{r_0^2} = \frac{U(r_0)}{\varepsilon}, \tag{2.15}$$

where $\varepsilon = \mu v^2/2$ is the kinetic energy of the particles in the center-of-mass frame of reference.

▶ **Problem 2.12** Determine the cross section of collision when colliding particles, e. g., an electron or a single charged ion and a spherical cluster of radius r_0 and charge Z, are contacted.

The colliding particles are contacted if the distance of the closest approach in their collision does not exceed the cluster radius. According to formula (2.15) this takes place if the impact parameter of collision does not exceed the value ρ_0, which follows from the relation

$$\rho_0^2 = r_0^2 \left[1 + \frac{U(r_0)}{\varepsilon}\right] = r_0^2 + \frac{Ze^2 r_0}{\varepsilon} ,$$

where we take the interaction potential of opposite charged particles to be $U(R) = -Ze^2/R$. This leads to the cross section of collisions when colliding particles are contacted, i. e., an electron or ion reaches the cluster surface,

$$\sigma = \pi\rho_0^2 + \pi r_0 \frac{Ze^2}{\varepsilon} . \tag{2.16}$$

▶ **Problem 2.13** For a spherical interaction potential $U(R)$ of classical particles determine the angle of particle scattering at a given impact parameter of collision.

Separating the velocity of a reduced particle moving in a central field with tangential and normal motion, we obtain from the conservation of the total energy of normal components

$$\frac{\mu v_R^2}{2} + \frac{\mu v_\tau^2}{2} = \varepsilon - U(R) ,$$

where $v_R = dR/dt$ and $v_\tau = v\rho/R = Rd\vartheta/dt$ are the normal and tangential velocity components. This relation allows us to determine evolution of the trajectory angle as

$$\frac{d\vartheta}{dR} = \frac{v_\tau}{v_R} = \frac{\rho}{R^2\sqrt{1 - \frac{\rho^2}{R^2} - \frac{U(R)}{\varepsilon}}} .$$

Solving this equation with taking into accounting the symmetry with respect the time reversal $R(t) = R(-t)$ and the boundary conditions $R(\infty) = \infty$, $R(0) = r_0$, we find the total scattering angle

$$\vartheta = \pi - 2\int_{r_0}^{\infty} \frac{\rho dR}{R^2\sqrt{1 - \frac{\rho^2}{R^2} - \frac{U(R)}{\varepsilon}}} . \tag{2.17}$$

According to this $\vartheta = 0$ at $U(R) = 0$. This formula gives the scattering angle in the classical case.

▶ **Problem 2.14** Estimate the cross section of scattering by a large angle.

Processes in ensembles of weakly interacting particles determine an equilibrium of this ensemble resulted from particle interaction that proceeds in particle collisions. For classical particles their behavior inside a gas is described by the trajectories of particles, and the curvature of trajectories is of importance for establishing an equilibrium and some properties of the ensemble. As a result, a remarkable change of the direction of a particle trajectory is determined by strong collisions with surrounding particles which are accompanied by scattering by a large angle in collisions of the particles. Thus, the typical cross section of large-angle scattering is of importance for some properties of an ensemble.

Let us estimate a typical cross section for elastic scattering by large angles. In this case the interaction potential at the distance of the closest approach is comparable to the kinetic energy of the colliding particles, and this cross section is given by the relation

$$\sigma = \pi\rho_o^2, \qquad \text{where} \qquad U(\rho_o) \sim \varepsilon \,. \tag{2.18}$$

In particular, for the interaction potential $U(R) = AR^{-k}$ the cross section of scattering by a large angle according to this formula is given by

$$\sigma \approx \pi \left(\frac{A}{\varepsilon}\right)^{2/k} . \tag{2.19}$$

► **Problem 2.15** Find the differential and diffusion cross section for collision of particles within the framework of the hard sphere model.

The interaction potential between particles for the hard sphere model has the form of a hard infinite wall (see Fig. 2.4), which is given by the relation

$$U(R) = 0,\ r > R_0,\ U(R) = \infty,\ r < R_0, \tag{2.20}$$

where R_0 is the hard sphere radius. The dependence of the distance of the closest approach on the impact parameter is given in Fig. 2.5, and Fig. 2.6 shows the character of scattering at such a collision. According to Fig. 2.6, the scattering angle in this collision is equal to $\vartheta = \pi - 2\alpha$, and $\sin\alpha = \rho/R_0$, i. e., $\rho = R_0\cos(\vartheta/2)$. This

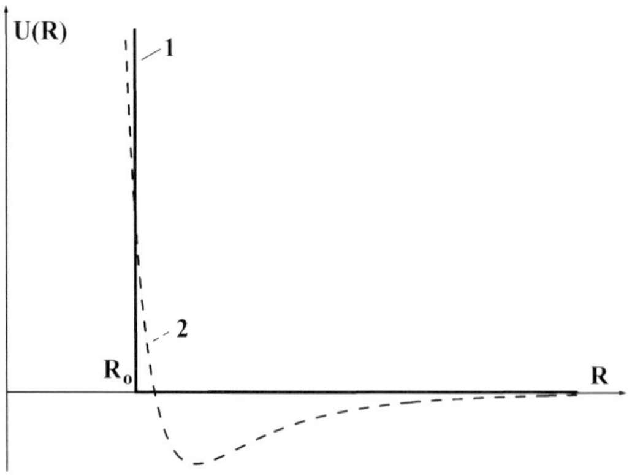

Fig. 2.4 The interaction potential of colliding particles as a function of a distance between them: dotted line corresponds to the interaction potential, a solid line is an approximation of this interaction potential for the hard sphere model.

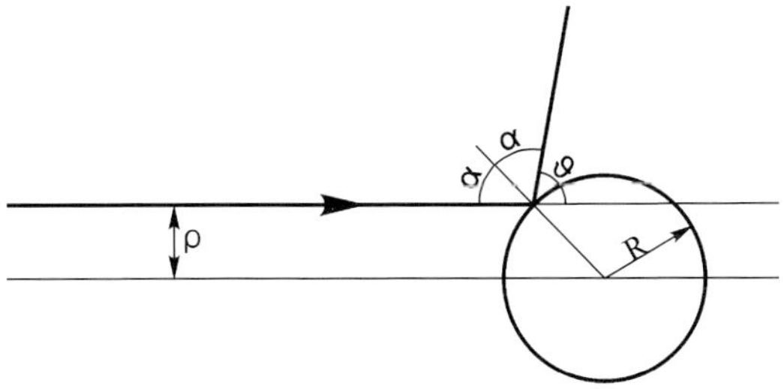

Fig. 2.5 The character of particle scattering for the hard sphere model: $R = r_0$ is the hard sphere radius, ρ is the impact parameter of collision, ϑ is the scattering angle.

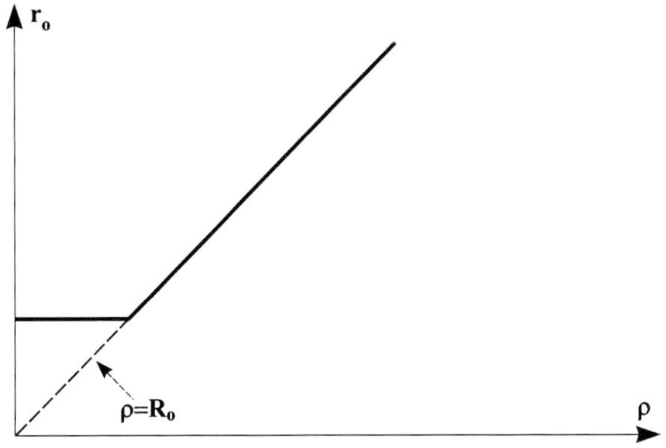

Fig. 2.6 The dependence of the distance of closest approach r_0 on the impact parameter of collision ρ for the hard sphere model.

leads to the following differential cross section:

$$d\sigma = 2\pi\rho d\rho = \frac{\pi R_0^2}{2} d\cos\vartheta\,, \tag{2.21}$$

and the diffusion cross section for the hard sphere model is equal to

$$\sigma^* = \int (1 - \cos\vartheta) d\sigma = \int_{-1}^{1} \frac{\pi R_0^2}{2} (1 - \cos\vartheta) d\cos\vartheta = \pi R_0^2. \tag{2.22}$$

▶ **Problem 2.16** Determine the capture cross section for collision of particles interacting through the potential $U(R) = -CR^{-k}$.

The dependence of the impact parameter on the distance of the closest approach for the above interaction potential is determined on the basis of formula (2.15) and is represented in Fig. 2.7. One can see two branches for this dependence $\rho(r_0)$ that are separated by the minimum. Region 2 is stable, whereas region 1 corresponds

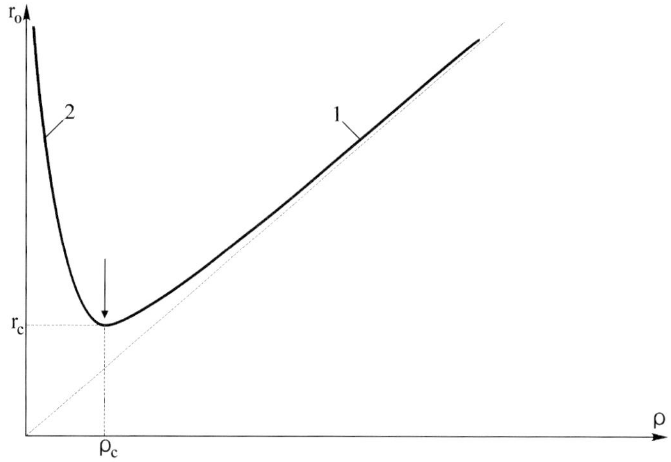

Fig. 2.7 The dependence of the distance of closest approach r_0 on the impact parameter of collision ρ for the the interaction potential $U(R) = -CR^{-k}$, so that this dependence for collision of free particles is given by curve 1, and motion of bound particles corresponds to curve 2. ρ_c is the impact parameter of collision below which the capture of particles takes place, R_c the distance of closest approach for this impact parameter.

to the bound state of particles, and then ρ characterizes the particle momentum in the bound state. Regions 1 and 2 are separated by the impact parameter ρ_c and the distance r_c of the closest approach. For collisions at the impact parameters below ρ_c the distance of the closest approach is zero. In reality the interaction potential at small distances between atomic particles corresponds to repulsion because of the exchange interaction, and therefore the distance of the closest approach is determined as $U(r_{\min}) = 0$, but at thermal collisions $r_{\min} \ll r_c$.

As for the interaction potential of particles under consideration $U(R) = -CR^{-k}$, the distance of the closest approach follows from formula (2.15) at $\rho \geq \rho_c$ and is zero at $\rho < \rho_c$. Therefore at small impact parameters the capture of particles takes place, which leads to the approach of particles up to zero distance, if the interaction potential corresponds to attraction. As follows from formula (2.15), the distance of the closest approach r_0 and the impact parameter ρ_c at this distance are given by

$$r_c = \left[\frac{C(k-2)}{2\varepsilon}\right]^{1/k}, \quad \rho_c = \left(\frac{k}{k-2}\right)^{1/2}\left[\frac{C(k-2)}{2\varepsilon}\right]^{1/k}. \tag{2.23}$$

From this it follows that the capture cross section

$$\sigma_c = \pi\rho_c^2 = \frac{\pi k}{k-2}\left[\frac{C(k-2)}{2\varepsilon}\right]^{2/k}. \tag{2.24}$$

Let us consider the case of polarization interaction of an ion and atom, when the interaction potential is $U(R) = -\alpha e^2/(2R^4)$, where α is the atomic polarizability.

Then the polarization cross section for capture of an atom by an ion is equal to

$$\sigma_c = 2\pi \left(\frac{\alpha e^2}{\mu v^2} \right)^{1/2} . \qquad (2.25)$$

▶ **Problem 2.17** Obtain the criterion of reality for the capture cross section determined in a real situation.

A real dependence for the pair interaction potential is given in Fig. 2.8. This interaction potential is characterized by the minimum

$$U(R_e) = -D$$

at a distance R_e between atomic particles. Evidently, the capture cross section (2.24) can be used in the case of this interaction potential if the collision energy ε satisfies the criterion

$$\varepsilon \ll D .$$

Then scattering of atomic particles is determined by a long-range part of the interaction potential that corresponds to distances between atomic particles $R > R_e$. Then, approximating the interaction potential by the dependence $U(R) = -CR^{-k}$, one can use formula (2.24) for the capture cross section. For this approximation we take R_0 from the relation $|U(R_0)| = \varepsilon$, and the parameter k of the interaction potential is

$$k = -\frac{R_0 U'(R_0)}{U(R_0)} . \qquad (2.26)$$

Figure 2.9 shows the dependence $r_0(\rho)$ for a real interaction potential, and capture of particles takes place at low collision energies.

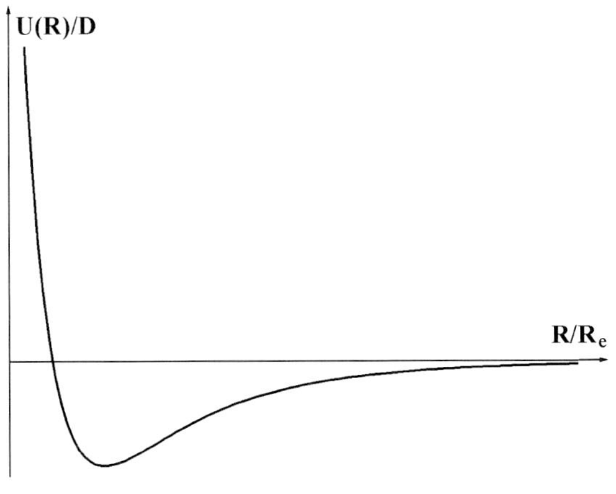

Fig. 2.8 A typical interaction potential of two atomic particles as a function of a distance between them.

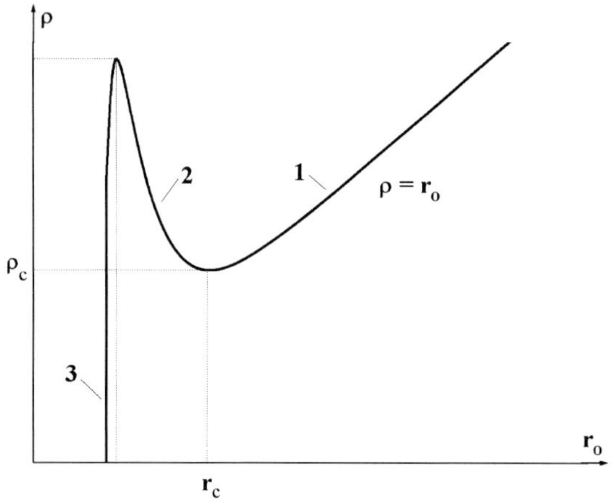

Fig. 2.9 The dependence of the distance of closest approach r_0 on the impact parameter of collision ρ for a typical interaction potential of two atomic particles. 1, 2 — free motion of colliding particles if the impact parameter is above and below ρ_c, so that particles are scattered or captured. Region 3 relates to bound particles.

▶ **Problem 2.18** Find the cross section of mutual neutralization in collisions of positive and negative ions within the framework of the model according to which the exchange event takes place if the distance between colliding particles is less than R_0.

This process of electron transition proceeds according to the scheme

$$A^+ + B^- \rightarrow A^* + B \tag{2.27}$$

and results in the electron transition from the atom field B to the field of the positive ion A^+. This process proceeds at enough strong interaction when the distance between colliding particles is not large. For the above model, the cross section of this process according to formula (2.15) is given by

$$\sigma = \pi R_0^2 + \frac{\pi R_0 e^2}{\varepsilon}, \tag{2.28}$$

where we take the interaction potential of colliding particles as $U(R) = e^2/R$.

▶ **Problem 2.19** Determine the scattering angle for collision of particles with a sharply varied repulsive interaction potential.

We use below a small parameter $1/k$, where $k = -d \ln U(R)/dR$. In the zero-th approximation we obtain the hard sphere model for an interaction potential. Then the distance of the closest approach is $r_0 = \rho$, $\rho \geq R_0$; $r_0 = R_0$, $\rho \leq R_0$, and according to formula (2.17) the scattering angle at $\rho \leq R_0$ is equal to

$$\vartheta = \pi - 2 \arcsin \frac{\rho}{r_0} .$$

Representing the first approximation in the form

$$\vartheta = \pi - 2\arcsin\frac{\rho}{r_0} + 2\Delta\vartheta\,,$$

we have

$$\Delta\vartheta = \int_{r_0}^{\infty}\frac{\rho dR}{R^2}\left(\frac{1}{\sqrt{1-\frac{\rho^2}{R^2}}} - \frac{1}{\sqrt{1-\frac{\rho^2}{R^2}-\frac{U(R)}{\varepsilon}}}\right).$$

The above expression for the scattering angle is exact, and $\Delta\vartheta \sim 1/k$, i. e., proportional to a small parameter of the problem. We determine the first expansion term of $\Delta\vartheta$ over $1/k$. In order to avoid the divergence in the integral for $\Delta\vartheta$, we use the relation

$$-\frac{d}{d\rho}\int_{r_0}^{\infty} dR\left(\sqrt{1-\frac{\rho^2}{R^2}-\frac{U(R)}{\varepsilon}} - \sqrt{1-\frac{\rho^2}{R^2}}\right)$$
$$= -\frac{dr_0}{d\rho}\sqrt{1-\frac{\rho^2}{R^2}} + \int_{r_0}^{\infty}\frac{\rho dR}{R^2}\left(\frac{1}{\sqrt{1-\frac{\rho^2}{R^2}}} - \frac{1}{\sqrt{1-\frac{\rho^2}{R^2}-\frac{U(R)}{\varepsilon}}}\right).$$

From this it follows that

$$\Delta\vartheta = \frac{dr_0}{d\rho}\sqrt{1-\frac{\rho^2}{R^2}} - \frac{d}{d\rho}\int_{r_0}^{\infty} dR\left(\sqrt{1-\frac{\rho^2}{R^2}-\frac{U(R)}{\varepsilon}} - \sqrt{1-\frac{\rho^2}{R^2}}\right).$$

Taking into account that the above integral converges near $R = r_0(R - r_0 \sim 1/k)$, one can calculate this integral with the accuracy of the order of $1/k$:

$$-\int_{r_0}^{\infty} dR\left(\sqrt{1-\frac{\rho^2}{R^2}-\frac{U(R)}{\varepsilon}} - \sqrt{1-\frac{\rho^2}{R^2}}\right) \approx \sqrt{1-\frac{\rho^2}{r_0^2}}\int_{r_0}^{\infty} dR\left[1-\sqrt{1-\frac{r_0^k}{R^k}}\right]$$
$$= \sqrt{r_0^2-\rho^2}\int_0^1 \frac{1-\sqrt{1-x}}{kx^{1+1/k}}dx$$
$$= \frac{2}{k}(1-\ln 2)\sqrt{r_0^2-\rho^2},$$

where

$$x = \frac{U(R)}{\varepsilon}\frac{1}{\sqrt{1-\frac{\rho^2}{R^2}}} = \frac{r_0^k}{R^k}\,.$$

From this we have

$$\Delta\vartheta = \frac{dr_0}{d\rho}\sqrt{1-\frac{\rho^2}{R^2}} + \frac{2}{k}(1-\ln 2)\frac{d}{d\rho}\sqrt{r_0^2-\rho^2}\,.$$

Let us introduce the value $u = U(r_0)/\varepsilon = 1 - \rho^2/r_0^2$. Since $k \gg 1$ and in the scattering region the value u is close to unity, we obtain from the above formulas

$$\vartheta = 2\arcsin\sqrt{u} + 2\left[\frac{2-(k-2)\ln 2}{k}\right]\frac{\sqrt{u(1-u)}}{1+(k-2)u/2}$$
$$\approx 2\arcsin\sqrt{u} - 2\ln 2\frac{\sqrt{u(1-u)}}{1+ku/2}, \qquad (2.29)$$

where $k = -d\ln u/dr_0$. We conserve 1 in the denominator of this formula in order to have the possibility for expansion of this formula at small scattering angles.

▶ **Problem 2.20** Determine the diffusion cross section of elastic scattering of particles as a result of expansion over a small parameter for scattering of particles with a sharply varied repulsion interaction potential.

According to the definition of the diffusion or transport cross section (2.6) and taking into account that collisions with $u \sim 1$ give the main contribution to the diffusion cross section, we have on the basis of formula (2.29) the scattering angle

$$\vartheta = 2\arcsin\sqrt{u} - \frac{4\ln 2}{k}\sqrt{\frac{1-u}{u}}\,.$$

From this we find the diffusion cross section

$$\sigma^* = 2\pi\int_0^1 u\left[(1-u)dr_0^2 - r_0^2du\right] + \pi\int_0^1 \frac{4\ln 2}{k}\sqrt{\frac{1-u}{u}}\cdot 2\sqrt{u(1-u)}r_0^2du\,.$$

Since $u = 1 - \rho^2/r_0^2$, we use the relation $d\rho^2 = (1-u)dr_0^2 - r_0^2du$. Taking into account that the first term is $\sim k$ times less than the second one, we take the second integral on parts. This gives

$$-2\pi\int_0^1 ur_0^2du = \pi\int_0^1 r_0^2du^2 = \pi r_0^2\Big|_0^1 - \pi\int_0^1 u^2dr_0^2$$
$$= \pi R_0^2 + \frac{2\pi}{k}\int_0^1 r_0^2udu = \pi R_0^2\left(1+\frac{1}{k}\right),$$

where $u(R_0) = 1$. We use $u \sim r_0^{-k}$, so that $dr_0/r_0 = -du/(ku)$. Note that $\rho = 0$ corresponds to $u = 1$, and $\rho = \infty$ corresponds to $u = 0$. Repeating these steps for the other integral and keeping only terms of the order of $1/k$, we finally find

$$\sigma^* = \pi R_0^2\left(1+\frac{3-4\ln 2}{k}\right).$$

If we represent the diffusion cross section in the form $\sigma^* = \pi R_1^2$, we have $R_1 = R_0 + (3-4\ln 2)/(2k)$. This gives

$$u(R_1) = (R_0/R_1)^k = \exp(-3/2 + 2\ln 2) = 4\exp(-3/2) = 0.89\,.$$

Thus, the diffusion cross section has the form

$$\sigma^* = \pi R_1^2, \quad \text{where} \quad \frac{U(R_1)}{\varepsilon} = 0.89. \tag{2.30}$$

► **Problem 2.21** Determine the diffusion cross section of collision for particles with the polarization potential of interaction $U(R) = -\alpha e^2/(2R^4)$.

Formula (2.29) for the scattering angle takes the form in this case

$$\vartheta = \pi - 2\int_{r_0}^{\infty} \frac{\rho dR}{R^2\sqrt{1 - \frac{\rho^2}{R^2} + \frac{\alpha e^2}{2R^4\varepsilon}}}.$$

Here the distance of the closest approach r_0 is zero if the impact parameter of collision $\rho \le \rho_c$, where $\rho_c = \sqrt{2}[\alpha e^2/(2\varepsilon)]^{1/4}$ according to formula (2.15) and it is given by formula (2.15) for $\rho > \rho_c$. We divide the range of impact parameters into two parts, $\rho \le \rho_c$ and $\rho > \rho_c$. Introducing the reduced variables $x = [\alpha e^2/(2\varepsilon R^4)]^{1/4}$ and $y = \rho/\rho_c$, we obtain for the diffusion cross section

$$\sigma^* = \sigma_c (1 + J_1 + J_2),$$

where $\sigma_c = \pi\rho_c^2 = 2\pi\sqrt{\alpha e^2/\varepsilon}$ according to formula (2.25) is the cross section of capture of an atom in the polarization interaction potential, the integrals are in the expression for the diffusion cross section

$$J_1 = \int_0^1 2y dy \cos\int_0^{\infty} \frac{2y\sqrt{2}dx}{\sqrt{1 + x^4 - 2x^2y^2}},$$

$$J_2 = \int_1^{\infty} 2y dy \left(1 + \cos\int_0^{x_0} \frac{2y\sqrt{2}dx}{\sqrt{1 + x^4 - 2x^2y^2}}\right),$$

and the distance of the closest approach is $x_0 = y^2 - \sqrt{y^4 - 1}$. These integrals are calculated by numerical methods and are equal to $J_1 = -0.101$, $J_2 = 0.207$. Thus the diffusion cross section of scattering in the polarization interaction potential is

$$\sigma^* = 1.10\sigma_c, \tag{2.31}$$

i. e., the contribution to the diffusion cross section from collisions with the impact parameters larger ρ_c is approximately 10 %.

► **Problem 2.22** Determine the diffusion cross section of two charged particles in an ideal plasma that satisfies the criterion (2.9).

The interaction potential of two charged particles of the same charge e according to formula (1.55) is given by

$$U = e\varphi = \frac{e^2}{r}\exp\left(-\frac{r}{r_D}\right), \tag{2.32}$$

where r is the distance between charged particles, r_D is the Debye–Hückel radius (1.55) and for an ideal plasma under consideration (2.9) it is large compared to an average distance between charged particles in a plasma. Therefore, a pair collision of charged particles in an ideal plasma is determined by the Coulomb interaction potential ($U = e^2/r$) between colliding particles.

The diffusion cross section of collision of two charged particles in a plasma is determined by small scattering angles, so that it is equal to

$$\sigma^* = \int (1 - \cos\vartheta) 2\pi\rho d\rho \approx \int \vartheta^2 \pi\rho d\rho \,, \tag{2.33}$$

where ρ is the impact parameter of the collision. The scattering angle is $\vartheta = \Delta p/p$, where Δp is the variation of the particle momentum, $p = \mu g$ is the momentum of the colliding particles in the center-of-mass frame, meaning that μ is the reduced mass of the particles and g is the relative velocity of the particles. We have for the variation Δp of the relative momentum of colliding particles at scattering at small angles when the trajectory of colliding particles in the center-of-mass frame of reference is straightforward,

$$\Delta \boldsymbol{p} = \int_{-\infty}^{\infty} \boldsymbol{F} dt \,,$$

where $\boldsymbol{F} = e^2\boldsymbol{n}/r^2$ is the force acted on the reduced particle moving in the Coulomb field of the force center, and $\boldsymbol{n}$ is the unit vector in the direction joining the force center and the reduced particle. We have

$$\Delta \boldsymbol{p} = \int_{-\infty}^{\infty} \frac{e^2 \boldsymbol{\rho}}{r^3} dt = \frac{2e^2 \boldsymbol{n}}{\rho v} \,, \tag{2.34}$$

where $\boldsymbol{n}$ is the unit vector directed along the impact parameter of collision $\boldsymbol{\rho}$, and we use the relation for a straightforward trajectory $r^2 = \rho^2 + v^2 t^2$. From this we obtain for the scattering angle

$$\vartheta = \frac{e^2}{\varepsilon\rho} \,, \tag{2.35}$$

where $\varepsilon = \mu g^2/2$ is the particle energy in the center-of-mass frame of reference.

Substituting this expression in formula (2.33) for the diffusion cross section of charged particles, we obtain

$$\sigma^* \approx \int \vartheta^2 \pi\rho d\rho = \frac{\pi e^4}{\varepsilon^2} \int \frac{d\rho}{\rho} \,. \tag{2.36}$$

This integral diverges in limits of both the small and large impact parameters of collision. The divergence at small impact parameters is due to violation of the assumption of small scattering angles. This limit should really correspond to $\vartheta \sim 1$, or $\rho_{\min} \sim e^2/\varepsilon$. The divergence at large impact parameters is caused by the infinite range of unscreened Coulomb interaction potential of charged particles in a vacuum, e^2/r. Replacing it by the interaction potential (2.32) in a plasma, we obtain the Debye–Hückel radius as the upper limit of integration. As a result, we have for the diffusion cross section for the scattering of two charged particles

$$\sigma^* = \frac{\pi e^4}{\varepsilon^2} \ln \Lambda, \quad \Lambda = r_D e^2/\varepsilon \,, \tag{2.37}$$

where the quantity $\ln \Lambda$ is the so-called Coulomb logarithm. According to its definition, $\Lambda \gg 1$, and the accuracy of $\ln \Lambda$ is determined by a factor of the order of 1. Thus, the accuracy of formula (2.37) improves with an increase in the Coulomb logarithm.

▶ **Problem 2.23** Within the framework of the liquid drop model determine the rate of atom attachment to the cluster surface from a surrounding vapor. Consider atom attachment to the cluster surface as a result of an atom–cluster contact.

Within the framework of the liquid drop model, the cluster M_n, which is a system of n bound atoms M, is considered as a bulk liquid drop of spherical form whose density coincides with the density of the bulk liquid. Then the radius r_n of such a cluster is connected with the number of cluster atoms n by the relation

$$r_n = r_W \cdot n^{1/3}, \quad r_W = \left(\frac{3m}{4\pi\rho}\right)^{1/3}, \tag{2.38}$$

where r_W is the Wigner–Seitz radius, m is the mass of an individual atom, and ρ is the density of the bulk liquid.

The process of atom attachment is

$$M_n + M \to M_{n+1} \,,$$

and we assume each contact of atom with the cluster surface leads to its sticking. Then the process of atom–cluster collision with atom attachment to the cluster surface is determined by the cross section $\sigma_n = \pi r_n^2 = \pi r_W^2 n^{2/3}$, and the rate of atom attachment, i. e., the atom flux to the cluster surface, is equal to

$$\nu_n = N v \sigma_n = N k_o n^{2/3}, \quad \text{where} \quad k_o = \sqrt{\frac{8T}{\pi m}} \pi r_W^2 . \tag{2.39}$$

Here v is the average atom velocity, T is the gaseous temperature, and m is the atom mass.

2.4 Phase Theory of Particle Elastic Scattering

▶ **Problem 2.24** A particle is scattered at a force center, and the force is spherically symmetric. Introduce the scattering phases as characteristics of scattering by expansion of the particle wave function over spherical harmonics.

Let us consider formula (2.11) for the asymptotic wave function of a scattering particle far from the scattered center, which has the form

$$\psi(\mathbf{R}) = C\left[e^{iqR} + f(\vartheta)\frac{e^{iqR}}{R}\right],$$

where $\mathbf{R}$ is the particle coordinate with the force center as the origin of the frame of reference, $\mathbf{q}$ is the particle wave vector, ϑ is the scattering angle, and $f(\vartheta)$ is the scattering amplitude.

Let us expand the particle wave function over spherical harmonics. For quantities of this formula we have

$$\psi(\mathbf{R}) = \frac{1}{R}\sum_{l=0}^{\infty} A_l \varphi_l(R) P_l(\cos\vartheta), \tag{2.40}$$

$$e^{iqR} = \sqrt{\frac{\pi}{2qR}}\sum_{l=0}^{\infty} i^l(2l+1) J_{2l+1/2}(qR) P_l(\cos\vartheta), \tag{2.41}$$

$$f(\vartheta) = \sum_{l=0}^{\infty} f_l P_l(\cos\vartheta). \tag{2.42}$$

Here $P_l(\cos\vartheta)$ is the Legendre polynomial, $J_{2l+1/2}(qR)$ is the Bessel function whose asymptotic form at large values of the argument has the following form:

$$J_{2l+1/2}(x) = \sqrt{\frac{2}{\pi x}}\sin(x - \pi l/2)$$

The particle radial wave functions are solutions of the Schrödinger equations

$$\frac{d^2\varphi_l}{dR^2} + \left[q^2 - \frac{2\mu U(R)}{\hbar^2} - \frac{l(l+1)}{R^2}\right]\varphi_l = 0, \tag{2.43}$$

and their asymptotic form is

$$\varphi_l(R) = \frac{1}{q}\sin(qR - \frac{\pi l}{2} + \delta_l). \tag{2.44}$$

The quantities δ_l are the scattering phases which describe scattering of particles. Expansions (2.41) and (2.42) are expressed through the scattering phases by the relations

$$A_l = e^{i\delta_l}(2l+1)i^l, \; f_l = \frac{1}{2iq}(2l+1)(e^{2i\delta_l} - 1). \tag{2.45}$$

Because the coefficients f_l characterize the scattering amplitudes, the scattering phases δ_l describe scattering of a particle in a given central field.

▶ **Problem 2.25** Within the framework of the phase theory express the diffusion and total cross section of scattering through scattering phases.

Let us use expression (2.12) for the differential cross section of elastic scattering, expansion (2.42) of the scattering amplitude over spherical harmonics and the diffusion σ^* and total σ_t cross sections of particle scattering. As a result, we obtain for the total and diffusion cross sections of scattering of atomic particles

$$\sigma^* = \int_0^1 |f(\vartheta)|^2 (1 - \cos\vartheta) 2\pi d\cos\vartheta = \frac{4\pi}{q^2} \sum_{l=0}^{\infty} (2l+1) \sin^2(\delta_l - \delta_{l+1}) \tag{2.46}$$

$$\sigma_t = \int_0^1 |f(\vartheta)|^2 2\pi d\cos\vartheta = \frac{4\pi}{q^2} \sum_{l=0}^{\infty} (2l+1) \sin^2\delta_l . \tag{2.47}$$

► **Problem 2.26** Obtain expressions for the scattering phases in the Born approximation.

Let us use relation (2.11) and expand it over spherical harmonics. Separating l-spherical harmonic, we obtain for the scattering phase the following equation:

$$\sin\delta_l = -\frac{\mu}{\hbar^2}\sqrt{2\pi q} \int_0^{\infty} \sqrt{R} J_{l+1/2}(qR) \varphi_l(R) U(R) dR . \tag{2.48}$$

Since $\varphi_l(R)$ is the correct particle wave function, we obtain an equation for the radial wave function. In the Born approximation one can change the radial wave function by its expression in the absence of interaction, when the radial wave function has the form

$$\varphi_l(R) = \sqrt{\frac{\pi R}{2q}} J_{l+1/2}(qR) .$$

This gives the following formula for the scattering phase in the Born approximation:

$$\delta_l = -\frac{\pi\mu}{\hbar^2} \int_0^{\infty} U(R) \left[J_{l+1/2}(qR)\right]^2 R dR . \tag{2.49}$$

In particular, in the case of electron–atom scattering and the polarization interaction potential between them, when $U(R) = -\alpha e^2/(2R^4)$, this formula gives

$$\delta_l = \frac{\pi\alpha q^2}{(2l-1)(2l+1)(2l+3)a_o} , \tag{2.50}$$

where $a_o = \hbar^2/(me^2)$ is the Bohr radius. This formula is valid if the main contribution to the integral gives large distances between an electron and atom, $R \sim 1/q \gg a_o$, where the polarization interaction takes place. One can see that it is not valid for the zero-th scattering phase, because the integral diverges in this case.

▶ **Problem 2.27** Derive the quasiclassical expressions for scattering phases.

We use the quasiclassical solution of the Schrödinger equation (2.43) for the radial wave function that has the form

$$\varphi_l(R) = C \sin \left[q \int_{r_0}^{R} dR' \sqrt{1 - \frac{U(R')}{\varepsilon} - \frac{(l+1/2)^2}{q^2 R^2}} + \frac{\pi}{4} \right] .$$

Here r_0 is the classical turning point, or the distance of the closest approach at which the integrand equals zero. Since the quasiclassical solution is valid at large momenta $l \gg 1$, we simplify this formula by replacing in it $l(l+1)$ by $(l+1/2)^2$.

Comparing at large distances this radial wave function with its asymptotic expression (12.3), we obtain from this formula for the scattering phase in the quasiclassical approximation:

$$\delta_l = \lim_{R \to \infty} \left[q \int_{r_0}^{R} dR' \sqrt{1 - \frac{U(R')}{\varepsilon} - \frac{(l+1/2)^2}{q^2 R^2}} + \frac{\pi}{2}(l + \frac{1}{2}) - qR \right] . \qquad (2.51)$$

▶ **Problem 2.28** Show the correspondence between the phase theory of scattering and the classical theory in the case when classical criterion is fulfilled.

Taking formula (2.12) for the differential cross section of elastic scattering and expansion (2.42) for the scattering amplitude for spherical harmonics, we obtain the following formula, which connects the differential cross section in an element of solid angle $d\Theta = 2\pi d \cos\vartheta$ with scattering phases:

$$\begin{aligned} d\sigma &= 2\pi d\cos\vartheta \left| f(\vartheta) \right|^2 \\ &= \frac{\pi d\cos\vartheta}{2q^2} \sum_{l=0}^{\infty} \sum_{n=0}^{\infty} (2l+1)(2n+1) P_l(\cos\vartheta) P_n(\cos\vartheta) \exp(2i\delta_l - 2i\delta_n). \end{aligned}$$

Being guided by the quasiclassical limit, we will consider large momenta l and not small scattering angles ϑ, so that $l\vartheta \gg 1$. In this limit and $l \gg 1$ the Legendre polynomials have the following asymptotic form:

$$P_l(\cos\vartheta) = \frac{\sqrt{2} \sin\left[(l+1/2)\,\vartheta + \pi/4\right]}{\sqrt{\pi(2l+1)\sin\vartheta}}, \quad l\vartheta \gg 1 .$$

From this we have for the differential cross section

$$\begin{aligned} d\sigma &= \frac{d\vartheta}{q^2} \sum_{l=0}^{\infty} \sum_{n=0}^{\infty} \sqrt{(2l+1)(2n+1)} \\ &\quad \left\{ \cos\left[(l-n)\vartheta\right] - \cos\left[\frac{l+n+1}{2}\vartheta\right] \right\} \exp(2i\delta_l - 2i\delta_n). \end{aligned}$$

Since the main contribution to these sums in the classical limit gives large collision momenta l, we replace the sums by integrals. Because of oscillations of cosines, the second term does not give a contribution to the result, while the integral for the first term converges near $l = n$. Therefore integrating over dn we obtain

$$d\sigma = \frac{\pi d\vartheta}{q^2} \int_0^\infty (2l+1)dl \left[\delta\left(2\frac{d\delta_l}{dl} - \vartheta\right) + \delta\left(2\frac{d\delta_l}{dl} + \vartheta\right) \right] .$$

Let us introduce an impact parameter of collision as $\rho = (l + 1/2)/q$ and the scattering angle in the classical limit from the relation

$$\vartheta_{\text{cl}} = \pm 2\frac{d\delta_l}{dl} , \quad (2.52)$$

where the plus sign corresponds to a repulsion interaction potential and minus relates to an attractive one. From this we obtain the classic cross section of elastic scattering (formula 2.6)

$$d\sigma = 2\pi\rho d\rho .$$

In this formula the angle of classical scattering ϑ_{cl} is connected with the impact parameter of collision ρ by the classical formula (2.17). One can obtain this formula on the basis of formula (2.52) by using the quasiclassical expression (2.51) for the scattering phase. This gives

$$\frac{\pi \pm \vartheta_{\text{cl}}}{2} = \int_{r_0}^\infty \frac{\rho dr}{\sqrt{1 - U(R)/\varepsilon - \rho^2/R^2}} ,$$

which is in accordance with formula (2.17) for a repulsed interaction.

▶ **Problem 2.29** Show the oscillation character of the differential cross section of elastic scattering for atomic particles interacted through a realistic interaction potential (see Fig. 2.8) due to scattering near the rainbow point.

Because of the correspondence between the classical and quantum formalism, one can include peculiarities of the quantum character of particle scattering in the range where the classical criterion is fulfilled. Indeed, in the classical approach the scattering amplitude is a real value, while in the quantum case it is a complex value. Hence, one can represent the scattering amplitude in the classic limit in the form

$$f(\vartheta) = f_{\text{cl}}(\vartheta) e^{i\eta_l} ,$$

where $f_{\text{cl}}(\vartheta)$ is the classical scattering amplitude, so its square is the classical differential cross section of scattering, and the quasiclassical phase η_l reflects the quantum nature of this quantity. This formula is valid in the classical limit for large momenta $l \gg 1$.

Figure 2.10 gives the dependence of the classical scattering angle on the impact parameter ρ or collision momentum $l = \mu\rho v/\hbar$ for the realistic interaction potential

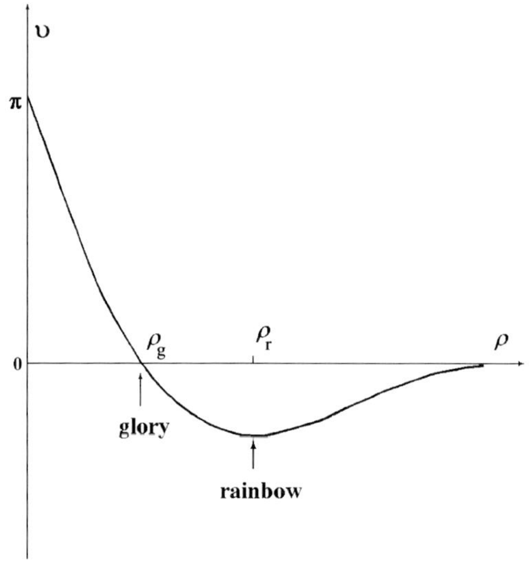

Fig. 2.10 The dependence of the classical scattering angle on an impact parameter of collision. The rainbow point corresponds to the maximum scattering angle in an attractive direction, and the glory point relates to the maximum of the scattering phase.

of particles that is given in Fig. 2.8 and corresponds to attraction of particles at large distances between them and their repulsion at not so large distances. Then frontal collisions are characterized by the scattering angle $\vartheta = \pi$; at large impact parameters the scattering angle tends to zero, and for an attractive interaction potential it is negative. Therefore, the scattering angle as a function of the impact parameter of collision has a negative minimum, the rainbow point. Near this point identical scattering angles correspond to two different impact parameters of collision in the region of negative angles of scattering. Hence, the scattering amplitude near the rainbow can be represented in the form

$$f(\vartheta) = f_1(\vartheta)e^{i\eta_1} + f_2(\vartheta)e^{i\eta_2} ,$$

where f_1, f_2 are the classical scattering amplitudes and are positive values. Correspondingly, the differential cross section for a given negative scattering angle is

$$\frac{d\sigma}{d\Omega} = f_1^2 + f_2^2 + 2f_1 f_2 \cos(\eta_1 - \eta_2) .$$

The phases η_1, η_2 are monotonic functions of both the impact parameter of collision and collision velocity. Therefore, the differential cross section of scattering has an oscillating structure in the range of negative scattering angles as a function of both the scattering angle and the collision velocity.

▶ **Problem 2.30** Find the behavior of scattering phases for scattering of a slow electron in an atom.

When the electron velocity v tends to zero, the scattering phases δ_l also tend to zero, and formula (2.48) allows us to find the threshold behavior of phases. In the case of a short-range electron–atom interaction, when $\varphi_l(R) \sim R^l$ near the atom, the scattering phase is $\delta_l \sim q^{2l+1}$ at low values of the electron wave vector q. Hence, at low electron energies the main expansion term gives $\delta_0 = -Lq$, and it is the definition of the electron–atom scattering length L. The cross section of scattering of a slow electron in an atom is equal to

$$\frac{d\sigma}{d\cos\vartheta} = 2\pi L^2, \quad \sigma_t = \sigma^* = 4\pi L^2. \tag{2.53}$$

As is seen, in contrast to collision of classical particles, the total cross section and the cross section of scattering by a large angle are equal. From this formula it follows that electron scattering is isotropic at zero energy, i. e., the cross section is determined by scattering of s-electron. Next, the scattering length may be expressed through the wave function of the scattered electron Ψ as

$$\left.\frac{d\ln\Psi}{dr}\right|_{r=0} = -\frac{1}{L},$$

where r is the distance of the scattering electron from the atom center. Note that this character of electron scattering takes place in the case of a short-range electron–atom interaction potential (Fermi potential)

$$U_{sh} = 2\pi L\frac{\hbar^2}{m_e}\delta(\boldsymbol{r}), \tag{2.54}$$

where $\boldsymbol{r}$ is the electron coordinate.

The above dependence of scattering phases on the electron wave vector q at its low values results for a weak penetration of an electron in an atom region. In the case of a long-range interaction potential the dependence is different because scattering at any electron momentum is determined by a region far from the atom center. In particular, for the polarization interaction potential the scattering phases are proportional to q^2 at small q according to formula (2.50).

► **Problem 2.31** Within the framework of the phase theory of scattering obtain the expression for the differential cross section of scattering of a slow electron by an atom by taking into account the short-range and polarization electron–atom interactions.

Along with a short-range electron–atom interaction, a long-range interaction can give a contribution to parameters of electron–atom scattering. In contrast to a short-range interaction, at low collision energies the contribution from a long-range interaction is determined by an electron region far from the atom where coordinates of scattered and atomic electrons may be separated. At low electron energies the main long-range interaction corresponds to the polarization electron–atom interaction, and the total electron–atom interaction potential for a slow electron has the form

$$U(r) = U_{\rm sh} - \frac{\alpha e^2}{2r^4} \, .$$

We treat this within the framework of the perturbation theory that is valid because scattering as a result of the polarization interaction is determined by large electron–atom distances where this interaction is small in comparison with the electron kinetic energy. Then we have for the scattering amplitude

$$f(\vartheta) = -L - \frac{m_{\rm e}}{2\pi\hbar^2} \int \frac{\alpha e^2}{2r^4} [1 - \exp(-i\boldsymbol{K}\,\boldsymbol{r})] \, d\boldsymbol{r} \, ,$$

where $K = |\boldsymbol{q} - \boldsymbol{q}'| = 2q \sin\vartheta/2$ is the variation of the electron wave vector as a result of scattering. This formula gives for the scattering amplitude at low electron velocities

$$f(\vartheta) = -L - \frac{\pi\alpha}{4a_{\rm o}} K = -L - \frac{\pi\alpha q}{2a_{\rm o}} \sin\frac{\vartheta}{2} \, , \tag{2.55}$$

where $a_{\rm o} = \hbar^2/(m_{\rm e}e^2)$ is the Bohr radius. This gives for the total and diffusion cross sections of electron–atom scattering at low electron energies

$$\sigma_{\rm t} = 4\pi L^2 \left(1 - \frac{4}{3}x + \frac{1}{2}x^2\right) , \quad \sigma^* = 4\pi L^2 \left(1 - \frac{8}{5}x + \frac{2}{3}x^2\right) , \quad x = -\frac{\pi\alpha q}{2La_{\rm o}} \, . \tag{2.56}$$

These formulas exhibit a sharp minimum for the cross sections of electron–atom scattering (see Fig. 2.11) at small collision energies if the scattering length L is negative (the Ramsauer effect), and the reason of this effect is such that the zero-th phase is zero δ_0 when the contribution of other phases to the cross section is relatively small because of a low electron energy. The Ramsauer effect is observed in elastic scattering of electrons in argon, krypton, and xenon atoms. As follows from formula (2.56), the total cross section in this approximation has the minimum $4\pi L^2/9$ at the electron wave number $q_{\rm min} = -8La_{\rm o}/(3\pi\alpha)$ $(x = 4/3)$. The

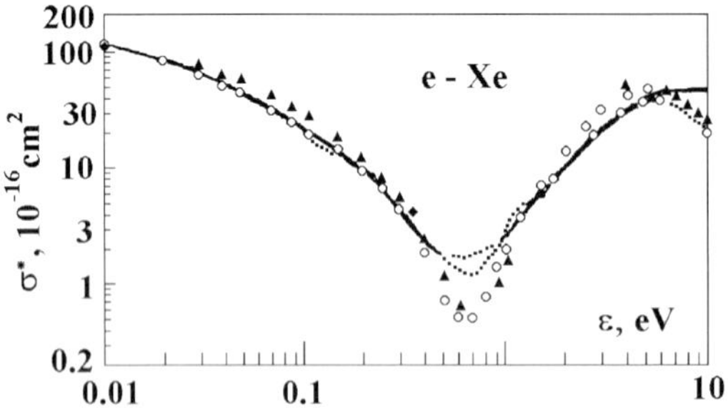

Fig. 2.11 Electron-atom xenon diffusion cross section as a function of the electron energy according to different experimental data and their approximation. The electron-atom scattering length in the xenon case is negative.

minimum of the diffusion cross section in this approximation is equal to $4\pi L^2/25$ and corresponds to the electron wave vector $q_{\rm min} = -12La_{\rm o}/(5\pi\alpha)$ $(x = 6/5)$ according to formula (2.56). Thus, within the framework of this approximation, the scattering cross section drops by an order of magnitude at low electron energies.

2.5
Total Cross Section of Elastic Collision

► **Problem 2.32** Show that the total cross section of elastic collision $\sigma_{\rm t} = \int d\sigma$, which results from integration of the differential cross sections over all solid angles, is infinity in the classical limit.

We assume that interaction of colliding particles takes place at any large distance between them. This means that scattering of particles takes place at any impact parameters of collision, and therefore the integral is taken for all impact parameters. As a result, we obtain the infinite total cross section of scattering in this case.

► **Problem 2.33** Estimate the total cross section of elastic collision for classical particles.

Our consideration for the total cross section of classical particles was based on an assumption that the classical character of scattering holds true at any impact parameters of particle collisions, but it is not valid at large impact parameters. Indeed, let us determine the variation of the particle's momentum when particles are moving along classical trajectories. This value is given by

$$\Delta \boldsymbol{p} = \int_{-\infty}^{\infty} \boldsymbol{F} dt \,,$$

where $\boldsymbol{F} = -\partial U/\partial \boldsymbol{R}$ is the force with which one particle acts upon the other, and U is the interaction potential between the particles. This leads to an estimate for the momentum variation $\Delta p \sim U(\rho)/v$.

In order to account for the classical description of scattering at small angles, we use the Heisenberg uncertainty principle, so that the value Δp can be determined up to an accuracy of $\hbar/\rho$. From this it follows that impact parameters which give the main contribution to the total scattering cross section satisfy the relation $\Delta p(\rho) \sim \hbar/\rho$. As a result, we obtain the following estimate:

$$\sigma_{\rm t} \sim \rho_{\rm t}^2 \,, \quad \text{where} \quad \frac{\rho_{\rm t} U(\rho_{\rm t})}{\hbar v} \sim 1 \tag{2.57}$$

for the total scattering cross section.

In particular, for the interaction potential of classical particles in the form $U(R) = C/R^k$, the total cross section is

$$\sigma_{\rm t} \sim \left(\frac{C}{\hbar v}\right)^{2/(k-1)} .$$

Since the scattering cross section is determined by quantum effects, it tends to infinity in the classical limit $\hbar \to 0$.

Particle motion will obey classical laws if the kinetic energy ε satisfies the condition

$$\varepsilon \gg \frac{\hbar}{\tau}, \tag{2.58}$$

where τ is a typical collision time. Since $\tau \sim \rho/v$ and $\varepsilon = \mu v^2/2$, it follows that the angular momentum of the particles $l = \mu\rho v/\hbar \gg 1$. If this criterion is valid, a description that the motion of the particles follows classical trajectories holds true.

▶ **Problem 2.34** Prove that for classical particles the total cross section of elastic collision exceeds significantly the cross section of particle scattering by a large angle.

Let us use the estimations (2.18) for the large-angle scattering cross section and (2.57) for the total cross section and assume the interaction potential of colliding particles $U(R)$ to be a monotonic function of the distance R between particles. From the indicated formulas we have

$$\frac{U(\rho_o)}{U(\rho_t)} \sim \frac{\mu\rho_t v}{\hbar} = l \gg 1\,,$$

where l is the orbital momentum of colliding particles, which is large for classical particles. Thus, for collision of classical particles we have $\rho_t \gg \rho_o$ because of a monotonic dependence $U(R)$, and this gives

$$\sigma_t \gg \sigma. \tag{2.59}$$

Note that if scattering of atomic particles has a quantum nature, the large-angle scattering cross section and the total scattering cross section have the same order of magnitude. This is typical for elastic scattering of electrons in atoms and molecules.

▶ **Problem 2.35** Express the total cross section of elastic scattering through the classical scattering phase in the classical limit.

In the classical limit large collision momenta l give the main contribution to the cross section. Introducing the impact parameter of collision $\rho = l/q$ and replacing the summation in formula (2.47) by integration, we find for the total cross section of elastic scattering from this formula,

$$\sigma_t = \int_0^\infty 8\pi\rho d\rho \sin^2\delta(\rho). \tag{2.60}$$

This formula allows us to evaluate the total cross section of elastic scattering, which is given in the classical limit

$$\sigma_t = 2\pi\rho_t^2, \quad \text{where} \quad \delta(\rho_t) \sim 1. \tag{2.61}$$

The parameter ρ_t is called the Weiskopf radius.

▶ **Problem 2.36** Find the total cross section for the interaction potential $U(R) = AR^n, n \gg 1$, in the classical limit.

Let us use formula (2.51) for the scattering phase in the classical limit. We use that the total cross section is determined by scattering at small angles where the interaction potential is small in comparison with the particle energy. Expanding formula (2.51) over this small parameter, we find the scattering phase for large collision momenta in the form

$$\delta_l = -\frac{1}{\hbar^2}\int_{r_0}^{\infty}\frac{U(R)dR}{\sqrt{q^2-\frac{(l+1/2)^2}{R^2}}},$$

where the distance of the closest approach corresponds to free particle motion $r_0 = \rho = (l+1/2)/q$. Introducing the classical time from the relation $R^2 = \rho^2 + v^2t^2$, we rewrite the above relation in the form

$$\delta_l = -\frac{1}{2\hbar}\int_{-\infty}^{\infty} U(R)dt. \tag{2.62}$$

Correspondingly, formula (2.61) can be rewritten in the form

$$\sigma_t = 2\pi\rho_t^2, \quad \text{where} \quad \frac{\rho_t U(\rho_t)}{\hbar v} \sim 1. \tag{2.63}$$

Formula (2.62) gives the interaction potential $U(R) = AR^n, n \gg 1$, in the classical limit $l \gg 1$,

$$\delta = -\frac{A}{2\hbar}\int_{-\infty}^{\infty}\frac{dt}{(\rho^2+v^2t^2)^n} = -\frac{A\sqrt{\pi}}{2\hbar v\rho^{n-1}}\frac{\Gamma\left(\frac{n+1}{2}\right)}{\Gamma\left(\frac{n}{2}\right)}.$$

Substituting this in formula (2.60) for the total cross section of elastic scattering, we obtain

$$\sigma_t = 2\pi\left(\frac{A}{\hbar v}\right)^{\frac{2}{n-1}}\left[\frac{\sqrt{\pi}\Gamma\left(\frac{n-1}{2}\right)}{\Gamma\left(\frac{n}{2}\right)}\right]^{\frac{2}{n-1}}\Gamma\left(\frac{n-3}{n-1}\right). \tag{2.64}$$

▶ **Problem 2.37** Show the oscillation character of the total cross section of elastic scattering in the classical limit for atomic particles interacted through a realistic (Fig. 2.8) interaction potential.

Using the correspondence (2.52) between the scattering angle and phase, we find that the phase δ_l as a momentum function has the maximum at the glory point (Fig. 2.10), where the classical scattering angle is zero. Near this point the scattering phase can be expanded near the glory point,

$$\delta_l = \delta_l^{(0)} + \alpha(l - l_g)^2,$$

where l_g is the collision momentum at the glory point, and $2\alpha = d^2\delta_l/dl^2$ at this point. The first term corresponds to a regular part of the scattering phase and the second one accounts for its stationary part near the glory point.

According to the general expression for the total cross section of scattering (2.47), this form of the scattering phase gives oscillations in the total cross section. The oscillation amplitude is $\sim 1/l_g$, and since $l_g \gg 1$, it is relatively small.

▶ **Problem 2.38** Find the ratio of the total cross section of elastic scattering to the diffusion one for the polarization interaction potential $U(R) = -\alpha e^2/(2R^4)$.

We use formula (2.64) for the total polarization cross section, taking in it $n = 4$. Dividing it by the cross section of polarization capture (2.25), we obtain

$$\frac{\sigma_t}{\sigma_c} = \left(\frac{\pi}{4}\right)^{2/3} \Gamma\left(\frac{1}{3}\right)\left(\frac{\alpha e^2 \mu^2 v^2}{\hbar^4}\right)^{1/6} = 2.3\, l_c^{2/3} ,$$

where $l_c = \mu\rho_c v/\hbar$ is the momentum of colliding particles that restrict the capture. Since in the classical limit $l_c \gg 1$, this ratio is large.

▶ **Problem 2.39** Find the total cross section for the sharply varied interaction potential.

Let us approximate the potential of the force center by the dependence $U(R) = A/R^n, n \gg 1$, and use formula (2.64) for the total cross section in the limit $n \gg 1$. We represent this formula in the form

$$\sigma_t = 2\pi\rho_t^2 , \quad \text{where} \quad \frac{\rho_t U(\rho_t)}{\hbar v} = g$$

and find in the limit of large n the parameter g, which is given by

$$g = \frac{\Gamma\left(\frac{n}{2}\right)}{\sqrt{\pi}\Gamma\left(\frac{n-1}{2}\right)} \left[\Gamma\left(\frac{n-3}{n-1}\right)\right]^{\frac{n-1}{2}} .$$

We have in this limit $n \gg 1$

$$\left[\Gamma\left(\frac{n-3}{n-1}\right)\right]^{\frac{n-1}{2}} = \left[\Gamma(1) - \frac{2}{n}\Gamma'(1)\right]^{\frac{n}{2}} = e^{\psi(1)}\Gamma(1) ,$$

where $\psi(1) = \Gamma'(1)/\Gamma(1) = -C + 1 = 0.423$. Applying the Stirling formula, we obtain

$$\frac{\Gamma\left(\frac{n}{2}\right)}{\Gamma\left(\frac{n-1}{2}\right)} = \frac{\left(\frac{n}{2}\right)^{\frac{n}{2}}}{\sqrt{e}\left(\frac{n-1}{2}\right)^{\frac{n-1}{2}}} = \sqrt{\frac{n}{2e}}\left(1 - \frac{2}{n}\right)^{\frac{n}{2}} = \sqrt{\frac{ne}{2}} .$$

Finally, we get $g = \sqrt{\frac{ne}{2\pi}} e^{\psi(1)} \approx \sqrt{n}$, and formula (2.64) for the total cross section of elastic scattering takes the form

$$\sigma_t = 2\pi\rho_t^2 , \quad \text{where} \quad \frac{\rho_t U(\rho_t)}{\hbar v} = \sqrt{n}. \tag{2.65}$$

3
Slow Atomic Collisions

3.1
Slow Collisions of Heavy Atomic Particles

► **Problem 3.1** Within the framework of the two-state model analyze slow collisions of heavy atomic particles.

We consider slow collisions of heavy atomic particles—ions, atoms, and molecules—when the relative velocity of their collision is small compared to a typical velocity of atom electrons, v_e. Under this condition, the electron distribution in each of the colliding atomic particles corresponds to the internal fields of the particles and differs slightly from their distributions at fixed nuclei. Then evolution of a system of colliding particles is described by parameters of a quasimolecule that consists of colliding atomic particles, and the electron energies for quasimolecule states as a function of the distance between the colliding particles are the electron terms of the quasimolecule, and their position determines the character of collision of atomic particles.

The transitions between states of colliding particles proceed at separations where some electron terms occur close or are intersected. We consider below the transition between two such states, and in considering this transition, one can ignore other quasimolecule states. Thus, we consider evolution of the system of colliding particles on the basis of the Schrödinger equation for the wave function Ψ of this system,

$$i\hbar\frac{\partial\Psi}{\partial t} = \hat{H}\Psi, \tag{3.1}$$

and we use the positions of the electron terms—the electron energies E_i as a function of the distance R between particles, which are the eigenvalues of the Hamiltonian at motionless nuclei

$$\hat{H}\Psi_i = E_i\,\Psi_i\,, \tag{3.2}$$

Plasma Processes and Plasma Kinetics. Boris M. Smirnov

ISBN: 978-3-527-40681-4

and Ψ_i are the eigenfunctions for these states. Thus, restricting by two quasimolecule states, we represent the wave function of colliding atoms in the form

$$\Psi = (c_1\psi_1 + c_2\psi_2)\exp[-\frac{i}{2}\int^t (H_{11} + H_{22})dt']. \qquad (3.3)$$

Here $H_{11} = \langle\psi_1|\hat{H}|\psi_1\rangle$, $H_{22} = \langle\psi_2|\hat{H}|\psi_2\rangle$, and we use the atomic units $m_e = e^2 = \hbar = 1$. Substituting this expansion into the Schrödinger equation, multiplying it by ψ_1^* or ψ_2^*, and integrating over electron coordinates, we obtain the following set of equations for the probability amplitudes

$$i\frac{dc_1}{dt} = \frac{\kappa}{2}c_1 + \frac{\Delta}{2}c_2, \quad i\frac{dc_2}{dt} = \frac{\Delta}{2}c_1 - \frac{\kappa}{2}c_2, \qquad (3.4)$$

where $\kappa(R) = H_{11} - H_{22}$, $\Delta(R) = 2H_{12} - (H_{11} + H_{22})\langle\psi_2 \mid \psi_1\rangle$. This is the diabatic set of equations for the amplitudes c_1, c_2 that describes evolution of the system of colliding atomic particles.

Note that the solution of the stationary Schrödinger equation (3.2) allows us find the eigenfunctions Ψ_1, Ψ_2 for the two-state basis ψ_1, ψ_2 that have the form

$$\Psi_1 = a_1\psi_1 + a_2\psi_2, \quad \Psi_2 = -a_2\psi_1 + a_1\psi_2, \quad a_{1,2} = \left[\frac{1}{2} \pm \frac{\kappa}{\sqrt{\kappa^2 + \Delta^2}}\right]^{1/2}, \qquad (3.5)$$

and the difference of the energies E_1, E_2 of the eigenstates is

$$E_1 - E_2 = \sqrt{\kappa^2 + \Delta^2}. \qquad (3.6)$$

In particular, this consideration is valid if the basis wave functions ψ_1, ψ_2 correspond to the location of an electron or excitation near the first or second atomic core, and at large distances R between colliding particles the interaction of the particles consists of two parts, the long-range interaction, $\kappa(R)$, and the exchange interaction $\Delta(R)$, which is determined by overlapping of the wave functions which belong to different cores.

▶ **Problem 3.2** Determine the velocity dependence of the probability of the inelastic transition between states of heavy slow particles at low collision velocities.

At low velocities it is convenient to expand the wave function of slow colliding particles over eigenfunctions of the quasimolecule $\psi_n(\mathbf{r}, R)$ at a fixed distance R between nuclei

$$\Psi = \sum_n b_n(t)\psi_n(\mathbf{r}, R)\exp\left[-\frac{i}{\hbar}\int^t \varepsilon_n(R)dt'\right], \qquad (3.7)$$

where $\varepsilon_n(R)$ are the eigenvalues of the Schrödinger equation (3.2), which has the form $\hat{H}\psi_n(\mathbf{r}, R) = \varepsilon_n(R)\psi_n(\mathbf{r}, R)$, and $\mathbf{r}$ is a sum of electron coordinates. Substituting expansion (3.7) in the Schrödinger equation (3.1) and multiplying this equation by the wave function $\psi_{\rm m}^*(\mathbf{r}, R)$, integrate it over electron coordinates. As a result, we obtain the adiabatic set of equations for the amplitudes $b_{\rm m}$,

$$i\frac{db_{\rm m}}{dt} = \sum_n b_n(t)\left(-i\frac{\partial}{\partial t}\right)_{mn} \psi_n(\mathbf{r}, R) \exp\left[-i\int^t \omega_{mn}(R)dt'\right] , \tag{3.8}$$

where $\omega_{mn} = [\varepsilon_n(R) - \varepsilon_n(R)]/\hbar$, and the matrix element is taken between eigenstates of the quasimolecule.

The operator in equation (3.8) consists of two parts and can be represented in the form

$$\frac{\partial}{\partial t} = \frac{dR}{dt}\frac{\partial}{\partial R} + \frac{d\theta}{dt}\frac{\partial}{\partial \theta} = \frac{dR}{dt}\frac{\partial}{\partial R} - i\frac{d\theta}{dt}\hat{l}_\theta ,$$

where $\hat{l}_\theta$ is the projection of the electron orbital momentum over the vector $d\boldsymbol{\theta}/dt$, which is perpendicular to the axis joining cores of colliding particles.

At low collision velocities one can solve the set of equations (3.8) on the basis of the perturbation theory taking in the zero-th approximation $b_n = \delta_{n0}$, so that the amplitude of the transition after collision is

$$b_{\rm m} = \int_{-\infty}^{\infty} \left(\frac{\partial}{\partial t}\right)_{0m} \exp\left[-i\int^t \omega_{m0}(R)dt'\right] dt ,$$

and this value is small because of fast oscillation of the integrand. One can estimate the probability of the transition as

$$P_{\rm m} = \left|b_{\rm m}(\infty)\right|^2 \sim \exp(-\xi) ,$$

and the Massey parameter ξ is

$$\xi = \frac{\Delta E \cdot a}{\hbar v} , \tag{3.9}$$

where a is a typical range of distances between nuclei associated with a significant change of the electron terms, v is the relative collision velocity, $\Delta E = \hbar\omega_{0m}$ is a minimal difference of energies for these two states of the quasimolecule. Thus, if the Massey parameter is large, the probability of the transition is adiabatically small, and transitions between electron states in slow atomic collisions can be a result of intersections or pseudointersections of corresponding electronic terms.

▶ **Problem 3.3** In the two-state approach express the probability of electron transfer in slow collisions of an ion and the parent atom with a valence *s*-electron through the energetic parameters of the quasimolecule consisting of colliding particles.

This resonant charge exchange process proceeds according to the scheme

$$A^+ + A \to A + A^+ . \tag{3.10}$$

In slow collisions electrons follow for the atomic fields, and their evolution is determined by the parameters of the quasimolecule that consists of colliding atomic particles with fixed nuclei. For a slow process (3.10) with s-valence electron the quasimolecule consisting of a colliding ion and a parent atom has a symmetry for reflection with respect of the symmetry plane that is perpendicular to the axis joining nuclei and divides it into two halves. Correspondingly, the wave function of the quasimolecule at large distances between nuclei can be even or odd with respect to the above operation and have the following form at large distances between nuclei:

$$\psi_{\rm g} = \frac{1}{\sqrt{2}}(\psi_1 + \psi_2), \quad \psi_{\rm u} = \frac{1}{\sqrt{2}}(\psi_1 - \psi_2), \tag{3.11}$$

where ψ_1, ψ_2 are the wave functions which are centered at indicated nuclei. These eigenfunctions of the electron Hamiltonian satisfy to the Schrödinger equation

$$\hat{H}\psi_{\rm g} = \varepsilon_{\rm g}\psi_{\rm g}, \qquad \hat{H}\psi_{\rm u} = \varepsilon_{\rm u}\psi_{\rm u} .$$

Let us consider the case when inelastic transitions are absent. The transitions of an electron from one core to other result from the interference of two eigenstates. Indeed, assume that at the beginning $t \to -\infty$ an electron is located near the first nucleus, and then the electron wave function is given by the formula

$$\begin{aligned} \Psi(\boldsymbol{r}, \boldsymbol{R}, t) &= \frac{1}{\sqrt{2}}\psi_{\rm g}(\boldsymbol{r}, \boldsymbol{R}) \exp\left[-i\int_{-\infty}^{t} \varepsilon_{\rm g}(t')dt'\right] \\ &+ \frac{1}{\sqrt{2}}\psi_{\rm u}(\boldsymbol{r}, \boldsymbol{R}) \exp\left[-i\int_{-\infty}^{t} \varepsilon_{\rm u}(t')dt'\right] . \end{aligned} \tag{3.12}$$

If we fix nuclei, the electron will oscillate between nuclei. In the case of ion–atom collision the transition probability $P_{\rm exc}$ in the end of collision is given by

$$P = \sin^2 \int_{-\infty}^{\infty} \frac{(\varepsilon_{\rm g} - \varepsilon_{\rm u})}{2} dt = \sin^2 \int_{-\infty}^{\infty} \frac{\Delta(R)}{2} dt , \tag{3.13}$$

where the exchange interaction potential between the ion and parent atom is introduced by $\Delta(R) = \varepsilon_{\rm g}(R) - \varepsilon_{\rm u}(R)$, and the distance between nuclei $R(t)$ corresponds to a certain law of particle collision. Usually, the colliding ion and atom move along straightforward trajectories, and therefore $R^2 = \varrho^2 + v^2t^2$. Thus, the nature of both the resonant charge exchange processes reduces to the interference of the states in absence of inelastic transitions between the states.

▶ **Problem 3.4** Determine the probability of the transition between two neighboring terms in the adiabatic approximation.

In expression (3.12) for the wave function of two colliding particles we neglect the transitions between states under consideration. Though this formula is obtained for a symmetric system of colliding particles, in neglecting of inelastic transitions one can write it for any states, and it has the form

$$\Psi(\mathbf{r}, \mathbf{R}, t) = \sum_k a_k(t)\psi_k(\mathbf{r}, \mathbf{R}) \exp\left[-i \int_{-\infty}^{t} \varepsilon_k(t')dt'\right],$$

and in the absence of transitions the values $|a_k(t)|$ are independent of time in this approximation, when the Massey parameter (3.9) is large. We determine below the transition probability between two states with nearby electron terms for which the transition probability is adiabatically small. In this case the Schrödinger equation gives the following equations for amplitudes of the quasimolecule location in these states,

$$\frac{d^2 a_{1,2}}{dt^2} + E_{1,2}^2 a_{1,2} = 0, \tag{3.14}$$

and these equations take into account the above dependence of the quasimolecule wave function on time. But because the adiabatic states of the quasimolecule are not stationary states exactly, this distinction causes weak transitions between the states when time t varies from $-\infty$ to $+\infty$. Therefore, if at $t = -\infty$ the quasimolecule is located in one state, a weak admixture of another state is absent in the wave function, and we will find it.

Let us take the energies of states under consideration in accordance with formula (3.6) as

$$E_{1,2} = \pm\frac{1}{2}\sqrt{\kappa^2 + \Delta^2};$$

we have formally the quasiclassical solution of equations (3.14) for the amplitude of the first state,

$$a_1(t) = \frac{1}{\sqrt{E_1(t)}} \exp\left[-\int_0^t E_1(t')dt'\right]$$

in the absence of transitions. Below we take this solution in the limit $t \to -\infty$ and find an admixture of another amplitude at $t \to +\infty$, which characterizes the transition probability of the quasimolecule transition into another state.

Evidently, this transition is possible when the energy difference is small. Therefore in order to find it, we move to the complex plane of time and then the transition will be determined by a time range where $E_1(t) - E_2(t)$ is close to zero. Introducing the time t_c of the intersection of these two electron terms on the complex plane of time, $E_1(t_c) = E_2(t_c)$, and from formula (3.6) we have near this point $E_1(t) - E_2(t) = A\sqrt{t_c - t}$. Indeed, formula (3.6) gives $E_1(R) = E_2(R) = \text{const}\sqrt{R - R_c}$ at distances R between nuclei near the distance R_c of the intersection of these electron terms, and since $R - R_c \sim (t - t_c)$, we obtain the above time dependence for the difference of the electron state energies in the complex plane of time.

The solution of equations (3.14) near the intersection of electron terms has the form

$$a_1(t) = B\sqrt{t_c - t}\, H^{(2)}_{1/3}\left(\frac{2A}{3}(t_c - t)^{2/3}\right) ,$$

where $H^{(2)}(x)$ is the Hankel function of the second type, and its asymptotic behavior at $t \to \infty$ is

$$a_1(t)_{t\to-\infty} = \frac{1}{\sqrt{E_1(t)}} \exp\left[-\int_{t_c}^{t} E_1(t')dt' - i\frac{5\pi}{12}\right] .$$

Note that if we find in the vicinity of the intersection point $t = t_c$, one can divide a space by three lines

$$\mathrm{Im}\left[-\int_{t_c}^{t} E_1(t')dt'\right] = -\frac{2A}{3}\mathrm{Im}(t - t_c)^{3/2} = 0 ,$$

which are represented in Fig. 3.1. Between two such lines the amplitude $a_1(t)$ is an analytical function of time, but intersecting any of the three lines, we move from one analytical region to another one, and this leads to a change of the amplitude phase. In our case, the three lines $\mathrm{Im}(t - t_c)^{3/2} = 0$ form the angle $2\pi/3$ with each other. When we move along a real time line from $t \to -\infty$ to $t \to \infty$, we intersect one line, i. e., transfer from one analytical region to another one. This leads to a phase change by $2\pi/3$ in the expression of the amplitude phase, i. e., changes the argument of the Hankel function by π. Using the relation for the Hankel functions

$$H^{(2)}_{1/3}(-z) = H^{(2)}_{1/3}(z) + e^{i\pi/3} H^{(1)}_{1/3}(z) ,$$

we obtain from this the form of the asymptotic amplitude in the limit $t \to \infty$,

$$a_1(t) \propto \frac{1}{\sqrt{E_1(t)}} \exp\left[-\int_{t_c}^{t} E_1(t')dt' - i\frac{5\pi}{12}\right] + \frac{e^{i\pi/3}}{\sqrt{E_1(t)}} \exp\left[\int_{t_c}^{t} E_1(t')dt' + i\frac{5\pi}{12}\right] .$$

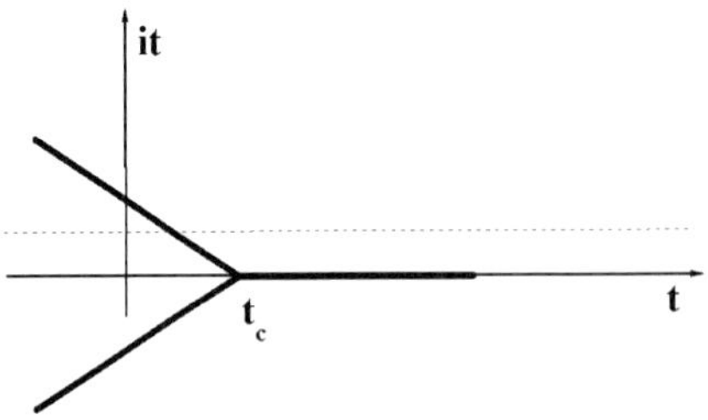

Fig. 3.1 Peculiarities of the function $\mathrm{Im}(t - t_c)^{3/2}$. Three analytical regions for this function are restricted by indicated lines, and transition between different analytical regions of this function leads to a stepwise change of its phase by $2\pi/3$. The line of integration for the amplitude $a_1(t)$ exponent is given by a dotted line.

The first term relates to the location of the quasimolecule in the first state, and the second term characterizes this for the second state. From the relation of these values we have the probability of the transition from the first state to the second one as the square of the amplitude relation,

$$P_{12} = \exp\left[-4\mathrm{Im}\int_0^{t_c} E_1(t')dt'\right] = \exp\left[-2\mathrm{Im}\int_0^{t_c} \sqrt{\kappa^2+\Delta^2}dt'\right], \tag{3.15}$$

where the energy difference is taken according to formula (3.6).

► **Problem 3.5** Determine the probability of the adiabatic transition between two quasimolecule states in the Landau–Zener case, if in the range of term intersection $\kappa(R) = F(R - R_o)$ and $\Delta(R) = \text{const}$, where R is the distance between colliding atomic particles, R_o is the distance of the term intersection in the absence of the exchange interaction, and Δ is the minimum energy difference for these terms (see Fig. 3.2).

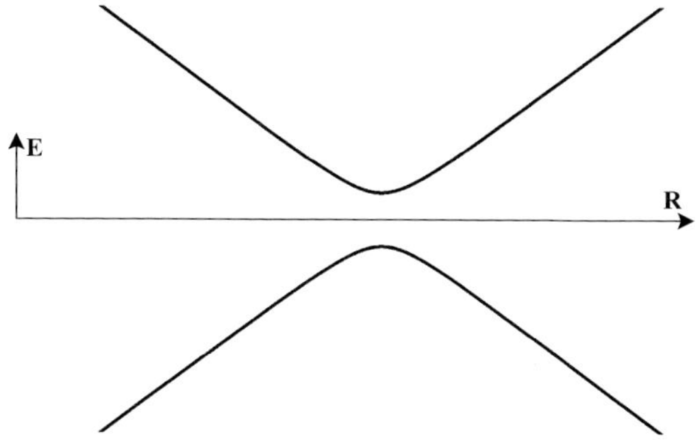

Fig. 3.2 The behavior of electron terms in the Landau-Zener case of transition.

Let us determine the probability of transition between two neighboring terms for the adiabatic approximation (3.15). We then have

$$2\mathrm{Im}\int_0^{t_c} \sqrt{-F^2 v_R^2 (t-t_c)^2 + \Delta^2}dt = \frac{2\Delta^2}{F v_R}\int_0^1 \sqrt{1-x^2}dx = \frac{\pi\Delta^2}{2F v_R}\,.$$

Here t_c is an imaginary time at which two electron terms are coincided, R_c is the complex distance of this intersection, so that $R - R_c = v_R(t - t_c)$, and v_R is the radial relative velocity of colliding atomic particles.

We also introduce the real distance between particles, R_0, where the diagonal energy part is zero, $\kappa(R) = F(R - R_0)$, and at this distance the energy difference has a minimum. Taking $R_0 > r_0$, where r_0 is the distance of the closest approach, we obtain that the quasimolecule passes twice the intersection distance. Ignoring the phase correlation for this transition, we have for the total probability P of transitions between states as a result of double passage of the distance of the electron term intersection in the collision,

$$P = 2P_{12}(1 - P_{12}) ,$$

where P_{12} is the probability of transition (3.15) as a result of a single passage of the intersection distance. Finally, we obtain for the probability of transition as a result of slow collisions in the Landau–Zener case,

$$P = 2\exp\left(-\frac{\pi\Delta^2}{2Fv_R}\right); \quad \frac{\Delta^2}{Fv_R} \gg 1 . \tag{3.16}$$

▶ **Problem 3.6** Determine the probability of the transition between two quasimolecule states in the Landau–Zener case (see Fig. 3.2) for any velocities of collision as a result of a single passage of the intersection distance.

Our task is to solve the set of equations (3.4) for the amplitudes $c_1(t)$, $c_2(t)$ in the Landau–Zener case, when $\kappa(t) = k(t - t_o), \Delta(t) = \text{const}$. It is convenient to introduce new amplitudes,

$$c_+ = \frac{c_1 + c_2}{\sqrt{2}}, \quad c_- = \frac{c_1 - c_2}{\sqrt{2}} ,$$

and the equations for these amplitudes have the following form instead of the set (3.4):

$$i\frac{dc_+}{dt} - \frac{k}{2}(t - t_o)c_+ = -\Delta c_-, \quad i\frac{dc_-}{dt} + \frac{k}{2}(t - t_o)c_- = -\Delta c_+ .$$

As is seen, the amplitudes c_+ , c_- relate to unperturbed energies $\pm k(t - t_o)/2$. One can reduce these equations to the form

$$i\frac{dc_+^2}{dt^2} + [\frac{k^2}{4}(t - t_o)^2 c_+ \Delta^2 + i\frac{k}{2}\Delta]c_+ = 0 \quad i\frac{dc_-^2}{dt^2} + [\frac{k^2}{4}(t - t_o)^2 c_+ \Delta^2 - \frac{ik}{2}\Delta]c_- = 0 .$$

Each of these equations is the equation of the quantum oscillator with a complex phase. Their solutions are expressed through the functions of the parabolic cylinder $E(x, y)$. In particular, for the symmetric amplitude we have

$$c_+(t) = cE\left(-\frac{i}{2} - \frac{\Delta^2}{k}, \sqrt{\frac{k(t - t_o)}{\Delta^2}}\right) .$$

The asymptotic expressions of these functions in the limit $t - t_o \to \pm\infty$ have the form

$$|c_+(\pm\infty)| \propto (t - t_o)^{-i\frac{\Delta^2}{k}} .$$

From this we obtain for the transition from $t = -\infty$ to $t = \infty$

$$\frac{|c_+(-\infty)|}{|c_+(-\infty)|} = [\exp(i\pi)]^{-i\frac{\Delta^2}{k}} = \exp\left(-\frac{\pi\Delta^2}{k}\right) .$$

This gives the expression for the probability of a single passage through the minimum energy difference,

$$P_{12} = \exp\left(-\frac{\pi\Delta^2}{2Fv_R}\right),$$

where we use $k = Fv_R$. As is seen, we obtain the Landau–Zener formula (3.16) for the probability of a single passage of the minimum energy distance. Above (previous Problem) we find this expression when the Massey criterion is large. As is seen, this expression is valid for any value of the Massey parameter.

► **Problem 3.7** Find the probability of the adiabatic transition between two electron terms, if the time dependence for diagonal and nondiagonal matrix elements of the quasimolecule energy in formula (3.6) is $\kappa(t) = \text{const}$, $\Delta(t) = \delta\exp(\gamma t)$ (the Rosen–Zener case, Fig. 3.3).

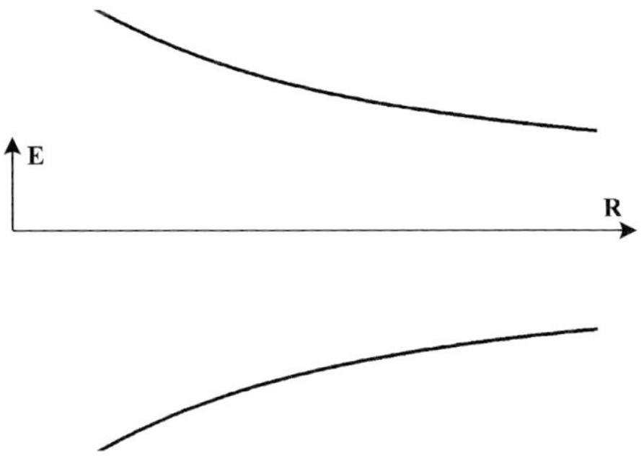

Fig. 3.3 The behavior of electron terms in the Rosen-Zener case of transition.

Formula (3.15) gives the transition probability of a single passage in the adiabatic limit,

$$P_{12} = \exp\left[-2\mathrm{Im}\int_0^{t_c}\sqrt{\kappa^2+\Delta^2 dt}\right]$$

$$= \exp\left[-2\kappa\mathrm{Im}\int_0^{t_c}\sqrt{1=\exp(2\gamma t)dt}\right] = \frac{\kappa}{\gamma}\int_0^{\pi}\sqrt{1+e^{i\varphi}d\varphi},$$

where $\varphi = 2i\gamma t$. Since

$$\mathrm{Im}\int_0^{\pi}\sqrt{1+e^{i\varphi}d\varphi} = \mathrm{Im}\sqrt{2\cos\frac{\varphi}{2}e^{i\varphi/4}d\varphi} = 4\sqrt{2}\int_{1/\sqrt{2}}^{1}\sqrt{2z^2-1dz} = \pi/2\,.$$

This gives the transition probability in the Rosen–Zener case

$$P_{12} = \exp\left(-\frac{\pi\kappa}{\gamma}\right). \tag{3.17}$$

▶ **Problem 3.8** On the basis of the set of diabatic equations (3.4) determine the probability of the transition between two states in a two-state approach, if the quasimolecule parameters in the range of transition can be approximated by the dependences $\kappa(t) = \text{const}$ and $\Delta(t) = 2\delta\exp(-t/\tau)$ (the Rosen–Zener–Demkov case, see Fig. 3.3).

A general method to solve the set of equations (3.4) is based on its solution in the limiting cases when $\Delta \gg \kappa$ or $\Delta \ll \kappa$ and the accurate solution of this set at $\Delta \sim \kappa$. In the range $\Delta \ll \kappa$ we have

$$c_1 = a\exp\left(i\int^t \frac{\kappa}{2}dt'\right), \quad c_2 = b\exp\left(-i\int^t \frac{\kappa}{2}dt'\right),$$

and the values $|c_1|$ and $|c_2|$ do not vary in time. In the other limiting case $\Delta \gg \kappa$ the solutions of the set of equations (3.4) are

$$c_1 = \cos\left(i\int^t \frac{\Delta}{2}dt' + \gamma\right),$$

$$c_2 = \sin\left(i\int^t \frac{\Delta}{2}dt' + \gamma\right),$$

where γ is the phase. In this range the values $|c_1 + c_2|$ and $|c_1 + c_2|$ do not vary in time. An intermediate range determines the transition probability between these two states.

Let us apply this to the Rosen–Zener–Demkov case and divide the time into some ranges, solve the set of equations (3.4) in each range and sew solutions on the boundary of the neighboring ranges. We have in the limit $t \to -\infty$ where $\Delta \ll \kappa$ that $|c_1| = 1$ and $c_2 = 0$. Next, in the intermediate range $\Delta \sim \kappa$ the set of equations (3.4) takes the form

$$i\frac{dc_1}{dt} = \frac{\kappa}{2}c_1 + \delta e^{t/\tau}c_2, \quad i\frac{dc_2}{dt} = \delta e^{t/\tau}c_1 - \frac{\kappa}{2}c_2, \tag{3.18}$$

where the values κ and δ do not depend on time. The solution of this set under boundary conditions $c_1(-\infty) = 1,\ c_2(-\infty) = 0$ is

$$c_1 = \sqrt{\frac{\pi\kappa\tau}{2\cosh\frac{\pi\kappa\tau}{2}}}e^{t/(2\tau)}J_{-1/2-i\kappa\tau/2}(\delta\tau e^{t/\tau}),$$

$$c_2 = \sqrt{\frac{\pi\kappa\tau}{2\cosh\frac{\pi\kappa\tau}{2}}}e^{t/(2\tau)}J_{1/2-i\kappa\tau/2}(\delta\tau e^{t/\tau}).$$

In the limit $t \to \infty$ where $\Delta \gg \kappa$ we have from this on the basis of the asymptotic expressions for the Bessel functions

$$c_1 = \left(\cosh\frac{\pi\kappa\tau}{2}\right)^{-1/2}\cos\left(\delta\kappa e^{t/\tau} + i\frac{\pi\kappa\tau}{4}\right),$$

$$c_2 = -i\left(\cosh\frac{\pi\kappa\tau}{2}\right)^{-1/2}\sin\left(\delta\kappa e^{t/\tau} + i\frac{\pi\kappa\tau}{4}\right).$$

One can reduce this to a general dependence for $\Delta(t)$ in a range $\Delta \gg \kappa$, and then these solutions in the limit $t \to \infty$ are

$$c_1 = \left(\cosh\frac{\pi\kappa\tau}{2}\right)^{-1/2}\cos\left(\int\limits_{-\infty}^{\infty}\Delta dt + i\frac{\pi\kappa\tau}{4}\right),$$

$$c_2 = -i\left(\cosh\frac{\pi\kappa\tau}{2}\right)^{-1/2}\sin\left(\int\limits_{-\infty}^{\infty}\Delta dt + i\frac{\pi\kappa\tau}{4}\right).$$

When the quasimolecule passes a range $\Delta \sim \kappa$ once more, its evolution is described by the diabatic set of equations (3.18) with change $t \to -t$. This set of equations is

$$i\frac{dc_1}{dt} = \frac{\kappa}{2}c_1 + \delta e^{-t/\tau}c_2, \quad i\frac{dc_2}{dt} = \delta e^{-t/\tau}c_1 - \frac{\kappa}{2}c_2.$$

Solving this set and transferring to the limit $t \to \infty$ as above, we finally obtain the probability of the transition of the Rosen–Zener–Demkov formula,

$$P = \left|c_2(t=\infty)\right|^2 = \frac{\sin^2\left(\int\limits_{-\infty}^{\infty}\Delta dt\right)}{\cosh^2\frac{\pi\kappa\tau}{2}}. \tag{3.19}$$

If we average over the phase, formula (3.19) gives the Rosen–Zener formula

$$P = \frac{1}{2\cosh^2\frac{\pi\kappa\tau}{2}}. \tag{3.20}$$

Note that the parameter τ in this formula is given by

$$\tau = \frac{a}{v_R}, \quad a = \left(\frac{d\ln\Delta(R)}{dR}\right)^{-1},$$

where v_R is the normal component of the relative velocity of colliding particles. We consider the adiabatic limit, i. e., the limit of small velocities when the Rosen–Zener formula gives the transition probability

$$P = 2\exp\left(-\pi\kappa\tau\right),$$

in accordance with formula (3.17) since $P = 2P_{12}$ in the adiabatic limit and $\gamma = 1/\tau$. The Massey parameter (3.9) is given by $\xi = \pi\kappa\tau$, and this coincides with formula (3.9) with the accuracy up to a constant factor. In the case when the Massey parameter (3.9) is small, the Rosen–Zener–Demkov formula (3.19) is transformed into formula (3.13), and the transition probability is 1/2 on average, as in the case of the resonant charge exchange at small impact collision parameters.

3.2 Resonant Charge Exchange and Similar Processes

► **Problem 3.9** Connect the cross section of the resonant charge exchange in slow collisions with the parameters of the problem, assuming the cross section to be large compared to an atomic value, i. e., electron transfer proceeds at large distances between a colliding ion and atom in comparison with their sizes.

Using expression (3.13) for the probability of the charge exchange process, we obtain the cross section of this process,

$$\sigma_{\rm res} = \int_0^\infty 2\pi\rho d\rho \sin^2\zeta(\rho), \qquad \zeta(\rho) = \int_\infty^\infty \frac{\Delta(R)}{2} dt. \tag{3.21}$$

Here $\zeta(\rho)$ is the charge exchange phase, and ρ is the impact parameter of collision. Taking into account that the main contribution to the cross section of electron transfer goes from the large impact parameter of collision, where the charge exchange phase depends strongly on the impact parameter, we approximate this quantity in the range of large impact parameters by the dependence $\zeta(\rho) = A\rho^{-n}$ with large n. Then we obtain the cross section of the resonant charge exchange,

$$\sigma_{\rm res} = \int_0^\infty 2\pi\rho d\rho \sin^2\zeta(\rho) = \frac{\pi}{2}(2A)^{2/n}\Gamma\left(1-\frac{2}{n}\right)\cos\frac{\pi}{n} .$$

Let us represent the cross section in the form

$$\sigma_{\rm res} = \frac{\pi R_o^2}{2} f_n ,$$

where R_o is determined by the relation $\zeta(R_o) = a$, and the function f_n is equal to

$$f_n = (2a)^{\frac{2}{n}}\Gamma(1-\frac{2}{n})\cos\frac{\pi}{n} .$$

We now expand the cross section over a small parameter $1/n$ and restrict by two expansion terms. For this representation it is convenient to take an arbitrary parameter a such that the expansion term of the function f_n that is proportional to $1/n$ would be zero. This gives

$$a = \frac{e^{-C}}{2} = 0.28 ,$$

where $C = 0.557$ is the Euler constant. Then the expansion of the function f_n at large n has the form

$$f_n = 1 - \frac{\pi^2}{6n^2} .$$

Correspondingly, the cross section of the resonant charge exchange is determined by the relation

$$\sigma_{\rm res} = \frac{\pi R_o^2}{2}, \quad \text{where} \quad \zeta(R_o) = \frac{e^{-C}}{2} = 0.28. \tag{3.22}$$

► **Problem 3.10** In the two-state approach express the exchange interaction potential for an ion and parent atom at large separations through the wave function of a valence s-electron.

The exchange interaction of atomic particles is determined by overlapping of the electron wave functions which belong to different atomic centers. The Hamiltonian of an electron when it is located in the field of two cores at large separations has the form

$$\hat{H} = -\frac{1}{2}\Delta + V(r_1) + V(r_2) + \frac{1}{R}\,.$$

Here we use the atomic units ($\hbar = m_e = e^2 = 1$), R is the distance between atomic cores, r_1, r_2 are the distances of the electron from the considering nucleus, $V(r)$ is the interaction potential of the electron with ion, and far from the ion this potential is the Coulomb one $V(r) = -1/r$. Due to the symmetry of the problem, the eigenfunctions are expressed by formulas (3.11) through the wave functions centered on a certain core and satisfy to the Schrödinger equations (3.2).

For determining the ion–atom exchange interaction potential we multiply the first equation by ψ_u^*, the second equation by ψ_g^*, take the difference of the obtained equations and integrate the result over the volume which is a half-space restricted by the symmetry plane. Since the distance between the nuclei is large, the wave function ψ_2 is zero inside this volume and the wave function ψ_1 is zero outside of this volume. Hence $\int_V \psi_u^* \psi_g d\mathbf{r} = 1/2$, and the relation obtained has the form

$$\frac{\Delta(R)}{2} = \frac{1}{2}\int_V \left(\psi_u \Delta\psi_g - \psi_g \Delta\psi_u\right) d\mathbf{r} = \frac{1}{2}\int_S \left(\psi_2 \frac{\partial}{\partial z}\psi_1 - \psi_1 \frac{\partial}{\partial z}\psi_2\right) ds\,,$$

where S is the symmetry plane which limits the integration range; we take the z axis that joins nuclei, and the origin of the reference frame is located in the center of the line joining nuclei. Since the electron is found in the s-state in the field of each atomic core, its wave functions in this coordinate system can be represented in the form

$$\psi_1 = \psi\left(\sqrt{(z+R/2)^2+\rho^2}\right), \quad \psi_2 = \psi\left(\sqrt{(z-R/2)^2+\rho^2}\right),$$

where ρ is the distance from the axis in the perpendicular direction to it. Since $ds = 2\pi\rho d\rho$, we have from the above relation

$$\frac{\Delta(R)}{2} = \int_0^\infty 2\pi\rho d\rho \left[\psi\left(\sqrt{(z-R/2)^2+\rho^2}\right)\frac{\partial}{\partial z}\psi\left(\sqrt{(z+R/2)^2+\rho^2}\right)\right.$$
$$\left. -\psi\left(\sqrt{(z+R/2)^2+\rho^2}\right)\frac{\partial}{\partial z}\psi\left(\sqrt{(z-R/2)^2+\rho^2}\right)\right]_{z=0}.$$

Let us use the obvious relation

$$\frac{\partial}{\partial z}\left[\psi\left(\sqrt{(z+R/2)^2+\rho^2}\right)\right]_{z=0} = R\frac{\partial}{\partial \rho^2}\psi\left(\sqrt{\frac{R^2}{4}+\rho^2}\right),$$

which reduces the above formula to the form

$$\frac{\Delta(R)}{2} = R\int_0^\infty d\rho^2 \frac{\partial}{\partial\rho^2}\psi^2\left(\sqrt{\frac{R^2}{4}+\rho^2}\right) = R\,\psi^2\left(\frac{R}{2}\right). \tag{3.23}$$

The wave function of this formula is centered on one core by taking into account the action of another core; this molecular wave function differs from the wave function in the field of one core by a numerical coefficient. At large distances r between the electron and core the atomic wave function of the electron is given by

$$\psi_{at}(r) = Ar^{1/\gamma-1}e^{-r\gamma},$$

where A is a numerical factor, and in atomic units $\gamma = \sqrt{2J}$, where J is the ionization potential of the atom. This leads to the following dependence on the separation R for the ion–atom exchange interaction potential:

$$\Delta(R) = CR^{2/\gamma-1}\exp(-R\gamma), \tag{3.24}$$

where C is a numerical coefficient.

▶ **Problem 3.11** Determine the dependence on the collision velocity for the cross section of the resonant charge exchange in slow collisions.

We consider the resonant charge exchange as a tunnel transition of a valence electron between two identical cores, and this transition proceeds at large impact parameters of collision and distances between nuclei of colliding atomic particles in comparison with typical atomic sizes. This leads to a sharp dependence of the exchange interaction potential (3.24) on a distance between nuclei, and correspondingly a sharp dependence of the charge exchange phase (3.21) on the impact parameter of collision, and we will approximate it by formula $\zeta(\rho) \sim e^{-\gamma\rho}/v$. Then we have on the basis of formula (3.22)

$$R_o = \frac{1}{\gamma}\ln\frac{v_o}{v},$$

and a parameter $v_o \gg 1$. This leads to the following formula for the cross section:

$$\sigma_{res} = \frac{\pi R_o^2}{2} = \frac{\pi}{2\gamma^2}\ln^2\frac{v_o}{v}. \tag{3.25}$$

Note that since transitions proceed at large distances between cores, the parameter γR_o is large (in reality in thermal collisions it exceeds 10). Therefore, the velocity dependence for the cross section is enough weak. Indeed, from this formula it follows

$$\frac{d\ln\sigma_{res}}{d\ln v} = -\frac{2}{\gamma R_o} \ll 1.$$

In particular, this allows us to represent the velocity dependence for the cross section in the form

$$\frac{\sigma_{res}(v)}{\sigma_{res}(v_1)} = \left(\frac{v_1}{v}\right)^\alpha, \quad \alpha = \frac{2}{\gamma R_o}. \tag{3.26}$$

► **Problem 3.12** Determine the cross section of the charge exchange process for collisions of a slow highly excited atom and ion, assuming a bound electron to be classical.

Under these conditions, a weakly bound electron is located in the field of two Coulomb centers and interacts with cores through the interaction potential

$$U = -\frac{1}{r_1} - \frac{1}{r_2} + \frac{1}{R},$$

where r_1, r_2 are the electron distances from the corresponding nucleus, and R is the distance between nuclei. We assume that the electron transition between cores proceeds in the under-barrier manner (see Fig. 3.4). Take the electron binding energy with its core to be $\varepsilon = -\gamma^2/2$, where $J = \gamma^2/2$ is the ionization potential for this atom state. Let us assume that an approach of the atom and ion proceeds enough fast so that the electron energy does not vary during motion of nuclei (the diabatic case). Then the condition of disappearance of the barrier $U(r_1 = r_2 = R_0/2) = \varepsilon$ gives $R_0 = 8/\gamma^2$. Assuming the under-barrier character of the electron transition to another core, we obtain the cross section of the resonant charge exchange process in this case:

$$\sigma_{res} = \frac{\pi R_0^2}{2} = \frac{32\pi}{\gamma^4} = \frac{8\pi}{J^2}. \tag{3.27}$$

In the other limiting case (the adiabatic case) the electron follows to the variation of core fields, and its binding energy in the fields of two cores becomes at a distance R between them $\varepsilon = -\gamma^2/2 + 1/R$. Correspondingly, the distance R_0 between nuclei when the barrier between cores disappears, which follows from the relation $U(r_1 = r_2 = R_0/2) = 4/R_0 = \varepsilon = -\gamma^2/2 + 1/R_0$, becomes $R_0 = 6/\gamma^2$. Hence, the cross section of the resonant charge exchange process in the adiabatic case is equal to

$$\sigma_{res} = \frac{\pi R_0^2}{2} = \frac{18\pi}{\gamma^4} = \frac{9\pi}{2J^2}. \tag{3.28}$$

As is seen, the cross sections differ significantly in the diabatic and adiabatic cases.

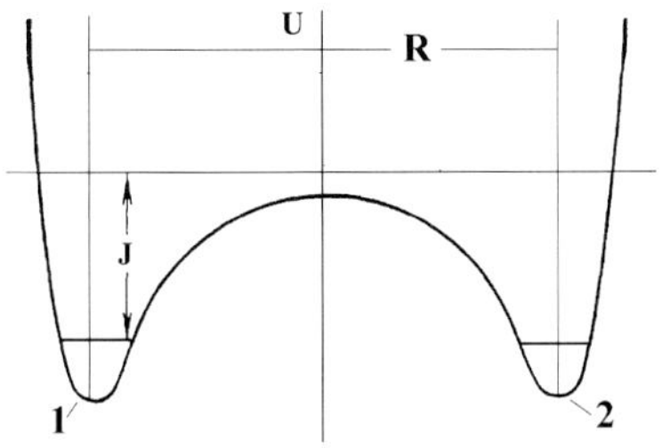

Fig. 3.4 Planar section for the potential well for the resonant charge exchange process when the electron transfers from the field of one core to another one.

▶ **Problem 3.13** Find the total cross section and the cross section of exchange excitation for collision of the S-atom in the ground state with the resonantly excited atom in the P-state.

Scattering of atoms in the ground and resonantly excited states results from the dipole–dipole interaction whose interaction operator V has the form

$$V = \frac{1}{R^3}\left[\boldsymbol{D}_1\boldsymbol{D}_2 - 3(\boldsymbol{D}_1\boldsymbol{k})(\boldsymbol{D}_2\boldsymbol{k})\right], \tag{3.29}$$

where R is the distance between atoms, $\boldsymbol{D}_1$, $\boldsymbol{D}_2$ are the operators of the dipole moment for the first and second atoms, and $\boldsymbol{k}$ is the unit vector along the axis that joins atoms.

The Hamiltonian of this atomic system consisting of two identical interacting atoms is conserved if electrons are reflected with respect to the symmetry plane that is perpendicular to the axis joining atoms and bisects it. Correspondingly, the states of this atomic system can be even or odd, so the corresponding wave function conserves or changes the sign as a result of electron reflection with respect to the symmetry plane. Let ψ be the wave function of the ground state and φ be the wave function of the resonantly excited state. Then the eigenfunctions Ψ_{g}, Ψ_{u} of the two interacting atoms are

$$\Psi_{\mathrm{g,u}} = \frac{1}{\sqrt{2}}(\psi_1\varphi_2 \pm \varphi_1\psi_2)\,,$$

where indices 1, 2 are the numbers of atoms, and the plus and minus signs correspond to the even and odd states, respectively. From this we obtain the interaction potential of atoms, $U(R)$, in the first order of the perturbation theory, taking the z-axis along the direction joining the atoms,

$$U(R) = \pm\frac{1}{R^3}\left(|\langle\psi|D_x|\varphi\rangle|^2 + |\langle\psi|D_y|\varphi\rangle|^2 - 2|\langle\psi|D_z|\varphi\rangle|^2\right),$$

where D_x, D_y, D_z are the operators of the corresponding projections of the atom dipole moment.

Let us introduce the parameter

$$d^2 = \frac{1}{3}\sum_m |\langle 00|\boldsymbol{D}|1m\rangle|^2, \tag{3.30}$$

where $\boldsymbol{D}$ is the operator of the atom dipole moment, the matrix element is taken between the ground S-state and the resonantly excited P-state, and m is the orbital momentum projection for the excited state. This value can be expressed in atomic units through the total oscillator strength f for the transition in the excited P-state in ignoring its spin–orbit splitting (S-P transition) as

$$d^2 = \frac{f}{2\Delta\varepsilon}, \tag{3.31}$$

where $\Delta\varepsilon$ is the transition energy. Table 3.1 contains the values of this parameter for atoms of the first and second groups of the periodical system of elements.

In ignoring spin–orbit splitting of levels in the interaction between atoms in the ground *S*-state and resonantly excited *P*-state, the interaction potentials are

$$U_{zz} = \pm\frac{d^2}{R^3}, \quad U_{xx} = \pm\frac{d^2}{R^3}(1 - 3\cos^2\theta), \qquad U_{yy} = \pm\frac{d^2}{R^3}(1 - 3\sin^2\theta)\,, \tag{3.32}$$

where the axis $\boldsymbol{R}$ joining atoms is located in the plane xy and forms an angle θ with the axis x; the plus and minus signs refer to the even and odd states, respectively, as above.

Table 3.1 Parameters of transitions between the ground and first excited states for atoms of alkaline metals and alkaline-earth metals.

Element	Transition	$\Delta\varepsilon$ (eV)	τ (ns)	$d^2, e^2a_o^2$	$v\sigma_t(10^{-7}\,\mathrm{cm}^3/\mathrm{s})$
H	$1^2S \to 2^2P$	10.20	1.60	0.555	0.516
He	$1^1S \to 2^1P$	21.22	0.56	0.177	0.164
Li	$2^2S \to 3^2P$	1.85	27	5.4	5.1
Be	$2^1S \to 2^1P$	5.28	1.9	3.4	9.6
Na	$3^2S_{1/2} \to 3^2P_{1/2}$	2.10	16	6.2	3.1
Na	$3^2S_{1/2} \to 3^2P_{3/2}$	2.104	16	6.2	4.8
Mg	$3^1S_0 \to 3^1P_1$	4.35	2.1	5.9	5.5
K	$4^2S_{1/2} \to 4^2P_{1/2}$	1.61	27	8.9	4.1
K	$4^2S_{1/2} \to 4^2P_{3/2}$	1.62	27	8.8	6.3
Ca	$4^1S_0 \to 4^1P_1$	2.93	4.6	7.9	7.3
Cu	$4^2S_{1/2} \to 4^2P_{1/2}$	3.79	7.0	2.4	1.1
Cu	$4^2S_{1/2} \to 4^2P_{3/2}$	3.82	7.2	2.4	1.7
Zn	$4^1S_0 \to 4^1P_1$	5.80	1.4	3.5	3.3
Rb	$5^2S_{1/2} \to 5^2P_{1/2}$	1.56	28	8.4	3.9
Rb	$5^2S_{1/2} \to 5^2P_{3/2}$	1.59	26	8.6	6.2
Sr	$5^1S_0 \to 5^1P_1$	2.69	6.2	10.1	9.4
Ag	$5^2S_{1/2} \to 5^2P_{1/2}$	3.66	7.9	2.4	1.1
Ag	$5^2S_{1/2} \to 5^2P_{3/2}$	3.78	6.7	2.4	1.7
Cd	$5^1S_0 \to 5^1P_1$	5.42	1.7	3.5	3.3
Cs	$6^2S_{1/2} \to 6^2P_{1/2}$	1.39	31	11.5	5.3
Cs	$6^2S_{1/2} \to 6^2P_{3/2}$	1.46	27	11.4	8.1
Ba	$6^1S_0 \to 6^1P_1$	2.24	8.5	9.7	9.0
Au	$6^2S_{1/2} \to 6^2P_{1/2}$	4.63	6.0	1.06	0.49
Au	$6^2S_{1/2} \to 6^2P_{3/2}$	5.10	4.6	1.04	0.75
Hg	$6^1S_0 \to 6^1P_1$	6.70	0.13	2.4	2.2

In order to analyze the dynamics of the development of the system of colliding atoms, we take as a basis the nonstationary Schrödinger equation for the wave function Ψ of colliding particles, $i\hbar\partial\Psi/\partial t = \hat{H}\Psi$, where $\hat{H}$ is the Hamiltonian of the system, which is a sum $\hat{H} = \hat{H}_o + V$ of the Hamiltonian $\hat{H}_o$ of noninteracting atoms and the operator V of their interaction. We take now the wave function in the form

$$\Psi = \sum_i [a_i^+\psi_i^+ + a_i^-\psi_i^-]\,.$$

Here i is the quantum number of the state, in this case it is the momentum projection onto a given direction, ψ_i^+ and ψ_i^- are the even and odd wave functions if an excited atom is found in the i-th state. If we substitute this expansion of the wave function in the nonstationary Schrödinger equation, after standard operations we obtain the equations for the expansion amplitudes:

$$i\hbar\dot{a}_z^+ = \frac{d^2}{R^3}a_z^+, \quad i\hbar\dot{a}_z^- = -\frac{d^2}{R^3}a_z^-$$

$$i\hbar\dot{a}_x^+ = \frac{d^2}{R^3}(1-3\cos^2\theta)a_x^+, \quad i\hbar\dot{a}_x^- = -\frac{d^2}{R^3}(1-3\cos^2\theta)a_x^-$$

$$i\hbar\dot{a}_y^+ = \frac{d^2}{R^3}(1-3\sin^2\theta)a_y^+, \quad i\hbar\dot{a}_y^- = -\frac{d^2}{R^3}(1-3\sin^2\theta)a_y^- .$$

We take as the quantum states for the P-atom the states with zero projection of the orbital momentum on the axes z, x, and y, respectively, and for collision of two atoms the z-axis is perpendicular to the motion plane, the axis x corresponds to the zero momentum projection onto the direction of the impact parameter of collision, and the axis y is directed along the relative velocity; atoms are moving along the straightforward trajectories. We use above the matrix elements (3.32) for the interaction operators, and θ is the angle between the axis $\mathbf{R}$ joining atoms and the impact parameter of collision $\boldsymbol{\varrho}$.

Note that the states x and y are bound through the angle θ, which vary in time such that $\dot{\theta} = -\rho v/R^2$, and a current distance R between atoms due to their free motion is expressed through the impact parameter ρ of collision and the collision velocity v as $R^2 = \rho^2 + v^2t^2$. Let us solve the equation for the z-state when the momentum projection is zero on the axis that is perpendicular to the motion plane and process of excitation transfer and momentum rotation are not entangled. We have

$$a_z^+ = \exp\left(-\frac{i}{\hbar}\int_{-\infty}^{\infty}\frac{d^2}{R^3}dt\right) = \exp\left(-\frac{2id^2}{\hbar v\rho^2}\right)$$

$$a_z^- = \exp\left(\frac{i}{\hbar}\int_{-\infty}^{\infty}\frac{d^2}{R^3}dt\right) = \exp\left(\frac{2id^2}{\hbar v\rho^2}\right) .$$

From this we obtain the cross section of the exchange excitation σ_{exc}, the cross section of the elastic scattering σ_{el}, and the total cross section in this collision $\sigma_{\text{t}} = \sigma_{\text{exc}} + \sigma_{\text{el}}$,

$$\sigma_{\text{exc}}^z = \int_0^{\infty}|\text{Im}\, a_z^+|^2 \cdot 2\pi\rho d\rho = \frac{\pi^2 d^2}{\hbar v}; \quad \sigma_{\text{el}}^z = \int_0^{\infty}|1-\text{Re}\, a_z^+|^2 \cdot 2\pi\rho d\rho = \frac{\pi^2 d^2}{\hbar v}$$

$$\sigma_{\text{t}}^z = \frac{2\pi^2 d^2}{\hbar v} .$$

In two other cases when at the beginning the axis onto which the momentum projection for an excited atom is zero is located in the motion plane, equations for the appropriate amplitudes are solved only numerically, but all the cross sections are proportional to $d^2/\hbar v$. Then the cross sections of the exchange transfer are

$$\sigma_{\rm exc}^x = \frac{2.65\pi d^2}{\hbar v}; \quad \sigma_{\rm exc}^y = \frac{\pi d^2}{\hbar v}; \quad \overline{\sigma_{\rm exc}} = 2.26\frac{\pi^2 d^2}{\hbar v}, \tag{3.33}$$

where an overline denotes an average over momentum directions. In the same manner we have for average cross sections

$$\overline{\sigma_{\rm el}} = \frac{2.58\pi d^2}{\hbar v}; \quad \overline{\sigma_{\rm t}} = \frac{4.84\pi d^2}{\hbar v}. \tag{3.34}$$

Table 3.1 contains the values $v\sigma_{\rm t}$ averaged over momentum projections.

► **Problem 3.14** In the case of collisions of alkali metal atoms in the ground *S*-state and resonantly excited ^2P_j-states ($j = 1/2,\ 3/2$) analyze the role of spin–orbit splitting in excitation transfer for such collisions.

We consider above the case when spin–orbit splitting of levels for excited states that we denote by $\delta_{\rm f}$, may be ignored. We now estimate when this consideration holds true for this consideration. Transitions due to the dipole–dipole interaction of atoms take place at typical distances

$$R_{\rm o} \sim \sqrt{\frac{d^2}{\hbar v}},$$

where transfer excitation proceeds. Transitions between fine structure states take place at distances $\sim R_{\rm f}$ where the potential of the dipole–dipole interaction of atoms $\sim d^2/R_{\rm f}^3$ is comparable with the fine splitting of levels $\delta_{\rm f}$, that is

$$R_{\rm f} \sim \left(\frac{d^2}{\delta_{\rm f}}\right)^{1/3}.$$

Evidently, fine splitting of levels can be ignored if $R_{\rm f} \ll R_{\rm o}$, or the parameter ζ, which is defined by

$$\zeta \equiv \frac{R_{\rm o}}{R_{\rm f}} = \frac{(\delta_{\rm f})^{1/3} d^{1/3}}{(\hbar v)^{1/2}}, \tag{3.35}$$

is large, $\zeta \gg 1$. Table 3.2 gives parameters for alkali atom collisions in the ground and first resonantly excited state if they proceed in a vapor of temperature 500 K.

The Massey parameter ξ for this process is estimated as

$$\xi \sim \frac{\delta_{\rm f} R_{\rm f}}{\hbar v} \sim \frac{(\delta_{\rm f})^{2/3} d^{2/3}}{\hbar v} \sim \zeta^2,$$

and according to the Table 3.2 data, the Massey parameter ξ is large for most cases of alkali metal atoms if an excited atom is found in the lowest state. In this case the probability of transitions between states of fine structure is small, and excitation transfer takes place in each state of fine structure independently. In addition, transitions between different fine states in collisions proceed only due to the intersection of electron terms of a different fine structure.

Table 3.2 Parameters of collision between alkali metal atoms in the ground and first resonantly excited state at the temperature of 500 K. $\overline{v}$ is the average collision velocity and the parameter ζ is given by formula (3.35).

Atom	Li	Na	K	Rb	Cs
$\delta_f\,(\text{cm}^{-1})$	0.34	17.2	57.6	238	554
d^2 (a.u.)	5.4	6.2	8.8	8.5	11.5
$\overline{v}\,(10^4\ \text{cm/s})$	17	9.6	7.4	5.0	4.0
ζ	0.44	2.3	4.2	9.6	13

▶ **Problem 3.15** Estimate the cross section and rate constant of excitation transfer in collisions of atoms in the ground S and resonantly excited P-states if the fine structure splitting is large.

As follows from the Table 3.2 data, the Massey criterion for thermal collisions of most alkali metal atoms in the ground ^2S and resonantly excited ^2P-states is large. This means that excitation transfer proceeds independently for each fine state of excited atoms. But excitation transfer for each collision channel is determined by dipole–dipole interaction atoms, as earlier, so the cross section of excitation transfer for each fine state is estimated as

$$\sigma \sim \frac{d^2}{\hbar v}\,,$$

and the depolarization cross section has the same order of magnitude. The numerical coefficients in this expression depend on the behavior of electron terms and are given in Table 3.3 for excitation transfer.

Table 3.3 The cross sections of excitation transfer for collisions of atoms in the ground S-state and resonantly excited P-states for two limiting cases with respect to the fine splitting of levels.

Colliding atoms	$\sigma_{\text{exc}}/\sigma_o$	$\sigma_{\text{el}}/\sigma_o$	σ_t/σ_o
$A(S) + A(P)$	2.26	2.58	4.84
$A(^2S_{1/2}) + A(^2P_{1/2})$	1.12	1.29	2.41
$A(^2S_{1/2}) + A(^2P_{3/2})$	1.66	2.06	3.72

Taken from Table 3.3 the values of the rate constants for the total elastic scattering and excitation transfer $\langle v\sigma_t\rangle$ for each state of fine structure are given in Table 3.1.

▶ **Problem 3.16** Determine the cross section and rate constant of transitions between fine structure states for collisions of atoms in the S- and P-states if the fine structure splitting may be ignored.

The process under consideration proceeds according to the scheme

$$M(^2S_{1/2}) + M(^2P_{1/2}) \longleftrightarrow M(^2S_{1/2}) + M(^2P_{3/2}). \tag{3.36}$$

As a result of this transition, the electron energy varies by δ_f, the energy of fine splitting of levels, so that this energy is transferred or is taken from the kinetic energy of colliding atoms. Since the electron states $^2P_{1/2}, ^2P_{3/2}$ are resonantly excited, transitions between these states in collisions vary the wavelength of subsequent radiation, and for this reason transitions between fine structure levels are of interest.

According to the data of Table 3.2, the condition of a smallness of a fine splitting energy is not fulfilled for the lowest resonantly excited states of alkali metals, but it is valid for more excited states with lower fine splitting of levels. The Massey parameter (3.9) is small for such collisions, and the system of colliding atoms does not "feel" fine splitting of atoms. Therefore, excitation transfer takes place independently on a fine state of an excited atom, so an excited electron "chooses" a final fine state in a random way. Hence, the cross sections of the transition between fine structure states in this case is equal to

$$\sigma\left(\frac{1}{2} \to \frac{3}{2}\right) = \frac{1}{2}\sigma\left(\frac{3}{2} \to \frac{1}{2}\right) = \frac{1}{3}\sigma_{\text{exc}} ,$$

where the argument indicates fine states of transition, and σ_{exc} is the cross section of the exchange transfer in accordance with formula (3.33), which is given by

$$\sigma_{\text{exc}} = 2.26\pi\frac{d^2}{\hbar v} ,$$

and correspondingly the cross sections for transitions between fine structure states are equal to

$$\sigma\left(\frac{1}{2} \to \frac{3}{2}\right) = 0.75\pi\frac{d^2}{\hbar v}, \quad \sigma\left(\frac{3}{2} \to \frac{1}{2}\right) = 1.5\pi\frac{d^2}{\hbar v} .$$

▶ **Problem 3.17** A resonantly excited atom of alkali metal in the state ^2P collides with the parent atom in the ground ^2S state. Find the cross section of the transition between states of fine structure if the Massey parameter (3.9) for the direct transition between fine states is large.

Because the Massey parameter (3.9) for the transition between fine structure states is large in this case, such transitions are possible only between intersection electron terms. This relates to a term for an electron state $^2P_{1/2}$ that is directed up and some electron term of the electron state $^2P_{3/2}$ that is directed down. these electron terms are intersected at a distance between colliding atoms,

$$R_0 \sim \left(\frac{d^2}{\delta_f}\right)^{1/3} .$$

The character of the transition under consideration is as follows. Because intersecting electron terms are characterized by different projections of the electron momentum onto the axis connected atoms, the transition takes place due to rotation of the axis and is realized at distances between nuclei when the energy difference for these electron terms is lower or is of the order of $\hbar\dot{\theta}$, where θ is the angle between the rotating axis and its initial direction. Thus, the transition between fine structure states proceeds in the range ΔR of the distances between atoms according to

$$\frac{d^2}{\hbar R_0^3}\frac{\Delta R}{R_0} \sim \dot{\theta} = \frac{v}{R_0} .$$

During this time the axis rotates by an angle $\Delta\theta \sim \Delta R/R_0$, and it is of the order of the transition probability

$$P \sim \frac{\Delta R}{R_0} \sim \frac{vR_0^2}{d^2} .$$

We now assume that the kinetic energy of colliding atoms ε exceeds significantly the fine structure splitting δ_f, which allows for the system of colliding atoms to remain on a new electron term up to the infinite distance between the atoms.

The cross section of the transition between fine structure states is estimated as

$$\sigma \sim PR_0^2 \sim \frac{vR_0^4}{d^2} \sim \frac{vd^{2/3}}{\delta_f^{4/3}} .$$

Using numerical coefficients in this expression by taking into account a certain behavior of electron terms for this electron system, we find for the transition cross section

$$\sigma\left(\frac{3}{2} \to \frac{1}{2}\right) = \frac{1}{2}\sigma\left(\frac{1}{2} \to \frac{3}{2}\right) = 1.4\frac{vd^{2/3}}{\delta_f^{4/3}} .$$

Correspondingly, the rate constant of the fine structure transition for the Maxwell distribution of atoms in this case $T/gg\delta_f$ is equal to

$$k\left(\frac{3}{2} \to \frac{1}{2}\right) = 3.9\frac{Td^{2/3}}{\mu\delta_f^{4/3}} ,$$

where μ is the reduced mass of colliding atoms.

▶ **Problem 3.18** Estimate the rate constant of the transition between the fine structure states in collisions between a resonantly excited atom of alkali metal in the state ^2P and the parent atom in the ground ^2S state if the distance between fine structure levels δ_f is small compared to a thermal energy T of atoms.

In this case the rate constants of the transition between fine structure levels are connected by the principle of detailed balance,

$$k\left(\frac{3}{2} \to \frac{1}{2}\right) = \frac{1}{2}\exp\left(-\frac{\delta_f}{T}\right) k\left(\frac{1}{2} \to \frac{3}{2}\right) ,$$

and this shows that the transition to an upper state $^2P_{3/2}$ is possible for a tail of the Maxwell distribution function of atoms. The transition to the lower electron terms results from the capture of atoms when colliding atoms are located enough long on the attractive electron term. The capture cross section for the dipole–dipole interaction of atoms ($U(R) \sim d^2/R^3$) is estimated as

$$\sigma_c \sim \left(\frac{d^2}{T}\right)^{2/3} .$$

Correspondingly, the rate constant of this process is

$$k \sim \frac{d^{4/3}}{T^{1/6}\mu^{1/2}} ,$$

where μ is the reduced mass of colliding atoms.

In this process the atom system is located enough long on an attractive electron term for an upper fine structure state at an infinite distance between the atoms, and during this time it has the possibility of transferring to the lower state near the distance of the intersection of these electron terms. Evidently, this character of the transition is valid if an atom capture proceeds at a large distance R_c between the atoms, i. e.,

$$\frac{d^{4/3}\mu^{1/2}T^{1/6}}{\hbar} \ll 1 ,$$

and this criterion along with the Massey criterion $\xi \gg 1$ holds true for excited states with a small value of d.

3.3 Processes Involving Negative Ions

► **Problem 3.19** On the basis of the delta-function model for an electron located in the field of two atoms find the distance between atoms when the electron term of the bound electron state intersects the boundary of the continuous spectrum. Use this in the analysis of the character of the negative ion detachment in collisions of the negative ion and atom.

The process

$$A + B^- \rightarrow A + B + e \tag{3.37}$$

is possible if the electron term of quasimolecules constructed from colliding particles crosses the boundary of the continuous spectrum, and then a weakly bound electron releases. We demonstrate this possibility on the basis of the delta-function model for a valence electron located in the field of two atoms. The delta-function model for the negative ion assumes that the radius of the action of the atom field in a negative ion to be small compared to the size of the negative ion. Hence, one can consider the atomic field as a boundary condition for the electron wave function at the point of atom location.

In considering the behavior of *s*-electron in the field of two atomic centers within the framework of the delta-function model, we take into account that outside the atoms fields are absent and the electron wave function satisfies the Schrödinger equation

$$-\frac{1}{2}\Delta\Psi = -\frac{1}{2}\alpha^2\Psi\,,$$

where $\frac{1}{2}\alpha^2$ is the electron binding energy, and we use below the atomic units. On the basis of this equation and accounting for a short-range atomic fields at points of atom location we construct the electron wave function as

$$\Psi = Ae^{-\alpha r_1}/r_1 + Be^{-\alpha r_2}/r_2\,,$$

where r_1, r_2 are electron distances from the corresponding nucleus. The boundary conditions for the electron wave function near each atom have the form

$$\frac{d\ln(r_1\Psi)}{dr_1}(r_1 = 0) = -\kappa_1, \qquad \frac{d\ln(r_2\Psi)}{dr_2}(r_2 = 0) = -\kappa_2,$$

where $1/\kappa$ is the electron scattering length for a given atom, or $\kappa^2/2$ is the electron binding energy in the negative ion formed on the basis of this atom.

The above boundary conditions give

$$-\kappa_1 = -\alpha + \frac{B}{A}e^{-\alpha R}/R; \qquad -\kappa_2 = -\alpha + \frac{A}{B}e^{-\alpha R}/R\,,$$

where R is the distance between the atoms. Excluding the parameters A and B from this equation, we obtain

$$(\alpha - \kappa_1)(\alpha - \kappa_2) - e^{-2\alpha R}/R^2 = 0. \tag{3.38}$$

The solution of this equation gives the dependence of the electron binding energy $\frac{1}{2}\alpha^2$ on the distance between the atoms. It is of importance that the electron term can intersect the boundary of the continuous spectrum ($\alpha = 0$). The electron bond is broken at $R_c = \sqrt{\alpha_1\alpha_2} = \sqrt{L_1L_2}$, where L_1, L_2 are the electron scattering lengths for the corresponding atom. Within the framework of the delta-function model, the electron–atom scattering length is $L = 1/\gamma$, where $\gamma^2/2$ is the electron binding energy in the negative ion (the electron affinity to the atom).

In slow collisions when the electron term is above the boundary of the continuous spectrum, i. e., the electron state is an autodetachment one and is characterized by a certain level width Γ, an electron can release as a result of decay of the autodetachment state. In any case, this takes place if $\Gamma\tau > 1$, where τ is the time of the location of the quasimolecule in the autodetachment state.

▶ **Problem 3.20** Estimate the rate constant of recombination of negative and positive ions in slow collisions.

In analyzing the pair process of mutual neutralization of positive and negative ions, which proceeds according to the scheme

$$A^+ + B^- \rightarrow A^* + B, \tag{3.39}$$

we assume the binding energy of an electron in the negative ion to be relatively small, and it can proceed in many excited states of a forming excited atom. Because of the tunnel character of this transition, its rate depends sharply (in the exponential way) on the distance between nuclei. Therefore, we use the model when a tunnel transition takes place and the transition probability is 1, if the distance of the closest approach between colliding ions is less than or equal to R_o, which is the parameter of the problem. If the distance of the closest approach exceeds R_o, the transition probability is zero. Then formula (2.15) gives the cross section of mutual neutralization of ions within the framework of this model,

$$\sigma_{rec} = \pi\rho_o^2 = \pi R_o^2\left(1 + \frac{e^2}{R_o\varepsilon}\right), \tag{3.40}$$

where ρ_o is the impact parameter at which the distance of the closest approach of the colliding ions is R_o, and ε is the kinetic energy of the ions in the center-of-mass frame of reference.

From this it follows that at low collision velocities the recombination cross section is inversely proportional to the collision energy,

$$\sigma_{rec} = \pi R_o e^2/\varepsilon. \tag{3.41}$$

This gives the recombination coefficient of positive and negative ions in an ionized gas, which equals $\alpha = v\sigma_{rec}$. Averaging over the Maxwell energy distribution function for the ions, we find

$$\alpha = \langle v\sigma_{rec}\rangle = \frac{2\sqrt{2\pi}R_o e^2}{\sqrt{\mu T}}, \tag{3.42}$$

where the angle brackets denote averaging over the relative velocities of the ions. Note that since the tunnel transition of a bound electron proceeds effectively at distances between nuclei which exceed significantly the atomic size, usually for thermal collisions $R_o \sim 10\, a_o$, where a_o is the Bohr radius.

► **Problem 3.21** Determine the rate of mutual neutralization of ions (3.39) considering it as the transition between two electron terms and assuming that the transition occurs only due to these two terms and proceeds under optimal conditions.

In contrast to the previous problem, we now consider the process of mutual neutralization of ions (3.39) as the transition between two electron terms, as is shown in Fig. 3.5. This transition is due to the intersection of two electron terms, and the transition process is described by the set of equations (3.4). Assuming that these terms are separated from others, we have for the probability of the transition, P, as a result of double passing of the distance R_c of term intersection,

$$P = 2P_{12}(1 - P_{12}),$$

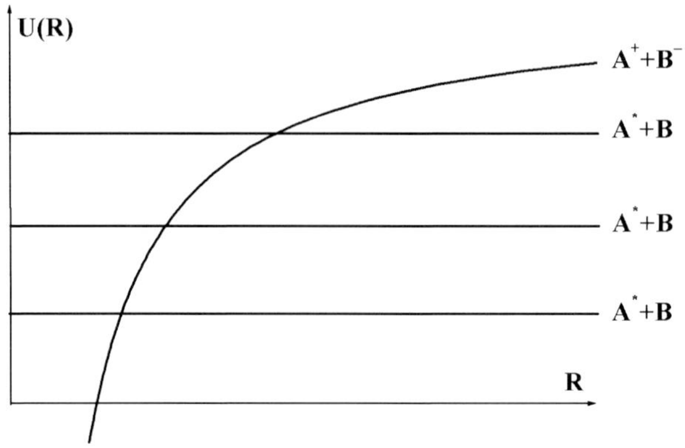

Fig. 3.5 Electron terms for the process of mutual neutralization where the electron term for interaction of positive and negative ions (Coulomb term) is intersected by a group of terms for interaction of neutral atoms in different excited electron states.

where P_{12} is the probability of the transition during one pass through the intersection point. As is seen, the optimal conditions correspond to $P_{12} = 1/2$ and $P = 1/2$, and according to the set of equations (3.4) this takes place if $\Delta \sim \kappa$, i. e., the Massey parameter (3.9) $\xi \sim 1$. Note that in the previous problem where the electron transition can proceed in many atom excited states, we take $P = 1$ under optimal conditions. Then under optimal conditions the cross section of this transition is $\sigma_{\text{rec}} = \pi R_c^2/2$. Correspondingly, the rate of transition is

$$k_{\text{rec}} = \sqrt{\frac{8T}{\pi\mu}}\,\sigma_{\text{rec}} = \sqrt{\frac{2\pi T}{\mu}}R_c^2\,, \tag{3.43}$$

where μ is the reduced mass of ions, and we assume here that the interaction potential of ions at the transition distance to be small compared to the thermal energy of ions $e^2/R_c \ll T$. Accounting for this interaction, we obtain the rate constant of this process,

$$k_{\text{rec}} = \sqrt{\frac{2\pi T}{\mu}}R_c^2\left(1 + \frac{e^2}{R_c T}\right). \tag{3.44}$$

▶ **Problem 3.22** Within the framework of the delta-function model for the electron in the field of two identical atoms with which an electron can form a negative ion, find the width of the autodetachment level at separations near the intersection of the electron term for a negative molecular ion and the boundary of the continuous spectrum.

When the electron term of the system A–A^- is located above the boundary of the continuous spectrum, the state becomes the autodetachment state that testifies the possibility of the atomic system to decay with electron release. On the other hand, one can consider this state as a bound state of the electron and molecule if its lifetime τ is enough large, i. e., this level is enough narrow, and its width $\Gamma = \hbar/\tau$ satisfies to the relation

$$\varepsilon \gg \Gamma , \tag{3.45}$$

where ε is the electron energy after the decay of the autodetachment state.

We now return to the problem under consideration when an electron is found in the field of two identical atoms with which it can form the negative ion. Then the electron terms of such bound states within the framework of the delta-function model for electron–atom interaction is given by equation (3.38), which for identical atoms has the form

$$(\alpha - \kappa)^2 - \frac{e^{-2\alpha R}}{R^2} = 0, \tag{3.46}$$

where for this model $\hbar^2\kappa^2/(2m_e)$ is the electron binding energy in the negative ion on the basis of one atom, and $\hbar^2\alpha^2/(2m_e)$ is the electron binding energy when it is located in the field of two atoms.

Note that because of the problem symmetry, the electron have two bound states with atoms, even and odd, depending on the property of the electron wave function to conserve or change its sign as a result of electron reflection with respect to the symmetry plane that bisects the axis joining nuclei is perpendicular to it. A decrease of the distance between nuclei leads to an increase in the electron binding energy in the even state and a decrease of it in the odd state. Therefore, considering the intersection of the electron term with the boundary of the continuous spectrum, we concentrate on the odd state only.

According to equation (3.46), the intersection takes place at the distance $R_c = 1/\kappa$ between nuclei, and we consider a distance range near the intersection, where equation (3.46) takes the form

$$\frac{1}{R} - \alpha + \frac{\alpha^2 R}{2} - \frac{\alpha^3 R^2}{6} + \alpha - \kappa = 0 ,$$

and we restrict by the expansion terms up to $\sim \alpha^3$. Let us denote $\Delta R = R_c - R$ and the electron energy above the continuous spectrum boundary $\varepsilon = \hbar^2 q^2/2m_e$, and a small parameter of the expansion is $qR_c \ll 1$. From the solution of the above equation we have

$$\varepsilon \equiv \frac{\hbar^2 q^2}{2m_e} = \frac{\hbar^2 \Delta R}{m_e R_c^3}; \quad \Gamma = \varepsilon \frac{qR_c}{3} . \tag{3.47}$$

One can see that criterion (3.45), which allows us to consider the autodetachment state as a bound electron state, holds true.

▶ **Problem 3.23** Determine the cross section of negative ion detachment in a slow collision with an identical atom based on formula (3.47) for the width of the autodetachment level.

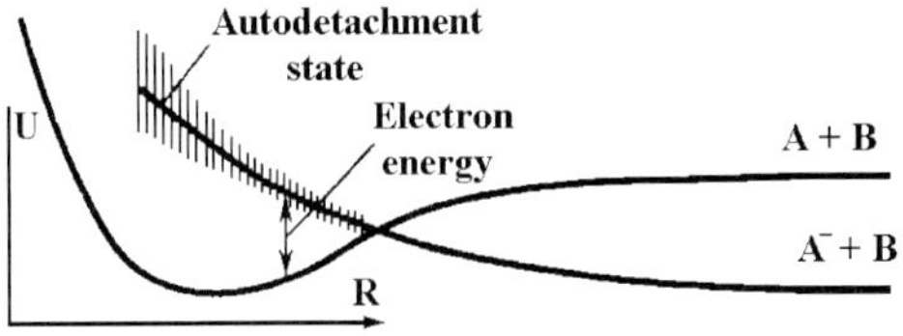

Fig. 3.6 Electron terms which determine the character of the negative ion detachment in collisions with an atom.

This process proceeds at distances between colliding particles where the negative ion term is over the continuous spectrum boundary, and this quasimolecule state becomes autodetaching one. Then the quasimolecule decays in a molecule and a free electron at distances below R_c, the distance of the term intersection (see Fig. 3.6). Correspondingly, the cross section of detachment of a negative ion in a slow collision with a parent atom is given by

$$\sigma_{\text{det}} = \frac{1}{2}\int_0^{R_c}\left[1 - \exp\left(-\frac{1}{\hbar}\int_{R<R_c} \Gamma dt\right)\right] 2\pi\rho d\rho \, .$$

A factor 1/2 accounts for the possibility of detachment of the odd state of the molecular negative only. We use expression (3.47) for the width of the detachment level in the form

$$\Gamma = \frac{\hbar^2\sqrt{2}}{3m_e}\frac{\Delta R^{3/2}}{R_c^{7/2}} \, . \tag{3.48}$$

Take into account that detachment of the negative ion proceeds mostly at distances between colliding particles near the intersection distance R_c, and therefore the probability of surviving for the negative ion is possible for impact parameters of collision which are close to R_c. Therefore, we first evaluate the integral $\hbar^{-1}\int_{R<R_c}\Gamma dt$ at impact parameters of collision near R_c under the assumption that colliding particles move along straightforward trajectories. We have

$$\frac{1}{\hbar}\int_{R<R_c} \Gamma dt = \frac{\pi\hbar\delta\rho^2}{4m_e v R_c^3} \, ,$$

where $\delta\rho = R_c - \rho \ll R_c$. From this we obtain for the detachment cross section as a result of collision of a negative ion with a parent atom

$$\sigma_c = \frac{\pi R_c^2}{2}\left(1 - \sqrt{\frac{m_e v R_c}{\hbar}}\right) . \tag{3.49}$$

Under the above assumptions, the second term in the parentheses is small compared to the first one because of a small collision velocity.

▶ **Problem 3.24** Determine the energy distribution function of electrons which are formed as a result of detachment of a negative ion in collision with an atom and results from the decay of a autodetachment state.

Detachment of a negative ion in collisions with atoms proceeds at distances between colliding particles below R_c where the electron term becomes autodetachment. The probability $P(t)$ of negative ion detachment is given by the equation

$$\frac{dP}{dt} = \frac{\Gamma}{\hbar}(1-P), \quad P = 1 - \exp\left(-\int_{t_c}^{t} \Gamma \frac{dt'}{\hbar}\right).$$

The energy of a released electron is equal to the difference of the energies for an autodetachment term and the boundary of continuous spectrum, so this energy is connected unambiguously with a time after the intersection of electron terms, and therefore the distribution function of electron energies is given by

$$f(\varepsilon) = \frac{dP}{d\varepsilon} = \frac{\Gamma}{\hbar}\frac{dt}{d\varepsilon}\exp\left(-\int_0^{\varepsilon} \Gamma \frac{d\varepsilon}{\hbar}\frac{dt}{d\varepsilon}\right).$$

If we take the energy dependence for the autodetachment level width as

$$\Gamma = a\varepsilon^k ,$$

we obtain the electron spectrum in the form

$$f(\varepsilon) = \frac{dP}{d\varepsilon} = C(k+1)\varepsilon^k \exp\left(-C\varepsilon^{k+1}\right) , \quad C = \frac{a}{v_R \hbar \frac{dE}{dR}} ,$$

where v_R is the normal component of the collision velocity, and E is the difference of the electron energies of the considering electron term and the boundary of the continuous spectrum. In particular, formula (3.47) gives $k = 3/2$, and in this case the distribution function is

$$f(\varepsilon) \sim \varepsilon^{3/2} \exp\left(-C\varepsilon^{5/2}\right). \tag{3.50}$$

3.4 Three-Body Processes

▶ **Problem 3.25** Determine the dependence of the rate of the three-body process on parameters of the problem.

We consider the three-body process that proceeds according to the scheme

$$A + B + C \to AB + C. \tag{3.51}$$

As a result of a three-body collision, particles A and B combine to form a bound system, and particle C carries away the energy released thereby. The balance equation for the number densities of particles of the process (3.51) has the form

$$d[AB]/dt = K[A][B][C] , \tag{3.52}$$

where $[X]$ is the number density of particles X, and K is the rate constant of the three-body process, with dimensionality cm^6/s. The balance equation (3.52) is the definition of the three-body rate constant.

The rate constant of the three-body process can be estimated on the basis of the Thomson theory. This theory takes into account the nature of the three-body process (3.51) and the fact that a binding energy of forming bound state of atomic particles is of the order of a typical thermal energy of particles, $\sim T$. The latter in turn is less significantly than the binding energy of particles in the final state AB. In addition, because the formation of a bound state corresponds to highly excited states of AB, the motion of the particles is governed by classical laws.

Let us find the rate of formation of a bound state AB in the three-body collision (3.51) on the basis of the character of this process. Indeed, formation of a bound state of particles A and B occurs in the following way. As particles A and B approach each other, their energy increases as the potential energy of interaction is converted into kinetic energy. If a third particle C interacts strongly with A or B when these particles are close to each other, then the third particle takes from A or B an energy in excess of the initial kinetic energy of these particles. The bound state of particles A and B is thus formed as a result of a collision with the third particle.

Being guided by the above character of formation of a bound state, we find the rate constant of a three-body process taking into account that typical kinetic energies of colliding particles are of the order of a thermal energy $\sim T$. Assume the mass of the third particle C to be comparable to the mass of either particle, A or B. Requiring the energy exchange to be more than the initial kinetic energy of particles A and B, we obtain that the interaction potential between these particles during collision with the third particle C must also be of the order of T. From this we define a critical radius b, i. e., a distance between particles C and A or B that provides the possibility of forming a bound state of AB in three-body collisions

$$U(b) \sim T . \tag{3.53}$$

Here $U(R)$ is the interaction potential for particle C with A or B.

On the basis of this character of bound state formation, we estimate the rate constant of the three-body process (3.51). The rate of conversion of particle B into particle AB is of the order of magnitude of the product of two factors. So the first factor is the probability of the location of particle B in the critical region near A, which is equal to $[A]b^3$. The second factor is the rate $[C]v\sigma$ of collisions with particle C, where v is the typical relative collision velocity, and σ is the cross section for the collision between C and either A or B, resulting in an energy exchange of the order of T. Assuming the masses of the colliding particles to be of the same order of magnitude, we find this cross section to be comparable to the cross section for elastic collision. This estimate for the rate of formation of particle AB is thus

$$d[AB]/dt \sim [A][B]b^3[C]v\sigma .$$

A comparison of this expression with the definition of the constant of the three-body process (3.52) gives the estimate for the rate constant,

$$K \sim vb^3\sigma . \tag{3.54}$$

Collision of three particles (3.51) is the three-body process if the number density of the third particle C is small, and therefore colliding particles A and B can simultaneously collide with one particle C only. Hence, the mean free path of particles A or B in a gas of particles C exceeds significantly the size b of the critical region. This leads to the criterion

$$[C]\sigma b \ll 1\,, \qquad (3.55)$$

where σ is the typical elastic cross section of a particle A or B with a particle C.

▶ **Problem 3.26** On the basis of the Thomson theory for three-body processes estimate the rate of three-body recombination of positive and negative ions if the third particle is a gas atom.

This process of three-body recombination of positive and negative ions proceeds according to the scheme

$$A^+ + B^- + C \rightarrow A^* + B + C\,. \qquad (3.56)$$

In this process a bound state of the positive and negative ions A^+ and B^- is formed first, and then a valence electron transfers from the field of the atom B to the field of the ion A^+, and the bound state decays into two atoms A^* and B. The second stage of the process occurs spontaneously when the three-body process finishes. We stop at the first stage of this process, and in evaluating the rate of this process we assume the atom C mass to be comparable to the mass of one of the ions A^+ or B^-.

Below we consider the Coulomb interaction between the ions and a polarization interaction between each ion and atom. Then the rate constant of the three-body recombination process according to formula (3.54) is given by

$$K = \frac{\alpha}{N_{\rm i}} \sim \frac{e^6}{T^3}\left(\frac{\alpha e^2}{m}\right)^{1/2}\,, \qquad (3.57)$$

where α is the atom polarizability, m is a particle mass, and σ is the cross section of the order of the diffusion cross section (2.31) in the case of the polarization interaction potential.

The criterion (3.55) of the three-body character of the recombination process has now the form

$$\frac{[C]e^2(\alpha e^2)^{1/2}}{T^{3/2}} \ll 1\,. \qquad (3.58)$$

In particular, if we estimate the typical polarizability of the atom to be several atomic units, the criterion (3.58) gives at room temperature the number density $[C]$ which must be much less than $10^{20}\,\text{cm}^{-3}$. Therefore, the three-body character of the recombination of positive and negative ions takes place at the gas pressure is of the order of one atmosphere, as it proceeds in atmospheric air.

▶ **Problem 3.27** On the basis of the Thomson theory for three-body processes estimate the rate of three-body process for formation of molecular ions in an atomic gas.

This three-body process of conversion of atomic ions into molecular ions develops according to the scheme

$$A^+ + 2A \to A_2^+ + A \tag{3.59}$$

and the rate constants of this process are given in Table 3.4 for inert gases. Assuming that this process be determined by the polarization interaction $U(R) = -\alpha e^2/(2R^4)$ of an ion A^+ with atoms A, where R is the ion–atom distance and α is the atom polarizability, we estimate on the basis of formula (3.54) the rate constant of three-body formation of molecular ions A_2^+,

$$K_{\text{ia}} \sim \frac{(\alpha e^2)^{5/4}}{m^{1/2}T^{3/4}} .$$

Table 3.4 The rate constant of the three-body process (3.59) given at room temperature and expressed in units 10^{-32} cm^6/s.

A	He	Ne	Ar	Kr	Xe
K_{ia}	11	6	3	2.3	3.6
C	37	20	2.4	1.6	1.7

Taking the rate constant of this process in the form

$$K_{\text{ia}} = C\frac{(\alpha e^2)^{5/4}}{m^{1/2}T^{3/4}} , \tag{3.60}$$

we give in Table 3.4 the values of factor C obtained on the basis of experimental data. The difference of this parameter for light and heavy inert gases testifies the role of the exchange ion–atom interaction in the three-body process for helium and neon along with the polarization interaction.

▶ **Problem 3.28** From the dimensionality consideration estimate the rate of three-body formation of molecules in an atomic gas, if the interaction potential between atoms is approximated by the dependence $U(R) = -AR^{-k}$, and $A > 0$.

The three-body process of formation of diatomic molecules in an atomic gas is described by the scheme (3.51). Assuming that the scattering cross section σ and the critical size b in formula (3.54) are determined by the polarization interaction potential, we have $\sigma \sim b^2$, and the rate constant of this process can be estimated as

$$K \sim vb^5 \sim (A)^{5/k} (T)^{1/2-5/k} m^{-1/2} , \tag{3.61}$$

where m is the atom mass, $v \sim (T/m)^{1/2}$ is the typical atom velocity, and the critical size (3.53) is estimated as $b \sim (A/T)^{1/k}$. In the case of the Coulomb interaction this formula gives $b \sim e^2/T$, whereas in the case of the polarization interaction this formula is transformed into formula (3.60).

► **Problem 3.29** Estimate the rate of three-body formation of molecular ions in an atomic gas from the dimensionality consideration.

In the case of process (3.59) we have three dimensional parameters, the interaction parameter αe^2, a typical thermal energy of the particles T, and the atom mass M. Calculating from these parameters the only quantity of the dimensionality of the three-body rate constant, we again obtain formula (3.60). The same dependence follows from formula (3.61).

Let us represent formula (3.60) in the form

$$K = K_0 \alpha^{5/4} m^{-1/2} (300/T)^{3/4} . \tag{3.62}$$

If the polarizability α is expressed in atomic units (a_0^3), the atom mass in units of atomic masses (1.66×10^{-24} g), and the temperature in K, we obtain from a comparison with the experimental data

$$K_0 = (8 \pm 4) \times 10^{-32}\,\text{cm}^6/\text{s} .$$

3.5 Principle of Detailed Balance

► **Problem 3.30** Establish the relation between the cross sections of inelastic collisions for direct and inverse processes (the principle of detailed balance).

Our task is to compare the cross sections of atom excitation and quenching for direct and inverse collision processes with other particles according to the scheme

$$A + B_i \longleftrightarrow A + B_f , \tag{3.63}$$

where B_i, B_f denote an atom in initial and final states, and the transition between these states results from collision with an atomic particle B. In order to establish the relation between the cross sections of these processes, we consider these atomic particles A and B to be located in a volume Ω, and the transitions of the atom B are possible between the states i and f only in collisions with an atomic particle A.

Let us introduce an interaction operator V between atomic particles A and B that is responsible for the above transitions. Then we have for the transition rates within the framework of the perturbation theory between states i and f

$$w_{if} = \frac{2\pi}{\hbar} |V_{if}|^2 \frac{dg_i}{d\varepsilon} , \qquad w_{fi} = \frac{2\pi}{\hbar} |V_{fi}|^2 \frac{dg_f}{d\varepsilon} .$$

Here $\frac{dg_i}{d\varepsilon}$, $\frac{dg_f}{d\varepsilon}$ are the statistical weights per unit energy for the appropriate channels of the process. On the basis of the definition, we have from this for the cross sections of these processes

$$\sigma_{\rm if} = \frac{w_{\rm if}}{N v_{\rm i}} = \Omega \frac{w_{\rm if}}{v_{\rm i}}, \quad \sigma_{\rm fi} = \frac{w_{\rm fi}}{N v_{\rm f}} = \Omega \frac{w_{\rm fi}}{v_{\rm f}},$$

where $N = 1/\Omega$ is the number density of particles, $v_{\rm i}, v_{\rm f}$ are the electron velocities for the transition channels, and an atom assumes to be motionless. The time reversal operation gives for the matrix elements of the interaction operator $V_{\rm if} = V_{\rm fi}$, which leads to the relation between the cross sections of direct and inverse processes in electron–atom collisions,

$$\sigma_{\rm if} v_{\rm i} \frac{dg_{\rm i}}{d\varepsilon} = \sigma_{\rm fi}\, v_{\rm f} \frac{dg_{\rm f}}{d\varepsilon} . \tag{3.64}$$

The statistical weights for the transition channels are equal to

$$dg_{\rm i} = \Omega \frac{d\boldsymbol{p}_{\rm i}}{(2\pi\hbar)^3} g_{\rm i}, \quad dg_{\rm f} = \Omega \frac{d\boldsymbol{p}_{\rm f}}{(2\pi\hbar)^3} g_{\rm f},$$

where $g_{\rm i}$, $g_{\rm f}$ are the statistical weights for given atom states. Then finally formula (3.64) takes the form

$$\sigma_{\rm if} = \sigma_{\rm fi} \frac{v_{\rm f}^2 g_{\rm f}}{v_{\rm i}^2 g_{\rm i}} . \tag{3.65}$$

▶ **Problem 3.31** Within the framework of the liquid drop model for a cluster establish the relation between the rate of atom evaporation from the cluster surface and the rate constant of atom attachment to the cluster surface.

We have now an equilibrium for the processes of cluster growth and evaporation according to the scheme

$$M + M_n \longleftrightarrow M_{n+1} . \tag{3.66}$$

Hence, the rate of atom evaporation $\nu_{n+1}^{\rm ev}$ for a cluster consisting of $n+1$ atoms is equal to the rate ν_n of atom attachment to a cluster

$$\nu_{n+1}^{\rm ev} = \nu_n = N k_{\rm 0} n^{2/3} ,$$

and the equilibrium number density of atoms is given by formula (1.80). As a result, we have for the rate of cluster evaporation

$$\nu_{n+1}^{\rm ev} = \nu_n = N_{\rm sat}(T) k_{\rm 0} n^{2/3} \exp\left(-\frac{\varepsilon_n - \varepsilon_{\rm o}}{T}\right) , \tag{3.67}$$

where ε_n is the binding energy of surface atoms of the cluster, and $\varepsilon_{\rm o}$ is the atom binding energy for a macroscopic surface.

▶ **Problem 3.32** On the basis of the principle of detailed balance determine the rate constant of formation of positive and negative ions as a result of collision of two atoms if we consider this process to be detailed inverse with respect to the process (3.39) of mutual neutralization of ions.

The process of formation of ions is the inverse process with respect to the mutual neutralization process (3.39) and we consider it as the transition between two electron terms in accordance with Fig. 3.5. Then the cross section is the same $\sigma_{\text{ion}} = \sigma_{\text{rec}}$ as for mutual neutralization of ions, and under optimal conditions it is equal to $\sigma_{\text{rec}} = \pi R_c^2/2$. Correspondingly, the rate constant of this process is

$$k_{\text{ion}} = \overline{v}\,\sigma_{\text{ion}}\,,$$

where $\overline{v}$ is the relative velocity of atoms at large distances between them. Then in the limit of a large thermal energy of atoms $e^2/R_c \ll T$ the rate constant of this process is given by formula (3.43). In a general case on the basis of the Maxwell distribution function on collision velocities we obtain

$$k_{\text{ion}} = \sqrt{\frac{2\pi T}{\mu}}\,R_c^2\left(1 + \frac{e^2}{R_c T}\right)\exp\left(-\frac{e^2}{R_c T}\right)\,. \tag{3.68}$$

4
Collisions Involving Electrons

4.1
Inelastic Electron–Atom Collisions

▶ **Problem 4.1** Find the asymptotic expansion for the wave function of the electron–atom under inelastic electron–atom collisions.

Let us start from the Schrödinger equation for the wave function $\Psi(\boldsymbol{r}, \boldsymbol{r}_e)$ of the electron–atom system, which has the form

$$\left[\widehat{H}_a(\boldsymbol{r}) - \frac{1}{2}\Delta_e - E\right]\Psi(\boldsymbol{r}, \boldsymbol{r}_e) = V(\boldsymbol{r}, \boldsymbol{r}_e)\Psi(\boldsymbol{r}, \boldsymbol{r}_e) ,$$

where $\boldsymbol{r}, \boldsymbol{r}_e$ are the coordinates of atomic electrons and the incident electron, respectively, $\widehat{H}_a(\boldsymbol{r})$ is the Hamiltonian of atomic electrons, $-\Delta_e/2$ is the operator of the kinetic energy of an incident electron, $V(\boldsymbol{r}, \boldsymbol{r}_e)$ is the interaction operator between incident and atomic electrons, and E is the total energy of the electron–atom system. Note that the vector $\boldsymbol{r}$ is many dimensional here. The atomic units are used, and the spin variables are included through the symmetry of the wave function. If we consider the right-hand side of the above equation as a nonuniformity, one can write a formal solution of this equation in the form

$$\Psi(\boldsymbol{r}, \boldsymbol{r}_e) = e^{i\boldsymbol{q}_0\boldsymbol{r}_e}\Phi_0(\boldsymbol{r}) + \int G(\boldsymbol{r}, \boldsymbol{r}_e; \boldsymbol{r}'; \boldsymbol{r}'_e)V(\boldsymbol{r}', \boldsymbol{r}'_e)\Psi(\boldsymbol{r}', \boldsymbol{r}'_e)d\boldsymbol{r}'d\boldsymbol{r}'_e ,$$

where $\Phi_0(\boldsymbol{r})$ is the atomic wave function for the initial state, $\boldsymbol{q}_0$ is the wave vector of an incident electron, and $G(\boldsymbol{r}, \boldsymbol{r}_e; \boldsymbol{r}'\boldsymbol{r}'_e)$ is the Green function of the Schrödinger equation, which satisfies the equation

$$\left[\widehat{H}_a(\boldsymbol{r}) - \frac{1}{2}\Delta_e - E\right]G(\boldsymbol{r}, \boldsymbol{r}_e; \boldsymbol{r}'; \boldsymbol{r}'_e) = \delta(\boldsymbol{r} - \boldsymbol{r}')\delta(\boldsymbol{r}_e - \boldsymbol{r}'_e) ,$$

and has the form

$$G(\boldsymbol{r}, \boldsymbol{r}_e; \boldsymbol{r}'; \boldsymbol{r}'_e) = \frac{1}{4\pi^3}\sum_k \Phi_k(\boldsymbol{r})\Phi_k(\boldsymbol{r}')\lim_{\epsilon\to 0}\int \frac{\exp[i\boldsymbol{q}(\boldsymbol{r}_e - \boldsymbol{r}'_e)]}{q_k^2 - q^2 + i\epsilon} .$$

Here $\{\Phi_k(\boldsymbol{r})\}$ is the system of eigen atomic wave functions, $\boldsymbol{q}_k$ is the wave vector of a scattering electron, and the summation is made over all the atomic states

Plasma Processes and Plasma Kinetics. Boris M. Smirnov

ISBN: 978-3-527-40681-4

including the states in which the transition is forbidden ($q_k < 0$). For the prohibited transitions we have

$$\lim_{\epsilon \to 0} \int \frac{\exp[i\boldsymbol{q}(\boldsymbol{r}_e - \boldsymbol{r}'_e)]}{q_k^2 - q^2 + i\epsilon} = \frac{2\pi^2 \exp[i\boldsymbol{q}_k(\boldsymbol{r}_e - \boldsymbol{r}'_e)]}{|\boldsymbol{r}_e - \boldsymbol{r}'_e|} .$$

This gives the following asymptotic expression for the wave function of the electron–atom system,

$$\Psi(\boldsymbol{r}, \boldsymbol{r}_e)\Big|_{r_e \to \infty} = e^{i\boldsymbol{q}_0 \boldsymbol{r}_e} \Phi_0(\boldsymbol{r}) + \sum_k f_{0k}(\vartheta) \Phi_k(\boldsymbol{r}) \frac{e^{iq_k r_e}}{r_e} . \tag{4.1}$$

This wave function of electrons is a sum of those for incident and scattering electrons, and $f_{0k}(\vartheta)$ is the scattering amplitude (the scattering angle ϑ is the angle between the vectors $\boldsymbol{r}_e$ and $\boldsymbol{q}_0$), which has the form

$$f_{0k}(\vartheta) = \frac{1}{2\pi} \int \Phi_k^*(\boldsymbol{r}') e^{-iq_k \boldsymbol{r}'_e \boldsymbol{n}} V(\boldsymbol{r}', \boldsymbol{r}'_e) \Psi(\boldsymbol{r}', \boldsymbol{r}'_e) d\boldsymbol{r}' d\boldsymbol{r}'_e , \tag{4.2}$$

where $\boldsymbol{n}$ is the unit vector along $\boldsymbol{r}_e$.

▶ **Problem 4.2** Express the differential cross section of the electron–atom inelastic scattering through the scattering amplitude.

We will use expression (4.1) for the wave function of the electron–atom system when the electron is located far from the atom. This wave function is normalized by the flux q_0 of an incident electron. The rate of an atom transition in a state k with electron scattering in a solid angle $d\Theta$ is equal to $q_k \left| f_{0k}(\vartheta) \right| d\Theta$ in this case, whereas the incident electron flux is q_0. Then according to the cross section definition as the ratio of the scattering rate to the flux of incident particles we have for the inelastic cross section now

$$d\sigma_{0k} = \frac{q_k}{q_0} \left| f_{0k}(\vartheta) \right| d\Theta . \tag{4.3}$$

▶ **Problem 4.3** Find the threshold behavior for the cross section of atom excitation by electron impact.

Let us use formula (4.2) for the scattering amplitude in the limit $q_k \to 0$. The electron–atom wave function $\Psi(\boldsymbol{r}, \boldsymbol{r}_e)$ in formula (4.2) does not depend on the wave vector of a scattering electron q_k in an atom region where scattering proceeds. This means that the scattering amplitude is independent of q_k at small q_k. Then taking the energy of an incident electron ε in the form $\varepsilon = \Delta E + q_k^2/2$, where ΔE is the excitation energy, and $q_k^2 \ll \Delta E$, we obtain the following energy dependence for the excitation cross section:

$$\sigma_{0k} = \text{const} \sqrt{\varepsilon - \Delta E} . \tag{4.4}$$

▶ **Problem 4.4** Determine the cross section of the inelastic ion–atom collision in the Born approximation.

Considering the electron–atom interaction potential $V(\boldsymbol{r}, \boldsymbol{r}_\mathrm{e})$ as a perturbation, we have for the wave function of the electron–atom system ignoring this interaction

$$\Psi(\boldsymbol{r}, \boldsymbol{r}_\mathrm{e}) = e^{i\boldsymbol{q}_0\boldsymbol{r}_\mathrm{e}}\Phi_0(\boldsymbol{r}) ,$$

and the scattering amplitude (4.2) takes the form in this limit

$$f_{0k}(\vartheta) = \frac{1}{2\pi}\int \Phi_k^*(\boldsymbol{r})e^{-i\boldsymbol{K}\boldsymbol{r}_\mathrm{e}}\Phi_0(\boldsymbol{r})\; V(\boldsymbol{r}, \boldsymbol{r}_\mathrm{e})d\boldsymbol{r}d\boldsymbol{r}_\mathrm{e} ,$$

where $\boldsymbol{K} = q_k\boldsymbol{n} - \boldsymbol{q}_0$ is the momentum (in atomic units) that is transferred to the incident electron.

Representing a general coordinate of atomic electrons $\boldsymbol{r}$ as a sum of coordinates $\boldsymbol{r}_i$ of individual atomic electrons, we have for the electron–atom interaction potential

$$V(\boldsymbol{r}, \boldsymbol{r}_\mathrm{e}) = -\frac{Z}{r_\mathrm{e}} + \sum_i \frac{1}{|\boldsymbol{r}_\mathrm{e} - \boldsymbol{r}_i|} ,$$

where Z is the core charge or the total number of valence electrons. Since

$$\int \frac{e^{-i\boldsymbol{K}\boldsymbol{r}_\mathrm{e}}}{|\boldsymbol{r}_\mathrm{e} - \boldsymbol{r}_i|}d\boldsymbol{r}_\mathrm{e} = -\frac{4\pi}{K^2}\left(\sum_i e^{-i\boldsymbol{K}\boldsymbol{r}_i}\right)_{0k} ,$$

we obtain for the scattering amplitude in the Born approximation

$$f_{0k}(\vartheta) = -\frac{2}{K^2}\left(\sum_i e^{-i\boldsymbol{K}\boldsymbol{r}_i}\right)_{0k} . \tag{4.5}$$

In determining the total cross section of atom excitation, we use the relation between the transferring electron momentum $\boldsymbol{K}$ and the scattering angle ϑ,

$$K^2 = q_0^2 + q_k^2 - 2q_0q_k\cos\vartheta, \qquad KdK = q_0q_kd\cos\vartheta .$$

This gives the total cross section of atom excitation by electron impact,

$$\sigma_{0k} = \frac{8\pi}{K^2}\int\limits_{|q_0-q_k|}^{q_0+q_k}\left|\left(\sum_i e^{-i\boldsymbol{K}\boldsymbol{r}_i}\right)_{0k}\right|^2\frac{dK}{K^3} . \tag{4.6}$$

The Born approximation is valid at high electron energies when the interaction electron–atom potential in a region that is responsible for scattering is small compared to the electron energy, and the wave function of the incident electron may be represented as a plane wave. At high electron velocities we have

$$q_0 - q_k = \sqrt{2\varepsilon_0} - \sqrt{2\varepsilon_k} = \frac{\Delta E}{\sqrt{2\varepsilon_0}} ,$$

where ε_0, ε_k are the electron energies for the corresponding channel, and ΔE is the excitation energy. Under these conditions we can expand the exponent in formula (4.5) over a small parameter $\boldsymbol{K}\boldsymbol{r}_i$. This gives the Bethe approximation for the atom excitation cross section

$$\sigma_{0k} = \frac{8\pi}{q_0^2}\left|(D_x)_{0k}\right|^2 \ln\frac{q_e}{|q_0 - q_k|}\,, \tag{4.7}$$

where $\boldsymbol{D} = \sum_i \boldsymbol{r}_i$ is the operator of the atom dipole moment, and q_e is a typical electron wave vector. It is also convenient to represent the atom excitation cross section by electron impact in the Born approximation in the form

$$\sigma_{0k} = \frac{4\pi}{\varepsilon}\left|(D_x)_{0k}\right|^2 \ln\left(\mathrm{const}\frac{\varepsilon}{\Delta E}\right), \qquad \varepsilon \gg \Delta E\,, \tag{4.8}$$

where const ~ 1. As is seen, resonantly excited atoms are excited in collisions with a fast electron, i. e., the transition takes place mostly between the states which are connected by a radiative transition.

▶ **Problem 4.5** Find the similarity law for the cross section of atom excitation by electron impact.

The cross section (4.8) of atom excitation in collision with a fast electron is proportional to the square of the matrix element of the atom dipole moment, as well as the rate of a radiative transition for an excited atom that has the form

$$\frac{1}{\tau} = \frac{2\Delta E^3}{c^3}\left|(D_x)_{0k}\right|^2 g_0\,. \tag{4.9}$$

Here $c = 137$ is the light velocity and g_0 is the statistical weight of the initial atom state. This formula allows us to transform formula (4.7) for the excitation cross section to the form

$$\sigma_{ex} = \frac{c^3}{\tau}\frac{g_k}{g_0}\frac{\pi}{\varepsilon\Delta E^3}\ln\left(\mathrm{const}\frac{\varepsilon}{\Delta E}\right), \qquad \varepsilon \gg \Delta E\,.$$

One can expand this formula in the total range of collision energies in the form

$$\sigma_{ex} = \frac{1}{\tau}\frac{g_k}{g_0}\varphi\left(\frac{\varepsilon}{\Delta E}\right), \tag{4.10}$$

where the universal function $\varphi(x) \sim \ln(\mathrm{const}\cdot x)/x$ at high argument values ($x \gg 1$), and at small x we have $\varphi(x) \sim \sqrt{x}$.

▶ **Problem 4.6** Determine the cross section of exchange scattering for collision of a classical electron with a one-electron atom at high electron energies.

In this problem we have two electrons, and an incident and atomic electrons are exchanged. The wave function in fast collisions is $\exp(i\boldsymbol{q}_0\boldsymbol{r}_e)\Phi_0(\boldsymbol{r})$ and

$\exp(-iq_k \boldsymbol{r}\,\boldsymbol{n})\Phi_k(\boldsymbol{r}_e)$ before and after collision, respectively, so the amplitude of the exchange electron scattering in fast collisions has the form

$$f_{0k}(\vartheta) = -\frac{1}{2\pi}\int \Phi_k^*(\boldsymbol{r}_e)\Phi_0(\boldsymbol{r})\exp(i\boldsymbol{q}_0\boldsymbol{r}_e - iq_k\boldsymbol{r}\,\boldsymbol{n})\left(\frac{1}{|\boldsymbol{r}_e-\boldsymbol{r}|} - \frac{1}{r_e}\right)d\boldsymbol{r}d\boldsymbol{r}_e\,. \qquad (4.11)$$

Using the relative electron coordinate $\boldsymbol{R} = \boldsymbol{r} - \boldsymbol{r}_e$, which is characterized by spherical components R, θ, φ, and the coordinate $\boldsymbol{r}_e$ of an incident electron will be described by spherical components r_e, θ_e, φ_e. Accounting for $q_0 \gg 1$ and integrating over $d\cos\theta$ in parts, we have

$$f_{0k}(\vartheta) = -\frac{1}{2\pi}\int \Phi_k^*(\boldsymbol{r}_e)\Phi_0(\boldsymbol{r}_e+\boldsymbol{R})\exp(-i\boldsymbol{K}\boldsymbol{r}_e + iq_0 R\cos\theta)\left(\frac{1}{R} - \frac{1}{r_e}\right)d\boldsymbol{R}d\boldsymbol{r}_e\,.$$

This gives

$$f_{0k}(\vartheta) = -\frac{2}{q_0^2}\left(e^{-i\boldsymbol{K}\,\boldsymbol{r}}\right)_{0k} + 0\left(\frac{1}{q_0^3}\right). \qquad (4.12)$$

One can see that the amplitude of exchange scattering (4.12) is less than that (4.5) for direct scattering in K^2/q_0^2 times.

Summarizing formulas (4.5) and (4.12), we get for the amplitude of the electron scattering of a fast electron on a one-electron atom depending on the total spin of two electrons

$$f_{0k}(\vartheta) = -2\left(\frac{1}{K^2} \pm \frac{1}{q_0^2}\right)\left(e^{-i\boldsymbol{K}\,\boldsymbol{r}}\right)_{0k}\,. \qquad (4.13)$$

Here the sign + relates to the zero total spin (the symmetric coordinate wave function of electrons with respect to their permutation), and the minus sign is taken if the total electron spin is 1 (the coordinate wave function of electrons changes the sign as a result of electron permutation). In addition, the second term is less than the first one when this expression holds true.

4.2 Atom Quenching by Electron Impact

▶ **Problem 4.7** Establish the relation between the cross section of the atom excitation by electron impact and the inverse process—quenching of an excited atom by electron impact (the principle of detailed balance).

Our task is to compare the cross sections of the atom excitation and quenching by electron impact if these processes are inversely opposite

$$e + A_0 \longleftrightarrow e + A_*\,, \qquad (4.14)$$

where A_0, A_* denote an atom in the ground and excited states, respectively. In order to establish the relation between the cross sections of these processes, we consider one electron and one atom in a volume Ω. The atom can find in the states 0 and $*$, and transitions between these states result from collisions with the electron. Introducing the interaction operator V that is responsible for these transitions, we have for the transition rates within the framework of the perturbation theory

$$w_{0*} = \frac{2\pi}{\hbar} |V_{0*}|^2 \frac{dg_*}{d\varepsilon}, \qquad w_{*0} = \frac{2\pi}{\hbar} |V_{*0}|^2 \frac{dg_0}{d\varepsilon} .$$

Here $dg_0/d\varepsilon$, $dg_*/d\varepsilon$ are the statistical weights per unit energy for the appropriate channels of the process. On the basis of the definition, we have from this for the cross sections of these processes

$$\sigma_{0*} = \frac{w_{0*}}{Nv_0} = \Omega \frac{w_{0*}}{v_0}, \qquad \sigma_{*0} = \frac{w_{*0}}{Nv_*} = \Omega \frac{w_{*0}}{v_*} ,$$

where $N = 1/\Omega$ is the number density of particles, v_0, v_* are the electron velocities for the transition channels, and an atom assumes to be motionless. The time reversal operation gives for the matrix elements of the interaction operator $V_{0*} = V_{*0}^*$, which leads to the relation between the cross sections of direct and inverse processes in electron–atom collisions

$$\sigma_{0*} \, v_0 \, \frac{dg_0}{d\varepsilon} = \sigma_{*0} \, v_* \, \frac{dg_*}{d\varepsilon} . \tag{4.15}$$

The statistical weights for the transition channels are equal to

$$dg_0 = \Omega \frac{d\mathbf{p}_0}{(2\pi\hbar)^3} g_0, \qquad dg_* = \Omega \frac{d\mathbf{p}_*}{(2\pi\hbar)^3} g_* ,$$

where g_0, g_* are the statistical weights for given atom states. Then finally formula (3.64) takes the form

$$\sigma_{ex} = \sigma_q \frac{v_*^2 g_*}{v_0^2 g_0} , \tag{4.16}$$

where we denote by $\sigma_{ex} = \sigma_{0*}$ the excitation cross section, and by $\sigma_q = \sigma_{*0}$ the quenching cross section.

▶ **Problem 4.8** Determine the dependence on the electron energy for the rate constant of atom quenching by a slow electron.

When a plasma contains excited atoms which influence equilibrium in a plasma, quenching of atoms by electron impact is responsible for the equilibrium of electrons in the plasma. The cross section of atom quenching in collisions with a slow electron is expressed through the threshold cross section of atom excitation by electron impact on the basis of the principle of detailed balance (4.4). Since the threshold excitation cross section has the form (3.64)

$$\sigma_{ex} = C\sqrt{\varepsilon - \Delta E}, \qquad \varepsilon - \Delta E \ll \Delta E ,$$

the cross section of atom quenching is

$$\sigma_q = C \frac{g_0 \Delta E}{g_* \sqrt{\varepsilon - \Delta E}} \,.$$

Here ε is the electron energy, ΔE is the atom excitation energy, C is a constant, and g_0, g_* are the statistical weights for the ground and excited states, respectively.

From this we obtain the rate constant of atom quenching by a slow electron, k_q,

$$k_q = v_* \sigma_q = C \frac{g_0 \Delta E \sqrt{2}}{g_* \sqrt{m_e}} \,, \tag{4.17}$$

where v_* is the electron velocity in the quenching channel and m_e is the electron mass. Correspondingly, the rate constant k_{ex} of atom excitation by electron impact is equal to

$$k_{ex} = k_q \frac{g_*}{g_0} \sqrt{\frac{\varepsilon - \Delta E}{\Delta E}} \,. \tag{4.18}$$

Thus, the rate constant of atom quenching by slow electrons k_q is independent of both the electron energy and the energy distribution function for slow electrons. Hence, it depends only on parameters of the transition atomic states, so that the quenching rate constant is a convenient parameter characterized also by excitation of atoms by electron impact near the threshold. The rate constant k_q is of the order of an atomic value, i. e., it has an order of 10^{-9}–10^{-8} cm^3/s. Table 4.1 contains the quenching rate constants for metastable atoms of inert gases, and these data confirm general conclusions.

Table 4.1 The rate constant of quenching of metastable states of inert gas atoms in collisions with slow electrons (N. B. Kolokolov. *Chemistry of Plasma*, **12**, 56, 1984).

Atom, transition	ΔE (eV)	k_q (10^{-10} cm^3/s)
$He(2^3S \to 1^1S)$	19.82	31
$Ne(2^3P_2 \to 2^1S)$	16.62	2.0
$Ar(3^3P_2 \to 3^2S)$	11.55	4.0
$Kr(4^3P_2 \to 4^2S_0)$	9.915	3.4
$Xe(5^2P_2 \to 5^2S_0)$	8.315	19

▶ **Problem 4.9** Determine the dependence of the rate constant for quenching of a resonantly excited atom state by a slow electron on the excitation energy.

In the case of a resonantly excited state, the quenching process is more effective than that for metastable states because the excitation cross section for a fast electron in the Born approximation for resonantly excited states exceeds significantly that for metastable states. Indeed, expressing the matrix element of the atom dipole moment through the oscillator strength f_{0*} for this transition, we obtain the excitation cross section of resonantly excited states near the threshold,

$$\sigma_{\mathrm{ex}}(\varepsilon) = \frac{2\pi e^4 f_{0*}}{\Delta E^{5/2}} a\sqrt{\varepsilon - \Delta E}\,,$$

where the numerical coefficient a in this formula, as follows from experimental data in atomic units, is equal to

$$a = 0.130 \pm 0.007\,.$$

This gives the quenching rate constant in atomic units,

$$k_{\mathrm{q}} = A\frac{g_0 f_{0*}}{g_* \Delta E^{3/2}}\,, \tag{4.19}$$

where $A = 1.16 \pm 0.06$. Table 4.2 gives the parameters of this formula for quenching of resonantly excited states for atoms of the first group of the periodical system of elements. As is seen, the quenching rate constant for resonantly excited states exceeds that for metastable states (Table 4.1) by one to two orders of magnitude.

Table 4.2 Parameters of resonantly excited states of some atoms and the rate constant of quenching of these states in collisions with a slow electron (f_{0*} is the oscillator strength for this transition, $\tau_{*\mathrm{o}}$ is the radiative lifetime of the resonantly excited state, and λ is the wavelength of the emitting photon).

Atom, transition	ΔE (eV)	λ (nm)	f	$\tau_{*\mathrm{o}}$ (ns)	k_{q} (10^{-8} cm^3/s)
$\mathrm{H}(2^1P \to 1^1S)$	10.20	121.6	0.416	1.60	0.79
$\mathrm{He}(2^1P \to 1^1S)$	21.22	58.43	0.276	0.555	0.18
$\mathrm{He}(2^1P \to 2^1S)$	0.602	2058	0.376	500	51
$\mathrm{He}(2^3P \to 2^3S)$	1.144	1083	0.539	98	27
$\mathrm{Li}(2^2P \to 2^2S)$	1.848	670.8	0.74	27	19
$\mathrm{Na}(3^2P \to 3^2S)$	2.104	589	0.955	16.3	20
$\mathrm{K}(4^2P_{1/2} \to 4^2S_{1/2})$	1.610	766.9	0.35	26	31
$\mathrm{K}(4^2P_{3/2} \to 4^2S_{1/2})$	1.616	766.5	0.70	25	32
$\mathrm{Rb}(5^2P_{1/2} \to 5^2S_{1/2})$	1.560	794.8	0.32	28	32
$\mathrm{Rb}(5^2P_{3/2} \to 5^2S_{1/2})$	1.589	780.0	0.67	26	33
$\mathrm{Cs}(6^2P_{1/2} \to 6^2S_{1/2})$	1.386	894.4	0.39	30	46
$\mathrm{Cs}(6^2P_{3/2} \to 6^2S_{1/2})$	1.455	852.1	0.81	27	43

▶ **Problem 4.10** Express the rate constant of quenching of resonantly excited atom states by a slow electron for atoms of the first group of the periodical system of elements through the radiative time of excited states and the excitation energy of these states.

Using the relation between the radiative lifetime $\tau_{*\mathrm{o}}$ of a resonantly excited state and the matrix element square for the operator of the atom dipole moment, one can rewrite formula (4.19) in the form

$$k_{\mathrm{q}} = \frac{k_{\mathrm{o}}}{(\Delta E)^{7/2}\, \tau_{*\mathrm{o}}} \,. \tag{4.20}$$

If the atom excitation energy ΔE is expressed in eV, and the radiative lifetime is given in ns, then the numerical coefficient in formula (4.20) is $k_{\mathrm{o}} = (4.4 \pm 0.7) \times 10^{-5}\,\mathrm{cm}^3/\mathrm{s}$, as follows from experimental data for the excitation cross sections of alkali metal atoms.

▶ **Problem 4.11** Based on the experimental data given in Table 4.2, express the dependence of the rate constant of quenching for resonantly excited atom states by a slow electron on the excitation energy.

According to formula (4.19), this dependence is $k_{\mathrm{q}} \sim \Delta E^{-3/2}$. Of course, one can treat the data of Table 4.2 for approximating the quenching rate constant k_{q} by a general power dependence $k_{\mathrm{q}} \sim \Delta E^{-\alpha}$ and finding the optimal value of the parameter α. But since k_{q} depends also on other parameters, the accuracy of this operation would not be sufficient. Hence we are restricted by the dependence (4.19) only. Then statistical treatment of the Table 4.2 data gives in atomic units

$$k_{\mathrm{q}} \Delta E^{-3/2} = 0.6 \pm 0.2 \,,$$

where k_{q} is taken in $10^{-7}\,\mathrm{cm}^3/\mathrm{s}$, and ΔE is expressed in eV. Note that the accuracy of this approximation is worse than the accuracy of the experimental data used in Table 4.2 by about 20 %.

4.3 Atom Ionization by Electron Impact

▶ **Problem 4.12** Determine the cross section of atom ionization by electron impact as a result of collision of the incident and valence electrons. Consider colliding electrons as classical particles and the valence electron to be the motionless particle (Thomson model).

The process of atom ionization by electron impact proceeds according to the scheme

$$e + A \to 2e + A^+ \,. \tag{4.21}$$

In this process the incident electron interacts with a valence electron and transfers to it a part of its kinetic energy, which can cause the detachment of the valence electron from the initially neutral atom. We analyze this process below in terms of a simple model developed by J. J. Thomson, in which it is assumed that electron collisions can be described on the basis of classical laws, and that the electrons do not interact with the atomic core during the collision. One can expect that the analysis of this quantum process on the basis of a classical consideration is incorrect. Nevertheless, because of the identity of the classical and quantum cross sections for elastic scattering of particles interacting through the Coulomb interaction potential this model can give a correct qualitative description of the ionization process.

An ionization event occurs within the framework of this model, if the energy transferred to the valence electron exceeds the ionization potential J of the atom. Hence, our task is to find the cross section for collisions of electrons in which the energy exchange between electrons exceeds a given value $\Delta\varepsilon$. We consider the limiting case when the kinetic energy of the incident electron, $\varepsilon = mv^2/2$, exceeds remarkably the energy exchange $\Delta\varepsilon$, and an incident electron is scattered in a small angle. An electron momentum variation is equal to according to formula (2.34)

$$\Delta p = \int_{-\infty}^{\infty} \frac{e^2\rho}{R^3} dt = \frac{2e^2}{\rho v} .$$

We assume that the valence electron is motionless before collision and the energy of an incident electron is relatively high. This gives the energy $\Delta\varepsilon$ of the valence electron after collision,

$$\Delta\varepsilon = \frac{\Delta p^2}{2m_e} = \frac{2e^4}{\rho^2 m_e v^2} = \frac{e^4}{\rho^2 \varepsilon} ,$$

where ε is the energy of the incident electron. The cross section for collisions accompanied by the exchange of energy $\Delta\varepsilon$ is

$$d\sigma = 2\pi\rho d\rho = \frac{\pi e^4 d\Delta\varepsilon}{\varepsilon\,(\Delta\varepsilon)^2} .$$

Though this formula was deduced for the case $\varepsilon \gg \Delta\varepsilon$, it is valid for any relative magnitudes of these parameters. The ionization cross section corresponds to $\varepsilon > \Delta\varepsilon > J$ and on the basis of formula (2.34) is given by

$$\sigma_{\text{ion}} = \int_J^{\varepsilon} d\sigma = \frac{\pi e^4}{\varepsilon}\left(\frac{1}{J} - \frac{1}{\varepsilon}\right) . \tag{4.22}$$

This expression is the Thomson formula for the atom ionization cross section by electron impact. The model may be generalized for atoms with several valence electrons by a simple multiplication of this cross section by the number of valence electrons since these electrons take part in the ionization process independently.

▶ **Problem 4.13** Determine the rate constant of atom ionization by electron impact within the framework of the Thomson model for the ionization cross section and for the Maxwell distribution function of electrons on energies.

On the basis of the Thomson formula (4.22) for the ionization cross section σ_{ion} for an atom with one valence electron by electron impact we have for the rate constant of ionization k_{ion} of atom ionization

$$k_{ion} = \frac{2}{\sqrt{\pi}} \int_{J/T_e}^{\infty} x^{1/2} e^{-x} dx \sqrt{\frac{2\varepsilon}{m_e}} \frac{\pi e^4}{\varepsilon} \left(\frac{1}{J} - \frac{1}{\varepsilon} \right) = k_J F(J/T_e) \,,$$

where $x = \varepsilon/T_e$, ε is the electron energy, and T_e is the electron temperature, J is the atom ionization potential. Next,

$$k_J = \sqrt{\frac{8J}{\pi m_e}} \frac{\pi e^4}{J^2} \,, \qquad F(y) = \sqrt{y} \int_{y}^{\infty} e^{-x} dx \left(1 - \frac{y}{x} \right) . \tag{4.23}$$

In the limiting cases we have from this for the rate constant of atom ionization

$$k_{ion} = k_J \sqrt{y}, \quad y \ll 1; \quad k_{ion} = \frac{k_J}{\sqrt{y}} e^{-y}, \quad y \gg 1 \,.$$

Figure 4.1 shows the dependence $F(y)$ in an intermediate range of the argument.

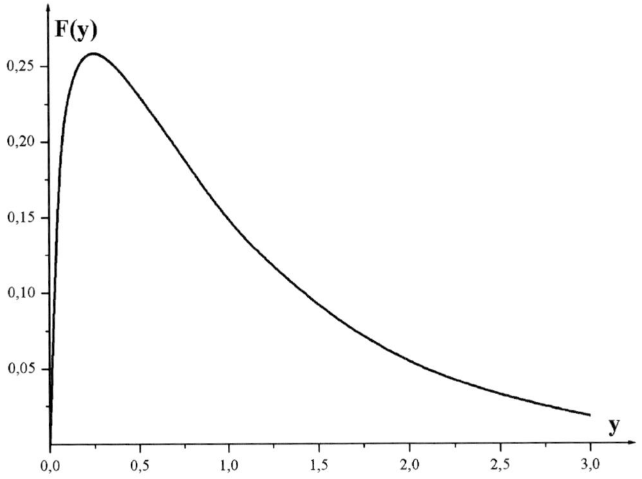

Fig. 4.1 Function $F(y)$ for the rate constant of atom ionization by electron impact according to the Thomson model.

▶ **Problem 4.14** On the basis of the dimensionality consideration find the dependence of the ionization cross section of an atom by electron impact on the parameters of this problem. Assume the character of collision of the incident and valence electrons to be classical and the distribution of a valence electron inside the atom to be described by one parameter—the ionization potential of the atom J.

Under the classical treatment, we characterize the ionization cross section by the following classical parameters of the problem: m_e (the electron mass), e^2 (the interaction parameter), ε (the electron energy), and J (the ionization potential). The most general form of the cross section expressed through these parameters is

$$\sigma_{\rm ion} = \frac{\pi e^4}{J^2} f\left(\frac{\varepsilon}{J}\right) , \qquad (4.24)$$

where $f(x)$ is a universal function that is identical for all atoms. For the Thomson model this function is given by

$$f(x) = 1/x - 1/x^2 .$$

In the case when the valence electron shell contains n electrons, this formula takes the form

$$\sigma_{\rm ion} = \frac{\pi e^4}{J^2} n f\left(\frac{\varepsilon}{J}\right) . \qquad (4.25)$$

▶ **Problem 4.15** Determine the atom ionization cross section by electron impact under the assumption of the Thomson model, but refuse from the assumption that a valence electron is motionless.

If the change of the momentum of a valence electron is $\Delta \boldsymbol{p}$ as a result of collision with the incident electron whose velocity is $\boldsymbol{u}$ before the collision, the change $\Delta\varepsilon$ of its energy is

$$\Delta\varepsilon = \frac{(m_e \boldsymbol{u} + \Delta\boldsymbol{p})^2 - m_e^2 u^2}{2m_e} ,$$

and the transferred momentum $\Delta \boldsymbol{p}$ is expressed through the energy change $\Delta\varepsilon$ as

$$\Delta \boldsymbol{p} = \sqrt{(m_e \boldsymbol{uk})^2 + 2m_e\Delta\varepsilon} - m\boldsymbol{uk} ,$$

where $\boldsymbol{k}$ is the unit vector directed along $\Delta\boldsymbol{p}$. From this we have for the ionization cross section

$$\sigma_{\rm ion} = \int_J^\varepsilon d\sigma = \frac{4\pi e^4}{v^2} \int_J^\varepsilon \left\langle \frac{m_e d\Delta\varepsilon}{\left(\sqrt{(m_e \boldsymbol{uk})^2 + 2m_e\Delta\varepsilon} - m\boldsymbol{uk}\right)^3 \sqrt{(m_e \boldsymbol{uk})^2 + 2m_e\Delta\varepsilon}} \right\rangle ,$$

where v is the velocity of the incident electron, and the angle brackets mean an average over an angle between the vectors $\boldsymbol{u}$ and $\boldsymbol{k}$, and over the velocities u of the valence electron. Averaging over the angle, we obtain

$$\sigma_{\rm ion} = \frac{\pi e^4}{\varepsilon} \int_J^\varepsilon \left\langle \frac{m_e d\Delta\varepsilon}{\Delta\varepsilon^3}\left(\Delta\varepsilon + \frac{2}{3} m_e u^2\right) \right\rangle = \frac{\pi e^4}{\varepsilon}\left[\frac{1}{J} - \frac{1}{\varepsilon} + \frac{2\overline{E}}{3}\left(\frac{1}{J^2} - \frac{1}{\varepsilon^2}\right)\right] ,$$

where $\overline{E} = \overline{m_e u^2/2}$, and averaging is made over the electron distribution inside the atom.

Under the assumption that the valence electron is motionless $\overline{E}$ we obtain from this the Thomson formula (4.22). If we assume that the valence electron is located mostly in the Coulomb field of the atomic core, we have from the virial theorem $\overline{E} = J$, and this formula gives for the ionization cross section

$$\sigma_{\text{ion}} = \frac{\pi e^4}{\varepsilon}\left(\frac{5}{3J} - \frac{1}{\varepsilon} - \frac{2J}{3\varepsilon^2}\right). \tag{4.26}$$

Let us compare this formula with the Thomson formula in order to understand to which changes leads accounting for the velocity distribution of the valence electron. At the threshold the Thomson formula gives

$$\sigma_{\text{ion}} = \frac{\pi e^4}{J^2}(\varepsilon - J), \qquad \varepsilon - J \ll J \tag{4.27}$$

while formula (4.26) gives near the threshold

$$\sigma_{\text{ion}} = \frac{7\pi e^4}{3J^2}(\varepsilon - J), \qquad \varepsilon - J \ll J\,. \tag{4.28}$$

The Thomson formula leads to the maximum of the cross section at $\varepsilon = 2J$ with the value $\sigma_{\max} = \pi e^4/(4J^2)$, whereas the maximum of the cross section according to formula (4.26) is approximately twice compared to that of the Thomson formula, and this maximum occurs at the electron energy $\varepsilon = 1.85J$. Next, the asymptotic cross section according to formula (4.26) exceeds $5/3$ times that of the Thomson formula. Thus, even large energies of the incident electron accounting for the nonzero velocity of a valence electron leads to a change in the ionization cross section.

▶ **Problem 4.16** Within the framework of the Thomson model for atom ionization by electron impact find the rate of atom ionization in a plasma for the ground and excited atom states which are in thermodynamic equilibrium with electrons. Assume a typical thermal energy of electrons to be small compared to the atom ionization potential.

We assume the Maxwell energy distribution function of electrons that is normalized to one,

$$f(\varepsilon)d\varepsilon = \frac{2\varepsilon^{1/2}d\varepsilon}{\pi^{1/2}T^{3/2}}\exp\left(-\frac{\varepsilon}{T_e}\right),$$

and the electron temperature T_e to be small compared to the ionization potential J of the atom. Then taking the Thomson formula for the ionization cross section near the threshold,

$$\sigma_{\text{ion}} = \int_J^{\varepsilon} d\sigma = \frac{\pi e^4 n}{J^2}(\varepsilon - J)\,,$$

where n is the number of valence electrons, we find the rate of atom ionization by electron impact,

$$\nu_{\text{ion}} = N_0\,\langle v_e\sigma_{\text{ion}}\rangle = N_0\frac{\pi e^4 n}{J^2}\sqrt{\frac{8T_e}{\pi m_e}}\exp\left(-\frac{J}{T_e}\right). \tag{4.29}$$

Here N_0 is the number density of atoms in the ground state, and this rate is taken per electron. The same formula can be obtained for excited atoms,

$$\nu_{ion}^{*} = N_{*}\frac{\pi e^4 n}{J_*^2}\sqrt{\frac{8T_e}{\pi m_e}}\exp\left(-\frac{J_*}{T_e}\right), \tag{4.30}$$

where N_* is the number density of atoms in a given excited state, and J_* is the ionization potential for an excited atom.

▶ **Problem 4.17** Within the framework of the Thomson model for atom ionization by electron impact compare the rates of atom ionization in a plasma from the ground and an excited atom state. Compare the contribution to the ionization rate from the ground and an excited state under the assumption of thermodynamic equilibrium between atoms and electrons. A thermal electron energy is small compared to the atom ionization potential.

Since excited atoms are found in thermodynamic equilibrium with atoms in the ground state, and this equilibrium is supported by atom collisions with electrons, the number densities of atoms in the ground and excited states are determined by the Boltzmann formula, so

$$\frac{N_*}{N_0} = \frac{g_*}{g_0}\exp\left(-\frac{\Delta\varepsilon}{T_e}\right),$$

where $\Delta\varepsilon = J - J_*$ is the excitation energy, g_0, g_* are the statistical weights for atoms in the ground and excited states.

So, we have the ratio of the ionization rates (4.29) and (4.30),

$$\frac{\nu_{ion}^{*}}{\nu_{ion}} = \frac{g_*}{n g_0}\left(\frac{J}{J_*}\right)^2. \tag{4.31}$$

The first factor is of the order of 1, but in the case of a highly excited state the second factor is large. Therefore, if atoms are excited in equilibrium with electrons, ionization of the ground state through excitation of this state is more effective than the direct ionization from the ground state. As a result, excited states give the main contribution to the ionization rate of atoms by electron impact in a dense plasma, where excited atom states are in thermodynamic equilibrium with the ground atom states due to collisions with plasma electrons. This character of atom ionization in an equilibrium plasma is stepwise ionization.

We consider as an example ionization of helium atoms by electron impact in an equilibrium plasma with the Maxwell distribution of electrons over energies. We assume that electrons can ionize both atoms in the ground $He(1^1S)$ and metastable $He(2^3S)$ states and due to collisions with electrons the Boltzmann equilibrium is established between these states. Then the ratio of ionization rates (4.31) through the metastable and ground atom states takes the form

$$\frac{\nu_{ion}^{*}}{\nu_{ion}} = \frac{3}{2}\left(\frac{J}{J_*}\right)^{3/2}\frac{F(J^*/T_e)}{F(J/T_e)},$$

where J = 24.6 eV and J* = 4.77 eV are the ionization potentials for helium atoms in the ground and metastable states, and $F(y)$ is given by formula (4.23). In the limit of low electron temperatures this formula is transformed to formula (4.31). Figure 4.2 gives this ratio as a function of the electron temperature.

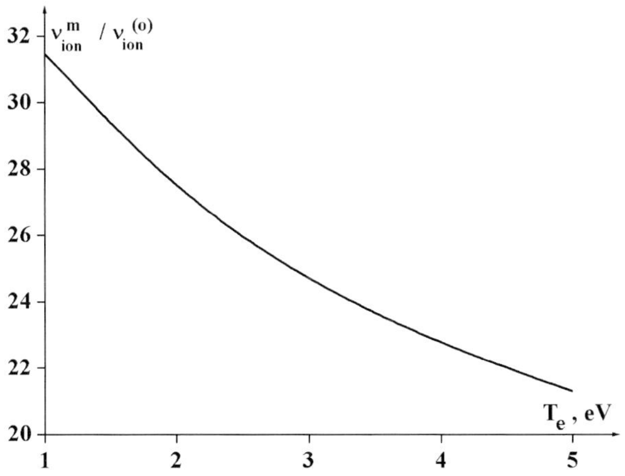

Fig. 4.2 The ratio of the ionization rates of the helium atom by electron impact if ionization proceeds through the metastable (ν_{ion}^{m}) and ground ($\nu_{ion}^{(0)}$) atom states.

► **Problem 4.18** Determine the threshold behavior of the atom ionization cross section by electron impact.

The peculiarities of this ionization threshold process are as follows. First, both electrons, incident and released ones, have small energies when they leave the atomic core during atom ionization. Hence, when these electrons are located far from the core, they can be considered to be classical. Second, both electrons leave the core in opposite directions. If these directions are not opposite, the electrons get an orbital momentum as a result of their interaction, and the total electron energy takes part in this motion. Third, current distances of electrons from the core during their removal are almost the same. If it is not so, a more removal electron is screened by the nearest one and carries the potential energy. Another electron remains to be bonded. These peculiarities of the threshold ionization process allow us to use the classical character of electron motion under simplified conditions, when in a space of electron coordinates $\mathbf{r}_1$ and $\mathbf{r}_2$ the region $r_1 \approx r_2$ is responsible for atom ionization. The motion equations for the coordinates $\mathbf{r}_1$ and $\mathbf{r}_2$ of electrons from the center have the following form in atomic units:

$$\frac{d^2\mathbf{r}_1}{dt^2} = -\frac{\mathbf{r}_1}{r_1^3} + \frac{\mathbf{r}_1 - \mathbf{r}_2}{|\mathbf{r}_1 - \mathbf{r}_2|}\,, \quad \frac{d^2\mathbf{r}_2}{dt^2} = -\frac{\mathbf{r}_2}{r_2^3} - \frac{\mathbf{r}_1 - \mathbf{r}_2}{|\mathbf{r}_1 - \mathbf{r}_2|}\,. \tag{4.32}$$

Let us introduce a new variable $\mathbf{r} = (\mathbf{r}_1 + \mathbf{r}_2)/2$. Since the electrons emit in opposite directions and are located at close distances from the center, we introduce

$$\mathbf{r}_1, \mathbf{r}_2 = \pm\mathbf{r} + (\Delta\mathbf{r} + \delta\mathbf{r})/2 \,,$$

where $\Delta\mathbf{r}$ is directed along of $\mathbf{r}$, and $\delta\mathbf{r}$ is directed perpendicular to $\mathbf{r}$. With an accuracy of up to $(\Delta\mathbf{r}/r)^2$ and $(\delta\mathbf{r}/r)^2$ we have from the energy conservation law

$$\left(\frac{dr}{dt}\right)^2 = \Delta\varepsilon + \frac{3}{2r} \,,$$

where $\Delta\varepsilon = \varepsilon - J$ is the electron energy over the threshold, so that ε is the energy of an incident electron, and J is the atom ionization potential. In the range $r \ll 1/\Delta\varepsilon$ this equation has the following solution:

$$r^3 = \frac{27}{8} t^2 \,. \tag{4.33}$$

As follows from the Newton equations (4.32) for electrons, the difference of the energies of releasing electrons is

$$\delta\varepsilon = \frac{1}{2}\left(\frac{dr_1}{dt}\right)^2 - \frac{1}{2}\left(\frac{dr_2}{dt}\right)^2 = 2\frac{dr}{dt}\frac{d\Delta r}{dt} \,.$$

The fulfillment of the criterion $\delta\varepsilon \leq \Delta\varepsilon$ is necessary for the ionization process. In order to find the range where it holds true, we analyze the equation for Δr that follows from formula (4.32) and has the form

$$\frac{d^2\Delta r}{dt^2} = \frac{2\Delta r}{r^3} \,.$$

Expressing t through r in the range $r \gg 1/\Delta\varepsilon$ in accordance with formula (4.33), we get

$$\frac{d^2\Delta r}{rdr^2} - \frac{1}{2r^2}\frac{d\Delta r}{dr} - \frac{4}{3}\frac{\Delta r}{r^3} = 0 \,.$$

The solution of this equation has the form

$$\frac{\Delta r}{r} = C_1 r^{-1/2-\alpha} + C_2 r^{\alpha} \,, \qquad \alpha = \frac{1}{4}\sqrt{\frac{100-9}{4-1}} = 1.127 \,.$$

The first term of this solution increases up to infinity in the limit $r \to \infty$. Since by the definition of Δr the value $\Delta r/r$ is restricted, we have $C_1 = 0$, which gives in the case of release of both electrons

$$\frac{\Delta r}{r} = C_2 r^{\alpha} \,, \qquad r \ll \Delta\varepsilon \,.$$

In some range of possible values C_2 we have $|\delta\varepsilon| \leq \Delta\varepsilon$, and release of both electrons takes place in this case. If the ratio $\Delta\varepsilon/J$ is small, the above relation is fulfilled in a relatively small range of possible value C_2. According to the dimension consideration, the reduced parameters $r_1\Delta\varepsilon$, $r_2\Delta\varepsilon$, $t\Delta\varepsilon^{3/2}$ may be conserved at small $\Delta\varepsilon/J$ when they are expressed through the parameter $\Delta\varepsilon$ only. Hence, from the dimension consideration it follows that the range of C_2 that corresponds to ionization varies as $C_2 \sim \Delta\varepsilon^\alpha$. From this it follows that the threshold energy dependence for the ionization cross section $\sigma_{\rm ion}$ has the form

$$\sigma_{\rm ion} \sim (\varepsilon - J)^\alpha\,, \tag{4.34}$$

and $\alpha = 1.127$ for atom ionization.

▶ **Problem 4.19** Determine the average energy of electrons formed in a plasma as a result of atom collisions with plasma electrons. Electrons are found in thermodynamic equilibrium, and their temperature $T_{\rm e}$ is small compared to the atom ionization potential J.

When collision of an electron of energy $\varepsilon > J$ with an atom leads to atom ionization, the total energy of two electrons (incident and released ones) after ionization is $\varepsilon - J$, i. e., the average energy per electron is $(\varepsilon - J)/2$, and the energy dependence for the ionization cross section is given by formula (4.34). Taking the rate constant of ionization in the form

$$k_{\rm ion} = \langle v\sigma_{\rm ion}\rangle\,,$$

where v is the electron velocity, an average is made over the Maxwell distribution function of electrons with the electron temperature $T_{\rm e}$. From this we find the average reduced electron energy after ionization,

$$\frac{\langle(\varepsilon - J)\rangle}{2T_{\rm e}} = \frac{\langle x\rangle}{2} = \frac{\langle x^{3/2+\alpha}\rangle}{\langle x^{3/2+\alpha}\rangle} = \frac{\int\limits_0^\infty x^{2+\alpha}e^{-x}dx}{\int\limits_0^\infty x^{1+\alpha}e^{-x}dx} = \frac{\alpha+1}{2}\,.$$

Hence, the energy that is consumed in the formation of one released electron is equal on average to ($\alpha = 1.127$)

$$\overline{E} = J + \frac{\langle(\varepsilon - J)\rangle}{2} = J + \frac{\alpha+1}{2}T_{\rm e} = J + 1.06\ T_{\rm e}, \tag{4.35}$$

and the second term is small compared to the first one.

▶ **Problem 4.20** Determine the dependence of the rate constant of stepwise ionization of atoms in a plasma on the plasma parameters.

We define the rate constant of stepwise ionization $k_{\rm ion}$ from the balance equation for the number density $N_{\rm e}$ of plasma electrons if ionization processes by electron impact are taken into account, and the balance equation has the form

$$\frac{dN_{\rm e}}{dt} = N_{\rm e}\sum_n k_n N_n \equiv N_{\rm e}k_{\rm ion}N_{\rm 0}\,. \tag{4.36}$$

Here N_0 is the number density of atoms in the ground state, N_n is the number density of atoms in an excited state, and k_n is the rate constant of atom ionization by electron impact from the n-th excited state. In a dense plasma the number density of excited atoms is determined by formula (4.22) if we use the Thomson formula for the ionization cross section. But in a reality the dependence of this formula from parameters of the problem is the same.

The main contribution to the rate constant of stepwise ionization is determined by the states with the ionization potential of the order of the thermal electron energy $J_n \sim T_e$ since a typical energy change between an incident and valence electrons is $\sim T_e$. We now estimate the rate constant of stepwise ionization on the basis of formula (4.36),

$$k_{ion} = \sum_n k_n \frac{N_n}{N_0} . \tag{4.37}$$

The statistical weight of the states with the dominant contribution to ionization is $g_n \sim g_i n^2 \sim g_i/J_n \sim g_i/T_e$, where n is the principal quantum number of these states, and g_i is the statistical weight of the ion that is a core for an excited atom. Since the number of terms of different n in this sum is $\sim n$, we obtain for the rate constant of stepwise ionization (4.37) by taking into accounting formula (4.30)

$$k_{ion} \sim \frac{g_i}{g_0 T_e^3} \exp\left(-\frac{J}{T_e}\right) .$$

Note that this ionization process is determined by highly excited states whose properties are determined by the Coulomb electron–core interaction, and hence other parameters of this formula correspond to the parameters of the hydrogen tom. Hence, reducing the above formula to the rate constant dimensionality, we obtain for the rate constant of stepwise ionization

$$k_{ion} = A \frac{g_i}{g_0} \frac{m_e e^{10}}{\hbar^3 T_e^3} \exp\left(-\frac{J}{T_e}\right) , \tag{4.38}$$

where the numerical coefficient $A \sim 1$ (in reality, $A \approx 3$).

▶ **Problem 4.21** Compare the rate constants of stepwise ionization in a dense plasma for two different kinds of atoms.

We consider the character of stepwise ionization of atoms by electron impact, so that the main contribution to its rate constant follows from highly excited states whose parameters are determined only by the Coulomb interaction between a bound electron and a core-ion. Hence, the individuality of atoms is lost for these states, i. e., the character of processes for these atom states proceeds in the same manner for atoms of different kinds. We therefore can use formula (4.38) for the rate constant of stepwise ionization, which gives the ratio of the rate constants of stepwise ionization in a dense plasma that consists of atoms of different kinds,

$$\frac{k_{ion}^{(1)}}{k_{ion}^{(2)}} = \frac{g_i^{(1)} g_0^{(2)}}{g_i^{(2)} g_0^{(1)}} \exp\left(\frac{J_2}{T_e} - \frac{J_1}{T_e}\right) , \tag{4.39}$$

where $k_{\rm ion}^{(1)}$, $k_{\rm ion}^{(1)}$ are the rate constants of stepwise ionization for indicated atoms, whose ground states are characterized by the ionization potentials J_1, J_2 and the statistical weights $g_0^{(1)}$ and $g_0^{(2)}$, respectively, and the ion statistical weights are $g_i^{(1)}$ and $g_i^{(2)}$, respectively.

4.4 Recombination of Electrons and Ions

► **Problem 4.22** On the basis of the Thomson theory for three-body processes estimate the rate of three-body recombination of electrons and ions with electrons as third particles.

We consider the three-body recombination process of electrons and ions that proceeds according to the scheme

$$2e + A^+ \rightarrow e + A^* , \tag{4.40}$$

and is of importance for a dense plasma. Evidently, the rate constant of this process is given by formula (3.54), and we find the parameters of this formula. In this process particles A and C are electrons, and particle B is ion, and these particles interact through the Coulomb interaction potential $U(R) = \pm e^2/r$, where r is the distance between them. Hence the cross section of the elastic scattering by large angles is $\sigma \sim e^4/T_{\rm e}^2$, and the size of the critical region for this process is $b \sim e^2/T_{\rm e}$. Then formula (3.54) gives for the rate constant of the process (4.40)

$$K_{\rm ei} = \frac{\alpha}{N_{\rm e}} = C\frac{e^{10}}{m^{1/2}T_{\rm e}^{9/2}} . \tag{4.41}$$

From the definition, we have in this case the recombination coefficient α,

$$\alpha = K_{\rm ei} N_{\rm e} = C N_{\rm e}\frac{e^{10}}{m^{1/2}T_{\rm e}^{9/2}} . \tag{4.42}$$

The numerical coefficient C in this expression is of the order of 1, $C \sim 1$. From treatment of numerical evaluation we have

$$C = 1.5 \times 10^{\pm 0.2} ,$$

and this value will be used below.

► **Problem 4.23** Estimate from dimensionality consideration the rate of three-body recombination electrons and ions with electrons as third particles.

While accounting for the character of the process (4.40), which leads to an electron capture in bound atom states with the ionization potential $\sim T_{\rm e}$, we have in this problem three-dimensional parameters, the interaction parameter e^2, the electron mass $m_{\rm e}$, and the thermal electron energy $T_{\rm e}$ (the electron temperature). We have only the combination of these parameters of the dimensionality cm^6/s, which corresponds to the rate constant of the three-body process,

$$K_{ei} \sim \frac{e^{10}}{m^{1/2}T^{9/2}} \,,$$

and this combination coincides with the rate constant (4.41) for the three-body recombination process (4.40).

▶ **Problem 4.24** Determine the rate constant of stepwise ionization of atoms in a plasma on the basis of the rate constant of three-body recombination of electrons and ions and the principle of detailed balance.

In considering this process in a dense low-temperature plasma (for example, in a gas-discharge plasma), we assume that a typical electron energy is usually considerably lower than atomic ionization potentials,

$$T_e \ll J \,. \tag{4.43}$$

Then ionization of atoms proceeds through excitation of atoms, as a result of stepwise ionization, and the last stage of this process is ionization of an excited atom with the electron binding energy of the order of the thermal electron energy in a plasma. Then three-body recombination of electrons and ions with an electron as a third body proceeds in the same succession, but in an inverse direction, and the rate coefficient of stepwise ionization is related to the rate of three-body electron–ion recombination by the principle of detailed balance. We consider this below.

Therefore, single ionization of atoms can occur only in collisions with high-energy electrons from the tail of the distribution function. Ionization can also occur as a result of the collision of an electron with an excited atom. For ionization by electrons that are not energetic enough to produce ionization directly, the atom must pass through the number of excited states, with transitions to these states caused by collisions with electrons. This mechanism for ionization of an atom is called stepwise ionization. We shall estimate the rate constant of stepwise ionization assuming the electron energy distribution to be the Maxwell one, and taking the electron temperature to be considerably lower than the atom ionization potential, so that stepwise ionization of atoms by electron impact can take place only with a high number density of electrons, and there are no competing channels for transitions between excited states. Then the stepwise ionization process is the detailed-balance inverse process to the three-body recombination of electrons and ions. In inverse processes, atoms undergo the same transformations but in opposite directions.

Assuming the electrons to be in thermodynamic equilibrium with the atoms in a plasma, where electrons are formed due to stepwise ionization, and their decay is due to the three-body recombination process, we have the balance equation

$$dN_e/dt = 0 = N_e N_a k_{ion} - \alpha N_e N_i \,.$$

Here N_e, N_i, and N_a are the number densities of electrons, ions, and atoms, respectively, and k_{ion} is the rate constant for stepwise ionization.

Because of thermodynamic equilibrium, the number densities of electrons, ions, and atoms are connected by the Saha distribution (1.69). From this we obtain the relationship between the rate constants of the competing processes as

$$k_{\text{ion}} = \frac{\alpha}{N_e} \frac{g_e g_i}{g_a} \left(\frac{m_e T_e}{2\pi\hbar^2} \right)^{3/2} \exp\left(-\frac{J}{T_e} \right) , \tag{4.44}$$

where g_e, g_i, and g_a are the statistical weights of electrons, ions, and atoms, respectively, and T_e is the electron temperature. Because the rate constants k_{ion} and α/N_e do not depend on the number densities, relation (4.44) is valid even if the Saha distribution does not hold. Thermodynamic equilibrium in the system is used here as a method that allows us to establish a relationship between the rate constants of direct and inverse processes.

The separate question relates to the real ratio of the statistical weights of particles. In considering the limiting case (4.43), we use expression (4.41) for the three-body rate constant of electron–ion recombination. In this limit the recombination process results in collision of two free electrons in the field of the Coulomb center with the formation a bound state for one of colliding electrons. In this process the statistical weights of colliding particles are equal to 1, in spite of a certain spin projection for each electron and the ion statistical weight. Correspondingly, during stepwise ionization, when an excited electron proceeds through some excited states, the statistical weight for this electron is also equal to 1. Therefore, according to the nature of this process when each valence electron is excited independently, it is necessary to change the ratio $g_i g_e / g_a$ by the number of valence electron n, and hence formula (4.44) for the rate constant of stepwise ionization of atoms in a plasma takes the form

$$k_{\text{ion}} = \frac{Cn}{(2\pi)^{3/2}} \frac{m_e e^{10}}{\hbar^3 T_e^3} \exp\left(-\frac{J}{T_e} \right) . \tag{4.45}$$

By taking into account the nature of the process, formula (4.45) coincides with formula (4.38) for the rate constant of stepwise ionization, which is derived from other method. Table 4.3 gives the rate constants for ionization of inert gas atoms evaluated by this formula and $C = 1.5$.

Table 4.3 Rate constants (in cm^3/s) of stepwise ionization of inert gas atoms by electron impact.

T_e (eV)	He	Ne	Ar	Kr	Xe
0.8	2.0×10^{-18}	2.7×10^{-16}	3.8×10^{-13}	3.3×10^{-12}	3.6×10^{-11}
1.0	4.9×10^{-16}	3.0×10^{-14}	1.0×10^{-11}	5.9×10^{-11}	3.8×10^{-10}
2.0	2.4×10^{-10}	1.8×10^{-10}	3.3×10^{-9}	8.0×10^{-9}	2.0×10^{-8}
3.0	7.8×10^{-10}	2.0×10^{-9}	1.4×10^{-8}	2.5×10^{-8}	4.6×10^{-8}
4.0	1.3×10^{-9}	5.0×10^{-9}	2.1×10^{-8}	3.3×10^{-8}	5.3×10^{-8}

▶ **Problem 4.25** Determine the recombination coefficient of a multicharged ion and electron through the formation of an autoionization state (dielectronic recombination).

Autoionization and autodetachment states are of importance as they are intermediate states in collision processes. In the previous chapter we considered the formation of autodetachment states during collisions involving a negative ion and an atom. The subsequent decay of these states during the collision leads to electron release. These states are intermediate states of collision processes involving electrons, especially in recombination and attachment processes, and therefore we consider now autoionization and autodetachment states in more detail.

An autoionization state is a bound state of an atom or a positive ion whose energy is above the boundary of the continuous spectrum. Hence an electron can be released in the decay of such a state. For example, the autoionization state $\mathrm{He}(2s^2,{}^1S)$ is the state of the helium atom where both electrons are located in the excited $2s$ state. This autoionization state can decay. As a result of such a decay one electron makes a transition to the ground state, and the other electron ionizes. The scheme of this process is

$$\mathrm{He}(2s^2,{}^1S) \rightarrow \mathrm{He}^+(1s,{}^2S) + e + 57.9\ \mathrm{eV}. \tag{4.46}$$

An autodetachment state is identical to the autoionization state, but occurs in a negative ion. The decay of such a state proceeds with the formation of a free electron and an atom or a molecule. An example of an autodetachment transition is

$$\mathrm{H}^-(2s^2,{}^1S) \rightarrow \mathrm{H}(1s,{}^2S) + e + 9.56\ \mathrm{eV}. \tag{4.47}$$

Autoionization and autodetachment states are characterized by a certain width Γ that is connected with the lifetime of the states τ with respect to their decay by the relation $\Gamma = \hbar/\tau$.

Dielectronic recombination of an electron and multicharged ion results from capture of the electron into an ion autoionization state and its subsequent decay by the radiative transition to a stable state. The total process proceeds according to the scheme

$$e + A^{+Z} \rightarrow \left[A^{+(Z-1)}\right]^{**}, \tag{4.48}$$

$$\left[A^{+(Z-1)}\right]^{**} \rightarrow A^{+Z} + e, \tag{4.49}$$

$$\left[A^{+(Z-1)}\right]^{**} \rightarrow A^{+(Z-1)} + \hbar\omega\,. \tag{4.50}$$

Since the radiative lifetime of the multicharged ion depends strongly ($\sim Z^{-4}$) on its charge Z, the process of dielectronic recombination is of importance for recombination involving multicharged ions. We evaluate below the rate of this process on the basis of the above scheme of these processes.

From the scheme of these processes we have the following balance equation for the number density N_{ai} of ions in a given autoionization state:

$$\frac{dN_{\mathrm{ai}}}{dt} = N_e N_Z k - N_{\mathrm{ai}}\frac{\Gamma}{\hbar} - N_{\mathrm{ai}}\frac{1}{\tau}\,,$$

where N_{ai} is the number density of ions in an autoionization states, N_e is the electron number density, N_Z is the number density of ions of a charge Z, k is the rate constant of the first process, Γ is the width of the autoionization level, and τ is the radiative lifetime of the autoionization state. From this equation we find the number density of ions in the autoionization state N_{ai}, and the rate of the recombination process $I = N_{ai}/\tau$ as

$$N_{ai} = \frac{N_e N_Z k}{\Gamma/\hbar + 1/\tau}, \qquad I = \frac{N_{ai}}{\tau} = \frac{N_e N_Z k}{\Gamma\tau/\hbar + 1}. \tag{4.51}$$

From this we find the recombination coefficient α as

$$\alpha = \frac{I}{N_e N_Z} = \frac{k}{\Gamma\tau/\hbar + 1}. \tag{4.52}$$

One can connect the rate constant k for electron capture in the autoionization state with the width Γ of the autoionization level. We use the Saha formula for the number density of ions in an autoionization state when electrons and ions are found in thermodynamic equilibrium,

$$\frac{N_Z N_e}{N_{ai}} = \frac{g_e g_Z}{g_{ai}} \left(\frac{m_e T_e}{2\pi\hbar^2}\right)^{3/2} \exp\left(\frac{E_a}{T_e}\right), \tag{4.53}$$

where g_e, g_Z, and g_{ai} are statistical weights for the atomic states, E_a is the excitation energy of the autoionization state, and T_e is the electron temperature. Since thermodynamic equilibrium corresponds to the limit $\tau \to \infty$, a comparison of the first expression in formula (4.51) with formula (4.53) gives for the rate constant of the electron capture in the autoionization state of the ion as

$$k = \frac{g_{ai}}{g_e g_Z} \left(\frac{2\pi\hbar^2}{m_e T_e}\right)^{3/2} \frac{\Gamma}{\hbar} \exp\left(-\frac{E_a}{T_e}\right). \tag{4.54}$$

On the basis of formulas (4.52) and (4.54) we finally obtain for the rate of dielectronic recombination

$$\alpha = \frac{g_{ai}}{g_e g_Z} \left(\frac{2\pi\hbar^2}{m_e T_e}\right)^{3/2} \frac{\Gamma/\hbar}{\Gamma\tau/\hbar + 1} \exp\left(-\frac{E_a}{T_e}\right). \tag{4.55}$$

► **Problem 4.26** Obtain the expression for the cross section of dissociative attachment of an electron to a diatomic molecule as a result of the electron capture on an autodetachment level or for the cross section of dissociative recombination of an electron and a diatomic molecular ion as a result of electron capture on an autoionization level. Then atomic particles fly away until an unstable state of atomic particles becomes a stable one (see Fig. 4.3).

The process of dissociative attachment of electrons to molecules proceeds through the formation of an autodetachment state:

$$e + AB \to (AB^-)^{**} \to A^- + B, \tag{4.56}$$

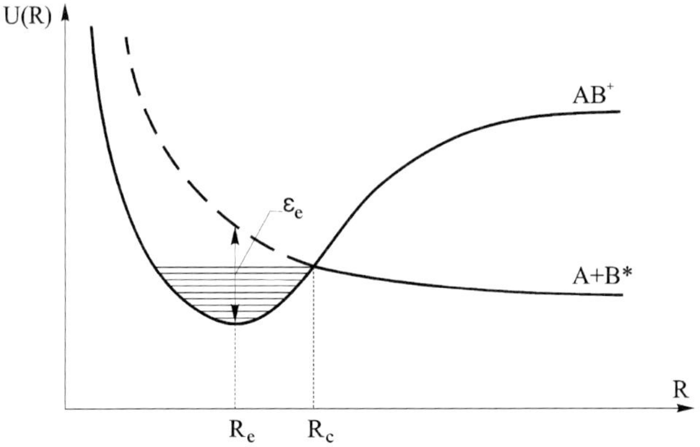

Fig. 4.3 Electron terms for the dissociative recombination process in collisions of an electron and molecular ion.

and in the same manner the dissociative recombination process proceeds,

$$e + AB \rightarrow (AB^-)^{**} \rightarrow A^- + B \,. \tag{4.57}$$

Formally, both processes are identical. Indeed, the first stage of the process is the electron capture on an autodetachment or autoionization level at a given distance between nuclei, and the second stage is flying away of atomic particles up to the formation of a stable state of atomic particles. The cross section for the first stage of the process, electron capture on an autodetachment or autoionization level, is given by the Breit–Wigner formula

$$\sigma_{\text{cap}} = \frac{\pi \hbar^2}{2m_e \varepsilon} \frac{\Gamma^2(R)}{[\varepsilon - \varepsilon_a(R)]^2 + \Gamma^2(R)/4} \,, \tag{4.58}$$

where ε is the electron energy, and $\varepsilon_a(R)$ is the position of the electron term above the energy of the initial molecule (or molecular ion) state at a given distance between nuclei where the electron capture takes place. In order to obtain from this the cross section of the process (4.56) or (4.57), it is necessary to average this over the initial distance between nuclei and multiply by the probability $\exp(-\int \Gamma dt/\hbar)$ of survival for the autodetachment or autoionization state. Thus, the cross section of this process is given by

$$\sigma_{\text{dis}} = \frac{\pi \hbar^2}{2m_e \varepsilon} \int dR \frac{\Gamma^2(R) \left|\varphi_0(R)\right|^2}{[\varepsilon - \varepsilon_a(R)]^2 + \Gamma^2(R)/4} \exp\left(-\int_R^{R_c} \frac{\Gamma(R')dR'}{\hbar v_R}\right) \,. \tag{4.59}$$

Here $\varphi_0(R)$ is the nuclear wave function, so $|\varphi_0(R)|^2 dR$ is the probability that the distance between nuclei in the initial state ranges from R up to $R + dR$, and we replace dt by dR'/v_R, where v_R is the radial velocity for a relative motion of nuclei. This formula is identical to both processes, dissociative attachment and dissociative recombination, and the difference for these processes is contained in positions of electron terms and dependence $\Gamma(R)$.

▶ **Problem 4.27** Obtain the integral relation for the cross section of dissociative attachment or dissociative recombination assuming the width of an autodetachment or autoionization level to be relatively small.

When the width of the autodetachment or autoionization level is small compared to the electron energy (only under this condition this unstable state can be considered as a bound electron state!), one can use the following change in scales of the electron energy:

$$\frac{1}{2\pi}\frac{\Gamma(R)}{[\varepsilon-\varepsilon_a(R)]^2+\Gamma^2(R)/4}=\delta[\varepsilon-\varepsilon_a(R)]\,.$$

Using this change in formula (4.59), we obtain the following integral relation:

$$\int_0^\infty \sigma_{dis}\varepsilon d\varepsilon=\frac{\pi^2\hbar^2}{m_e}\left\langle\Gamma(R)\exp\left(-\int_R^{R_c}\frac{\Gamma(R')dR'}{\hbar v_R}\right)\right\rangle, \qquad (4.60)$$

where the angle brackets mean an average over the distance between nuclei.

▶ **Problem 4.28** Determine the rate constant of dissociative attachment of an electron to a diatomic molecule for the Maxwell velocity distribution function of electrons.

According to the definition, the rate constant of dissociative attachment or dissociative recombination is given by

$$k_{dis}=\frac{2}{\sqrt{\pi}T_e^{3/2}}\int_0^\infty\sqrt{\frac{2\varepsilon}{m_e}}\sigma_{dis}(\varepsilon)\sqrt{\varepsilon}\exp\left(-\frac{\varepsilon}{T_e}\right)d\varepsilon\,.$$

On the basis of formula (4.59) we reduce it to the form

$$k_{dis}=\left(\frac{2\pi}{m_eT_e}\right)^{3/2}\hbar^2\left\langle\Gamma(R)\exp\left[-\frac{\varepsilon_a(R)}{T_e}-\int_R^{R_c}\frac{\Gamma(R')dR'}{\hbar v_R}\right]\right\rangle. \qquad (4.61)$$

In derivation this formula we assume $\Gamma \ll T_e$ and use the integral relation (4.60).

▶ **Problem 4.29** Determine the rate constant of dissociative recombination of an electron and molecular ion assuming the width of an autoionization electron term to be independent of the excitation energy of this term, and this term varies linearly with the distance between nuclei.

According to its nature, dissociative recombination is similar to dissociative attachment and proceeds according to the scheme

$$e+AB^+\to A+B^+\,.$$

Because of many excited states, this process usually is more effective than the electron attachment process and chooses optimal excited atom states in the end. As is shown, Fig. 4.4 represents the behavior of electron terms for this process.

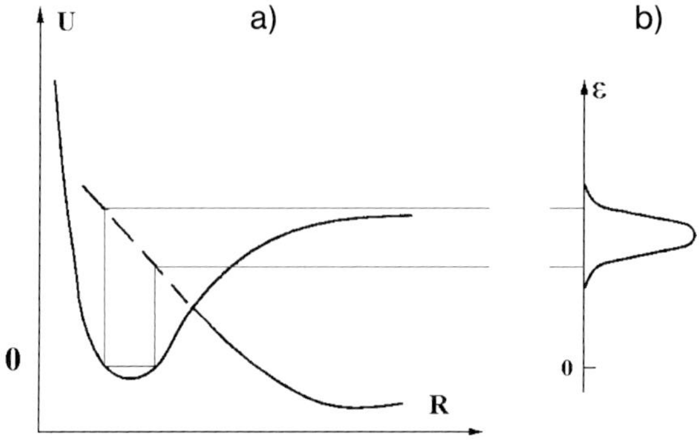

Fig. 4.4 Electron terms which are responsible for electron attachment to a molecule (a) and the spectrum of captured electrons (b).

The rate constant of this process is determined by formula (4.61) with $\Gamma(R) =$ const,

$$k_{\text{dis}} = \hbar^2 \Gamma \left(\frac{2\pi}{m_e T_e} \right)^{3/2} \int |\varphi_0(R)|^2 dR \exp\left[-\frac{\varepsilon_a(R)}{T_e} - \frac{\Gamma}{\hbar} \sqrt{\frac{2\mu(R_c - R)}{E_R}} \right],$$

where $E_R = d\varepsilon_a(R)/dR$. Introducing the excitation energy of the autoionization term E_a at the distance R_c of the term intersection, i. e., $\varepsilon_a(R) = E_a + E_R(R_c - R)$, we consider two limiting cases with respect to the electron temperature and the width of the autoionization term. In the case

$$\Gamma \ll \frac{\hbar E_R}{\sqrt{\mu T_e}}$$

the integral for the rate constant of dissociative recombination converges near the intersection distance due to the energy dependence for the electron distribution function, which gives in this case

$$k_{\text{dis}} = \frac{\hbar^2 \Gamma T_e}{E_R} \left(\frac{2\pi}{m_e T_e} \right)^{3/2} |\varphi_0(R_c)|^2 \exp\left(-\frac{E_a}{T_e} \right) . \tag{4.62}$$

In the opposite limiting case, the range of distances which gives the contribution to the rate constant is determined by survival of the autoionization term, and we have in this limiting case

$$k_{\text{dis}} = \frac{2\hbar^4 \Gamma}{E_R \mu} \left(\frac{2\pi}{m_e T_e} \right)^{3/2} |\varphi_0(R_c)|^2 \exp\left(-\frac{E_a}{T_e} \right) . \tag{4.63}$$

In both cases we assume that the range of distances that determines dissociative recombination is narrow and is concentrated in the vicinity of the intersection distance R_c.

▶ **Problem 4.30** Determine the cross section of dissociative attachment of an electron to a molecule if the distances between nuclei in the molecule are concentrated in a narrow range near R_0, and this distance is separated significantly from the intersection distance R_c of the electron terms.

We use formula (4.61) for the cross section of dissociative attachment of an electron to a molecule

$$\sigma_{dis} = \frac{\pi\hbar^2}{2m_e\varepsilon}\int dR \frac{\Gamma^2(R)\left|\varphi_0(R)\right|^2}{[\varepsilon-\varepsilon_a(R)]^2+\Gamma^2(R)/4}\exp\left(-\int_R^{R_c}\frac{\Gamma(R')dR'}{\hbar v_R}\right). \tag{4.64}$$

Let us take the nuclear wave function of the initial (ground) vibrational state in the form

$$\left|\varphi_0(R)\right|^2 = \frac{1}{\sqrt{\pi}\Delta R}\exp\left[-\frac{(R-R_0)^2}{\Delta R^2}\right],$$

where R_0 is the equilibrium distance between nuclei in the molecule, and ΔR is the amplitude of nuclear oscillations. At small widths of an autoionization electron term

$$\Gamma \ll \frac{dE}{dR}\Delta R,$$

electron capture proceeds in a narrow range of distances that allows us to use the change

$$\frac{1}{2\pi}\frac{\Gamma(R)}{[\varepsilon-\varepsilon_a(R)]^2+\Gamma^2(R)/4} \longrightarrow \delta[\varepsilon-\varepsilon_a(R)]$$

as we have done above. This gives the cross section

$$\sigma_{dis}(\varepsilon) = \frac{\pi^2\hbar^2\Gamma(R_\varepsilon)}{m_e\varepsilon}\frac{\left|\varphi_o(R_\varepsilon)\right|^2}{E_R}\exp\left(-\int_{R_\varepsilon}^{R_c}\frac{\Gamma(R')dR'}{\hbar v_R}\right),$$

where from the resonance of the electron energy we have $\varepsilon = \varepsilon_a(R_\varepsilon)$, and $\varepsilon_a(R) = \varepsilon_a(R_0) + E_R(R-R_0)$.

From this we obtain the resonance energy dependence for the cross section of dissociative attachment

$$\sigma_{dis}(\varepsilon) = \sigma_{max}\exp\left[-\frac{(\varepsilon-\varepsilon_0)^2}{(\Delta R\cdot E_R)^2}\right], \tag{4.65}$$

where we denote $\varepsilon_0 = \varepsilon_a(R_0)$, and the maximum cross section is equal to

$$\sigma_{max} = \frac{\pi^{3/2}\hbar^2}{m_e\varepsilon_0}\frac{\Gamma(R_0)}{\Delta R\cdot E_R}\exp\left(-\int_{R_0}^{R_c}\frac{\Gamma(R)dR}{\hbar v_R}\right). \tag{4.66}$$

As is seen, the width of resonance in the cross section of dissociative attachment is $\Delta\varepsilon \sim \Delta R\cdot E_R$, and we assume it to be small compared to the electron energy. Note that the formulas obtained relate by an equal degree to the process of dissociative recombination of an electron and molecular ion under the used conditions.

5
Elementary Radiative Processes in Excited Gases

5.1
Broadening of Spectral Lines

► **Problem 5.1** Find the broadening of spectral lines due to the Doppler effect owing to thermal motion of atoms.

The distribution of emitting photons over frequencies is characterized by the distribution function a_ω, so that $a_\omega d\omega$ is the probability that the photon frequency ranges from ω up to $\omega + d\omega$. As the probability, the frequency distribution function of photons is normalized as

$$\int a_\omega d\omega = 1. \tag{5.1}$$

Because spectral lines are narrow, in scales of emitting photons, the distribution function of photons is

$$a_\omega = \delta(\omega - \omega_0), \tag{5.2}$$

where ω_0 is the frequency of an emitting photon.

If a radiating particle is moving with a velocity v_x in the direction to a receiver and emit a photon of frequency ω_0, according to the Doppler effect it is conceived by the receiver as having a frequency

$$\omega = \omega_0 \left(1 + \frac{v_x}{c}\right), \tag{5.3}$$

where c is the velocity of light.

If radiating atoms have the Maxwell distribution over velocities, they are characterized by the distribution function,

$$f(v_x) = C \exp\left(-\frac{mv_x^2}{2T}\right),$$

where T is the atom temperature expressed in energetic units, m is the radiating particle mass, C is the normalized constant. Evidently,

$$a_\omega d\omega = f(v_x) dv_x,$$

Plasma Processes and Plasma Kinetics. Boris M. Smirnov

ISBN: 978-3-527-40681-4

if we take the normalization constant such that the distribution function is normalized to 1. This gives

$$a_\omega = \frac{1}{\omega_0} \frac{mc^2}{2\pi T}^{1/2} \exp\left[-\frac{mc^2(\omega-\omega_0)^2}{2T\omega_0^2}\right] . \qquad (5.4)$$

From this it follows that the typical width of a spectral line due to Doppler broadening $\Delta\omega_D$ is relatively small,

$$\frac{\Delta\omega_D}{\omega_0} \sim \sqrt{\frac{T}{mc^2}} .$$

For example, for hydrogen atoms at room temperature this ratio is approximately 10^{-5}.

▶ **Problem 5.2** Determine the spectral line shape due to a finite lifetime τ of the excited state.

The stationary wave function is characterized by a time factor $\exp(-iEt/\hbar)$, where E is the energy of a given state, so the amplitude of transition with emission of a photon of frequency ω_0 is given by a factor $\exp(-i\omega_0 t)$. If an upper state of transition has the lifetime τ, the time factor of its wave function is $\exp(-iE_* t/\hbar) - t/(2\tau)$, where E_* is the energy of the upper state of transition. Correspondingly, the amplitude of transition with emission of a photon of frequency ω_0 is given now by a factor $c(t) \sim \exp[-i\omega_0 t - t/(2\tau)]$.

Taking the distribution function $a_\omega \sim |c_\omega|^2$, where c_ω is the Fourier component of the amplitude of the radiation transition, we obtain by using the normalization condition (5.1)

$$a_\omega = \frac{1}{2\pi\tau} \frac{1}{(\omega-\omega_0)^2 + [1/(2\tau)]^2} . \qquad (5.5)$$

The spectral line for this character of broadening has the Lorentz form in contrast to the Gauss form of the spectral line for the Doppler broadening mechanism (5.4). In addition, one can prove that the relative width of spectral lines is low when it is determined by the radiation lifetime. Indeed, the radiation time of resonantly excited atoms is $\sim 10^{-9}$ s, while the frequency of this radiation is $\sim 10^{-16}$ s. Note that while considering broadening of spectral lines, we take their relatively small widths.

▶ **Problem 5.3** Determine the spectral line shape if the Doppler and Lorentz mechanisms act simultaneously, but the width of a spectral line is given by the Doppler mechanism.

When we have two independent broadening mechanisms, which are characterized by the photon distribution functions $a_\omega^{(1)}$ and $a_\omega^{(2)}$, the resultant distribution function a_ω is given by

$$a_\omega = \int a_{\omega'}^{(1)} a_{\omega-\omega'}^{(2)} d\omega' .$$

Using in this formula the photon distribution functions for the Doppler (5.4) and Lorentz (5.5) broadening mechanisms, we obtain for the resultant photon distribution function

$$a_\omega = \frac{c}{2\pi\tau\omega_0}\int_{-\infty}^{\infty} \exp\left[-\frac{mc^2(\omega'-\omega_0)^2}{2T\omega_0^2}\right]\left[(\omega'-\omega)^2+[1/(2\tau)^2]\right]^{-1}d\omega' .$$

We have under the considered condition

$$\Delta\omega_D \sim \omega_0\sqrt{\frac{T}{mc^2}} \ll \frac{1}{\tau} .$$

Hence for frequencies $|\omega-\omega_0| \ll 1/\tau$ one can replace in this formula the Lorentz factor by the delta function, and we obtain the Doppler broadening (5.4) for the photon distribution function.

It is of importance for this case that the Doppler function gives a more sharp frequency dependence than the Lorentz one. Let us introduce the boundary frequency ω_b as

$$\exp\left[\frac{mc^2(\omega_b-\omega_0)^2}{2T\omega_0^2}\right] = \Delta\omega_D\tau ,$$

or

$$|\omega_b-\omega_0| = \Delta\omega_D \ln(\Delta\omega_D\tau) . \qquad (5.6)$$

Because of a small width of the spectral line due the Doppler broadening, we have $|\omega_b-\omega_0| > \Delta\omega_D$ and $|\omega_b-\omega_0|\tau > 1$. One can see that due to a sharp frequency dependence for the Doppler photon distribution function it is possible that the trunk of the distribution function with $|\omega-\omega_0| < |\omega_b-\omega_0|$ corresponds to the Doppler broadening, whereas for the tail of the distribution function $|\omega-\omega_0| > |\omega_b-\omega_0|$ we have the Lorentz shape of the spectral line. Figure 5.1 gives examples of the photon distribution functions under consideration.

▶ **Problem 5.4** Express the frequency distribution function of photons through the correlation function.

We use a general frequency dependence for the frequency distribution function of emitting photons as

$$a_\omega = |c_\omega|^2 \sim \left|\int \exp(-i\omega t)\langle\Psi_0|\boldsymbol{D}|\Psi_*\rangle dt\right|^2 .$$

Here c_ω is the transition amplitude, $\boldsymbol{D}$ is the operator of the atom dipole moment that determines the intensity of the radiative transition, the wave function of the lower Ψ_0 and upper Ψ_* states of transition depends on time as $\Psi_0 \sim \exp(-iE_0t/\hbar)$, $\Psi_* \sim \exp(-iE_*t/\hbar)$, where E_0, E_* are the energies of these states, and angle brackets denote an average over internal atom coordinates.

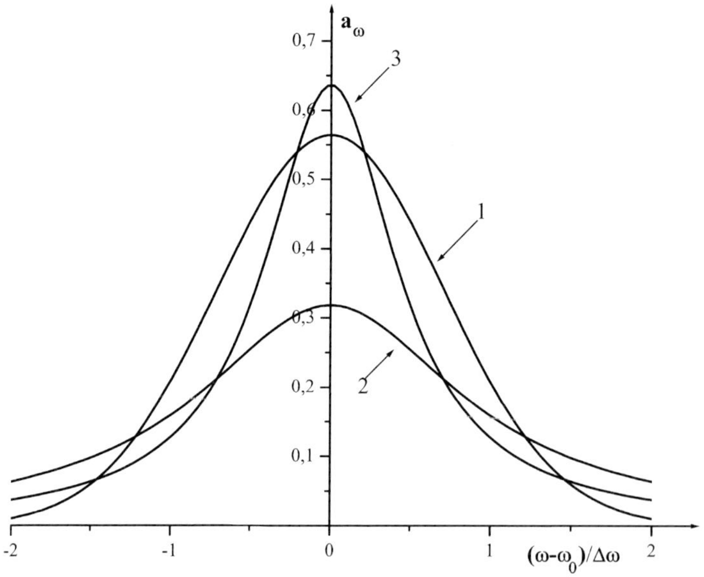

Fig. 5.1 Profiles of spectral lines for the Doppler broadening (1), and Lorenz broadening (2) for identical line width, and for Lorenz broadening (3) with double width of the spectral line. The photon distribution function with respect to the reduced frequency is $a_x = \frac{1}{\sqrt{\pi}} \exp(-x^2)$, $\frac{1}{\pi(1+x^2)}$ and $\frac{2}{\pi(1+4x^2)}$ in these cases.

Because the operator $\boldsymbol{D}$ does not depend on time, we obtain from this, if a radiating atom is isolated,

$$a_\omega \sim \int_{-\infty}^{\infty} dt \int_{-\infty}^{\infty} dt' \exp[i(\omega - \omega_0)(t - t')] = \delta(\omega - \omega_0),$$

where $\omega_0 = (E_* - E_0)/\hbar$, and we use the normalization condition (5.1).

In a general case, if a radiating atom interacts with an environment, we denote the matrix element as $\varphi(t) = \langle \Psi_0 | \boldsymbol{D} | \Psi_* \rangle$, and then we obtain the photon distribution function in the form

$$a_\omega = |c_\omega|^2 \sim \int \exp(-i\omega t)\varphi(t)dt \int \exp(i\omega t')\varphi^*(t')dt' .$$

Introducing the variable $\tau = t - t',\ t'$, we obtain

$$a_\omega = \frac{1}{2\pi} \int \Phi(\tau) \exp(-i\omega\tau) d\tau ,$$

where we denote the correlation function as

$$\Phi(\tau) = \lim_{T\to\infty} \frac{1}{T} \int_{-T/2}^{T/2} \varphi^*(t)\varphi(t+\tau)dt = \overline{\varphi^*(t)\varphi(t+\tau)} , \tag{5.7}$$

where the overline denotes averaging over time. We also chose the numerical factor such that $a_\omega = \delta(\omega - \omega_0)$ if $\varphi(t) = \exp(-i\omega_0 t)$.

► **Problem 5.5** Determine the correlation function $\Phi(\tau)$ considcring collisions of an excited atom with surrounding ones and assuming a random character of such collisions.

Collisions of a radiating atom with surrounding atoms shifts a phase of the wave function. We take into account such a shift for an excited atom state where the interaction with surrounding atoms is more than that for the atom in a lower state of radiation transition. Let us replace the energy of this state E_* by $E_* + U(R_k)$, where $U(R_k)$ is the interaction potential of a given atom with k-th atom from an environment, and R_k is the distance between these atoms. Excluding simultaneous collision of a test atom with others, we obtain a shift of phase $\chi_k = \int_{-\infty}^{\infty} U(R_k)dt$ as a result of collision with k-th atom.

This gives for the amplitude $\varphi(t)$

$$\varphi(t) = \exp(-i\omega_0 t) - i\sum_k \chi_k \eta(t - t_k) ,$$

where t_k is a time of k-th collision, χ_k is a phase shift due to this collision, and $\eta(t - t_k)$ is the unit function, i. e., $\eta(x) = 1,\ x > 0$, and $\eta(x) = 0,\ x < 0$.

For determining the correlation function $\Phi(\tau)$, let us compose the combination

$$\begin{aligned} \Delta\Phi(\tau) &= \Phi(\tau) - \exp(-i\omega_0\Delta\tau)\Phi(\tau + \Delta\tau) \\ &= \overline{\varphi^*(t)[\varphi(t+\tau) - e^{-i\omega_0\Delta\tau}\varphi(t + \tau + \Delta\tau)]} . \end{aligned}$$

Let us take the value $\Delta\tau$ to be large compared to a time of a strong atom interaction during collision, but small compared to a time between neighboring collisions. Since during this time only one collision proceeds, we obtain

$$\begin{aligned} \varphi(t+\tau) - e^{-i\omega_0\Delta\tau}\varphi(t+\tau+\Delta\tau) &= \varphi(t+\tau) \\ \left[1 - \exp\left(i\sum_k \chi_k\eta(t+\tau+\Delta\tau - t_k) - \chi_k\eta(t+\tau - t_k)\right)\right] &= \varphi(t+\tau)(1 - e^{i\chi_k\Delta\tau}) , \end{aligned}$$

where we assume that the k-th collision proceeds in the time interval $\Delta\tau$. As a result, we have for the correlation function change

$$\Delta\Phi(\tau) = \Phi(\tau)\overline{(1 - e^{i\chi})} ,$$

where an average is made over a time interval $\Delta\tau$. This average over collisions gives

$$\overline{1 - e^{i\chi}} = \Delta\tau N v \int 2\pi\rho d\rho[1 - e^{i\chi(\rho)}] = \Delta\tau(\nu' + i\nu'') ,$$

where $\nu' = Nv\sigma'$, $\nu'' = Nv\sigma''$, and $\sigma' = \int_0^\infty 2\pi\rho d\rho[1 - \cos\chi(\rho)]$, $\sigma'' = -\int_0^\infty 2\pi\rho d\rho \sin\chi(\rho)$.

Note that the phase of the elastic collision of particles at large collision momenta according to formula (2.62) is given by

$$\delta_l = -\frac{1}{2\hbar}\int_{-\infty}^{\infty} U dt = -\frac{\chi(\rho)}{2},$$

and the cross section of the spectral line broadening as a result of the collision is

$$\sigma' = \int_0^{\infty} 2\pi\rho d\rho[1 - \cos\chi(\rho)] = \int_0^{\infty} 4\pi\rho d\rho \sin^2\frac{\chi(\rho)}{2} = \frac{1}{2}\sigma_t,$$

where σ_t is the total cross section of elastic scattering that is given by formula (2.60).

Let us return to the equation for the change of the correlation function $\Delta\Phi(\tau)$

$$\Delta\Phi(\tau) = \Phi(\tau)\overline{(1 - e^{i\chi})} = \Phi(\tau)\Delta\tau(\nu' + i\nu'')$$

Using the definition of the function $\Delta\Phi(\tau)$ and turn the value $\Delta\tau$ to zero, we have the following equation for $\Delta\Phi(\tau)$:

$$\begin{aligned}\Delta\Phi(\tau) &= \Phi(\tau) - \exp(-i\omega_o\Delta\tau)\Phi(\tau + \Delta\tau)\\ &= \Phi(\tau) - (1 - i\omega_o\Delta\tau)(\Phi + \frac{d\Phi}{d\tau}\Delta\tau) = \Delta\tau(i\omega_o\Phi + \frac{d\Phi}{d\tau}).\end{aligned}$$

Equalizing it to the above expression for $\Delta\Phi(\tau)$, we obtain the equation for the correlation function $\Phi(\tau)$,

$$\frac{d\Phi(\tau)}{d\tau} = \Phi(\tau)[i(\omega_o - \nu'')t - \nu' t].$$

Solving this equation by using $\Phi(0) = 1$, we obtain finally for the correlation function

$$\Phi(\tau) = \exp[i(\omega_o - \nu'')t - \nu' t]. \tag{5.8}$$

▶ **Problem 5.6** On the basis of the above correlation function determine the frequency distribution function a_ω due to collisions with surrounding particles.

Using the relation between the frequency distribution function a_ω and the correlation function $\Phi(\tau)$, we obtain for the distribution function

$$a_\omega = \frac{\nu}{2\pi\left[(\omega - \omega_o + \Delta\nu)^2 + \left(\frac{\nu}{2}\right)^2\right]}, \tag{5.9}$$

where $\nu = Nv\sigma_t$ is responsible for broadening of the spectral line, and $\Delta\nu = Nv\sigma''$ gives a shift of the spectral line. One can see that collision broadening gives the Lorentz form of the spectral line. In addition, due to the structure of the cross sections we have $\sigma_t \gg \sigma''$, if the total cross section is determined by large collision momenta. Therefore, usually one can ignore a collision shift of the spectral line in comparison with its width.

The criterion of validity of formula (5.9) is based on the assumption that the probability of locating for two and more surrounding particles in a region of a strong interaction with a radiating atom is small. A typical size of this region corresponds to the Weiskopf radius $\rho_t \sim \sqrt{\sigma_t}$, and the criterion of the collision broadening of a spectral line (the impact theory) for a typical frequency shift is

$$N\sigma_t^{3/2} \ll 1. \tag{5.10}$$

▶ **Problem 5.7** Compare the broadening of a spectral line for the transition between the ground and resonantly excited states of an alkali metal atom due to the Doppler and Lorenz broadening mechanisms. Determine the number density N_{tr} of atoms when the line widths for these mechanisms are coincided.

Let us represent the distribution function of emitting photons for the Doppler line broadening (5.4) in the form

$$a_\omega = \frac{1}{\sqrt{\pi}\Delta\omega_D} \cdot \exp\left[-\frac{(\omega-\omega_0)^2}{\Delta\omega_D^2}\right]$$

where

$$\Delta\omega_D = \omega_0\sqrt{\frac{T}{mc^2}}$$

is the width of a spectral line for the Doppler broadening mechanism. The values of $\Delta\omega_D$ for resonant transitions of alkali metal atoms are given in Table 5.1.

Table 5.1 Broadening parameters for spectral lines of radiative transitions between the ground and first resonantly excited states of alkali metal atoms. λ is the photon wavelength for the resonant transition, τ is the radiative lifetime of the resonantly excited state, $\Delta\omega_D$, $\Delta\omega_L$ are the widths of these spectral lines due to the Doppler and Lorenz mechanisms of the broadening, N_{tr} is the number density of atoms at which these widths become identical. The temperature of alkali metal atoms is 500 K.

Element	Transition	λ (nm)	τ (ns)	$\Delta\omega_D$ (10^9 s^{-1})	$\Delta\omega_L/N$ (10^{-7}cm^3/s)	N_{tr} (10^{16}cm^{-3})
Li	$2^2S \to 3^2P$	670.8	27	8.2	2.6	16
Na	$3^2S_{1/2} \to 3^2P_{1/2}$	589.59	16	4.5	1.6	15
Na	$3^2S_{1/2} \to 3^2P_{3/2}$	589.0	16	4.5	2.4	9.4
K	$4^2S_{1/2} \to 4^2P_{1/2}$	769.0	27	2.7	2.0	6.5
K	$4^2S_{1/2} \to 4^2P_{3/2}$	766.49	27	2.7	3.2	4.2
Rb	$5^2S_{1/2} \to 5^2P_{1/2}$	794.76	28	1.7	2.0	4.5
Rb	$5^2S_{1/2} \to 5^2P_{3/2}$	780.03	26	1.8	3.1	2.9
Cs	$6^2S_{1/2} \to 6^2P_{1/2}$	894.35	31	1.2	2.6	2.3
Cs	$6^2S_{1/2} \to 6^2P_{3/2}$	852.11	27	1.3	4.0	1.6

For the Lorenz broadening mechanism the width of a spectral line $\Delta\omega_L$ is given by formula (5.9),

$$\Delta\omega_L = \frac{1}{2} N\overline{v\sigma_t} ,$$

where N is the number density of atoms. The total cross section σ_t for collision of atoms in the ground and resonantly excited states is given by formula (3.34), and its values for alkali metal atoms are given in Table 3.1. Along with the widths of spectral lines due to the Doppler and Lorenz mechanisms of line broadening, we show in Table 5.1 the values of the atom number density N_{tr} at which these widths are identical, i. e.,

$$N_{tr} = \frac{2\Delta\omega_D}{\overline{v\sigma_t}} .$$

▶ **Problem 5.8** Within the framework of the quasistatic theory of spectral line broadening determine the frequency distribution function of photons for a spectral line wing, if the broadening occurs mostly due to the interaction of an upper state of the radiation transition. Assume the interaction potential of a radiating atom with surrounding ones to be pairwise and isotropic.

The quasistatic theory of the broadening of spectral lines assumes interacting atoms to be motionless, and the shift of the spectral line of a radiating atom owing to its interaction with surrounding atoms is equal, for a given configuration of surrounding atoms, to

$$\Delta\omega \equiv \omega - \omega_0 = \frac{1}{\hbar}\sum_k U(R_k) , \tag{5.11}$$

where R_k is the coordinate of k-th atom in the frame of reference where the radiating atom is the origin.

The wing of a spectral line is created by nearby atoms at distances where the probability of atom location is small. Therefore, we have for the frequency distribution function

$$a_\omega d\omega = w(R)dR = N \cdot 4\pi R^2 dR' ,$$

where $w(R)dR$ is the probability of atom location in a range from R up to $R + dR$ for the origin. Assuming a monotonic dependence $U(R)$, we obtain from this

$$a_\omega = 4\pi R^2 N\hbar (dU/dR)^{-1} . \tag{5.12}$$

▶ **Problem 5.9** Determine the frequency distribution function for the quasistatic character of the broadening of spectral lines for the pair interaction of an upper state of a radiating atom with surrounding gas atoms.

Because of a pair interaction of a radiating atom with surrounding ones, an interaction of a radiating atom with a given gas atom is independent of the interaction with other gas atoms. Hence the distribution function is

$$a_\omega = \prod_k \int p(U_k)dU_k \,,$$

where $p(U_k)dU_k$ is the probability that the interaction potential with the k-th atom ranges from U_k up to $U_k + dU_k$, and the spectral line shift for a given configuration of surrounding atoms is equal to

$$\hbar(\omega - \omega_0) = \sum_k U_k$$

Let us introduce the characteristic function

$$\Lambda(t) = \int_{-\infty}^{\infty} e^{i(\omega-\omega_0)t} a_\omega d\omega \,,$$

and the frequency distribution function results from the inverse transformation

$$a_\omega = \frac{1}{2\pi} \int_{-\infty}^{\infty} e^{-i(\omega-\omega_0)t} \Lambda(t) dt \,.$$

Correspondingly, the characteristic function is the product of the characteristic functions $g_k(t)$ of all the atoms,

$$\Lambda(t) = \prod_k g_k(t), \qquad g_k(t) = \int \exp\left(\frac{iU_k t}{\hbar}\right) p(U_k)dU_k \,.$$

Let n gas atoms ($n \gg 1$) be located in a large volume Ω, so that $p(U_k)dU_k = d\boldsymbol{R}_k/\Omega$ is the probability for the k-th atom to be found in this volume element. We obtain

$$g_k(t) = \frac{1}{\Omega} \int \exp\left(\frac{iU(\boldsymbol{R}_k)t}{\hbar}\right) d\boldsymbol{R}_k = 1 + \frac{1}{\Omega} \int \left[\exp\left(\frac{iU(\boldsymbol{R}_k)t}{\hbar}\right) - 1\right] d\boldsymbol{R}_k$$

and

$$\ln g_k(t) = \frac{1}{\Omega} \int \left[\exp\left(\frac{iU(\boldsymbol{R}_k)t}{\hbar}\right) - 1\right] d\boldsymbol{R}_k \,.$$

Because of an independent and random distribution of n gas atoms, we have

$$\ln \Lambda(t) = \sum_k \ln g_k(t) = n g_k(t) = N \int \left[\exp\left(\frac{iU(\boldsymbol{R}_k)t}{\hbar}\right) - 1\right] d\boldsymbol{R}_k \,,$$

N is the number density of atoms. This gives for the frequency distribution function

$$a_\omega = \frac{1}{2\pi} \int_{-\infty}^{\infty} e^{-i(\omega-\omega_0)t} \Lambda(t) dt$$

$$= \frac{1}{2\pi} \int_{-\infty}^{\infty} \exp\left(-i(\omega-\omega_0)t + N \int \left[\exp\left(\frac{iU(\boldsymbol{R})t}{\hbar}\right) - 1\right] d\boldsymbol{R}\right) dt \,, \qquad (5.13)$$

where $U(\boldsymbol{R})$ is the pair interaction potential.

▶ **Problem 5.10** Determine the mean shift of the spectral line for the polarization interaction of the core of a radiating atom in an upper state of the radiative transition with surrounding atoms for the quasistatic character of the broadening of spectral lines.

In this case the upper state of a radiating atom is strongly excited, and a size of its orbit exceeds remarkably an average distance between gas atoms. Therefore, the nearest atoms are located closer to a core of the radiating atom than its excited electron, and the interaction potential with these atoms results from the interaction of the core charge with an induced atom dipole of a gas atom. As a result, the energy of a transferred photon differs from that of a free radiating atom by an energy of the polarization interaction between a charged core and surrounding atoms located inside an orbit of the upper state atom.

According to formula (5.11) a shift of the photon frequency as a result of the polarization interaction with gas atoms is

$$\Delta\omega = \omega - \omega_o = -\frac{1}{\hbar}\sum_k \frac{\alpha e^2}{2R_k^4} ,$$

where α is the atom polarizability, R_k is the distance from the core to the k-th atom. One can replace a sum by the integral for large distances from the core,

$$\Delta\omega = -\frac{\alpha e^2}{2\hbar}\int_R^\infty \frac{N \cdot 4\pi R^2 dR}{R^4} ,$$

where N is the number density of atoms. If for the average shift, we take the lower limit R_o such that one atom is located in the sphere of this radius, we obtain for the average shift of a spectral line

$$\overline{\Delta\omega} = -\frac{10\alpha e^2 N^{4/3}}{\hbar} .$$

Let us determine the average shift of a spectral line more accurately, on the basis of the frequency distribution function (5.12), which in the case of the polarization interaction has the form

$$a_\omega = \frac{1}{2\pi}\int_{-\infty}^{\infty} \exp[-it(\omega - \omega_o) - (it\nu)^{3/4}]dt ,$$

$$\nu = \frac{\alpha e^2}{2\hbar}\left[\frac{4\pi N}{3}\Gamma\left(\frac{1}{4}\right)\right] \approx \frac{19\alpha e^2 N^{4/3}}{\hbar} .$$

The maximum of the frequency distribution function a_ω determines the average frequency shift $\overline{\Delta\omega}$, which is given by

$$\overline{\Delta\omega} = -0.52\nu = -\frac{10\alpha e^2 N^{4/3}}{\hbar} . \tag{5.14}$$

As is seen, this coincides with that obtained in a rough manner. The frequency distribution function a_ω for the case of the polarization interaction is given in Fig. 5.2.

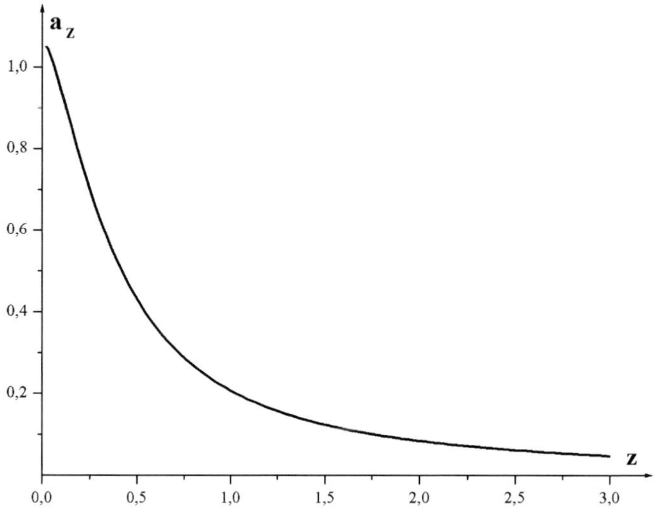

Fig. 5.2 Profile of a spectral line for static polarization broadening, and the reduced frequency is $x = (\omega - \omega_o)/\nu^{3/4}$.

▶ **Problem 5.11** Give the criterion of the validity of the quasistatic theory of spectral line broadening and compare it with that of the impact theory of broadening (formula 5.10).

The quasistatic theory of spectral line broadening requires atoms to be motionless during a typical time ($\sim 1/\Delta\omega$) of creation of this broadening, and this holds true under the condition

$$\frac{1}{\Delta\omega} \ll \frac{R}{v} \,,$$

where R is the typical distance from a nearest atom and v is the typical velocity of their relative motion. For the average shift we have $\Delta\omega \sim U(R_o)/\hbar$, where $R_o \sim N^{-1/3}$ is the average distance between nearest neighbors, and N is the number density of gas atoms. Substituting this in the above criterion, we get

$$\frac{R_o U(R_o)}{\hbar v} \gg 1$$

Assuming the interaction potential $U(R)$ to be a monotonic function of the distance R between atoms, we have from this

$$R_o < \rho_t \,,$$

where ρ_t is the Weiskopf radius. Thus, the quasistatic theory of spectral line broadening is valid at high densities of gas atoms. One can see that the criterion of the validity of the quasistatic theory is opposite with respect the criterion of the collision broadening theory (5.10). Thus, the quasistatic and impact theories of broadening of spectral lines relate to opposite physical cases of broadening.

▶ **Problem 5.12** Determine the distribution function of electric field strengths at a given point of a plasma if the resultant electric field is created by motionless ions.

Nearest ions create an electric field at a point of location of a radiating atom, and we first find the distribution on the electric field strengths. The electric field strength is $\boldsymbol{e}_k = e\boldsymbol{R}_k/R_k^3$ from an individual ion located at point $\boldsymbol{R}_k$ in the frame of reference where the origin is a radiating atom. Then one can obtain the distribution on electric field strengths assuming a random distribution of ions in a space and using the derivation of formula (5.12) spreading it from a scalar argument to a vector one. We obtain by analogy with formula (5.12) for the probability $P(\boldsymbol{E})$ that the total electric field strength that is summed from individual ions is equal to a given value $\boldsymbol{E}$,

$$P(\boldsymbol{E}) = \frac{1}{(2\pi)^3}\int \exp\left[-i\boldsymbol{r}\boldsymbol{E} + N_\mathrm{i}\int\left(e^{i\boldsymbol{r}\boldsymbol{e}} - 1\right)d\boldsymbol{R}\right]d\boldsymbol{r}, \tag{5.15}$$

where $\boldsymbol{e}(\boldsymbol{R}) = e\boldsymbol{R}/R^3$ we take into account $\boldsymbol{E} = \sum_k \boldsymbol{e}_k$, and N_i is the number density of ions. Using the isotropic character of the distribution on the electric field directions, one can reduce this formula to the form

$$P(\boldsymbol{E})dE = H(z)dz, \qquad z = \frac{E}{2\pi e}\left(\frac{15}{4}\right)^{2/3} N_\mathrm{i}^{-2/3},$$

and the Holtzmark function $H(z)$ is given by

$$H(z) = \frac{2}{\pi z}\int_0^\infty x\sin x\exp\left[-\left(\frac{x}{z}\right)^{3/2}\right]dx\,.$$

In the limiting cases the Holtzmark function is equal to

$$H(z) = \frac{4}{3\pi}z^2, \quad z \ll 1, \qquad H(z) = \frac{15}{8}\sqrt{\frac{2}{\pi}}z^{-5/2}, \quad z \gg 1\,.$$

The Holtzmark function is represented in Fig. 5.3.

▶ **Problem 5.13** Determine the frequency distribution function in a hydrogen plasma if the shift of spectral lines for radiative transitions involving an excited hydrogen atoms when the broadening results from the interaction of excited hydrogen atoms in degenerate states with electric fields which are created by plasma ions (the Holtzmark broadening).

The distribution function on electric field strengths (5.15) allows us to determine the frequency distribution of emitting photons. Indeed, if an excited hydrogen atom is characterized by the parabolic quantum numbers n, n_1, n_2, a shift of the electron energy $\Delta\varepsilon$ under the action of an electric field of a strength E is

$$\Delta\varepsilon = \frac{3\hbar^2 E}{2m_e e}n(n_2 - n_1)\,.$$

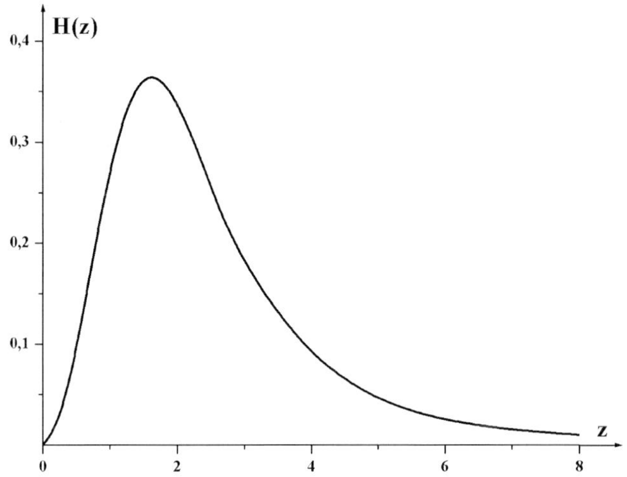

Fig. 5.3 The Holzmark function $H(z) = \int_0^\infty x \sin x \exp\left[-\left(\frac{x}{z}\right)^{3/2}\right] dx$.

Let us assume identically the probability of the location of a hydrogen atom with a given principal quantum number n in states with certain parabolic quantum numbers n_1, n_2. Ignoring a shift or splitting the lower transition state, we then obtain the frequency distribution function of emitting photons

$$a_\omega d\omega = \frac{1}{n^2} \sum_{n_1, n_2} H(z_{12}) dz_{12}, \qquad z_{12} = \frac{1}{3\pi} \left(\frac{15}{4}\right)^{2/3} \frac{m_e(\omega - \omega_0)}{\hbar n(n_2 - n_1)} N_i^{-2/3}$$

▶ **Problem 5.14** Show that in an intermediate case of broadening, when the quasistatic theory of broadening is valid for the central part of a spectral line, whereas the wing of the spectral line is described by the impact theory of broadening, an estimate for the width of the spectral line is identical for both methods.

On the basis of the impact theory, we have for the width of a spectral line

$$\Delta\omega \equiv |\omega - \omega_0| \sim v/R \sim U(R)/\hbar ,$$

where R is the typical distance between interacting atoms that determines a given shift of the spectral line. Comparing the last relation ($v/R \sim U(R)/\hbar$) with the estimation of the total cross section of particle scattering σ_t by formula (2.57), we obtain $R \sim \sqrt{\sigma_t}$. On the other hand, on the basis of the quasistatic theory of broadening we obtain the frequency distribution function, if the broadening for a given frequency is created by distances form a radiating atom $\sim R$,

$$a_\omega = 4\pi R^2 N\hbar \left|dU/dR\right|^{-1} \sim \hbar N R^3 / \left|U(R)\right| \sim N\sigma^{3/2} / |\omega - \omega_0| \sim N\sigma^2/v .$$

As follows from the impact theory of broadening, the frequency distribution function is

$$a_\omega \sim v / |\omega - \omega_0|^2 \sim N v\sigma/(v^2\sigma) \sim N\sigma^2/v .$$

One can see that both approaches exhibits the same behavior in the transition region. This demonstrates the relation between the impact and quasistatic theories of the broadening of spectral lines as opposite limiting cases of the interaction between the emitting and the surrounding atoms.

5.2
Cross Sections of Radiative Tansitions

▶ **Problem 5.15** Determine the rate of the radiative transition between two atom states with nonzero matrix element of the atom dipole moment.

The resonantly excited atom states can be formed as a result of absorption of the dipole radiation by atoms in the ground states, and below we will concentrate on such excited states because they are important in the interaction between atoms and a radiation field. We consider below the radiative process

$$A_0 + \hbar\omega \to A_f\,, \tag{5.16}$$

where the subscripts denote atom states, and we have in mind that index 0 relates to the ground atom state, while f denotes a resonantly excited state. The transition in an excited state under the action of an electromagnetic wave results from the interaction between between the radiation and atomic fields, and the simplest and strongest operator of this interaction has the form

$$V = -\boldsymbol{E}\boldsymbol{D}\,,$$

where $\boldsymbol{D}$ is the operator of the atom dipole moment and $\boldsymbol{E}$ is the strength of the radiation electromagnetic field.

We now find the rate of the process (5.16) on the basis of the nonstationary perturbation theory, taking the electromagnetic wave strength as $\boldsymbol{E} = \boldsymbol{E}_\omega \cos\omega t$, where ω is the frequency of the electromagnetic field, and considering the atom–field interaction as a perturbation. Then the atom wave function has the form

$$\Psi = \psi_0 e^{-i\varepsilon_j t/\hbar} + c_f \cdot \psi_f e^{-i\varepsilon_f t/\hbar}\,,$$

where ψ_0 and ψ_f are the eigenfunctions of the unperturbed Hamiltonian $\hat{H}_0$ for the noninteracting atom and radiation field, and the energies ε_0, ε_f correspond to these states. In addition, this is valid if $c_f \ll 1$. The Schrödinger equation for this system is

$$i\hbar\frac{\partial\Psi}{\partial t} = (\hat{H}_0 + V)\Psi\,,$$

and for the transition amplitude this equation can be rewritten as

$$i\hbar\dot{c}_f = V_{0f}\exp(-i\omega_0 t) = \mathrm{Re}\,\langle 0\,|\boldsymbol{E}_\omega\boldsymbol{D}|\,f\rangle\, e^{i(\omega-\omega_0)t}\,.$$

Here $\hbar\omega_0 = \varepsilon_f - \varepsilon_0$, and the angle brackets denote the matrix element taken between the initial (0) and final (f) atomic states. Solving this equation in the limit

$\omega \approx \omega_o$, we ignore in the expression the c_f terms proportional to $(\omega + \omega_o)^{-1}$, since they are small compared to terms proportional to the $(\omega - \omega_o)^{-1}$. This gives the probability $|c_f|^2$ of the transition for large times $t \gg \omega^{-1}$,

$$|c_f|^2 = \frac{1}{\hbar^2} \left|\langle 0 | \boldsymbol{E}_\omega \boldsymbol{D} | f \rangle\right|^2 \frac{\sin^2 [(\omega - \omega_o) t/2]}{(\omega - \omega_o)^2} .$$

In the limit $\omega \to \omega_o$ we use the delta function by replacing

$$\frac{\sin^2 [(\omega - \omega_o) t/2]}{(\omega - \omega_o)^2} \to \frac{\pi \hbar}{2} \delta(\hbar\omega - \hbar\omega_o) .$$

This gives the probability of the radiative transition per unit time,

$$w_{0f} = \frac{|c_f|^2}{t} = \frac{\pi}{2\hbar} |\langle 0 | \boldsymbol{E}_\omega \boldsymbol{D} | f \rangle|^2 \delta[\hbar\omega - \hbar\omega_o] .$$

We now connect the strength of an electromagnetic wave E_ω with the number photons of the radiation field n_ω per state. We have for the radiation energy per unit time

$$\left\langle \frac{E^2}{8\pi} \right\rangle + \left\langle \frac{H^2}{8\pi} \right\rangle = \left\langle \frac{E^2}{4\pi} \right\rangle = \frac{E_\omega^2}{8\pi} ,$$

where brackets indicate an average over time; we account for the equality of the average of the electric and magnetic energies of electromagnetic wave and the definition of E_ω, so that $E_\omega^2 = E^2/2$. Note that in this way we introduce E_ω to be the electric field strength for a given mode of the radiation field. We now consider the number of photons n_ω per state on the basis of relation (1.63) for the spectral density of radiation U_ω,

$$U_\omega = \frac{E_\omega^2}{8\pi} = \frac{\hbar\omega^3}{\pi^2 c^3} \cdot n_\omega .$$

Replacing the electric field strength by the number of photons per state in the expression for the rate of photon absorption, we obtain

$$w_{0f} = \frac{4\omega^3}{\hbar c^3} \cdot n_\omega \cdot \left|\langle 0 | \boldsymbol{D} \boldsymbol{s} | f \rangle\right|^2 ,$$

where $\boldsymbol{s}$ is a unit vector of photon polarization ($\boldsymbol{E}_\omega = \boldsymbol{s} E_\omega$). An average over polarizations and summation over final atom state give the following expression for the rate of photon absorption with atom transition from the ground state to a resonantly excited one,

$$w_{0f} = \frac{4\omega^3}{3\hbar c^3} \cdot \left|\langle 0 | \boldsymbol{D} | f \rangle\right|^2 g_f \cdot n_\omega , \tag{5.17}$$

where g_f is the statistical weight of the final state.

▶ **Problem 5.16** Express the rates of radiative transitions between the ground and excited states through Einstein coefficients.

Let us introduce the Einstein coefficients A and B on the basis of relations for the rates of radiative transitions,

$$w(0, n_\omega \to f, n_\omega - 1) = A \cdot n_\omega; \quad w(f, n_\omega \to 0, n_\omega + 1) = \frac{1}{\tau_f} + B \cdot n_\omega \,, \tag{5.18}$$

where τ_f is the radiative lifetime of an excited state f with respect to the spontaneous radiative transition to a state 0. Formula (5.17) gives the expression for the Einstein coefficient A. We find from the balance of transitions under thermodynamic equilibrium the relation between the Einstein coefficients A and B and the expression of the spontaneous lifetime τ_f for the transition in the ground state.

Indeed, because of an equilibrium, the number of emissions per unit time is equal to the number of absorptions per unit time, which gives

$$N_0\, w(i, \overline{n}_\omega \to f, \bar{n}_\omega - 1) = N_f\, w(f, \overline{n}_\omega - 1 \to 0, \overline{n}_\omega) \,.$$

Using expressions (5.18) of the transition rates, we get

$$N_0\, A\overline{n}_\omega = N_f\,(1/\tau + B\overline{n}_\omega) \tag{5.19}$$

In the case of thermodynamic equilibrium, we have the Boltzmann formula (1.12) connecting the number densities of atoms in the ground N_0 and excited N_f states and the Planck formula (1.62) for the average number of photons $\overline{n}_\omega$ per state which are given by

$$N_f = \frac{g_f}{g_0} N_0 \exp\left(-\frac{\hbar\omega}{T}\right), \quad \overline{n}_\omega = \left[\exp\left(-\frac{\hbar\omega}{T}\right) + 1\right]^{-1},$$

where g_0 and g_f are the statistical weights for the ground and excited states, and the photon energy $\hbar\omega$ coincides with the energy difference between the two states. Substituting these expressions in formula (5.19), we obtain the Einstein coefficients

$$A = \frac{g_f}{g_0 \tau_f} \qquad B = \frac{1}{\tau_f} = \frac{4\omega^3}{3\hbar c^3} \cdot \left|\langle j|\boldsymbol{D}|f\rangle\right|^2 g_0 \,. \tag{5.20}$$

In addition, the rates of radiative processes are

$$w(0, n_\omega \to f, n_\omega - 1) = \frac{g_f}{g_0 \tau_f} n_\omega \qquad w(f, n_\omega \to 0, n_\omega - 1) = \frac{1}{\tau_f} + \frac{n_\omega}{\tau_f} \,. \tag{5.21}$$

This analysis leads to the conclusion of the presence of stimulated radiation, which is described by the last term and is of fundamental importance.

▶ **Problem 5.17** Derive the expressions for the cross section of absorption and the cross section of stimulated radiation.

Using the above expressions (5.21) for the rates of absorption and stimulate radiation, one can find the cross section of these processes on the basis of the cross section definition (2.2), as the ratio of the process rate to the flux of incident particles. Indeed, the photon flux is cdN_ω, where c is the velocity of light, and the number density of photons is $dN_\omega = 2n_\omega d\mathbf{k}/(2\pi)^3$, where $\mathbf{k}$ is the photon wave vector and n_ω is the number of photons in a given state; the factor 2 accounts for the two independent polarization states, and $d\mathbf{k}/(2\pi)^3$ is the number of states in an element $d\mathbf{k}$ of the wave vector values. From this we obtain the photon flux

$$j_\omega d\omega = n_\omega \frac{\omega^2 d\omega}{\pi^2 c^2} ,$$

by using the dispersion relation $\omega = kc$ for photons. Because the absorption probability per unit time in an interval $d\omega$ of photon frequencies is $An_\omega a_\omega d\omega$, we get the absorption cross section from its definition,

$$\sigma_{\text{abs}} = \frac{\pi^2 c^2}{\omega^2} A a_\omega = \frac{\pi^2 c^2}{\omega^2} \frac{g_f}{g_0} \frac{a_\omega}{\tau_f} \tag{5.22}$$

In the same manner we obtain the stimulated photon emission cross section

$$\sigma_{\text{em}} = \frac{\pi^2 c^2}{\omega^2} B a_\omega = \frac{\pi^2 c^2}{\omega^2} \frac{a_\omega}{\tau_f} = \frac{g_0}{g_f} \sigma_{\text{abs}} . \tag{5.23}$$

► **Problem 5.18** Determine the maximum of the absorption cross section.

Evidently, maximum absorption corresponds to the center of the spectral line under conditions when the width is determined by a finite lifetime τ_f of an excited state due to spontaneous radiation. In this case we have $a_{\omega_0} \equiv a(\omega_0) = 2\tau_f/\pi$, and the maximum absorption cross section is given by

$$\sigma_{\text{abs}} = 2\pi \frac{g_f}{g_0} \frac{c^2}{\omega^2} = \frac{g_f}{g_0} \frac{\lambda^2}{2\pi} , \tag{5.24}$$

where $\lambda = 2\pi c/\omega$ is the photon wavelength. Thus, the maximum absorption cross section is of the order of the square of the photon wave length. This is a large value. In particular, for photons in the optical region of the spectrum, this value is of the order of 10^{-10}–$10^{-9}\,\text{cm}^2$, which exceeds a typical gas-kinetic cross section by approximately six order of magnitude.

► **Problem 5.19** Give the integral relations for the radiative cross sections.

These relations use the normalization condition (5.1) for the frequency distribution function. On the basis of this relation, using the normalization condition $\int a_\omega d\omega = 1$, and recognizing that the integral converges in a narrow region of photon frequencies, we obtain

$$\int \sigma_{\text{abs}}(\omega) d\omega = \frac{\pi^2 c^2}{\omega^2} \frac{g_f}{g_0} \frac{1}{\tau_f} , \tag{5.25}$$

and

$$\int \sigma_{em}(\omega)d\omega = \frac{\pi^2 c^2}{\omega^2}\frac{1}{\tau_f} . \tag{5.26}$$

5.3 Absorption Coefficient for Resonant Photons

▶ **Problem 5.20** Derive the expression for the absorption coefficient of resonant photons.

According to the definition of the absorption coefficient k_ω for radiation propagated in a gas, the intensity of radiation I_ω of frequency ω is given by

$$\frac{dI_\omega}{dx} = -k_\omega I_\omega , \tag{5.27}$$

where x is the direction of radiation propagation. Then taking into account both absorption and stimulated emission, we can represent the absorption coefficient as

$$k_\omega = N_i\sigma_{abs} - N_f\sigma_{em} = N_i\sigma_{abs}\left(1 - \frac{N_f}{N_i}\frac{g_i}{g_f}\right) . \tag{5.28}$$

Here the absorption cross section σ_{abs} is given by formula (5.22), and the cross section of stimulated emission σ_{em} is determined by formula (5.23), N_0, N_f are the number densities of atoms in the ground and excited states, and g_0 and g_f are the statistical weights of these states.

▶ **Problem 5.21** Derive the expression for the absorption coefficient of resonant photons in the case of thermodynamic equilibrium for atoms in the ground and excited states.

In this case the relation between the number densities in the ground and excited state is determined by the Boltzmann formula (1.12). Substituting this formula in expression (5.28) for the absorption coefficient, we get this expression in the form

$$k_\omega = N_0\sigma_{abs}\left[1 - \exp\left(-\frac{\hbar\omega}{T}\right)\right] . \tag{5.29}$$

▶ **Problem 5.22** Obtain the condition of laser operation.

In this case the absorption coefficient is negative, which takes place under the condition of an inverted population of levels

$$\frac{N_0}{N_f} < \frac{g_0}{g_f} \tag{5.30}$$

If this condition is fulfilled, the photon flux passing through the gas is amplified, and this is a basis of operation of lasers, which are generators of monochromatic radiation. A medium in which condition (5.30) holds true is an active medium.

▶ **Problem 5.23** Show that the absorption coefficient does not depend on the number density of the atom number density in the center of the spectral line for resonant radiation, if the line width is determined by collision processes involving these atoms.

Use formula (5.29) for the absorption coefficient and formula (5.22) for the absorption cross section. Assuming that the broadening of the spectral line result from the collision of a resonantly excited atom with its atom in the ground state, we obtain according to formula (5.9) the frequency distribution function in the line center, $a_\omega = 2/(\pi\nu) = 2/(\pi N_0)\langle v\sigma_t\rangle$, where N_0 is the number density of atoms in the ground state, σ_t is the total cross section of the collision of atoms in the ground and resonantly excited states that is averaged over momentum projections, v is the relative collision velocity, and the angle brackets denote an average over collision velocities. As a result, we obtain the absorption coefficient in the line center assuming the temperature to be relatively small

$$k_o = \frac{\pi c^2}{2\omega^2}\frac{1}{\langle v\sigma_t\rangle\tau}\frac{g_f}{g_0}\left[1-\exp\left(-\frac{\hbar\omega}{T}\right)\right]. \tag{5.31}$$

As is seen, the maximum absorption coefficient under the above conditions does not depend on the number density of atoms. Moreover, because $\sigma_t \sim \frac{1}{v}$, the absorption coefficient does not depend on the temperature if $T \ll \hbar\omega$, i. e., the number density of excited atoms is less than that for atoms in the ground state.

Table 5.2 gives the parameters of the first resonantly excited states for atoms of alkali metals and of alkaline-earth metals whose valence electron shell contains one or two electrons. Here λ is the photon wavelength for the resonant transition, τ is the radiative lifetime of the resonantly excited state, and k_o is the absorption coefficient in the line center when the number density of excited atoms is relatively small.

▶ **Problem 5.24** Determine the dependence of the absorption coefficient for the resonant spectral line center in an alkali metal vapor on the atom number density.

The width of a spectral line for a resonant transition between the ground and resonantly excited states of alkali metal atoms is created by both the Doppler effect and collisions of the emitting atom with surrounding ones. The absorption coefficient for a given frequency ω is inversely proportional to the frequency distribution function of emitting photons a_ω, which is given by the formula

$$a_\omega = \frac{1}{\sqrt{\pi}\Delta\omega_D}\int_{-\infty}^{\infty} d\omega' \exp\left[-\frac{(\omega-\omega')^2}{\Delta\omega_D^2}\right]\frac{1}{(\omega_0-\omega')^2+(\Delta\omega_L)^2},$$

Table 5.2 Parameters of radiative transitions between the ground and first resonantly excited states for atoms of alkali metals and alkaline-earth metals.

Element	Transition	λ nm	τ (ns)	g_*/g_0	$v\sigma_t$ (10^{-7} cm^3/s)	k_0 (10^5 cm^{-1})
H	$1^2S \to 2^2P$	121.57	1.60	3	0.516	8.6
He	$1^1S \to 2^1P$	58.433	0.56	3	0.164	18
Li	$2^2S \to 3^2P$	670.8	27	3	5.1	1.6
Be	$2^1S \to 2^1P$	234.86	1.9	3	9.6	1.4
Na	$3^2S_{1/2} \to 3^2P_{1/2}$	589.59	16	1	3.1	1.1
Na	$3^2S_{1/2} \to 3^2P_{3/2}$	589.0	16	2	4.8	1.4
Mg	$3^1S \to 3^1P$	285.21	2.1	3	5.5	3.4
K	$4^2S_{1/2} \to 4^2P_{1/2}$	769.0	27	1	4.1	0.85
K	$4^2S_{1/2} \to 4^2P_{3/2}$	766.49	27	2	6.3	1.1
Ca	$4^1S \to 4^1P$	422.67	4.6	3	7.3	2.5
Cu	$4^2S_{1/2} \to 4^2P_{1/2}$	327.40	7.0	1	1.1	2.1
Cu	$4^2S_{1/2} \to 4^2P_{3/2}$	324.75	7.2	2	1.7	2.8
Zn	$4^1S \to 4^1P$	213.86	1.4	3	3.3	4.8
Rb	$5^2S_{1/2} \to 5^2P_{1/2}$	794.76	28	1	3.9	0.91
Rb	$5^2S_{1/2} \to 5^2P_{3/2}$	780.03	26	2	6.2	1.2
Sr	$5^1S \to 5^1P$	460.73	6.2	3	9.4	1.7
Ag	$5^2S_{1/2} \to 5^2P_{1/2}$	338.29	7.9	1	1.1	2.0
Ag	$5^2S_{1/2} \to 5^2P_{3/2}$	328.07	6.7	2	1.7	2.9
Cd	$5^1S \to 5^1P$	228.80	1.7	3	3.3	4.5
Cs	$6^2S_{1/2} \to 6^2P_{1/2}$	894.35	31	1	5.3	0.77
Cs	$6^2S_{1/2} \to 6^2P_{3/2}$	852.11	27	2	8.1	1.0
Ba	$6^1S \to 6^1P$	553.55	8.5	3	9.0	1.9
Au	$6^2S_{1/2} \to 6^2P_{1/2}$	267.60	6.0	1	0.49	1.9
Au	$6^2S_{1/2} \to 6^2P_{3/2}$	242.80	4.6	2	0.75	4.2
Hg	$6^1S \to 6^1P$	184.95	1.3	3	2.2	5.7

where ω_0 is the frequency of the line center and ω is a current frequency. From this we have for the distribution function for the line center

$$a_0 = \frac{1}{\sqrt{\pi}\Delta\omega_D}\int_{-\infty}^{\infty} d\omega' \exp\left[-\frac{(\omega_0-\omega')^2}{\Delta\omega_D^2}\right]\frac{1}{(\omega_0-\omega')^2+(\Delta\omega_L)^2}$$
$$= \frac{1}{\sqrt{\pi}\Delta\omega_D}F\left(\frac{\Delta\omega_D}{\Delta\omega_L}\right),$$

and the function $F(z)$ is given by

$$F(z) = \frac{1}{\sqrt{\pi}}\int_{\infty}^{\infty}\frac{dx}{x^2z^2+1}\exp(-x^2)\,.$$

In the limiting cases we have

$$F(z) = 1, \quad z \to 0 \qquad F(z) = \frac{\sqrt{\pi}}{z}, \quad z \to \infty\,.$$

Figure 5.4 gives the function $F(z)$. The dependence of the absorption coefficient

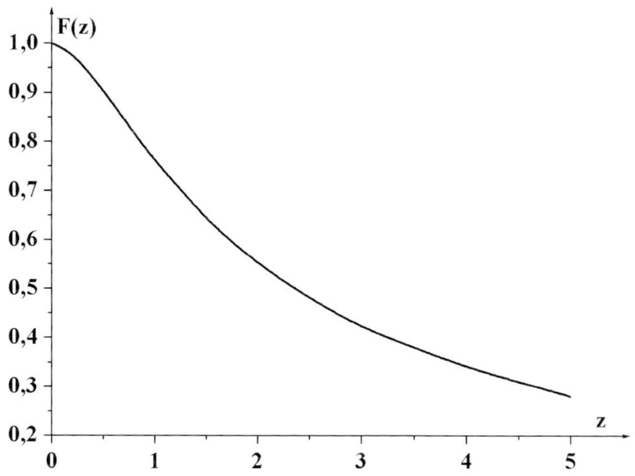

Fig. 5.4 Function $F(z) = \frac{1}{\sqrt{\pi}} \int_{\infty}^{\infty} \frac{dx}{x^2 z^2 + 1} \exp(-x^2)$.

on the number density of atoms is approximated by the dependence

$$k(N) = \frac{k_o}{\sqrt{1 + A(N_{tr}/N)^2}}, \tag{5.32}$$

where k_o is the absorption coefficient in the limit of a large number density N of atoms as is given in Table 5.2, N_{tr} is the atom number density when the Doppler $\Delta\omega_D$ and Lorenz $\Delta\omega_L$ widths of the spectral line are coincided, and the values N_{tr} are given in Table 5.1 for alkali metal vapors. The optimal value of the numerical coefficient in this formula is $A \approx 0.3$.

5.4
Absorption Coefficient in Molecular Gases

► **Problem 5.25** Introduce parameters which determine passing of infrared radiation through a molecular gas.

A spectrum of molecular transitions includes many spectral lines due to rotation-vibration transitions. If the width of an individual spectral line is less than the difference of frequencies for neighboring transitions, the absorption coefficient of a molecular gas is an oscillation frequency function with minima for frequencies between two neighboring lines and maxima for centers of spectral lines for corresponding transitions. Evaluating the total radiation flux, it is necessary to average over such oscillations. We have considered before some parameters of emitting radiation where an average on oscillations is fulfilled.

The absorption function $A(\omega_0, \Delta\omega)$ is defined as

$$A(\omega_0, \Delta\omega) = \frac{1}{\Delta\omega} \int_{\omega_0-\Delta\omega/2}^{\omega_0+\Delta\omega/2} \left(1 - e^{-u_\omega}\right) d\omega \tag{5.33}$$

and is the average probability that a photon is absorbed in a given gas layer. Here u_ω is the optical thickness of this layer, and $\Delta\omega$ is the width of the frequency range considered. If $\Delta\omega$ is large compared to the distance between neighboring lines, the absorption function $A(\omega_0, \Delta\omega)$ depends weakly on $\Delta\omega$, and the absorption function is a convenient characteristic of radiation passing through a given layer, as well as the transmission function that is defined as

$$1 - A(\omega_0, \Delta\omega) = \frac{1}{\Delta\omega} \int_{\omega_0-\Delta\omega/2}^{\omega_0+\Delta\omega/2} e^{-u_\omega} d\omega \,. \tag{5.34}$$

If a spectral band is narrow compared to the frequency of emitting photons, one can represent the total photon flux from a flat layer as

$$j = j_\omega^{(0)} \Delta\omega, \qquad \Delta\omega = 2 \int_{-\infty}^{\infty} d\omega \int_0^1 d\cos\theta \left(1 - e^{-u_\omega/\cos\theta}\right) \cos\theta \,, \tag{5.35}$$

and $j_\omega^{(0)}$ is the flux from the black body surface at a given temperature, and $\Delta\omega$ is the effective width of a spectral band.

One can represent the absorption coefficient of a molecular gas in the infrared spectral range due to the vibration–rotation and rotation transitions as

$$k_\omega = \sum_k S_k a_{\omega-\omega_k} \,. \tag{5.36}$$

Here the frequency ω_k corresponds to a center of the transition, and the intensity S_k of this transition is given by

$$S_k = \frac{\pi^2 c^2}{\omega_k^2 \tau_k} N_0 \left[1 - \exp\left(-\frac{\hbar\omega_k}{T}\right)\right] , \tag{5.37}$$

where N_0 is the number density of molecules for the lower state of transition.

▶ **Problem 5.26** Consider simple models for description of photon absorption in a molecular layer as a result of vibration–rotation and rotation transitions.

If the optical thickness of an emitting layer is small, and photons of an intermediate frequency leave a molecular layer freely, the emission spectrum of the layer is separated in independent spectral lines, and we call this consideration as a model of single lines. For this model one can use formulas for transport of resonance radiation. In particular, the width of an individual spectral line follows from the relation $u_o a_{\omega_o \pm \Delta\omega} / a_{\omega_o} \sim 1$, where u_o is the optical thickness of a gaseous volume for the line center. Correspondingly, the absorption function for this model is equal to $A \sim \Delta\omega/\delta$, where δ is the difference of frequencies of neighboring transitions.

Within the framework of the regular model the spectral lines are regularly spaced, and their intensities vary slightly with variation of the transition number. Then formula (5.36) takes the form

$$k_\omega = S(\omega) \sum_k a(\omega - \omega_o \pm k\delta) , \tag{5.38}$$

where δ is the frequency difference for neighboring transitions. This model describes linear molecules.

For a random model the average difference between neighboring transition frequencies δ is given, while spectral lines are spaced in a random way, and the probability $p(S)dS$ is that the transition intensity is concentrated in a given range.

► **Problem 5.27** Find the average absorption coefficient of a spectral band if the width of an individual spectral line is small in comparison with the distance between neighboring lines.

In this case neighboring spectral lines are not overlapped, and the sum (5.36) is divided into independent ranges. Correspondingly, formula (5.36) gives in this case, if we take into consideration n individual lines,

$$k_\omega = \sum_{k=1}^{n} \int S_k a_{\omega - \omega_k} \frac{d\omega}{n\delta} = \frac{1}{n\delta} \sum_{k=1}^{n} S_k = \frac{\overline{S}}{\delta} . \tag{5.39}$$

► **Problem 5.28** Evaluate the absorption coefficient for a molecular gas, which is found in thermodynamic equilibrium and consists of linear molecules, for the Lorentz shape of the spectral line. Absorption results from rotation-vibration transitions and the spectral line width is small compared to the difference of frequencies for neighboring lines.

Under thermodynamic equilibrium the number density of molecules in vibration and rotation states is given by formulas (1.14) and (1.16), which give

$$N_{vJ} = N_0 e^{-\hbar\omega_o v/T} \frac{B}{T} e^{-BJ(J+1)/T} ,$$

where N_0 is the number density of molecules in the ground vibration and rotation states, $\hbar\omega_o$ is the energy of excitation of the lowest vibration state, v is the number of vibration level, J is the rotation momentum, B is the rotation constant, T is the temperature expressed in energetic units, and we considered that it is usually valid that $B \ll T$. In addition, we restrict by only transition from the ground to the first vibration state. The absorption coefficient (5.29) with the Lorentz shape (5.5) of the spectral line is given by

$$k_\omega = \frac{\pi^2 c^2}{\omega^2} \sum_J w_J N_0 e^{-\hbar\omega_o v/T} \frac{B}{T} e^{-BJ(J+1)/T} \frac{\nu}{(\omega - \omega_J)^2 + \nu^2} ,$$

where w_J is the rate of the radiative transition from the excited vibration state, and ν is the width of a spectral line, the transition energy is $\hbar\omega_J = \hbar\omega_o + B(J+1)$ for the absorption transition $J \to J+1$ (P-branch) and $\hbar\omega_J = \hbar\omega_o - BJ$ for the transition $J \to J-1$ (R-branch). Taking into account $B \ll \hbar\omega_o$, we ignore the difference in the transition rate $(w(J \to J+1) \approx w(J \to J-1) = 1/2\tau)$. Therefore, the absorption coefficient has the form

$$k_\omega = \frac{\pi^2 c^2 \nu}{2\omega^2 \tau}\frac{B}{T} N_0 \sum_{J=-\infty}^{\infty} e^{-BJ(J+1)/T} \frac{(J+1/2)}{\left[\omega - \omega_o - \frac{2B}{\hbar}(J+1/2)\right]^2 + \nu^2} .$$

On the basis of the formula

$$\sum_{k=-\infty}^{\infty} \frac{1}{(x-k)^2 + y^2} = \frac{\pi \sinh(2\pi y)}{y(\cosh 2\pi y - \cos 2\pi x)}$$

and using $B \ll T$, we reduce the expression for the absorption coefficient to the form

$$k_\omega = \frac{\pi^2 c^2 \nu}{4\omega^2 \tau}\frac{\hbar^2 |\omega - \omega_o|}{TB} N_0 \left(1 - e^{-\hbar\omega_o/T}\right) \frac{\sinh \frac{\pi\hbar\nu}{B} \exp\left[-\frac{\hbar^2(\omega-\omega_o)^2}{4BT}\right]}{\left(\cosh \frac{\pi\hbar\nu}{B} - \cos \frac{\pi\hbar|\omega-\omega_o|}{B}\right)} \tag{5.40}$$

One can see that the absorption coefficient is almost a periodic function of frequency with the period of $2B/\hbar$ ($2B$ is the difference of the energies for neighboring transitions). This frequency dependence is named the Elsasser function. The ratio of the values for the absorption coefficient in the neighboring maximum and minimum is equal to

$$\frac{1 + \cosh \frac{\pi\hbar\nu}{B}}{\cosh \frac{\pi\hbar\nu}{B} - 1}$$

This ratio is large in the limit $\hbar\nu \ll B$ and equals $[2B/(\pi\hbar\nu)]^2$. The maximum absorption coefficient corresponds to excitation of the state with the maximum population and the transition frequency $\omega = \omega_o \pm \sqrt{2BT}/\hbar$. The maximum absorption coefficient is equal to

$$k_\omega = \frac{\pi^2 c^2 \nu}{\omega^2}\frac{\hbar\sqrt{2}}{\tau\sqrt{eTB}} N_0 \left(1 - e^{-\hbar\omega_o/T}\right) \frac{\sinh \frac{\pi\hbar\nu}{B}}{\left(\cosh \frac{\pi\hbar\nu}{B} - 1\right)} . \tag{5.41}$$

Figure 5.5 gives an example of the frequency dependence for the absorption coefficient of linear molecules that shows the oscillation character of this function.

► **Problem 5.29** Determine the criterion of the validity of formula (5.40).

In derivation of formula (5.40) for the absorption coefficient we neglect the population change if the rotation quantum number for the molecular transition varies

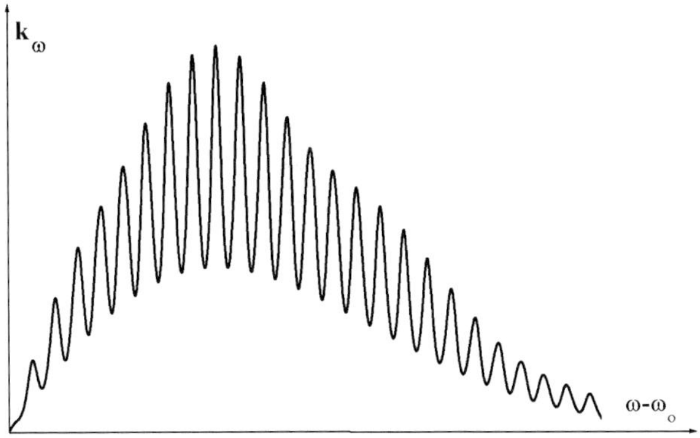

Fig. 5.5 The Elsasser form for the absorption coefficient in a gas of linear molecules as a frequency function due to vibration-rotation transitions. $k_\omega = \frac{x\exp(-\alpha x^2)}{\cosh(\pi a)-\cos(\pi x)}$ in accordance with formula (5.40) and parameters $x = \hbar(\omega - \omega_0)/B$, $\alpha = B/(4T) \ll 1$, $a = \hbar\nu/B$.

by 1. This can be valid for basic rotation numbers of molecules if they are large, i. e., if the rotation constant is less than a thermal energy ($B \ll T$). But it is not fulfilled for relatively small rotation numbers J for which $BJ \sim T$, but simultaneously $J \sim 1$. Under these conditions we cannot assume the intensities of neighboring lines to be identical, as it was used in the derivation formula (5.40). Hence, this formula holds true for transitions from rotation states for which $J \gg \sqrt{T/B}$.

▶ **Problem 5.30** Within the framework of the regular model find the absorption coefficient of a gas consisting of linear molecules for the Lorentz shape of the spectral line.

On the basis of formula (5.36) and the Lorentz shape (5.9) of the spectral line, we have for the absorption coefficient

$$k_\omega = \sum_{k=-\infty}^{\infty} \frac{S\nu}{\pi}\left[(\omega - \omega_0 - k\delta)^2 + \nu^2\right]^{-1},$$

where ν is the width of an individual spectral line, ω_0 is the band center, δ is the distance between neighboring lines, S is the intensity of a spectral line which assumed to be independent of the frequency. On the basis of the Mittag–Leffler theorem, we have

$$\sum_{k=-\infty}^{\infty}\left[(x-n)^2 + y^2\right]^{-1} = \frac{\pi\sinh 2\pi y}{y(\cosh 2\pi y - \cos 2\pi x)},$$

which gives for the absorption coefficient

$$k_\omega = \frac{S}{\delta}\sinh\frac{2\pi\nu}{\delta}\left[\cosh\frac{2\pi\nu}{\delta} - \cos\frac{2\pi(\omega-\omega_0)}{\delta}\right]^{-1}.$$

The ratio of the absorption coefficient in the center of an individual line, k_{max}, to that between two neighboring lines, k_{min}, is equal to

$$\frac{k_{max}}{k_{min}} = \frac{\cosh \frac{2\pi\nu}{\delta} + 1}{\cosh \frac{2\pi\nu}{\delta} - 1} .$$

In the limit $\nu \gg \delta$, when spectral lines are overlapped, this ratio is maximum $[\delta/(\pi\nu)]^2$.

6
Boltzmann Kinetic Equation

6.1
Boltzmann Equation for a Gas

► **Problem 6.1** Derive the kinetic equation for the distribution function of atomic particles of a gas consisting of identical atomic particles.

The concept of the kinetic equation is based on the fact that each particle of a gas is free and moves along a straightforward trajectory during a basic time, and during a small time this particle takes part in a pair collision with other gas particles. According to the definition of cross section, a relative time when a particle interacts strongly with surrounding particles is $\sim N\sigma^{3/2} \ll 1$, where N is the particle number density and σ is the cross section of scattering by a large angle.

The distribution function $f(\boldsymbol{v}, \mathrm{J}, \boldsymbol{r}, t)$ of particles is defined such that $f(\boldsymbol{v}, \mathrm{J}, \boldsymbol{r}, t)d\boldsymbol{v}$ is the number density of particles at the point $\boldsymbol{r}$ at time t, the particle velocity ranges from $\boldsymbol{v}$ to $\boldsymbol{v} + d\boldsymbol{v}$, and J describes all internal quantum numbers. The normalization condition has the form

$$N(\boldsymbol{r}, t) = \sum_{\mathrm{J}} \int f(\boldsymbol{v}, \mathrm{J}, \boldsymbol{r}, t) d\boldsymbol{v} .$$

The variation of the distribution function allows one to analyze the evolution of the gas and has the form

$$\frac{df}{dt} = I_{\mathrm{col}}(f) , \tag{6.1}$$

where $I_{\mathrm{col}}(f)$ is the collision integral that takes into account pair collisions of particles in evolution of the gas.

The left-hand side of this equation, which describes the motion of particles in external fields, has the form

$$\frac{df}{dt} = \frac{f(\boldsymbol{v} + d\boldsymbol{v}, \mathrm{J}, \boldsymbol{r} + d\boldsymbol{r}, t + dt) - f(\boldsymbol{v}, \mathrm{J}, \boldsymbol{r}, t)}{dt} .$$

Since $d\boldsymbol{v}/dt = \boldsymbol{F}/m$, where $\boldsymbol{F}$ is the external force acted on a single particle, m is the particle mass, and $d\boldsymbol{r}/dt = \boldsymbol{v}$, we have

$$\frac{df}{dt} = \frac{\partial f}{\partial t} + \boldsymbol{v} \cdot \frac{\partial f}{\partial \boldsymbol{r}} + \frac{\boldsymbol{F}}{m} \cdot \frac{\partial f}{\partial \boldsymbol{v}} ,$$

Plasma Processes and Plasma Kinetics. Boris M. Smirnov

ISBN: 978-3-527-40681-4

and the kinetic equation (6.1) takes the form

$$\frac{\partial f}{\partial t} + \boldsymbol{v} \cdot \frac{\partial f}{\partial \boldsymbol{r}} + \frac{\boldsymbol{F}}{m} \cdot \frac{\partial f}{\partial \boldsymbol{v}} = I_{\text{col}}(f) . \tag{6.2}$$

This is also called the Boltzmann kinetic equation.

▶ **Problem 6.2** Give a simple form of the kinetic equation expressing the collision integral through a typical time of change of a particle trajectory (tau-approximation).

The collision integral contained in the kinetic equation characterizes the evolution of the system as a result of pairwise collisions of particles. A typical relaxation time of the distribution function in a gas can be estimated as $\tau \sim (N\sigma v)^{-1}$, where v is a typical collision velocity, σ is the collision cross section. This value suggests the simple approximation, tau-approximation,

$$I_{\text{col}}(f) = -\frac{f - f_0}{\tau}$$

for the collision integral, where f_0 is the equilibrium distribution function. The nature of this approximation can be illustrated by a simple example. If we disturb an equilibrium state of the system described by the distribution function f_0 so that the distribution function at the initial time is $f(0)$, then the subsequent evolution of the system is described by the equation

$$\frac{df}{dt} = -\frac{f - f_0}{\tau} , \tag{6.3}$$

and its solution has the form

$$f = f_0 + [f(0) - f_0] \exp\left(-\frac{t}{\tau}\right) . \tag{6.4}$$

Thus, the relaxation time τ of the system is of the order of the time between consecutive collisions of a test particle with others. It can be dependent on the collision velocity.

▶ **Problem 6.3** Obtain the expression for the collision integral in the case of elastic collisions.

The collision integral accounts for changes in the distribution function as a result of pairwise collisions of particles. We analyze first the collision integral in the case of an atomic gas, where it is expressed in terms of the elastic scattering cross section of atoms. The transition probability per unit time per unit volume is denoted by $W(\boldsymbol{v}_1, \boldsymbol{v}_2 \to \boldsymbol{v}'_1, \boldsymbol{v}'_2)$, so the rate $W d\boldsymbol{v}'_1 d\boldsymbol{v}'_2$ is the probability per unit time and per unit volume for collision of two atoms with velocities $\boldsymbol{v}_1$ and $\boldsymbol{v}_2$, if their final velocities are in an interval from $\boldsymbol{v}'_1$ to $\boldsymbol{v}'_1 + d\boldsymbol{v}'_1$ and from $\boldsymbol{v}'_2$ to $\boldsymbol{v}'_2 + d\boldsymbol{v}'_2$, respectively. By definition, the collision integral is

$$I_{\text{col}}(f) = \int (f'_1 f'_2 W' - f_1 f_2 W) d\boldsymbol{v}'_1 d\boldsymbol{v}'_2 d\boldsymbol{v}_2 , \tag{6.5}$$

where we use the notations that $f_1 = f(\boldsymbol{v}_1)$, $W = W(\boldsymbol{v}_1, \boldsymbol{v}_2 \to \boldsymbol{v}_1', \boldsymbol{v}_2')$, and other quantities take subscripts according to the same rule. The principle of detailed balance, which accounts for reversed evolution of a system as would occur in a system in the case of a physical time reversal $t \to -t$, yields $W = W'$. Taking it into account reduces the collision integral to the form

$$I_{\text{col}}(f) = \int (f_1' f_2' - f_1 f_2) W d\boldsymbol{v}_1' d\boldsymbol{v}_2' d\boldsymbol{v}_2 \,. \tag{6.6}$$

▶ **Problem 6.4** Express the collision integral through the differential cross section of elastic scattering of particles.

The elastic scattering cross section follows from its definition as the ratio of the number of scattering events per unit time to the flux of incident particles. The differential cross section for elastic scattering is

$$d\sigma = \frac{f_1 f_2 W d\boldsymbol{v}_1 d\boldsymbol{v}_2 \boldsymbol{v}_1' \boldsymbol{v}_2'}{f_1 d\boldsymbol{v}_1 f_2 d\boldsymbol{v}_2} = \frac{W d\boldsymbol{v}_1' d\boldsymbol{v}_2'}{|\boldsymbol{v}_1 - \boldsymbol{v}_2|} \,.$$

Substitution of this expression into formula (6.6) gives the collision integral

$$I_{\text{col}}(f) = \int (f_1' f_2' - f_1 f_2) |\boldsymbol{v}_1 - \boldsymbol{v}_2| d\sigma d\boldsymbol{v}_1' d\boldsymbol{v}_2' d\boldsymbol{v}_2 \,. \tag{6.7}$$

As is seen, the nine integrations inferred in formula (6.6) are replaced by the five integrations in formula (6.7). This is a consequence of accounting for the conservation of momentum for the colliding particles (three integrations) and the conservation of their total energy (one more integration).

▶ **Problem 6.5** Derive the collision integral for electrons located in a gas if this collision integral is determined by elastic collisions of electrons with gas atoms.

Because of a low number density, electrons do not influence the distribution function of atoms that is the Maxwell one $\varphi(v_a)$. Here $\boldsymbol{v}_a$ is the atom velocity, and the velocity distribution function of atoms is independent of the velocity direction. The collision integral for electrons with atoms (6.7) takes the form

$$I_{\text{col}}(f) = \int (f' \varphi' - f \varphi) v d\sigma d\boldsymbol{v}_a \,, \tag{6.8}$$

where $\boldsymbol{v}$ is the electron velocity, $f(\boldsymbol{v})$ is the velocity distribution function of electrons, and $d\sigma$ is the differential cross section of elastic electron–atom scattering. We account for a large velocity of electrons compared to atom velocities, $v \gg v_a$. As is seen, in this case the collision integral is linear with respect to the electron distribution function.

6.2 Peculiarities of Statistical Description of Gas Evolution

▶ **Problem 6.6** On the basis of the Boltzmann kinetic equation for atoms of a gas find the equilibrium distribution function of atoms on velocities.

If external fields do not act on gas atoms and the gas state is uniform and stationary, the left-hand side of the kinetic equation (6.1) is zero, and it has the form $I_{col}(f) = 0$. Then on the basis of formula (6.6) for the collision integral for atoms we obtain

$$f_1 f_2 = f_1' f_2'$$

for any pair of colliding atoms. It is convenient to rewrite this relation in the form

$$\ln f(\boldsymbol{v}_1) + \ln f(\boldsymbol{v}_2) = \ln f(\boldsymbol{v}_1') + \ln f(\boldsymbol{v}_2') .$$

From this it follows that the value $\ln f(\boldsymbol{v}_1) + \ln f(\boldsymbol{v}_2)$ is conserved as a result of any collisions of two particles. This means that $\ln f(\boldsymbol{v})$ must be an additive function of the integrals of motion. One can find three such values for elastically collided atoms: a constant, the atom momentum $\boldsymbol{p}$, and the atom energy ε. This leads to the following general form of the distribution function:

$$\ln f(\boldsymbol{v}) = C_1 + \boldsymbol{C}_2 \boldsymbol{p} + C_3 \varepsilon ,$$

where C_1, C_2, C_3 are constants. This gives the distribution function

$$f(\boldsymbol{v}) = A \exp\left[-\alpha(\boldsymbol{v} - \boldsymbol{w})^2\right] .$$

This expression is identical to formula (1.20) for the Maxwell distribution function in the frame of reference, where atoms are motionless as a whole. Here A is the normalization constant, $\boldsymbol{w}$ is the average velocity of the distribution, and $\alpha = m/(2T)$, where m is the particle mass and T is the gaseous temperature.

▶ **Problem 6.7** Show for a uniform gas and in the absence of an external field that according to the Boltzmann equation, the functional $H(t) = (\boldsymbol{v}, t) \ln f(\boldsymbol{v}, t) d\boldsymbol{v}$ cannot increase during the evolution of an atom (the Boltzmann H-theorem).

The kinetic equation (6.1) has the following form in this case:

$$\frac{\partial f_1}{\partial t} = \int W(\boldsymbol{v}_1, \boldsymbol{v}_2 \rightarrow \boldsymbol{v}_1', \boldsymbol{v}_2')(f_1' f_2' - f_1 f_2) d\boldsymbol{v}_1' d\boldsymbol{v}_2' d\boldsymbol{v}_2 .$$

From this it follows that

$$\frac{dH}{dt} = \int W(\boldsymbol{v}_1, \boldsymbol{v}_2 \rightarrow \boldsymbol{v}_1', \boldsymbol{v}_2')(f_1' f_2' - f_1 f_2) \ln f_1 d\boldsymbol{v}_1' d\boldsymbol{v}_2' d\boldsymbol{v}_1 d\boldsymbol{v}_2 + \frac{d}{dt}\int f_1 d\boldsymbol{v}_1 .$$

Because of the conservation of the total number of particles, the second term on the right-hand side of this equation is zero. From the symmetry for the rate $W(\boldsymbol{v}_1, \boldsymbol{v}_2 \rightarrow \boldsymbol{v}_1', \boldsymbol{v}_2')$ with respect to the changes $\boldsymbol{v}_1 \longleftrightarrow \boldsymbol{v}_2$ and $\boldsymbol{v}_1 \longleftrightarrow \boldsymbol{v}_1'$, $\boldsymbol{v}_2 \longleftrightarrow \boldsymbol{v}_2'$, we have

$$\frac{dH}{dt} = \frac{1}{4}\int W(\boldsymbol{v}_1, \boldsymbol{v}_2 \rightarrow \boldsymbol{v}_1', \boldsymbol{v}_2')(f_1' f_2' - f_1 f_2) \ln\left(\frac{f_1 f_2}{f_1' f_2'}\right) d\boldsymbol{v}_1' d\boldsymbol{v}_2' d\boldsymbol{v}_1 d\boldsymbol{v}_2 .$$

Since $W \geq 0$, the function $(y - x) \ln \frac{x}{y}$ is negative at any positive values of variables x and y, and is zero if $x = y$. From this it follows that

$$\frac{dH}{dt} \leq 0 \,. \tag{6.9}$$

Note that $\frac{dH}{dt} = 0$ at $f_1 f_2 = f_1' f_2'$, and this corresponds to the equilibrium. From the Boltzmann H-theorem (6.9) it follows that in the absence of external fields evolution of a gas leads to the equilibrium distribution of its particles.

The Boltzmann H-theorem proves that entropy increase when a gas tends to an equilibrium and is of principle for the statistical physics.

▶ **Problem 6.8** A gas is found in a nonequilibrium state and tends to an equilibrium. Show that the statistical description of this process on the basis of the Boltzmann kinetic equation and its dynamic description on the basis of the Newton equations for particles (or the Schrödinger equations in the quantum case) have a different nature.

According the Boltzmann H-theorem, an ensemble of weakly interacting particles that is out of an equilibrium at initial time, will tend to the equilibrium in time. Let an initial distribution function of particles be $\varphi(\boldsymbol{v}, \boldsymbol{r})$ and let the ensemble relax to an equilibrium distribution function f_o, but at time t the distribution function does not reach the equilibrium function. If we reverse time $t \to -t$, the distribution function does not reach the initial distribution function, but will tend to the equilibrium function, as before.

Now let us analyze this scenario from another standpoint. Let the initial distribution function of particles be $\varphi(\boldsymbol{v}, \boldsymbol{r})$, as before, and we describe each particle of an ensemble by the Newton equation that for the i-th particle has the form

$$m \frac{d^2 \boldsymbol{r}_i}{dt^2} = \sum_k \boldsymbol{F}_{ik} \,,$$

where $\boldsymbol{r}_i$ is the coordinate of the i-th particle. The force that acts on this particle from the k-th particle is

$$\boldsymbol{F}_{ik} = -\frac{\partial U(|\boldsymbol{r}_i - \boldsymbol{r}_k|)}{\partial \boldsymbol{r}_i} \,,$$

where U is the interaction potential between two particles, and $\boldsymbol{r}_k$ is the coordinate of the k-th particle.

Let us stop the development of this ensemble at time t and reverse the time $t \to -t$. As is seen, the ensemble will develop in the reversed time, and at time t all the particles will have the initial positions and will move with opposite velocity. So, if we again reverse the time, we obtain the distribution function of particles to be $f = \varphi(\boldsymbol{v}, \boldsymbol{r})$, which is the same as at the beginning.

Thus, two methods of description of a particle ensemble, the statistical method through the kinetic equation and the dynamic method through the set of Newton equations for particles, lead to different results. Note that in the quantum case when we describe the particle state by the Schrödinger equation, we also obtain the inverse development of the particle ensemble at time reversal, as we have above in the classical case on the basis of Newton equations.

▶ **Problem 6.9** Show that the entropy of a uniform ensemble of weakly interacting classical particles is different for the statistical and dynamical description of this ensemble.

The entropy of an ensemble of independent particles is proportional to a number of particles of this ensemble, and below we will consider the entropy S of a test particle, which is introduced as

$$S = -P_i \sum_i \ln P_i \,, \tag{6.10}$$

where P_i is the probability for a particle to be located in a given state i. Because the definition of a state of a classical particle is arbitrary, the entropy is defined with the accuracy up to constant. One can see that H-functional is the entropy taking into account the velocity space.

Take an ensemble of particles to be restricted by a volume V and a typical velocity of particles to be $v_{\rm T}$. Assume that the particle coordinate at a given time is known with the accuracy Δx, Δy, and Δz, and the accuracy of a given velocity component is known with the accuracy Δv_x, Δv_y, and Δv_z for the corresponding component. Then according to the above formula for the entropy, the difference of the entropies per particle for the statistical $S_{\rm st}$ and dynamic $S_{\rm dyn}$ description of the ensemble is

$$S_{\rm st} - S_{\rm dyn} = \ln\left(\frac{V \cdot v_{\rm T}^3}{\Delta x \cdot \Delta y \cdot \Delta z \cdot \Delta v_x \cdot \Delta v_y \cdot \Delta v_z}\right) . \tag{6.11}$$

Thus, the statistical description is irreversible and particles are characterized by a more high entropy at this description than that at the dynamical reversible description of this ensemble. The transition to the statistical description may be a result of the action of a random force on a particle. Random fields may be very weak, but their action on an ensemble of many particles is capable to make their distribution to be random during a not large time. Then the statistical description of this ensemble becomes correct, and the state of a particle becomes uncertain.

▶ **Problem 6.10** Consider the Poincare instability that determines the divergence of particle trajectories at their small uncertainty.

A weak interaction of a particle with random external fields leads to a displacement of its trajectory that is intensified in time. Let us consider two neighboring trajectories of a test particle and ascertain the divergence of these trajectories in time in collisions with other particles. If the distance between the trajectories exceeds the action radius of forces between particles, the trajectories become different after the first collision, and as a result they redispersed after the first collision through the time $\sim L/v_T$, where L is the typical size of a volume occupied by particles, and v_T is a typical thermal velocity of particles.

We consider another limiting case when the trajectory divergence is small compared to the action radius of interparticle forces, but the motion of a particle along two trajectories remains classical. Moreover, for definiteness we assume that scattering of colliding particles is described by the hard-sphere model, so that the impact parameter of collision ρ is connected with the scattering angle θ and the sphere radius R_o by the relation $\rho = R_o \cos\theta$. Assuming the initial displacement of the trajectory δ to be small, we have the difference of impact parameters for two trajectories $\Delta\rho_1 \sim \delta$ and the difference of scattering angles $\Delta\theta_1 \sim \delta/R_o$. From this it follows that the divergence of the impact parameters at the second collision is $\Delta\rho_2 \sim \lambda\Delta\theta_1 \sim \lambda\delta/R_o$, where λ is the mean free path, i. e., the distance which the particle travels between two collisions. Since for a gas $\lambda \gg R_o$, the divergence of trajectories increases between each neighboring collision.

From this it follows that a small deviation of the particle trajectory is enforced with time, i. e., any particle trajectory is instable with respect to its small displacement. Evidently, if a significant change of the trajectory takes place after k collisions, when $\Delta\theta_k \sim 1$, from the above formula we have for an effective number of collisions k, which lead to *randomization*,

$$k = \frac{\ln \frac{\lambda}{\delta}}{\ln \frac{\lambda}{R_o}} \,. \tag{6.12}$$

Let us make an estimate for randomization in an atmospheric air. Then the number density of molecules $N \sim 3 \times 10^{19}\,\mathrm{cm}^{-3}$ and the gas-kinetic cross section $\sigma \sim 3 \times 10^{-15}\,\mathrm{cm}^2$, so that $\lambda = (N\sigma)^{-1} \sim 10^{-5}$ cm, $R_o = \sqrt{\sigma/\pi} \sim 3 \times 10^{-8}$ cm. If the initial displacement of the trajectory is of the order of the nuclear size $\delta \sim 10^{-13}$ cm and the violation of classical laws on such distances is neglected, we obtain from formula (6.12) $k \approx 4$, i. e., several collisions of molecules lead to a random distribution of molecules. Thus, the chaotization process proceeds fast and leads to irreversibility of this system.

Thus, the above Poincare instability results in intensification of a random displacement of the particles' trajectory due to the random weak external fields' divergence due to collisions with other particles. Collisions lead to an exponential growth of the particle entropy in time, and therefore under several conditions the transition proceeds from a deterministic system to a random one described in the statistical terms. Thus, the irreversibility of a statistical ensemble of particles results from randomization inside this particle ensemble due to a weak external action and collisions between particles.

► **Problem 6.11** Analyze the collapse of wave functions for entangled processes and its role in the irreversibility of evolution of a closed system of particles.

We now consider one more mechanism of randomization in the distribution of particles by states that follows from the nature of quantum processes and relates to entangle states. Indeed, the reversibility of the Newton or Schrödinger equations belongs to continuous processes. This can be violated if the ground state of particles is degenerated, so two particles can form an Einstein–Podolsky–Rosen pair, and a subsequent collision of one of these particles with a particle of the system chooses the state of this particle of the pair. This chooses automatically the state of the second particle of the pair. Such an action, the collapse of wave functions, is added to the Schrödinger equation as an additional condition, and since it is a prompt transition of the second particle in a certain state, and this transition has a random character, all this causes an irreversibility of the particle ensemble. It cannot return to the initial state as a result of time reversal.

► **Problem 6.12** An atomic ion is moving in a parent gas and can transfer a charge to atoms as a result of the resonant charge exchange process. Consider a colliding ion and an atom as an example of the Einstein–Podolsky–Rosen pair.

We have the wave function of a colliding atom and its ion according to formula (3.3),

$$\Psi = \frac{1}{\sqrt{2}}\left[\psi_1 \cos\zeta(t) + i\psi_2 \sin\zeta(t)\right] \exp\left(-\frac{i}{\hbar}\int^t \varepsilon dt\right) , \tag{6.13}$$

where the wave functions ψ_1, ψ_2 correspond to ion location near the first or second nuclei, respectively, $\zeta(t) = \int^t (\varepsilon_g - \varepsilon_u) dt/\hbar$ is the charge exchange phase, $\varepsilon = (\varepsilon_g + \varepsilon_u)/2$, ε_g, ε_u are the energies of the even and odd states for the quasimolecule consisting of the ion and atom. The value $\varepsilon_g - \varepsilon_u$ drops exponentially with an increase of the distance between nuclei at large distances. Hence, the transition of a valence electron between fields of two ions finishes at some distances between nuclei, so the probability for the first nucleus to belong to a charged particle after collision equals $\cos^2\zeta(\infty)$, and the probability for the second nuclei to belong to the charged particle after collision is $\sin^2\zeta(\infty)$. Let us measure the charge of the first particle, for example, using a mass spectrometer. This measurement chooses automatically the charge of the second particle. Indeed, if the first particle is charged, the other is neutral, and vice versa. Such a measurement leads to the collapse of wave functions. The collapse of the wave function chooses only one term in the above formula for the wave function of the quasimolecule.

This measurement in a gas proceeds automatically as a result of collision of one of these atomic particles with a gas atom, and this collision establishes that whether a colliding particle is an ion or atom. Therefore, each collision of an atom and ion leads to the collapse of their wave functions through some time after it. Because the collapse of wave functions has a random character, it leads to irreversible evolution of this system. The collapse of wave functions is possible for any system of atomic particles with degenerated states of individual particles.

6.3
Integral Relations from the Boltzmann Equation

▶ **Problem 6.13** Derive the macroscopic transport equations for a gas from the Boltzmann kinetic equation.

To this end we multiply the kinetic equation by an appropriate velocity function $\Psi(\boldsymbol{v})$ and integrate over the particle velocities. We have, taking the integral for parts, for each term on the left-hand side of the kinetic equation

$$\int \Psi \frac{\partial f}{\partial t} d\boldsymbol{v} = \frac{\partial}{\partial t} \int \Psi f d\boldsymbol{v} - \int f \frac{\partial \Psi}{\partial t} d\boldsymbol{v} = \frac{\partial}{\partial t}(N\overline{\Psi}) - N\overline{\frac{\partial \Psi}{\partial t}} ,$$

where an overline means an average over particle velocities, and N is the number density of particles.

In the same manner we have for the second term on the left-hand side of the kinetic equation

$$\int \Psi v_x \frac{\partial f}{\partial x} d\boldsymbol{v} = \frac{\partial}{\partial x} \int \Psi v_x f d\boldsymbol{v} - \int f v_x \frac{\partial \Psi}{\partial x} d\boldsymbol{v} = \frac{\partial}{\partial t}(N\,\overline{\Psi v_x}) - N\,\overline{v_x \frac{\partial \Psi}{\partial x}} .$$

The above operations for the third term on the left-hand side of the kinetic equation give

$$\int \Psi \frac{\partial f}{\partial v_x} d\boldsymbol{v} = \int \frac{\partial}{\partial v_x}(\Psi f) d\boldsymbol{v} - \int f \frac{\partial \Psi}{\partial v_x} d\boldsymbol{v} = -N\overline{\frac{\partial \Psi}{\partial v_x}} .$$

Hence, we obtain the macroscopic equation in the form

$$\frac{\partial}{\partial t}(N\overline{\Psi}) + \mathrm{div}\,(N\,\overline{\Psi \boldsymbol{v}}) - N\left[\overline{\frac{\partial \Psi}{\partial t}} + \overline{\boldsymbol{v}\nabla\Psi} + \overline{\frac{\boldsymbol{F}}{m}\frac{\partial \Psi}{\partial \boldsymbol{v}}}\right] = \overline{\Psi(\boldsymbol{v})\,I_{\mathrm{col}}(f)} .$$

Let us prove that if $\Psi(\boldsymbol{v})$ is the integral of motion, i. e., if the sum of these values for colliding particles is identical before and after collision $\Psi(\boldsymbol{v}) + \Psi(\boldsymbol{v}_1) = \Psi(\boldsymbol{v}') + \Psi(\boldsymbol{v}_1')$, the right-hand side of the last equation is zero. Indeed, since particles are identical, we have

$$\overline{\Psi(\boldsymbol{v})\,I_{\mathrm{col}}(f)} = \frac{1}{2}\int \left[\Psi(\boldsymbol{v}) + \Psi(\boldsymbol{v}_1)\right] W(\boldsymbol{v}, \boldsymbol{v}_1 \to \boldsymbol{v}', \boldsymbol{v}_1') \left[f(\boldsymbol{v}')f(\boldsymbol{v}_1') - f(\boldsymbol{v})f(\boldsymbol{v}_1)\right] d\boldsymbol{v} d\boldsymbol{v}_1 d\boldsymbol{v}' d\boldsymbol{v}_1' .$$

The transition rate satisfies the principle of detailed balance according to which the time reversal $t \to -t$ leads to the reversal transition that is analogous to the change $\boldsymbol{v} \leftrightarrow \boldsymbol{v}'$, $\boldsymbol{v}_1 \leftrightarrow \boldsymbol{v}_1'$. This allows us to represent this value in the form

$$\overline{\Psi(\boldsymbol{v})\,I_{\mathrm{col}}(f)} = \frac{1}{4}\int \left[\Psi(\boldsymbol{v}) + \Psi(\boldsymbol{v}_1) - \Psi(\boldsymbol{v}') + \Psi(\boldsymbol{v}_1')\right] W \left[f(\boldsymbol{v}')f(\boldsymbol{v}_1') - f(\boldsymbol{v})f(\boldsymbol{v}_1)\right] d\boldsymbol{v} d\boldsymbol{v}_1 d\boldsymbol{v}' d\boldsymbol{v}_1' ,$$

and since $\Psi(\boldsymbol{v})$ is the integral of motion, this average value is zero. Finally, we obtain the macroscopic equation in the form

$$\frac{\partial}{\partial t}(N\overline{\Psi}) + \mathrm{div}\,(N\,\overline{\Psi \boldsymbol{v}}) - N\left[\overline{\frac{\partial \Psi}{\partial t}} + \overline{\boldsymbol{v}\nabla\Psi} + \overline{\frac{\boldsymbol{F}}{m}\frac{\partial \Psi}{\partial \boldsymbol{v}}}\right] = 0 , \qquad (6.14)$$

where $\boldsymbol{F}$ is a force from an external field that acts on one gas particle, and m is the particle mass.

▶ **Problem 6.14** Derive the macroscopic transport equations of a gas for the cases $\Psi(\boldsymbol{v}) = \text{const}$, $\Psi(\boldsymbol{v}) = m\boldsymbol{v}$, and $\Psi(\boldsymbol{v}) = \varepsilon = mv^2/2$.

Taking $\Psi(\boldsymbol{v})$ to be constant, we obtain the continuity equation

$$\frac{\partial N}{\partial t} + \operatorname{div}(N\boldsymbol{w}) = 0\,, \tag{6.15}$$

where $\boldsymbol{w}$ is the average atom velocity at this time and coordinate. Applying the obtained equation for $\Psi(\boldsymbol{v}) = m\boldsymbol{v}$, we have for vector components of the momentum transport equation

$$\frac{\partial}{\partial t}(Nmw_i) + \frac{\partial}{\partial x_k}(Nmw_i w_k) + \frac{\partial}{\partial x_k}P_{ik} = 0\,, \tag{6.16}$$

where

$$P_{ik} = Nm\overline{(v_i - w_i)(v_k - w_k)} \tag{6.17}$$

is the pressure tensor. Taking it in the form

$$P_{ik} = Nm\delta_{ik}\overline{(v_i - w_i)^2} = p\delta_{ik}\,, \tag{6.18}$$

where p is the gas pressure, we obtain the Euler equation

$$\frac{\partial \boldsymbol{w}}{\partial t} + (\boldsymbol{w}\nabla)\boldsymbol{w} + \frac{1}{Nm}\nabla p - \frac{\boldsymbol{F}}{m} = 0\,. \tag{6.19}$$

Let us define the heat flux intensity as

$$q_i = \frac{1}{2}mN\,\overline{(\boldsymbol{v} - \boldsymbol{w})^2(v_i - w_i)}\,, \tag{6.20}$$

and the gas temperature as

$$T = \frac{\overline{m(\boldsymbol{v} - \boldsymbol{w})^2}}{3}\,.$$

Taking $\Psi(\boldsymbol{v}) = \varepsilon = mv^2/2$, substituting it into the macroscopic equation (6.14), and excluding from this the continuity equation and the momentum transport equation, we finally obtain the equation of energy transport or the heat conduction equation for a gas in the form

$$\frac{3}{2}N\left(\frac{\partial T}{\partial t} + \boldsymbol{w}\nabla T\right) = -P_{ik}\frac{\partial w_i}{\partial x_k} - \frac{\partial q_k}{\partial x_k}\,. \tag{6.21}$$

▶ **Problem 6.15** Define the kinetic coefficients in a gas with fluxes of particles, momentum, and heat.

Under thermodynamic equilibrium such gas parameters as the number density of atoms or molecules of each species, the mean velocity of atoms or molecules, and the temperature are constants in a region occupied by a gas. If some of these values vary in this region, appropriate fluxes arise in order to equalize these parameters over the total volume of a gas or plasma. The fluxes are small if variations of the parameters are small on distances of the order of the mean free path for atoms or molecules. Then a stationary state of the system with fluxes exists, and such states are conserved during times much longer than typical times of collisions between particles. In other words, the inequality

$$\lambda \ll L \tag{6.22}$$

is satisfied for the systems under discussion, where λ is the mean free path for particles, and L is the typical size of the system or a distance over which a parameter varies noticeably. If this criterion is fulfilled, the system is in a stationary state to the first approximation, and transport of particles, heat, or momentum occurs in the second approximation in terms of an expansion over a small parameter λ/L. Below we consider various types of such transport.

Let us introduce the kinetic coefficients or transport coefficients as the coefficients of proportionality between fluxes and corresponding gradients. We start from the diffusion coefficient D that is introduced as the proportionality factor between the particle flux $\boldsymbol{j}$ and the gradient of concentration c of a given species, or

$$\boldsymbol{j} = -DN\nabla c \,. \tag{6.23}$$

Here N is the total particle number density. If the concentration of a given species is low ($c_i \ll 1$), that is, this species is an admixture to the gas, the flux of particles of this species can be written as

$$\boldsymbol{j} = -D_i \nabla N_i \,, \tag{6.24}$$

where N_i is the number density of particles of the given species.

The thermal conductivity coefficient κ is defined as the proportionality factor between the heat flux $\boldsymbol{q}$ and the temperature gradient ∇T on the basis of the relation

$$\boldsymbol{q} = -\kappa \nabla T \,. \tag{6.25}$$

The viscosity coefficient η is the proportionality factor between the frictional force acting on a unit area of a moving gas, and the gradient of the mean gas velocity in the direction perpendicular to the surface of a gas element (see Fig. 6.1). If the mean gas velocity $\boldsymbol{w}$ is parallel to the x-axis and varies in the z direction, the frictional force is proportional to $\partial w_x/\partial z$ and acts on an xy surface in the gas. Thus the force F per unit area is

$$F = -\eta \frac{\partial w_x}{\partial z} \,. \tag{6.26}$$

This definition and the previous ones are valid not only for gases, but also for liquids.

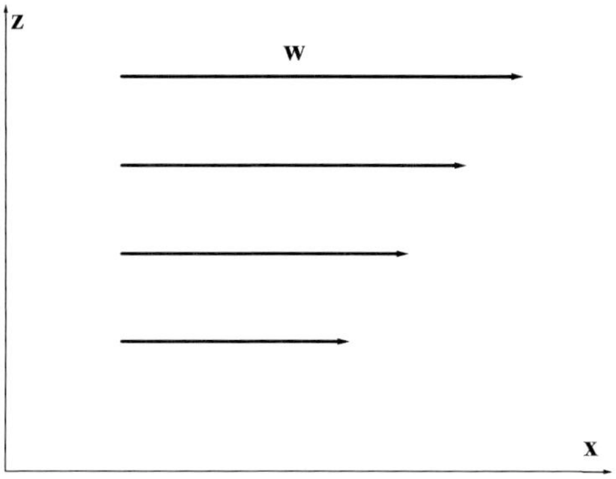

Fig. 6.1 The geometry of a viscous flux of a gas.

▶ **Problem 6.16** Give a general expression for the pressure tensor (6.17) of a gas by taking into account the gas viscosity.

Our goal is to generalize expression (6.18) for the pressure tensor by including in it the gas viscosity. This is a slow process because a constant pressure $p = \text{const}$ is established with the sound speed. Since the gas temperature is constant in a space, from the state equation $p = NT$ we have that the number density N of atoms does not change in the space. Hence, from the continuity equation (6.15) in the stationary case we have $\operatorname{div}(N\boldsymbol{w}) = 0$, and if the average gas velocity $\boldsymbol{w}$ is directed along the axis x, we have from this

$$\frac{\partial w_x}{\partial x} = 0 \,.$$

From this, accounting for expression (6.26) for the force per unit area and symmetrizing it by taking into account the above formula, we write the pressure tensor (6.17) by adding to expression (6.18) the symmetrized expression (6.26). Then accounting for $\operatorname{div}\boldsymbol{w} = 0$, we require the trace of the viscosity part of the pressure tensor to be zero. This leads to the following expression for the pressure tensor:

$$P_{ik} = p\delta_{ik} - \eta\left(\frac{\partial w_i}{\partial x_k} + \frac{\partial w_k}{\partial x_i} - \frac{2}{3}\delta_{ik}\frac{\partial w_i}{\partial x_k}\right) , \qquad (6.27)$$

where as usual x_i, x_k are the coordinates x, y, z, and the summation takes place over twice repeated indices.

▶ **Problem 6.17** Generalize the Euler equation (6.19) taking into account the gas viscosity.

We generalize the equation for momentum variation (6.16) by inserting in it the pressure tensor in the form (6.27). Dividing the initial equation by the atom mass, we obtain the Navier–Stokes equation

$$\frac{\partial \boldsymbol{w}}{\partial t} + (\boldsymbol{w}\cdot\nabla)\boldsymbol{w} = -\frac{\nabla p}{\rho} + \nu\Delta\boldsymbol{w} + \frac{\boldsymbol{F}}{m} , \qquad (6.28)$$

where $\nu = \eta/\rho$ is the kinematic viscosity that has the dimensionality of the diffusion coefficient (cm^2/s), and we account for in this equation that $\nabla \boldsymbol{w} = 0$.

▶ **Problem 6.18** Derive the heat transport equation for a motionless and weakly nonuniform gas where heat transport results from the gas thermal conductivity.

We use the equation of energy transport (6.21) taking there the zero gas velocity $\boldsymbol{w} = 0$ and introducing the heat flux intensity (6.20) according to formula (6.25). We also assume that a nonuniform gas is supported at constant pressure. Therefore instead of the heat capacity of an atom at the constant volume $c_V = 3/2$, which was used in equation (6.21), we introduce in this equation the heat capacity c_p per atom or molecule of this gas or constant pressure. As a result, we obtain the energy balance equation (6.21) in the form

$$c_p N \frac{\partial T}{\partial t} = \kappa \Delta T \,, \tag{6.29}$$

where κ is the thermal conductivity coefficient of a gas. It is convenient to include in this equation the thermal diffusivity coefficient χ for this gas, which is defined as

$$\chi = \frac{\kappa}{c_p N} \,. \tag{6.30}$$

Then the heat balance equation under the considered conditions takes the form

$$\frac{\partial T}{\partial t} = \chi \Delta T \,. \tag{6.31}$$

▶ **Problem 6.19** Determine the contribution to the heat balance of gas particles due to internal degrees of freedom.

The heat balance equation (6.29) includes the heat capacity c_p per gas particle. This value is equal to

$$c_p = \frac{i+2}{2} \,,$$

where i is the number of degrees of freedom for gas particles. In particular, for an atomic gas ($i = 3$) we have $c_p = 5/2$, and for a molecular gas consisting of diatomic molecules ($i = 5$) we have $c_p = 7/2$. In the latter case we assume vibration degrees of freedom to be in thermodynamic equilibrium with translation degrees of freedom. In addition, the energy per degree of freedom is taken to be $T/2$, where T is the temperature expressed in energy units. This is not fulfilled for the transition in a continuous spectrum, as in the cases of dissociation of molecules or ionization of atomic particles. In these cases the binding energy of such transitions usually exceeds significantly the thermal energy of particles. Hence, we extract heat transport for such degrees of freedom separately, representing the heat flux in a gas as a sum of two terms as

$$\boldsymbol{q} = -\kappa_t \nabla T - \kappa_i \nabla T \,,$$

where κ_t is the thermal conductivity coefficient due to the transport of translational energy, and the second term is due to transport of energy in the internal degrees of freedom. The thermal conductivity coefficient is then just the sum of these two terms

$$\kappa = \kappa_t + \kappa_i \ ,$$

and we now analyze the second term, where an internal state of the gas particle is denoted by the subscript i.

Because of a temperature gradient, the number density of particles in this state is not constant in space, and the diffusion flux is given by

$$\boldsymbol{j}_i = -D_i \nabla N_i = -D_i \frac{\partial N_i}{\partial T} \nabla T \ .$$

From this we obtain the heat flux

$$\boldsymbol{q} = \sum_i \varepsilon_i \boldsymbol{j}_i = -\sum_i \varepsilon_i D_i \frac{\partial N_i}{\partial T} \nabla T \ ,$$

where ε_i is the excitation energy for the i-th state. Assuming the diffusion coefficient to be the same for excited and nonexcited particles, we obtain the thermal conductivity coefficient due to internal degrees of freedom,

$$\kappa_i = \sum_i \varepsilon_i D_i \frac{\partial N_i}{\partial T} = D \frac{\partial}{\partial T} \sum_i \varepsilon_i N_i = D \frac{\partial}{\partial T} \bar{\varepsilon} N = D c_p \ , \qquad (6.32)$$

where $\bar{\varepsilon} = \sum_i \varepsilon_i N_i / N$ is the mean excitation energy of the particles, $N = \sum_i N_i$ is the total number density, and $c_p = \partial \bar{\varepsilon} / \partial T$ is the heat capacity per particle at constant pressure.

► **Problem 6.20** Charged particles are moving in a gas in an external electric field and collide elastically with gas atoms. Based on the tau-approximation for the collision integral of charged particles and gas atoms, derive the macroscopic equation for momentum transport and compare it with the Newton equation for an individual charged particle.

Using the kinetic equation in the form (6.2), we take the collision integral in the tau-approximation (6.3). The left-hand side of the equation for the average momentum of charged particles is given by equation (6.16) with $\boldsymbol{F} = e\boldsymbol{E}$, where $\boldsymbol{E}$ is the electric field strength. The right-hand side is $m w / \tau$. So, the equation for the average momentum of a charged particles followed from the Boltzmann equation for a uniform gas ($N = \text{const}$) takes the form

$$M \frac{dw}{dt} + M \frac{w}{\tau} = e\boldsymbol{E} \ ,$$

where M is the charged particle mass. This equation coincides with the Newton equation for a charged particle moving in a gas if we introduce in this equation the friction force $M w / \tau$ resulted from collisions of this charge particle with gas atoms.

▶ **Problem 6.21** Derive the momentum integral equation of ions in a gas in the stationary regime by taking into account elastic ion–atom collisions.

We now derive the stationary equation of the previous problem taking into account ion–atom collisions correctly. Then multiplying equation (6.2) with the collision integral (6.7) by $m\boldsymbol{v}_1$, and integrating over $d\boldsymbol{v}_1$, we obtain

$$e\boldsymbol{E}N_\mathrm{i} = \int M(\boldsymbol{v}_\mathrm{i}' - \boldsymbol{v}_\mathrm{i})gd\sigma f_\mathrm{i} f_\mathrm{a} d\boldsymbol{v}_\mathrm{i} d\boldsymbol{v}_\mathrm{a} \,.$$

Here the quantities $\boldsymbol{v}_\mathrm{i}$ and $\boldsymbol{v}_\mathrm{a}$ are the initial velocities of the ion and atom before collision, and $\boldsymbol{v}_\mathrm{i}'$ and $\boldsymbol{v}_\mathrm{a}'$ are their values after collision, respectively. The distribution functions of ions $f_\mathrm{i}(\boldsymbol{v}_\mathrm{i})$ and atoms $f_\mathrm{a}(\boldsymbol{v}_\mathrm{a})$ are normalized to the ion N_i and atom N_a number densities; M and m are ion and atom masses, and g is their relative velocity that is conserved in the collision. We have used the principle of detailed balance, which assures the invariance under time reversal of the evolution of the system and yields $\int \boldsymbol{v}_1 f_1' f_2' d\sigma d\boldsymbol{v}_1 d\boldsymbol{v}_2 = \int \boldsymbol{v}_1' f_1 f_2 d\sigma d\boldsymbol{v}_1 d\boldsymbol{v}_2$.

Let us express the ion velocity $\boldsymbol{v}_\mathrm{i}$ before and after collision in terms of the relative ion–atom velocity $\mathbf{g}$ and the center-of-mass velocity $\boldsymbol{V}$ by the relation $\boldsymbol{v}_\mathrm{i} = \boldsymbol{V} + m\mathbf{g}/(m+M)$. This gives $M(\boldsymbol{v}_\mathrm{i} - \boldsymbol{v}_\mathrm{i}') = \mu(\mathbf{g} - \mathbf{g}')$. The relative ion–atom velocity after collision has the form $\mathbf{g}' = \mathbf{g}\cos\vartheta + \boldsymbol{k}g\sin\vartheta$, where ϑ is the scattering angle and $\boldsymbol{k}$ is a unit vector directed perpendicular to $\mathbf{g}$. Because of the random character of scattering, the second term disappears after averaging, i. e., the integration over scattering angles gives $\int(\mathbf{g} - \mathbf{g}')d\sigma = \mathbf{g}\sigma^*(g)$, where $\sigma^*(g) = \int(1 - \cos\vartheta)d\sigma$ is the diffusion cross section of ion–atom scattering. As a result, the above equation takes the form

$$e\boldsymbol{E}N_\mathrm{i} = \int \mu \mathbf{g} g \sigma^*(g) f_\mathrm{i} f_\mathrm{a} d\boldsymbol{v}_\mathrm{i} d\boldsymbol{v}_\mathrm{a} \,. \tag{6.33}$$

This also means that as a result of ion collisions with gas atoms, a force is generated,

$$\boldsymbol{F} = -\frac{1}{N_\mathrm{i}} \int \mu \mathbf{g} g \sigma^*(g) f_\mathrm{i} f_\mathrm{a} d\boldsymbol{v}_\mathrm{i} d\boldsymbol{v}_\mathrm{a} \,,$$

which acts on a moving ion in a gas.

▶ **Problem 6.22** Find the drift velocity in an atomic gas for polarization ion–atom interaction.

In the case of polarization interaction between an ion and atom, the diffusion cross section is close to the cross section of polarization capture (2.25), and is inversely proportional to the relative velocity g of collision. Then we have

$$\int \mathbf{g} f_\mathrm{i} f_\mathrm{a} d\boldsymbol{v}_\mathrm{i} d\boldsymbol{v}_\mathrm{a} = (\boldsymbol{w}_\mathrm{i} - \boldsymbol{w}_\mathrm{a}) N_\mathrm{i} N_\mathrm{a} = \boldsymbol{w}_\mathrm{i} N_\mathrm{i} N_\mathrm{a} \,,$$

where $\boldsymbol{w}_\mathrm{i}$ is the average ion velocity, and a gas is motionless, $\boldsymbol{w}_\mathrm{a} = 0$. From this we obtain

$$\boldsymbol{w}_\mathrm{i} = \frac{e\boldsymbol{E}}{\mu N_\mathrm{a} k_\mathrm{c}} \,, \tag{6.34}$$

where $k_c = 2\pi\sqrt{\alpha e^2/\mu}$ is the rate constant of the polarization capture process, and α is the polarizability of the atom. Note that this formula is valid at any electric field strength including very strong fields, when the ion distribution function is very different from the Maxwell distribution dependence on the gas temperature. Thus, the integral equation (6.33) can be a basis for the analysis of ion behavior in a gas in any external electric field.

▶ **Problem 6.23** Derive the macroscopic equation for the power that transfers to ions from an external electric field and then to a gas as a result of collisions of ions with gas atoms.

This operation is similar to that during the derivation of equation (6.33). Indeed, we multiply the kinetic equation (6.2) with the collision integral (6.7) by the ion kinetic energy $Mv_i^2/2$ and integrate it over ion velocities. This gives the following equation:

$$e\boldsymbol{E}\boldsymbol{w}N_i N_a = \int M[(v_i')^2 - v_i^2]gd\sigma f_i f_a d\boldsymbol{v}_i d\boldsymbol{v}_a ,$$

where we used the symmetry with respect to the time inversion $t \leftrightarrow -t$ and use the same notations as in the derivation of equation (6.33). Introducing the velocity $\boldsymbol{V} = (M\boldsymbol{v}_i + m\boldsymbol{v}_a)/(m+M)$ of the center of mass for the colliding ion and atom and the relative ion–atom velocity $\mathbf{g} = \boldsymbol{v}_i - \boldsymbol{v}_a$, we obtain

$$\frac{M}{2}[(v_i')^2 - v_i^2] = \mu\boldsymbol{V}(\mathbf{g} - \mathbf{g}') ,$$

where $\mathbf{g}$, $\mathbf{g}'$ are the relative ion–atom velocities before and after collision. This leads to the relation

$$e\boldsymbol{E}\boldsymbol{w}N_i N_a = \mu\int \boldsymbol{V}\mathbf{g}g\sigma^*(g)f_i f_a d\boldsymbol{v}_i d\boldsymbol{v}_a . \tag{6.35}$$

This is the balance equation for the ion energy if an ion is moving in a gas in an external electric field and is scattering on gas atoms elastically. The left-hand side of this equation is the specific power that an ion acquires from an electric field, and the right-hand side of this equation is the specific power transmitted to gas atoms.

▶ **Problem 6.24** Find the average ion energy in the case of elastic ion–atom collisions in a gas in an external electric field, if the rate of ion–atom collisions is independent of the collision velocity.

Under these conditions, an ion moves in a constant electric field with a drift velocity that is given by formula (6.34) and is $w = eE/(\mu\nu)$. Here the rate of elastic ion–atom collisions is $\nu = N_a g\sigma^*(g)$ and does not depend on the relative ion–atom velocity g. Correspondingly, formula (6.35) gives in this case

$$w^2 = \langle\boldsymbol{V}\mathbf{g}\rangle = \frac{M}{M+m}\langle v_i^2\rangle - \frac{m}{M+m}\langle v_a^2\rangle + \frac{M-m}{M+m}\langle\boldsymbol{v}_i\rangle\langle\boldsymbol{v}_a\rangle ,$$

where brackets mean an average over ion or atom velocities. Assuming that a gas is motionless, i. e., $\langle \boldsymbol{v}_a \rangle = 0$, we introduce the average ion kinetic energy as $\bar{\varepsilon} = M\langle v_i^2 \rangle/2$ and the gas temperature as $T = m\langle v_a^2 \rangle/3$. As a result, we obtain from the above equation for the average ion energy

$$\bar{\varepsilon} = \frac{(m+M)w^2}{2} + \frac{3}{2}T\,. \tag{6.36}$$

▶ **Problem 6.25** In the case when an ion is moving in a gas in an external electric field derive the macroscopic equation for the variation of the average ion kinetic energy per unit time for an ion motion along the field.

This equation as well as the method of its derivation is similar to the derivation of equation (6.35). To this end we multiply the kinetic equation (6.2) with the collision integral (6.7) by the ion kinetic energy along the field $Mv_{ix}^2/2$ (x is the field direction), and integrate it over ion velocities, which gives

$$e\boldsymbol{E}\boldsymbol{w} N_i N_a = \int M[(v'_{ix})^2 - v_{ix}^2] g d\sigma f_i f_a d\boldsymbol{v}_i d\boldsymbol{v}_a\,.$$

As earlier, we introduce the velocity of the center of mass for the colliding ion and atom, $\boldsymbol{V} = (M\boldsymbol{v}_i + m\boldsymbol{v}_a)/(m+M)$, and the relative ion–atom velocity $\boldsymbol{g} = \boldsymbol{v}_i - \boldsymbol{v}_a$. Then we obtain

$$(v_{ix}^2 - v'_{ix})^2 = \frac{2m}{M+m} V_x(g_x - g'_x) + \left(\frac{m}{M+m}\right)^2 [g_x^2 - (g'_x)^2]\,.$$

Take the relative ion–atom velocity after collision as

$$(g)' = (g)\cos\vartheta + \boldsymbol{n} g \sin\vartheta\,,$$

where $\boldsymbol{g}$, $\boldsymbol{g}'$ are the relative ion–atom velocities before and after collision, respectively, and $\boldsymbol{n}$ is the unit vector in the scattering plane that is perpendicular to the vector $\boldsymbol{g}$. We have

$$\int (g_x - g'_x) d\sigma = \int (\boldsymbol{g} - \boldsymbol{g}')_x d\sigma = g_x \int (1 - \cos\vartheta) d\sigma - \overline{n_x} g \int \sin\vartheta d\sigma = g_x v^*(g)\,,$$

since $\overline{n_x} = \overline{\cos\theta\cos\vartheta + \sin\theta\sin\vartheta\cos\Phi} = 0$, so that θ, Φ are the polar angles of the vector $\boldsymbol{g}$ with respect of the polar axis x, and ϑ is the scattering angle.

Next, taking

$$g'_x = g(\cos\theta\cos\vartheta + \sin\theta\sin\vartheta\cos\Phi) = g_x\cos\theta + g_\perp \sin\theta\cos\Phi\,,$$

where $g_\perp$ is the projection of the relative velocity on the plane that is perpendicular to $\boldsymbol{E}$, we have, averaging over the azimuthal angle Φ,

$$\begin{aligned} g_x^2 - (g'_x)^2 &= g_x^2 - g_x^2\cos^2\vartheta - g_\perp^2 \sin^2\vartheta\overline{\cos^2\Phi} - 2g_x g_\perp \sin\vartheta\cos\vartheta\overline{\cos\Phi} \\ &= \left(g_x^2 - \frac{g_\perp^2}{2}\right)\sin^2\vartheta = \frac{3}{2}\left(g_x^2 - \frac{g^2}{3}\right)\sin^2\vartheta\,, \end{aligned}$$

where we use $g^2 = g_x^2 + g_\perp^2$. Introducing the average cross section $\sigma^{(2)} = \int(1-\cos^2\vartheta)d\sigma$, we have

$$\int [g_x^2 - (g_x')^2]d\sigma = \frac{3}{2}(g_x^2 - \frac{g^2}{3})\sigma^{(2)} .$$

As a result, we obtain the following integral relation:

$$eEw = \mu\langle V_x g_x \nu_1\rangle + \frac{3\mu m}{4(m+M)}\left\langle (g_x^2 - \frac{g^2}{3})\nu_2 \right\rangle . \tag{6.37}$$

Here

$$\nu_1 = N_a g\sigma^*(g), \quad \nu_2 = N_a g\sigma^{(2)(g)} .$$

This is the balance equation for the ion kinetic energy in the field direction. The left-hand side of this equation is the specific power that an ion acquires from an electric field and is transformed into the kinetic energy of ions in the field direction. The right-hand side of this equation is the specific power that is transferred from the ion kinetic energy in the field direction to atoms as a result of ion–atom elastic collisions.

▶ **Problem 6.26** An ion is moving in a gas in a constant electric field. Assuming the ion–atom cross section of elastic scattering to be inversely proportional to the collision velocity, determine the ion kinetic energy in the field direction.

Under these conditions the rates ν_1 and ν_2 in equation (6.37) are independent of the collision velocity, and this equation takes the form

$$eEw = \mu\nu_1 \frac{\langle Mv_{ix}^2\rangle - \langle mv_{ax}^2\rangle}{4(m+M)} + \frac{3\mu m\nu_2}{4(m+M)}\left\langle (g_x^2 - \frac{g^2}{3})\right\rangle . \tag{6.38}$$

While deriving this relation, we take into account that a gas is motionless, $\langle v_{ax}\rangle = 0$, and the distribution of atoms on velocities is isotropic $\langle v_{ax}^2\rangle = \langle v_a^2\rangle/3$. Since the ion drift velocity in this case is given by formula (6.34), $w = eE/(\mu\nu_1)$, and the average kinetic ion energy is determined by formula (6.36), we have from this for the average kinetic energy of the ion in the field direction

$$\left\langle \frac{Mv_{ix}^2}{2}\right\rangle = \frac{T}{2} + \frac{(m+M)w^2}{2}\frac{\nu_1 + m\nu_2/(4M)}{\nu_1 + 3m\nu_2/(4M)} . \tag{6.39}$$

Correspondingly, formulas (6.36) and (6.39) give the average ion kinetic energy in directions that are perpendicular to the field

$$\left\langle \frac{Mv_{iy}^2}{2}\right\rangle = \left\langle \frac{Mv_{iz}^2}{2}\right\rangle = \frac{1}{2}\left(\left\langle \frac{Mv_i^2}{2}\right\rangle - \left\langle \frac{Mv_{ix}^2}{2}\right\rangle\right) = \frac{T}{2} + \frac{(m+M)w^2}{2}\frac{m\nu_2/(4M)}{\nu_1 + 3m\nu_2/(4M)} . \tag{6.40}$$

Let us analyze the results. In the case $M \ll m$, which corresponds, in particular, to electron motion in a gas, the average kinetic energy of the ion for motion in different directions is identical. Indeed, the velocity distribution function is isotropic in this case. In another limiting case, when the ion mass is large compared the atom mass, the ion kinetic energy in the field direction exceeds that in other directions at high electric field strengths.

▶ **Problem 6.27** Prove the virial theorem for a uniform ensemble of classical particles that is the relation between the average kinetic energy of particles and their potential energy.

The virial theorem establishes the relation between the mean kinetic energy of a particle and averaged parameters of its interaction with surrounding particles or fields. We deduce it in the classical case when the motion equation for the particle is given by the Newton equation

$$m\frac{d^2\boldsymbol{r}}{dt^2} = \boldsymbol{F} \,,$$

where $\boldsymbol{r}$ is the particle coordinate, m is its mass, and $\boldsymbol{F}$ is the force which acts on this particle from other particles of this ensemble or fields.

Let us multiply this equation by x, the particle coordinate along the x axis, and integrate this over a large time range. We have

$$mx\frac{d^2x}{dt^2} = m\frac{d}{dt}\left(x\frac{dx}{dt}\right) - m\left(\frac{dx}{dt}\right)^2 .$$

Averaging over a large time τ, we obtain for the second term

$$\frac{1}{\tau}\int_0^\tau m\frac{d}{dt}\left(x\frac{dx}{dt}\right) = \frac{m}{\tau}x\frac{dx}{dt}\bigg|_0^\tau \,,$$

and in the limit $\tau \to \infty$ this term tends to zero. Thus, we get from the above equation

$$\frac{m}{2}\overline{\left(\frac{dx}{dt}\right)^2} = -\frac{1}{2}\overline{xF_x} = \frac{1}{2}\overline{x\frac{\partial U}{\partial x}} \,. \tag{6.41}$$

This is the force virial equation that is useful for the analysis of uniform ensembles of interacting particles. Here the overline means averaging over time, F_x is the force component, and U is the interaction potential for this particle and others in the presence of external fields. The term on the left-hand side of this equation is the average kinetic energy of this particle, and the right-hand side of this equation corresponds to interaction of this particle with an environment.

▶ **Problem 6.28** Derive the state equation of a uniform ensemble of particles on the basis of the virial theorem.

The state equation for an ensemble of particles under a constant pressure establishes the relation between the temperature, pressure, and other macroscopic parameters of the ensemble. To create a constant pressure in this system, we place the particles in a vessel where a certain pressure is supported. From the virial theorem we have for the particle ensemble

$$\sum_i \frac{m}{2}\overline{\left(\frac{dx_i}{dt}\right)^2} = -\frac{1}{2}\sum_i \overline{x_i(F_x)_i} - \frac{1}{2}\left[\sum_i \overline{x_i(F_x)_i}\right]_{\text{walls}} ,$$

where an index i indicates the particle number, an overline means averaging over time, and we divide the force virial for an individual particle into two parts, so that the first one relates to other particles of the ensemble, and the second one refers to the vessel walls. Because of equilibrium, one can change the second term on the right-hand side of this equation by a force acting from the walls on the particle ensemble. The force from a surface element ds acting on particles of the ensemble is pds, where p is the pressure. Hence, we have

$$\frac{1}{2}\left[\sum_i \overline{x_i(F_x)_i}\right]_{\text{walls}} = \frac{1}{2}\int_S pxds = \frac{p}{2}\int_\Omega d\Omega = \frac{1}{2}p\Omega ,$$

where Ω is the volume restricted by the walls, and the particle ensemble is located in this volume.

According to the temperature definition, the average kinetic energy of an individual particle for motion in a given direction is

$$\frac{m}{2}\overline{\left(\frac{dx_i}{dt}\right)^2} = \frac{T}{2} ,$$

and the force virial averaged over a large time range is identical for different particles, which is

$$\sum_i \overline{x_i(F_x)_i} = n\,\overline{x_i(F_x)_i} ,$$

where n is the total number of particles of the ensemble under consideration. Thus, we obtain

$$nT = -n\overline{x_i(F_x)_i} + p\Omega .$$

Introducing a specific volume per atom, $V = \Omega/n$, we reduce this equation to the form

$$T = -\overline{x_i(F_x)_i} + pV . \qquad (6.42)$$

This is the state equation for a particle ensemble, and the left-hand side of this equation is the kinetic energy of a particle. The first term on the right-hand side of this equation accounts for the force virial which acts on an individual particle from other particles, and the second term takes into account the action of walls. In particular, for a gas, i. e., for a system of weakly interacting particles, one can neglect by the virial force, and the state equation takes the form

$$T = pV \,. \tag{6.43}$$

6.4 Stepwise Quantities and Processes

▶ **Problem 6.29** Find the probability that some variable x gets a given value after $n \gg 1$ steps if each step is random.

We now consider a group of systems and processes when a parameter under consideration is a sum of many small elements which have a random nature. An example of this is the Brownian motion of a particle whose displacement after many collisions with an environment is a sum of many random displacements.

Let us introduce the probability $f(x,n)$ that the variable has a given value after n steps, and $\varphi(x_k)dz_k$ is the probability that the variable change ranges from x_k to $x_k + dz_k$ after the k-th step. Since the functions $f(x)$, $\varphi(x)$ are the probabilities, they are normalized by the condition

$$\int_{-\infty}^{\infty} f(x,n)dz = \int_{-\infty}^{\infty} \varphi(x)dx = 1 \,.$$

From the definition of the above functions we have

$$f(x,n) = \int_{-\infty}^{\infty} dx_1 \cdots \int_{-\infty}^{\infty} dx_n \prod_{k=1}^{n} \varphi(x_k) \,,$$

and

$$x = \sum_{k=1}^{n} x_k \,.$$

Let us use the characteristic functions

$$G(p) = \int_{-\infty}^{\infty} f(x)\exp(-ipx)dx, \quad g(p) = \int_{-\infty}^{\infty} \varphi(x)\exp(-ipx)dx \,.$$

On the basis of the inverse operation, we have

$$f(x) = \frac{1}{2\pi}\int_{-\infty}^{\infty} G(p)\exp(ipx)dp, \quad \varphi(x) = \frac{1}{2\pi}\int_{-\infty}^{\infty} g(p)\exp(ipx)dp \,.$$

Using the definition of the characteristic functions, we obtain

$$g(0) = \int_{-\infty}^{\infty} \varphi(x)dx = 1; \quad g'(0) = i\int_{-\infty}^{\infty} z\varphi(x)dx = i\overline{x_k}; \quad g''(0) = -\overline{x_k^2}\,,$$

where $\overline{x_k}$, $\overline{x_k^2}$ are the average shift and the average square shift of the variable after one step. Taking the displacement of the variable after n steps as a sum of those from individual steps, we have for the characteristic function

$$G(p) = \int_{-\infty}^{\infty} \exp(-ip\sum_{k=1}^{n} x_k)\prod_{k=1}^{n}\varphi(x_k)dx_k = g^n(p)\,,$$

and this gives

$$f(x) = \frac{1}{2\pi}\int_{-\infty}^{\infty} g^n(p)\exp(ipx)dp = \frac{1}{2\pi}\int_{-\infty}^{\infty}\exp(n\ln g\, +\, ipx)dp\,.$$

For a large number of steps $n \gg 1$ the integral converges at small p. Expanding $\ln g$ in a series over small p, we have

$$\ln g = \ln\left(1 + i\overline{x_k}p - \frac{1}{2}\overline{x_k^2}\,p^2\right) = i\overline{x_k}p - \frac{1}{2}\left(\overline{x_k^2} - \overline{x_k}^2\right)p^2\,.$$

From this it follows that

$$\begin{aligned} f(x) &= \frac{1}{2\pi}\int_{-\infty}^{\infty} dp \exp\left[ip(n\overline{x_k} - x) - \frac{n}{2}\left(\overline{x_k^2} - \overline{x_k}^2\right)p^2\right] \\ &= \frac{1}{\sqrt{2\pi\Delta^2}}\exp\left[-\frac{(x-\overline{x})^2}{2\Delta^2}\right]\,. \end{aligned} \tag{6.44}$$

This formula is the Gauss distribution. Here $\overline{x} = n\overline{x_k}$ is the average shift of the variable after n steps, and $n\Delta^2 = n\left[\overline{x_k^2} - (\overline{x_k})^2\right]$ is the average square displacement of this quantity. Δ is the fluctuation of this quantity for a system of many identical elements. The Gauss distribution holds true if small p gives the main contribution to the characteristic function, i. e., $\overline{x_k}p \ll 1$, $\overline{x_k^2}p^2 \ll 1$. Since this integral is determined by $n\overline{x_k^2}\,p^2 \sim 1$, the Gauss distribution is valid for a large number of steps or elements, $n \gg 1$.

▶ **Problem 6.30** N free particles are located in a closed volume Ω. Find the probability that in a small part Ω_0 of this volume n particles are located, if the average number of particles $\overline{n} = N\Omega_0/\Omega$ is large there.

The probability P_n of finding n particles in a given volume is the product of the probability of locating n particles in this volume $(\Omega_0/\Omega)^n$, the probability of locating other $N - n$ particles outside this volume $(1 - \Omega_0/\Omega)^{N-n}$, and a number of ways C_N^n of doing it. Hence this probability is given by the formula

$$P_n = C_N^n \left(\frac{\Omega_0}{\Omega}\right)^n \left(1 - \frac{\Omega_0}{\Omega}\right)^{N-n} .$$

This probability satisfies the normalization condition $\sum_n P_n = 1$.

Take the limiting case $n \gg 1$. Then the average number of particles in a given volume is $\overline{n} = N\frac{\Omega_0}{\Omega} \gg 1$. In addition, we have $n \ll N$, $n^2 \ll N$. Using the above small parameters, we finally obtain the Poisson formula

$$P_n = \frac{\overline{n}^n}{n!} \exp(-\overline{n}) . \qquad (6.45)$$

▶ **Problem 6.31** Expanding the Poisson formula for a large number of particles in a given volume, find the fluctuations in the particle distribution and compare them with an average particle number.

In the considered limiting case $n \gg 1$, $\overline{n} \gg 1$, the probability P_n has a sharp maximum at $n = \overline{n}$. On the basis of the Stirling formula

$$n! = \frac{1}{\sqrt{2\pi n}} \left(\frac{n}{e}\right)^n , \quad n \gg 1 ,$$

we obtain the expansion of the probability P_n near its maximum $\overline{n}$, which has the form

$$\ln P_n = \ln P_0 - \frac{(n - \overline{n})^2}{2\overline{n}} ,$$

where $P_0 = (2\pi\overline{n})^{-1/2}$, and the fluctuation of a number of particles in a given volume is equal to

$$\Delta = \sqrt{\overline{n^2} - (\overline{n})^2} = \sqrt{\overline{n}} \ll \overline{n} . \qquad (6.46)$$

This allows us to demonstrate the role of fluctuations for a system of many identical particles. Let us divide the total volume, where an ensemble of particles is located, into some cells, so that the average number of particles is equal to $\overline{n_i} = N\frac{\Omega_i}{\Omega}$ in the i-th cell of a volume Ω_i. Here N is the total number of particles in the total volume Ω. Then, ignoring the fluctuations, we deal with mean values $\overline{n_i}$ of particles in cells, and the distribution of the number of particles in a given cell is concentrated near its average number. One can see that the fluctuations are small, and the above statement is valid if the number of particles in cells is enough large, $\overline{n_i} \gg 1$.

Note that the distribution of particles in cells by neglecting the fluctuations can be obtained by two methods. In the first case, we measure the distribution in cells and find n_i particles in the i-th cell. This value coincides with its average value $\overline{n_i}$ with the accuracy up to fluctuations. In the second case, we follow a test particle which is found in a cell i during a time t_i from the total observation time t. Then the number of particles in the i-th cell is equal to Nt_i/t and it coincides with $\overline{n_i}$ with the accuracy up to fluctuations. Thus, when we find average values in the statistical physics, in the first approximation we neglect fluctuations.

▶ **Problem 6.32** Derive the Fokker–Planck equation for the probability density $W(x_0, t_0; x, t)$ to realize the variable x at time t if at time t_0 this variable has the value x_0, assuming that the variable x varies by small steps.

We use the Smoluchowski equation for the probability density

$$W(x_0, t_0; x, t+\tau) = \int W(x_0, t_0, ; z, t) W(z, t; x, \tau) dz \,, \tag{6.47}$$

which uses the definition of this quantity, and the normalization condition that has the form

$$\int W(x_0, t_0, ; x, t) dx = 1 \,. \tag{6.48}$$

Because the variable x changes by a small increment in each individual event, the system is diffusive in nature in the variable x, and the Smoluchowski equation may be reduced to the Fokker–Planck equation in the limit $\tau \to 0$.

For this operation we multiply the Smoluchowski equation by an arbitrary continuous function $g(x)$, which is zero at the boundary of the x-range together with their derivatives, and integrate over dx. This gives

$$\int g(x) W(x_0, t_0; x, t+\tau) dx = \int W(x_0, t_0, ; z, t) dz \int g(x) W(z, t; x, t+\tau) dx \,.$$

We now consider the limit $\tau \to \; 0$ and expand the function $g(x)$ over a small parameter $x - z$. Transferring the first term on the right-hand side to the left-hand side and dividing the result by τ, we obtain

$$\int g(x) dx \frac{W(x_0, t_0, ; x, t+\tau) - W(x_0, t_0, ; x, t)}{\tau} = \int W(x_0, t_0, ; z, t) dz$$

$$\left[g'(z) \frac{\overline{x - z}}{\tau} + g''(z) \frac{\overline{(x-z)^2}}{2\tau} + g'''(z) \frac{\overline{(x-z)^3}}{6\tau} \right] .$$

Transferring to the limit $\tau \to \; 0$ and ignoring the cube term and others of more high degree, we obtain

$$\int g(x) dx \frac{\partial W(x_0, t_0, ; x, t)}{\partial t} = \int W(x_0, t_0, ; z, t) [g'(z) A(z, t) + g''(z) B(z, t)] dz \,,$$

where

$$\begin{aligned} A(x, t) &= \lim_{\tau \to 0} \frac{1}{\tau} \int (z - x) W(z, t; x, t+\tau) dz, \\ B(x, t) &= \lim_{\tau \to 0} \frac{1}{2\tau} \int (z - x)^2 W(z, t; x, t+\tau) dz \,. \end{aligned} \tag{6.49}$$

Integrating by parts and accounting for $g = g' = 0$ at the boundary, we obtain

$$\int g(x) dx \left[\frac{\partial W}{\partial t} + \frac{\partial (AW)}{\partial x} - \frac{\partial^2 (BW)}{\partial x^2} \right] = 0 \,.$$

Since it is valid for an arbitrary function $g(x)$, the expression in the square brackets is zero, which corresponds to the Fokker–Planck equation

$$\frac{\partial W}{\partial t} = -\frac{\partial (AW)}{\partial z} + \frac{\partial^2 (BW)}{\partial z^2} . \tag{6.50}$$

Note that the limit of small τ nevertheless means that there are many events at this τ.

► **Problem 6.33** Consider the Fokker–Planck equation for the probability density $W(x_0, t_0; x, t)$ as the continuity equation for this quantity and establish the physical sense of the equation components.

Due to a small change of the variable in one event, the evolution of the system has the continuous character, so that the probability W satisfies the continuity equation

$$\frac{\partial W}{\partial t} + \frac{\partial j}{\partial x} = 0$$

and the flux j can be represented in the form

$$j = AW - B\frac{\partial W}{\partial x} . \tag{6.51}$$

Here the first term is the hydrodynamic flux, and the second corresponds to the diffusion flux. Therefore, the Fokker–Planck equation accounts for drift and diffusion in a space of the variable x.

► **Problem 6.34** Generalize the Fokker–Planck equation for the case of curvilinear coordinates.

In the case of curvilinear coordinates, if a state of a given parameter x is characterized by the state density $\rho(x)$, the normalization condition for the probability of transition after many events takes the form

$$\int \rho(x) W(x_0, t_0; x, t) dx = 1$$

instead of equation (6.48). Correspondingly, the probability density W is replaced by ρW, so we obtain the Fokker–Planck equation in the form

$$\rho\frac{\partial W}{\partial t} = -\frac{\partial (\rho AW)}{\partial x} + \frac{\partial^2 (\rho BW)}{\partial x^2} \tag{6.52}$$

instead of equation (6.51).

► **Problem 6.35** Derive the Fokker–Planck equation as a function of the initial state.

First we simplify the notation of the density probability $W(x, t_0; z, t) \equiv W(x, z, t)$ by taking into account the uniformity of the transition processes. Let us start from the Fokker–Planck equation (6.52),

$$\rho(z)\frac{\partial W(x,z,t)}{\partial t} = -\frac{\partial[\rho(z)A(z,t)W(x,z,t)]}{\partial z} + \frac{\partial^2[\rho(z)B(z,t)W(x,z,t)]}{\partial x^2} ,$$

and the Smoluchowski equation (13.29) has the form

$$W(x, z, t+\tau) = \int W(x, \xi, t)\rho(\xi)W(\xi, z, \tau)dz\ .$$

From the Smoluchowski equation it follows that

$$\frac{\partial W(x, z, t+\tau)}{\partial t} = \int \frac{\partial W(x, \xi, t)}{\partial t}\rho(\xi)W(\xi, z, \tau)dz\ .$$

Using the Fokker–Planck equation (6.52) and integrating this equation by parts, we get

$$\begin{aligned}\frac{\partial W(x, z, t+\tau)}{\partial t} &= \int \frac{\partial}{\partial \xi}[A(\xi)\rho(\xi)W(x, \xi, t)]W(\xi, z, \tau)d\xi \\ &+ \int \frac{\partial^2}{\partial \xi^2}[B(\xi)\rho(\xi)W(x, \xi, t)] \\ W(\xi, z, \tau)d\xi &= -\int A(\xi)\rho(\xi)W(x, \xi, t)\frac{\partial W(\xi, z, \tau)}{\partial \xi}d\xi \\ &+ \int B(\xi)\rho(\xi)W(x, \xi, t)\frac{\partial^2 W(\xi, z, \tau)}{\partial^2 \xi}d\xi\ .\end{aligned}$$

Along with this, we have

$$\frac{\partial W(x, z, t+\tau)}{\partial t} = \frac{\partial W(x, z, t+\tau)}{\partial \tau} = \int W(x, \xi, t)\frac{\partial W(\xi, z, \tau)}{\partial \tau}\rho(\xi)d\xi\ .$$

From comparison of the equations obtained, we finally have

$$\frac{\partial W(\xi, z, \tau)}{\partial \tau} = -A(\xi)\frac{\partial W(\xi, z, \tau)}{\partial \xi} + B(\xi)\frac{\partial^2 W(\xi, z, \tau)}{\partial \xi^2}\ . \tag{6.53}$$

▶ **Problem 6.36** Find the relation between the quantities A and B which characterize the hydrodynamic and diffusion fluxes in formula (6.51).

The relation between the quantities A and B can be found from the condition that if the distribution function coincides with the Maxwell distribution function ($f_0 \sim \exp(-\varepsilon/T)$), the collision integral will be zero. This condition yields

$$A = \left(-\frac{B}{T} + \frac{dB}{d\varepsilon} + \frac{B}{\rho}\frac{d\rho}{d\varepsilon}\right)\ . \tag{6.54}$$

Relation (6.54) between the quantities A and B which are characteristics of the hydrodynamic and diffusion fluxes (6.51) is a general relation between transport coefficients in an arbitrary space if the system is found in the thermodynamic equilibrium. In the case when a particle is moving in a space in a gas under an action of a weak external field these connect the mobility and diffusion coefficient of the particle and have the form of the Einstein relation for a neutral (1.49) or charged (1.50) particle.

► **Problem 6.37** Determine the recombination coefficient involving electrons if the energy of a recombining electron varies by a small amount in an elementary process of collision with other particles.

We now consider the Fokker–Planck equation for evolution of the recombining electron in the energy space, and the recombination process results in many elementary events. The Fokker–Planck equation for the probability $W(\varepsilon, \varepsilon', t)$ of transition from an initial state of energy ε to a state of energy ε' during a time t is governed by the Fokker–Planck equation (6.53). One can define the boundary of a continuous spectrum for electrons and assume that approaching this boundary leads to recombination–formation of a bound state for these electrons. On the basis of these considerations we introduce the probability $v(\varepsilon, t)$ for the electron to approach a bound state, which is equal to

$$v(\varepsilon, t) = 1 - \int W(\varepsilon, \varepsilon', t) d\varepsilon \,,$$

where the integral is taken over all continuous spectrum of electron states. Then from the Fokker–Planck equation (6.53) for the probability $W(\varepsilon, \varepsilon', t)$ of transition from an initial state we obtain

$$\frac{\partial(1-v)}{\partial\tau} = -A(\varepsilon)\frac{\partial(1-v)}{\partial\varepsilon} + B(\varepsilon)\frac{\partial^2(1-v)}{\partial\varepsilon^2} \,.$$

Let us introduce the average recombination time

$$\tau(\varepsilon) = \int_0^\infty [1 - v(\varepsilon, t)]dt = \int_0^\infty t\frac{v(\partial\varepsilon, t)}{\partial t}dt \,,$$

which satisfies the equation

$$B(\varepsilon)\tau'(\varepsilon) - A(\varepsilon)\tau'' = -1 \,. \tag{6.55}$$

Note that in reality states of a given energy are characterized by a certain density of states $\rho(\varepsilon)$ and to account for this fact, it is necessary to replace the probability W by the quantity ρW. But because we are dealing with an integral value $\tau(\varepsilon)$, this replacement does not change equation (6.55).

Taking into account the relation (6.54) between the coefficients of equation (6.55), we reduce it to the form

$$B\tau'' - \left(\frac{B}{T} - B' - B\frac{\rho'}{\rho}\right)\tau' = -1 \,,$$

and the solution of this equation is

$$\tau(\varepsilon) = \int_{\varepsilon_0}^{\varepsilon} \frac{\exp(x/T)dx}{\rho(x)B(x)} \int_x^C \exp(-\frac{y}{T})\rho(y)dy \,,$$

where ε_0 and C are constants of integration.

Being guided by the recombination of an electron and a positive ion, we will consider the recombination process as a result of electron drift over free and bound electron states. Then we take as ε_0 the energy of the ground state of a forming atom, i. e., $\varepsilon_0 = -J$, where J is the atom ionization potential. Assuming this value to be large compared to the thermal electron energy $J \gg T$, one can replace ε_0 by $-\infty$. Next, if the electron energy ε is large compared to its thermal energy T, an electron loses energy, so one can neglect the term $B\tau''$ in equation (6.55). This is equivalent to replace the integration constant C by infinity. As a result, the expression for the quantity τ takes the form

$$\tau(\varepsilon) = \int_{-\infty}^{\varepsilon} dx \int_{x}^{\infty} dy \frac{\exp\left(\frac{x}{T} - \frac{y}{T}\right)\rho(y)}{\rho(x)B(x)} .$$

The integration in the above formula is taken over a range that is shown in Fig. 6.2. This range may be narrowed under real conditions when this process takes place in an ideal plasma. We find the number density $\rho(\varepsilon)$ in this case. This is given by the relation

$$\rho(\varepsilon) = \frac{dn}{d\varepsilon} = \int \delta\left(\varepsilon - \frac{p^2}{2m_e} + \sum_i \frac{e^2}{r_i}\right) \frac{d\mathbf{p}\mathbf{r}}{(2\pi\hbar)^3} ,$$

where n is the number of states, p is the electron momentum, r_i is the coordinate of the i-th ion with respect to the electron position. From this we have for the number density of states of a continuous spectrum $\rho_{\text{cont}}(\varepsilon)$ if the criterion $\varepsilon \gg e^2/N_i$ holds true,

$$\rho_{\text{cont}}(\varepsilon) = \int \delta\left(\varepsilon - \frac{p^2}{2m_e}\right) \frac{d\mathbf{p}\mathbf{r}}{(2\pi\hbar)^3} = \frac{4\pi\sqrt{2m_e\varepsilon}m_e V}{(2\pi\hbar)^3} ,$$

where V is the plasma volume. For bound states $\varepsilon < 0$ under the condition $|\varepsilon| \gg e^2 N_i$ the integral for the state density is divided in a sum of integral near each ion, that is

$$\rho_{\text{bound}}(\varepsilon) = N_i V \int \delta\left(-|\varepsilon| - \frac{p^2}{2m_e} + \frac{e^2}{r}\right) \frac{d\mathbf{p}\mathbf{r}}{(2\pi\hbar)^3} = \frac{\sqrt{2}\pi^3 e^6 m_e^{3/2} N_i V}{(2\pi\hbar)^3} |\varepsilon|^{5/2} .$$

One can see for an ideal plasma $\rho_{\text{bound}} \ll \rho_{\text{cont}}$. Hence, the basic time of recombination a test electron is located in bound states, i. e., $\tau(\varepsilon) \approx \tau(0)$. Therefore during

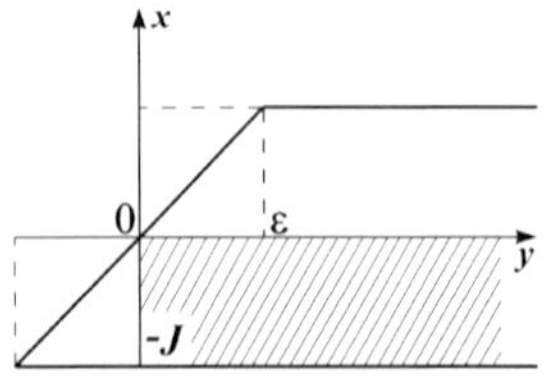

Fig. 6.2 The range of integration in evaluation of the recombination time.

the integration of the formula for $\tau(\varepsilon)$ over the range of Fig. 6.2, one can restrict by a shaded range only. This gives the following formula for the recombination time,

$$\tau(\varepsilon) \approx \tau(0) = \frac{2T_e^{3/2}}{|\pi^{3/2}e^6 N_i} \int_0^\infty \frac{\exp(-|\varepsilon|/T_e)|\varepsilon|^{5/2} d|\varepsilon|}{B(|\varepsilon|)} , \quad (6.56)$$

and we include in this expression the electron temperature T_e, which determines the above processes. Note that along with the criterion of an ideal plasma (1.56) we assume above the thermal electron energy to be relatively small ($T_e \ll J$).

6.5 Collision Integral for Electrons

► **Problem 6.38** Find the collision integral for the spherically symmetric part of the electron distribution function in the energy space if electrons take part in elastic collisions with gas atoms.

In a uniform gas without external fields the kinetic equation for the spherically symmetric part f_o of the electron energy distribution function has the form

$$\frac{\partial f_o}{\partial t} = I_{ea}(f_o) ,$$

where $I_{ea}(f_o)$ is the collision integral taking into account elastic electron–atom collisions, and the normalization of the distribution function has the form

$$\int_0^\infty f_o \, \varepsilon^{1/2} d\varepsilon = 1 .$$

We account for the analogy between the above kinetic equation and the Fokker–Plank equation that allows us to represent the collision integral for electrons as the right-hand side of the Fokker–Planck equation. So, we have

$$I_{ea}(f_o) = \frac{1}{\rho(\varepsilon)} \frac{\partial}{\partial \varepsilon} \left[-A\rho f_o + \frac{\partial}{\partial \varepsilon}(B\rho f_o) \right] ,$$

where $\rho(\varepsilon) \sim \varepsilon^{1/2}$.

On the basis of relation (6.54) between the quantities A and B, we have

$$I_{ea}(f_o) = \frac{1}{\rho(\varepsilon)} \frac{\partial}{\partial \varepsilon} \left[\rho(\varepsilon) B(\varepsilon) \left(\frac{\partial f_o}{\partial \varepsilon} + \frac{f_o}{T} \right) \right] , \quad (6.57)$$

where T is the gas temperature.

We now determine the quantity $B(\varepsilon)$. By definition, it is given by

$$B(\varepsilon) = \frac{1}{2} \left\langle \int (\varepsilon - \varepsilon')^2 N_a v d\sigma(\varepsilon \to \varepsilon') \right\rangle ,$$

where the angle brackets signify an average over atomic energies, and $d\sigma$ is the electron–atom cross section corresponding to a given variation of the electron energy. Note that the relative electron–atom velocity does not change in the collision process, that is $|\boldsymbol{v}-\boldsymbol{v}_a| = |\boldsymbol{v}'-\boldsymbol{v}_a|$, where $\boldsymbol{v}$ and $\boldsymbol{v}'$ are the electron velocities before and after the collision, respectively, and $\boldsymbol{v}_a$ is the velocity of the atom, unvarying in a collision with an electron. From this it follows that $v^2-(v')^2=2\boldsymbol{v}_a(\boldsymbol{v}-\boldsymbol{v}')$, which leads to the following expression:

$$B(\varepsilon) = \frac{m_e^2}{2}\langle\frac{v_a^2}{3}\rangle \int (\boldsymbol{v}-\boldsymbol{v}')^2 N_a v d\sigma = \frac{m_e}{M} T \frac{m_e v^2}{2} N_a v \sigma^*(v)\,. \tag{6.58}$$

In this equation, $\langle v_a^2/3\rangle = T/M$, T is the gas temperature, m_e and M are the electron and atom masses, $|\boldsymbol{v}-\boldsymbol{v}'| = 2v\sin(\vartheta/2)$, where ϑ is the scattering angle, and $\sigma^*(v) = \int(1-\cos\vartheta)d\sigma$ is the diffusion cross section of electron–atom scattering.

Thus the collision integral from the spherical part of the electron distribution function takes the form

$$I_{ea}(f_0) = \frac{m_e}{M}\frac{\partial}{v^2\partial v}\left[v^3\nu\left(\frac{\partial f_0}{m_e v\partial v}+\frac{f_0}{T}\right)\right], \tag{6.59}$$

where $\nu = N_a v\sigma^*(v)$ is the rate of electron–atom collisions.

► **Problem 6.39** Determine the recombination coefficient for electrons and ions as a result of three-body collisions of electrons with gas atoms.

The recombination coefficient α of electrons and ions in three-body collisions with atoms is as follows:

$$\alpha = \frac{1}{\tau N_i},$$

where the time τ of electron passage through its discrete spectrum is given by formula (6.56) with expression (6.58) for the quantity $B(|\varepsilon|)$. Evaluating the average value of the electron velocity cube $\overline{v_e^3}$ in the bound space of electron energies, we have

$$\overline{v_e^3} = \frac{\int v_e^3\delta\left(\varepsilon-\frac{m_e^2}{2}+\frac{e^2}{r}\right)d\boldsymbol{v}_e d\boldsymbol{r}}{\int\delta\left(\varepsilon-\frac{m_e^2}{2}+\frac{e^2}{r}\right)d\boldsymbol{v}_e d\boldsymbol{r}} = \frac{16}{3\pi}\left(\frac{2|\varepsilon|}{m_e}\right)^{3/2}.$$

We also take into account that the average atom velocity square in the frame of reference where an ion is motionless is

$$\overline{v_a^2} = \frac{3T}{m_a} + v_i^2,$$

where T is the gas temperature, m_a is the atom mass, and v_i is the velocity of an ion that takes part in the recombination event with a test electron. Substituting this in formula (6.58), we obtain for $\varepsilon < 0$

$$B(|\varepsilon|) = \frac{T}{m_a}\left(1+\frac{m_i v_i^2}{3}\right)N_a\sigma^*\frac{16\sqrt{m_a}}{3\pi}(2|\varepsilon|)^{3/2}.$$

Here N_a is the number density of atoms, σ^* is the diffusion cross section of electron–atom scattering, and we assume it to be independent of the electron velocity. We also assume that an ion velocity remains constant during several electron–atom collisions. As a result, we obtain from this on the basis of formula (6.56) for a recombination time

$$\tau = \frac{3m_a T_e^{7/2}}{16\sqrt{2\pi m_e} e^6 N_i (1 + m_a v_i^2/3T)} N_a \sigma^* .$$

Averaging over the ion velocities, we obtain for the recombination electron–ion coefficient due to three-body electron–atom collisions

$$\alpha = \frac{1}{\tau N_i} = \frac{16\sqrt{2\pi}}{3} \frac{N_a \sigma^* e^6 T \sqrt{m_e}}{T_e^{7/2} \mu} , \tag{6.60}$$

where μ is the reduced ion–atom mass.

► **Problem 6.40** Find the collision integral with participation of free electrons in a plasma (the Landau integral).

Since the diffusion cross section of electron–electron collisions (2.37) is determined mostly by scattering on small angles, a remarkable change of the momentum of a test electron results from many collisions with other electrons, and therefore the collision integral can be expressed through a flux j_α in the momentum space,

$$I_{ee}(f) = -\frac{\partial j_a}{\partial v_a} .$$

But in contrast to the Fokker–Planck equation, now the collision integral is a nonlinear function of the electron distribution function. In deriving the expression for the collision integral for electron–electron collisions, we consider expression (6.6). Using the symmetry of the rate $W(\mathbf{v}_1, \mathbf{v}_2 \to \mathbf{v}_1', \mathbf{v}_2')$ in this formula, we write it in the symmetric form as

$$W = W\left(\frac{\mathbf{v}_1 + \mathbf{v}_1'}{2}, \frac{\mathbf{v}_2 + \mathbf{v}_2'}{2}, \Delta\mathbf{v}\right) = W\left(\mathbf{v}_1 + \frac{\Delta\mathbf{v}}{2}, \mathbf{v}_2 - \frac{\Delta\mathbf{v}}{2}, \Delta\mathbf{v}\right) ,$$

where $\mathbf{v}_1, \mathbf{v}_2$ are the velocities of colliding electrons, and $\Delta\mathbf{v}$ is a variation of the electron velocity as a result of collisions that is small compared with the velocity of each colliding electron. From the principle of the detailed balance it follows that W is an even function of $\Delta\mathbf{v}$, i. e., $W(\Delta\mathbf{v}) = W(-\Delta\mathbf{v})$.

Our task is to expand the collision integral over a small parameter that is proportional to $\Delta\mathbf{v}$. The first order of the expansion of the collision integral over the small parameter gives

$$I_{ee}(f) = -\int \left[f(\mathbf{v}_2)\frac{\partial f(\mathbf{v}_1)}{\partial \mathbf{v}_1} - f(\mathbf{v}_1)\frac{\partial f(\mathbf{v}_2)}{\partial \mathbf{v}_2}\right] \Delta\mathbf{v} W d(\Delta\mathbf{v}) d\mathbf{v}_2 .$$

Since W is the even function of $\Delta\mathbf{v}$, this approximation gives zero.

From the second-order approximation in $\Delta\boldsymbol{v}$ we have

$$I_{ee}(f) = -\int d\Delta\boldsymbol{v}d\boldsymbol{v}_2 W\left(\frac{1}{2}\Delta_\alpha\Delta_\beta\frac{\partial^2 f_1}{\partial v_{1\alpha}\partial v_{1\beta}}f_2 - \Delta_\alpha\Delta_\beta\frac{\partial f_1}{\partial v_{1\alpha}}\frac{\partial f_2}{\partial v_{2\beta}} + \frac{1}{2}\Delta_\alpha\Delta_\beta f_1\frac{\partial^2 f_2}{\partial v_{2\alpha}\partial v_{2\beta}}\right) - \int \Delta\boldsymbol{v}d\boldsymbol{v}_2\frac{1}{2}\Delta_\alpha\left(\frac{\partial W}{\partial v_{1\alpha}} - \frac{\partial W}{\partial v_{2\alpha}}\right)\Delta_\beta\left(\frac{\partial f_1}{\partial v_{1\beta}}f_2 - f_1\frac{\partial f_2}{\partial v_{2\beta}}\right),$$

where indices $\alpha, \beta \equiv x, y, z$, we use notations $f_1 \equiv f(\boldsymbol{v}_1)$, $f_2 \equiv f(\boldsymbol{v}_2)$, $\Delta_\alpha \equiv \Delta v_\alpha$, and the summation is made over twice repeating indices. One can evaluate some of the terms of the above expression by their integration by parts. We have

$$\frac{1}{2}\int d\Delta\boldsymbol{v}d\boldsymbol{v}_2 W\cdot\Delta_\alpha\Delta_\beta\frac{\partial f_1}{\partial v_{1\alpha}}\frac{\partial f_2}{\partial v_{2\beta}} + \frac{1}{2}\int d\Delta\boldsymbol{v}d\boldsymbol{v}_2\Delta_\alpha\Delta_\beta\frac{\partial W}{\partial v_{2\alpha}}\frac{\partial f_1}{\partial v_{1\beta}}f_2 = \frac{1}{2}\int d\Delta\boldsymbol{v}d\boldsymbol{v}_2\Delta_\alpha\Delta_\beta\frac{\partial f_1}{\partial v_{1\alpha}}\frac{\partial}{\partial v_{2\beta}}(Wf_2) = 0,$$

$$\frac{1}{2}\int d\Delta\boldsymbol{v}d\boldsymbol{v}_2 W\cdot\Delta_\alpha\Delta_\beta f_1\frac{\partial^2 f_2}{\partial v_{2\alpha}\partial v_{2\beta}} + \frac{1}{2}\int d\Delta\boldsymbol{v}d\boldsymbol{v}_2\Delta_\alpha\Delta_\beta\frac{\partial W}{\partial v_{2\alpha}}f_1\frac{\partial f_2}{\partial v_{2\beta}} = \frac{1}{2}\int d\Delta\boldsymbol{v}d\boldsymbol{v}_2\cdot\Delta_\alpha\Delta_\beta f_1\frac{\partial}{\partial v_{2\alpha}}\left(W\frac{\partial f_2}{\partial v_{2\beta}}\right) = 0,$$

since the distribution function is zero at $v_{2\beta} \to \pm\infty$. After eliminating these terms we find

$$I_{ee}(f) = -\frac{1}{2}\int d\Delta\boldsymbol{v}d\boldsymbol{v}_2\Delta_\alpha\Delta_\beta\left(W\frac{\partial^2 f_1}{\partial v_{1\alpha}\partial v_{1\beta}}f_2 - W\frac{\partial f_1}{\partial v_{1\alpha}}\frac{\partial f_2}{\partial v_{2\beta}} + \frac{\partial W}{\partial v_{1\alpha}}\frac{\partial f_1}{\partial v_{1\beta}}f_2 - \frac{\partial W}{\partial v_{2\alpha}}f_1\frac{\partial f_2}{\partial v_{2\beta}}\right).$$

We represent the collision integral in the form

$$I_{ee}(f) = -\frac{\partial j_\beta}{\partial v_{1\beta}}, \tag{6.61}$$

and the flux in the space of electron velocities is equal to

$$j_\beta = \int d\boldsymbol{v}_2\left(f_1\frac{\partial f_2}{\partial v_{2\beta}} - \frac{\partial f_1}{\partial v_{1\beta}}f_2\right)D_{\alpha\beta}, \quad D_{\alpha\beta} = \frac{1}{2}\int \Delta_\alpha\Delta_\beta W d\Delta\boldsymbol{v}. \tag{6.62}$$

Thus we represent the collision integral for electrons as the term of the continuity equation in the space of the electron velocities, and the quantity $D_{\alpha\beta}$ in this equation is an analog of the diffusion coefficient for motion in usual space.

▶ **Problem 6.41** Determine the tensor $D_{\alpha\beta}$ of electron diffusion in the velocity space for electrons located in an ideal plasma.

To evaluate the diffusion coefficient $D_{\alpha\beta}$ in the velocity space we consider a small variation of the electron energy in collisions with surrounding electrons. Indeed, according to formula (2.34) we have for variation of the electron velocity after a weak collision with other electron

$$\Delta_\alpha = \frac{2e^2 \rho_\alpha}{\rho^2 g m_e} ,$$

where we denote $\boldsymbol{\Delta} = \Delta \boldsymbol{v}$: here ρ is the impact parameter of collision, ρ_α is its projection on a given direction, and g is the relative velocity of electrons. From this we have

$$D_{\alpha\beta} = \frac{1}{2}\int \Delta_\alpha \Delta_\beta W d\Delta \boldsymbol{v} = \frac{1}{2}\int \Delta_\alpha \Delta_\beta g d\sigma = \frac{2e^4}{m_e^2 g}\int \frac{\rho_\alpha \rho_\beta}{\rho^4} d\sigma = \frac{4\pi e^4}{m_e^2 g} n_\alpha n_\beta \ln \Lambda . \tag{6.63}$$

Here n_α, n_β are the components of the unit vector $\boldsymbol{n}$ directed along $\boldsymbol{\rho}$, and $\ln \Lambda$ is the Coulomb logarithm that is given by

$$\ln \Lambda = \int_{\rho_<}^{\rho_>} \frac{d\rho}{\rho} . \tag{6.64}$$

According to formula (2.37) the Coulomb logarithm is equal to

$$\ln \Lambda = \ln \frac{r_D \varepsilon}{e^2} .$$

If we take the frame of reference in such way that the direction of the relative velocity of electrons $\mathbf{g}$ is directed along the x-axis, and xy is the motion plane, we obtain that only Δ_y is nonzero. Correspondingly, only the tensor component D_{yy} is nonzero and is equal to

$$D_{yy} = \frac{2e^4}{m_e g}\int \frac{1}{\rho^2} 2\pi\rho d\rho = \frac{4\pi e^4}{m_e g} \ln \Lambda . \tag{6.65}$$

For an arbitrary position of the frame of reference, one can construct the tensor $D_{\alpha\beta}$ on the basis of two symmetric tensors $\delta_{\alpha\beta}$ and $g_\alpha g_\beta$. Evidently, it has the form

$$D_{\alpha\beta} = \frac{4\pi e^4}{m_e g^3} g_\alpha g_\beta \ln \Lambda . \tag{6.66}$$

Thus, finally one can represent the Landau collision integral that accounts for collisions between electrons in the form

$$I_{ee}(f) = -\frac{\partial j_\beta}{\partial v_{1\beta}}, \quad j_\beta = \int d\boldsymbol{v}_2 \left(f_1 \frac{\partial f_2}{\partial v_{2\alpha}} - \frac{\partial f_1}{\partial v_{1\alpha}} f_2 \right) D_{\alpha\beta}, \quad D_{\alpha\beta} = \frac{4\pi e^4}{m_e^2 g^3} g_\alpha g_\beta \ln \Lambda . \tag{6.67}$$

In contrast to the right-hand side of the Fokker–Planck equation in a velocity space, the Landau integral is nonlinear with respect to the distribution function of electrons.

▶ **Problem 6.42** Find the expression for the Landau integral in the case of fast electrons in a plasma.

In this case one can separate a group of fast electrons from others. Because a number of fast electrons is relatively small, we obtain a two-component system with a small concentration of a subsystem under consideration. Correspondingly, the collision integral $I_{ee}(f)$ becomes linear with respect to the distribution function of fast electrons, and we can use formula (6.57) for the collision integral with expression (6.58) for the diffusion coefficient of fast electrons in the energy space. Thus, we have now

$$I_{ee}(f_0) = \frac{1}{m_e v^2}\frac{\partial}{\partial v}\left[v B_{ee}(\varepsilon)\left(\frac{\partial f_0}{m_e v \partial v} + \frac{f_0}{T_e}\right)\right] . \tag{6.68}$$

Thus, this collision integral is linear with respect to the distribution function of fast electrons and the index o indicates that the distribution function of electrons is spherically symmetric. The diffusion coefficient $B(\varepsilon)$ for a fast electron with an energy $\varepsilon = m_e v^2/2$ is equal to

$$\begin{aligned} B(\varepsilon) &= \frac{1}{2}\int \overline{(\varepsilon - \varepsilon')^2 N_e v d\sigma(\varepsilon \to \varepsilon')} = \frac{N_e}{2}\int \overline{(\varepsilon - \varepsilon')^2 W d\Delta \mathbf{v}} \\ &= \frac{N_e}{2}\int \overline{m_e v_\alpha (v_\alpha - v'_\alpha)\cdot m_e v_\beta (v_\beta - v'_\beta) W d\Delta \mathbf{v}} = \frac{N_e}{2} m_e^2 v_\alpha v_\beta D_{\alpha\beta} . \end{aligned}$$

Here an overline means an average over scattering angles, N_e is the electron number density, $\varepsilon, \varepsilon'$ are the energies of a fast electron before and after collision, and the summation takes place over repeating indices. Use a small change of the energy of a test electron as a result of collision, $\varepsilon - \varepsilon' = m_e v_\alpha (v_\alpha - v'_\alpha)$. This allows us to use the diffusion cross section for two charged particles (2.37) that follows after integration over scattering angles or final electron velocities. Then on the basis of the diffusion cross section (2.37) or the diffusion coefficient for a fast electron in the velocity space (6.67) we obtain

$$B_{ee}(\varepsilon) = 2\pi e^4 v N_e \ln \Lambda . \tag{6.69}$$

▶ **Problem 6.43** Find the collision integral $I_{ea}(f_0)$ for the spherically symmetric part of the electron distribution function in the energy space if the loss of the energy of electrons is determined by excitation of diatomic molecules to the first vibrational level, and the average electron energy exceeds remarkably the vibrational excitation energy $\hbar\omega$.

The balance of electrons in the energy space has the form

$$4\pi v^2 dv \cdot I_{ea}(f_0) = -4\pi v^2 dv \cdot \nu_{ex}(v) f_0(v) + 4\pi (v^2 + \frac{2\hbar\omega}{m_e}) \cdot \nu_{ex}\left(\sqrt{v^2 + \frac{2\hbar\omega}{m_e}}\right) f_0\left(\sqrt{v^2 + \frac{2\hbar\omega}{m_e}}\right) d\sqrt{v^2 + \frac{2\hbar\omega}{m_e}} ,$$

where $\nu_{ex}(v)$ is the rate of molecule excitation at the electron velocity v, and m_e is the electron mass. We take into account that molecules of the gas are found in the ground vibrational state. The left-hand side of this equation is the number of electrons which go into a velocity range dv as a result of collisions with molecules, and the terms on the right-hand side of this equation indicate the character of these transitions. Expanding the right-hand side of the equation over a small parameter $\hbar\omega/\varepsilon$, we obtain

$$I_{ea}(f_0) = \frac{\hbar\omega}{m_e}\frac{1}{v^2}\frac{\partial}{\partial v}(v\nu_{ex}(v)f_0) \,. \tag{6.70}$$

In derivation of this formula, we account for excitation of the first vibrational level only in electron–molecule collisions. In terms of the Fokker–Planck equation, this expression accounts for the hydrodynamic flux only with $A = \hbar\omega\nu_{ex}$. The excitation rate can be represented in the form $\nu_{ex} = N_m v\sigma_{ex}$, where N_m is the number density of molecules, and σ_{ex} is the cross section of excitation of the vibrational state of the molecule.

► **Problem 6.44** Find the equilibrium number density of clusters of a given size in a vapor.

Growth and evaporation of clusters of a given size result from the processes of atom attachment and cluster evaporation whose rates are given by formulas (2.39) and (3.67). The kinetic equation for the size distribution function f_n of clusters, i. e., number density of clusters consisting of n atoms, has the form

$$\frac{\partial f_n}{\partial t} = Nk_{n-1}f_{n-1} - Nk_nf_n - \nu_nf_n + \nu_{n+1}f_{n+1} \,. \tag{6.71}$$

From this it follows that if an equilibrium between clusters of a given size is supported by the processes of atom attachment and cluster evaporation, we have the size distribution functions

$$f_{n+1}\nu_{n+1} = f_nNk_n \,.$$

Using formulas (2.39) and (3.67) for the rates of these processes, we obtain

$$\frac{f_{n+1}}{f_n} = S\exp\left(\frac{\varepsilon_{n+1} - \varepsilon_0}{T}\right) ,$$

with the supersaturation degree S defined as

$$S = \frac{N_{sat}(T)}{N} \,. \tag{6.72}$$

Within the framework of the liquid drop model for large clusters, we have for the total binding energy of cluster atoms E_n,

$$E_n = \varepsilon_0 n - An^{2/3} , \quad n \gg 1 , \tag{6.73}$$

where ε_o is the atom binding energy for a macroscopic system, and the second term accounts for the cluster surface energy with the specific surface energy A. This gives for the atom binding energy ε_n in a large cluster $n \gg 1$,

$$\varepsilon_n = \frac{dE_n}{dn} = \varepsilon_o - \frac{2A}{3n^{1/3}} \, .$$

From this we have for the ratio of the equilibrium size distribution functions for clusters of neighboring sizes

$$\frac{f_{n+1}}{f_n} = S \exp\left(-\frac{2A}{3n^{1/3}T}\right) . \tag{6.74}$$

From this it follows that for a supersaturated vapor $S > 1$ the size distribution function of large clusters grows with an increase of a cluster size. Next, the size distribution function of clusters as a function of their sizes has a minimum at the critical number of cluster atoms n_{cr}, which is according to the above formula

$$n_{cr} = \left(\frac{\Delta\varepsilon}{T \ln S}\right)^3 . \tag{6.75}$$

▶ **Problem 6.45** Determine the collision integral for clusters in a space of their sizes.

For large cluster sizes we represent the collision integral of clusters in the form

$$I_{col}(f_n) = -\frac{\partial j_n}{\partial n}$$

and according to the kinetic equation (6.71) we have in the limit $n \gg 1$

$$j_n = k_o(T)\xi n^{2/3} \left[N f_n - N_{sat}(T) f_{n+1} \exp(\frac{2A}{3Tn^{1/3}})\right] .$$

Thus, the collision integral has the form of a flux in a space of cluster size. Taking $f_{n+1} = f_n + \partial f_n/\partial n$, one can represent the collision integral in the form of the sum of two fluxes, so that the first one is, the hydrodynamic flux, is expressed through the first derivative over n, and the second flux, the diffusion one, includes the second derivative over n. The diffusion flux is small compared to the hydrodynamic one, but it is responsible for the width of the distribution function of clusters on sizes. Neglecting the diffusion flux for large n, we get

$$I_{col}(f_n) = -\frac{\partial}{\partial n}\left\{k_o(T)\xi n^{2/3} f_n \left[N - N_{sat}(T) \exp\left(\frac{\Delta\varepsilon}{Tn^{1/3}}\right)\right]\right\} \tag{6.76}$$

and the expression inside the square brackets is zero at the critical cluster size.

7
Transport and Kinetics of Electrons in Gases in External Fields

7.1
Electron Drift in a Gas in an Electric Field

▶ **Problem 7.1** Derive the kinetic equation for electrons moving in a gas in an external electric field, when elastic electron–atom collisions dominate, whereas electron–electron collisions are not essential.

The Boltzmann kinetic equation for the velocity distribution function $f(\boldsymbol{v})$ of electrons according to equation (6.2) has the form

$$\frac{e\boldsymbol{E}}{m_e}\frac{\partial f}{\partial \boldsymbol{v}} = I_{ea}(f) , \quad (7.1)$$

where I_{ea} is the electron–atom collision integral which accounts for electron–atom collisions.

In analyzing this kinetic equation, we take into account a small energy exchange in electron–atom collisions, whereas the direction of electron motion varies significantly at each collision. Hence the velocity distribution of electrons moving in a gas in an external electric field is nearly symmetrical with respect to directions of electron motion and has therefore the form

$$f(\boldsymbol{v}) = f_0(v) + v_x f_1(v) , \quad (7.2)$$

where x is the direction of the electric field $\boldsymbol{E}$. The electron–atom collision integral has a linear dependence on the distribution function $f(\boldsymbol{v})$ according to expression (6.8), which gives

$$I_{ea}(f) = I_{ea}(f_0) + I_{ea}(v_x f_1) .$$

The first term of this formula is given by equation (6.59). The second term has the form

$$I_{ea}(v_x f_1) = \int (\boldsymbol{v}' - \boldsymbol{v})_x v d\sigma\, f_1(v)\varphi(v_a) d\boldsymbol{v}_a ,$$

where $\boldsymbol{v}, \boldsymbol{v}'$ are the electron velocities before and after collision, respectively, $\boldsymbol{v}_a$ is the atom velocity. Because of a small atom velocity, the character of collision does

Plasma Processes and Plasma Kinetics. Boris M. Smirnov

ISBN: 978-3-527-40681-4

not depend on $\boldsymbol{v}_a$, and the integration over atomic velocities gives $\int \varphi(\boldsymbol{v}_a) d\boldsymbol{v}_a = N_a$, where N_a is the atom number density. Next, we represent the electron velocity after collision as

$$\boldsymbol{v}' = \boldsymbol{v} \cos \vartheta + v\boldsymbol{k} \sin \vartheta \ ,$$

where ϑ is the scattering angle, and $\boldsymbol{k}$ is the unit vector located in the plane that is perpendicular to the initial electron velocity $\boldsymbol{v}$. Since this vector has an arbitrary direction in a given plane, $\int \boldsymbol{k} d\sigma = 0$, we obtain $\int (\boldsymbol{v}' - \boldsymbol{v})_x d\sigma = -v_x \sigma^*(v)$, where $\sigma^*(v) = \int (1 - \cos \vartheta) d\sigma$ is the diffusion cross section of electron–atom scattering. This gives finally

$$I_{ea}(v_x f_1) = -\nu v_x f_1(v) \ ,$$

where $\nu = N_a v \sigma^*(v)$ is the rate of electron–atom collisions.

Thus, the kinetic equation (7.1) by taking into account for the expansion (7.2) and the expressions for the collision integrals takes the form

$$\frac{eE}{m_e} \left(\frac{v_x}{v} \frac{df_0}{dv} + f_1 + v_x^2 \frac{df_1}{dv} \right) = -\nu v_x f_1 + I_{ea}(f_0) \ .$$

It is convenient to represent it in terms of spherical harmonics. For this goal we integrate this equation over $d(\cos \theta)$, where θ is the angle between the vectors $\boldsymbol{v}$ and $\boldsymbol{E}$, and multiplying this equation by $\cos \theta$, integrate it over angles. Then we obtain instead of this equation the set of the following equations:

$$a \frac{df_0}{dv} = -\nu v f_1 , \qquad \frac{a}{3v^2} \frac{d}{dv} \left(v^3 f_1 \right) = I_{ea}(f_0) \ , \tag{7.3}$$

where $a = eE/m_e$. This set of equations establishes the relation between the spherical and nonspherical parts of the electron distribution function.

▶ **Problem 7.2** Derive the expression for the electron drift velocity if it is moving in an atomic gas in an external electric field.

From the set of equations (7.3) for the electron distribution function, we obtain the electron drift velocity in a gas—the average electron velocity in the field direction,

$$w_e = \int v_x^2 f_1 d\boldsymbol{v} = \frac{eE}{3m_e} \left\langle \frac{1}{v^2} \frac{d}{dv} \left(\frac{v^3}{\nu} \right) \right\rangle \tag{7.4}$$

with averaging over the spherical distribution function of the electrons. In particular, if the rate ν of electron–atom collisions is independent of the collision velocity, the electron drift velocity w_e and its mean energy $\bar{\varepsilon}$ are given by

$$w_e = \frac{eE}{m_e \nu} , \qquad \bar{\varepsilon} = \frac{3}{2} T + \frac{M}{2} w_e^2 \ . \tag{7.5}$$

► **Problem 7.3** Find the criterion in the case of electron drift in an atomic gas in an external electric field when one can ignore electron–electron collisions.

Since electron–atom and electron–electron collisions are not correlated in the course of electron drift in a gas, we separate these processes and represent the collision integral for electrons as a sum of the integral I_{ea} of electron–atom collisions and the integral I_{ee} of electron–electron collisions. As a result, the kinetic equation (7.1) takes the form

$$\frac{e\mathbf{E}}{m_e}\frac{\partial f}{\partial \mathbf{v}} = I_{ee}(f) + I_{ea}(f) \,. \tag{7.6}$$

The collision integrals in equation (7.6) are estimated as $I_{ea} \sim \frac{m_e}{M} v\sigma_{ea} N_a f$ and $I_{ee} \sim v\sigma_{ee} N_e f$, where v is a typical electron velocity, m_e, M are masses of electrons and atoms, N_e, N_a are the number densities of electrons and atoms, and σ_{ea}, σ_{ee} are typical cross sections for collisions of electrons with atoms and electrons, respectively. Above we consider the limiting case $I_{ee} \ll I_{ea}$. This limiting case requires the validity of the criterion

$$N_e \ll \frac{m_e}{M}\frac{\sigma_{ea}}{\sigma_{ee}} N_a \tag{7.7}$$

In order to present a more correct criterion (7.7), we compare the kinetic equation (7.6) terms due to electron–electron and electron–atom collisions. Evidently, the criterion (7.7) has then the form $B_{ee} \ll B_{ea}$, where B_{ee} is given by formula (6.69) and B_{ea} by formula (6.58). As a result, we obtain instead of criterion (7.7)

$$\frac{N_e}{N_a} \ll \xi = \frac{m_e}{M}\frac{1}{\ln\Lambda}\frac{T\varepsilon\sigma^*_{ea}}{2\pi e^4} \,, \tag{7.8}$$

where ε is the electron energy. We now obtain that a comparison between electron–electron and electron–atom rates depends on the electron energy.

Since $m_e \ll M$ and $\sigma_{ea} \ll \sigma_{ee}$, this criterion can be valid for an ionized gas with a weak degree of ionization. As an example, Fig. 7.1 contains the parameter ξ of formula (7.8) for helium if $\ln\Lambda = 10$ and $\sigma^*_{ea} = 6$ Å^2. As is seen, electron–electron collisions dominate in establishment of an electron equilibrium even at very low degree of ionization.

► **Problem 7.4** Determine the electron distribution function in the case when electrons are moving in an atomic gas in an external electric field, and elastic electron–atom collisions dominate for establishment of an electron-gas equilibrium.

In determination of the electron distribution function we take into account that both distribution functions f_0 and f_1 tend to zero at large electron energies. Then, integrating the second equation of the set (7.3) with the use of expression (6.59) for the collision integral, we obtain the second equation of the set (7.3) in the form

$$f_1 = \frac{3\nu}{a}\frac{m_e}{M}\left(f_0 + T\frac{df_0}{d\varepsilon}\right) \,,$$

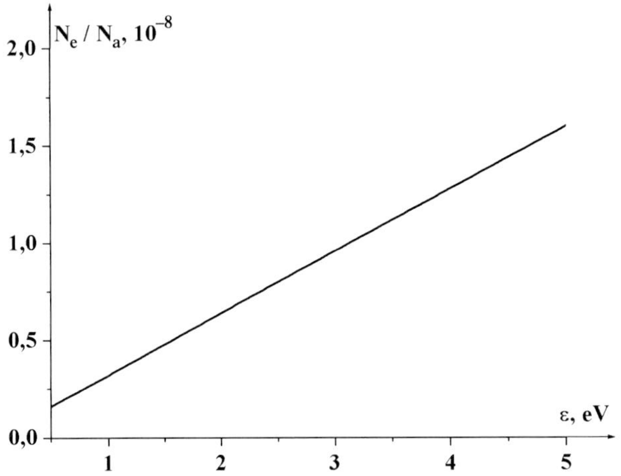

Fig. 7.1 The boundary concentration of electrons when the transition takes place between two regimes of electron equilibrium in a plasma in an external electric field, so that the parameter (7.8) $\xi = 1$.

where ε is the electron energy. Substituting this expression for f_1 in the first equation of the set (7.3), we reduce this set to the equation

$$f_0 + \left(T + \frac{Mu^2}{3}\frac{df_0}{d\varepsilon}\right) = 0 \,,$$

where $u = a/v = eE/(m_e v)$.

Solving this set, we obtain by taking into account $f_0 \to 0$ at large electron energies

$$f_0(v) = A \exp\left(-\int_0^v \frac{m_e v' dv'}{T + Mu^2/3}\right) . \tag{7.9}$$

One can see that the spherically symmetric function is transformed into the Maxwell distribution function in the absence of an external electric field with the gas temperature. Note that this distribution function is normalized by

$$\int_0^\infty 4\pi v^2 f_0 dv = N_e \,.$$

Correspondingly, for the asymmetric distribution function f_1 we have

$$f_1(v) = \frac{m_e u f_0}{T + Mu^2/3} . \tag{7.10}$$

From expressions (7.9) and (7.10) it follows that the ratio vf_1/f_0 of terms of formula (7.2) is small. Nevertheless, taking into account for the asymmetric part f_1 in the electron distribution function is of principle, because through this one the interaction with an electric field is realized.

▶ **Problem 7.5** Determine the drift velocity and the average energy for an electron that moves in a gas in an external electric field. Consider the case when the rate of electron–atom collisions is independent of the electron velocity or is proportional to it.

On the basis of expressions (7.9) and (7.10) we obtain the electron drift velocity in a gas,

$$w_e = \frac{1}{3}\int_0^\infty v^2 f_1 4\pi v^2 dv = \frac{4\pi}{3}\int_0^\infty \frac{m_e u f_0}{T + Mu^2/3} v^4 dv \,. \tag{7.11}$$

From this we have for the case when the rate of electron–atom collisions is independent of the electron velocity ($\nu = \text{const}$)

$$w_e = \frac{eE}{m_e \nu}\,, \qquad \bar{\varepsilon} = \frac{3T}{2} + \frac{Mw_e^2}{2}\,, \tag{7.12}$$

where w_e is the electron drift velocity and $\bar{\varepsilon}$ is its mean energy.

In the case when the electron–atom diffusion cross section is independent of the electron velocity ($\sigma^*(v) = \text{const}$), we obtain in the limit $\bar{\varepsilon} \gg T$

$$w_e = 0.857\left(\frac{m_e}{M}\right)^{1/4}\sqrt{\frac{eE\lambda}{m_e}}\,, \qquad \bar{\varepsilon} = 0.427eE\lambda\sqrt{\frac{M}{m_e}} = 0.530Mw_e^2\,, \tag{7.13}$$

where $\lambda = 1/(N_a\sigma^*)$ is the electron mean free path in a gas.

▶ **Problem 7.6** Determine the rate constant of atom ionization by electron impact if electron evolution is determined by elastic electron–atom collisions and the rate of such collisions does not depend on the collision velocity.

The electron distribution in this case according to formula (7.9) has the form

$$f_0(v) = \frac{2}{\sqrt{\pi}}(T + Mu^2/3)^{3/2}\exp\left[-\frac{m_e v^2}{2(T + Mu^2/3)}\right]\,,$$

where $u = eE/(m_e\nu)$, the electron energy is $\varepsilon = m_e v^2$, the effective temperature is $T_e = T + Mu^2/3$, and the distribution function f_0 is normalized by the condition

$$\int_0^\infty f_0 \varepsilon^{1/2} d\varepsilon = 1\,.$$

The threshold ionization cross section σ_{ion} is given by formula (4.34) and we take for simplicity $\alpha = 1$, i. e., $\sigma_{ion}(\varepsilon) = \sigma_0(\varepsilon - J)/J$. In particular, for the Thomson model (4.27) we have

$$\sigma_0 = \frac{\pi e^4}{J^2}$$

Within the framework of the Thomson model, we have for the rate constant of atom ionization by electron impact when the rate of elastic electron–atom collisions does not depend on the electron velocity

$$k_{\rm ion} = \exp\left(-\frac{J}{T_e}\right)\sqrt{\frac{8\pi T_e}{m_e}\frac{e^4}{J^2}} \,. \qquad (7.14)$$

▶ **Problem 7.7** An electron is moving in helium in a constant electric field. Assuming the cross section of elastic electron–atom scattering to be independent of the collision velocity, determine the rate of atom ionization by electron impact within the framework of the Thomson model for this process.

The elastic transport cross section for collision between an electron and helium atom lies between 6 Å^2 and 7 Å^2 if the electron energy varies from 0.3 eV to 6 eV. Hence we will consider below this cross section to be independent of the electron velocity. In this case the distribution function (7.9) has the form

$$f_0(\varepsilon) = A\exp\left(-\frac{\varepsilon^2}{\varepsilon_0^2}\right), \qquad \varepsilon_0 = \sqrt{\frac{M}{3m_e}}eE\lambda, \quad \lambda = (N_a\sigma^*)^{-1} \qquad (7.15)$$

if the average electron energy ε_0 exceeds significantly the thermal atom energy T. From this expression we find the rate constant of atom ionization by electron impact

$$k_{\rm ion} = \frac{\int_J^\infty v\sigma_{\rm ion}(v) f_0(v)\varepsilon^{1/2}d\varepsilon}{\int_0^\infty f_0(v)\varepsilon^{1/2}d\varepsilon} \,,$$

where the electron energy is $\varepsilon = m_e v^2/2$, and we use the Thomson formula (4.27) for the ionization cross section. Assuming the ionization potential J to be large compared with the average electron energy ε_0, we obtain from this

$$k_{\rm ion} = k_o \exp\left(-\frac{J^2}{\varepsilon_0^2}\right) \qquad k_o = 1.81\sqrt{\frac{\varepsilon_0}{m_e}\frac{e^4}{J^2}}\left(\frac{\varepsilon_0}{J}\right)^2 \,. \qquad (7.16)$$

▶ **Problem 7.8** Determine the population of metastable helium atoms He(2^3S) in a plasma that is supported by an electric field of strength E if electron–electron collisions are not essential.

Assuming in this case the electron–atom transport cross section to be independent of the electron energy, we take the energy distribution of electrons according to formula (7.15). We use the principle of detailed balance (3.64) for the excitation $k_{\rm ex}$ and quenching $k_{\rm q}$ rate constants. Then the number density of metastable atoms $N_{\rm m}$ in comparison with that in the ground state $N_{\rm o}$ follows from the balance of excitation and quenching events,

$$N_{\rm m}k_{\rm q} = \int f_0(\varepsilon)k_{\rm ex}(\varepsilon)\sqrt{\varepsilon}d\varepsilon \,,$$

where the electron distribution is normalized by the condition

$$\int f_0(\varepsilon)\sqrt{\varepsilon}d\varepsilon = 1$$

and is equal to

$$f_0(\varepsilon) = \frac{1.63}{\varepsilon_0^{3/2}}\exp\left(-\frac{\varepsilon^2}{\varepsilon_0^2}\right), \qquad \varepsilon_0 = \sqrt{\frac{M}{3m_e}}eE\lambda, \quad \lambda = (N_a\sigma^*)^{-1} .$$

From the principle of detailed balance (3.65) we obtain in the helium case ($g_*/g_0 = 3$) taking the quenching cross section to be independent of the electron energy

$$\frac{N_m}{N_0} = 3\int f_0(\varepsilon)\sqrt{\frac{\varepsilon(\varepsilon-\Delta E)}{\Delta E}}d\varepsilon ,$$

where ΔE is the atom excitation energy. Evaluating the above integral in the limit $\Delta E \gg \varepsilon_0$, we obtain

$$\frac{N_m}{N_0} = 1.53\left(\frac{\varepsilon_0}{\Delta E}\right)^{3/2}\exp\left(-\frac{\Delta E^2}{\varepsilon_0^2}\right) . \tag{7.17}$$

One can see that in the limit under consideration $\Delta E \gg \varepsilon_0$ the population of metastable states is small. In particular, even if $\varepsilon_0 = \Delta E/2$, the population of metastable states is 1%. Figure 7.2 gives the population of helium metastable states in a plasma under the above conditions.

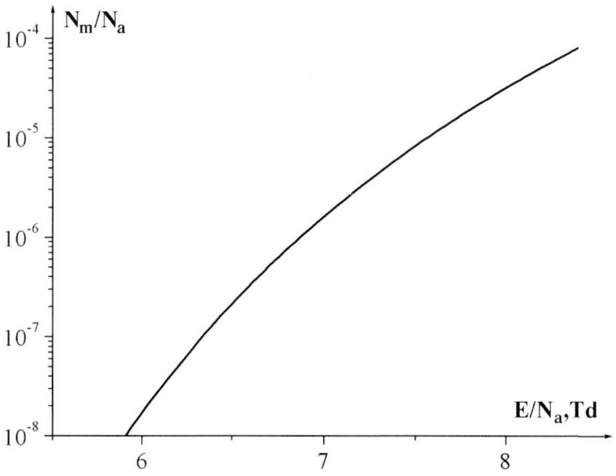

Fig. 7.2 The relative population of metastable state He(2^3S) for a helium plasma which is located in a constant electric field. Excitation and quenching of metastable atoms results from collisions with electrons.

▶ **Problem 7.9** Find the electron distribution function in the limiting case when electron–electron collisions dominate.

This case is opposite with respect to the limiting case to equation (7.7) and corresponds to the criterion

$$N_e \gg \frac{m_e}{M} \frac{\sigma_{ea}}{\sigma_{ee}} N_a \, . \tag{7.18}$$

Expanding the kinetic equation (7.6) over this small parameter, we obtain in zero approximation

$$I_{ee}(f) = 0$$

The solution of this equation leads to the Maxwell distribution function which has now the form

$$\varphi(v) = N_e \left(\frac{m_e}{2\pi T_e} \right)^{3/2} \exp\left(-\frac{m_e v^2}{2T_e} \right) , \tag{7.19}$$

and the electron temperature T_e here can differ from the gaseous one T and is determined from both interaction of electrons with an electric field and collisions with atoms. The electron temperature is determined by the character of the energy transfer from an external electric field to a gas. Then the energy transfers first from an external field to electrons, and later it goes from electrons to atoms as a result of collisions.

▶ **Problem 7.10** Derive the set of kinetic equations for the electron distribution function, if electrons are moving in a gas in an external electric field, and electron–electron collisions dominate.

Let us multiply the kinetic equation equation (7.6) by the electron energy $m_e v^2/2$ and integrate the result over electron velocities. We consider the relation

$$\int \frac{m_e v^2}{2} I_{ee} d\boldsymbol{v} = 0 \, ,$$

which reflects the physical sense of the collision integral and use conservation of the total energy in the electron subsystem. Hence, we have the integral relation

$$eEw_e = \int \frac{m_e v^2}{2} I_{ea} d\boldsymbol{v} \, . \tag{7.20}$$

This is the energy balance equation for electrons; the left-hand side of this relationship is the power which an electron obtains from the electric field and the right-hand side is the power transmitted from an electron to atoms as a result of their collisions. From equation (7.6) it follows that ions give a small contribution to the energy transfer between an external field and a gas in comparison with electrons, because the electron drift velocity exceeds remarkably the ion drift velocity.

Thus, the character of the energy transfer in a weakly ionized gas from an external electric field to electrons, and from electrons to atoms does not depend if the criterion (7.7) or (7.18) holds true. If the criterion (7.18) is valid, one can consider electrons as a subsystem. If the criterion (7.7) holds true, another character of equilibrium takes place for the electron–atom system.

Accounting for formula the spherical distribution function of electrons and the character of electron collisions, we obtain instead of the set of kinetic equations (7.3) the following one

$$f_1 = \frac{eE}{\nu T_e} f_0 , \qquad \frac{a}{3v^2}\frac{d}{dv}\left(v^3 f_1\right) = I_{ea}(f_0) . \tag{7.21}$$

▶ **Problem 7.11** Determine the electron diffusion coefficient in a weakly ionized gas.

The diffusion coefficient of electrons D_e follows from the equation for the electron flux $\boldsymbol{j}_e = -D_e \boldsymbol{\nabla} N_e$. When the electron number density varies in a space, the kinetic equation for the electron distribution function has the form

$$v_x \nabla f = I_{ea}(f) ,$$

and the standard expansion of the distribution function is

$$f = f_0(v) + v_x f_1(v) ,$$

where the x-axis is directed along the gradient of the electron number density. Because the distribution function is normalized to the electron number density $f \sim N_e$, we have $\nabla f = f \nabla N_e / N_e$. Hence the kinetic equation for the electron distribution function has the form

$$v_x f_0 \nabla N_e / N_e = -\nu v_x f_1 ,$$

which gives

$$f_1 = -\frac{f_0 \nabla N_e}{\nu N_e} .$$

From this we obtain the electron flux

$$\boldsymbol{j}_e = \int \boldsymbol{v} f d\boldsymbol{v} = \int v_x^2 f_1 d\boldsymbol{v} = -\frac{\boldsymbol{\nabla} N_e \int v_x^2 f_0 d\boldsymbol{v}}{\nu N_e} = -\boldsymbol{\nabla} N_e \left\langle \frac{v_x^2}{\nu} \right\rangle , \tag{7.22}$$

where brackets mean averaging over the electron distribution function. Comparing this formula with the definition of the diffusion coefficient according to the equation $\boldsymbol{j}_e = -D_e \boldsymbol{\nabla} N_e$, we find for the electron diffusion coefficient

$$D_e = \left\langle \frac{v^2}{3\nu} \right\rangle . \tag{7.23}$$

When an electron is moving in an external electric field, this formula is correct for transverse diffusion, because only in this case one can separate corrections to the spherical electron distribution function due to the electric field and those due to the gradient of the electron number density.

7.2
Energy Balance for Electrons Moving in a Gas in an Electric Field

▶ **Problem 7.12** Analyze times of equilibria establishment if an electron is moving in a gas in an external electric field.

There are two typical times for electron–atom collisions when electrons are moving in an atomic gas in an external field. The first one is $\tau = 1/\nu$ (ν is the typical rate of electron–atom collisions) and characterizes an electron momentum variation in the course of electron–atom collisions, and the second time $\sim M/(m_e\nu) \sim \tau M/m_e$ is a typical time of variation of the electron energy as a result of collisions with atoms. As is seen, the ratio of these times is of the order of M/m, i. e., these times differ very significantly. Indeed, the change of the electron momentum or the direction of electron motion results from one strong electron–atom collision, whereas a remarkable variation of the electron energy proceeds after $\sim M/m$ such collisions.

In terms of these times, the first equation of the set (7.3) is established for times $t \sim 1/\nu$, while the second equation is established for times $t \sim M/(m_e\nu)$. Hence, at times $1/\nu \ll t \ll M/(m_e\nu)$ we have

$$f_1(v) = -\frac{a}{\nu v}\frac{df_0(v)}{dv} ,$$

while the symmetric distribution function differs from the solution (7.9) or (7.19) of the equation set (7.3) under equilibrium conditions. Moreover, because the equilibrium for the electron momentum is established fast, the above relation between $f_0(v)$ and $f_1(v)$ is independent of the character of establishment of the electron energy.

▶ **Problem 7.13** Determine the electron drift velocity and the temperature of electrons if electrons are moving in a gas in an external electric field, and electron–electron collisions dominate.

This case is described by the set (7.21) of kinetic equations, and therefore the electron drift velocity according to its definition is equal to

$$w_e = \int \frac{v_x^2}{3} f_1 d\boldsymbol{v} = \frac{eE}{3T_e}\left\langle \frac{v^2}{\nu} \right\rangle . \tag{7.24}$$

The electron temperature T_e is a parameter which can be found from the balance equation (7.20) for the power transferred from an external field to electrons and from electrons to atoms of a gas. This equation by using formula (6.59) for the spherical part of the electron–atom collision integral of the electron distribution function takes the form

$$eEw_e = \int \frac{m_e v^2}{2} I_{ea}(f_0) d\mathbf{v} = \frac{m_e^2}{M} \left(1 - \frac{T}{T_e}\right) \langle v^2 \nu \rangle \,. \tag{7.25}$$

Substituting formula (7.24) for the electron drift velocity in a gas in (7.25), we obtain

$$T_e - T = \frac{Ma^2}{3} \frac{\langle v^2/\nu \rangle}{\langle v^2 \nu \rangle} \,. \tag{7.26}$$

where $a = eE/m_e$. In particular, in the case $\nu = const$, we have from this

$$w_e = \frac{eE}{m_e \nu}, \qquad T_e - T = \frac{Mw_e^2}{3}. \tag{7.27}$$

If the diffusion cross section is independent of the electron velocity ($\sigma^*(v) = \text{const}$), we obtain, introducing the electron mean free path $\lambda = (N_a \sigma^*)^{-1}$ in a gas,

$$w_e = \frac{eE\lambda}{3T_e} \langle v \rangle = \frac{eE\lambda}{3T_e} \sqrt{\frac{8T_e}{\pi m_e}} = 0.532 \frac{eE\lambda}{\sqrt{m_e T_e}}, \qquad T_e - T = \frac{3\pi M w_e^2}{32} = \frac{Mw_e^2}{3.4} \tag{7.28}$$

in accordance with formulas (7.12). The last relation between the electron temperature and drift velocity depends weakly on the relation between the rate of electron–atom elastic collision and collision velocity. In the limit when the electron temperature is large compared to the gas temperature formulas (7.28) give

$$w_e = 0.99 \frac{(eE\lambda)^{1/2}}{(m_e M)^{1/4}}, \qquad T_e = \frac{1}{2\sqrt{3}} \sqrt{\frac{M}{m_e}} eE\lambda \,. \tag{7.29}$$

▶ **Problem 7.14** Since the cross section of electrons with ions at small collision energies exceeds significantly that for electron–atom collisions, the presence of ions in a weakly ionized gas may be remarkable in the electron mobility even at low ion number densities. Assuming that a not low number density of electrons establishes the Maxwell distribution function of electron with the electron temperature, find the electron mobility in a weakly ionized gas in a low electric field the presence of ions.

On the basis of formula (7.24) for the electron drift velocity in a weak field, we represent the rate of electron collisions in this formula in the form

$$\nu = \nu_{ea} + \nu_{ei} = N_a v \sigma_{ea}^*(v) + N_i v \sigma_{ei}^*(v) \,.$$

Here v is the electron velocity that is large compared to the atom and ion velocity, ν_{ea}, ν_{ei} are the rates of electron–atom and electron–ion collisions, N_a, N_i are the atom and ion number densities, σ_{ea}^* is the transport electron–atom cross section, and σ_{ei}^* is the transport electron–ion cross section. Below for definiteness we take the transport electron–atom cross section to be independent of an electron velocity and use formula (2.37) for the electron–ion cross section because of the Coulomb character of their interaction in a plasma with Debye screening. Substituting this expression for the rate of collisions in formula (7.24), we obtain the electron drift velocity in a weak field,

$$w_e = w_0 \varphi(\lambda N_i \frac{\pi e^4}{T_e^2} \ln \Lambda) ,$$

where w_0 is the electron drift velocity in accordance with formula (7.28), when ions are absent, $\lambda = (N_a \sigma_{ea})^{-1}$ and the function $\varphi(y)$ is given by

$$\varphi(y) = \int_0^\infty \frac{x e^{-x} dx}{1 + y/x^4}$$

Figure 7.3 shows the dependence $\varphi(y)$.

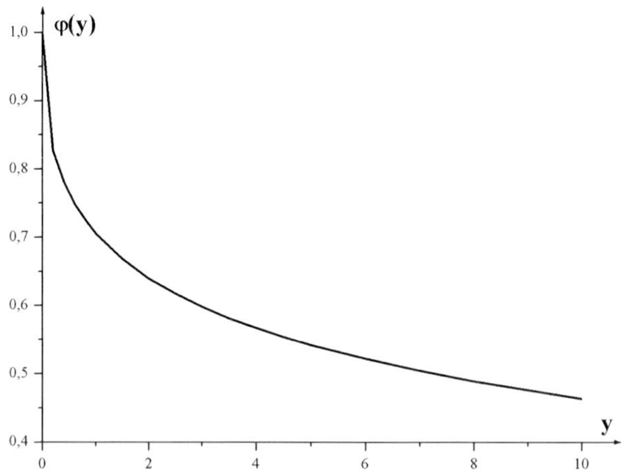

Fig. 7.3 The function $\varphi(y) = \int_0^\infty \frac{x e^{-x} dx}{1+y/x^4}$.

Note that at small y this function is $\varphi(y) = 1 - \pi\sqrt{y}/4$. Hence the electron drift velocity at small degree of gas ionization is equal to

$$w_e = w_0 \left(1 - \frac{\pi}{4} \sqrt{\frac{N_i \pi e^4 \ln \Lambda}{N_a \sigma_{ea} T_e^2}} \right) .$$

In particular, let us consider a certain case, if electrons are located in helium, $N_i/N_a = 0.01$, and $T_e = 2$ eV. Taking $\ln \Lambda = 10$, we obtain in this case ($\sigma_{ea} = 2 \cdot 10^{-15}$ cm^2) $w_e/w_0 = 0.8$.

▶ **Problem 7.15** In the case when electron–electron collisions dominate in establishment of equilibrium for electrons, moving in a gas in an external electric field, determine the rate for relaxation of the electron temperature.

If we start from the nonstationary kinetic equation, we obtain instead of formula (7.20) the following balance equation:

$$\frac{d\bar{\varepsilon}}{dt} = eEw_e - \int \frac{m_e v^2}{2} I_{ea} d\boldsymbol{v} \,. \tag{7.30}$$

This leads instead of formula (7.25) to the following for establishment of the electron average energy $\bar{\varepsilon}$:

$$\frac{d\bar{\varepsilon}}{dt} = eEw_e - \frac{m_e^2}{M}\left(1 - \frac{T}{T_e}\right)\langle v^2 \nu\rangle \,. \tag{7.31}$$

One can rewrite this equation for relaxation of the electron temperature

$$\frac{dT_e}{dt} = -\nu_\varepsilon (T_e - T_o) \,, \tag{7.32}$$

where T_o is the equilibrium electron temperature, T_e is the current electron temperature, and ν_ε is the rate of relaxation of the electron temperature. We have for this rate in the limiting cases

$$\nu_\varepsilon = 2\frac{m_e}{M}\nu, \quad \nu = \text{const} \,, \tag{7.33}$$

and

$$\nu_\varepsilon = \frac{16}{3\sqrt{\pi}} \frac{m_e}{M} \frac{1}{\lambda} \sqrt{\frac{2T_e}{m_e}}, \quad \sigma^*(v) = const \tag{7.34}$$

Thus, the typical time of variation of the electron momentum, $\tau \sim 1/\nu$ is approximately M/m_e times less than the typical time of variation of the electron energy, $\tau_\varepsilon \sim M/(m_e \nu)$.

▶ **Problem 7.16** Analyze the effect of runaway of electrons that corresponds to acceleration of fast electrons in a plasma in an external electric field because the cross section of the Coulomb scattering decreases significantly with an increase of the relative velocity.

In considering fast electrons on a tail of the distribution function, we analyze the momentum balance equation for the momentum $m_e v_x$ of a test electron when it is moving in the electrical field of strength E,

$$m_e \frac{dv_x}{dt} = eE - \frac{1}{v_x}\frac{d\varepsilon}{dt} \,.$$

Here $d\varepsilon/dt$ is the variation of the electron energy per unit time in collisions with plasma electrons. We take into account that an individual collision leads to scattering at small angles, and an individual act of collision is accompanied by a small energy variation. This derivative is equal to

$$\frac{d\varepsilon}{dt} = \int N_e v \cdot 2\pi\rho d\rho \cdot \frac{\Delta p^2}{2m_e} \,.$$

Here v is the velocity of a test electron, and $v \approx v_x$, N_e is the electron number density, ρ is the impact parameter of collision, and Δp is the momentum which is transferred from a test electron to a plasma one during their collision. According to formula (2.35) we have

$$\Delta p = \frac{2e^2}{\rho v} ,$$

which gives

$$\frac{d\varepsilon}{dt} = N_e \cdot \frac{4\pi e^4}{m_e v} \ln \Lambda .$$

Here the Coulomb logarithm $\ln \Lambda$ is given by formula (2.37). Finally, we obtain the balance equation for the momentum of a test fast electron which moves in an electric field in a plasma,

$$m_e \frac{dv_x}{dt} = eE - N_e \frac{4\pi e^4}{m_e v^2} \ln \Lambda . \tag{7.35}$$

From this balance equation it follows that fast electrons are accelerated in the electric field starting from energies

$$\varepsilon \geq \varepsilon_{cr} = N_e \frac{2\pi e^4}{eE} \ln \Lambda . \tag{7.36}$$

In particular, if the electric field strength E is measured in V/cm, the number density of electrons N_e is given in $10^{13}\ \mathrm{cm}^{-3}$, and the electron energy ε is measured in eV, the criterion (7.36) has the following form, if we take $\ln \Lambda = 10$,

$$\varepsilon_{cr} = 13 \frac{N_e}{E} . \tag{7.37}$$

7.3
Dynamics of Electrons in a Gas in Electric and in Magnetic Fields

▶ **Problem 7.17** Determine the drift velocity of electrons moving in a gas in an alternative electric field.

When an electron is moving in a gas in a harmonic electric field of strength $E \cos \omega t$ under the condition $\omega\tau \gg m_e/M$, i. e., the electron energy does not vary during the field period, one can represent the electron distribution function by analogy with formula (7.2) in the form

$$f(v, t) = f_0(v) + v_x f_1 \exp(i\omega t) + v_x f_{-1} \exp(-i\omega t) ,$$

where the axis x is directed along the field. Substituting this expansion into the kinetic equation (7.1) and separating the corresponding harmonics by the standard method, we obtain the following set of equations instead of equations (7.3):

$$\begin{aligned}\frac{a}{2}\frac{df_0}{dv} + (\nu + i\omega)v f_1 &= 0,\\ \frac{a}{2}\frac{df_0}{dv} + (\nu - i\omega)v f_1 &= 0,\\ \frac{a}{6v^2}\left[v^3\left(f_1 + f_{-1}\right)\right] &= I_{ea}(f_0)\,.\end{aligned} \tag{7.38}$$

Ignoring electron–electron collisions and using the electron–atom collision integral, one can represent the solution of the set of equations (7.38) in the form

$$f_0 = C\exp\left\{-\int_0^v \left[T + \frac{Ma^2}{6(\omega^2 + \nu^2)}\right]^{-1} m_e v dv\right\}. \tag{7.39}$$

Other harmonics of the electron distribution function follow from this on the basis of the set of equations (7.38). This gives the electron drift velocity instead of formula (7.4),

$$w_e(t) = \int v_x^2 \left[f_1\, e^{i\omega t} + f_{-1}e^{-i\omega t}\right] d\boldsymbol{v} = \frac{eE}{3m_e}\left\langle \frac{1}{v^2}\frac{d}{dv}\left(v^3 \frac{\nu\cos\omega t + \omega\sin\omega t}{\omega^2 + \nu^2}\right)\right\rangle. \tag{7.40}$$

This expression corresponds to expansion over a small parameter $m_e/(M\omega\tau) \ll 1$ that allows us ignore other terms of expansion over the spherical harmonics and time harmonics. Note that the relation between the parameters ω and ν may be arbitrary, and just the relation between these parameters determines the phase shift for a drift electron with respect to an external field. Formula (7.40) is transformed into formula (7.4) in the limit $\omega t = 0$.

▶ **Problem 7.18** Determine the difference between the electron and gas temperatures if electrons are located in a gas in an alternative electric field and electron–electron collisions dominate.

Formula (7.40) is determined by the equilibrium for the electron momentum and holds true under both criteria (7.7) and (7.18). We now consider electron motion in an atomic gas in an alternating field when the criterion (7.18) is valid, and the symmetric electron distribution function is the Maxwell one (7.19). Then formula (7.40) for the electron drift velocity takes the form

$$w_e(t) = \frac{eE}{3T_e}\left\langle v^2\left(\frac{\nu\cos\omega t + \omega\sin\omega t}{\omega^2 + \nu^2}\right)\right\rangle.$$

This formula is transformed into formula (7.24) in the limit $\omega t = 0$.

The electron temperature can be found by analogy with the case of a constant electric field (formula 7.26) on the basis of an analysis of the balance equation for the electron energy that has the form

$$\overline{eEw\cos\omega t} = \int \frac{m_e v^2}{2} I_{ea}(f_0) d\boldsymbol{v}\,,$$

where an overline denotes an average over time. Note that the expression for the electron–atom collision integral does not depend on external fields. If electron–electron collisions dominate, and f_0 is the Maxwell distribution function (7.19), on the basis of the above expression for the electron drift velocity we reduce the balance equation to the form

$$\overline{eEw\cos\omega t} = \frac{eE}{6T_e}\left\langle \frac{v^2\nu}{\omega^2+\nu^2} \right\rangle .$$

From this we obtain on the basis of the above expression for the electron drift velocity by taking into account the time averages $\overline{\cos^2\omega t} = 1/2$ and $\overline{\cos\omega t \sin\omega t} = 0$,

$$T_e - T = \frac{Ma^2}{6\langle v^2\nu\rangle}\left\langle v^2 \frac{\nu}{\omega^2+\nu^2} \right\rangle . \qquad (7.41)$$

In the limit $\omega \ll \nu$ this formula coincides with formula (7.26) if we employ in that the effective value $E/\sqrt{2}$ of the electric field strength instead of its amplitude E.

▶ **Problem 7.19** Analyze motion of electrons in a gas in mutually perpendicular constant electric and magnetic fields.

Let us take the directions of the electric and magnetic fields along the x and z, respectively. Then electrons are accelerated along the electric field, but the magnetic field compels them to rotate and in this manner stop them. In addition, a circular motion of electrons in a magnetic field creates electron currents in the direction perpendicular to the electric and magnetic fields—the Hall effect. In order to find the velocity electron distribution function, we solve the kinetic equation for electrons that has the form

$$\left(e\mathbf{E} + \frac{e}{c}[\mathbf{v}\mathbf{H}]\right)\frac{\partial f}{\partial \mathbf{v}} = I_{ea}(f),$$

where $\mathbf{E}$ is the electric field strength, and $\mathbf{H}$ is the magnetic field strength.

Using the standard method for the solution of this kinetic equation, we represent the electron distribution function in the form

$$f(\mathbf{v}) = f_0(v) + v_x f_1(v) + v_y f_2(v) .$$

Extracting spherical harmonics in the standard method, we obtain now instead of the first equation of the set (7.3) the following equations:

$$v f_1 = \frac{a\nu}{(\nu^2+\omega_H^2)}\frac{df_0}{dv}, \qquad v f_2 = \frac{a\omega_H}{(\nu^2+\omega_H^2)}\frac{df_0}{dv},$$

where $a = eE/m_e$, $\omega_H = eH/(m_e c)$ is the Larmor frequency, and $\nu = N_a v\sigma^*_{ea}$ is the rate of electron–atom collisions. These equations give the following expressions for the components of the electron drift velocity:

$$w_x = \frac{eE}{3m_e}\left\langle \frac{1}{v^2}\frac{d}{dv}\left(\frac{\nu v^3}{\nu^2+\omega_H^2}\right)\right\rangle, \qquad w_y = \frac{eE}{3m_e}\left\langle \frac{1}{v^2}\frac{d}{dv}\left(\frac{\omega_H v^3}{\nu^2+\omega_H^2}\right)\right\rangle . \qquad (7.42)$$

In the limit $\omega_H \ll \nu$ the first formula is transformed into formula (7.4).

Let us consider the case ν = const when the rate of electron–atom collisions does not depend on the collision velocity. Then formulas (7.42) can be represented in the vector form

$$w = -\frac{e\mathbf{E}\nu}{m_e(\omega_H^2 + \nu^2)} + \frac{e[\mathbf{E} \times \boldsymbol{\omega}_H]}{m_e(\omega_H^2 + \nu^2)}, \tag{7.43}$$

where the vector $\boldsymbol{\omega}_H = e\mathbf{H}/(m_e c)$ is parallel to the magnetic field.

► **Problem 7.20** Determine the difference between the electron and gas temperatures if a weakly ionized gas is located in mutually perpendicular constant electric and magnetic fields.

We assume the criterion (7.18) to be fulfilled, i. e., due to fast electron–electron collisions one can introduce the electron temperature. The balance equation for the electron energy has the form

$$eEw_x = \int \frac{m_e v^2}{2} I_{ea}(f_0) d\mathbf{v} \, .$$

Using formula (7.42) for the electron drift velocity and formula (6.59) for the electron–atom collision integral, we obtain

$$T_e - T = \frac{Ma^2}{3} \frac{\left\langle \frac{v^2\nu}{\nu^2+\omega_H^2} \right\rangle}{\langle v^2 \nu \rangle} \tag{7.44}$$

In the limiting case ν = const, this formula gives

$$T_e - T = \frac{Ma^2}{3(\nu^2 + \omega_H^2)} \, ,$$

and in the limiting case $\omega_H \gg \nu$ from formula (7.44) it follows

$$T_e - T = \frac{Ma^2}{3\omega_H^2} = \frac{Mc^2 E^2}{3H^2} \, . \tag{7.45}$$

► **Problem 7.21** Determine an increase of the electron temperature in comparison with the gas temperature if a weakly ionized gas is flowing with a velocity u in a transversal magnetic field of strength H.

Under these conditions in the frame of axes where this gas is motionless, the electric field of strength $E' = Hu/c$ occurs, where c is the light velocity. This field creates an electric current which can be used for transformation of the kinetic energy of the plasma flux into the electric energy. In this way the flow energy of a gas is converted into the electric and heat energy. Correspondingly, this process leads to a deceleration of the gas flow and to a decrease of its average velocity. Along with this, an origin of an electric field causes an increase in the electron temperature that is given by formula (7.45). From this formula it follows that the maximum increase of the electron temperature corresponds to the limit $\omega_H \gg \nu$ when formula (7.45) takes the form

$$T_e - T = \frac{Ma^2}{3\omega_H^2} = M\frac{u^2}{3} \,. \tag{7.46}$$

▶ **Problem 7.22** Find the transversal diffusion coefficient of electrons in a strong magnetic field.

A high magnetic field corresponds to the condition $\omega_H \gg \nu$, where $\omega_H = eH/(m_e c)$ is the Larmor electron frequency and ν is the electron–atom collision rate. The electric field, if it exists, directs along the magnetic field. Under these conditions, the projection of the electron trajectory on a plane perpendicular to the magnetic field consists of circles whose centers and radii vary after each collision. By definition, the diffusion coefficient in the transversal direction is given by $D_\perp = \langle x^2 \rangle /t$, where $\langle x^2 \rangle$ is the square of the displacement after time t in the direction x perpendicular to the field.

We have $x - x_o = r_H \cos \omega_H t$, where x_o is the x-coordinate of the center of the electron rotational motion, and $r_H = v_\rho/\omega_H$ is the Larmor radius; and v_ρ is the electron velocity in the direction perpendicular to the field. From this it follows that

$$\left\langle x^2 \right\rangle = n\left\langle (x - x_o)^2 \right\rangle = \frac{n v_\rho^2}{2\omega_H^2} \,,$$

where n is the number of collisions. Since $t = n/\nu$, where ν is the rate of electron–atom collisions, we obtain

$$D_\perp = \left\langle \frac{v_\rho^2 \nu}{2\omega_H^2} \right\rangle = \left\langle \frac{v^2 \nu}{3\omega_H^2} \right\rangle , \qquad \omega_H \gg \nu \,, \tag{7.47}$$

where brackets mean averaging over electron velocities. Combining formula (7.47) with the expression for the diffusion coefficient of electrons (7.23) in the absence of a magnetic field, we find for the transverse diffusion coefficient of electrons in a gas at an arbitrary relation between the Larmor frequency and the rate of electron–atom collisions

$$D_\perp = \frac{1}{3}\left\langle \frac{v^2 \nu}{\omega_H^2 + \nu^2} \right\rangle \,. \tag{7.48}$$

▶ **Problem 7.23** Find the drift velocity of an electron that is moving in a gas in an external electric and magnetic field within the framework of tau-approximation.

The kinetic equation for electrons in terms of tau-approximation (6.3) has the form

$$\frac{\partial f}{\partial t} + \frac{\mathbf{F}}{m_e} \cdot \frac{\partial f}{\partial \mathbf{v}} = -\frac{f - f_0}{\tau} \,.$$

Here $\mathbf{F}$ is a force acting on the electron from external fields, m_e is the electron mass, $1/\tau = N_a v \sigma_{ea}$ is the frequency of electron–atom collisions, N_a is the number density of atoms, and σ_{ea} is the cross section for electron–atom collisions. For simplicity, we assume τ to be independent of the collision velocity. Note that in this consideration we neglect electron–electron collisions, i. e., it relates to one electron moving in a gas.

We take a general form for an external force $\boldsymbol{F}$ that allows us to analyze a variety of aspects of the electron behavior and has the form

$$\boldsymbol{F} = -e\boldsymbol{E}\exp(-i\omega t) - \frac{e}{c}\left[\boldsymbol{v} \times \boldsymbol{H}\right] ,$$

where $\boldsymbol{E}$ and $\boldsymbol{H}$ are the electric and magnetic field strengths, ω is the frequency of the electric field, and $\boldsymbol{v}$ is the electron velocity. We assume the magnetic field to be constant, and the electric field to be harmonic. We define a coordinate system such that the vector $\boldsymbol{H}$ is directed along the z-axis and the vector $\boldsymbol{E}$ lies in the plane xz.

Let us multiply the above kinetic equation by $m_e\boldsymbol{v}$ and integrate over the electron velocities, which finally gives the equation of electron motion in the form

$$m_e\frac{d\boldsymbol{w}_e}{dt} + m_e\frac{\boldsymbol{w}_e}{\tau} = -e\boldsymbol{E}\exp\left(-i\omega t\right) - \frac{e}{c}\left[\boldsymbol{w}_e \times \boldsymbol{H}\right] , \tag{7.49}$$

where $\boldsymbol{w}_e$ is the electron drift velocity. From this equation it follows that electron–atom collisions create a frictional force $m_e\boldsymbol{w}_e/\tau$. Thus, this problem is reduced to the problem of motion of an individual electron in external fields in a friction matter.

Equation (7.49) is separated into three scalar equations having the form

$$\frac{dw_x}{dt} + \frac{w_x}{\tau} = a_x e^{i\omega t} + \omega_H w_y, \quad \frac{dw_y}{dt} + \frac{w_y}{\tau} = -\omega_H w_x, \quad \frac{dw_z}{dt} + \frac{w_z}{\tau} = a_z e^{i\omega t} .$$

Here $\omega_H = eH/(m_e c)$ is the Larmor frequency, $a_x = -eE_x/m_e$ and $a_z = -eE_z/m_e$. The stationary solution of these equations for the components of the electron drift velocity is

$$\begin{aligned} w_x &= \frac{\tau(1 + i\omega\tau)a_x e^{i\omega t}}{1 + (\omega_H^2 - \omega^2)\tau^2 + 2i\omega\tau}, \\ w_y &= \frac{\omega_H \tau^2 a_x e^{i\omega t}}{1 + (\omega_H^2 - \omega^2)\tau^2 + 2i\omega\tau}, \\ w_z &= \frac{\tau a_z e^{i\omega t}}{1 + i\omega\tau} . \end{aligned} \tag{7.50}$$

▶ **Problem 7.24** Find the drift of electrons in crossed electric and magnetic constant fields.

Let the electric field be directed along the x-axis, and the magnetic field be directed along the z-axis. According to formula (7.50), we have ($\omega = 0$, $\omega_H\tau \gg 1$)

$$w_y = -\frac{eE_x}{m_e\omega_H} = -c\frac{E_x}{H} , \quad w_x = \frac{w_y}{\omega_H\tau} . \tag{7.51}$$

As is seen, an electron (or a charged particle) in a crossed field is moving in the direction perpendicular to field directions, and the drift velocity is proportional to the electric field strength and inversely proportional to the magnetic field strength. Finally, the drift velocity $\boldsymbol{w}$ of a charged particle in the crossed electric $\boldsymbol{E}$ and magnetic $\boldsymbol{H}$ fields is equal to

$$\boldsymbol{w} = c\frac{[\boldsymbol{E} \times \boldsymbol{H}]}{H^2} . \tag{7.52}$$

▶ **Problem 7.25** Prove that the magnetic moment of a charged particle is conserved in the course of its motion in a nonuniform magnetic field. The magnetic moment of a charged particle μ is the ratio of the particle kinetic energy along the magnetic field $mv_\tau^2/2$ to the magnetic field strength H.

Assuming that the presence of charged particles does not influence the space distribution of the magnetic field, we have from the equation div $\boldsymbol{H} = 0$, when in the case of the axial symmetry for the magnetic field,

$$\frac{1}{\rho}\frac{\partial}{\partial\rho}\left(\rho H_\rho\right) + \frac{\partial H_z}{\partial z} = 0 .$$

Because of the axial symmetry, on the axis, H_ρ is zero. According to the above equation near the axis we have

$$H_\rho = -\frac{\rho}{2}\left(\frac{\partial H_z}{\partial z}\right)_{\rho=0} .$$

The force along the symmetry axis that acts on a moving charged particle from the magnetic field is

$$F_z = -\frac{e}{c}v_\tau H_\rho = \frac{v_\tau \rho}{2c}\frac{\partial H_z}{\partial z} . \tag{7.53}$$

Let us analyze these equations from the standpoint of the particle magnetic moment μ, which is defined as $\mu = IS/c$, where I is the particle current, S is the area enclosed by its trajectory, and c is the velocity of light. In this case we have $I = e\omega_H/(2\pi)$, and $S = \pi r_L^2$, where $r_L = v_\tau/\omega_H$ is the particle Larmor radius, and the Larmor frequency is $\omega_H = eH/(mc)$. In terms of the particle magnetic moment μ we have for the force acting on the particle in the magnetic field

$$F_z = -\mu\frac{\partial H_z}{\partial z} . \tag{7.54}$$

The minus sign indicates the force is in the direction of decreasing magnetic field.

One can prove that the magnetic moment of the particle is an integral of the motion; that is, it is a conserved quantity. We can analyze the particle motion along a magnetic line of force when averaged over gyrations. The equation of motion along the magnetic field gives $m dv_z/dt = F_z = -\mu dH_z/dz$, and since $v_z = dz/dt$, it follows from this that $d(mv_z^2/2) = -\mu dH_z$. From the law of conservation of the particle energy we have

$$\frac{d}{dt}\left(\frac{mv_\tau^2}{2} + \frac{mv_z^2}{2}\right) = \frac{d}{dt}\left(\mu H_z + \frac{mv_z^2}{2}\right)$$

$$\frac{d}{dt}(\mu H_z) - \mu\frac{dH_z}{dt} = H_z\frac{d\mu}{dt} = 0\,,$$

and the latter equation gives for the magnetic momentum of the particle motion

$$\frac{d\mu}{dt} = 0\,, \tag{7.55}$$

so the magnetic moment is conserved during the motion of the particle.

► **Problem 7.26** An electron is moving along an axial weakly varied magnetic field in the direction of increasing magnetic field. Find the character of variation of the kinetic energy ε_τ of transversal motion with variation of the magnetic field.

Let the magnetic field be directed along the z-axis and because of the equation div $\boldsymbol{H} = 0$ the radial component of the magnetic field occurs that follows from this equation,

$$\frac{1}{\rho}\frac{\partial(\rho H_\rho)}{\partial\rho} + \frac{\partial H_z}{\partial z} = 0\,.$$

From this it follows that

$$H_\rho = -\frac{\rho}{2}\frac{\partial H_z}{\partial z}\,, \tag{7.56}$$

and the transversal component of the magnetic field is small compared to the longitudinal one ($H_\rho \ll H_z$). The equation of motion along the axis z is

$$m_e\frac{dv_z}{dt} = \frac{e}{c}[\boldsymbol{v}\times\boldsymbol{H}]\,.$$

Under these conditions, the transversal electron motion proceeds along circles in the plane xy, and the force (7.53) acts on the electron in the longitudinal direction, so the electron motion in this direction is described by the equation

$$m_e\frac{dv_z}{dt} = F_z\,.$$

From this we find the longitudinal force (7.53) on the basis of formula (7.56)

$$F_z = \frac{v_\tau\rho}{2c}\frac{\partial H_z}{\partial z} = -\frac{e}{2c}\frac{v_\tau^2}{\omega_H}\frac{\partial H_z}{\partial z} = \frac{\varepsilon_\tau}{H}\frac{\partial H_z}{\partial z}$$

where v_τ is the transversal electron velocity and ε_τ is the transversal part of the kinetic electron energy. Multiplying the motion equation by v_z, we reduce it to the form

$$\frac{d\varepsilon_z}{dt} = -\frac{\varepsilon_\tau}{H}\frac{dH_z}{dt}\,.$$

Because of the conservation of the total electron kinetic energy $\varepsilon_z + \varepsilon_\tau = \text{const}$, solution of this equation is

$$\frac{\varepsilon_\tau}{H} = \text{const} \tag{7.57}$$

Thus, the magnetic field gradient leads to redistribution between the parts of the electron kinetic energy in the longitudinal and transversal directions, so that if the electron is moving in the direction of a magnetic field increase, the Larmor radius increases in the course of this motion, as is shown in Fig. 7.4. Note that relation (7.57) is analogous to equation (7.55). Indeed, the magnetic moment of an electron that is moving along a circular trajectory is equal to

$$\mu = \frac{IS}{c} = \frac{\varepsilon_\tau}{H},$$

where the electric current is $I = e\omega_H/(2\pi)$, and the square area for this current is $S = \pi r_H^2$, r_H is the Larmor radius. Therefore, relation (7.57) corresponds to the conservation of the electron magnetic moment (7.55) in the course of redistribution between the transversal and longitudinal electron kinetic energies.

▶ **Problem 7.27** Analyze the drift of a charged particle in a magnetic field that varies weakly in a space.

Let us take the magnetic field direction to be z and the gradient magnetic field direction to be y. Then a charged particle rotates in the plane xy. If the magnetic field decreases along y and a particle moves clockwise, the upper Larmor radius of particle rotation is larger than the lower one. As a result the particle shifts in the x-direction (see Fig. 7.5).

We find the drift velocity on the basis of formula (7.52) where the force acted per charged particle is given by formula (7.54). The magnetic moment for a rotating

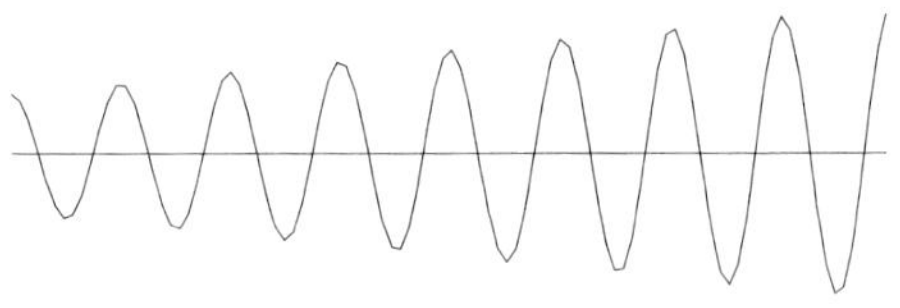

Fig. 7.4 Trajectory of a charged particle moved in an slowly varied magnetic field along the magnetic field (the axis z) in the direction of its decrease.

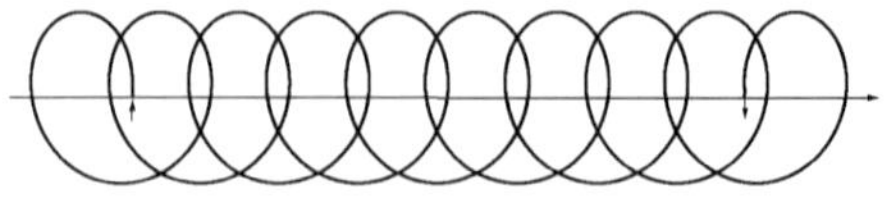

Fig. 7.5 Drift of a charged particle in an slowly varied magnetic field that proceeds in the direction of the magnetic field gradient.

particle of charge e is

$$\mu = \frac{1}{c}\pi r_L^2 e\omega_H ,$$

and according to formula (7.52) the drift velocity of a charged particle in a varied magnetic field is

$$w = \frac{v^2}{2\omega_H}\frac{[\nabla H \times \boldsymbol{H}]}{H^2} , \tag{7.58}$$

where $v = r_L\omega_H$ is the particle velocity on the Larmor orbit.

▶ **Problem 7.28** The behavior of magnetic lines of force in the magnetic trap is given in Fig. 7.6. Find the angle of motion of a charged particle in the minimum magnetic field region when it is captured by the magnetic trap.

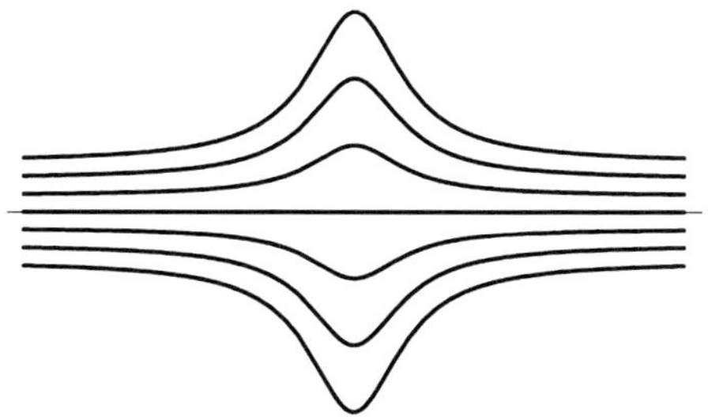

Fig. 7.6 Magnetic lines of force in a simple magnetic trap with the axial symmetry. Charged particles may be closed in this trap.

When the magnetic field increases, the transversal part of the particle kinetic energy increases. If the longitudinal part of the particle kinetic energy becomes zero, the particle reflects from this point and returns back to a region of less fields. Let us find conditions when it is fulfilled. Let us denote by $H_{\min}$ the minimal magnetic field in the trap, and by ε_z the longitudinal kinetic energy of a charged particle, and by θ the angle between the particle velocity in a region of the minimum magnetic field and the magnetic field direction z, so $\varepsilon_z = \varepsilon\cos^2\theta$, where ε is the total kinetic energy of this particle.

According to formula (7.57) the longitudinal part of the particle kinetic energy ε'_z in a region with a magnetic field H is

$$\varepsilon'_z = \varepsilon_z - (\varepsilon - \varepsilon_z)\frac{H}{H_{\min}} .$$

From this it follows that the condition $\varepsilon'_z > 0$ is violated in the region of strong magnetic fields if the angle θ exceeds θ_m, which is given by the relation

$$\sin^2\theta_m = \frac{H_{\min}}{H_{\max}} . \tag{7.59}$$

Then the charged particle is reflected from the region of a strong magnetic field and is trapped in a bounded space. This formula is a basis of various magnetic traps and relates for both positively and negatively charged particles, i. e., both electrons and positive ions. The particle trajectory is a helix wound around a magnetic line of force.

In particular, radiation belts of the Earth act according to this principle. Fast electrons and protons can be captured in a region of a weak magnetic field of the Earth and are reflected in vicinity of the Earth where the magnetic field is more strong. Collisions of these particles with others allow them to escape from the magnetic trap. The number of captured protons is greater than that of electrons because protons have a longer lifetime due to a larger mass. In addition, due to the influence of the solar wind, this magnetic trap acts more effectively on the side of the Earth opposite to the Sun.

7.4 Conductivity of a Weakly Ionized Gas

▶ **Problem 7.29** Determine the of a weakly ionized gas located in mutually perpendicular constant electric and magnetic fields.

In the absence of a magnetic field, the plasma conductivity is scalar. The presence of a magnetic field transforms the conductivity of a weakly ionized gas into a tensor. We use Ohm's law that takes the form

$$i_\alpha = \Sigma_{\alpha\beta} E_\beta \,, \tag{7.60}$$

where i_α is a component of the current density, and the summation takes place over twice repeating indices. In the case when the collision rate ν does not depend on the electron velocity, components of the are given by

$$\Sigma_{xx} = \Sigma_{yy} = \Sigma_0 \cdot \frac{1}{1+\omega_H^2 \tau^2}\,, \qquad \Sigma_{yx} = -\Sigma_{xy} = \Sigma_0 \cdot \frac{\omega_H \tau}{1+\omega_H^2 \tau^2}\,, \tag{7.61}$$

where $\tau = 1/\nu$, $\Sigma_0 = N_e e^2 m_e/\tau$.

In the limiting case $\omega_H \tau \gg 1$ the electric current is directed perpendicular to both the electric and magnetic fields. In this case the plasma conductivity and electric current do not depend on the collision rate because the change of the electron motion direction is determined by electron rotation in a magnetic field. Then we have

$$i_y = ecN_e \frac{E_x}{H} = \frac{E_x}{R_H H}\,,$$

where $R_H = 1/(ecN_e)$ is the Hall constant.

▶ **Problem 7.30** Show that the conductivity of a weakly ionized gas is determined mostly by electron transport.

The conductivity of a weakly ionized gas, Σ, is created by free electrons and is defined as the proportionality factor between the electric current density $\boldsymbol{i}$ and the electric field strength $\boldsymbol{E}$ in the Ohm law (7.60). The electric current is a sum of two components, the electron and ion current,

$$\boldsymbol{i} = -eN_e\boldsymbol{w}_e + eN_i\boldsymbol{w}_i ,$$

where N_e, N_i are the electron and ion number density, and $\boldsymbol{w}_e, \boldsymbol{w}_i$ are the electron and ion drift velocities, respectively, which are expressed through the electron and ion mobilities K_e, K_i,

$$\boldsymbol{w}_e = K_e\boldsymbol{E} , \quad \boldsymbol{w}_i = K_i\boldsymbol{E} ,$$

which gives for the conductivity of a quasineutral ionized gas

$$\Sigma = e(K_e + K_i) .$$

Estimating the mobility of a charged particle in a gas, we have

$$K \sim \frac{e}{N_a\sigma\sqrt{\mu T}} ,$$

where N_a is the number density of atoms, σ is the typical cross section of collision of charged particles with atoms, and the temperature T is their typical relative energy in such collisions. From this it follows that $K_e \gg K_e$, i. e., electrons give the main contribution to the gas conductivity. This gives the following estimation for the gas conductivity:

$$\Sigma \sim \frac{N_e e^2}{N_a\sigma_{ea}^*\sqrt{m_e T_e}} ,$$

where σ_{ea} is a typical cross section of electron–atom scattering. Next, from formula (7.4) we obtain the plasma conductivity, if it is determined by electron–atom collisions,

$$\Sigma = \frac{N_e e^2}{3m_e}\left\langle \frac{1}{v^2}\frac{d}{dv}\left(\frac{v^3}{\nu}\right)\right\rangle , \tag{7.62}$$

where $\nu = N_a v\sigma_{ea}^*$ is the rate of electron–atom collisions. In particular, introducing a collision time $\tau = 1/\nu$ and assuming it to be independent of the collision velocity v, we obtain the plasma conductivity in the traditional form

$$\Sigma_0 = \frac{N_e e^2\tau}{m_e} . \tag{7.63}$$

▶ **Problem 7.31** Find the conductivity of an ionized gas at high ionization degrees.

Note that electron–electron collisions do not change the total electron momentum and therefore do not influence the plasma conductivity. Hence, for the determination of the plasma conductivity of a strongly ionized plasma, where electron–ion collisions are prevailed over electron–atom collisions, it is necessary to use the rate ν_{ei} of electron–ion collisions in formula (7.62) instead of electron–atom collisions, and this formula takes the form

$$\Sigma = \frac{N_e e^2}{3m_e} \left\langle \frac{1}{v^2} \frac{d}{dv} \left(\frac{v^3}{\nu_{ei}} \right) \right\rangle . \tag{7.64}$$

Here $\nu_{ei} = N_i v \sigma^*$, where v is the electron velocity, and the transport cross section σ^* of electron–ion scattering is given by formula (2.37),

$$\sigma^* = \frac{\pi e^4 \ln \Lambda}{\varepsilon^2} ,$$

where ε is the electron energy, and the Coulomb logarithm is equal to $\ln \Lambda = \ln[e^2/(r_D T)]$, and r_D is the Debye-Hückel radius. Taking the Maxwell distribution function for electrons considering $\nu_{ei} \sim v^{-3}$, and $N_e = N_i$ we finally obtain the Spitzer formula for the plasma conductivity,

$$\Sigma = \frac{2^{5/2} T_e^{3/2}}{\pi^{3/2} m_e^{1/2} e^2 \ln \Lambda} . \tag{7.65}$$

▶ **Problem 7.32** Within the framework of tau-approximation for electron–atom collisions determine the conductivity of a weakly ionized gas with a small electron number density where collisions between electrons are negligible. This gas is located in external electric and magnetic fields.

We consider the Ohm law for the electric current that is determined by electrons,

$$j_\alpha = \Sigma_{\alpha\beta} E_\beta ,$$

where j_α is the current component. On the other hand, one can use the direct expression for the electron current

$$j_\alpha = -e N_e w_\alpha ,$$

where the drift velocity is given by expressions (7.50). In particular, if an electron is moving in a constant electric field, these formulas give for the conductivity of this plasma

$$\Sigma_{\alpha\beta} = \Sigma_0 \delta_{\alpha\beta}, \qquad \Sigma_0 = \frac{N_e e^2 \tau}{m_e} , \tag{7.66}$$

where N_e is the electron number density.

In a general case, directing the magnetic field along the z-axis, we obtain the components of the tensor of the reduced plasma conductivity $\sigma_{\alpha\beta} = \Sigma_{\alpha\beta}/\Sigma_o$,

$$\begin{aligned} \sigma_{xx} &= \sigma_{yy} = \frac{1 + i\omega\tau}{1 + (\omega_H^2 - \omega^2)\tau^2 + 2i\omega\tau} , \\ \sigma_{xy} &= \sigma_{yx} = \frac{\omega_H \tau}{1 + (\omega_H^2 - \omega^2)\tau^2 + 2i\omega\tau} , \\ \sigma_{zz} &= \frac{1}{1 + i\omega\tau} . \end{aligned} \tag{7.67}$$

In addition to this we have

$$\sigma_{xz} = \sigma_{yz} = \sigma_{zx} = \sigma_{zy} = 0 \, .$$

▶ **Problem 7.33** Establish the relation between the dielectric constant tensor for a weakly ionized gas located in external fields and the conductivity tensor.

The dielectric tensor of a plasma $\varepsilon_{\alpha\beta}$ is defined on the basis of the relation

$$D_\alpha = \varepsilon_{\alpha\beta} E_\beta , \quad D_\alpha = E_\alpha + 4\pi P_\alpha , \tag{7.68}$$

where $\boldsymbol{D}$ is the electric displacement vector and $\boldsymbol{P}$ is the polarization per unit volume of the plasma, i. e., $\boldsymbol{P}$ is the dipole moment per plasma unit volume, and this moment is generated by an external electric field. In order to extract an electron motion that is induced by the action of an external field, we represent the time dependence of the electron coordinate as $\boldsymbol{r} = \boldsymbol{r}_0 + \boldsymbol{r}' \exp(i\omega t)$, where $\boldsymbol{r}_0$ is independent of an external field and $\boldsymbol{r}'$ is determined by electron motion under the field action. Hence, the electron velocity induced by the external field is

$$\boldsymbol{w}_\mathrm{e} = \frac{d\boldsymbol{r}}{dt} = i\omega \boldsymbol{r}' \exp(i\omega t) \, .$$

The plasma polarization is

$$\boldsymbol{P} = -e \sum_k \boldsymbol{r}'_k \exp(i\omega t) = -i N_\mathrm{e} \boldsymbol{w}_\mathrm{e} / \omega \, ,$$

where the index k denotes an individual electron, and the sum is taken over electrons per unit plasma. Expressing the electron current through an average electron drift velocity $\boldsymbol{i} = -e N_\mathrm{e} \boldsymbol{w}_\mathrm{e}$, we have $\boldsymbol{D} = \boldsymbol{E} + 4\pi i \boldsymbol{i}/\omega$. From this it follows for the dielectric constant tensor that

$$\varepsilon_{\alpha\beta} = \delta_{\alpha\beta} + \frac{4\pi i}{\omega} \Sigma_{\alpha\beta} \, , \tag{7.69}$$

where $\delta_{\alpha\beta}$ is the Kronecker symbol. This relation establishes the connection between the dielectric constant tensor of a plasma and its conductivity tensor.

▶ **Problem 7.34** Determine the dielectric constant tensor for a weakly ionized isotropic gas located in external fields within the framework of the tau-approximation for electron–atom collisions.

We are based on formula (7.69), which allows us to express the dielectric constant through the conductivity of an ionized gas that according formulas (7.66) and (7.67) is given by

$$\Sigma = \frac{N_\mathrm{e} e^2 \tau}{m_\mathrm{e}} \frac{1}{1 + i\omega\tau} \, .$$

This formula describes the case when the magnetic field is absent; N_e is the electron number density, and ω is the frequency of an electric field. From this we obtain the dielectric constant of an ionized gas,

$$\varepsilon = 1 - \frac{i\omega_{\mathrm{p}}^2 \tau}{\omega(1 + i\omega\tau)}, \tag{7.70}$$

and $\omega_{\mathrm{p}} = \sqrt{4\pi N_{\mathrm{e}} e^2 / m_{\mathrm{e}}}$ is the plasma frequency. In the limit of high frequencies $\omega\tau \gg 1$ this formula gives

$$\varepsilon = 1 - \frac{\omega_{\mathrm{p}}^2}{\omega^2}. \tag{7.71}$$

▶ **Problem 7.35** Determine the conductivity of an ionized gas near the cyclotron resonance for electrons moved in a gas located in a magnetic field and alternating electric field that is perpendicular to the magnetic field, and the frequency of an electric field is close to the Larmor frequency.

According to formula (7.67) the conductivities $\Sigma_{\parallel}$ and $\Sigma_{\perp}$ in the directions parallel and perpendicular to the magnetic field are

$$\Sigma_{\parallel} \equiv \Sigma_{xx} = \frac{\Sigma_0(1 + i\omega\tau)}{1 + (\omega_H^2 - \omega^2)\tau^2 + 2i\omega\tau}, \quad \Sigma_{\perp} \equiv \Sigma_{xy} = \frac{\Sigma_0 \omega_H \tau}{1 + (\omega_H^2 - \omega^2)\tau^2 + 2i\omega\tau}. \tag{7.72}$$

One can see a resonance at $\omega = \omega_H$, where these components of the conductivity tensor are $\Sigma_{\parallel} = i\Sigma_{\perp} = \Sigma_0/2$, and the cyclotron resonance width is $\Delta\omega \sim 1/\tau$.

The cyclotron resonance has a simple physical explanation. In the absence of an electric field, an electron travels in a circular orbit with the Larmor frequency ω_H. If an electric field is applied in the plane of the circular orbit, and if this field varies such that its direction remains parallel to the electron velocity, the electron continuously receives energy from the field. Then like the motion in a constant electric field, the electron is accelerated until it collides with atoms. Hence, in both cases the conductivities are of the same order of magnitude and are expressed in terms of the rate $1/\tau$ of electron–atom collisions. If the field frequency ω differs from the cyclotron frequency ω_H, the conductivity is considerably lower, since the conditions of the interaction between the electron and field are not optimal.

▶ **Problem 7.36** Find the power of absorbed energy in a weakly ionized plasma located in an alternating electric and constant magnetic fields. Show that the conditions of the cyclotron resonance corresponds to the maximum absorption.

If a plasma is found in an external field, it transforms the field energy into heat, and the power of this process per unit plasma volume is equal to $p = \boldsymbol{j} \cdot \boldsymbol{E}$. Taking the x-axis as a direction of the electric field, we obtain the specific power absorbed by the plasma,

$$p = (\Sigma_{xx} + \Sigma_{xx}^*)\frac{E^2}{2}.$$

Taking the magnetic field along the axis z, we obtain the specific absorbed power on the basis of formula (7.72),

$$p = \frac{p_0}{2}\left[\frac{1 - i\omega\tau}{1 + (\omega_H^2 - \omega^2)\tau^2 - 2i\omega\tau} + \frac{1 + i\omega\tau}{1 + (\omega_H^2 - \omega^2)\tau^2 + 2i\omega\tau}\right] , \tag{7.73}$$

where $p_0 = \Sigma_0 E^2$ is the specific absorbed power for the constant electric field. Note that absorption of energy by a plasma results from electron–atom collisions that lead to transfer of energy to atoms obtained by the electrons from an external field.

Let us consider the limiting cases of formula (7.73). In the absence of a magnetic field ($\omega_H = 0$) it gives

$$p = p_0 \frac{1}{1 + \omega^2\tau^2} .$$

From formula (7.73) it follows for a constant electric field ($\omega = 0$)

$$p = p_0 \frac{1}{1 + \omega_H^2\tau^2} .$$

In the region of the cyclotron resonance, with $\omega\tau \gg 1$, $\omega_H\tau \gg 1$, and $|(\omega - \omega_H)\tau| \sim 1$, formula (7.73) yields

$$p = \frac{p_0}{2} \frac{1}{1 + (\omega_H - \omega)^2\tau^2} \tag{7.74}$$

One can see that the resonant absorbed power is less than half of what is absorbed in a constant electric field. The fact that the same order of magnitude is obtained for these values is explained by the related character of the electron motion in these cases.

7.5 Thermal Conductivity and Thermal Diffusion of Electrons in a Gas

► **Problem 7.37** Determine the electron thermal conductivity of a weakly ionized gas if the criterion (7.18) holds true, which allows us to introduce the electron temperature T_e, and a weak gradient of the electron temperature ∇T_e exists in a space that causes a thermal flux.

Because of a small mass of electrons, their transport can give a contribution to the thermal conductivity of a weakly ionized gas. We introduce the thermal conductivity coefficient κ_e due to electrons on the basis of the formula

$$q_e = -\kappa_e \nabla T_e , \tag{7.75}$$

and below we evaluate the coefficient of the thermal conductivity of electrons.

In this case the velocity electron distribution function at each space point is the Maxwell one $\varphi(v)$ and is given by formula (7.19). By taking into account the electron temperature gradient, this distribution function can be represented in the standard form

$$f(\boldsymbol{v}) = \varphi(v) + (\boldsymbol{v}\nabla \ln T_e) f_1(v) \,.$$

Then the kinetic equation $\boldsymbol{v}\nabla f = I(f)$ takes the following form:

$$\varphi(v)\left(\frac{m_e v^2}{2T_e} - \frac{5}{2}\right)\boldsymbol{v}\nabla T_e = I_{ea}(f)$$

Here we take the direction of the temperature gradient along the x-axis and assume the electron pressure $p_e = N_e T_e$ to be constant in a space. From this on the basis of the formula $I_{ea}(v_x f_1) = -\nu v_x f_1$ we obtain the nonsymmetric part of the distribution function,

$$f_1(v) = -\frac{\varphi(v)}{\nu}\left(\frac{m_e v^2}{2T_e} - \frac{5}{2}\right) .$$

Let us determine the electron heat flux that is given by

$$q_e = \int \frac{m_e v^2}{2} v_x f(\boldsymbol{v}) d\boldsymbol{v} = \int \frac{m_e v^2}{2} v_x^2 \nabla \ln T_e f_1(v) d\boldsymbol{v} \,.$$

Then on the basis of formula (7.75) we obtain the thermal conductivity coefficient of electrons

$$\kappa_e = N_e \left\langle \frac{v^2}{3\nu}\frac{m_e v^2}{2T_e}\left(\frac{m_e v^2}{2T_e} - \frac{5}{2}\right)\right\rangle , \qquad (7.76)$$

where brackets mean averaging over the Maxwell electron distribution function.

Assuming $\nu \sim v^n$, i. e., $\nu(v) = \nu_0 z^{n/2}$, where $z = m_e v^2/(2T_e)$, we have from formula (7.76)

$$\kappa_e = \frac{4}{3\sqrt{\pi}}\frac{T_e N_e}{\nu_0 m_e}\left(1 - \frac{n}{2}\right)\Gamma\left(\frac{7-n}{2}\right)$$

In particular, if $\nu = \text{const}$, this formula gives

$$\kappa_e = \frac{5 T_e N_e}{2\nu_0 m_e}$$

If $n = 1$, i. e., $\nu = v/\lambda$ (λ is the mean free path), we have from the above formula

$$\kappa_e = \frac{2}{3\sqrt{\pi}} N_e \lambda \sqrt{\frac{2T_e}{m_e}} \,.$$

▶ **Problem 7.38** Connect the coefficient of the electron thermal conductivity of a weakly ionized gas with the total coefficient of the thermal conductivity of this gas.

In order to determine the contribution of electron transport to the electron thermal conductivity, it is necessary to reduce the heat flux due to electrons to the gradient of the gas temperature. For this goal we must connect the gradients of the electron T_e and gas T temperatures that follows from the relation (7.26). Take for definiteness $T_e \gg T$ and use the velocity dependence $d \ln \nu dv$ for the rate ν of electron–atom collisions. This gives from formula (7.26) $\nabla T_e = \nabla T/(1+n)$ and leads to the following expression for the total thermal conductivity coefficient of a weakly ionized gas:

$$\kappa = \kappa_a + \kappa_e \frac{\nabla T_e}{\nabla T} = \kappa_a + \frac{\kappa_e}{1+n}, \qquad (7.77)$$

where κ_a is the thermal conductivity coefficient of the atomic gas. Estimating the contribution of electrons to the thermal conductivity of a weakly ionized gas, one can find this contribution to be remarkable at low values of N_e/N_a due to a low electron mass in comparison with the atomic one.

▶ **Problem 7.39** Find the thermodiffusion coefficient for a weakly ionized gas.

In electron transport in a weakly ionized gas cross-fluxes may be of importance. In particular, an external electric field and the gradient of the electron temperature cause both the electron flux $\boldsymbol{j}$ and the heat flux $\boldsymbol{q}$, which are given by the equations

$$\boldsymbol{j} = N_e K \boldsymbol{E} - D_T N \nabla \ln T_e , \qquad \boldsymbol{q} = -\kappa_e \boldsymbol{q} \nabla T_e + \alpha e \boldsymbol{E} . \qquad (7.78)$$

In particular, the electron flux under a gradient of the electron temperature, which will be analyzed below, is equal to

$$j_e = -D_T N \nabla \ln T_e , \qquad (7.79)$$

where D_T is the thermodiffusion coefficient, and we will evaluate this value in the case when the criterion (7.18) holds true, which is the electron distribution function (7.19) that is characterized by the electron temperature T_e. Correspondingly, the variation of the electron distribution function owing to the electron temperature gradient is described by the following equation for the asymmetric part $f_1(v)$ of the distribution function:

$$v_x \frac{\partial f_0}{\partial x} = -\nu v_x f_1,$$

where ν is the rate of electron–atom elastic collisions. This gives for the flux of electrons (the temperature gradient is directed along the x-axis)

$$j_x = \int v_x f dv = \int v_x^2 f_1 dv = -\frac{1}{3}\int \frac{v^2}{\nu}\frac{\partial f_0}{\partial x} dv = -\frac{d}{dx}\left[N_e \left\langle \frac{v^2}{3\nu} \right\rangle\right],$$

where brackets mean an average over electron velocities. Since the x-dependence occurs due to a gradient of the electron temperature, we obtain from this formula

$$j_x = -\nabla T_e \frac{d}{dT_e}\left(N_e \left\langle \frac{v^2}{3\nu} \right\rangle\right)$$

Comparing this expression with formula (7.79), we find for the thermodiffusion coefficient,

$$D_T = T_e \frac{d}{dT_e}\left[\frac{N_e}{N}\langle\frac{v^2}{3\nu}\rangle\right] = T_e \frac{d}{dT_e}\left(\frac{N_e}{N}D_e\right),$$

where the diffusion coefficient of electrons in a gas D_e is given by formula (7.23). If the electron pressure $p_e = N_e T_e$ is constant, this expression takes the form

$$D_T = T_e^2 \frac{N_e}{N}\frac{d(D_e/T_e)}{dT_e}. \tag{7.80}$$

In particular, if $\nu = \text{const}$, this formula gives $D_T = 0$. In the case of the power velocity dependence for the rate of electron–atom collisions $\nu \sim v^n$, we have

$$D_T = -n\frac{N_e}{N}D,$$

and the direction of the electron flux with respect to the temperature gradient depends on the sign of n.

▶ **Problem 7.40** Determine the cross-flux coefficient α for the heat flux under the action of an electric field.

An external electric field acted on a weakly ionized gas creates an electric current and also a heat flux. Since it is determined by an asymmetric part of the electron distribution function that is given by the set of equations (7.3), we obtain this distribution function part $f_1 = eEf_o/(\nu T_e)$, and the coefficient α of the set (7.78) of equations is equal to

$$\alpha = \frac{m_e N_e}{6T_e}\cdot\left\langle\frac{v^4}{\nu}\right\rangle = \frac{4T_e N_e}{3\sqrt{\pi}m_e\nu_o}\Gamma\left(\frac{7}{2}-\frac{n}{2}\right),$$

where we take as usually $\nu = \nu_o\left(v/\sqrt{\frac{2T_e}{m_e}}\right)^n$. This gives for $n = 0$

$$\alpha = \frac{5T_e N_e}{2m_e\nu}$$

and for $n = 1$, when $\nu = v/\lambda$, this formula yields

$$\alpha = \sqrt{\frac{2T_e}{m_e}}\frac{\lambda}{3\sqrt{\pi}} = \frac{2\lambda N_e}{3v_T},$$

where $v_T = \sqrt{\frac{8T_e}{\pi m_e}}$ is the mean electron velocity.

▶ **Problem 7.41** Determine the heat flux due to electrons when a plasma is located in a metallic enclosure and in a dielectric enclose.

Depending on the enclosure character, different boundary conditions are realized. Namely, if the walls are dielectric, $\boldsymbol{j} = 0$, which corresponds to the regime of ambipolar diffusion when electrons travel together with ions. In the case of a metallic enclosure, the transversal electric field is absent $E = 0$. Hence, the heat flux due to electrons is different depending on the wall type. We represent the heat flux in the form

$$\boldsymbol{q} = -C\kappa_e \boldsymbol{\nabla} T_e ,$$

and determine the coefficient C for different enclosures. Evidently, for a metallic enclosure $C = 1$. If walls are dielectric, $\boldsymbol{j} = 0$, and that leads to origin of an electric field of a strength $\boldsymbol{E} = D_T \boldsymbol{\nabla} \ln T_e/(N_e K_e)$. This gives $C = 1 - \alpha N D_T e/(\kappa_e T_e N_e K_e)$. Using the Einstein relation for the electron mobility in a gas $D = KT_e/e$ and formula (7.80) for the electron thermodiffusion coefficient gives

$$C = 1 + \frac{\alpha n}{\kappa_e} ,$$

where we use as usually the velocity dependence for the rate of electron–atom collisions in the form $d \ln \nu / d \ln v = n$. This gives

$$C = \frac{n+2}{2-n}$$

As is seen, the effective thermal conductivity coefficient for electrons in the two considering cases of metallic and dielectric walls depends on n. For $n = 0$ this value is identical for both cases, while for $n = 1$ it is more in the second case in three times than that in the first case.

8
Transport of Ions and Atoms in Gases and Plasmas

8.1
General Peculiarities of Transport Processes

▶ **Problem 8.1** Define the kinetic coefficients in a gas connected with fluxes of particles, momentum and heat with the corresponding gradients.

Under thermodynamic equilibrium such gas parameters as the number density of atoms or molecules of each species, the mean velocity of atoms or molecules and the temperature are constants in a region occupied by a gas. If some of these values should vary in this region, appropriate fluxes arise in order to equalize these parameters over the total volume of a gas or plasma. The fluxes are small if variations of the parameters are small at distances of the order of the mean free path for atoms or molecules. Then a stationary state of the system with fluxes exists, and such states are conserved during times much longer than typical times between particle collisions. In other words, the inequality

$$\lambda \ll L \tag{8.1}$$

is satisfied for the systems under discussion, where λ is the mean free path for particles, and L is a typical size of the system or a distance over which a parameter varies noticeably. If this criterion is fulfilled, the system is in a stationary state to first approximation, and the transport of particles, heat or momentum occurs in the second approximation in terms of an expansion over a small parameter (8.1). Below we consider various types of such transport.

Let us introduce the kinetic coefficients or transport coefficients as the coefficients of proportionality between fluxes and corresponding gradients. We start from the diffusion coefficient D that is introduced as the proportionality factor between the particle flux $\boldsymbol{j}$ and the gradient of concentration c of a given species, or

$$\boldsymbol{j} = -DN\nabla c\,. \tag{8.2}$$

Here N is the total particle number density. If the concentration of a given species is low ($c_i \ll 1$), that is, this species is an admixture to the gas, the flux of particles of this species can be written as (6.24)

Plasma Processes and Plasma Kinetics. Boris M. Smirnov

ISBN: 978-3-527-40681-4

$$j = -D_i \nabla N_i \,, \tag{8.3}$$

where N_i is the number density of particles of the given species.

The thermal conductivity coefficient κ is defined as the proportionality factor between the heat flux $\boldsymbol{q}$ and the temperature gradient ∇T on the basis of the relation (6.25)

$$\boldsymbol{q} = -\kappa \nabla T \tag{8.4}$$

The viscosity coefficient η is the proportionality factor between the frictional force acting on unit area of a moving gas, and the gradient of the mean gas velocity in the direction perpendicular to the surface of a gas element (see Fig. 6.1). If the mean gas velocity $\boldsymbol{w}$ is parallel to the x-axis and varies in the z direction, the frictional force is proportional to $\partial w_x/\partial z$ and acts on the xy surface in the gas. Thus the force F per unit area is in accordance with formula (6.26)

$$F = -\eta \frac{\partial w_x}{\partial z} \,. \tag{8.5}$$

This definition as well as the previous ones refer both to liquids as well as gases.

▶ **Problem 8.2** Estimate the diffusion coefficient of atomic particles in a gas based on the character of particle transport in a gas.

We have a group of atomic particles in a gas and will find the kinetic coefficients for test particles taking into account their collisions with the atomic particles of a gas. For this purpose we consider the definition of a kinetic coefficient and use the character of transport processes. This allows us to find the dependence of kinetic coefficients on the parameters of a gas and to estimate these values.

We start with the diffusion coefficient, considering a group of identical test particles in a gas and assuming particle transport to occur in both opposite directions, which establishes a certain equilibrium in a gas. For the i-th types of particles their fluxes in both directions, in order of magnitude, are $N_i v$, where N_i is the number density of the particles of a given species, and v is a typical velocity of these particles. Therefore, the net current behaves as $j \sim \Delta N_i v$, where ΔN_i is the difference in the number densities of oppositely directed particles participating in the transport. Particles, which reach a given point without collisions, have distances from this point of the order of the mean free path $\lambda \sim (N\sigma)^{-1}$, where σ is a typical cross section for elastic scattering in large angles, and N is the total number density of the gas particles. Hence, $\Delta N_i \sim \lambda \nabla N_i$ and the diffusive flux behaves as $j \sim \lambda v \nabla N_k$. Comparing this with the definition of the diffusion coefficient (6.24), we obtain

$$D_i \sim v\lambda \sim \frac{\sqrt{T}}{N\sigma\sqrt{m}} \,. \tag{8.6}$$

Here T is the gas temperature and m is the mass of particles of a given species, assumed to be of the same order of magnitude as the masses of other particles comprising the gas. In this analysis we do not need to account for the sign of the flux because it is simply opposite to that of the number density gradient, and tends to equalize the particle number densities at neighboring points. The same can be said about the signs of the fluxes and gradients for the other transport phenomena.

▶ **Problem 8.3** Estimate the thermal conductivity coefficient for a one-component gas, using the character of heat transport in a gas.

For this purpose, we use the same procedure as in the case of estimation of the diffusion coefficient. Taking any gas point, we find the heat flux in each of the opposite directions to be $q \sim NvT$ for a given space point, where Nv is a flux of particles, and T is the gas temperature that is expressed in energetic units. The difference of these heat fluxes in terms of a temperature gradient is determined by a difference in the transferred thermal energy or by the difference in temperatures. Hence, the resultant heat flux can be estimated as $q \sim Nv\Delta T$. Since individual particles propagate without collisions at a distance of the order of the mean free path $\lambda \sim (N\sigma)^{-1}$, we obtain a typical temperature difference $\Delta T \sim \lambda \nabla T$. Substituting this in the estimation for the heat flux, we obtain the thermal conductivity coefficient

$$\kappa \sim Nv\lambda \sim \frac{v}{\sigma} \sim \frac{\sqrt{T}}{\sigma\sqrt{m}} . \tag{8.7}$$

The thermal conductivity coefficient is independent of the particle number density. Indeed, an increase in the particle number density causes an increase in the number of particles that transfer heat, but this then leads to a decrease in the mean free path of particles, i. e., a distance on which heat is transported. These two effects mutually cancel each other out.

▶ **Problem 8.4** Estimate the viscosity coefficient of a gas, based on the character of momentum transport in a gas.

In considering transport of momentum in a moving gas where the gas mean velocity varies in the direction perpendicular to the mean velocity, we base ourselves on the character of friction that is determined, in this case, by transport of particles in a moving gas with a gradient of the average velocity, as shown in Fig. 6.1. Indeed, taking two neighboring gas elements which have different average gas velocities, we obtain an exchange of particle momenta between these elements due to particle transport between them. This creates a frictional force that slows down particles of gas elements with higher velocities and accelerates particles of gas elements having lower velocities.

Let us estimate the viscosity coefficient on the basis of its definition (8.5) by analogy with the procedures employed for the diffusion and thermal conductivity coefficients by using the character of momentum transport in this case. The force acting per unit area of a gas as a result of the momentum transport is $F \sim vm\Delta w_x$, where Nv is the particle flux, and $m\Delta w_x$ is the difference of the mean momentum carried by particles which are moving in opposite directions at a given point. Since particles reaching this point without collisions are located from it at distances of the order of the mean free path λ, we have $m\Delta w \sim m\lambda\partial w_x/\partial z$. Hence, the force acting per unit area of a gas is $F \sim N v m\lambda\partial w_x/\partial z$. Comparing this with the definition of the viscosity coefficient (8.5), and using $(T/m)^{1/2}$ instead of v and $(N\sigma)^{-1}$ instead of λ, we obtain the estimate

$$\eta \sim \frac{\sqrt{mT}}{\sigma} \tag{8.8}$$

for the viscosity coefficient η. One can see that the viscosity coefficient is independent of the particle number density. By analogy with the thermal conductivity coefficient, this independence results from the compensation of opposite effects occurring with the momentum transport. Indeed, the number of momentum carriers is proportional to the number density of particles, while a typical transport distance, the mean free path, is inversely proportional to it. These effects cancel each other out mutually.

▶ **Problem 8.5** Estimate the mobility of particles in a gas in weak external fields.

We use the definition of the mobility b of a particle in a gas according to the relation $\boldsymbol{w} = b\boldsymbol{F}$ between the mean velocity $\boldsymbol{w}$ of the particle and the force $\boldsymbol{F}$ acts on the particle from an external field. Next, we use the Einstein relation $b = D/T$ (1.49) between the particle mobility b in a weak field and its diffusion coefficient D in a gas. Then using the estimate (8.6) for the diffusion coefficient, we obtain the following estimate for particle mobility

$$b \sim \frac{1}{N\sigma\sqrt{mT}} . \tag{8.9}$$

▶ **Problem 8.6** Determine the diffusion coefficient of a test particle in a gas if the cross section of collision between a the test and the gas particles is inversely proportional to their relative velocity.

We have an admixture of test particles of a small concentration in a gas or extract a small group of test atomic particles in the parent gas for determination of the self-diffusion coefficient. In both cases the diffusion coefficient D is defined by formula (8.2) from the relation for the flux of test atoms $\boldsymbol{j} = -D\nabla N_1$, where N_1 is the number density of the admixture or test particles. The kinetic equation (6.2) for the distribution function of these atoms with the collision integral (6.7) has the form

$$\boldsymbol{v}_1 f_1^{(0)} \frac{\nabla N_1}{N_1} = \int (f_1' f_2' - f_1 f_2) |\boldsymbol{v}_1 - \boldsymbol{v}_2| \, d\sigma d\boldsymbol{v}_2 . \tag{8.10}$$

Here f_1 is the distribution function of test particles that takes into account a density gradient, f_2 is the Maxwell distribution function of the gas atoms, and the Maxwell distribution function $f_1^{(0)}$ of test particles is normalized by the condition $\int f_1^{(0)} d\boldsymbol{v}_1 = N_1$ and has the form

$$f^{(0)}(v) = N(r)\left[\frac{m}{2\pi T(r)}\right]^{3/2}\exp\left[-\frac{m[v-w(r)]^2}{2T(r)}\right] , \tag{8.11}$$

where $N(r)$ is the number density of particles, $T(r)$ is the temperature, and $w(r)$ is the average velocity of particles at a point r.

Taking the density gradient to be directed along the axis x, we multiply the kinetic equation by the test particle momentum along the density gradient Mv_{1x}, where M is the mass of a test atom, and integrate over velocities of test atoms. On the left-side of the integral relation we obtain $T\partial N_1/\partial x$. Introducing the relative velocity $\mathbf{g} = v_1 - v_2$ and the velocity of the center of mass $V = (Mv_1 + mv_2)(M + m)$, we reduce this relation to the form

$$T\frac{\partial N_1}{\partial x} = \mu \int f_1 f_2 (g'_x - g_x) g d\sigma dv_1 dv_2 . \tag{8.12}$$

Here M, m are the masses of the test and the gas particles, and $\mu = mM/(m + M)$ is the reduced mass of these particles. We above used the symmetry with respect to the operation $t \to -t$ that gives

$$\int f'_1 f'_2 g_x dv_1 dv_2 dv'_1 dv'_2 = \int f_1 f_2 g'_x dv_1 dv_2 dv'_1 dv'_2 .$$

Let us represent the relative velocity of particles $\mathbf{g}'$ after collisions as

$$\mathbf{g}' = \mathbf{g}\cos\vartheta + \mathbf{k}g\sin\vartheta ,$$

where ϑ is the scattering angle and $\mathbf{k}$ is the unit vector of a random direction in the plane that is perpendicular to $\mathbf{g}$. Integrating over scattering angles, we reduce equation (8.12) to the form

$$T\frac{\partial N_1}{\partial x} = -\mu \int f_1 f_2 g_x g\sigma^*(g) dv_1 dv_2 , \tag{8.13}$$

where $\sigma^*(g) = \int (1 - \cos\vartheta) d\sigma$ is the transport cross section of atom collisions.

Accounting for the rate constant $k^{(1)} = g\sigma^*(g)$ to be independent of the collision velocity and the normalization condition $\int f_1 dv_1 = N_1$, we find from the above equation the drift velocity of particles

$$w_x = \frac{\int f_1 v_{1x} dv_1}{\int f_1 dv_1} = -\frac{T}{\mu k^{(1)} N_1 N_2}\frac{\partial N_1}{\partial x} .$$

Defining the particle flux as

$$j = w_x N_1 = -D\frac{\partial N_1}{\partial x} ,$$

we obtain the diffusion coefficient

$$D = -\frac{w_x N_1}{\partial N_1/\partial x} = \frac{T}{\mu k^{(1)} N_2} . \tag{8.14}$$

▶ **Problem 8.7** Determine the mobility of atomic ions in a foreign gas assuming the polarization interaction between a drifting ion and gas atoms.

We take the diffusion cross section of ion–atom collision at low collision energies to be equal to the polarization cross section of capture according to formula (2.25), and this is valid to an accuracy of 10 % (see formula 2.31). Substituting the polarization cross section (2.25) into expression (8.14) for the diffusion coefficient, we obtain the latter formula in the form

$$D = \frac{T\sqrt{\mu}}{2\pi N\sqrt{\alpha e^2}} , \tag{8.15}$$

where N is the number density of gas atoms, α is the atom polarizability. On the basis of the Einstein relation (8.66) this gives for the ion mobility

$$K = \frac{\sqrt{\mu}}{2\pi N\sqrt{\alpha}} .. \tag{8.16}$$

▶ **Problem 8.8** Give a general expression for the pressure tensor (6.17) by taking into account a gas viscosity.

Our goal is to generalize expression (6.18) for the pressure tensor including in it the gas viscosity. This is a slow phenomenon because a constant pressure $p = \text{const}$ is established with the sound speed. Since the gas temperature is constant in a space, from the state equation $p = NT$ we have that the number density N of atoms does not change in a space. Hence, from the continuity equation (6.15) in the stationary case we have $\operatorname{div}(N\boldsymbol{w}) = 0$, and if the average gas velocity $\boldsymbol{w}$ is directed along the axis x, we have from this

$$\frac{dw_x}{dx} = 0 .$$

Thus, accounting for expression (8.5) for the force per unit area and symmetrizing it with taking into account the above formula, we write the pressure tensor (6.17) by adding to expression (6.18) the symmetrized expression (8.5)

$$P_{ik} = p\delta_{ik} - \eta\left(\frac{\partial w_i}{\partial x_k} + \frac{\partial w_k}{\partial x_i}\right) , \tag{8.17}$$

where as usual x_i, x_k are the coordinates x, y, z.

▶ **Problem 8.9** Derive the kinetic equation for transport in a gas if thermodynamic equilibrium results from the elastic collisions of gas particles.

Since the mean free path of gas particles is small compared to the typical sizes of a system, we have a local thermodynamic equilibrium in each point, i. e., in each point the Maxwell distribution function (8.11) of particles is established. In reality, this point occupies a region whose size is comparable to the mean free path λ of particles. For large systems comparable to λ we change this region by a point.

Expanding the distribution function over a small parameter λ/L, we represent it in the form

$$f(\boldsymbol{v}) = f^{(0)}(\boldsymbol{v}) + f^{(1)}(\boldsymbol{v}) , \tag{8.18}$$

if we restrict ourselves to the first expansion term. We take the number density of particles $N(\boldsymbol{r})$, the average velocity of particles $\boldsymbol{w}(\boldsymbol{r})$ and the temperature $T(\boldsymbol{r})$ to be determined by zero-approximation for the distribution function, and then we have the average momenta from the first-order distribution function equal to zero, that is

$$\int f^{(1)}(\boldsymbol{v})d\boldsymbol{v} = 0, \quad \int f^{(1)}(\boldsymbol{v})\boldsymbol{v}d\boldsymbol{v} = 0, \quad \int f^{(1)}(\boldsymbol{v})\,(\boldsymbol{v}-\boldsymbol{w})^2\,d\boldsymbol{v} = 0 . \tag{8.19}$$

Let us represent the correction term for the distribution function due to gradients of corresponding values that has the form

$$f^{(1)}(\boldsymbol{v}) = f^{(0)}(\boldsymbol{v})\theta , \tag{8.20}$$

where the function θ characterizes the deviation of the distribution function from the Maxwell one. Since in the zero-th approximation we neglect nonuniformities of the gas, we have $I_{\rm col}(f^{(0)}) = 0$, that leads to formula (8.11) for the distribution function. The next expansion terms over a small parameter λ/L in the kinetic equation (6.2) and the integral collision in the form (8.20) gives

$$\boldsymbol{v}_1\nabla f_1^{(0)} + \frac{\boldsymbol{F}}{m}\frac{\partial f_1^{(0)}}{\partial \boldsymbol{v}_1} = \int f_1^{(0)} f_2^{(0)}\left(\theta_1' + \theta_2' - \theta_1 - \theta_2\right)|\boldsymbol{v}_1 - \boldsymbol{v}_2|\,d\sigma d\boldsymbol{v}_2 . \tag{8.21}$$

Here subscripts 1, 2 give the number of colliding particles, and a superscript $'$ characterizes their parameters after collision, while the initial parameters do not have a superscript. This equation is useful for determining the transport parameters of a gas.

8.2 Thermal Conductivity and Viscosity of Atomic Gases

► **Problem 8.10** Determine the thermal conductivity coefficient of an atomic gas in the tau-approximation.

The kinetic equation (8.21) by using the collision integral in the form of tau-approximation (6.3), has the form

$$f^{(0)}(\boldsymbol{v})\left(\frac{mv^2}{2T} - \frac{5}{2}\right)\boldsymbol{v}\nabla \ln T = -\frac{f^{(0)}(\boldsymbol{v})\theta}{\tau} ,$$

where we use expression (8.11) for unperturbed distribution function of gas atoms and accounts for the temperature gradient in the gas. From this we find the correction to the distribution function,

$$\theta = -\tau\left(\frac{mv^2}{2T} - \frac{5}{2}\right)\boldsymbol{v}\nabla \ln T ,$$

and from the definition of the heat flux $\boldsymbol{q} = \int (mv^2/2)\boldsymbol{v} f(\boldsymbol{v}) d\boldsymbol{v}$ we obtain the heat flux

$$\boldsymbol{q} = \int \tau \left(\frac{mv^2}{2T} - \frac{5}{2} \right) \boldsymbol{v} \nabla \ln T \cdot \frac{mv^2}{2} \boldsymbol{v} f^{(0)}(\boldsymbol{v}) d\boldsymbol{v} .$$

Comparing this with the definition of the thermal conductivity coefficient $\boldsymbol{q} = -\kappa \nabla T$, we find the thermal conductivity coefficient

$$\kappa = \int \tau \left(\frac{mv^2}{2T} - \frac{5}{2} \right) \cdot \frac{mv^2}{2T} \cdot \frac{v^2}{3} f^{(0)}(\boldsymbol{v}) d\boldsymbol{v} .$$

We above have take into account the spherical symmetry of the distribution function. If we assume the relaxation time τ to be independent of the collision velocity, we obtain

$$\kappa = \frac{5T\tau N}{2m} . \tag{8.22}$$

This corresponds to the above estimate (formula 8.7) for the thermal conductivity coefficient.

▶ **Problem 8.11** Determine the viscosity coefficient of an atomic gas in the tau-approximation.

When a gas is moving in the direction x with the mean velocity w_x and the gradient of the mean gas velocity occurs in the direction z, the kinetic equation (8.21) with the collision integral in tau-approximation (6.3) has the form

$$v_z \frac{m(v_x - w)}{T} \frac{\partial w_x}{\partial z} f^{(0)}(\boldsymbol{v}) = -\frac{f^{(0)}(\boldsymbol{v})\theta}{\tau} , \tag{8.23}$$

that gives for the correction to the distribution function if it is represented in the form (8.18) and (8.20)

$$\theta = -\tau v_z \frac{m(v_x - w)}{T} \frac{\partial w_x}{\partial z} .$$

We now determine the force per unit area that is equal to the product of the particle flux Nv_z and the momentum transferred $m(v_x - w)$. Hence, this specific force is equal to

$$F = \int v_z f(\boldsymbol{v}) \cdot m(v_x - w) d\boldsymbol{v} = -\frac{\partial w_x}{\partial z} \int \frac{mv_z^2}{T} m(v_x - w)^2 \tau f^{(0)}(\boldsymbol{v}) d\boldsymbol{v} .$$

Comparing this equation with formula (8.5) that defines the viscosity coefficient, we obtain the viscosity coefficient

$$\eta = \int \frac{mv_z^2}{T} m(v_x - w)^2 \tau f^{(0)}(\boldsymbol{v}) d\boldsymbol{v} .$$

In particular, if the relaxation time τ is independent of the collision velocity, we obtain

$$\eta = T\tau N . \tag{8.24}$$

This result corresponds to estimate (8.5).

▶ **Problem 8.12** Determine the thermal conductivity coefficient of an atomic gas if the differential cross section of atom elastic scattering does not depend on the collision velocity.

We consider the kinetic equation (8.21) that under the condition $p = NT = \text{const}$ at a given temperature gradient, dT/dx has the form

$$f(\boldsymbol{v}_1)\left(\frac{mv_1^2}{2T} - \frac{5}{2}\right) v_{1x}\frac{d\ln T}{dx} = \int \left(f_1' f_2' - f_1 f_2\right)|\boldsymbol{v}_1 - \boldsymbol{v}_2|\, d\sigma d\boldsymbol{v}_2 \,.$$

Let us multiply this equation by $v_{1x}v_1^2$ and integrate over atom velocities $d\boldsymbol{v}_1$. On the left-hand side of this equation we have, using expression (8.11), for the atom distribution function

$$\int v_{1x}v_1^2 f(\boldsymbol{v}_1)\left(\frac{mv_1^2}{2T} - \frac{5}{2}\right) v_{1x}\frac{d\ln T}{dx} = \frac{5NT}{m^2}\frac{dT}{dx} \,.$$

On the right-hand side of this integral relation we use the symmetry of the integrand is conserved at time reversal, i. e., as a result of transformation $\boldsymbol{v}_{1,2} \leftrightarrow \boldsymbol{v}_{1,2}'$ and also as a result of exchange by colliding atoms $\boldsymbol{v}_{1,2} \leftrightarrow \boldsymbol{v}_{2,1}$. We obtain

$$\frac{5NT}{m^2}\frac{dT}{dx} = \frac{1}{2}\int \left(v_{1x}'v_1'^2 + v_{2x}'v_2'^2 - v_{1x}v_1^2 - v_{2x}v_2^2\right) f_1 f_2\, |\boldsymbol{v}_1 - \boldsymbol{v}_2|\, d\sigma d\boldsymbol{v}_1 d\boldsymbol{v}_2 \,. \quad (8.25)$$

Introducing the relative velocity of atoms $\mathbf{g} = \boldsymbol{v}_1 - \boldsymbol{v}_2$ and the center-of-mass velocity $\boldsymbol{V} = (\boldsymbol{v}_1 + \boldsymbol{v}_2)/2$, we reduce this equation to the form

$$\frac{5NT}{m^2}\frac{dT}{dx} = \frac{1}{2}\int [g_x'(\mathbf{Vg}') - g_x(\mathbf{Vg})] f_1 f_2 g d\sigma d\boldsymbol{v}_1 d\boldsymbol{v}_2 \,. \quad (8.26)$$

We now integrate the integrand on the right-hand side of the equation over angles of vectors and scattering angles, taking the relative velocity of atoms after collision in the form $\mathbf{g}' = \mathbf{g}\cos\vartheta + \boldsymbol{k}g\sin\vartheta$, where ϑ is the scattering angle, and $\boldsymbol{k}$ is the unit vector distributed randomly in the plane that is perpendicular to the vector $\mathbf{g}$. Taking the direction of the axis x in an arbitrary manner, we evaluate the integral

$$\int \left\langle [\mathbf{g}'(\mathbf{Vg}') - \mathbf{g}(\mathbf{Vg})] d\sigma \right\rangle = -\sigma^{(2)}(g)\mathbf{g}(\mathbf{Vg}) +$$

$$\int \left\langle [\boldsymbol{k}g(\mathbf{Vg})\sin\vartheta\cos\vartheta + \mathbf{g}(\boldsymbol{V}\boldsymbol{k})g\sin\vartheta\cos\vartheta + \boldsymbol{k}g^2(\boldsymbol{V}\boldsymbol{k})\sin^2\vartheta] d\sigma \right\rangle \,.$$

Here the angle brackets denote an average over angles between the vectors $\mathbf{g}$ and $\boldsymbol{V}$, and

$$\sigma^{(2)}(g) = \int (1 - \cos^2\vartheta) d\sigma. \quad (8.27)$$

The second and third terms give zero after integration over directions of the vector $\boldsymbol{k}$. In the last term we expand the vector $\boldsymbol{k}$ in components $\boldsymbol{k} = \boldsymbol{l}\cos\varphi + \boldsymbol{m}\sin\varphi$, where $\boldsymbol{l}, \boldsymbol{m}$ are the unit vectors, so that the vector $\boldsymbol{l}$ is located in the plane of the vectors $\mathbf{g}$ and $\boldsymbol{V}$, and the vector $\boldsymbol{m}$ is perpendicular to these vectors, and φ is the polar angle in the plane perpendicular to $\mathbf{g}$. Averaging over the polar angle φ, we obtain

$$\int \left\langle [\mathbf{g}'(\mathbf{V}\mathbf{g}') - \mathbf{g}(\mathbf{V}\mathbf{g})] d\sigma \right\rangle = -\sigma^{(2)}(g)[\mathbf{g}(\mathbf{V}\mathbf{g}) - \frac{\boldsymbol{l}}{2}(\mathbf{V}\boldsymbol{l})g^2] .$$

According to the definition of the vector $\boldsymbol{l}$ we have

$$g^2\boldsymbol{l}(\boldsymbol{V}\boldsymbol{l}) = g^2\boldsymbol{V} - \mathbf{g}(\boldsymbol{V}\mathbf{g}) ,$$

and hence we obtain equation (8.26) in the form

$$\frac{5NT}{m^2}\frac{dT}{dx} = -\frac{1}{4}\int [3g_x(\mathbf{V}\mathbf{g}) - V_x g^2]\, f_1 f_2 g\sigma^{(2)}(g) d\boldsymbol{v}_1 d\boldsymbol{v}_2 . \tag{8.28}$$

Up to now the equation has been deduced without any assumptions. At this stage we assume that the value $k^{(2)} = g\sigma^{(2)}(g)$ does not depend on the relative collision velocity g. Then, returning to the velocities $\boldsymbol{v}_1, \boldsymbol{v}_2$ of the individual particles and using the symmetry of the integrand with respect to the exchange $\boldsymbol{v}_1 \leftrightarrow \boldsymbol{v}_2$, we obtain this equation in the form

$$\frac{5NT}{m^2}\frac{dT}{dx} = -\frac{k^{(2)}}{2}\int [v_{1x}v_1^2 - 2v_{1x}v_2^2 + v_{1x}(\boldsymbol{v}_1\boldsymbol{v}_2)]\, f_1 f_2 d\boldsymbol{v}_1 d\boldsymbol{v}_2 .$$

We consider the distribution functions of colliding atoms are almost spherically symmetric, and the deviation from this symmetry is determined by a small temperature gradient. One can see that the third term is proportional to the square of this small parameter, and therefore we ignore this term. The second term corresponds to the atom flux under the action of a temperature gradient, i. e., it describes the thermodiffusion process. Focusing on the thermal conductivity process, we are restricted by the first term only, and this gives

$$\frac{5NT}{m^2}\frac{dT}{dx} = -\frac{k^{(2)}N}{m}q_x , \tag{8.29}$$

where according to its definition, the heat flux is $\boldsymbol{q} = \int \boldsymbol{v}(mv^2/2)f(\boldsymbol{v})d\boldsymbol{v}$. From this, based on definition (8.4) of the thermal conductivity coefficient κ, we obtain

$$\kappa = \frac{5T}{mk^{(2)}} . \tag{8.30}$$

▶ **Problem 8.13** Determine the viscosity coefficient of an atomic gas if the differential cross section of atom elastic scattering is inversely proportional to the collision velocity.

We now consider the kinetic equation (6.2) for the distribution function of atoms and the collision integral (6.7), and assuming that the distribution function is close to equation (8.11), we reduce this kinetic equation to the form

$$v_{1z}\frac{\partial f_1}{\partial z} \equiv v_{1z}\frac{m(v_{1x} - w)}{T}\frac{\partial w}{\partial z} f_1 = \int (f_1' f_2' - f_1 f_2)\, |\boldsymbol{v}_1 - \boldsymbol{v}_2|\, d\sigma d\boldsymbol{v}_2 .$$

We also use the definition of the viscosity coefficient η according to formula (8.5) for the force F due to a gradient of the drift velocity

$$F = \int v_z f d\boldsymbol{v} \cdot m(v_x - w) = -\eta \frac{\partial w_x}{\partial z} ,$$

where the first factor of the integrand is the particle flux, the second factor is the transferring momentum.

For determination of the viscosity coefficient we multiply the above kinetic equation by $mv_{1z}(v_{1x} - w)$ and integrate over the velocities. This gives the right-hand side of the resultant equation as

$$\int mv_{1z}(v_{1x} - w)\boldsymbol{v}_1 \nabla f_1 d\boldsymbol{v}_1 = \int mv_{1z}^2 \frac{m(v_{1x} - w)^2}{T} \frac{\partial w_x}{\partial z} = NT \frac{\partial w_x}{\partial z} .$$

The right-hand side of the resultant equation is

$$\int mv_{1z}(v_{1x} - w)(f_1' f_2' - f_1 f_2)\, |\boldsymbol{v}_1 - \boldsymbol{v}_2|\, d\sigma d\boldsymbol{v}_1 d\boldsymbol{v}_2 =$$

$$\int \left[mv_{1z}'(v_{1x}' - w) - mv_{1z}(v_{1x} - w)\right] f_1 f_2 |\boldsymbol{v}_1 - \boldsymbol{v}_2| d\sigma d\boldsymbol{v}_1 d\boldsymbol{v}_2 .$$

Using the symmetry with respect to time reversal $t \to -t$ in the above equation allows us to change parameters after collision by parameters before collision and vice versa. On the basis of the relative velocity of atoms $\boldsymbol{g} = \boldsymbol{v}_1 - \boldsymbol{v}_2$ and the center-of-mass velocity $\boldsymbol{V} = (\boldsymbol{v}_1 + \boldsymbol{v}_2)/2$, we reduce this resultant equation to the form $(\boldsymbol{V}' = \boldsymbol{V})$

$$NT \frac{\partial w_x}{\partial z} = m \int \left[\frac{V_z}{2}(g_x' - g_x) + \frac{V_x}{2}(g_z' - g_z) + \frac{1}{4}(g_x' g_z' - g_x g_z)\right] f_1 f_2 g d\sigma d\boldsymbol{V} d\boldsymbol{g} .$$

Let us integrate the integrand on the right-hand side of the equation over scattering angles, taking the atom relative velocity after collision in the form $\boldsymbol{g}' = \boldsymbol{g}\cos\vartheta + \boldsymbol{k} g \sin\vartheta$, where ϑ is the scattering angle, and $\boldsymbol{k}$ is the unit vector that is located in the plane perpendicular to the vector $\boldsymbol{g}$ and is distributed randomly in this plane. Integrating over directions of the vector $\boldsymbol{k}$ for the first and second terms in square bracket gives zero, and after integration over the scattering angles we obtain

$$NT \frac{\partial w_x}{\partial z} = -\frac{m}{4} \int g_x g_z g \sigma^{(2)}(g) f_1 f_2 d\boldsymbol{V} d\boldsymbol{g} , \tag{8.31}$$

where the averaged cross section $\sigma^{(2)}(g)$ is given by formula (8.27).

We now take into account that the value $k^{(2)} = g\sigma^{(2)}(g)$ is independent of the collision velocity g. Then returning to the velocities of colliding particles $\boldsymbol{v}_1$ and $\boldsymbol{v}_2$ and ignoring cross terms in the integrand, because they relate to other transport processes, we obtain

$$NT \frac{\partial w_x}{\partial z} = -\frac{mk^{(2)}}{4} \int g_x g_z f_1 f_2 d\boldsymbol{V} d\boldsymbol{g}$$

$$= -\frac{mk^{(2)}}{4} \int (v_{1x} v_{1z} + v_{2x} v_{2z}) f_1 f_2 d\boldsymbol{v}_1 d\boldsymbol{v}_2 = -\frac{Nk^{(2)}}{2} F ,$$

where the force F per unit area is connected by formula (8.5) with the viscosity coefficient. Hence, the latter is equal to

$$\eta = \frac{2T}{k^{(2)}} \,. \tag{8.32}$$

▶ **Problem 8.14** For a dissociating molecular gas, estimate the contribution to heat transport due to the dissociation degree of freedom.

In a dissociating gas under consideration, along with thermodynamic equilibrium between translation and vibrational degrees of freedom of molecules, the dissociation equilibrium

$$A + A \longleftrightarrow A_2$$

also occurs.

Under thermodynamic equilibrium, the number densities of atoms N_a and molecules N_m are connected by the Saha formula (1.69), so that $N_a^2/N_m = F(T)\exp(-D/T)$, where D is the dissociation energy of the molecule, and a function $F(T)$ is characterized by a weak temperature dependence in comparison to the exponential one. Taking such a dissociation regime that $N_a \gg N_m$, and since $D \gg T$, we have $\partial N_m/\partial T = (D/T^2)N_m$. We hence obtain from relation (6.32)

$$\kappa_i = \left(\frac{D}{T}\right)^2 D_m N_m \,,$$

where D_m is the diffusion coefficient of molecules in an atomic gas. Comparing this expression with the thermal conductivity coefficient (8.7) due to translational heat transport, we have

$$\frac{\kappa_i}{\kappa_t} \sim \left(\frac{D}{T}\right)^2 \frac{N_m}{N_a} \,.$$

In the regime under consideration the number density of molecules is relatively small $N_a \gg N_m$, while $D \gg T$. Therefore, heat transport due the dissociation equilibrium may determine heat transport in a dissociating molecular gas. In this case a molecule moves from a cold region to a hot one and is dissociated there. Though the number of such molecules is small compared to that of atoms, each molecule carries a more energy than an individual atom. Therefore, the contribution from molecules can be dominant even at a weak dissociation degree of this gas.

8.3 Diffusion and Drift Character of Particle Motion

▶ **Problem 8.15** Derive the macroscopic equation for the number density of particles due to diffusion of particles.

Let us use the continuity equation (6.15) that has the form $\partial N_i/\partial t + \mathrm{div}\,\boldsymbol{j} = 0$ for the i-th gas component, and the expression for the diffusion flux $\boldsymbol{j} = -D\nabla N_i$ From this we obtain the diffusion equation

$$\frac{\partial N_i}{\partial t} = D\Delta N_i; \tag{8.33}$$

In this case a typical time τ_L of particle transport on typical distances L for this gas is large compared with a typical time $\tau_o \sim \lambda/v$ between successive collisions. Indeed, according to this equation $\tau_L \sim L^2/D$, and using the estimate for the diffusion coefficient $D \sim \lambda v$, we have $\tau_L \sim \tau_o(L/\lambda)^2$, i. e., $\tau_L \gg \tau_o$. This allows us to consider the diffusion process as time-independent.

▶ **Problem 8.16** Determine the average displacement of a test particle in a gas for a time t due to its diffusion in the gas. This time is large in comparison to the time between successive collisions.

Let us introduce the probability $P(\boldsymbol{r}, t)$ that a test particle is located at point $\boldsymbol{r}$ at moment t, if it was found at the spatial origin in the beginning. Evidently, this probability is spherically symmetrical and satisfies the normalization condition

$$\int\limits_0^\infty P(r,t)4\pi r^2 dr = 1\,. \tag{8.34}$$

The probability P is described by the diffusion equation (8.33), which in the spherically symmetrical case takes the form

$$\frac{\partial P}{\partial t} = \frac{D}{r}\frac{\partial^2}{\partial r^2}(rP)\,.$$

To find mean values for the diffusion parameters, we multiply this equation by $4\pi r^4 dr$ and integrate the result over all r. The left-hand side of the equation yields

$$\int_0^\infty 4\pi r^4 dr \frac{\partial P}{\partial t} = \frac{d}{dt}\int_0^\infty r^2 P 4\pi r^2 dr = \frac{d}{dt}\overline{r^2}\,,$$

where $\overline{r^2}$ is the mean square of the distance from the origin. Integrating twice by parts and using the normalization condition (8.34), we transform the right-hand side of the equation into

$$D\int_0^\infty 4\pi r^4 dr \frac{1}{r}\frac{\partial^2}{\partial r^2}(rP) = -3D\int_0^\infty 4\pi r^2 dr \frac{\partial}{\partial r}(rP) = 6D\int_0^\infty P\cdot 4\pi r^2 dr = 6D\,.$$

The resulting equation has the forms $d\overline{r^2} = 6Ddt$. Since at time zero the particle is located at the origin, the solution of this equation has the form

$$\overline{r^2} = 6Dt\,. \tag{8.35}$$

Because the motion in different directions is independent and has a random character, it follows from this that

$$\overline{x^2} = \overline{y^2} = \overline{z^2} = 2Dt\,. \tag{8.36}$$

▶ **Problem 8.17** Find the solution of the one-dimensional diffusion equation and also the solution of the three-dimensional equation in the spherically symmetric case.

The diffusion process is an example of random processes, and hence the solution of the diffusion equation is given by the Gauss distribution (6.44). In the one-dimensional case we have $\Delta = \overline{x^2} = 2Dt$, and the solution of the diffusion equation

$$\frac{\partial P(x,t)}{\partial t} = D\frac{\partial^2 P(x,t)}{\partial x^2} \tag{8.37}$$

has the form

$$P(x,t) = (4\pi Dt)^{-1/2} \exp\left[-\frac{x^2}{4Dt}\right] \tag{8.38}$$

This formula relates to the initial condition $P(x,0) = \delta(x)$, i. e., the particle is located in the origin of the frame of reference in the beginning. In the spherically symmetric three-dimensional case we use the symmetry of the distribution

$$P(r,t) = P(x,t)P(y,t)P(y,t) ,$$

and the density probability in the right-hand side of this relation is given by formula (8.38). This leads to the following probability density:

$$P(r,t) = \frac{1}{(4\pi Dt)^{3/2}} \exp\left(-\frac{r^2}{4Dt}\right) . \tag{8.39}$$

If the particle is not located in the origin in the beginning, this distribution function may be used as the Green function of the diffusion equation. Indeed, if the initial space distribution of the particle is described by a distribution function $\varphi(\mathbf{r})$ that normalized by the condition $\int \varphi(\mathbf{r})d\mathbf{r} = 1$, the probability $p(\mathbf{r},t)$ of particle location at point $\mathbf{r}$ at time t is given by

$$p(\mathbf{r},t) = \int P(\mathbf{r}-\mathbf{r}',t)\varphi(\mathbf{r}')d\mathbf{r}' ,$$

where this probability is normalized by the condition $\int p(\mathbf{r})d\mathbf{r} = 1$.

▶ **Problem 8.18** Ions are injected in a drift chamber and are moving there in a gas under a constant electric field. Taking the drift velocity of ions in the drift chamber to be w, and the diffusion coefficient to be D, find the distribution function of ions through a time t at a distance x from the drift chamber entrance.

Taking into account the drift of ions in a gas along with their diffusion, one can change the coordinate x in formula (8.37) by $x - wt$. This leads to the following expression for the distribution function, i. e., the probability of finding the particle at a distance x from the origin at time t,

$$P(x,t) = (4\pi Dt)^{-1/2} \exp\left[-\frac{(x-wt)^2}{4Dt}\right] . \tag{8.40}$$

► **Problem 8.19** Ions of two kinds are injected in a drift chamber whose length is L. Determine the time dependence for the ion current on the collector, if the drift velocities of ions are w_1 and w_2, respectively, and they are characterized by the identical diffusion coefficients D in a gas. Find the criterion when signals corresponding to different ions can be separated.

On the basis of formula (8.40) we obtain the current on the collector at time t

$$I(t) = \frac{J_1}{\sqrt{4\pi D_1 t}} \exp\left[-\frac{(L - w_1 t)^2}{4D_1 t}\right] + \frac{J_2}{\sqrt{4\pi D_2 t}} \exp\left[-\frac{(L - w_2 t)^2}{4D_2 t}\right], \tag{8.41}$$

where indices 1 and 2 refer to the two types of ions, and J_1, J_2 are the total number of ions of a given type at the drift chamber entrance. Evidently, signals from these two ion types can be resolved, if the time dependence $I(t)$ has a minimum between two maxima. Let us find the criterion when it is realized.

Denote $w = (w_1 + w_2)/2$, $\Delta w = w_1 - w_2$ $(w_1 > w_2)$ and take $D_1 = D_2 = D$ in accordance with the conditions of this problem. We assume the transport parameters of different ions to be close to each other, so then $\Delta w \ll w$. In this case both maxima and minimum are located near $t_0 = L/w$. In order to find the minimum of the current (8.41) under this condition, we introduce the reduced variable x and the reduced constant A according to the relations

$$x = \frac{2w}{\Delta w}\left(\frac{w}{L}t - 1\right), \quad A = \frac{eEL}{8T}\left(\frac{\Delta w}{w}\right)^2.$$

We consider the case of low electric field strengths E in the drift camera $w = KE$ and the Einstein relation $D = eK/T$. In new variables the relation (8.41) takes the form

$$I(t) = C\left\{J_1 \exp\left[-A(1+x)^2\right] + J_2 \exp\left[-A(1-x)^2\right]\right\}, \qquad C = \sqrt{\frac{eE}{4\pi LT}}.$$

Let us analyze this expression from the standpoint of resolution of two signals. Evidently, these signals are resolved if the dependence $I(t)$ has two maxima with a minimum between them. Then one can relate to an appropriate ion a signal part from the corresponding side of the minimum. Hence, the possibility of resolving ions of different sorts connects with the existence of the minimum for the dependence $I(t)$. As it follows from formula for $I(t)$, the extremum condition for the function $I(t)$ is given by the relation

$$F(x) = \frac{J_2}{J_1}\frac{1-x}{1+x}e^{2Ax} = 1,$$

and the extremum ion currents are located in a range

$$1 \geq x \geq -1.$$

We now use the condition of this problem $J_1 = J_2 = J$. Then due to the symmetric form of $F(x)$ we obtain the relation

$$F(x)F(-x) = 1.$$

This gives the solution of this equation $x = 0$, and this solution is a minimum of $I(t)$ if $d^2 I/dt^2 > 0$. The latter is fulfilled if $A > 1/2$ or

$$U = EL > U_0 = \frac{4T}{e}\left(\frac{w}{\Delta w}\right)^2 , \tag{8.42}$$

where U is the voltage in the drift chamber. As is seen, the limiting voltage does not depend on the gas pressure.

▶ **Problem 8.20** A charged cluster is located in a gas near a charged wall that repulses the cluster. The diffusion coefficient of the cluster in a gas D, and under the drift velocity w of the cluster is created by the wall electric field strength. Find the probability of cluster attachment to walls if in the beginning it is located at a distance x_0 from walls that is small compared to the radius of the wall curvature.

According to the problem conditions, the wall surface is plane for the cluster, and the problem is one-dimensional. Taking the boundary condition $P(0, t) = 0$, where $P(x, t)$ is the probability for cluster location at a distance x from the walls at time t, and this probability can be composed on the basis of formula (8.39) in the form

$$P(x, t) = (4\pi Dt)^{-1/2}\left\{\exp\left[-\frac{(x - x_0 - wt)^2}{4Dt}\right] - \exp\left[-\frac{(x + x_0 + wt)^2}{4Dt}\right]\right\} . \tag{8.43}$$

This gives for the cluster flux to the boundary

$$j = -D\frac{\partial P(x, t)}{\partial x} = \frac{x_0 + wt}{(4\pi D)^{1/2} t^{3/2}} \exp\left[-\frac{(x_0 + wt)^2}{4Dt}\right] . \tag{8.44}$$

From this we find the probability of cluster attachment to walls (we assume the attachment takes place if the cluster coordinate reaches the value $x = 0$)

$$W = \int_0^\infty \frac{x_0 + wt}{(4\pi D)^{1/2}}\frac{dt}{t^{3/2}} \exp\left[-\frac{(x_0 + wt)^2}{4Dt}\right] . \tag{8.45}$$

Introducing the reduced parameter $\eta = x_0 w/(4D)$ and the reduced variable $\tau = wt/x_0$, we represent this expression in the form

$$W(\eta) = \sqrt{\frac{\eta}{\pi}} \int_0^\infty \frac{(1 + \tau)d\tau}{\tau^{3/2}} \exp\left[-\frac{\eta(1 + \tau)^2}{\tau}\right] . \tag{8.46}$$

This integral is approximated by formula

$$W(\eta) = \exp(-4\eta) = \exp\left(-\frac{x_0 w}{D}\right) . \tag{8.47}$$

8.4 Chapman–Enskog Approximation

▶ **Problem 8.21** Evaluate the thermal conductivity coefficient of an atomic gas in the first Chapman–Enskog approximation by taking into consideration elastic collisions of atoms.

The Chapman–Enskog method corresponds to expansion of the kinetic coefficients over a numerical parameter. In considering above the ion mobility in the first Chapman–Enskog approximation, we find the term which is proportional to a strength of a weak field. Now we use this method directly, based on the kinetic equation in the form (8.21) that now has the form

$$f_1^{(0)}(\boldsymbol{v})\left(\frac{mv_1^2}{2T}-\frac{5}{2}\right)\boldsymbol{v}_1\nabla\ln T=\int f_1^{(0)}f_2^{(0)}\left(\theta_1'+\theta_2'-\theta_1-\theta_2\right)|\boldsymbol{v}_1-\boldsymbol{v}_2|\,d\sigma d\boldsymbol{v}_2\,, \tag{8.48}$$

and the correction to the distribution function θ is proportional to the temperature gradient that is relatively small, and we represent this correction as $\theta=-\boldsymbol{A}\nabla\ln T$. A vector $\boldsymbol{A}$ can be constructed on vectors which we dispose in zero approximation when gradients are absent. The only such vector is the atom velocity, and hence we have $\boldsymbol{A}=A(v)\boldsymbol{v}$. Introducing the reduced velocity of particles $\boldsymbol{u}=\sqrt{m/(2T)}\boldsymbol{v}$, we rewrite the above equation as

$$f_1^{(0)}\left(u_1^2-\frac{5}{2}\right)\boldsymbol{u}_1\left(\frac{2T}{m}\right)^{1/2}=$$
$$\int f_1^{(0)}f_2^{(0)}\left(A_1'\boldsymbol{u}_1'+A_2'\boldsymbol{u}_2'-A_1\boldsymbol{u}_1-A_2\boldsymbol{u}_2\right)|\boldsymbol{u}_1-\boldsymbol{u}_2|\,d\sigma d\boldsymbol{u}_2\,, \tag{8.49}$$

and the normalization condition for the distribution function is $\int f^{(0)}d\boldsymbol{u}=1$.

Our goal is to find the value $A(u)$ that we can get by solving the nonlinear integral equation. We use a numerical solution for this that is based on the expansion of $A(u)$ over the Sonin polynomials

$$A(u)=\sum_{i=0}^{\infty}a_i S_n^i(u^2)\,.$$

The Sonin polynomials are defined as

$$S_n^i(x)=\sum_{k=0}^{i}(-x)^k\frac{\Gamma(n+i+1)}{\Gamma(n+k+1)k!(i-k)!}\,, \tag{8.50}$$

and the first Sonin polynomials are $S_n^0=1$, $S_n^1=n+1-x$. The Sonin polynomials satisfy the orthogonality condition,

$$\int_0^{\infty}S_n^m(x)S_n^k(x)e^{-x}x^n dx=\delta_{mk}\,.$$

The orthogonality condition for the Sonin polynomials allows us to satisfy conditions (8.19) for each expansion term. In the case of heat transport we take $n = 3/2$, i. e.,

$$A(u) = \sum_{i=0}^{\infty} a_i S^i_{3/2}(u^2) ,$$

which gives $a_0 = 0$ from the first condition (8.19), and the second condition of (8.19) is fulfilled separately for each expansion term. Using this expansion, we obtain the heat flux

$$\begin{aligned} \boldsymbol{q} &= \int \frac{mv^2}{2}\boldsymbol{v} f d\boldsymbol{v} = \int \frac{mv^2}{2}\boldsymbol{v} f^{(0)}\theta d\boldsymbol{v} = -\frac{2T^2}{m}\int u^2 \boldsymbol{u}(\boldsymbol{u}\boldsymbol{\nabla})\ln(T) A f^{(0)} d\boldsymbol{u} \\ &= -\frac{2T^2}{3m}\int u^4 \boldsymbol{\nabla} \ln T \sum_{k=1}^{\infty} a_k S^k_{3/2}(u^2) f^{(0)} d\boldsymbol{u} = -\kappa \boldsymbol{\nabla} T . \end{aligned}$$

Using the equilibrium Maxwell distribution function $f^{(0)}$ according to formula (8.11), we obtain the thermal conductivity coefficient

$$\kappa = \frac{4TN}{3\sqrt{\pi}m}\int_0^{\infty} e^{-z} z^5 dz \sum_{k=1}^{\infty} a_k S^k_{3/2}(z) = \frac{5TN}{2m} a_1 , \tag{8.51}$$

and our task is to evaluate a_1. Let us use the notations

$$b_{km} = \int f_1^{(0)} f_2^{(0)} \left[\boldsymbol{u}_1 S^m_{3/2}(u_1^2) + \boldsymbol{u}_2 S^m_{3/2}(u_2^2) - \boldsymbol{u}'_1 S^m_{3/2}(u_1'^2) - \boldsymbol{u}'_2 S^m_{3/2}(u_2'^2) \right] |\boldsymbol{u}_1 - \boldsymbol{u}_2| \, d\sigma d\boldsymbol{u}_2 .$$

Multiplying the kinetic equation (8.49) by $\boldsymbol{u}_1 S^m_{3/2}(u_1^2)$ and integrating it over $d\boldsymbol{u}_1$, we obtain the following set of equations for the coefficients a_k:

$$\sum_{k=1}^{\infty} a_k b_{km} = \frac{15}{4} N \delta_{m1} , \quad m = 1, 2, \ldots \tag{8.52}$$

From the solution of this set of equations one can determine the values a_k.

For the coefficient a_1 we have from this set of equations

$$a_1 = \frac{15N}{4} \lim_{n\to\infty} \frac{\begin{vmatrix} b_{22} & b_{23} & \ldots & b_{2n} \\ b_{32} & b_{33} & \ldots & b_{3n} \\ \ldots & \ldots & \ldots & \ldots \\ b_{n2} & b_{n3} & \ldots & b_{nn} \end{vmatrix}}{\begin{vmatrix} b_{11} & b_{12} & \ldots & b_{1n} \\ b_{22} & b_{22} & \ldots & b_{2n} \\ \ldots & \ldots & \ldots & \ldots \\ b_{n1} & b_{n2} & \ldots & b_{nn} \end{vmatrix}} = \frac{\frac{15N}{4}}{b_{11} + b_{12}\frac{D_2}{D + b_{13}}\frac{D_3}{D+} \cdots} ,$$

where D is the upper determinant, and D_i is the lower determinant in which the i-th columns and lines are absent. The Chapman–Enskog approximation allows us to restrict ourselves to the first terms of expansion over a numerical parameter. In the first Chapman–Enskog approximation we use only one term in the denominator. In this approximation we have

$$a_1 = \frac{15N}{4b_{11}}, \qquad \kappa = \frac{75TN^2}{8mb_{11}},$$

and our task is to evaluate the integral

$$b_{11} = \sqrt{\frac{2T}{m}} \int f_1^{(0)} f_2^{(0)} \boldsymbol{u}_1 (u_1^2 - \frac{5}{2}) \left[\boldsymbol{u}_1 (u_1^2 - \frac{5}{2}) + \boldsymbol{u}_2 (u_2^2 - \frac{5}{2}) - \boldsymbol{u}_1' \left(u_1'^2 - \frac{5}{2} \right) - \boldsymbol{u}_2' (u_2'^2 - \frac{5}{2}) \right] |\boldsymbol{u}_1 - \boldsymbol{u}_2| \, d\sigma d\boldsymbol{u}_1 d\boldsymbol{u}_2 \,.$$

This integral conserves if we change the atoms 1 and 2, and because of the symmetry with respect to time reversal $t \to -t$, the integral conserves if $\boldsymbol{u} \to -\boldsymbol{u}', \boldsymbol{u}' \to -\boldsymbol{u}$. This allows us to reduce the integral to the form

$$b_{11} = \sqrt{\frac{2T}{m}} \int f_1^{(0)} f_2^{(0)} \left[\boldsymbol{u}_1 (u_1^2 - \frac{5}{2}) + \boldsymbol{u}_2 (u_2^2 - \frac{5}{2}) - \boldsymbol{u}_1' (u_1'^2 - \frac{5}{2}) - \boldsymbol{u}_2' (u_2'^2 - \frac{5}{2}) \right]^2 |\boldsymbol{u}_1 - \boldsymbol{u}_2| \, d\sigma d\boldsymbol{u}_1 d\boldsymbol{u}_2 \,.$$

Let us introduce the reduced velocity of the mass center $\boldsymbol{U}$ and the reduced relative velocity $\boldsymbol{g}$. Since the velocity of the mass center is conserved in a collision of particles, we have $\boldsymbol{U} = \boldsymbol{U}'$, and from the energy conservation law it follows $g = g'$. From this we obtain the reduced velocities of particles before and after collisions

$$\boldsymbol{u}_1 = \boldsymbol{U} + \boldsymbol{g}/2, \ \boldsymbol{u}_2 = \boldsymbol{U} - \boldsymbol{g}/2, \ \boldsymbol{u}_1' = \boldsymbol{U} + \boldsymbol{g}'/2, \ \boldsymbol{u}_2' = \boldsymbol{U} - \boldsymbol{g}'/2 \,,$$

and this gives

$$\boldsymbol{u}_1 (u_1^2 - \frac{5}{2}) + \boldsymbol{u}_2 (u_2^2 - \frac{5}{2}) - \boldsymbol{u}_1' (u_1'^2 - \frac{5}{2}) - \boldsymbol{u}_2' (u_2'^2 - \frac{5}{2}) = \boldsymbol{g}(\boldsymbol{g}\boldsymbol{U}) - \boldsymbol{g}'(\boldsymbol{g}'\boldsymbol{U}) \,.$$

Using the expression for the Maxwell distribution functions $f_1^{(0)}, f_2^{(0)}$ of particles in accordance with formula (8.11), we obtain

$$b_{11} = N^2 \sqrt{\frac{2T}{m}} \int e^{-2U^2 - g^2/2} \left[\boldsymbol{g}(\boldsymbol{g}\boldsymbol{U}) - \boldsymbol{g}'(\boldsymbol{g}'\boldsymbol{U}) \right]^2 g d\sigma \frac{d\boldsymbol{U} d\boldsymbol{g}}{4\pi^3} \,.$$

Let us introduce the scattering angle ϑ, so that $\boldsymbol{g}\boldsymbol{g}' = \boldsymbol{g}^2 \cos\vartheta$. Take the direction $\boldsymbol{U}$ as the direction of the polar axis, so that the vector $\boldsymbol{g}$ is characterized by the polar angle θ, and $\boldsymbol{g}\boldsymbol{U} = gU\cos\theta$. Correspondingly, $\boldsymbol{g}'\boldsymbol{U} = gU(\cos\vartheta\cos\theta + \sin\vartheta\sin\theta\cos\varphi)$, where φ is the azimuthal angle. Next, $d\boldsymbol{g} = g^2 d\cos\theta d\varphi$. After integration we get

$$b_{11} = 12\sqrt{\frac{T}{\pi m}} N^2 \overline{\sigma_2} \,,$$

where the average cross section is

$$\overline{\sigma_2} = \int_0^\infty t^2 \exp(-t) \sigma^{(2)}(t) dt, \ t = \frac{mg^2}{4T}, \ \sigma^{(2)}(t) = \int (1 - \cos^2\vartheta) d\sigma \qquad (8.53)$$

Correspondingly, the thermal conductivity coefficient is equal in the first Chapman–Enskog approximation

$$\kappa = \frac{25\sqrt{\pi T}}{32\overline{\sigma_2}\sqrt{m}} . \tag{8.54}$$

If the scattering process is described within the framework of the hard sphere model, so that $d\sigma = \frac{\pi R_0^2}{2} d\cos\vartheta$, we have $\sigma_2 = \int(1-\cos^2\vartheta)d\sigma = 2\pi R_0^2/3$, and the first Chapman–Enskog approximation gives for the gas thermal conductivity

$$\kappa = \frac{75\sqrt{T}}{64R_0^2\sqrt{\pi m}} . \tag{8.55}$$

▶ **Problem 8.22** Find the mobility of ions in a gas at low strengths of an external electric field in the first Chapman–Enskog approximation.

In weak electric fields the energy acquired by an ion from the field in a time interval between two successive collisions is of the order of $eE\lambda \sim eE/N\sigma$ ($\lambda \sim 1/(N\sigma)$ is the mean free path of ions in a gas) and much lower than the mean thermal energy of ions T, i. e.,

$$\frac{eE\lambda}{T} \equiv \frac{eE}{TN\sigma} \ll 1 . \tag{8.56}$$

Under condition (8.56), the drift velocity of an ion is much lower than its thermal velocity. Therefore, the velocity distribution function for ions differs slightly from the Maxwell distribution. In accordance with the apparatus of the previous problem, the Chapman–Enskog approximation represents an addition to the Maxwell distribution function due to a weak electric field as an expansion over Sonin polynomials, and the ion mobility is an expansion over a numerical parameter. Restricting ourselves to the first Chapman–Enskog approximation, we use only one constant in this approximation. Then one can use a more simple method in comparison to the analysis of the set of equations as that (8.52) while determining the thermal conductivity coefficient. In particular, we represent the ion distribution function in this case as

$$f(\boldsymbol{v}) = \varphi(v)[1 + v\cos\theta \cdot \psi(v)] , \tag{8.57}$$

where θ is the angle between the vectors $\boldsymbol{v}$ and $\boldsymbol{E}$ and where $\psi(v)$ is assumed to be independent of the collision velocity.

Substituting this expansion into the integral relation (6.33) for ions, one can find ψ. As a result, we obtain the following expression for the ion mobility $K_{\rm I}$ that corresponds to the first Chapman–Enskog approximation

$$K_{\rm I} = \frac{3\sqrt{\pi}e}{8N\overline{\sigma}\sqrt{2T\mu}}, \tag{8.58}$$

where

$$\overline{\sigma} = \int_0^\infty \sigma^*(x)\exp(-x^2)x^2 dx, \qquad x = \frac{\mu g^2}{2T}. \tag{8.59}$$

Formula (8.58) relates to small field strengths according to criterion (8.56) when the ion drift velocity is small compared to the thermal velocity of ions.

▶ **Problem 8.23** Express the ion mobility in a gas using the average cross section of ion–atom collision in usual units.

Usually the mobility is reduced to the normal number density of the gas atoms $N = 2.69 \cdot 10^{19}$ cm^{-3}, which corresponds to the gas temperature of 0 °C and the pressure of 1 atm. Expressing the gas temperature in formula (8.58) in Kelvin, the reduced mass in atomic mass units, and the mean cross section in the units $\pi a_0^2 = 0.879 \cdot 10^{-16}$ cm^{-2}, we reduce this equation to the form

$$K_\mathrm{I} = \frac{2.1 \cdot 10^4}{\overline{\sigma}\sqrt{\mu T}} \frac{\mathrm{cm}^2}{\mathrm{V \cdot s}} \tag{8.60}$$

▶ **Problem 8.24** Determine the diffusion coefficient of an atomic gas with elastic collisions of atoms in the first Chapman–Enskog approximation.

In the Chapman–Enskog approximation for diffusion of test atoms in a gas we represent the distribution function of test atoms as

$$f(\boldsymbol{v}_1) = f^{(0)}(v_1)\left[1 - v_{1x}\frac{\partial \ln N_1}{\partial x} h(\boldsymbol{v}_1)\right],$$

and we take $h(v) = \text{const}$ in the first Chapman–Enskog approximation. Then we find from equation (8.13)

$$h = \frac{3TM}{\mu^2 N \langle g^3 \sigma^*(g)\rangle},$$

where the angle brackets denote an average over the Maxwell distribution for relative velocities of test atoms and gas atoms. The flux of test atoms in the gradient direction is equal to

$$j = \int v_{1x} f(\boldsymbol{v}_1) d\boldsymbol{v}_1 = -\frac{\partial \ln N_1}{\partial x} h \int v_{1x}^2 f_1^{(0)} d\boldsymbol{v}_1 = -\frac{hT}{M}\frac{\partial N_1}{\partial x}.$$

Comparing this with the definition of the diffusion coefficient (8.2), we find for the diffusion coefficient of a test atom in an atomic gas

$$D_{12} = \frac{3T^2}{\mu^2 \langle g^3 \sigma^*(g)\rangle} = \frac{3\sqrt{\pi T}}{8N\sqrt{2\mu}\overline{\sigma}}, \qquad \overline{\sigma} = \frac{1}{2}\int_0^\infty e^{-t} t^2 \sigma^*(t) dt, \qquad t = \frac{\mu g^2}{2T}. \tag{8.61}$$

▶ **Problem 8.25** Determine the viscosity coefficient for an atomic gas in the first Chapman–Enskog approximation while accounting for elastic collisions of atoms.

We take, initially, the flow geometry according to Fig. 6.1, when the average gas velocity $\boldsymbol{w}$ is directed along x and its gradient along z, and star from the integral equation (8.31). The distribution function of atoms is close to the Maxwell distribution function (8.11), and the correction to it is given by formula (8.20), so that in the case of the viscosity transport process, we take the factor of formula (8.20) that is responsible for the viscosity process, and it has the form

$$\theta = -C(v_x - w)v_z \frac{\partial w_x}{\partial z} .$$

Thus, within the framework of the first Chapman–Enskog approximation we use one free parameter C which we will find it from equation (8.5).

Substituting this expression in equation (8.5) and ignoring cross terms with respect to atom velocities, since these terms correspond to other transport processes, we find the parameter C from this equation

$$C = \frac{8NT}{m\int g_x^2 g_z^2 g\sigma^{(2)}(g) f_1^{(0)} f_2^{(0)} d\mathbf{V} dg} = \frac{120T}{mN\langle g^5\sigma^{(2)}(g)\rangle} ,$$

where an average is made with the Maxwell distribution function over relative particle velocities. The force per unit area for this distribution function is

$$F = \int v_z f(\boldsymbol{v}) d\boldsymbol{v} \cdot m(v_x - w) = -C\int m v_z^2 (v_x - w)^2 f^{(0)}(\boldsymbol{v}) d\boldsymbol{v} \frac{\partial w_x}{\partial z} .$$

Since

$$\overline{(v_x - w)^2 v_z^2} = \frac{T^2}{m^2} ,$$

where an average is made over the Maxwell distribution function (8.11), we have from this

$$F = -C\frac{NT^2}{m}\frac{\partial w_x}{\partial z}$$

Thus, we have in the first Chapman–Enskog approximation

$$\eta = C\frac{NT^2}{m} = \frac{120T^3}{m^2\langle g^5\sigma^{(2)}(g)\rangle} .$$

Finally, we obtain the viscosity coefficient

$$\eta = \frac{5\sqrt{\pi T m}}{16\overline{\sigma}_2} ,$$

and the averaged cross section is given by formula (8.53).

▶ **Problem 8.26** Find the viscosity coefficient for assuming atomic gas at an isotropic character of atom scattering within the framework of the hard sphere model.

The differential cross section of atom scattering for the hard sphere model according to formula (2.21) is $d\sigma = \frac{\pi R_o^2}{2} d\cos\vartheta$, where R_o is the radius of the scattering sphere. This gives $\sigma_2 = \int(1-\cos^2\vartheta)d\sigma = 2\pi R_o^2/3$, and from the first Chapman–Enskog approximation it follows for the viscosity coefficient

$$\eta = \frac{15\sqrt{Tm}}{32\sqrt{\pi}R_o^2} . \tag{8.62}$$

▶ **Problem 8.27** Determine the ratio of the thermal conductivity and viscosity coefficients.

We evaluated above the thermal conductivity and viscosity coefficients in three cases: in the tau-approximation, in the first Chapman–Enskog approximation, and also in the case when the collision cross section of gas atoms is inversely proportional to the collision velocity. One can see that averaging over scattering angles proceeds in the same manner for these transport coefficients, and their ratio is

$$\frac{\kappa}{\eta} = \frac{5}{2m} = \frac{c_p}{m} , \tag{8.63}$$

where m is the atom mass, and c_p is the heat capacity per atom for an atomic ideal gas at constant pressure.

8.5
Diffusion of Ions in Gas in an External Electric Field

▶ **Problem 8.28** Derive the expression for the diffusion coefficient of ions, moving in a gas in an external electric field, on the basis of the kinetic equation.

We evaluated above the diffusion coefficient for ions in a gas when external fields are absent and the average drift velocity of ions is zero. We now consider a general case when the drift velocity of ions is nonzero, and the gradient of the number density N_i of ions causes the diffusion flux

$$\boldsymbol{j} = -D\nabla N_i .$$

We use the kinetic equation for the distribution function $f(\boldsymbol{v}, \boldsymbol{r})$ of ions which is

$$\frac{\partial f}{\partial t} + \boldsymbol{v}\nabla f + \frac{e\boldsymbol{E}}{M}\frac{\partial f}{\partial \boldsymbol{v}} = I_{col}(f) ,$$

where M is the ion mass. In the stationary regime the ion distribution function depends on the parameter $\boldsymbol{r} - \boldsymbol{w}t$ instead of $\boldsymbol{r}$, and taking this into account, we reduce the kinetic equation to the form

$$(\boldsymbol{v} - \boldsymbol{w}) f \nabla \ln N_i + \frac{e\boldsymbol{E}}{M}\frac{\partial f}{\partial \boldsymbol{v}} = I_{col}(f) .$$

We now expand this equation over a small parameter $\lambda\partial \ln N_i/\partial x$, where x is the gradient direction, and find the correction to the unperturbed distribution function due to this small parameter. Then we represent the expansion of the distribution function in the form

$$f(\boldsymbol{v},\boldsymbol{r}) = f^{(0)}(\boldsymbol{v}) - \frac{\partial \ln N_i}{\partial z}\Phi(\boldsymbol{v}) ,$$

where the unperturbed distribution function satisfies the equation

$$\frac{e\boldsymbol{E}}{M}\frac{\partial f^{(0)}}{\partial \boldsymbol{v}} = I_{col}(f^{(0)}) ,$$

and from the normalization condition for the distribution function we have

$$\int \Phi(\boldsymbol{v})d\boldsymbol{v} = 0 . \tag{8.64}$$

As it follows from the kinetic equation for the distribution functions $f(\boldsymbol{v},\boldsymbol{r})$ and $f^{(0)}(\boldsymbol{v})$, the equation for the function $\Phi(\boldsymbol{v})$ has the form

$$(v_x - w_x)f^{(0)} + \frac{e\boldsymbol{E}}{M}\frac{\partial \Phi}{\partial \boldsymbol{v}} = I_{col}(\Phi) . \tag{8.65}$$

Here we account for collisions of ions with gas atoms only since the number density of ions is respectively small. Therefore, the collision integral is the linear functional with respect to the function $\Phi(\boldsymbol{v})$. The ion flux is $\boldsymbol{j} = \int \boldsymbol{v} f(\boldsymbol{v})d\boldsymbol{v}$. Taking into consideration that this flux is created due to the ion density gradient and using expression (8.65) for the ion distribution function, we obtain the diffusion coefficient from the comparison of this flux with the diffusion one

$$D_i = \int v_x \Phi(\boldsymbol{v})d\boldsymbol{v} .$$

Using the condition of normalization (8.64) for $\Phi(\boldsymbol{v})$, it is convenient to represent the ion diffusion coefficient in the symmetric form

$$D_i = \int (v_x - w_x)\Phi(\boldsymbol{v})d\boldsymbol{v} .$$

One can use this expression for evaluation of the ion diffusion coefficient.

▶ **Problem 8.29** Prove the Einstein relation for ions on the basis of the kinetic equation.

For small electric field strength we represent the distribution function of ions as $f = \varphi_0 + v_x\varphi_1$, where φ_0 is the Maxwell distribution function and the x-axis is directed along the electric field. Expanding the kinetic equation for the ion distribution function

$$\frac{e\boldsymbol{E}}{M}\frac{\partial f}{\partial \boldsymbol{v}} = I_{col}(f)$$

over a small parameter, we have in the first approximation (the zero-th approximation gives $I_{\rm col}(\varphi_{\rm o}) = 0$)

$$\frac{e\mathbf{E}}{M}\frac{d\varphi_{\rm o}}{d\mathbf{v}} \equiv \frac{eEv_x}{T}\varphi_{\rm 0} = I_{\rm col}(v_x\varphi_1)\,,$$

and since the collision integral is linear with respect to the argument, we have from this

$$\frac{eE}{T}\varphi_{\rm 0} = I_{\rm col}(\varphi_1)\,.$$

This equation coincides with equation (8.65) in the limit $E \to 0$,when $w_x = 0$. From the analogy of these equations it follows the analogy of the values $v_x\varphi_1$ and $\frac{eE}{T}\Phi$. Therefore, according to formula (8.65) the diffusion coefficient is

$$D_{\rm i} = \frac{T}{eE}\int v_x^2\varphi_1 d\mathbf{v}\,,$$

where the gradient of the ion number density is directed along x. On the other hand, the ion drift velocity according to its definition is

$$w_x = \int v_x^2\varphi_1 d\mathbf{v} = EK_{\rm i}\,,$$

where $K_{\rm i}$ is the ion mobility. Comparing these expressions, we find the Einstein relation

$$K_{\rm i} = \frac{eD_{\rm i}}{T}\,. \tag{8.66}$$

▶ **Problem 8.30** Find the drift velocity of ions whose mass is much greater than the mass of the gas atoms.

The width of the ion velocity distribution is determined either by the thermal velocity of ions or by the mean variation of the ion velocity in one collision with a gas atom. The thermal velocity of ions is low compared with the thermal velocity of atoms and the mean ion velocity variation is of the order of mg/M. Hence the width of the ion velocity distribution is always smaller than the relative velocity $\mathbf{g}$ of the ion–atom collision. Therefore, if we use the distribution function of ions in formula (6.33) in the form $f(\mathbf{v}) = \delta(\mathbf{v} - \mathbf{w})$, we obtain

$$\frac{eE}{\mu N} = \left(\frac{m}{2\pi T}\right)^{3/2}\int \exp\left(-\frac{mv_{\rm a}^2}{2T}\right) g_x g\sigma^*(g)d\mathbf{v}_{\rm a}\,,$$

where m is the atom mass, $\mathbf{v}_{\rm a}$ is the atom velocity. This gives

$$\frac{eE}{\mu N} = \frac{1}{w^2}\exp\left(-\frac{mw^2}{2T}\right)\left(\frac{2T}{\pi m}\right)^{1/2}\int_0^\infty \exp\left(-\frac{mg^2}{2T}\right) g^2\sigma^*(g)(y\cosh y - \sinh y)dg\,,$$

$$y = \frac{mwg}{T}\,. \tag{8.67}$$

When the ion drift velocity is small compared to a thermal velocity of gas atoms, it follows from the above

$$w = \frac{3\sqrt{\pi}eE}{8N\overline{\sigma}(2Tm)^{1/2}}\,, \tag{8.68}$$

where

$$\overline{\sigma} = \int_0^\infty \sigma^*(x)\exp(-x^2)x^2dx\,, \quad x = \frac{mg^2}{2T}\,.$$

This result corresponds with the first Chapman–Enskog approximation, but it holds true in a wider velocity range ($w \ll \sqrt{T/m}$) than the Chapman–Enskog approximation is valid ($w \ll \sqrt{T/M}$). Note that the method used of replacing the distribution function with the delta function is applicable if $w \gg \sqrt{T/M}$. However, the above result is valid for all drift velocities, since for $w \ll \sqrt{T/m}$ the drift velocity is proportional to the electric field strength, and the proportionality factor can be found on the basis of this method.

At high drift velocities $w \gg \sqrt{T/m}$, the above formula gives

$$\frac{eE}{mN} = w^2\sigma^*(w)\,.$$

This relation can be obtained directly for the motion equation for a test ion

$$\frac{dP_x}{dt} = eE - \int \Delta P_x Nw d\sigma = eE - mNw^2\sigma^*(w) = 0\,,$$

where P_x is the projection of the ion momentum onto the electric field direction, and $\Delta \mathbf{P} = m\mathbf{w}(1-\cos\vartheta)$ is the momentum variation as a result of ion–atom collision, and ϑ is the scattering angle.

▶ **Problem 8.31** Compare the exact formula for the mobility of electrons in a gas in weak electric fields with the Chapman–Enskog approximation. Take the dependence of the cross section of electron–atom elastic scattering on the collision velocity to be $\sigma^*(v) = Cv^{-k}$.

The mobility of an electron in a weak electric field is given by

$$K = \frac{w}{E} = \frac{8e}{3\sqrt{\pi}m}\int_0^\infty \frac{t^4\exp(-t^2)}{\nu_{ea}}dt\,, \qquad t = \sqrt{\frac{mv^2}{2T}}\,,$$

where ν_{ea} is the rate of electron–atom elastic collisions.

Comparing this expression with that based on the first Chapman–Enskog approximation (8.9), we obtain for a given dependence of the diffusion cross section of electron–atom scattering $\sigma^*(v)$ on the electron velocity v

$$\frac{K}{K_I} = \frac{16}{9\pi}\Gamma\left(3-\frac{k}{2}\right)\Gamma\left(2+\frac{k}{2}\right)\,. \tag{8.69}$$

Note that in most of the actual cases $0 < k < 2$, so that this relation is between 1 and 1.13. In the case of the Coulomb interaction between particles ($k = 4$) collisions involve many particles, while the above equations have been derived for two-particle collisions.

► **Problem 8.32** Find the dependence of the ion drift velocity on the electric field strengths in strong fields. Assuming that the ion–atom scattering has the classical nature and the potential of ion–atom interaction is approximated by the dependence $U(R) \sim R^{-k}$, where R is the distance between the nuclei.

In strong electric fields, which satisfy a criterion that is opposite to (8.56), the ion drift velocity is much higher than the thermal velocity of ions and atoms. In this case the relative velocity of collision between an ion and a gas particle is equal to the ion velocity, and according to equation (6.33) we have

$$\frac{eE}{\mu} = \int f(\boldsymbol{v}) v v_x \sigma^*(v) d\boldsymbol{v} .$$

Since in this limiting case the only parameter with the dimension of velocity, determining the distribution function, is the ion drift velocity, we obtain the above equation $eE/(\mu N) \sim w^2 \sigma^*(w)$. For the interaction potential $U(R) \sim R^{-k}$, the cross section according to formula (2.19) is given by $\sigma \sim \mu^{-2/k} v^{-4/k}$. From this it follows

$$w \sim \frac{1}{\sqrt{\mu}} \left(\frac{eE}{N} \right)^{\frac{k}{2(k-2)}} .$$

► **Problem 8.33** Determine the distribution function of ions which are moving in a gas in an external electric field, if the ion mass is greater than the atom mass. Find the parameters of the velocity distribution function if the ion–atom cross section of elastic scattering is inversely proportional to the collision velocity.

Since the velocity of large ions varies weakly in the course of one ion–atom collision, the velocity distribution function of ions is delta-function. The width of this distribution function is of the order of the ion thermal velocity which is realized in the absence of an electric field. Based on the Gauss distribution, one can represent the ion distribution function in the form

$$f(\boldsymbol{v}) = \text{const} \cdot \exp\left[-\frac{M(v_x - w)^2}{2T_\parallel} - \frac{Mv_\perp^2}{2T_\perp} \right] . \tag{8.70}$$

Here Const is the normalized coefficient, v_x, $v_\perp$ are the ion velocities in the electric field direction and perpendicular to it. In the absence of the electric field the longitudinal $T_\parallel$ and transversal $T_\perp$ temperatures coincide with the gas temperature T. In the limit of small longitudinal and transversal temperatures this distribution function is

$$f(\boldsymbol{v}) = \delta(\boldsymbol{v}) .$$

The longitudinal and transversal temperatures in formula (8.70) can be found from equations (6.39) and (6.40). With an accuracy up to the first expansion term over a small parameter m/M we have

$$T_{\|} \equiv \langle M(v_x - w)^2 \rangle = T + mw^2 \left(1 - \frac{\nu_2}{2\nu_1}\right),$$
$$T_{\perp} \equiv \langle Mv_y^2 \rangle = \langle Mv_z^2 \rangle = T + mw^2 \frac{\nu_2}{4\nu_1} . \qquad (8.71)$$

In particular, it follows from this that if the ion drift velocity is small compared to the thermal atom velocity, the ion distribution function is the Maxwell one and its temperature coincides with the gas, i. e.,

$$f(\boldsymbol{v}) = \exp\left[-\frac{M(\boldsymbol{v} - \boldsymbol{w})^2}{2T}\right] .$$

▶ **Problem 8.34** Determine the diffusion coefficient of ions in a gas in an external electric field if the rate of ion–atom collisions is independent of the collision velocity.

Let us multiply equation (8.65) with the velocity v_z in the direction of the density gradient and integrate the result over velocities. Accounting for the normalization condition (8.64), we obtain

$$\left(\langle v_z^2 \rangle - w_z^2\right) N_i = -\int v_z I_{col}(\Phi) d\boldsymbol{v} ,$$

where brackets mean an average over ion velocities and N_i is the number density of ions. The right-hand side of this equation is

$$-\int v_z I_{col}(\Phi) d\boldsymbol{v} = \int v_z [\Phi(\boldsymbol{v})\varphi(\boldsymbol{v}_a) - \Phi(\boldsymbol{v}')\varphi(\boldsymbol{v}'_a)] |\boldsymbol{v} - \boldsymbol{v}'| d\sigma d\boldsymbol{v} d\boldsymbol{v}_a$$
$$\int (v_z - v'_z) \Phi(\boldsymbol{v})\varphi(\boldsymbol{v}_a) d\sigma d\boldsymbol{v} d\boldsymbol{v}_a .$$

Here $\boldsymbol{v}, \boldsymbol{v}'$ are the ion velocities before and after collision, $\boldsymbol{v}_a, \boldsymbol{v}'_a$ are the atom velocities before and after collision. We used above the detailed balance principle for elastic collisions that is based on the symmetry with respect to the operation $t \leftrightarrow -t$ and conserves the relative collision velocity of colliding particles.

Integrating over scattering angles, as was done for deriving equation (6.33), transforms the above equation to the form

$$\left(\langle v_z^2 \rangle - w_z^2\right) N_i = \frac{m}{m+M} \int g_z \Phi(\boldsymbol{v})\varphi(\boldsymbol{v}_a) g\sigma^* d\boldsymbol{v} d\boldsymbol{v}_a . \qquad (8.72)$$

Assuming that the rate of ion–atom collisions $\nu_1 = N_a g \sigma^*(g)$ is independent of the collision velocity. This transforms equation (8.72) to the form

$$\left(\langle v_z^2 \rangle - w_z^2\right) N_i = \frac{\nu_1 m}{N_a(m+M)} \int (v_z - V_{az}) \Phi(\boldsymbol{v})\varphi(\boldsymbol{v}_a) d\boldsymbol{v} d\boldsymbol{v}_a .$$

Using it in formula (8.41), we obtain the ion diffusion coefficient in the z direction

$$D_z = \frac{M(\langle v_z^2\rangle - w_z^2)}{\mu \nu_1} . \tag{8.73}$$

On the basis of formulas (6.39) and (6.40) for the average ion kinetic energy along the field and perpendicular to it, we obtain the longitudinal

$$D_\parallel = \frac{1}{|mu\nu_1|}\left[T + mw^2 \frac{\nu_1 + (m-2M)\nu_2/(4M)}{\nu_1 + 3m\nu_2/(4M)}\right]$$

and transversal ion diffusion coefficients

$$D_\perp = \frac{1}{|mu\nu_1|}\left[T + mw^2 \frac{(1+m/M)\nu_2}{4(\nu_1 + 3m\nu_2/(4M))}\right] .$$

▶ **Problem 8.35** Determine the transversal diffusion coefficient for heavy ions located in a gas in an external electric field.

Using formula (8.72), we take into account that a typical transversal ion velocity is small compared to the thermal atom velocity at low fields or when compared to the drift ion velocity. This allows us to integrate over atom velocities that transforms formula (8.72) to the form

$$\langle v_z^2\rangle N_\mathrm{i} = \frac{m}{m+M} \int v_z \Phi(v) dv \overline{\nu_1} ,$$

where $\overline{\nu_1}$ is the rate of ion–atom collisions averaged over ion and atom velocities. In the limit of low fields this value is

$$\overline{\nu_1} = N_\mathrm{a} \langle v_\mathrm{a} \sigma^*(v_\mathrm{a})\rangle ,$$

where an average is made over atom velocities. In the other limiting case of high fields

$$\overline{\nu_1} = N_\mathrm{a} w \sigma^*(w) .$$

In any case $\overline{\nu_1}$ does not depend on the transversal ion velocity. This gives

$$D_\perp = \frac{m+M}{m} \frac{\langle v_z^2\rangle}{\overline{\nu_1}} = \frac{T_\perp}{\mu \overline{\nu_1}} ,$$

where $\mu \approx m$ is the reduced ion–atom mass.

▶ **Problem 8.36** Determine the longitudinal diffusion coefficient for heavy ions located in a gas in an external electric field.

If the ion drift velocity is small compared to a typical thermal velocity of gas atoms, the longitudinal ion diffusion coefficient is determined in the same manner as the transversal diffusion coefficient and according to formula (8.73) is equal to

$$D_\parallel = \frac{M(\langle v_x^2\rangle - w_x^2)}{\mu \overline{\nu_1}} = \frac{T_\parallel}{\mu \overline{\nu_1}} ,$$

since $\overline{\nu_1}$ does not depend on the longitudinal ion velocities.

In the other limiting case

$$w \gg \sqrt{\frac{T}{m}} ,$$

we have from formula (8.72)

$$\left(\langle v_x^2 \rangle - w_x^2\right) N_i = \frac{m}{m+M} \int v_z \Phi(\boldsymbol{v}) \nu(\boldsymbol{v}) d\boldsymbol{v} .$$

Let us introduce the parameter

$$\gamma = -\left(\frac{d \ln \nu}{d \ln v}\right)\bigg|_{v=w} ,$$

so that $\nu(v) \sim v^{-\gamma}$. Because the ion velocities are concentrated near w, we have

$$\nu(v) \approx \nu(v) \left[1 + \frac{(v_x - w)}{w}\right] .$$

Substituting this into equation for $\Phi(\boldsymbol{v})$, we obtain that the first expansion term gives zero because of the normalization condition (8.64). As a result, we obtain the longitudinal diffusion coefficient of an ion for strong fields

$$D_{\|} = \frac{(m+M)\left(\langle v_x^2 \rangle - w_x^2\right)}{m(1+\gamma)\nu_1(w)} = \frac{T_{\|}}{\mu(1+\gamma)\nu_1(w)} .$$

▶ **Problem 8.37** Express the ion mobility in a mixture of gases in weak electric fields in terms of the ion mobilities in each of the constituent gases.

The ion mobility in a mixture of gases K in the first Chapman–Enskog approximation (8.58) can be represented as

$$\frac{1}{K} = A\nu_{\text{eff}} .$$

Here A is a function of gas parameters, and ν_{eff} is the effective rate of collisions between ions and atoms of the gas. Since the gas is a mixture of different gases, the total scattering rate is an additive function of the rates of collisions with the particles of each of the gas species. Since we evaluate the ion mobility for the constant total number density of gas atomic particles and the rate of collisions with the particles of the i-th gas species is $\nu_i r = N_i \langle v \sigma_i \rangle$, where the number density of particles of the i-th component is $N_i = c_i N$, where N is the total number density of atomic particles of the gas, and c_i is the concentration of the i-th component, we have $\nu_{\text{eff}} = \sum_i c_i \nu_i$. Introducing the ion mobility K_i in a gas consisting of a pure i-th component according to formula $1/K_i = A\nu_i$, we obtain the ion mobility in a mixture of gases

$$\frac{1}{K} = \sum_i \frac{c_i}{K_i} . \tag{8.74}$$

This relation is known as the Blanck law and it has been derived for weak electric fields only when it has the same accuracy as the Chapman–Enskog approximation.

► **Problem 8.38** Determine the recombination of positive and negative ions in a dense gas in the limit of a large density.

The case of a large gas density is opposite to the case (3.55) of the three-body process, and the following criterion is required for this case

$$\lambda \ll b , \tag{8.75}$$

where $b = e^2/T$ is the critical radius and λ is the mean free path of ions in a gas. Under such circumstances, frequent collisions of the ions with gas particles prevent them from approaching one another, and thus the typical recombination time is essentially the time required for the ions to approach each other. If the distance between ions is R, each ion is subjected to the field produced by the other ion, and the electric field strength is $E = e/R^2$. This field causes oppositely charged ions to move towards each other with the velocity $w = e(K_+ + K_-)/R^2$, where K_+ and K_- are the mobilities of the positive and negative ions in the gas. Such a consideration holds true if $R \gg \lambda$.

The rate of the process when negative ions intersect a sphere around a positive ion of radius R is the product of the surface area of the sphere, $4\pi R^2$, and the negative ion flux $N_- w = N_- e(K_+ + K_-)/R^2$. This leads to the balance equation for the number density of positive ions as

$$\frac{dN_+}{dt} = -N_+ N_- 4\pi e(K_+ + K_-) ,$$

and comparing this with the definition of the recombination coefficient, we obtain the Langevin formula for the recombination coefficient to be

$$\alpha = 4\pi e(K_+ + K_-) . \tag{8.76}$$

We note that from criterion (8.75) it follows that $eE\lambda \ll T$, and the electric field strength E from a positive ion is relatively small until the distance from the positive ion is large compared to the critical distance b. This means that at distances between ions which mainly contribute to the recombination coefficient, the ions move in a weak electric field. Therefore the ion mobilities K_+ and K_- in the Langevin formula (8.76) correspond to small fields. Next, because these mobilities are inversely proportional to the number density of gas atoms, the recombination coefficient of ions has the same dependence on the number density of gas atoms. This means that an increase in gas density leads to an increase in the frictional force for ions that slows the approach of the ions.

► **Problem 8.39** Analyze the dependence of the recombination coefficient for positive and negative ions on gas density in a wide range of gas densities.

At very low number density of gas atoms the exchange cross section is estimated by formula (2.28), and the rate of this process does not depend on the number density of gas atoms. At higher number densities which satisfy to the criterion (3.58) this process has a three-body character, its rate is proportional to the number density of gas atoms and is given by formula (3.57). At high number densities the rate of mutual neutralization is given by formula (8.76) and is inversely proportional to the number density of gas atoms.

Figure 8.1 shows the dependence of the rate of the charge exchange process on the number density of gas atoms that is deduced on the basis of the above consideration. In the limit of the low number density of gas atoms (range 1), the recombination coefficient of positive and negative ions at room temperature is estimated as $\alpha_1 \geq \hbar^2/(m_e^2 \mu T)^{1/2}$, if we take $R_o \gg a_o$ in formula (2.28), where μ is the reduced mass of ions, and $a_o = \hbar^2/(m_e e^2)$ is the Bohr radius. At higher number densities N of atoms (range 2 of Fig. 8.1) the recombination coefficient is given by formula (3.57) $\alpha_2 \sim N(e^6/T^3)(\beta e^2/\mu)^{1/2}$, where β is the polarizability of the particle C. In the range 3 of Fig. 8.1 according to formulas (8.76) and (8.16) the recombination coefficient is $\alpha_3 \sim e\sqrt{\mu}/(N\sqrt{\beta})$, where β is the atom polarizability.

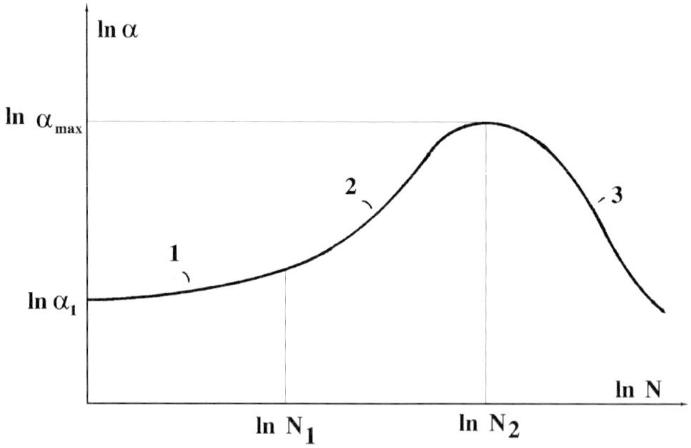

Fig. 8.1 The dependence of the recombination coefficient α of positive and negative ions in a gas on the number density of gas atoms or molecules: 1 — pair recombination, 2 — three body recombination, 3 — the Lanzhevin mechanism when approach of positive and negative ions is hampered by collisions with gas atoms.

We estimate the number densities of transient regions, so that the number density of gas atoms N_{12} between ranges 1 and 2, and also the number density N_{23} between ranges 2 and 3 are equal, if we take the atom polarizability $\beta \sim a_o^3$ and $R_o \sim a_o$

$$N_{12} \sim \frac{T^{5/2}}{e^5\sqrt{a_o}}, \qquad N_{23} \sim \left(\frac{T}{a_o e^2}\right)^{3/2}.$$

The recombination coefficient on the boundaries of these regions is

$$\alpha_{12} \sim \alpha_1 \sim \frac{e^2 a_o}{\sqrt{T\mu}}, \qquad \alpha_{23} \sim \frac{e^4}{\mu^{1/2} T^{3/2}}.$$

One can see that the maximum recombination coefficient in number density corresponds to the largest recombination α_{23}, which corresponds to the Coulomb cross section of ion scattering.

We make the above estimations for air at room temperature. According to the above estimations, we have $N_{12} \sim 10^{17}\ \mathrm{cm}^{-3}$, $N_{23} \sim 10^{20}\ \mathrm{cm}^{-3}$, $\alpha_{12} \sim 10^{-9}\ \mathrm{cm}^3/\mathrm{s}$, and $\alpha_{23} \sim 10^{-6}\ \mathrm{cm}^3/\mathrm{s}$. As is seen, the maximum recombination coefficient corresponds to approximately atmospheric pressure. Figure 8.2 gives experimental values for the recombination coefficients of positive and negative ions in air at room temperature as a function of pressure. This confirms the above estimations.

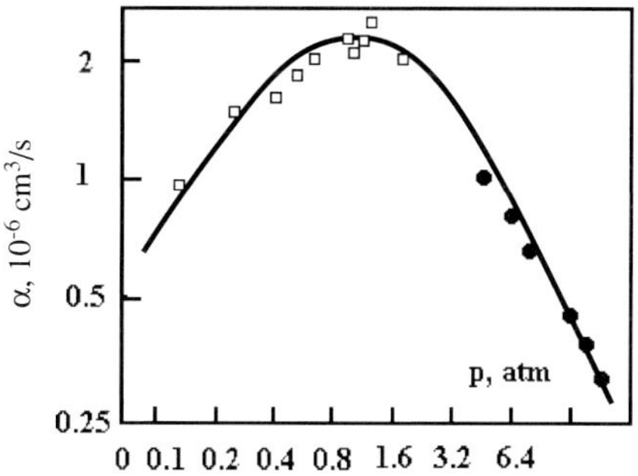

Fig. 8.2 The recombination coefficient α of positive and negative ions in air at room temperature depending on its pressure. Signs are experimental data, solid curve is their approximation.

8.6
Transport of Atomic Ions in the Parent Gas in an External Electric Field

▶ **Problem 8.40** Determine the mobility of atomic ions in the parent gas in the first Chapman–Enskog approximation at low electric field strengths.

In this case the ion–atom scattering is determined by the resonant charge exchange process, and if elastic ion–atom scattering is weak, the character of ion–atom scattering is represented in Fig. 8.3 (the Sena effect). Then as a result of resonant charge exchange, an incident atom is transformed into an ion, and therefore the ion acquires the energy of a former atom.

The diffusion ion–atom cross section (2.6) is equal in this case

$$\sigma^* = \int_0^\infty [1 - P(\rho)](1 - \cos\vartheta)2\pi\rho d\rho + \int_0^\infty P(\rho)(1 - \cos\vartheta')2\pi\rho d\rho,$$

where $P(\rho)$ is the probability of resonant charge exchange for an impact parameter ρ of collision, ϑ is the scattering angle for the nucleus which initially belonged to

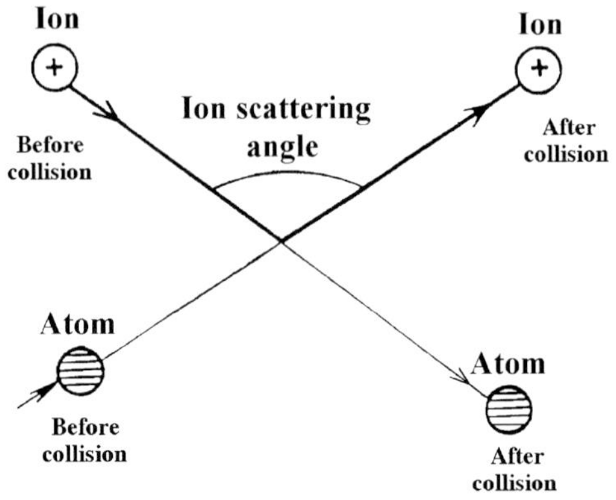

Fig. 8.3 The Sena effect for ion scattering as a result of resonant charge exchange if atom and ion are moving along straightforward trajectories for capture particles.

the ion in the center-of-mass reference frame, ϑ' is the angle between the velocity of this nucleus before collision and the velocity of another nucleus after collision in the center-of-mass reference frame; this angle describes the ion scattering if resonant charge transfer occurs. Since $\vartheta + \vartheta' = \pi$, we have

$$\sigma^* = \int_0^\infty (1 - \cos\vartheta) \cdot 2\pi\rho d\rho + \int_0^\infty P(\rho)\cos\vartheta \cdot 2\pi\rho d\rho \,. \tag{8.77}$$

In the absence of ion–atom elastic scattering ($\vartheta = 0$), the diffusion cross section of ion scattering is

$$\sigma^* = \int_0^\infty P(\rho) \cdot 2\pi\rho d\rho = 2\sigma_{\text{res}} \,, \tag{8.78}$$

where σ_{res} is the cross section of resonant charge transfer.

Substituting this diffusion cross section into the mobility that corresponds to the first Chapman–Enskog approximation, we obtain the ion mobility in the parent gas after accounting for a weak dependence of the resonant charge exchange cross section on the collision velocity

$$K_{\text{I}} = \frac{3\sqrt{\pi}e}{16N\sqrt{TM}\sigma_{\text{res}}} \,, \tag{8.79}$$

where $M = 2\mu$ is the mass of the ion or atom and μ is their reduced mass. The main contribution to the mobility corresponds to the relative collision velocity $g = \sqrt{5T/\mu} = \sqrt{10T/M}$, and the cross section of resonant charge exchange in formula (8.79) is taken at this collision velocity.

Formula (8.79) is valid at low field strengths in accordance with the criterion (8.56) when an energy ($\sim eE\lambda$) that an ion obtains from the field between two nearest exchange events is small compared to the thermal ion energy T.

▶ **Problem 8.41** Determine the drift velocity of ions in the parent gas at low electric fields in the second Chapman–Enskog approximation while ignoring ion–atom elastic scattering.

Since the charge exchange process in the absence of ion–atom elastic scattering leads to a change in velocities between an ion and atoms, the Boltzmann kinetic equation (6.2) with the collision integral (6.7) takes the following form in this case:

$$\frac{\partial f}{\partial v_x} = \int \left[f(\boldsymbol{v}')\varphi(v) - f(\boldsymbol{v})\varphi(v)'\right] |\boldsymbol{v} - \boldsymbol{v}'| N_a \sigma_{res} d\boldsymbol{v}' .$$

Here the electric field is directed along the x-axis, $\boldsymbol{v}$, $\boldsymbol{v}'$ are the ion and atom velocity before collision, and N_a is the number density of atoms. The ion distribution function $f(\boldsymbol{v})$ of ions and the Maxwell distribution function $\varphi(v)$ of atoms are normalized to unity.

At low electric field strengths we represent the ion distribution function in the form

$$f(\boldsymbol{v}) = \varphi(v)\left[1 + \frac{eE\lambda}{T} v_x \psi(v)\right] ,$$

and then the kinetic equation is reduced to the following equation

$$v_x \varphi(v) = \int |\boldsymbol{v} - \boldsymbol{v}'| \varphi(v)\varphi(v)' \left[v_x \psi(v) - v'_x \psi(v')\right] d\boldsymbol{v}' , \qquad (8.80)$$

and we use in this derivation $\partial\varphi(v)\partial v_x = -M v_x \varphi(v)/T$. It is convenient to expand $\psi(v)$ over Sonin polynomials

$$\psi(v) = \sum_{n=0}^{n=\infty} a_n S^n_{3/2}\left(\frac{Mv^2}{2T}\right) ,$$

and the ion mobility is

$$K = \frac{\langle v_x \rangle}{E} = \frac{e\lambda}{M} a_o .$$

We use a general method to solve the kinetic equation (8.80), if we multiply it by $S^m_{3/2}(Mv^2/2T)$ and integrate over ion velocities. This leads to the following set of equations:

$$\delta_{m0} = \int \frac{Mv_x}{T} S^m_{3/2}\left(\frac{Mv^2}{2T}\right) |\boldsymbol{v} - \boldsymbol{v}'| \varphi(v)\varphi(v)' \cdot \sum_n a_n \left[v_x S^n_{3/2}\left(\frac{Mv^2}{2T}\right) - v'_x S^n_{3/2}\left(\frac{M(v')^2}{2T}\right)\right] d\boldsymbol{v} d\boldsymbol{v}' .$$

It is convenient to rewrite this equation in the symmetric form after taking into account quantities a_n, which have the dimensionality $\sqrt{M/T}$. Thus, we have

$$\sum_n a_n \frac{M}{2T} \int \left[v_x S^m_{3/2}\left(\frac{Mv^2}{2T}\right) - v'_x S^m_{3/2}\left(\frac{M(v')^2}{2T}\right)\right] \cdot \left[v_x S^n_{3/2}\left(\frac{Mv^2}{2T}\right) - v'_x S^n_{3/2}\left(\frac{M(v')^2}{2T}\right)\right] |\boldsymbol{v} - \boldsymbol{v}'| \varphi(v)\varphi(v)' d\boldsymbol{v} d\boldsymbol{v}' = \delta_{m0} .$$

Let us solve this set of equations in the second Chapman–Enskog approximation, using the reduced velocities $\boldsymbol{u} = \sqrt{M/(2T)}\boldsymbol{v}$ and $\boldsymbol{u}' = \sqrt{M/(2T)}\boldsymbol{v}'$. Restricting two expansion terms of $\psi(v)$ over Sonin polynomials, we have

$$a_0 \left\langle (u_x - u'_x)^2 |\boldsymbol{u} - \boldsymbol{u}'| \right\rangle$$
$$+ a_1 \left\langle (u_x - u'_x) \left[u_x \left(\frac{5}{2} - u^2 \right) - u'_x \left(\frac{5}{2} - (u')^2 \right) \right] |\boldsymbol{v} - \boldsymbol{v}'| \right\rangle = \sqrt{\frac{M}{2T}} ,$$

where brackets mean an average on the basis of the Maxwell distribution function of ions and atoms.

Introducing the reduced ion–atom velocity $\boldsymbol{s} = \boldsymbol{u} - \boldsymbol{u}'$ and the reduced velocity of the ion-atom center of mass $\boldsymbol{S} = (\boldsymbol{u} - \boldsymbol{u}')/2$, we have

$$u_x \left(\frac{5}{2} - u^2 \right) - u'_x \left(\frac{5}{2} - (u')^2 \right) = s_x \left(\frac{5}{2} - S^2 \frac{1}{4} s^2 \right) - 2S_x \boldsymbol{s}\boldsymbol{S} ,$$

and the above set of two equations is reduced to the form

$$a_0 \langle s_x^2 s \rangle + a_1 \left\langle s_x s \left[s_x \left(\frac{5}{2} - S^2 \frac{1}{4} s^2 \right) - 2S_x \boldsymbol{s}\boldsymbol{S} \right] \right\rangle = \sqrt{\frac{M}{2T}}$$
$$a_0 \left\langle s_x s \left[s_x \left(\frac{5}{2} - S^2 \frac{1}{4} s^2 \right) - 2S_x \boldsymbol{s}\boldsymbol{S} \right] \right\rangle$$
$$+ a_1 \left\langle s \left[s_x \left(\frac{5}{2} - S^2 \frac{1}{4} s^2 \right) - 2S_x \boldsymbol{s}\boldsymbol{S} \right] \right\rangle = \sqrt{\frac{M}{2T}} = 0 .$$

In particular, in the first Chapman–Enskog approximation we restrict ourselves only to the first equation and assume $a_1 = 0$. This gives,

$$a_0 = \frac{3}{\langle s^2 \rangle} \sqrt{\frac{M}{2T}} = \frac{3\sqrt{\pi M}}{16\sqrt{T}} ,$$

which gives for the mobility

$$K = \frac{e\lambda a_0}{M} = \frac{3\lambda\sqrt{\pi}}{16M\sqrt{T}} ,$$

that corresponds to formula (8.58).

Integrating the above set of equations over angles, we reduce it to the form

$$a_0 \frac{\langle s^2 \rangle}{3} + a_1 \left(\frac{5}{6} \langle s^3 \rangle - \frac{5}{2} \langle s^3 \rangle \langle S^2 \rangle - \frac{\langle s^5 \rangle}{12} \right) = \sqrt{\frac{M}{2T}} ;$$
$$a_0 \left(\frac{5}{6} \langle s^3 \rangle - \frac{5}{9} \langle s^3 \langle S^2 \rangle - \frac{\langle s^5 \rangle}{12} \right)$$
$$+ a_1 \left[\langle s^3 \rangle \left(\frac{25}{12} - \frac{25}{3} \langle S^2 \rangle + \frac{10}{3} \langle S^4 \rangle \right) + \langle s^5 \rangle \left(\frac{5}{18} \langle S^2 \rangle - \frac{5}{12} \right) + \frac{\langle s^7 \rangle}{48} \right] = 0 .$$

After evaluation of the average values of the reduced center-of-mass velocities, we reduce this set to the form

$$\frac{\langle s^3\rangle}{3}\left(a_0-\frac{a_1}{4}\right)=\sqrt{\frac{M}{2T}}\,,\ \frac{\langle s^3\rangle}{12}\left(a_0-\frac{31a_1}{2}\right)=0\,.$$

From this, we have

$$a_0^{(2)}=\frac{62}{63}a^{(1)}=\frac{31M\sqrt{\pi}}{168\sqrt{T}}\,,$$

and the superscript means the Chapman–Enskog approximation has been used. From this it follows that the result of the second Chapman–Enskog approximation differs from that of the first by approximately 2 %. This confirms an excellent convergence for the Chapman–Enskog method. Finally, in the second approximation we obtain the ion mobility in a parent gas

$$K=\frac{e\lambda a_0}{M}=\frac{31\lambda\sqrt{\pi}}{168\sqrt{TM}}=\frac{0.327}{N_a\sigma_{res}\sqrt{TM}}\,,$$

instead of formula (8.58) for the first Chapman–Enskog approximation.

▶ **Problem 8.42** Determine the field correction for the atomic ion mobility in a parent gas in the limit of low fields if ion–atom elastic scattering is ignored.

A small parameter is proportional to the electric field strength, but since a small parameter is a scalar value, it contains a combination $\boldsymbol{E}\boldsymbol{v}$, where $\boldsymbol{v}$ is the ion velocity. Hence, we represent the ion distribution function as the following expansion

$$F(\boldsymbol{v})=\varphi(v)\left[1+\sum_{n=1}^{\infty}\xi^n v_x^n\chi_n(v)\right]\,,$$

where $\varphi(v)$ is the Maxwell distribution function, and a small reduced parameter is

$$\xi=\frac{eE\lambda}{T}\,.$$

Substituting this expansion in the kinetic equation and extracting in it terms of an identical degree with respect to a small parameter ξ, one can find values χ_n. Then the ion mobility is

$$K=\frac{\langle v_x\rangle}{E}=\frac{e\lambda}{T}\left(\langle v_x^2\chi_1\rangle+\xi^2\langle v_x^4\chi_3\rangle+\cdots\right)\,,$$

and extracting terms with identical degrees E^N in the kinetic equation, we get

$$\frac{T}{M}\frac{d[v_x^{n-1}\chi_{n-1}(v)\varphi(v)]}{dv_x}$$
$$=\int|\boldsymbol{v}-\boldsymbol{v}'|\varphi(v)\varphi(v')[(v_x')^n\chi_n(v')-v_x^n\chi_n(v)]d\boldsymbol{v}'[(v_x')^n-v_x^n]d\boldsymbol{v}'\,.$$

Taking χ_n for small n to be independent of the velocity, we reduce this equation to the form

$$\chi_{n-1}\frac{T}{M}\frac{d[v_x^{n-1}\varphi(v)]}{dv_x} = \chi_n \int |\boldsymbol{v}-\boldsymbol{v}'|\varphi(v)\varphi(v')[(v'_x)^n - v_x^n]d\boldsymbol{v}' \, .$$

For the solution of this equation we transform it into an integral equation by multiplying v_x^n and integrating over ion velocities. Integrating the left-hand side of the equation in parts and reducing the right-hand side to the symmetric form, we obtain

$$\chi_{n-1}\frac{T}{M}\left\langle v_x^{2n-2}\right\rangle = \frac{1}{2}\chi_n \left\langle |\boldsymbol{v}-\boldsymbol{v}'|[(v'_x)^n - v_x^n]\right\rangle d\boldsymbol{v}' \, .$$

This recurrent relation allows us to connect the previous and subsequent expansion terms. In particular, for first terms we have

$$\chi_2 = \frac{4T}{M}\chi_1\frac{\langle v_x^2\rangle}{\langle |\boldsymbol{v}-\boldsymbol{v}'|[(v'_x)^2 - v_x^2]^2\rangle}$$

$$\chi_3 = \frac{6T}{M}\chi_2\frac{\langle v_x^4\rangle}{\langle |\boldsymbol{v}-\boldsymbol{v}'|[(v'_x)^3 - v_x^3]^2\rangle} \, .$$

From this we find the first two terms of the mobility expansion over a small parameter

$$K = \frac{e\lambda}{M}(\chi_1\langle v_x^2\rangle + \xi^2\chi_3\langle v_x^4\rangle) = K_o\left(1+\xi^2\frac{\chi_3\langle v_x^4\rangle}{\chi_1\langle v_x^2\rangle}\right) =$$
$$K_o\left[1+\xi^2\frac{24T^2\langle v_x^4\rangle^2}{M^2\langle |\boldsymbol{v}-\boldsymbol{v}'|[v_x^2-(v'_x)^2]^2\rangle\langle |\boldsymbol{v}-\boldsymbol{v}'|[v_x^3-(v'_x)^3]^2\rangle}\right] ,$$

where K_o is the ion mobility in the limit of zero electric field strength.

We now evaluate the average values which are inserted in the above formula. Thus, we have

$$\langle v_x^4\rangle = \frac{15}{4}\left(\frac{2T}{M}\right)^2 = \frac{15T^2}{M^2} \, .$$

For evaluation of the denominator terms, we introduce the reduced relative velocity $\boldsymbol{s} = \sqrt{M/(2T)}(\boldsymbol{v}-\boldsymbol{v}')$ and the reduced velocity of the center of mass $\boldsymbol{S} = \sqrt{M/(2T)}(\boldsymbol{v}+\boldsymbol{v}')/2$. We have

$$\left(\frac{M}{2T}\right)^{5/2}\langle |\boldsymbol{v}-\boldsymbol{v}'|[v_x^2-(v'_x)^2]^2\rangle = 4\langle s s_x^2 S_x^2\rangle = \frac{4}{9}\langle s^2\rangle\langle S^2\rangle = \frac{8\sqrt{2}}{3\sqrt{\pi}}$$

$$\left(\frac{M}{2T}\right)^{7/2}\langle |\boldsymbol{v}-\boldsymbol{v}'|[v_x^3-(v'_x)^3]^2\rangle = \left\langle s s_x^2\left(3S_x^2+\frac{s_x^2}{4}\right)^2\right\rangle$$
$$= \frac{3}{5}\langle s^3\rangle\langle S^4\rangle + \frac{1}{10}\langle s^5\rangle\langle S^2\rangle + \frac{1}{112}\langle s^7\rangle = \frac{3\cdot 169\sqrt{2}}{70\sqrt{\pi}} \, .$$

From this it follows for the first two terms of expansion for the ion mobility

$$K = K_o(1+6.9\xi^2) \, . \tag{8.81}$$

► **Problem 8.43** Determine the drift velocity of ions in the parent gas in the limit of strong electric fields.

The criterion of strong electric fields is inverse with respect to the criterion (8.56) and has the form

$$\frac{eE}{TN\sigma} \gg 1\,, \tag{8.82}$$

so the drift velocity of ions is much higher than the thermal velocity of atoms. Therefore, the atoms can be assumed to be motionless. Hence the ions will stop after each charge transfer event, and subsequently will be accelerated by the action of the electric field. The probability $P_1(t)$ that the last charge exchange event for a test ion occurs at a time t earlier, is given by the equation

$$\frac{dP_1}{dt} = -\nu P_1, \quad P_1 = \exp(-\int_0^t \nu dt')\,.$$

Here $\nu = Nv_x\sigma_{\rm res}$, v_x is the ion velocity that is directed along the field, N is the number density of atoms, and $\sigma_{\rm res}$ is the cross section of resonant charge exchange. The ion velocity at time t after the charge exchange event follows from the Newton equation

$$\frac{Mdv_x}{dt} = eE\,,$$

that gives $v_x = eEt/M$. Taking into consideration that the velocity distribution function $f(v_x)$ is proportional to the probability $P(t)$ and connecting a time after the last charge transfer with the ion velocity, we obtain the distribution function

$$f(v_x) = C\exp(-\int_0^{v_x} \nu\frac{M}{eE}dv'_x)\,, \quad v_x > 0\,.$$

This is the velocity distribution function of ions, and C is the normalization constant. Assuming the cross section of resonant charge exchange to be independent of the collision velocity, we obtain the distribution function and the ion drift velocity

$$f(v_x) = C\exp\left(-\frac{Mv_x^2}{2eE\lambda}\right)\eta(v_x), \quad C = \sqrt{\frac{2M}{\pi eE\lambda}}, \quad w = \sqrt{\frac{2eE\lambda}{\pi M}}\,. \tag{8.83}$$

► **Problem 8.44** Determine the velocity distribution function for atomic ions moving in the parent gases under an external electric field.

We use the fact that kinetics of atomic ions in the parent gas is determined by the resonant charge exchange process, and assume that elastic ion–atom scattering may be ignored. Then the collision of an ion with the velocity $\boldsymbol{v}$ and an atom with a velocity $\boldsymbol{v}'$ leads after the resonant charge exchange event to formation of an atom with velocity $\boldsymbol{v}$ and an ion with velocity $\boldsymbol{v}'$. Next, we assume that the ion distribution function $f(\boldsymbol{v})$ can be expanded in the longitudinal and transversal directions

$$f(\boldsymbol{v}) = f(v_x)F(v_\perp) .$$

In this case the kinetic equation can also be expanded in the transversal and longitudinal directions. Because of the absence of an electric field in the transversal direction, the kinetic equation for this distribution function is $I_{\text{col}}(F) = 0$. Using expression (6.7) for the collision integral, we have this equation in the form

$$F(v_\perp)\int \sigma_{\text{res}}|\boldsymbol{v}-\boldsymbol{v}'|f(v_x)\varphi(v'_x)\varphi(v'_\perp)dv_x d\boldsymbol{v}' =$$

$$\varphi(v_\perp)\int \sigma_{\text{res}}F(v'_\perp)|\boldsymbol{v}-\boldsymbol{v}'|f(v'_x)\varphi(v_x)dv_x d\boldsymbol{v}' .$$

Here $\varphi(v_x), \varphi(v_\perp)$ are the Maxwell distribution functions of atoms for the transversal and longitudinal directions of atom motion. One can see that the solution of this equation is the Maxwell distribution function for the transversal motion of ions

$$F(v_\perp) = \varphi(v_\perp) = \frac{M}{T}\exp\left(-\frac{Mv_\perp^2}{2T}\right) . \tag{8.84}$$

In deriving this formula we assume that the transversal ion velocity is independent of the longitudinal one. This is justified since an external electric field does not act on the transversal velocity component, and also the character of ion–atom scattering does not depend on it. Hence, the transversal ion velocity is taken from an atom after the charge exchange event, and therefore the ion distribution function on transversal velocity is the Maxwell one independently on the electric field strength.

► **Problem 8.45** Determine the average kinetic energy of atomic ions moving in the parent gases under an external electric field for different directions of ion motion ignoring elastic ion–atom scattering in comparison with the charge exchange process.

We assume transport of atomic ions in the parent gas to be resulted from the resonant charge exchange process, as it is given in Fig. 8.3. This means that colliding ion and atom move along straightforward trajectories, i. e., the cross section of their elastic scattering is small compared to the resonant charge exchange cross section. From this it follows that the ion–atom scattering cross section has the form

$$d\sigma = \sigma_{\text{res}}\delta(1+\cos\vartheta)d\cos\vartheta .$$

This means that the scattering angle is $\vartheta = \pi$ if the charge exchange events proceeds and gives formula (8.78) for the transport cross section of ion–atom scattering. In addition, $\sigma^{(2)} = \int(1-\cos^2\vartheta)d\sigma = 0$ in this case.

We now use formulas (6.39) and (6.40) for the ion kinetic energies in different directions. From these formulas it follows

$$\left\langle \frac{Mv_x^2}{2} \right\rangle = \frac{T}{2} + \frac{Mw^2}{2}, \qquad \left\langle \frac{Mv_y^2}{2} \right\rangle = \left\langle \frac{Mv_z^2}{2} \right\rangle = \frac{T}{2}. \tag{8.85}$$

Here v_x is the ion velocity component along the field and v_y, v_z are the velocity components in the perpendicular direction to the field, M is the ion mass, w is the ion drift velocity. As it follows from formula (8.85), atomic ions have a thermal energy in the transversal directions even at large fields.

► **Problem 8.46** Determine the longitudinal diffusion coefficient for atomic ions moving in the parent atomic gas under a large electric field. The resonant charge exchange cross section is independent of the collision velocity.

Denoting the initial ion and atom velocities as $\boldsymbol{v}$, $\boldsymbol{v}'$, we found that their velocities are changed after collision. Therefore, the kinetic equation (6.2) with the collision integral (6.7) has the following form in this case:

$$\frac{eE}{M}\frac{\partial f}{\partial v_x} = \int [f(v'_x)\varphi(v'_x) - f(v_x)\varphi(v_x)]|v_x - v'_x|\sigma_{\text{res}} dv'_x .$$

Here $f(v_x)$ is the ion distribution function and $\varphi(v_x)$ is the Maxwell distribution function. We consider that the transversal ion velocities are of the order of thermal atom velocities, and since a typical atom velocity is small compared to the ion drift velocity, we obtain the atom distribution function as

$$\varphi(v_x) = N_a\delta(v_x) ,$$

and the above kinetic equation is transformed to the form

$$\frac{eE}{M}\frac{\partial f}{\partial v_x} = \frac{\delta(v_x)}{\lambda}\int v'_x f(v'_x) dv'_x - \frac{v_x}{\lambda} f(v_x) .$$

Here N_a is the number density of atoms, $\lambda = (N_a\sigma_{\text{res}})^{-1}$. From this we derive equation (8.65) for the function $\Phi(v_x)$ through which the ion diffusion coefficient is expressed as

$$(v_x - w)f(v_x) = \frac{eE}{M}\frac{d\Phi}{dv_x} + \frac{v_x}{\lambda}\Phi(v_x) - \frac{\delta(v_x)}{\lambda}\int v'_x\Phi(v'_x)dv'_x .$$

It is convenient to use the reduced variable

$$u = \left(\frac{Mv_x^2}{eE\lambda}\right)^{1/2} = v_x\left(\frac{MN_a\sigma_{\text{res}}}{eE}\right)^{1/2} = \frac{2}{\sqrt{\pi}}\frac{v_x}{w} .$$

Using the ion distribution function (9.44), we reduce this equation to the form

$$\frac{2N_i\lambda}{\pi w}\left(u - \sqrt{\frac{2}{\pi}}\right)\exp(-u^2)\eta(u) = \frac{d\Phi}{du} + u\Phi - \delta(u)\int u'\Phi(u')du' ,$$

where N_i is the ion number density and $\eta(u)$ is the unit function. The solution of this equation is

$$\Phi(u) = \frac{2\lambda}{\pi w} N_i \exp(-u^2)\left(\frac{u^2}{2} - \sqrt{\frac{2}{\pi}}u + A\right)\eta(u) .$$

The integration constant A follows from the normalization condition (8.64) and is equal to $A = 2/\pi - 1/2$, which gives

$$\Phi(u) = \frac{2\lambda}{\pi w} N_i \exp(-u^2)\left(\frac{u^2}{2} - \sqrt{\frac{2}{\pi}}u + \frac{2}{\pi} - \frac{1}{2}\right)\eta(u) .$$

From this we find the diffusion coefficient of ions in the parent gas along the field

$$D_{\parallel} = \frac{1}{N_i}\int (v_x - w)\Phi dv_x = \frac{1}{N_i}\frac{eE\lambda}{M}\int \Phi(u) u du = \lambda w\left(\frac{2}{\pi} - \frac{1}{2}\right) = 0.137 w\lambda . \tag{8.86}$$

▶ **Problem 8.47** Determine the transversal diffusion coefficient for atomic ions moving in the parent atomic gas under a large electric field. The resonant charge exchange cross section is independent of the collision velocity.

According to formula (8.84) the ion distribution function on transversal velocity components is the Maxwell one in this case, and hence equation (8.65) for the transversal direction z takes the form

$$v_z \varphi(v_z) = \frac{w}{\lambda}\Phi(v_z) - \frac{w}{\lambda}\varphi(v_z)\int \Phi(v_z')v_z' .$$

The last integral in the right-hand side of this equation is zero according to the normalization condition (8.64). Hence, the solution of this equation is

$$\Phi(v_z) = \frac{\lambda v_z \varphi(v_z)}{w} .$$

This gives for the transversal diffusion coefficient of ions

$$D_{\perp} = \frac{\lambda \langle v_z^2 \rangle}{w} = \frac{\lambda T}{M w} . \tag{8.87}$$

As it follows from this formula, at high fields the transversal diffusion coefficient of ions diminishes in $\sim \sqrt{eE\lambda/T}$ times compared to that without fields.

9
Kinetics and Radiative Transport of Excitations in Gases

9.1
Resonant Radiation of Optically Thick Layer of Excited Gas

► **Problem 9.1** For the Lorentz and Doppler shape of a spectral line find the probability $P(r)$ that a resonant photon crosses a distance r that exceeds significantly the mean free path of the photon for the center of a spectral line. Assume a gas to be uniform.

This probability is the product of the probability $a_\omega d\omega$ for emission at a given frequency and the probability $e^{-k_\omega r}$ of photon surviving, i. e., this quantity is equal to

$$P(r) = \int a_\omega d\omega e^{-k_\omega r} . \tag{9.1}$$

For the Lorentz line shape (5.5) we obtain, using a new variable $s = (\omega - \omega_o)/\nu$,

$$a_\omega d\omega = \frac{ds}{\pi(1+s^2)}, \qquad k_\omega = \frac{k_0}{1+s^2} ,$$

where k_0 is the absorption coefficient for the spectral line center. Let us define $u = k_0 r$ the optical thickness in this way for the center line frequency; according to the problem condition, $u \gg 1$. Formula (9.1) gives the probability for the photon to cross a given distance

$$P(r) = \frac{1}{\pi}\int_{-\infty}^{\infty} \frac{ds}{(1+s^2)} \exp\left(-\frac{u}{1+s^2}\right) = \int_0^{\pi} d\varphi \exp(-u\cos^2\frac{\varphi}{2}) = e^{-u/2} I_0\left(\frac{u}{2}\right) ,$$

where $\varphi = \arctan s$ and $I_0(x)$ is the Bessel function. In the limit under consideration we have

$$P(r) = \frac{1}{\sqrt{\pi u}}, \qquad u \gg 1 . \tag{9.2}$$

For the Doppler shape of the spectral line we use the variable

$$t = u \exp\left[-\frac{mc^2}{2T}\left(\frac{\omega - \omega_0}{\omega_0}\right)^2\right]$$

Plasma Processes and Plasma Kinetics. Boris M. Smirnov

ISBN: 978-3-527-40681-4

that gives the probability to pass a distance r for the resonance photon

$$P(r) = \frac{1}{\sqrt{\pi u}} \int_0^u e^{-t} dt \left(\ln \frac{u}{t}\right)^{-1} = \frac{1}{\sqrt{\pi u}\sqrt{\ln u + C}}, \quad u \gg 1, \tag{9.3}$$

where $C = 0.577$ is the Euler constant.

Because wings of the Doppler spectral line drop sharper than that in the case of the Lorentz spectral line, the probability to propagate on large distances for the Lorentz spectral line is larger than that the Doppler one.

▶ **Problem 9.2** Derive an expression for the flux of resonant photons of a given frequency and the total flux of photons from a given gas volume. Resonant photons originate from radiative transitions of excited atoms.

Let us take a point from the surface of this volume and evaluate the flux that results from all the points inside the gas volume and therefore is the integral from the following factors:

$$j_\omega d\omega = \int \frac{N_*(\boldsymbol{r})d\boldsymbol{r}}{\tau} a_\omega d\omega \frac{1}{4\pi r^2} \exp\left(-\int_0^r k_\omega dr'\right) \cos\theta. \tag{9.4}$$

Here the first term is the rate of generation of photons in a volume element, the second is the probability of photon generation in a given frequency range, the third converts the rate in the flux under the assumption that photons are generated isotropically, the fourth is the probability that a photon generated in a given point reaches the surface, and the last term projects the flux from a given point on the direction of the total flux. In addition, $N_*(\boldsymbol{r})$ is the number density of resonantly excited atoms in a given point, τ is the radiative lifetime of the excited state, r is a distance from the surface point, which is taken as an origin from a current volume point, and θ is an angle between the direction of the total flux from all the points and the direction that joins the surface and a current point. In the case of the axial symmetry, the total flux is directed perpendicular to the surface.

This formula is transformed into the total flux of photons j in the following way:

$$j = \int j_\omega d\omega = \int \frac{N_*(\boldsymbol{r})d\boldsymbol{r}}{\tau} \frac{P(r)}{4\pi r^2} \cos\theta. \tag{9.5}$$

We above integrate formula (9.4) over frequencies and assume for simplicity that all the partial fluxes have the same direction. In principle, the flux direction may be different for different photon frequencies. In the case of axial symmetry all partial fluxes of any frequency are directed perpendicular to the surface.

▶ **Problem 9.3** Evaluate the radiation flux of a given frequency outside a plane layer of a thickness L, if this layer is infinite in transverse directions and gas parameters depend on a distance from the surface only.

Let us introduce the optical thickness u of a given layer located at a distance x from the surface as $u = \int_0^x k_\omega dx'$ and the total optical thickness of the layer as $u_\omega = \int_0^x k_\omega dx$. We have $du = k_\omega dx$, and the volume element is $d\mathbf{r} = 2\pi r^2 dr d\cos\theta$. Using the relations (5.24) and (5.28) between the absorption coefficient k_ω and the frequency distribution function a_ω, we obtain the photon flux given by (9.4),

$$j_\omega = \int d\cos\theta du \frac{N_* a_\omega}{2\tau k_\omega} e^{-u/\cos\theta} = \frac{\omega^2}{2\pi^2 c^2} \int_0^1 d\cos\theta \int_0^{u_\omega} due^{-u/\cos\theta} \left(\frac{g_* N_0}{g_0 N_*} - 1\right)^{-1} . \tag{9.6}$$

In the case of thermodynamic equilibrium inside the layer when the temperature T does not depend on a layer we have according to the Boltzmann formula (1.12)

$$\frac{g_* N_0}{g_0 N_*} = \exp\left(\frac{\hbar\omega}{T}\right) ,$$

and formula (9.6) takes the form

$$j_\omega = j_\omega^{(0)} \int \cos\theta d\cos\theta \left[1 - \exp\left(-\frac{u_\omega}{\cos\theta}\right)\right] , \tag{9.7}$$

where $j_\omega^{(0)}$ is the flux of photons that goes outside the black body and is given by

$$j_\omega^{(0)} = \frac{\omega^2}{4\pi^2 c^2} \left[\exp\left(\frac{\hbar\omega}{T}\right) - 1\right]^{-1} . \tag{9.8}$$

In the limiting cases the photon flux is

$$j_\omega = j_\omega^{(0)} , \quad u_\omega \gg 1; \quad j_\omega = 2 j_\omega^{(0)} u_\omega , \quad u_\omega \ll 1 . \tag{9.9}$$

As is seen, the optical thickness is the layer parameter that characterizes departure of radiation outside the system.

▶ **Problem 9.4** Estimate the flux of resonant photons of a given frequency which live a gaseous volume with the constant number density of atoms in the ground and resonantly excited states. The mean free path of a given photon is smaller than a dimension of the gaseous volume and the curvature of its surface.

In this case the number density of atoms in the ground and resonantly excited states inside the gaseous volume is supported by processes which are not connected with radiation transfer, and there is an equilibrium inside the volume with respect to absorption and emission processes. Denoting by i_ω the photon flux of a given frequency that is propagated inside the volume, we obtain the rate of absorption acts per unit volume and in a frequency range from ω to $\omega + d\omega$ to be $i_\omega k_\omega d\omega$, where k_ω is the absorption coefficient given by formula (5.28). On the other hand, the number of emitting photons is given by $N_* a_\omega d\omega/\tau$, where N_* is the number density of excited atoms, τ is its radiative lifetime. From the equality of these rates it follows

$$i_\omega = \frac{a_\omega N_*}{k_\omega \tau} . \tag{9.10}$$

On the basis of formula (5.28) for the absorption coefficient we obtain from this

$$i_\omega = \frac{\omega^2}{\pi^2 c^2}\left(\frac{N_0}{N_*}\frac{g_*}{g_0} - 1\right)^{-1} . \tag{9.11}$$

This photon flux is isotropic inside the gaseous volume until the distance from its boundary is large compared with its mean free path $(N_0 k_\omega)^{-1}$. Assuming the curvature of the surface for photons of a given frequency to be small compared to the surface curvature, we consider the volume boundary to be flat for these photons. Since the photon flux to the boundary goes from one side and is directed randomly, the resultant photon flux outside the volume is

$$j_\omega = \int_0^{\pi/2} i_\omega \cos\theta d(\cos\theta)\left(\int_{-\pi/2}^{\pi/2} d(\cos\theta)\right)^{-1} = \frac{i_\omega}{4} .$$

Here θ is an angle between the normal to the gas surface and the direction of photon propagation; and we have taken into account that the total photon flux outside the system is normal to the system surface. Finally, the flux of photons of frequency ω outside the gaseous system is

$$j_\omega = \frac{\omega^2}{4\pi^2 c^2}\left(\frac{N_0}{N_*}\frac{g_*}{g_0} - 1\right)^{-1} , \quad k_\omega L \gg 1 ,$$

and this formula coincides with formula (9.6).

► **Problem 9.5** Analyze the conditions of self-reversal of a spectral line and give the criterion for this phenomenon.

The above analysis of propagation of resonant radiation through gas layer that is large compared to mean free path of photons in the center of a spectral line is not diffusive. The surface of a region occupied by a gas is crosses by photons which are originated on a distance $\sim 1/k_\omega$ from the surface. Moreover, under thermodynamic equilibrium in the gas region when the gas temperature is constant over all the volume, the equilibrium flux of photons (9.8) is reached for frequencies for which $k_\omega L \geq 1$, where L is a dimension of the gas region.

Let us now consider spread conditions when the temperature in the region, which is responsible for emission in the center of the spectral line and hence is located on a distance $\sim 1/k_0$ from the surface, is lower that that for more deep regions which are responsible for emission on wings of the spectral line. As a result, the flux for the central part of the spectral line becomes lower than that its center, which leads to self-reversal of the spectral line as it is shown in Fig. 9.1.

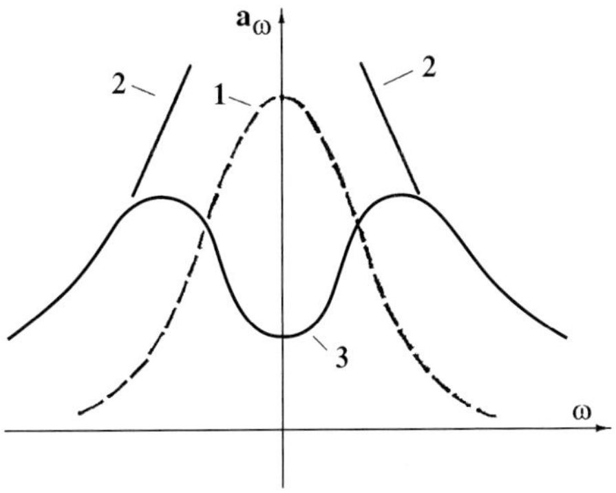

Fig. 9.1 Self-reversal of spectral lines. (1) shape of individual spectral line, (2) spectral power of emitting radiation at constant temperature, (3) spectral power of emitting radiation if the temperature drops to the boundary.

Let us give the criterion when self-reversal of the spectral line will be remarkable. Assuming a local thermodynamic equilibrium in each point, so that the number density of atoms is given by Boltzmann formula (1.12) at each point. In addition, the temperature is small $T \ll \hbar\omega$, and the number density of excited atoms is less than that of excited ones. Then, connecting photons of a given frequency ω with a distance $\sim 1/k_\omega$ from the surface, one can represent the criterion of self-reversal of spectral lines in the form

$$\frac{1}{k_0}\frac{\hbar\omega}{T^2}\left|\frac{dT}{dx}\right| \gg 1\,. \tag{9.12}$$

► **Problem 9.6** Analyze the self-reversal of a spectral line for the plane layer with infinite transversal sizes, if the temperature varies inside the layer as $T(x) = T_0 + \alpha x$, where x is the distance from the surface. In addition, $\alpha L \ll 1$, where L is the layer thickness, and the spectral line has the Lorentz shape.

Self-reversal of spectral lines results from radiation of a gas of a variable temperature when emission for the center of a spectral line is determined by a gas region of a more low temperature, and Fig. 9.1 gives a typical spectral power emission in this case. Let us represent formula (9.6) for the photon flux in the form

$$j_\omega = \frac{\omega^2}{2\pi^2 c^2}\int_0^1 d\cos\theta \int_0^{u_\omega} du e^{-u/\cos\theta} F(u), \quad F(u) = \left(e^{\hbar\omega/T} - 1\right). \tag{9.13}$$

We suppose local thermodynamic equilibrium and a weak dependence on the argument for the function $F[u(T)]$. When it is expanded in a series

$$F = F(T_0) + \alpha x \frac{dT}{dx}\,,$$

the second term is less than the first one. Under these conditions, the optical thickness for a layer on a distance x from the surface is $u = u_\omega x/L$. Next, for the

Lorentz shape of the spectral line we have $k_\omega = k_0(1+s^2)^{-1}$, where the variable $s = (\omega - \omega_o)/\nu$, so that ν is the line width, k_0 is the absorption coefficient for the line center. From this we have the photon flux with accounting for two expansion terms

$$j_\omega = j_\omega^{(0)} \int_0^1 d\cos\theta \int_0^{u_\omega} du e^{-u/\cos\theta} \left[1 + \frac{d\ln F}{dT} \frac{\alpha u(1+s^2)}{k_0}\right] , \tag{9.14}$$

where the equilibrium flux $j_\omega^{(0)}$ corresponds to the surface temperature T_0 and the derivative $d\ln F/dT$ is taken at the surface also.

Let us evaluate this integral for frequencies of a large optical thickness $u_\omega \gg 1$. Then for the first term we also determine the correction with respect to a small parameter $1/u_\omega$, while for the second term we restrict by the main term. As a result, we obtain

$$j_\omega = j_\omega^{(0)} \left(1 - \frac{e^{-u_\omega}}{u_\omega} + \frac{a}{u_\omega}\right) , \quad u_\omega = \frac{k_0 L}{1+s^2} , \quad a = \frac{2\alpha L}{3} \frac{d\ln F}{dT} , \tag{9.15}$$

and $a \ll 1$.

From the analysis of formula (9.15) it follows that if $u_0 \equiv k_0 L \gg \ln(1/a)$, the photon flux increases at removal from the center of the spectral line. The maximum photon flux corresponds to the frequency where

$$(u_\omega + 1)\exp(-u_\omega) = a . \tag{9.16}$$

One can see that since $a \ll 1$, at this frequency $u_\omega \gg 1$, i. e., the above formulas hold true. In addition, the maximum photon flux corresponding to this frequency is equal to

$$j_\omega = j_\omega^{(0)} \left(1 + \frac{a}{u_\omega + 1}\right) = j_\omega^{(0)} [1 + \exp(-u_\omega)] , \tag{9.17}$$

and because $u_\omega \gg 1$, $a \ll 1$, the second term in this formula is small compared to the first one.

▶ **Problem 9.7** A region of a gas located in thermodynamic equilibrium is a source of resonance radiation, and the mean free path of photons in the center of a resonance spectral line is small compared to dimensions of this region. Estimate the yield flux of photons for an arbitrary form of this region and for the Lorentz shape of the spectral line. Find the photon flux if a gas region is located between two parallel infinite planes with a distance d between them, inside a cylinder tube of a diameter d, and inside a spherical cavity of a diameter d.

First we estimate the photon flux. In the absence of absorption inside the system this flux is estimated as

$$j \sim \frac{N_*}{\tau} d ,$$

where d is the minimal dimension of the gas region. But according to formula (9.2) the probability to reach for a photon the surface is $P(d) \sim (k_0 d)^{-1/2}$, which gives the photon flux

$$j \sim \frac{N_* d^{1/2}}{\tau k_0^{1/2}} . \tag{9.18}$$

This estimate relates to an arbitrary geometry of the system.

We now consider the symmetric cases under consideration when the resultant flux is directed perpendicular to the surface. Then we use formula (9.5) for the photon flux and the asymptotic expression (9.2) for the probability for a photon to reach the boundary. This gives

$$j = \frac{N_*}{4\pi^{3/2} k_0^{1/2} \tau} \int \frac{d\mathbf{r} \cos\theta}{r^{5/2}} . \tag{9.19}$$

Taking the direction x to be perpendicular to the surface and introducing in accordance with formula (9.18)

$$j_0 = \frac{N_* d^{1/2}}{k_0^{1/2} \tau} , \tag{9.20}$$

we obtain a symmetric case

$$j = j_0 \frac{1}{4\pi^{3/2} d^{1/2}} \int \frac{x dx dy dz}{(x^2 + y^2 + z^2)^{7/4}} . \tag{9.21}$$

We now consider separately the cases under consideration. If the gas is located between two parallel infinite planes, we obtain, introducing new variables $\rho = \sqrt{y^2 + z^2}$, $\varphi = \arctan(y/z)$

$$j = \frac{j_0}{4\pi^{3/2} d^{1/2}} \int_0^d x dx \int_0^\infty \frac{2\pi \rho d\rho}{(\rho^2 + x^2)^{7/4}} = \frac{2 j_0}{3\pi^{1/2}} \approx 0.376 j_0 . \tag{9.22}$$

In the other case when a gas is located inside an infinite cylinder tube of a diameter d, on the basis of new variables $\rho = \sqrt{x^2 + y^2}$, $\varphi = \arctan(y/x)$ and formula (9.21) we obtain

$$\begin{aligned} j &= \frac{j_0}{4\pi^{3/2} d^{1/2}} \int \rho \cos\varphi \rho d\rho d\varphi \frac{\sqrt{\pi}\,\Gamma(5/4)}{\rho^{5/2}\Gamma(7/4)} \\ &= \frac{j_0}{4\pi d^{1/2}} \frac{\Gamma(5/4)}{\Gamma(7/4)} \int_0^d \cos\varphi d\varphi \int_0^{d\cos\varphi} \frac{d\rho}{\sqrt{\rho}} \approx 0.274 j_o . \end{aligned} \tag{9.23}$$

If an excited gas occupies the volume inside the sphere of a diameter d, we use spherical variables r, θ, φ and the origin to be located on the sphere surface, so that $0 \le r \le d\cos\theta$, $0 \le \theta \le \pi/2$, $0 \le \varphi \le 2\pi$. Integrating (9.21) over the region inside the sphere, we obtain

$$j = \frac{j_0}{4\pi^{3/2}d^{1/2}} \int_0^1 2\pi d\cos\theta \int_0^{d\cos\theta} r^2 dr \frac{r\cos\theta}{r^{7/2}}$$

$$= \frac{j_0}{\sqrt{\pi}} \int_0^1 \cos^{3/2}\theta d\cos\theta \approx 0.226 j_0\,. \tag{9.24}$$

▶ **Problem 9.8** Combine formulas (9.22), (9.23), and (9.24) to find the dependence of the total flux of photons on the volume and surface of the region that is occupied by an excited gas. Compare this flux with that in the absence of absorption inside the volume.

On the basis of formulas (9.22), (9.23), and (9.24) one can suggest a general formula for the total photon flux in the form

$$j = \frac{0.54 N_*}{\tau k_0^{1/2}} \sqrt{\frac{V}{S}}\,. \tag{9.25}$$

A numerical factor 0.54 differs from the accurate expressions in the above three cases less than by 3%. Indeed, for emission from a flat layer this formula gives $0.382 j_0$ for the total photon flux, the value $0.272 j_0$ for emission of the cylinder tube, and the value $0.220 j_0$, if an excited gas is located inside a sphere.

All these results relate to the case when the mean free path k_0^{-1} for resonance photons in the center of a spectral line is small compared to dimensions of this region occupied by an excited gas. If absorption inside this volume is absent and photons leave this volume freely, the photon flux is equal to

$$j = \frac{N_*}{\tau}\frac{V}{S}\,.$$

One can see that the probability for a photon to leave the volume is

$$P = 0.5\sqrt{\frac{S}{k_0 V}} \tag{9.26}$$

for the Lorentz shape of the spectral line. This corresponds to formula (9.2).

▶ **Problem 9.9** For the Lorentz and Doppler shape of a spectral line estimate the width of a resonance spectral line for radiation leaving a volume of dimension L occupied by an excited gas. The mean free path of photons in the center of the spectral line $1/k_0$ is small compared to L.

The above analysis shows that under these conditions the width of a spectral line for radiation leaving a volume of an excited gas is more than the width of the spectral line of an individual atoms, because photons in the central part of the spectral line are absorbed inside the volume, while photons on wings of the spectral line passes through this volume freely. Hence, the boundaries of the spectral line can be estimated from the condition that the optical thickness for the boundary frequencies is of the order of one

$$u_\omega = \int_0^L k_\omega dx \sim k_\omega L \sim 1 . \qquad (9.27)$$

In the case of Lorentz broadening of the spectral line, if at wings $k_\omega = k_o \nu^2/(\omega - \omega_0)^2$ according to formula (5.5), the width of the spectral line for the total radiation flux is given by

$$\Delta\omega \sim \nu\sqrt{k_o L}, \qquad k_o L \gg 1, \qquad (9.28)$$

where ν is the width of an individual line. In the same manner, when the spectral line has the Doppler shape according to formula (5.3), the width of the spectral line is

$$\Delta\omega \sim \Delta\omega_D \sqrt{\ln(k_o L)}, \qquad k_o L \gg 1 , \qquad (9.29)$$

where $\Delta\omega_D$ is the width of the Doppler-broadened spectral line in the case of small optical thickness of the plasma system. Thus resonant radiation emitting an excited gas is characterized by broader spectral lines than is radiation from individual atoms, because the main contribution to the emergent radiation arises largely from the wings of the spectra of individual atoms.

9.2
Radiation Transport in Optically Thick Medium

▶ **Problem 9.10** Evaluate the photon flux for a given frequency from a semi-infinite layer with a weak temperature change that depend on a layer depth only.

We now consider formula (9.13) for the emission flux that has the form

$$j_\omega = \frac{\omega^2}{2\pi^2 c^2} \int_0^1 d\cos\theta \int_0^{u_\omega} du e^{-u/\cos\theta} F(u), \qquad F(u) = \left(e^{\hbar\omega/T} - 1\right) .$$

In the case of a constant temperature, the optically thick layer emits as a black body with its temperature. If the layer temperature weakly varies, one can use the flux of a black body with a temperature of a layer whose optical depth is of the order of one.

This is an estimate, and to obtain a correct result, we expand $F(u)$ near a point u_1

$$F(u) = F(u_1) + (u - u_1)F'(u_1) + \frac{(u_1 - u)^2}{2} F''(u_1) .$$

Substituting this expansion in the expression for the photon flux, we obtain

$$j_\omega = \frac{\omega^2}{2\pi^2 c^2}\left[\frac{1}{2}F(u_1) + \left(\frac{1}{3} + \frac{u_1}{2}\right)F'(u_1) + \left(\frac{1}{4} - \frac{u_1}{3} + \frac{u_1^2}{4}F''(u_1)\right)\right] .$$

Taking $u_1 = 1/3$ such that the second term would be zero, we obtain the photon flux

$$j_\omega = \frac{\omega^2}{4\pi^2 c^2}\left[F\left(u = \frac{2}{3}\right) + \frac{8}{5}F''\left(u = \frac{2}{3}\right)\right] . \tag{9.30}$$

Ignoring the last term, we find that the radiation flux coincides with that from the surface of a black body whose temperature is equal to the temperature of a layer with the optical thickness of 2/3.

Let us estimate a small parameter used for deriving formula (9.30). This is equal to

$$\frac{1}{F}\frac{dF}{du} = \frac{\frac{dF}{dT}\frac{dT}{dx}}{\frac{du}{dx}F} = \frac{d\ln T}{k_\omega dx}\frac{\exp(\hbar\omega/T)}{\exp(\hbar\omega/T)-1} \sim \frac{d\ln T}{k_\omega dx} \ll 1 .$$

Thus, this expression for the photon flux holds true if the temperature varies weakly on distances compared to the optical thickness.

▶ **Problem 9.11** Determine the coefficient of radiative heat transfer in an optically thick gas if a small temperature gradient occurs in a longitudinal direction, while all the gas parameters are constant in transversal directions.

Let us determine the radiation flux j_ω that is equal at a given point to

$$j_\omega = j_\omega^+ - j_\omega^- ,$$

where j_ω^+, j_ω^- are the photon fluxes at this point which are created by spaces from different sizes from this point. Taking formula (9.30) for the photon flux from each side, we obtain the total photon flux at a given frequency

$$j_\omega = \frac{\omega^2}{4\pi^2 c^2}\left[F(u = \frac{2}{3}) - F(u = -\frac{2}{3})\right] .$$

Using the expression

$$F(u) = \frac{1}{\exp\left(\frac{\hbar\omega}{T}\right) - 1} ,$$

and assuming the temperature gradient dT/dx to be relatively small (x is the longitudinal direction), we obtain the photon flux at a given frequency

$$j_\omega = \frac{\omega^2}{3\pi^2 c^2}\frac{dF}{du} = \frac{\omega^2}{3\pi^2 c^2}\frac{dF/dx}{du/dx} = \frac{\omega^2}{3\pi^2 c^2}\frac{\hbar\omega\exp\left(\frac{\hbar\omega}{T}\right)}{k_\omega T^2\left[\exp\left(\frac{\hbar\omega}{T}\right) - 1\right]^2}\left|\frac{dT}{dx}\right| .$$

From this we find the heat flux q due to radiation transfer

$$q = \int \hbar\omega j_\omega d\omega \, .$$

On the basis of the above expression for the partial photon flux we obtain the coefficient $\kappa_{\rm rad}$ of radiative heat transfer due to its definition $q = -\kappa dT/dx$

$$\kappa = \int \frac{\omega^2}{3\pi^2 c^2}\left(\frac{\hbar\omega}{T}\right)^2 \frac{\exp\left(\frac{\hbar\omega}{T}\right) d\omega}{k_\omega\left[\exp\left(\frac{\hbar\omega}{T}\right) - 1\right]^2} \, . \tag{9.31}$$

One can see that the main contribution to the heat transfer coefficient follows from the photon energies $\hbar\omega \sim T$. Next, this formula is valid for a medium with a wide absorption spectrum where the mean free path of photons k_ω^{-1} ia small compared to typical medium dimensions L. This is not valid for transport of resonance radiation where for each medium dimension L one can find such frequencies for which $k_\omega L \sim 1$. Then radiation transfer occurs with these frequencies which correspond to wings of the resonance spectral line. Hence in this case the diffusion character of heat transport will be violated.

9.3
Emission of Infrared Radiation from Molecular Layer

▶ **Problem 9.12** Within the framework of a random model determine the average absorption function $\overline{A}$ in a given frequency range where the distribution of the transition intensities S is given by the probability $p(S)$.

If the spectral band contains n spectral lines and the average distance between neighboring spectral lines is δ, the absorption function according to formula (5.33) is given by

$$\overline{A} = 1 - \frac{1}{n\delta}\int\limits_{\omega_o - n\delta/2}^{\omega_o + n\delta/2} \exp\left[-\sum_k u_\omega^{(k)}\right] d\omega \, ,$$

where ω_o is the center of this frequency range, and $u_\omega^{(k)}$ is the optical thickness at a given frequency due to the k-th transition. The random model gives that the probability of location of the k-th line center in the frequency range between ω_k and $\omega_k + d\omega_k$ is proportional to $d\omega_k$, and we have

$$\overline{A} = 1 - \frac{\exp\left[-\sum\limits_k u_\omega^{(k)}\right]\prod\limits_k d\omega_k}{\prod\limits_k \int d\omega_k} = 1 - \frac{\prod\limits_k \int\limits_{\omega_o - n\delta/2}^{\omega_o + n\delta/2} d\omega_k \exp\left[-u_\omega^{(k)}\right]}{\prod\limits_k \int\limits_{\omega_o - n\delta/2}^{\omega_o + n\delta/2} d\omega_k} \, .$$

This average absorption function is the probability that a photon of a given frequency range will be absorbed by the molecular layer. Because of the identity of the frequency ranges, this integral takes the form

$$\overline{A} = 1 - \left[\frac{1}{n\delta} \int\limits_{\omega_0 - n\delta/2}^{\omega_0 + n\delta/2} \exp(-u_\omega) d\omega \right]^n ,$$

where u_ω is the optical thickness for an individual spectral line. We assume the frequency range to be large compared to the width of an individual line, which allows us to rewrite the above expression in the form

$$\overline{A} = 1 - \left[1 - \frac{1}{n\delta} \int\limits_{\omega_0 - n\delta/2}^{\omega_0 + n\delta/2} \left(1 - e^{-u_\omega}\right) d\omega \right]^n$$

In the limit of an infinite number of spectral lines in a given frequency range this formula gives

$$\overline{A} = 1 - \exp\left[- \int \left(1 - e^{-u_\omega}\right) \frac{d\omega}{\delta} \right] . \tag{9.32}$$

Using the distribution on intensities of spectral lines, one can represent this expression in the form

$$\overline{A} = 1 - \exp\left[- \int \left(1 - e^{-u_\omega(S)}\right) \frac{d\omega}{\delta} p(S) dS \right] . \tag{9.33}$$

▶ **Problem 9.13** Determine the distribution function on optical thicknesses for a given frequency range on the basis of a random model.

The parameters of this problem are average frequency difference between neighboring spectral lines δ and the distribution $p(S)$ on the intensities in a given frequency range. Our task is to find the distribution function $f(u)$ on optical thicknesses. It is convenient to operate with the Fourier component of the distribution function that is given by

$$\chi(t) = \int\limits_0^\infty e^{-itu} f(u) du = \prod_k \int e^{-itu} w(u_k) du_k = \prod_k \chi_k(t) ,$$

where u_k is the optical thickness from a given spectral line, and the total optical thickness $u = \sum_k u_k$; since within the framework of a random model distributions for each line are independent, $f(u)du = \prod_k w(u_k) du_k t$. On the basis of the inverse operation we have

$$f(u) = \frac{1}{\pi} \int\limits_{-\infty}^{\infty} e^{itu} \chi(t) dt = \frac{1}{\pi} \int\limits_{-\infty}^{\infty} \prod_k \chi_k(t) e^{itu_k} dt .$$

Assuming the width of an individual spectral line to be small compared to the width of a taken frequency range $\Delta\omega$, we have the probability that the center of a given spectral to be located in a range between ω_k and $\omega_k + d\omega_k$ is $d\omega_k/\Delta\omega$. Then we have for the partial characteristic function

$$\chi_k(t) = \int e^{-itu_k} \frac{d\omega_k}{\Delta\omega} = 1 - \frac{1}{\Delta\omega} \int \left(1 - e^{-itu_k}\right) d\omega_k ,$$

and

$$\ln \chi_k(t) = \frac{1}{\Delta\omega} \int \left(1 - e^{-itu_k(\omega_k)}\right) d\omega_k .$$

This gives

$$\ln \chi(t) = \frac{1}{\Delta\omega} \sum_k \int \left(1 - e^{-itu_k(\omega_k)}\right) d\omega_k .$$

In the limit $\Delta\omega/\delta \to \infty$ we obtain from this

$$\ln \chi(t) = \int \left(1 - e^{-itu_k(\omega-\omega_k,S)}\right) \frac{d\omega}{\delta} p(S) dS .$$

Here ω_k is the center of the k-th spectral line, u_k is the optical thickness due to this line, and S is its intensity. From this it follows the distribution function x

$$\begin{aligned} f(u) &= \frac{1}{\pi} \int_{-\infty}^{\infty} e^{itu} \chi(t) dt \\ &= \frac{1}{\pi} \int_{-\infty}^{\infty} e^{itu} \chi(t) dt \exp\left[itu - \int \frac{d\omega}{\delta} p(S) dS \left(1 - e^{-itu_k(\omega-\omega_k,S)}\right)\right] . \end{aligned} \qquad (9.34)$$

Let us determine from this the average absorption function that according to its definition (5.33) is equal to

$$\overline{A} = \int_0^{\infty} (1 - e^{-u}) f(u) du = 1 - \int_0^{\infty} e^{-u} f(u) du .$$

Using the above expression for the distribution function du and evaluating the integral over du, we obtain

$$\overline{A} = 1 - \frac{1}{\pi} \int_0^{\infty} \frac{dt}{(1 - it)} \exp\left[-\int \frac{d\omega}{\delta} p(S) dS \left(1 - e^{-itu_k(\omega-\omega_k,S)}\right)\right] .$$

The integrand has a pole at $t = -i$. Expressing the integral through a residue of the integrand, we get finally

$$\overline{A} = 1 - \exp\left[-\int \frac{d\omega}{\delta} p(S) dS \left(1 - e^{-itu_k}\right)\right] .$$

This formula coincides with formula (9.33).

▶ **Problem 9.14** For a random model of spectral lines evaluate the effective frequency width for emission of a molecular layer.

The definition of the effective width $\Delta\omega$ of an emission band gives

$$\Delta\omega = 2\int d\omega \int \cos\theta d\cos\theta \left[1 - e^{-u(\omega)/\cos\theta}\right] . \tag{9.35}$$

Let us average this expression over optical thicknesses by using the distribution function (9.34) on optical thicknesses. We have

$$\overline{1-\exp\left(-\frac{u}{\cos\theta}\right)} = 1 - \frac{1}{\pi}\int_{-\infty}^{\infty} dt \int_{0}^{\infty} du \exp\left[-\frac{u}{\cos\theta} + itu - \int \frac{d\omega_k}{\delta}\left(1 - e^{-itu_k}\right)\right] ,$$

where u_k is the optical thickness that is created by an individual spectral line. Integrating this expression over du and then over dt, as well as in the previous problem, we obtain a residue of the integrand. As a result, we find

$$\overline{1-\exp\left(-\frac{u}{\cos\theta}\right)} = 1 - \exp\left[-\int \frac{d\omega_k}{\delta}\left(1 - e^{-u_k/\cos\theta}\right)\right] .$$

This gives for the effective frequency width of the emission band

$$\Delta\omega = 2\int_{-\infty}^{\infty} d\omega \int_{0}^{1} \cos\theta d\cos\theta \left(1 - \exp\left[-\int \frac{d\omega_k}{\delta}\left(1 - e^{-u_k/\cos\theta}\right)\right]\right) .$$

On the basis of the distribution function $p(S)$ on frequencies for an individual spectral line one can rewrite this expression in the form

$$\Delta\omega = 2\int_{-\infty}^{\infty} d\omega \int_{0}^{1} \cos\theta d\cos\theta \left(1 - \exp\left[-\int p(S)dS\frac{d\omega_k}{\delta}\left(1 - e^{-u_k/\cos\theta}\right)\right]\right) . \tag{9.36}$$

▶ **Problem 9.15** For the Lorentz shape of the spectral line within the framework of the random model of spectral lines evaluate the effective frequency width for emission of a molecular layer.

The optical thickness of a layer due to an individual line according to formula (5.5) is given by

$$u = u_{\max}\frac{\nu}{(\omega-\omega_k)^2 + \nu^2} = \frac{u_{\max}}{1+s^2} ,$$

where ω_k is the frequency at the center of a spectral line, $u_{\max}$ is the optical thickness of the molecular layer for the line center, ν is the width of an individual spectral line, and the variable $s = (\omega - \omega_k)/\nu$. Substituting this expression in formula (9.36), we get

$$\Delta\omega = 2\int_{-\infty}^{\infty} d\omega \int_0^1 \cos\theta d\cos\theta \left(1 - \exp\left[-\frac{\nu}{\delta}\int_{-\infty}^{\infty} ds \left(1 - \exp\left[-\frac{u_{\max}(\omega)}{(1+s^2)\cos\theta}\right]\right)\right]\right). \tag{9.37}$$

Here the optical thickness of the molecular layer at the line center $u_{\max}(\omega)$ is proportional to the intensity of this spectral line, and the intensity in turn depends on the transition frequency. The main contribution to this integral give frequencies for which

$$\frac{\nu}{\delta}\int_{-\infty}^{\infty} ds \left(1 - \exp\left[-\frac{u_{\max}(\omega)}{(1+s^2)\cos\theta}\right]\right) \sim 1\,.$$

We first consider the case when the width ν of an individual spectral line is small compared to the mean difference of frequencies for neighboring transitions δ, i. e., $\nu \ll \delta$. Then the main contribution to integral (9.37) follows from the frequencies whose difference with the frequency of the line center is larger than the line width ν, i. e., $s \gg 1$. This gives

$$\frac{\nu}{\delta}\int_{-\infty}^{\infty} ds \left(1 - \exp\left[-\frac{u_{\max}}{(1+s^2)\cos\theta}\right]\right) = 2\frac{\nu}{\delta}\sqrt{\frac{\pi u_{\max}}{\cos\theta}}\,.$$

Correspondingly, the effective width of an emission range is equal to

$$\Delta\omega = 2\int_{-\infty}^{\infty} d\omega \int_0^1 \cos\theta d\cos\theta \left[1 - \exp\left(-\frac{2\nu}{\delta}\sqrt{\frac{\pi u_{\max}(\omega)}{\cos\theta}}\right)\right].$$

For evaluating this integral, we take ω' such that this frequency is close to the central frequency ω_0 of the band, but the exponent at this frequency is large. Taking into account a sharp frequency dependence of the optical thickness $u_{\max}(\omega)$ in centers of lines, we approximate it by the dependence

$$u_{\max}(\omega) = u_{\max}(\omega')\exp[-\alpha(\omega - \omega')], \qquad \alpha = \left.\frac{d\ln u_{\max}}{d\omega}\right|_{\omega=\omega'}.$$

Introducing a new variable, we reduce this expression to the form

$$\Delta\omega = 2(\omega' - \omega_0) + \frac{8}{\alpha}\int_0^1 \cos\theta d\cos\theta \int_0^{B/\sqrt{\cos\theta}} (1 - e^{-z})\frac{dz}{z}\,,$$

where

$$z = \frac{B}{\sqrt{\cos\theta}}\exp\left[-\frac{\alpha}{2}(\omega - \omega')\right], \quad B = \frac{2\nu}{\delta}\sqrt{\pi u_{\max}(\omega')}\,.$$

Using the asymptotic expression for the integral in $\Delta\omega$, we reduce it to the form

$$\Delta\omega = 2(\omega' - \omega_0) + \frac{8}{\alpha}\int_0^1 \cos\theta d\cos\theta \frac{Be^C}{\sqrt{\cos\theta}}$$
$$= 2(\omega' - \omega_0) + \frac{4}{\alpha}\ln(Be^{C+1/2}) = 2(\omega_1 - \omega_0),$$

where $C = 0.577$ is the Euler constant, and the frequency ω_1 is given by

$$u_{max}(\omega_1) = \frac{\delta^2}{4\pi e^{2C+1/2}\nu^2} = 0.015\frac{\delta^2}{\nu^2}. \tag{9.38}$$

In the other limiting case $\nu \gg \delta$ at frequencies, which give the main contribution to the integral, $u_{max} \ll 1$. From this we have

$$\frac{\nu}{\delta}\int_{-\infty}^{\infty} ds\left(1 - \exp\left[-\frac{u_{max}(\omega)}{(1+s^2)\cos\theta}\right]\right) = \frac{\pi\nu u_{max}(\omega)}{\delta\cos\theta}.$$

Repeating the operations of the previous limiting case, we obtain

$$\Delta\omega = 2(\omega' - \omega_0) + \frac{8}{\alpha}\int_0^1 \cos\theta d\cos\theta \int_0^{B/\sqrt{\cos\theta}} (1 - e^{-z})\frac{dz}{z},$$

and now

$$z = \frac{B}{\cos\theta}\exp[-\alpha(\omega - \omega')], \quad B = \frac{\pi\nu}{\delta}u_{max}(\omega') \gg 1.$$

Evaluating the effective band width in the same method as early, we obtain

$$\Delta\omega = 2(\omega' - \omega_0) + \frac{4}{\alpha}\int_0^1 \cos\theta d\cos\theta \frac{Be^C}{\cos\theta}$$
$$= 2(\omega' - \omega_0) + \frac{2}{\alpha}\ln(Be^{C+1/2}) = 2(\omega_1 - \omega_0),$$

where $C = 0.577$ is the Euler constant and the frequency ω_1 is given by

$$u_{max}(\omega_1) = \frac{\delta}{\pi e^{C+1/2}\nu} = 0.11\frac{\delta}{\nu}. \tag{9.39}$$

▶ **Problem 9.16** For the Lorentz shape of the spectral line evaluate the effective width of an emission band for a layer of a molecular gas whose vibration and transversal temperatures coincide and are constant over the layer. Absorption results from rotation–vibration transitions of linear molecules, and the intensities of centers of spectral lines of these transitions varyslightly for neighboring lines, but vary sharply in the limits of all the band.

Formula (5.40) gives the optical thickness of a layer for a given frequency

$$u_\omega = u_{\max}(\omega)\frac{\cosh\frac{2\pi\nu}{\delta} - 1}{\left(\cosh\frac{2\pi\nu}{\delta} - \cos\frac{2\pi|\omega-\omega_0|}{\delta}\right)}\,. \tag{9.40}$$

Here $u_{\max}(\omega)$ is the optical thickness in the centers of spectral lines, the difference of frequencies of neighboring lines is $\delta = 2B/\hbar$, ν is the width of an individual spectral line, and ω_0 is the frequency of the band center. The quantity $u_{\max}(\omega)$ varies weakly, if the frequency ω varies by δ, but it varies strongly, if the frequency ω varies by a value compared to the band width, i. e., at band edges it is small compared to its value at the band center ω_0. The band is symmetric with respect to its center.

According to a general formula (5.35), the effective width of the spectral band is now given by

$$\Delta\omega = 2\int_{-\infty}^{\infty} d\omega \int_0^1 \cos\theta\, d\cos\theta \left(1 - \exp\left[-\frac{u_{\max}(\omega)}{\cos\theta}\frac{\cosh\frac{2\pi\nu}{\delta} - 1}{\left(\cosh\frac{2\pi\nu}{\delta} - \cos\frac{2\pi|\omega-\omega_0|}{\delta}\right)}\right]\right)\,. \tag{9.41}$$

We consider below the limiting cases of this formula. If the width of an individual spectral line ν is small compared to the distance δ between neighboring lines, this formula takes the form

$$\Delta\omega = 2\int_{-\infty}^{\infty} d\omega \int_0^1 \cos\theta\, d\cos\theta \left(1 - \exp\left[-\frac{u_{\max}(\omega)}{\cos\theta}\right]\right)\,. \tag{9.42}$$

Because of a sharp dependence $u_{\max}(\omega)$ in scales of the band width we use the standard method of evaluation of this integral. Let us take a frequency ω' such that $u_{\max}(\omega') \gg 1$, but ω' is close to a frequency ω_1 for which $u_{\max}(\omega_1) \sim 1$. Using in this range the dependence $u_{\max}(\omega) = u_{\max}(\omega')\exp\left[-\alpha(\omega-\omega')\right]$, where α is the logarithm derivative of the function $u_{\max}(\omega)$ at $\omega = \omega'$, we divide the integral over frequencies in two parts with the boundary $\omega = \omega'$. This gives

$$\Delta\omega = 2(\omega - \omega') + \frac{4}{\alpha}\int_0^1 \cos\theta\, d\cos\theta \int_0^{\frac{u_{\max}(\omega')}{\cos\theta}} \left(1 - e^{-z}\right)\frac{dz}{z}\,,$$

where $z = u_{\max}(\omega)/\cos\theta$. Taking into account $u_{\max}(\omega') \gg 1$ on the basis of the asymptotic expression of the last integral we find

$$\Delta\omega = 2(\omega - \omega') + \frac{4}{\alpha}\ln\left[u_{\max}(\omega')e^{C+1/2}\right] \equiv 2(\omega_1 - \omega_0)\,,$$
$$u_{\max}(\omega_1) = e^{-C-1/2} = 0.34\,.$$

Here $C = 0.577$ is the Euler constant.

In the other limiting case, when the width of an individual line ν is small compared to the distance δ between neighboring lines, we divide the integral (9.41) in integrals near the centers of individual lines

$$\Delta\omega = \frac{\delta}{\pi}\int_0^1 \cos\theta \, d\cos\theta \sum_k \int_0^{2\pi} d\varphi \left(1 - \exp\left[-\frac{u_{max}(\omega)}{\cos\theta}\frac{\cosh\frac{2\pi\nu}{\delta} - 1}{\left(\cosh\frac{2\pi\nu}{\delta} - \cos\varphi\right)}\right]\right),$$

where a new variable is introduced as

$$2\pi(\omega - \omega_0) = \varphi + 2\pi k .$$

Assuming that the main contribution to the sum give many values of k, we replace summation by integration over frequencies that gives

$$\Delta\omega = \frac{1}{\pi}\int_0^1 \cos\theta \, d\cos\theta \int_{-\infty}^{\infty} d\omega \int_0^{2\pi} d\varphi \left(1 - \exp\left[-\frac{u_{max}(\omega)}{\cos\theta}\frac{\cosh\frac{2\pi\nu}{\delta} - 1}{\left(\cosh\frac{2\pi\nu}{\delta} - \cos\varphi\right)}\right]\right).$$

This operation allows us to separate the oscillating part from the part that varies slightly for an oscillation period.

Let us expand the integrand over a small parameter ν/δ. The main contribution to the integral give $\varphi \sim 1$, which allows us to expand $\cosh\frac{2\pi\nu}{\delta}$ over a small parameter. This gives

$$\Delta\omega = \frac{1}{\pi}\int_0^1 \cos\theta \, d\cos\theta \int_{-\infty}^{\infty} d\omega \int_0^{2\pi} d\varphi \left(1 - \exp\left[-\frac{u_{max}(\omega)}{\cos\theta}\frac{2\pi^2\nu^2}{\delta^2(1 - \cos\varphi)}\right]\right).$$

As before, we introduce a frequency ω' such that $u_{max}(\omega') \gg 1$, but ω' is close to a frequency ω_1 for which $u_{max}(\omega_1) \sim 1$. Using in this frequency range the dependence $u_{max}(\omega) = u_{max}(\omega')\exp[-\alpha(\omega - \omega')]$, where α is the logarithm derivative of the function $u_{max}(\omega)$ at $\omega = \omega'$, we divide the integral over frequencies in two parts with the boundary $\omega = \omega'$. We get

$$\Delta\omega = 2(\omega' - \omega_0) + \frac{2}{\pi}\int_0^1 \cos\theta \, d\cos\theta \int_0^{2\pi} d\varphi \int_0^{\frac{B}{\cos\theta(1-\cos\varphi)}} \left(1 - e^{-z}\right)\frac{dz}{z},$$

where

$$z = \frac{B}{\cos\theta(1 - \cos\varphi)}\exp\left[-\alpha(\omega - \omega')\right], \qquad B = 2\pi^2\frac{\nu^2}{\delta^2}u_{max}(\omega') \gg 1 .$$

Using the asymptotic expression for the last integral at large values of the upper limit, we obtain

$$\Delta\omega = 2(\omega' - \omega_0) + \frac{2}{\pi\alpha}\int_0^1 \cos\theta \, d\cos\theta \int_0^{2\pi} d\varphi \ln\frac{Be^C}{\cos\theta(1 - \cos\varphi)} .$$

We reduce this integral to the form

$$\Delta\omega = 2(\omega - \omega') + \frac{2}{\alpha}\ln\left[2Be^{C+1/2}\right] \equiv 2(\omega_1 - \omega_0),$$

$$u_{\max}(\omega_1) = \frac{\delta^2}{4\pi^2\nu^2}e^{-C-1/2} = 0.0086\frac{\delta^2}{\nu^2}\,.$$

Note that in the last limiting case $\nu \ll \delta$ the optical thickness in the center of a spectral line is created and is determined by one individual line. In the opposite limiting case $\nu \gg \delta$ when many individual spectral lines take part in absorption at a given frequency, the total optical thickness in the center of an individual spectral line increases compared to the optical thickness for this spectral line in $\pi\nu/\delta$ times.

▶ **Problem 9.17** Within the framework of a model of single spectral lines and their Lorentz shape find the average absorption function for a given frequency and the effective width of an emission band for a layer of a molecular gas located under thermodynamic conditions.

According to formula (5.5), the optical thickness of a layer due to an individual line in the case of the Lorentz shape of spectral lines is given by

$$u = u_{\max}\frac{\nu}{(\omega - \omega_k)^2 + \nu^2} = \frac{u_{\max}}{1 + s^2}\,,$$

where ω_k is the frequency at the center of a spectral line, $u_{\max}$ is the optical thickness of the molecular layer for the line center, ν is the width of an individual spectral line, and the variable $s = (\omega - \omega_k)/\nu$ is used. In the case under consideration in the center of an individual spectral line the optical thickness $u_{\max}(\omega) \gg 1$, whereas between two individual lines the optical thickness is less than one. The absorption function is equal to

$$A = \frac{\nu}{\delta}\int_{-\infty}^{\infty} ds\left[1 - \exp\left(-\frac{u_{\max}}{1+s^2}\right)\right]\,,$$

where δ is the average difference of frequencies for neighboring lines. Since the main contribution to this integral follows from $s \gg 1$ $(u_{\max} \gg 1)$, we have from this

$$A = \frac{\nu}{\delta}\int_{-\infty}^{\infty} ds\left[1 - \exp\left(-\frac{u_{\max}}{s^2}\right)\right] = \frac{\nu}{\delta}\sqrt{\pi u_{\max}} \quad \delta \gg \nu\,,$$

and the average absorption function is

$$\overline{A} = \frac{\nu\sqrt{\pi}}{\delta}u_{\max}^{1/2}\,.$$

The width of an emission spectral band is equal to

$$\Delta\omega = \nu\sum_k \int_0^1 \cos\theta\, d\cos\theta \int_{-\infty}^{\infty} ds\left(1-\exp\left[-\frac{u_{\max}^{(k)}(\omega)}{s^2\cos\theta}\right]\right)$$

$$= 2\nu\sum_k \int_0^1 \cos\theta\, d\cos\theta\sqrt{\pi u_{\max}^{(k)}\cos\theta} = \frac{4\nu\sqrt{\pi}}{5}\sum_k\sqrt{u_{\max}^{(k)}}\,, \tag{9.43}$$

where $u_{\max}^{(k)}$ is the optical thickness in the center of k-th spectral line. Assuming that the effective band width is created by many individual spectral lines. Then according to the definition of the line intensity

$$u_{\max}^{(k)}(\omega) = \frac{1}{2\pi\nu}\int S(\omega)dx\,,$$

where the integral is taken over the depth dx of a molecular layer. Assuming a certain frequency dependence for the intensity of spectral lines $S(\omega)$ and introducing the distribution function $p(S)$ on intensities of spectral lines, we obtain

$$\Delta\omega = \frac{2\sqrt{2\nu}}{5\delta}\int\sqrt{\int S(\omega)dx}\;\; p[S(\omega)]d\omega\,.$$

Assuming the intensity S is connected with the frequency ω unambiguously, we reduce this formula to the form

$$\Delta\omega = \frac{2\sqrt{2\nu}}{5\delta}\int d\omega\sqrt{\int S(\omega)dx}\,.$$

► **Problem 9.18** For the Lorentz shape of a spectral line determine the distribution function $f(u)$ on optical thicknesses u on the basis of the random model under conditions when the regular model holds true.

On the basis of formula (5.31), we have the optical thickness of a layer u if we express it through the minimal optical thickness $u_{\min}$ in the middle between centers of neighboring lines

$$u = u_{\min}\frac{\cosh\frac{2\pi\nu}{\delta}+1}{\cosh\frac{2\pi\nu}{\delta}-\cos\varphi}\,,$$

where δ is the difference of frequencies of neighboring lines, $\varphi = 2\pi(\omega-\omega_k)/\delta$, and we assume intensities of individual lines to be identical. The probability for a given frequency ω to be located on the distance from the center of a nearest line between $\omega-\omega_k$ and $\omega-\omega_k+d\omega$ is proportional to $d\omega$ or $d\varphi$. From this it follows for the distribution function

$$f(u)du = \frac{1}{\pi}d\varphi\,,$$

where φ ranges from 0 to π and the normalization of the distribution function has the form $\int f(u)du = 1$. From this it follows

$$f(u) = \frac{1}{\pi}\left(\frac{du}{d\varphi}\right)^{-1} . \tag{9.44}$$

We consider below the limiting case $\delta \gg \nu$, and in this case the regular model gives the following relation between the optical thickness u and the parameter φ:

$$u = \frac{2u_{\min}}{1-\cos\varphi}, \qquad \sin\frac{\varphi}{2} = \sqrt{\frac{u_{\min}}{u}} .$$

This gives for the distribution function according to formula (9.44)

$$f(u)du = \frac{\sqrt{u_{\min}}du}{\pi u\sqrt{u-u_{\min}}} . \tag{9.45}$$

We now use the random model when the distribution function is given by formula (9.34)

$$f(u)du = \frac{1}{2\pi}\int_{-\infty}^{\infty} dt \exp\left(itu - \int_{-\infty}^{\infty}\frac{d\omega}{\delta}\left[1-\exp\left(-\frac{itu_{\max}}{1+s^2}\right)\right]\right) ,$$

where we consider identical intensities of lines. Here $u_{\max}$ is the optical thickness for the center of some line, $s = (\omega - \omega)/\delta$, so that ω is a current frequency, ω_k is the frequency of a center of an individual line. Introducing as early $u_{\min} = u_{\max}(\pi\nu/\delta)^2$ and using a new variable $y = \pi\nu s/\delta = \pi(\omega-\omega_k)/\delta$, transform the above expression to the form

$$f(u) = \frac{1}{2\pi}\int_{-\infty}^{\infty} dt \exp\left(itu - \frac{1}{\pi}\int_{-\infty}^{\infty} dy\left[1-\exp\left(-\frac{itu_{\min}}{y^2}\right)\right]\right) .$$

Evaluating the integral over dy, we obtain

$$f(u) = \frac{1}{2\pi}\int_{-\infty}^{\infty} dt \exp\left(itu - \sqrt{\frac{2}{\pi}tu_{\min}} - i\sqrt{\frac{2}{\pi}tu_{\min}}\right) .$$

In the limit $u \gg u_{\min}$ we obtain from this

$$f(u) = \frac{1}{\pi}\frac{u_{\min}^{1/2}}{u^{3/2}} ,$$

i.e., this distribution function coincides with the accurate one. When $u \sim u_{\min}$, the random model gives the result that does not coincides with the accurate one.

▶ **Problem 9.19** Determine the absorption function of a flat layer within the framework of the model of single spectral lines if the optical thickness sharply drops along an individual spectral line and the optical thickness is small in the middle between neighboring lines.

We have for the absorption function within the framework of the single line model as

$$A = \int \frac{d\omega}{\delta}\left(1 - e^{-u}\right) .$$

In fact, the absorption function is now a part of the spectral frequencies near the centers of individual lines in which photons are absorbed by a gas layer. On the basis of this formula, the absorption function may be estimated as

$$A = 2\frac{|\omega_1 - \omega_k|}{\delta} , \qquad u(\omega_1) \sim 1 , \tag{9.46}$$

where ω_k is the line center and δ is the average difference of frequencies for neighboring lines. Let us evaluate this value more precise by using the standard method that takes into account a sharp dependence $u(\omega)$. Then we introduce a typical frequency ω' such that $u(\omega') \gg 1$, but ω' is close to a frequency ω_1 for which $u(\omega_1) \sim 1$. Using this range of frequencies the approximation $u(\omega) = u(\omega')\exp\left[-\alpha(\omega - \omega')\right]$, where α is the logarithm derivative of the function $u(\omega)$ at $\omega = \omega'$, we obtain $d\omega = -du/(\alpha u)$. This gives

$$A = \frac{2}{\delta\alpha}\int_0^{u(\omega_0)} \frac{du}{u}\left(1 - e^{-u}\right) = \frac{2(\omega' - \omega_k)}{\delta} + \frac{2}{\delta\alpha}\int_0^{u(\omega')} \frac{du}{u}\left(1 - e^{-u}\right) .$$

The asymptotic expression for this integral for $u(\omega') \gg 1$ is

$$\int_0^{u(\omega')} \frac{du}{u}\left(1 - e^{-u}\right) = \ln u(\omega') + C ,$$

where $C = 0.577$ is the Euler constant. Thus, we have

$$A = \frac{2(\omega' - \omega_k)}{\delta} + \frac{2}{\delta\alpha}\left(\ln u(\omega') + C\right) = \frac{2(\omega_1 - \omega_k)}{\delta} , \quad u(\omega_1) = e^{-C} = 0.56 . \tag{9.47}$$

▶ **Problem 9.20** Determine the effective width of the emission band within the framework of the model of single spectral lines for the Doppler form of the spectral line.

The width of an individual line is small compared to the average distance δ between neighboring lines, which leads to an independent contribution to the effective band width from each individual spectral line. Therefore, we restrict by the contribution $\Delta\omega$ from an individual spectral line, which is given by

$$\Delta\omega_k = 2\int_{-\infty}^{\infty} d\omega \int_0^1 \cos\theta d\cos\theta \left[1 - e^{-u(\omega)/\cos\theta}\right] ,$$

where $u(\omega)$ is a sharply decreased function with removal from the central frequency; in addition, the function $u(\omega)$ is symmetric with respect to the line center. Taking the integral on parts, we have

$$\Delta\omega_k = 2\int_{-\infty}^{\infty} d\omega\left[1 - e^{-u(\omega)}\right] + \int_{-\infty}^{\infty} u(\omega)d\omega \int_0^1 d\cos\theta e^{-u(\omega)/\cos\theta}$$
$$= A\delta + \int_{-\infty}^{\infty} u(\omega)d\omega \int_0^1 d\cos\theta e^{-u(\omega)/\cos\theta} ,$$

where A is the absorption function that is given by formula (9.47). The second integral can be evaluated by the standard method for a sharply varied function $u(\omega)$. In this method we take the frequency ω' such that $u(\omega') \gg 1$, but ω' is close to a frequency ω_1 for which $u(\omega_1) \sim 1$. Using this range of frequencies the approximation $u(\omega) = u(\omega')\exp\left[-\alpha(\omega - \omega')\right]$, where α is the logarithm derivative of the function $u(\omega)$ at $\omega = \omega'$, we have $d\omega = -du/(\alpha u)$, and the second integral takes the form

$$2\int_0^{\infty} u(\omega)d\omega \int_0^1 d\cos\theta e^{-u(\omega)/\cos\theta} = \frac{2}{\alpha}\int_0^1 \cos\theta d\cos\theta = \frac{1}{\alpha} ,$$

where we use $u(\omega_k) \gg 1$. As a result, we obtain the effective line width

$$\Delta\omega = 2(\omega_1 - \omega_k) + \frac{1}{\alpha} = 2(\omega_2 - \omega_k) , \quad u(\omega_2) = e^{-C-1/2} = 0.34 . \tag{9.48}$$

We now use this result for the Doppler form of a spectral line (5.4), so that

$$u(\omega) = u(\omega_k)\exp\left[-\frac{mc^2}{2T}\frac{(\omega - \omega_k)^2}{\omega_k^2}\right] .$$

Assuming $\Delta\omega \ll \omega_k$, we obtain the average absorption function on the basis of formula (9.47)

$$\overline{A} = \frac{2\omega_k}{\delta}\left(\frac{T}{mc^2}\right)^{1/2} \overline{(\ln u(\omega_k) + C)^{1/2}} ,$$

where an average is made over centers of individual lines, and the effective width of the emission band due to an individual line is

$$\Delta\omega_k = 2\omega_k\left(\frac{T}{mc^2}\right)^{1/2}(\ln u(\omega_k) + C + 1/2)^{1/2} .$$

The total effective width of the emission band is equal to

$$\Delta\omega = \sum_k \Delta\omega_k . \tag{9.49}$$

▶ **Problem 9.21** The absorption coefficient for an individual spectral line varies by $(\omega - \omega_k)^{-n}$ as the frequency ω removes from the line center ω_k, and the width of an individual spectral line is less than the distance between neighboring lines. The emission band is symmetric with respect to the band center ω_o, and the layer is also optically thick near ω_o between neighboring lines, whereas the optical depth at centers of spectral lines drops sharply with removal from the central frequency. Determine the effective width of a spectral band for such a layer, if the effective width of an individual spectral line is relatively small, i. e., the optical thickness at frequencies between two line centers is low.

Using the variable $s = |\omega - \omega_k| / \nu$, where ν is of the order of the width of an individual spectral line, so that the layer optical width due to an individual spectral line can be approximated as

$$u(\omega) = u(\omega_k)\left(\frac{\nu}{\omega - \omega_k}\right)^n = u(\omega_k)s^{-n} ,$$

where $u(\omega_k)$ is the optical thickness of the layer for the center of a given spectral line.

On the basis of formula (9.48) we obtain the contribution to the effective band width due to an individual spectral line

$$\Delta\omega_k = \nu\left[u(\omega_k)e^{C+1/2}\right]^{1/n} .$$

The total effective width of the spectral band is equal according to formula (9.49)

$$\Delta\omega = \sum_k \Delta\omega_k = \frac{\nu}{\delta}\int_{-\infty}^{\infty} d\omega_k \left[u(\omega_k)e^{C+1/2}\right]^{1/n} ,$$

where we assume that many transitions determine the effective band width that allows us to replace summation by integration; δ is the average difference of frequencies for neighboring transitions. Next, we assume a sharp decrease of the optical thickness in centers of spectral lines when we remove from the center band. Let us take

$$u(\omega_k) = u(\omega_o)\exp[-\alpha(\omega_k - \omega_o)] ,$$

where ω_o is the frequency of the band center. This gives for the total effective width of the spectral band

$$\Delta\omega = \frac{2\nu n}{\alpha\delta}\left[u\left(\omega_o e^{C+1/2}\right)\right]^{1/n} . \tag{9.50}$$

9.4
Propagation of Resonant Radiation in Optically Thick Gas

▶ **Problem 9.22** Derive the kinetic equation for the number density of resonantly excited atoms by taking into account re-emission of resonance photons.

The balance equation (9.58) and its solution (9.59) relate to an average number density of resonantly excited atoms, and this average is made over all the plasma regions. We will refer now the number density of excited atoms to a certain point, rather than to the total plasma. Then the balance equation for the number density $N_*(\boldsymbol{r}, t)$ has the form

$$\frac{\partial N_*}{\partial t} = N_e N_o k_{ex} - N_e N_* k_q - \frac{N_*}{\tau} + \frac{1}{\tau} \int N_*(\boldsymbol{r}') G(\boldsymbol{r}', \boldsymbol{r}) d\boldsymbol{r}' . \tag{9.51}$$

This equation is based on equation (9.58), but differs from it. First, it contains a partial time derivative because the number density of excited atoms also depends on coordinates. Second, τ is the radiative lifetime for an individual atom, rather than a radiative time outside a plasma region. And the principal difference these balance equations consists in the last term that accounts for reabsorption. Indeed, $G(\boldsymbol{r}', \boldsymbol{r}) d\boldsymbol{r}'$ is the probability that a photon, which is originated at point $\boldsymbol{r}$, will be absorbed in a volume $d\boldsymbol{r}'$ near a point $\boldsymbol{r}'$.

▶ **Problem 9.23** Obtain the expression for the Green function $G(\boldsymbol{r}', \boldsymbol{r})$ of equation (9.51).

The Green function $G(\boldsymbol{r}', \boldsymbol{r})$ of equation (9.51) that describes the reabsorption of resonance photons is symmetric with respect to its coordinates

$$G(\boldsymbol{r}', \boldsymbol{r}) = G(\boldsymbol{r}, \boldsymbol{r}') ,$$

as it follows from its physical nature. If a plasma occupies infinite space,

$$\int G(\boldsymbol{r}', \boldsymbol{r}) d\boldsymbol{r}' = \int G(\boldsymbol{r}', \boldsymbol{r}) d\boldsymbol{r} = 1 .$$

In order to write the expression for the Green function, we determine the probability $G(\boldsymbol{r}', \boldsymbol{r}) d\boldsymbol{r}$ that a photon that is originated at point $\boldsymbol{r}'$ will be absorbed in a volume element $d\boldsymbol{r}$ near a point $\boldsymbol{r}$. Let us represent this volume element $d\boldsymbol{r} = dx ds$, where x is directed along the photon propagation, nd the area element ds is perpendicular to this direction. One can represent the probability $G(\boldsymbol{r}', \boldsymbol{r}) d\boldsymbol{r}$ as a product of some probabilities. We have

$$G(\boldsymbol{r}', \boldsymbol{r}) d\boldsymbol{r} = \int a_\omega d\omega \frac{ds}{4\pi |\boldsymbol{r} - \boldsymbol{r}'|^2} k_\omega dx \exp\left(-\int k_\omega dx\right) .$$

The first term is the probability that a photon is emitted in a given frequency range, the second means the probability that a photon emitted isotropically will intersect a given area element. The third term is the probability that the photon will be absorbed in a range dx, and the fourth is the probability to survive for the photon on the way to a point $\boldsymbol{r}$. Thus, we obtain the Green function

$$G(\boldsymbol{r}', \boldsymbol{r}) = \int \frac{a_\omega k_\omega d\omega}{4\pi |\boldsymbol{r} - \boldsymbol{r}'|^2} \exp\left(-\int k_\omega dx\right) . \tag{9.52}$$

Correspondingly, equation (9.51) now takes the form

$$\frac{\partial N_*}{\partial t} = N_e N_0 k_{ex} - N_e N_* k_q - \frac{N_*}{\tau} + \frac{1}{\tau}\int N_*(\mathbf{r}')d\mathbf{r}' \int \frac{a_\omega k_\omega d\omega}{4\pi |\mathbf{r}-\mathbf{r}'|^2} \exp\left(-\int k_\omega dx\right), \tag{9.53}$$

This equation is named the Biberman–Holstein equation.

► **Problem 9.24** Determine the Green function $G(\mathbf{r}', \mathbf{r})$ of the Biberman–Holstein equation (9.53) for a uniform optically thick gas or plasma for photons in the center of the spectral line and both forms of a spectral line, the Lorentz and Doppler ones.

The Green function (9.52) for a uniform gas has the form

$$G(R) = \int \frac{a_\omega k_\omega d\omega}{4\pi R^2} \exp(-k_\omega R). \tag{9.54}$$

In the case of the Lorentz form of the spectral line (5.9) use the variable $s = 2(\omega - \omega_0)/\nu$, and the Green function takes the form

$$G(R) = \frac{k_0}{4\pi R^2} \int \frac{ds}{(1+s)^2} \exp\left(-\frac{k_0 R}{1+s^2}\right) = \frac{k_0}{4\pi R^2} \frac{d}{dk_0 R}\left[e^{-k_0 R/2} I_0\left(\frac{k_0 R}{2}\right)\right],$$

where k_0 is the absorption coefficient in the line center. In deriving of this expression we used the variable $\varphi = 2\arctan s$, and the definition of the Bessel function $I_0(z) = \frac{1}{\pi}\int_0^\pi \exp(-z\cos\varphi)d\varphi$. In the limiting case $k_0 R \gg 1$ this formula gives

$$G(R) = \frac{1}{(4\pi k_0)^{1/2} R^{7/2}}. \tag{9.55}$$

In the case of the Doppler form of the spectral line (5.4) use the variable $s = \frac{(\omega-\omega_0)}{\omega_0}(mc^2/T)^{1/2}$, so that $a_\omega d\omega = \pi^{-1/2}\exp(-s^2)ds$ and $k_\omega = k_0 \exp(-s^2)$. Substituting this in formula for the Green function and introducing the variable $t = k_0 R \exp(-s^2)$, we obtain

$$G(R) = \frac{k_0}{4(\pi)^{3/2} R^2} \int_{-\infty}^{\infty} ds \exp(-s^2 - t) = \frac{1}{4(\pi)^{3/2} k_0 R^4} \int_0^{k_0 R} \frac{t e^{-t} dt}{\sqrt{\ln(k_0 R/t)}}.$$

In the limiting case under consideration $k_0 R \gg 1$, replacing the upper limit by infinity and considering $\ln k_0 R \gg 1$, we get

$$G(R) = \frac{1}{4(\pi)^{3/2} k_0 R^4 \sqrt{\ln(k_0 R/t_0)}},$$

where t_0 is the solution of the equation $\int_0^\infty t e^{-t} dt \sqrt{\ln(t/t_0)} = 0$. Solving this equation, we obtain

$$G(R) = \frac{1}{4(\pi)^{3/2} k_0 R^4 \sqrt{\ln(k_0 R) - 0.42}}. \tag{9.56}$$

▶ **Problem 9.25** Obtain the criterion of thermodynamic equilibrium of resonantly excited atoms in a plasma, if this equilibrium is established as a result of electron collisions with atoms in the ground and excited states.

We use the following processes which determine the formation and decay of resonantly excited atoms:

$$e + A \longleftrightarrow e + A^* \; ; \; A^* \rightarrow A + \hbar\omega . \qquad (9.57)$$

Here A and A^* are the atoms in the ground and excited states, respectively, and $\hbar\omega$ denotes a resonance photon. On the basis of this scheme of processes we have the following balance equation for the number density of resonantly excited atoms:

$$\frac{dN_*}{dt} = N_e N_0 k_{ex} - N_e N_* k_q - \frac{N_*}{\tau} , \qquad (9.58)$$

where N_0, $N_*$are the number densities of atoms in the ground and resonantly excited states, N_e is the number density of electrons, k_{ex} is the rate constant of atom excitation by electron impact, k_q is the rate constant of quenching of a resonantly excited atom by electron impact, and τ is a lifetime of the resonantly excited state with respect to radiation. Note that if an emitting photon is absorbed in a plasma again, an excitation is conserved in a plasma. Therefore τ is the lifetime with respect to the departure of the photon outside a plasma region. Solving this balance equation under stationary conditions, we obtain

$$N_* = N_0 \frac{k_{ex}}{k_q} \left(1 + \frac{1}{N_e k_q \tau} \right)^{-1} . \qquad (9.59)$$

Let us analyze this formula. If the number density of electrons is large, i. e., the criterion

$$N_e k_q \tau \gg 1 \qquad (9.60)$$

holds true, departure of radiation does not violate equilibrium that is established by electrons. We assume the energy distribution of electrons to be the Maxwell one with a certain temperature. Then in electron–atom collisions (9.57) an equilibrium is established between the ground and resonantly excited atom states, i. e., the relation between the number density of atoms in the ground and resonantly excited states is given by the Boltzmann formula (1.12) with the electron temperature.

Thus, criterion (9.60) characterizes thermodynamic equilibrium for excited atoms. Note that the quenching rate constant is included to this criterion, rather than the excitation rate constant.

▶ **Problem 9.26** Derive the criterion when resonant radiation does not violate thermodynamic equilibrium between the ground and resonantly excited atom states that is established in electron–atom collisions (9.57).

We roughly obtain this criterion (9.60) above considering the plasma region as a whole. We now obtain this criterion from the Biberman–Holstein (9.53) equation.

It is convenient to rewrite this equation for the function $y(\boldsymbol{r}) = N_*/N_*^B$, where the number density of resonantly excited atoms N_*^B is given by the Boltzmann formula (1.12) and corresponds to thermodynamic equilibrium. Introducing a reduced parameter $\beta = N_e k_q \tau$ and the reduced time $t' = t/\tau$, where τ is the radiative lifetime of an individual atom, we reduced the Biberman–Holstein equation (9.53) to the form

$$\frac{\partial y(\boldsymbol{r})}{\partial t} = \int y(\boldsymbol{r}')G(\boldsymbol{r}, \boldsymbol{r}')d\boldsymbol{r}' + \beta - (1+\beta)y(\boldsymbol{r}) . \tag{9.61}$$

We consider the stationary regime and from this equation it follows that if $\beta \gg 1$, the solution of this equation is $y(\boldsymbol{r}) = 1$, i. e., thermodynamic equilibrium is supported. Let us introduce the probability $W(\boldsymbol{r}) = \int G(\boldsymbol{r}, \boldsymbol{r}')d\boldsymbol{r}'$ that a photon arising at a point $\boldsymbol{r}$ will be absorbed in a plasma region. We will operate below with an average over the plasma region quantities $\overline{y}$ and $\overline{W}$. Then the approximated solution of equation (9.61) is

$$\overline{y} = \frac{\beta}{1+\beta-\overline{W}} . \tag{9.62}$$

Introducing the mean probability for a photon to go outside a plasma region $\overline{P} = 1 - \overline{W}$, we obtain the criterion of thermodynamic equilibrium for resonantly excited atoms in the form $\beta \gg \overline{P}$, or

$$N_e k_q \tau \gg \overline{P} .$$

This coincides with criterion (9.60) if we take into account that the quantity τ in this criterion is a typical time of excitation location inside a plasma region, i. e., the value $\tau/\overline{P}$ in these notations.

► **Problem 9.27** Assuming the number density of excited atoms N_* and other parameters of a radiating gas to be constant inside a gas region, determine N_* and the flux of emitting photons for the Lorentz line form and an optically thick gaseous region.

Let us introduced the reduced parameter

$$\beta_{ef} = \frac{\beta}{\overline{P}} = N_e k_q \tau/\overline{P} ,$$

and the criterion (9.60) has the form $\beta_{ef} \gg 1$. Then the solution (9.62) has the form

$$\overline{y} = \frac{\beta_{ef}}{\beta_{ef} + \overline{P}}$$

gives the number density of resonantly excited atoms.

In the limiting case $\beta_{ef} \ll 1$ when the number density of excited atoms N_* is smaller than that under thermodynamic equilibrium and is equal to

$$N_* = N_*^B N_e k_q \tau/\overline{P} .$$

The rate J of emitting photons (a number of photons per unit time that leave a gas region) is equal to

$$J = \frac{N_* V}{\tau} P ,$$

where V is the gas region volume. For the Lorentz line form a typical probability for an emitting photon to leave a gas region according to formula (9.2) is estimated as $\overline{P} \sim (k_o R)^{-1/2}$, here R is a dimension of the gas region.

In the other limiting case $\beta_{ef} \gg 1$ the number density of excited atoms is given by the Boltzmann formula (1.12), and the rate of emitting photons in accordance with formula (9.26) is given by

$$J = 0.5\sqrt{\frac{S}{k_o V}} N_*^B V = \frac{0.5 N_*^B}{\tau \sqrt{k_o}} \sqrt{VS} ,$$

where S is the area of a surface that restricts the region occupied by an excited gas.

9.5 Kinetics of Atom Excitation by Electron Impact in a Gas in Electric Field

▶ **Problem 9.28** Electrons are moving in a gas in a constant electric field. Electrons obtain energy from an electric field and lost it in elastic collisions with atoms and due to excitation of atoms. If the electron energy exceeds the threshold energy $\Delta\varepsilon$ of atom excitation, electrons excite atoms. Find the rate of atom excitation.

For description of this process, we use expansion (7.2) for a nonstationary distribution function of electrons, and using the standard procedure, as for deduction of the set of equations (7.3), we instead obtain this set of equations

$$\frac{\partial f_o}{\partial t} + \frac{a}{3v^2}\frac{\partial (v^3 f_1)}{\partial v} = I_{ea}(f_o), \qquad \frac{\partial f_1}{\partial t} + a\frac{\partial f_o}{\partial v} = -\nu v f_1 . \tag{9.63}$$

Assuming the excitation flux to be relatively small, we ignore a nonstationarity overall, except the first term, which corresponds to a small flux in the energy space. As a result, we obtain

$$\frac{\partial f_o}{\partial t} = I_{ea}(f_o) + \frac{a}{3v^2}\frac{d}{dv}\left(\frac{v^2}{\nu}\frac{df_o}{dv}\right) . \tag{9.64}$$

The nonstationarity of the distribution function is only due to atom excitation. Hence, the rate of excitation is

$$\frac{dN_*}{dt} = -\frac{dN_e}{dt} = -\int 4\pi v^2 dv \frac{\partial f_o}{\partial t} ,$$

where N_* is the number density of excited atoms. Using the collision integral (6.59) for the electron distribution function f_o, we obtain from this

$$\frac{dN_*}{dt} = 4\pi \frac{m_e}{M} v^3 \nu \left[\left(T + \frac{Ma^2}{3\nu^2}\right)\frac{df_o}{d\varepsilon} + f_o\right]\Bigg|_{\varepsilon=\Delta\varepsilon} , \tag{9.65}$$

where $\varepsilon = m_e v^2/2$ is the electron energy and $\Delta\varepsilon$ is the energy of atom excitation.

▶ **Problem 9.29** Determine the energy distribution function of electrons under the conditions of the previous problem.

We use the boundary condition $f_0(\Delta\varepsilon) = 0$ for the distribution function which satisfies to the following equation under stationary conditions and below the excitation threshold far from it

$$\left(T + \frac{Ma^2}{3\nu^2}\right)\frac{df_0}{d\varepsilon} + f_0 = 0\,.$$

This means a fast absorption of electrons above the excitation threshold and gives the distribution function

$$f_0(\varepsilon) = C\left[\varphi_0(\varepsilon) - \varphi_0(\Delta\varepsilon)\right] = C\left[\exp\left(-\int_0^{\varepsilon}\frac{d\varepsilon'}{T + \frac{Ma^2}{3\nu^2}}\right) - \exp\left(-\int_0^{\Delta\varepsilon}\frac{d\varepsilon'}{T + \frac{Ma^2}{3\nu^2}}\right)\right], \tag{9.66}$$

and $\varphi_0(\varepsilon)$ is the distribution function if we ignore absorption of fast electrons due to the excitation process, so that far from the excitation threshold $\varphi_0(\varepsilon) = f_0(\varepsilon)$. The constant C follows from the normalization condition

$$C = N_e\left[4\pi\int_0^{v_0} v^2 dv \exp\left(-\int_0^{\varepsilon}\frac{d\varepsilon'}{T + \frac{Ma^2}{3\nu^2}}\right)\right]^{-1}. \tag{9.67}$$

Here N_e is the number density of electrons and the electron threshold velocity is $v_0 = \sqrt{2\Delta\varepsilon/m_e}$.

▶ **Problem 9.30** Determine the efficiency of atom excitation by considering energy losses due to elastic electron–atom collisions under the conditions of the previous two problems. Take the rate of electron–atom elastic collisions to be independent of the electron velocity.

Thus, we obtain the rate of atom excitation by individual electrons in a gas in an external electric field

$$\frac{dN_*}{dt} = 4\pi v_0^3\frac{m_e}{M}\nu(v_0)\varphi_0(v_0) = N_e\frac{m_e}{M}\nu(v_0)\cdot\frac{\exp\left(-\int_0^{\Delta\varepsilon}\frac{d\varepsilon'}{T + \frac{Ma^2}{3\nu^2}}\right)}{\int_0^{v_0}\left(\frac{v}{v_0}\right)^2\frac{dv}{v_0}\exp\left(-\int_0^{\varepsilon}\frac{d\varepsilon'}{T + \frac{Ma^2}{3\nu^2}}\right)}, \tag{9.68}$$

where $\varphi_0(v_0) = \varphi_0(\Delta\varepsilon)$ is the electron distribution function at the excitation threshold if we neglect the excitation process. In the case $\nu(v_0) = \text{const}$ this formula takes the form

$$\frac{dN_*}{dt} = \frac{4}{\sqrt{\pi}}\left(\frac{\Delta\varepsilon}{T + \frac{Ma^2}{3\nu^2}}\right)^{3/2} N_e\frac{m_e}{M}\nu(v_0)\exp\left(-\frac{\Delta\varepsilon}{T + \frac{Ma^2}{3\nu^2}}\right). \tag{9.69}$$

It is of interest to find which part ξ of the power, taken by electrons from an external electric field, is consumed on atom excitation. We assume that the power obtained by electrons from the electric field is transformed mostly into the atom thermal energy as a result of elastic collisions between electrons and atoms, and this power per electron is eFw, where w is the electron drift velocity. In the case $\nu = \text{const}$ we have from formula (9.69) neglecting the atom thermal energy ($T \ll Mw^2$)

$$\xi = \frac{\Delta\varepsilon \frac{dN_*}{dt}}{eFwN_e} = \frac{4}{3\sqrt{\pi}} \left(\frac{\Delta\varepsilon}{\overline{\varepsilon}}\right)^{3/2} \exp\left(-\frac{\Delta\varepsilon}{\overline{\varepsilon}}\right), \tag{9.70}$$

where $\overline{\varepsilon} = Ma^2/(3\nu^2) = Mw^2/3$ is the average electron energy. Figure 9.2 contains the dependence of the efficiency of atom excitation ξ on the electron energy $\overline{\varepsilon}$ under these conditions.

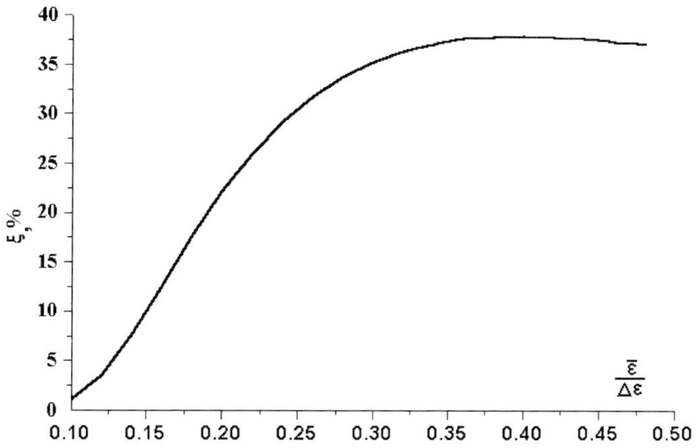

Fig. 9.2 The efficiency of atom excitation in a plasma located in a constant electric field as a function of the reduced average electron energy in the case when the rate of electron-atom elastic collisions is independent of the electron velocity.

▶ **Problem 9.31** Determine the energy distribution function of electrons near the excitation threshold if the excitation process is weak and corresponds to a tail of the distribution function. Find the criterion of this.

The above formulas are based on the assumption that the rate of atom excitation is determined mostly by diffusion of electrons in an energy space from small energies up to the atom excitation energy. We now consider another limiting case when excitation on the tail of the energy distribution function proceeds weakly and the efficiency of atom excitation near the threshold is determined by individual electrons which move in a gas in an external electric field. We first evaluate the electron distribution function above the excitation threshold in the energy range $\varepsilon \geq \Delta\varepsilon$, including the kinetic equation for electrons the term of inelastic electron–atom collisions. We assume that quenching of the excited atom proceeds by not electron impact because of a small number density of electrons. Then the second equation of set (7.3) takes the form

$$\frac{a}{3v^2}\frac{d}{dv}(v^3 f_1) = I_{ea}(f_o) - \nu_{ex} f_0 \,, \tag{9.71}$$

where $\nu_{ex} = N_a k_{ex}$, N_a is the number density of atoms, and k_{ex} is the excitation rate constant of the atom by electron impact. The collision integral I_{ea} (6.59) takes into account elastic electron–atom collisions. Using relation (7.3) between f_0 and f_1, we obtain the following equation for f_0:

$$\frac{a}{3v^2}\frac{d}{dv}\left(\frac{v^2}{\nu}\frac{df_0}{dv}\right) + I_{ea}(f_o) - \nu_{ex} f_0 = 0 \,. \tag{9.72}$$

Based on expression (6.59) for the electron–atom collision integral and neglecting the atom kinetic energy ($\sim T$) compared to the electron energy, we have

$$\frac{a}{3v^2}\frac{d}{dv}\left(\frac{v^2}{\nu}\frac{df_0}{dv}\right) + \frac{m_e}{M}\frac{1}{v^2}\frac{d}{dv}(v^3 \nu f_o) - \nu_{ex} f_0 = 0 \,. \tag{9.73}$$

We assume the average electron energy $\bar{\varepsilon}$ to be small compared to the atom excitation energy $\Delta\varepsilon$. Then, as it follows from formula (7.26), the mean electron energy is $\bar{\varepsilon} \sim Ma^2/\nu^2$. In addition, we assume that atom excitation does not influence on the electron distribution function below the excitation threshold, i. e.,

$$\nu \gg \nu_{ex} \gg \nu \frac{m_e}{M}\frac{\Delta\varepsilon}{\bar{\varepsilon}} \,. \tag{9.74}$$

This allows us to neglect the second term of the kinetic equation (9.73). Let us solve the resultant kinetic equation for the tail of the distribution function on the basis of the quasiclassical method accepting $f_0 = A\exp(S)$, where $S(v)$ is a smooth function, i. e., $(S')^2 \gg S''$. This gives $S' = \sqrt{3\nu_{ex}\nu}/a$, $a = eF/m_e$, and the distribution function for $\varepsilon \gg \bar{\varepsilon}$ has the form

$$f_0(v) = f_0(v_0)\exp(-S) = f_0(v_0)\exp\left(-\int_{v_0}^{v}\frac{dv}{a}\sqrt{3\nu_{ex}\nu}\right) , \tag{9.75}$$

where $v_0 = \sqrt{2\Delta\varepsilon/m_e}$ and $f_0(v_0)$ is determined by elastic electron–atom collisions. Near the threshold of atom excitation this formula gives

$$S = \frac{2v_0}{5a}\sqrt{3\frac{g_*}{g_o}\nu_q\nu_o}\left(\frac{\varepsilon - \Delta\varepsilon}{\Delta\varepsilon}\right)^{5/4} , \tag{9.76}$$

where the rate of elastic electron–atom collisions at the excitation threshold is $\nu_o = \nu(v_o)$, $\nu_q = N_a k_q$, k_q is the rate constant of quenching of the excited atom by electron impact, g_0, g_* are the statistical weights of the ground and excited atom states, $a = eF/m_e$, and we use formula (4.18) for the rate constant of atom excitation by electron impact that connects this rate constant and the rate of quenching of an excited atom by a slow electron. Using formula (9.75) for the electron distribution function, we assume the logarithm derivative of the distribution function to be determined by the excitation process not far from the threshold.

► **Problem 9.32** Determine the efficiency of atom excitation under the conditions of the previous problem.

Formula (9.75) gives the rate of atom excitation by electrons if this process proceeds mostly near the excitation threshold

$$\frac{dN_*}{dt} = \int 4\pi v^2 dv f_0(v_o) e^{-S} \nu_{ex}(v) = 4.30 a v_o^2 \left(\frac{a}{v_o \nu_o}\right)^{1/5} \left(\frac{\nu_q g_*}{\nu_o g_o}\right)^{2/5} f_0(v_o) , \tag{9.77}$$

and the distribution function is normalized by condition (2.30).

Comparing formulas (9.69) and (9.77) for the rate of atom excitation by individual electrons moving in a gas in an external electric field, one can do a choice between these two limiting cases. Indeed, in the case

$$\left(\frac{a}{v_o \nu_o}\right)^{6/5} \left(\frac{\nu_q g_*}{\nu_o g_o}\right)^{2/5} \gg 1 \tag{9.78}$$

the excitation process is restricted by diffusion of electron in an energy space to the excitation threshold, and the rate of this process is determined by formulas (9.68) and (9.69). In the other limiting case the excitation rate is determined by formula (9.77). Note that formula (9.77) is valid at low electric field strengths, whereas formula (9.69) holds true at high field strengths.

► **Problem 9.33** Determine the efficiency of atom excitation if the number density of electrons is not small and electron–electron collisions establish the Maxwell distribution function of electrons.

When electrons are located in a plasma, the energy distribution function of electrons drops strongly on the tail due to excitation of atoms and can be restored owing to collisions between electrons. Analyzing the character of atom excitation in a plasma, we assume for simplicity that excited states are destroyed as a result of radiation, i. e., quenching by electron impact is absent, and the excitation energy does not return to electrons. We assume criterion (7.18) to be valid, so that we have the Maxwell distribution function of electrons on velocities. In the first limiting case we assume the Maxwell distribution function is restored at energies $\varepsilon \geq \Delta\varepsilon$ ($\Delta\varepsilon$ is the atom excitation energy) which are responsible for excitation of atoms. Then the rate of atom excitation is equal to

$$\frac{dN_*}{dt} = N_a \int 4\pi v^2 dv \varphi(v) k_{ex}(v) , \tag{9.79}$$

where N_* is the number density of excited atoms, N_a is the number density of atoms in the ground state, $\varphi(v)$ is the Maxwell distribution function of electrons, k_{ex} is the rate constant of atom excitation by electron impact which is given by formula (4.18). Averaging over the Maxwell distribution function of electrons, we have

$$\frac{dN_*}{dt} = N_a N_e \overline{k_{ex}} = N_a N_e k_q \frac{g_*}{g_0} \exp\left(-\frac{\Delta\varepsilon}{T_e}\right) , \tag{9.80}$$

where the average rate constant of atom excitation in the limit $\Delta\varepsilon \gg T_e$ (T_e is the electron temperature) is equal to

$$\bar{k}_{ex} = \frac{1}{N_e} \int 4\pi v^2 dv \varphi(v) k_{ex}(v) = \frac{g_*}{g_0} k_q \exp\left(-\frac{\Delta\varepsilon}{T_e}\right) . \tag{9.81}$$

▶ **Problem 9.34** Determine the energy distribution function of electrons near the atom excitation threshold and the efficiency of atom excitation if the Maxwell distribution function of electrons is violated near the threshold because of excitation of atoms, while far from the excitation threshold it is the Maxwell one.

Let us consider the other limiting case of excitation of atoms by electrons in a plasma when criterion (7.18) is valid, but the Maxwell distribution function of electrons is not restored due to electron–electron collisions above the excitation limit because of absorption of fast electrons as a result of the excitation process. Then the excitation rate of atoms is determined by the rate of formation of fast electrons with the energy $\varepsilon > \Delta\varepsilon$ as a result of elastic collisions of electrons. Then on the basis of the kinetic equation (7.6), by using expression (6.68) for the electron–electron collision integral, we obtain the excitation rate per unit volume as

$$\begin{aligned} \frac{dN_*}{dt} &= -\int_{v_0}^{\infty} 4\pi v^2 dv \frac{\partial f}{\partial t} \\ &= -\int_{v_0}^{\infty} 4\pi v^2 dv\, I_{ee}(f_0) = -\frac{4\pi v_0}{m_e} B_{ee}(v_0) \left(\frac{f_0}{T_e} + \frac{df_0}{d\varepsilon}\right) , \end{aligned} \tag{9.82}$$

where the distribution function f_0 is taken at the excitation energy $\varepsilon = \Delta\varepsilon$. The electron distribution function in this case is the solution of the equation $I_{ee}(f_0) = 0$ under the boundary condition $f_0(v_0) = 0$ which accounts for an effective absorption of electrons above the excitation threshold. Then we have for the distribution function

$$f_0(v) = N_e \left(\frac{m_e}{2\pi T_e}\right)^{3/2} \left[\exp\left(-\frac{\varepsilon}{T_e}\right) - \exp\left(-\frac{\Delta\varepsilon}{T_e}\right)\right] , \qquad \varepsilon \le \Delta\varepsilon . \tag{9.83}$$

From this it follows that the electron distribution function is the Maxwell one far from the excitation threshold, while near the threshold the distribution function tends to zero because of absorption of electrons due to excitation of atoms. Using this distribution function and expression (6.69) for $B_{ee}(v)$, we obtain in this case the rate of excitation as

$$\frac{dN_*}{dt} = 4\sqrt{2\pi}\cdot\frac{N_e^2 e^4 \Delta\varepsilon \ln\Lambda}{m_e^{1/2} T_e^{5/2}} \exp\left(-\frac{\Delta\varepsilon}{T_e}\right). \tag{9.84}$$

Formula (9.84) is valid at high number densities of electrons when fast establishment of the equilibrium takes place for the electron distribution function on velocities. The corresponding criterion has the form

$$\frac{N_e}{N_a} \gg \frac{k_q}{k_{ee}}, \tag{9.85}$$

where the effective rate constant of elastic collisions of electrons k_{ee} due to the Coulomb interaction of electrons follows from comparison of formulas (9.80) and (9.84) and has the form

$$k_{ee} = 4\sqrt{2\pi}\cdot\frac{g_o}{g_*}\frac{e^4 \Delta\varepsilon \ln\Lambda}{m_e^{1/2} T_e^{5/2}}. \tag{9.86}$$

Formula (9.84) is valid under the opposite condition with respect to the criterion (9.85). As it is seen, the criterion (9.85) is much stronger than the criterion (7.18) because $m_e \ll M$. Thus both considering regimes of atom excitation in a plasma are possible. At relatively small number densities of electrons the distribution function is given by formula (9.76), while the Maxwell distribution function of electrons is valid at not low degrees of ionization. Correspondingly, the rate of atom excitation in a plasma varies from that by formula (9.80) to that by formula (9.84), as the electron number density increases.

As a demonstration of these results, Table 9.1 contains values of the rate constants (9.86) for rare gas atoms under typical conditions $T_e = 1\text{eV}$, $\ln\Lambda = 10$, and the boundary ionization degree is given by the relation

$$\left(\frac{N_e}{N_a}\right)_b = \frac{k_q}{k_{ee}} \tag{9.87}$$

for these parameters.

Table 9.1 The parameters of the criterion (9.86) for metastable inert gas atoms.

Metastable atom	**$\Delta\varepsilon$ (eV)**	**$k_{ee}(10^{-4}\text{cm}^3/\text{s})$**	**$\left(\frac{N_e}{N_a}\right)_b (10^{-6})$**
$He(2^3S)$	19.82	5.8	5.4
$Ne(2^3P_2)$	16.62	2.9	0.69
$Ar(3^3P_2)$	11.55	2.0	2.0
$Kr(4^3P_2)$	9.915	1.7	2.0
$Xe(5^2P_2)$	8.315	1.4	13

Note that in the case of high electron densities when the electron distribution function is the Maxwell one in the basic range of electron energies, this value is represented in the form $f = f(v_o)\exp(-S)$, which is characterized by the following exponent:

$$S = \frac{\varepsilon - \Delta\varepsilon}{T_e} = 3v_o^2\frac{(\varepsilon - \Delta\varepsilon)}{Ma^2} . \tag{9.88}$$

Here for simplicity we assume $\nu(v) = const.$ Because of criterion (9.85), formula (9.83) gives a more slight decrease of the distribution function with an increase in electron energy than that follows from formula (9.76), which holds true in the limit when collisions between electrons are not significant.

▶ **Problem 9.35** Determine the energy distribution function of electrons near the atom excitation threshold if quenching of excited atoms is determined by electron impact.

Above we assume that quenching of excited atoms is determined by other processes than electron impact. We now consider the other case, when quenching of excited atoms is determined by electron–atom collisions. Then, based on criterion (9.85), we found that fast electrons are generated and destroyed as a result of inelastic collisions between electrons and atoms. Because of equilibrium between the considering atomic states, this gives

$$\nu_{ex} f_o(v) v^2 dv = \nu_q\, f_o(v') v'^2 . \tag{9.89}$$

Here $v^2 = 2\Delta\varepsilon/m_e + v'^2$, and v, v' are the velocities of fast and slow electrons, $\nu_{ex} = N_a k_{ex}$, $\nu_q = N_i k_q$ are the rates of excitation and quenching of atomic states by electron impact, so that N_a, N_i are the number densities of atoms in the ground and excited states correspondingly, k_{ex}, k_q are the rate constants of the corresponding processes which are connected by the principle of detailed balance (4.18). From this we have

$$\frac{N_a}{g_o} f_o(v) = \frac{N_*}{g_*} f_o(\sqrt{v^2 - v^2}), \quad v > \sqrt{2\Delta\varepsilon/m} . \tag{9.90}$$

This relation establishes the relation between the distribution functions of slow and fast electrons. The relation can be written in the form

$$f_o(v) = \frac{f_o(v_o) f_o(\sqrt{v^2 - v_o^2})}{f_o(0)} . \tag{9.91}$$

In particular, for the Maxwell distribution function of slow electrons $[f_o \sim \exp(-\varepsilon/T_e)]$ this formula gives

$$f_o(v) = f_o(v_o)\exp\left(\frac{\varepsilon - \Delta\varepsilon}{T_e}\right) , \tag{9.92}$$

where T_e is the electron temperature, $\varepsilon = m_e v^2/2$ is the electron energy. Thus, inelastic collisions of electrons with excited atoms restore the Maxwell distribution function above the threshold of the atom excitation.

The above cases of atom excitation by electrons in a plasma show that this process depends on the character of establishment of the electron distribution function near the threshold of excitation. The result depends both on the rate of restoring of the electron distribution function in electron–electron or electron–atom collisions and on the character of quenching of excited atoms. Competition of these processes yields different ways of establishment of the electron distribution function and different expressions for the effective rate of excitation of atoms in a gas and plasma. Thus, the excitation rates depend on the collision processes which establish the electron distribution function below and above the excitation threshold, and the equilibrium for excited atoms.

10 Processes in Photoresonant Plasma

10.1 Interaction of Resonant Radiation and Gas

► **Problem 10.1** A weak beam of resonant photons of a flux j_+ is moving perpendicular to a semi-infinite gas layer, so that processes of absorption and reabsorption determine the number density of excited atoms in a gas. Find the number density of excited atoms near the gas boundary.

We have two photon fluxes inside the gas region, the incident flux j_+ and an isotropic flux $i(x)$ that is created by radiation of excited atoms. Here x is the coordinate inside the gas region along the photon beam. When the stationary regime is established, we have the equality of an incident and reflecting photon fluxes near the boundary, that gives $j_+(0) = i(0)/4$, i. e., under equilibrium between incident resonant radiation and an absorbed gas we have $i(0) = 4j_+(0)$.

The number density of excited atoms N_* follows from the balance equation between the rate of emission events N_*/τ per unit volume and the rate of absorption events that is equal to $[j_+ + i(0)/2]k_\omega$ near the gas boundary, where k_ω is the absorption coefficient. We consider distances from the boundary exceeded the mean free path of photons $1/k_\omega$ and emission of spontaneous radiation as the channel of a loss of excited atoms. From this we obtain the number density of excited atoms

$$N_* = 3k_\omega j_+\tau \,. \tag{10.1}$$

We above assume the flux of incident radiation to be small that allows us to ignore stimulated radiation. In addition, collision processes involving excited atoms assume to be weak compared with reabsorption processes.

► **Problem 10.2** A narrow beam of resonant radiation whose frequency corresponds to the center of a spectral line of a resonant transition, passes through a gas and is absorbed by it. Introducing the temperature of excitation according to the Boltzmann formula (1.12), find its connection with the incident radiation flux.

The balance equation for the number density of resonantly excited atoms N_*, by taking into account the processes of absorption and emission of resonant photons

Plasma Processes and Plasma Kinetics. Boris M. Smirnov

ISBN: 978-3-527-40681-4

in accordance with equations (5.27) and (5.28) has the form

$$\frac{dN_*}{dt} = j_\omega \sigma_{abs} N_0 - j_\omega \sigma_{em} N_* - \frac{N_*}{\tau} ,$$

where $j_\omega = I_\omega/(\hbar\omega)$ is the photon flux, $\sigma_{\rm abs}$ is the absorption cross section, $\sigma_{\rm em} = \sigma_{\rm abs} g_0/g_*$ is the cross section of stimulated emission, τ is the radiative lifetime of excited atoms in a plasma, which in the absence of reabsorption processes is equal to the radiative lifetime of an individual atom.

We introduce the temperature T_* of excitation on the basis of the Boltzmann formula (1.12)

$$\frac{N_*}{N_0} = \frac{g_*}{g_0} \exp\left(-\frac{\hbar\omega}{T_*}\right) .$$

Then the balance equation for the number density of excited atoms may be represented in the form

$$\frac{dN_*}{dt} = j_\omega k_o - j_\omega k_o \frac{N_*}{N_0}\frac{g_0}{g_*} - \frac{N_*}{\tau} . \tag{10.2}$$

The stationary form of this equation is

$$j_\omega k_o \left[1 - \exp\left(-\frac{\hbar\omega}{T_*}\right)\right] = \frac{N_*}{\tau} .$$

It is convenient to rewrite this relation in the form

$$j_\omega = \frac{j_0}{\exp\left(\frac{\hbar\omega}{T_*}\right) - 1} ; \qquad j_0 = \frac{N_0 g_*}{g_0 k_o \tau} . \tag{10.3}$$

This relation connects the flux of resonant photons in the spectral line center and the temperature of excitation. This relation can be represented in the form

$$T_* = \frac{\hbar\omega}{\ln\frac{1+\eta}{\eta}} ; \qquad \eta = \frac{j_\omega}{j_0} = j_\omega \sigma_{\rm abs} \tau \frac{g_0}{g_*} . \tag{10.4}$$

In particular, if $j_\omega = j_0$, we have $T_* = 1.44\hbar\omega$. Table 10.1 gives values of this excitation temperature T_*, of the specific photon flux j_0/N_0 for $g_*/g_0 = 1$ and the specific intensity of incident radiation $I_0/N_0 = \hbar\omega \cdot j_0/N_0$ for alkali metal vapors at the Lorenz shape of a spectral line. Figure 10.1 shows the excitation temperature as a function of an incident radiation flux in accordance with formula (10.4).

▶ **Problem 10.3** The frequency of a narrow beam of resonant radiation is shifted by $\omega - \omega_o$ from the center of a spectral line of a resonant transition with the Lorenz character of broadening of the spectral line. Give the relation between the temperature of excitation and the incident radiation flux.

The balance equation for resonantly excited atoms is given by formula (10.2) in which the absorption coefficient in the center of a spectral line k_o for incident

Table 10.1 Parameters of interaction of resonant radiation with vapors of alkali metals.

	k_0 (10^5 cm^{-3})	T_* (eV)	j_0/N_0 (100 cm/s)	I_0/N_0 (10^{-17} W · cm)
Li(2^2P)	1.6	2.61	2.3	6.8
Na($3^2P_{1/2}$)	1.1	3.04	5.6	19
Na($3^2P_{3/2}$)	1.4	3.04	4.4	15
K($4^2P_{1/2}$)	0.85	2.32	4.5	15
K($4^2P_{3/2}$)	1.1	2.33	3.6	9.4
Rb($5^2P_{1/2}$)	0.91	2.25	3.9	10
Rb($5^2P_{1/2}$)	1.2	2.29	3.2	8.0
Cs($6^2P_{1/2}$)	0.77	2.00	4.3	9.6
Cs($6^2P_{3/2}$)	1.0	2.10	3.7	8.6

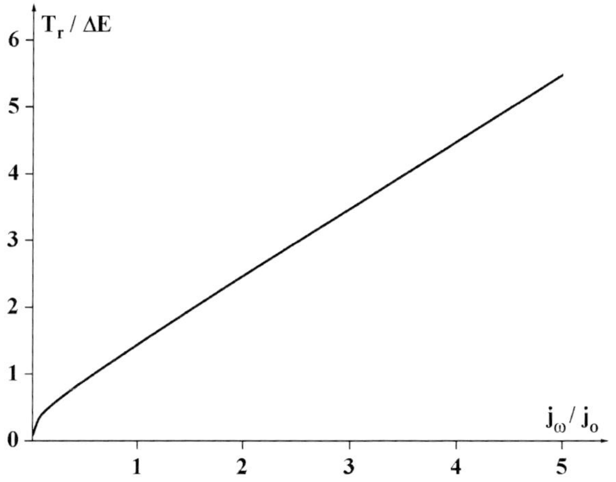

Fig. 10.1 The relative excitation temperature T_* for for resonantly excited atoms versus the radiation flux according to formula (10.4).

radiation must be replaced by the value k_ω, and for the Lorenz form of a spectral line these values are connected by the relation

$$k_\omega = k_0 \frac{(\omega - \omega_0)^2}{\Delta\omega^2} ,$$

where $\Delta\omega$ is the width of the spectral line. Correspondingly, it is necessary to replace the value k_0 by k_ω in formula (10.3).

► **Problem 10.4** A broad beam of resonant radiation of a weak intensity ($j_\omega \ll j_0$) has the cylinder symmetry and the photon frequency of this beam corresponds to the center of a spectral line for a resonance transition. Find the effective radiative time for the resonantly excited state.

The condition of a broad radiation beam corresponds to a large beam radius ρ_0 in comparison with the mean free path of resonant photons $1/k_0$

$$\rho_0 k_0 \gg 1 ,$$

where k_0 is the absorption coefficient at the spectral line center. In this case the radiative lifetime of resonantly excited atoms τ_{ef} exceeds that τ of an individual atom. In particular, for the Lorenz shape of the spectral line in accordance with formula (9.23) we have

$$\tau_{ef} \sim \tau\sqrt{\rho_0 k_0} .$$

▶ **Problem 10.5** For a beam of incident photoresonant radiation of a small intensity propagated in a broad gas region find the connection between the number density of excited atoms of a photoresonant plasma and the intensity of incident radiation. Broadening of spectral lines results from collisions of an excited atom with atoms in the ground state.

The following chain of successive processes takes place in a forming photoresonant plasma

$$A^* + A^* \rightarrow A_2^+ + e; \quad A^* \rightarrow A + \hbar\omega; \quad e + A^* \rightarrow e + A, \tag{10.5}$$

and these processes are realized in certain gases and vapors and under certain conditions. We will keep below this scheme for a photoresonant plasma.

One can see that an equilibrium of a photoresonant plasma with incident resonant radiation is established by photons which leave this plasma. This allows us to estimate the number density of excited atoms. Indeed, if I_ω is the intensity of an incident radiation, we obtain the balance equation between an absorbed and emitting power as

$$\frac{N_*}{\tau} P(R) V \sim \frac{I_\omega}{\hbar\omega} ,$$

where N_* is the number density of resonantly excited atoms, τ is the radiative lifetime of an excited atom, R is a dimension of the plasma region, $V \sim R^3$ is its volume, and $P(R)$ is the probability for an emitting photon to leave the plasma region.

Since according to formula (9.2) $P(R) \sim (k_0 R)^{-1/2}$ for the Lorenz shape of a spectral line, we obtain the number density of excited atoms from the balance equation

$$N_* \sim \frac{I_\omega \tau k_0^{1/2}}{\hbar\omega R^{5/2}} . \tag{10.6}$$

If we represent this formula in the form

$$N_* = C \frac{I_\omega}{R^{5/2}}$$

and according to parameters of Table 5.1 take $k_0 \sim 10^4 \text{cm}^{-1}$, $\hbar\omega \sim 1$ eV, $\tau \sim 10$ ns, we estimate the proportionality coefficient as $C = \frac{\tau k_0^{1/2}}{\hbar\omega} \sim 10^{13}\ \text{W}^{-1}\text{cm}^{-1/2}$.

▶ **Problem 10.6** Consider the distribution of atoms of a photoresonant plasma on excited sublevels depending on the number density of atoms and the intensity of polarized incident radiation.

Incident radiation of a given polarization transfers atoms in excited states of a certain polarization. Subsequent collisions of excited atoms with atoms in the ground states lead to change of an atom polarization. Therefore, we have two possibilities depending on gas and beam parameters. In the first case excited atoms return in the ground state under the action of stimulated and spontaneous radiation of excited atoms. Then we have excited atoms of one polarization only, and the statistical weight of excited atoms which take part in the radiation processes is equal to the statistical weight of atoms in the ground state, i. e., $g_*/g_0 = 1$ in this case.

In the other case forming excited atoms as a result of collisions with atoms in the ground state can change their polarization, and excited atoms are redistributed over polarizations in this case, where the ratio of the statistical weight of resonantly excited and ground states of alkali metal atoms is $g_*/g_0 = 3$ in this case.

The total cross section of collision of an atom in the ground S-state and excited P-states due to dipole–dipole interaction of colliding atoms is given by formula (3.34). In such a collision, the cross section of variation of the atom momentum projection is $0.602\pi d^2/(\hbar v)$ without exchange excitation and is $0.56\pi d^2/(\hbar v)$ in collisions with exchange excitation (the notations are given in formula (3.34)). Hence, the total cross section of depolarization of an excited atom in these collisions is $1.16\pi d^2/(\hbar v) = 0.24\sigma_t$, where the total cross section of this collision σ_t is given by formula (3.34). The values of the depolarization rate constants k_{dep} in accordance with this formula are given in Table 10.2. Note that these formulas relate to the case when spin-orbit splitting is relatively small. Though this is not fulfilled for the most cases, these data may be used for estimations.

Table 10.2 The rate constant of depolarization k_{dep} for resonantly excited atoms of alkali metals in collisions with the same atoms in the ground state.

	k_{dep} (10^{-7}cm^3/s)	N_{dep} (10^{14}cm^{-3})
$Li(2^2P)$	1.2	3.1
$Na(3^2P_{1/2})$	1.5	4.1
$Na(3^2P_{3/2})$	1.5	4.1
$K(4^2P_{1/2})$	2.1	1.8
$K(4^2P_{3/2})$	2.1	1.9
$Rb(5^2P_{1/2})$	2.0	1.8
$Rb(5^2P_{1/2})$	2.1	1.7
$Cs(6^2P_{1/2})$	2.8	1.2
$Cs(6^2P_{3/2})$	2.7	1.4

On the basis of the above values, we have the criterion that only one excited state takes part in interaction with incident resonant radiation

$$N_0 \ll k_{dep}\tau\frac{j_\omega + j_0}{j_\omega} = N_{dep}\left(1 + \frac{j_0}{j_\omega}\right) . \tag{10.7}$$

The values of the transient number density of atoms N_{dep} for alkali metal vapors are given in Table 10.2.

It is convenient to characterize mixing of excited atoms over polarizations by the parameter

$$\zeta = \frac{k_{dep} \tau N_o}{(1 + \frac{j_\omega}{j_o})^2} ,$$

where N_o is the total number density of atoms which interact with resonant radiation. Then the ratio of the effective statistical weights of the excited g_* and ground g_0 states is equal to

$$\frac{g_*}{g_0} = \frac{1 + 3\zeta}{1 + \zeta}$$

for atoms of alkali and alkali earth metals with the ground *S*-state and resonantly excited *P*-state.

This formula means that at small rates of mixing one polarization of an excited state is occupied only, whereas at high number densities of atoms redistribution takes place between sublevels of excited states as a result of atom collision with rotation of the momentum of an excited atom. Due to this process, the probabilities for an excited atom to have arbitrary momentum projections are identical.

10.2
Excited Atoms in Photoresonant Plasma

► **Problem 10.7** Estimate a typical time of establishment of an equilibrium for incident radiation of a weak intensity in an optically thick photoresonant plasma.

This equilibrium results from reabsorption processes. In particular, if an incident radiation corresponds to the center of the spectral line, the number density of excited atoms near the surface is given by formula (10.1), that is higher than that of formula (10.6). And this number density of excited atoms is supported near the surface during plasma radiation, i. e., a photoresonant plasma is not uniform. Forming excited atoms emit resonant radiation, and for the Lorenz form of the spectral line this radiation reaches any plasma surface for a time of the order of $\tau\sqrt{k_o R}$, where k_o is the absorption coefficient for the spectral line center, and R is a plasma dimension. According to the nature of the reabsorption process, only this time is responsible for establishment of an equilibrium for resonantly excited atoms in a photoresonant plasma.

► **Problem 10.8** Find the dependence of the mean free path for resonant photons in a photoresonant plasma depending on the intensity of a beam of resonant radiation and a shift of its frequency with respect to the center of a spectral line of the resonant transition.

Let us use the balance equation (10.2) for the photon flux of incident resonant radiation that has the form

$$\frac{dj_\omega}{dx} = -j_\omega \sigma_{abs} N_0 + j_\omega \sigma_{em} N_* ,$$

where the axis x is directed along an incident beam. For a weak beam intensity, when the second term may be ignored and when the radiation frequency coincides with the center of the corresponding spectral line, this equation takes the form

$$\frac{dj_\omega}{dx} = -\frac{j_\omega}{\lambda_o}, \qquad \lambda_o = (\sigma_o N_0)^{-1} = \frac{1}{k_o} ,$$

where σ_o is the absorption cross section and k_o is the absorption coefficient at the line center.

Taking into account stimulated radiation, we obtain the mean free path of photons

$$\lambda_\omega = \frac{\lambda_o}{1 - \exp\left(-\frac{\hbar\omega}{T_*}\right)} . \tag{10.8}$$

As is seen, transition of a part of atoms in an excited state causes blooming of a gas with respect to an incident radiation that occurs in decrease of its absorption by the gas. If the frequency of incident radiation does not coincide with the center of the corresponding spectral line, in accordance with the result of the previous problem this expression must be multiplied by the factor $(\omega - \omega_o)^2/\Delta\omega^2$ in the case of the Lorenz shape of the spectral line.

▶ **Problem 10.9** Estimate the penetration depth of a narrow beam of incident resonant radiation propagated inside a photoresonant plasma.

Evidently, the penetration depth coincides with the mean free path of photons in a photoresonant plasma that follows from the balance equation (10.2) for the photon flux of incident resonant radiation

$$\frac{dj_\omega}{dx} = -j_\omega k_o \left[1 - \exp\left(-\frac{\hbar\omega}{T_*}\right)\right] ,$$

and for simplicity we take the frequency of incident photons to be coincided with the center of a spectral line of atom transition. This gives the connection of the mean free path of photons λ with the excitation temperature T_*

$$\lambda = k_o^{-1} \left[1 - \exp\left(-\frac{\hbar\omega}{T_*}\right)\right]^{-1} .$$

One can find this value from the balance equation for absorbed and reabsorbed photons

$$j_\omega k_o \left[1 - \exp\left(-\frac{\hbar\omega}{T_*}\right)\right] = \frac{N_*}{\tau} ,$$

which gives

$$\lambda = \frac{j_\omega \tau}{N_*} .$$

In particular, if $j_\omega \sim j_o \sim N_0/(k_o \tau)$ and $N_* \sim N_0$, from this it follows the above relations $T_* \sim \hbar\omega$ and $\lambda \sim k_o^{-1}$.

Since the absorption coefficient for the Lorenz form of a spectral line is $k_o \sim 10^5 \text{cm}^{-1}$, the mean free path of a resonant radiation beam photons of a low or intermediate intensity is small ($\sim 10^{-5}$cm) for the center of a spectral line. In order to increase this value, it is necessary to increase the intensity of an incident radiation beam or to shift its frequency from the spectral line center.

▶ **Problem 10.10** Give the saturation criterion for a narrow beam of resonant radiation as the criterion to ignore spontaneous radiation.

This is a strictly criterion of saturation for resonant radiation when only processes of absorption and stimulated emission for this radiation are of importance. Based on the balance equation (10.2) for the number density of resonantly excited atoms, one can represent the possibility to neglect there spontaneous radiation. As a result, we have the criterion of saturation $j_\omega \gg j_o$ or

$$T_* \gg \hbar\omega .$$

If the opposite criterion is fulfilled, spontaneous radiation of resonantly excited atoms is of importance for the balance of excited atoms.

Note the principal peculiarity of interaction of a high intensity incident beam of radiation with a gas when excited atoms are quenched mostly due to generation of stimulated radiation. These excited atoms have a certain polarization that is determined by the polarization of incident radiation, and therefore the ratio of the statistical weights of the ground g_0 and excited g_* states in the Boltzmann formula (1.12) is $g_0/g_* = 1$. In addition, the region of excitation is restricted mostly by the region of propagation of an incident radiation beam.

▶ **Problem 10.11** Estimate a typical time of establishment of an equilibrium for resonantly excited atoms in an optically thin photoresonant plasma created by a beam of resonant radiation of a weak intensity.

We now consider the balance equation (10.2) for the resonantly excited atoms that is represented in the form

$$\frac{dN_*}{dt} = -\frac{[N_* - N_*^{(0)}]}{\tau} \left(1 + \frac{j_\omega}{j_o}\right) .$$

Here $N_*^{(0)}$ is the equilibrium number density of excited atoms. From this it follows for a strong intensity of incident radiation $j_\omega \gg j_o$ that a typical time τ_{exc} of establishment of the equilibrium number density of excited atoms is

$$\tau_{exc} = \tau \frac{j_o}{j_o + j_\omega} ,$$

where τ is the radiative lifetime of an excited state. As is seen, a typical time of establishment of the equilibrium number density of excited atoms decreases with an increase of the intensity of incident radiation, and for a strong intensity of incident resonant radiation is equal to

$$\tau_{\text{exc}} = \tau \left[1 - \exp\left(-\frac{\hbar\omega}{T_*}\right)\right] , \tag{10.9}$$

where T_* is the excitation temperature.

In this case of a high intensity of an incident radiation beam reabsorption processes are not essential because of importance of stimulated radiation for establishment of an equilibrium for excited atoms.

▶ **Problem 10.12** A photoresonant plasma is formed under the action of a narrow laser beam of resonant radiation whose radius r_0 is small compared to the mean free path of excited atoms in a gas. Find the condition when the process of transport of absorbed atoms influences the properties of a forming photoresonant plasma.

Let us write the balance equation for the number density of excited atoms based on equation (10.2) and adding to this equation the term that accounts for transport of excited atoms. This balance equation has the form

$$\frac{dN_*}{dt} = j_\omega k_0 - j_\omega k_0 \frac{N_*}{N_0}\frac{g_*}{g_0} - \frac{N_*}{\tau} - \frac{N_*}{\tau_{\text{tr}}} ,$$

where the last term accounts for passage of atoms outside the laser beam. One can see that the last term is negligible if $\tau_{\text{tr}} \gg \tau$.

We now evaluate a transport time τ_{tr} for a cylinder tube of a small radius, and this is a time of atom renewal inside this tube. This time is the ratio of the total number of atoms inside a tube of a length l, which is $\pi r_0^2 l N$ to the total number of atoms $2\pi r_0 l j$ that intersect the tube boundary per unit time. Here N is the number density of atoms, the atom flux through the tube boundary is $j = \overline{v} N$, and $\overline{v}$ is the average atom velocity. From this we have

$$\tau_{\text{tr}} = \frac{r_0}{2\overline{v}} ,$$

and the possibility to neglect the atom renewal $\tau_{\text{tr}} \gg \tau$ has the form

$$r_0 \ll 2\overline{v}\tau .$$

The parameter of the right-hand side of this criterion for alkali metal vapors (Na, K, Rb, Cs) at temperature of 500 K is given in Table 10.3. Since simultaneously the mean free path of atoms is large in comparison with the beam radius r_0, transport of excited atoms may be of importance for interaction of a laser beam with a gas at low gas pressure or a narrow laser beam. Note that in this consideration we assume the mean free path of photons to be large compared to the width of the laser beam. But since for an intense laser beam the mean free path of transversal photons is

Table 10.3 Parameter $2\bar{v}\tau$ of transport of alkali metal atoms in the field of a resonant radiation beam.

Element	$r = 2\bar{v}\tau$ (μm)
Li	76
Na	22
K	27
Rb	19
Cs	16

determined by stimulated radiation, the above condition does not influence this analysis.

10.3
Processes in Photoresonant Plasma Involving Electrons

▶ **Problem 10.13** Obtain the criterion when associative ionization in collisions between resonantly excited atoms in a photoresonant plasma do not influence on the number density of excited atoms.

Collisions of two excited atoms are of importance for formation of highly excited atoms or ions as it follows from scheme (10.5), i. e., these processes lead to plasma formation. These processes may be responsible for the balance of excited atoms. Analyzing ionization of an excited gas according to scheme (10.5), we consider collision processes involving two excited atoms to be secondary ones, i. e., they are weak in comparison to radiative processes of decay of excited atoms, and hence these processes do not determine the number density of excited atoms. But these processes lead to the formation of free electrons on the first stage of plasma evolution according to the scheme

$$2A^* \rightarrow A_2^+ + e - \Delta\varepsilon_i, \qquad (10.10)$$

where $\Delta\varepsilon_i$ is the energy for this process. Table 10.4 gives values of the rate constant k_{as} of this process for resonantly excited atoms of alkali metals at the temperature of 500 K.

Table 10.4 Parameters of process (10.10) at the temperature of 500 K, the number density N_{ef} of excited atoms according to formula (10.11).

A^*	$\Delta\varepsilon_i$ (eV)	k_{as} (cm^3/s)	N_{ef} (10^{17} cm^{-3})
Na(3^2P)	< 0	$4 \cdot 10^{-11}$	0.14
K(4^2P)	0.1	$9 \cdot 10^{-13}$	4.5
Rb(5^2P)	0.2	$4 \cdot 10^{-13}$	4.2
Cs (6^2P)	0.33	$7 \cdot 10^{-13}$	4.1

The criterion of the validity of scheme (10.10) has the form

$$\frac{N_*}{\tau} P(R) \gg N_*^2 k_{as} ,$$

where k_{as} is the rate constant of the process of associative ionization of two excited atoms (10.10). Thus, the criterion that associative ionization is the secondary process in a photoresonant plasma created by a weak incident resonant radiation ($T_* \ll \hbar\omega$) has the form

$$N_* \ll N_{ef} = \frac{1}{\tau k_{ion} \sqrt{k_0 R}} , \qquad (10.11)$$

and under typical parameters ($\tau \sim 10^{-8}$s, $k_0 \sim 10^5$cm^{-1}, $R \sim 1$ cm, $k_{as} \sim 10^{-12}$cm^3/s) this gives $N_* \ll 3 \cdot 10^{18}$cm^{-3}. Table 10.3 contains values of the quantity N_{ef} if $R = 1$ mm. Criterion (10.11) holds true for a weakly excited photoresonant plasma.

► **Problem 10.14** Analyze the character of ionization in a photoresonant plasma where associative ionization leads to the formation of free electrons on the first stage of plasma evolution, and a subsequent growth of the electron number density results from ionization of excited atoms by electron impact.

When electrons are formed in a photoresonant plasma, they collide with excited atoms that leads to the equilibrium

$$e + A_* \leftrightarrow e + A \qquad (10.12)$$

where A, and A_* are the atoms in the ground and resonantly excited states, respectively. As a result, the electron temperature is established that corresponds to the temperature of excited atoms in accordance with the Boltzmann formula (1.12)

$$T_* = \frac{\Delta\varepsilon}{\ln \frac{N_0 g_*}{N_* g_0}} ,$$

and this formula is based on definition (1.12) of the excitation temperature. Here $\Delta\varepsilon = \hbar\omega$ is the excitation energy for a resonantly excited state, N_0, N_* are the number densities of atoms in the ground and excited states, and g_0, g_* are their statistical weights. This temperature of excitation is equal to the electron temperature $T_e = T_*$ if the above process for the balance of the electron energy dominates. Therefore, finally the electron number density is established in accordance with the ionization equilibrium that is described by the Saha formula (1.69).

We now consider the character of evolution of a gas with resonantly excited atoms to ionization equilibrium. We include in this scheme (10.5) ionization of excited atoms by electron impact

$$e + A_* \leftrightarrow 2e + A^+ . \qquad (10.13)$$

Evidently, this process is stepwise. As a result, electrons are formed both in collisions of two excited atoms according to scheme (10.5) and as a result of ionization of excited atoms by electron impact. Correspondingly, the balance equation for the number density of electrons on the first stage of evolution of a photoresonant plasma takes the form

$$\frac{dN_e}{dt} = M + N_e N_* k_{ion} ,$$

where N_e, N_* are the number densities of electrons and excited atoms, k_{ion} is the rate constant of ionization of an excited atom by electron impact, and $M = k_{as} N_*^2$ in accordance with scheme (10.5). The solution of this balance equation is

$$N_e = \frac{M}{N_* k_{ion}} \left[\exp(N_* k_{ion} t) - 1 \right] . \tag{10.14}$$

It is seen that the process of associative ionization is of importance in growth of the electron number density on the first stage of evolution of a photoresonant plasma. Subsequent growth of the electron number density is determined by ionization of atoms by electron impact.

▶ **Problem 10.15** Determine a typical time of establishment of the average electron energy.

We have the balance equation for the average energy of electrons per unit volume that results from processes (10.12)

$$\frac{d(N_e \varepsilon_e)}{dt} = \hbar\omega N_e (k_q N_* - k_{exc} N_0) ,$$

where $\varepsilon_e = 3T_e/2$ is the average electron energy, k_q is the rate constant of quenching of excited atoms by electron impact, and k_{exc} is the rate constant of atom excitation by electron impact. We assume the energy distribution function of electrons to be the Maxwell one, whereas the electron temperature T_e varies in time. Introducing the excitation temperature T_* in accordance with formula (10.4) and using the principle of the detailed balance that connects the rate constants k_q and k_{exc}, we reduce this balance equation to the form

$$\frac{dT_e}{dt} = \frac{2\hbar\omega}{3} k_q N_* \left[1 - \exp\left(\frac{\hbar\omega}{T_*} - \frac{\hbar\omega}{T_e} \right) \right] .$$

This balance equation in the case when the electron temperature T_e and the excitation temperature T_* are nearby takes the form

$$\frac{dT_e}{dt} = \frac{2(\hbar\omega)^2}{3T_*^2} k_q N_* (T_* - T_e) = -\frac{(T_* - T_e)}{\tau_T} .$$

From this we find a typical time for establishment of the electron temperature τ_T in a photoresonant plasma

$$\tau_T = \frac{3T_*^2}{2(\hbar\omega)^2 k_q N_*} \tag{10.15}$$

▶ **Problem 10.16** Compare two regimes of establishment of the electron temperature in a photoresonant plasma, so that in both electrons obtain energy as a result of quenching of excited atoms, while the energy loss results from excitation of atoms in the ground state in the first regime and from ionization of excited atoms in the second regime.

In the first regime the evolution of a photoresonant plasma is governed by processes (10.12)

$$e + A_* \leftrightarrow e + A \,,$$

whereas in the second regime of electron equilibrium the following processes are dominant:

$$e + A_* \rightarrow e + A \,, \; e + A^* \rightarrow 2e + A_+ \,. \tag{10.16}$$

In the first regime the average electron energy is determined by processes of excitation and quenching of atoms in collisions with electrons, whereas in the second regime of evolution of the photoresonant plasma the electrons obtain energy by quenching of excited atoms and lose the energy as a result of ionization processes. In this manner an average electron energy is established.

Let us analyze the transition between these two regimes that corresponds to the following balance equation for the electron energy:

$$\hbar\omega N_* N_e k_q = J_* N_* k_{ion} N_e \,,$$

where k_q is the rate constant of quenching of a resonantly excited atom by electron impact, k_{ion} is the rate constant of ionization of an excited atom in collision with an electron that assumes to have the stepwise character, and J_* is the ionization potential of a resonantly excited atom. From this we obtain the electron temperature T_e^* at transition between two regimes

$$k_{ion}(T_e^*) = \frac{\hbar\omega}{J_*} k_q(T_e^*) \,. \tag{10.17}$$

One can see this electron temperature to be independent of the number density of atoms.

Table 10.4 gives the value of the parameter T_e^* for excited alkali metal vapors, if the rate constant of quenching of a resonantly excited atom by electron impact is taken from Table 4.2, and formula (4.38) is used for the rate constant of stepwise ionization of excited atoms in collisions with electrons. Table 10.4 also contains values of the equilibrium constant $K(T_e^*)$ at this temperature that is given by the relation

$$K_{ion} = \frac{N_e^2}{N_*} \,,$$

where N_e and N_* are the number densities of electrons and excited atoms, respectively, in a quasineutral plasma under thermodynamic equilibrium.

Along with the above parameters, we give in Table 10.5 the intensities of incident beam of photoresonant radiation I_ω^* at the temperature $T_* = T_e^*$, which is given by

$$I_\omega^*(T_e^*) = \frac{\hbar\omega j_0}{\exp\left(\frac{\hbar\omega}{T_*}\right) - 1},$$

and j_0 is determined by formula (10.3).

Table 10.5 The transition temperature T_e^* between two regimes of evolution of a photoresonant plasma, the equilibrium constant K_{ion} for ionization equilibrium at this temperature, and the intensity of the incident radiation I_ω^* if the excitation temperature is equal to the transition temperature T_e^*.

	T_e^* (eV)	K_{ion} (10^{17} cm^{-3})	I_ω^* (W/cm^2)
$Li(2^2P)$	0.55	6.4	0.07
$Na(3^2P_{1/2})$	0.45	3.8	0.02
$Na(3^2P_{3/2})$	0.45	3.9	0.02
$K(4^2P_{1/2})$	0.40	3.0	0.04
$K(4^2P_{3/2})$	0.41	3.2	0.04
$Rb(5^2P_{1/2})$	0.38	2.3	0.03
$Rb(5^2P_{1/2})$	0.38	2.4	0.03
$Cs(6^2P_{1/2})$	0.38	3.1	0.05
$Cs(6^2P_{3/2})$	0.36	2.5	0.03

Thus, in the second regime of evolution of a photoresonant plasma, when the electrons lose energy as a result of ionization of excited atoms, the electron temperature does not increase the intensity of incident resonant radiation and correspondingly the excitation temperature. As it follows from the data of Table 10.5, the equilibrium constant K_{ion} exceeds a typical number density of atoms, and this corresponds to total ionization of atoms under stationary conditions. Hence, this regime of evolution of a photoresonant plasma is not stationary. When the number density of electrons becomes of the order of the initial number density of atoms, absorption of incident radiation decreases. A typical time when this will be attained is the lifetime of the photoresonant plasma.

▶ **Problem 10.17** Assuming that the equilibrium for the electron temperature results from electron collisions with atoms in the ground and resonantly excited states, find the critical electron temperature T_{cr} when the equilibrium electron number density under stationary conditions corresponds to the number density of atoms.

In the stationary regime the electron temperature is established as a result of processes (10.12). If this regime is realized, the electron temperature T_e^{cr} coincides with the excitation temperature. Hence, the equilibrium constant for the ionization process $K_{ion}(T_e^{cr}) = N_e^2/N_0$ coincides with the number density of atoms N_0, i. e., under this temperature

$$N_e(T_e^{cr}) = N_0(T_e^{cr}) . \tag{10.18}$$

Table 10.6 gives values of the critical electron temperature T_e^{cr} for $N_0 = 1 \cdot 10^{16} \mathrm{cm}^{-3}$. This table also coincides with the total number density of resonantly excited atoms at this temperature according to the Boltzmann formula (1.12), and a typical time of establishment of this equilibrium is given by

$$\tau_{ion} = \left(k_{ion} N_0\right)^{-1} ,$$

where the rate constant of stepwise ionization of atoms by electron impact is given by formula (4.38). As is seen, a typical time of establishment of ionization equilibrium is enough large, so that electron temperatures above the critical one are available in reality.

Table 10.6 The critical electron temperature T_e^{cr} at which the equilibrium electron number density in a stationary photoresonant plasma is equal to the number density of atoms, the number density of excited atoms at this temperature N_*, a time of establishment of the ionization equilibrium, and the electron number density $(N_e)_0$ starting from which the Maxwell distribution function is established for electrons.

	T_e^{cr} (eV)	N_* (10^{13} cm^{-3})	τ_{ion} s	$(N_e)_0$ (10^9 cm^{-3})
$Li(2^2P)$	0.26	2.7	1200	3.4
$Na(3^2P)$	0.25	0.70	990	0.72
$K(4^2P)$	0.22	1.7	490	2.4
$Rb(5^2P)$	0.21	1.6	420	2.2
$Cs(6^2P)$	0.19	2.4	310	4.2

▶ **Problem 10.18** Find the electron number density $(N_e)_0$ in a photoresonant plasma in the course of growth of the number density of electrons starting from which one can use the electron temperature T_e as a characteristic of the energy distribution function of electrons.

Using the electron temperature as a characteristic of the average electron energy, we assume the Maxwell distribution function for electrons. In reality the Maxwell distribution of electrons requires that the electron equilibrium is established by electron–electron collisions. Hence, the rate of electron–electron collisions $k_{ee} N_e$ exceeds that for electron–atom collisions that is $k_q N_*$. In this case electrons obtain energy in quenching collisions with excited atoms, and mixing of electrons in the energy space results from electron–electron collisions. Hence, the Maxwell distribution function of electrons starts from the electron number density $(N_e)_0$ that is given by the relation

$$(N_e)_0 = \frac{k_q}{k_{ee}} N_* .$$

In particular, one can use formula (9.86) for the rate constant k_{ee} of electron–electron collisions with variation of the electron energy by the value of the order of the atom excitation energy ΔE. The corresponding electron number densities $(N_e)_0$ are given in Table 10.5 for $N_0 = 1 \cdot 10^{16}$ cm^{-3}.

▶ **Problem 10.19** For a photoresonant plasma created by resonant radiation of an intermediate intensity find typical times of its evolution: a typical time τ_{exc} for establishment of the equilibrium for excited atoms, a typical time τ_T for establishment of the average electron energy, the time τ_{ion} for total ionization of atoms.

Let us consider the transition from small intensities of incident radiation to high ones, so that the radiation flux is $j_\omega = j_0$. The corresponding values of the excitation temperature T_* for alkali photoresonant plasma are given in Table 10.1, and a typical time of establishment of equilibrium between atoms in the ground and resonantly excited states is given by formula (10.9) that under these conditions ($j_\omega = j_0$) is $\tau_{exc} = \tau/2$, where τ is the radiative lifetime of an excited state. Values of τ_{exc} are given in Table 10.6.

A typical time of establishment of the average electron energy is given by formula (10.15)

$$\tau_T = \frac{3T_*^2}{2(\hbar\omega)^2 k_q N_*} .$$

Note that under used conditions ($j_\omega = j_0$), we have $T_* = \hbar\omega/\ln 2$, and one third part of atoms are transferred in an excite state. Hence, this formula takes the form

$$\tau_T = \frac{9.4}{k_q N_0} ,$$

where N_0 is the initial number density of atoms. Values of this time are given in Table 10.7 for an alkali metal photoresonant plasma at $N_0 = 1 \cdot 10^{15} \text{cm}^{-3}$. As is seen, a time of establishment of the electron average energy is less than that for the excitation temperature, and hence the average electron energy follows for excitation of atoms.

Table 10.7 Typical times of establishment of equilibria in a photoresonant plasma of alkali metals if the flux of radiation $j_\omega = j_0$ and the initial number density of alkali metal atoms is $N_0 = 1 \cdot 10^{15}$ cm^{-3}.

	τ_{exc} (10^{-9} s)	τ_T (10^{-8} s)	τ_{ion} (10^{-7} s)	k_{ion} (10^{-7} cm^3/s)
Li(2^2P)	14	4.9	3.6	1.0
Na(3^2P)	8	4.7	1.8	1.4
K(4^2P)	13	3.0	2.0	1.8
Rb(5^2P)	14	2.9	2.1	1.8
Cs(6^2P)	14	2.1	1.5	2.5

In considering the growth of the electron number density in a photoresonant plasma, we take into account that formation of free electrons results from the associative ionization process (10.10) on the first stage of plasma evolution, and subsequently the process of ionization of excited atoms by electron impact (10.16) is responsible for this. Then in accordance with scheme (10.5) of processes involving electrons, the balance equation for the electron number density N_e has the form

$$\frac{dN_e}{dt} = M + k_{ion} N_e N_* ,$$

where $M = k_{as} N_*^2$, k_{as} is the rate constant of associative ionization in collisions of two excited atoms, k_{ion} is the ionization rate constant of an excited atom by electron impact. From this we have the variation of the electron number density in time in accordance with formula (10.14)

$$N_e(t) = \frac{M}{k_{ion} N_*} \exp(k_{ion} N_* t) ,$$

and this dependence is violated when $N_e \sim N_o$, where N_o is the initial number density of atoms (or their total number density in the course of atom excitation). We assume $\tau_{ion} \gg \tau_{exc}$, i. e., during a basic time of ionization the number density of excited atoms does not vary in time. We have then for a typical lifetime of this photoresonant plasma

$$\tau_{ion} = \frac{1}{k_{ion} N_*} \ln \frac{N_o k_{ion} N_*}{M} .$$

We take C as an initial electron number density in this formula such that the process of associative ionization of excited atoms gives the same contribution to growth of the electron number density as the process of ionization of excited atoms by electron impact, that gives $k_{as} N_*^2 \sim k_{ion} C N_*$, where k_{as} is the rate constant of associative ionization in collisions of excited atoms (see Table 10.2). Since $N_* \sim N_o$, where N_o is the initial number density of atoms, we have for the lifetime of a photoresonant plasma with respect to ionization of its atoms

$$\tau_{ion} = \frac{\exp\left(\frac{\hbar\omega}{T_*}\right) + 1}{k_{ion} N_o} \ln \frac{k_{ion}}{k_{as}} . \tag{10.19}$$

Here N_o is the total number density of nuclei, i. e., $N_o = N_0 + N_* + N_e$, and according to the Boltzmann formula (1.12) $N_* / N_0 = g_* \exp(-\hbar\omega/T_*)/g_0$, where T_* is the excitation temperature. In particular, in the case $j_\omega = j_o$ and the criterion (10.7) holds true, one third part of atoms is found in excited states, and then formula (10.19) gives

$$\tau_{ion} = \frac{3}{k_{ion} N_*} \ln \frac{k_{ion}}{k_{as}} .$$

Table 10.7 contains the values obtained on the basis of this formula for the lifetime τ_{ion} of a photoresonant plasma of alkali metal vapors with respect to total ionization of atoms. In the regime under consideration the ionization rate constant is given by relation (10.17). The rate constants of associative ionization are given from Table 10.3, and the logarithm in formula (10.19) is equal in average to 12 ± 2. This value is used in the lithium case where the rate constant of associative ionization is absent. As it follows from Table 10.7, the lifetime of this photoresonant plasma exceeds by one order of magnitude of typical times of establishment of equilibria.

▶ **Problem 10.20** Find the rate constants of atom ionization by electron impact in a photoresonant plasma of alkali metals in the limit of high intensities of incident radiation.

These values follow from formula (10.17) and correspond to the largest possible electron temperatures. Assuming the rate constant of atom quenching by electron impact to be independent of the electron energy and using for them the data of Table 4.2, we evaluate the values of these rate constants which are given in Table 10.7. Since these values are concentrated in a restricted range, we make statistical averaging over different alkali metals. This gives for the average rate constant of ionization of an excited atom by electron impact at maximal electron temperatures

$$k_{ion} = 1.6 \cdot 10^{-7} \text{cm}^3/\text{s} \cdot 10^{\pm 0.15} \, .$$

▶ **Problem 10.21** Determine a typical time of ionization τ_{ion} for a photoresonant plasma if the ratio between a typical time between a time of equilibrium establishment for excited atoms τ_{exc} and the ionization time τ_{ion} is arbitrary.

As above, we take the first stage of electron formation due to collisions of excited atoms including associative ionization (10.10), and subsequently generation of electrons results from ionization of excited atoms by electron impact (10.16). Correspondingly, as early the balance equation for the electron number density N_e has the form

$$\frac{dN_e}{dt} = M + k_{ion}(T_e) N_e N_*(T_*) \, ,$$

but the electron temperature T_e and the excitation temperature T_* are not constants now in time. Here $M = k_{as} N_*^2$, so that k_{as} is the rate constant of associative ionization in collisions of two excited atoms, k_{ion} is the ionization rate constant of an excited atom by electron impact. From the solution of this equation and taking into consideration dependences $T_e(t)$ and $T_*(t)$, one can find the time of full plasma ionization.

In the course of equilibria establishment the excitation temperature T_e grows up to its equilibrium value, as well as the electron temperature T_*. Correspondingly, ionization processes are accelerated as the equilibrium is established for the electron temperature T_e and for the excitation temperature T_*. Subsequent growth of the electron number density proceeds according to the above balance equation with constant parameters k_{ion} and N_*. From this one can conclude that the increase of the electron number density starts practically when the equilibrium electron T_e and excitation T_* temperatures are established. This gives that the typical time of full ionization of the photoresonant plasma is

$$\tau_{exc} + \tau_T + \tau_{ion} \, ,$$

where we use formula (10.9) for a time τ_{exc} of equilibrium establishment for the number density of excited atoms, formula (10.15) for a time τ_T of establishment of the equilibrium electron temperature T_e, and formula (10.19) for a time of full ionization of a photoresonant plasma, if the electron T_e and excitation T_* temperatures are constants in the course of this process.

▶ **Problem 10.22** Obtain the relation between the electron temperature T_e of a photoresonant plasma and the excitation temperature T_*.

Based on processes (10.12) and (10.13) in a photoresonant plasma, we obtain the following balance equation for the average energy of electrons per unit volume:

$$\frac{d(N_e \bar{\varepsilon})}{dt} = \hbar\omega N_e(k_q N_* - k_{exc} N_0) - j_* k_{ion} N_e N_* ,$$

where $\bar{\varepsilon} = 3T_e/2$ is the average electron energy, k_q is the rate constant of quenching of excited atoms by electron impact, k_{exc} is the rate constant of atom excitation by electron impact, and k_{ion} is the rate constant of ionization of an excited atom by electron impact.

Let us use the Boltzmann formula (1.12) that connects the number densities of atoms in the ground and excited states through the excitation temperature T_{exc} and the principle of detailed balance that connects the rate constants of excitation k_{exc} and quenching k_q through the electron temperature T_e. This gives the following relation between the excitation temperature and the electron temperature in the stationary case

$$\hbar\omega k_q \left[1 - \exp\left(\frac{\hbar\omega}{T_*} - \frac{\hbar\omega}{T_e}\right)\right] = j_* k_{ion}(T_e) . \quad (10.20)$$

Ignoring the second term in the left-hand side of this equation, we obtain from this formula (10.17).

Taking the quenching rate constant k_q in relation (10.20) to be independent of the electron energy, we obtain the connection between the excitation T_* and electron T_e temperatures, and is convenient to represent this relation in the form

$$T_* = \left[\frac{1}{T_e} + \frac{1}{\hbar\omega} \ln\left(1 - \frac{j_* k_{ion}}{\hbar\omega k_q}\right)\right]^{-1} . \quad (10.21)$$

The solution of this equation gives $T_* = T_e$ in the limit of low temperatures, and in the limit of high excitation temperatures the electron temperature tends to the limit T_e^* (formula (10.17)). Figure 10.2 shows the dependence $T_e(T_*)$ for a sodium photoresonant plasma on the basis of equation (10.21).

▶ **Problem 10.23** Find the criterion when the excitation temperature is determined by radiative processes in accordance with the balance equation (10.2), and processes involving electrons do not influence on the excitation temperature.

We take the balance equation (10.2), as the basis, for the number density N_* of excited atoms and add in this equation the terms connected with electron processes (10.10) and (10.16). This gives

$$\frac{dN_*}{dt} = j_\omega k_o - j_\omega k_o \frac{N_* g_0}{N_0 g_*} - \frac{N_*}{\tau} - k_q N_* N_e + k_{exc} N_0 N_e - k_{ion} N_* N_e .$$

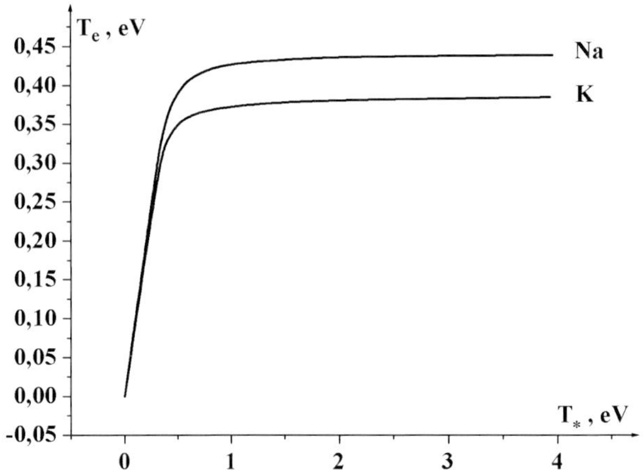

Fig. 10.2 The electron temperature T_e of a sodium (1) and potassium (2) photoresonant plasma as a function of the excitation temperature T_* according to equation (10.21).

We consider the electron temperature to be small in comparison to the excitation energy, i. e., ignore the process of formation of excited atoms by electron impact. Using relation (10.20) between the rate constants of ionization and quenching of excited atoms by electron impact under the equilibrium for electrons, reduce the balance equation for the number density of excited atoms to the form

$$\frac{dN_*}{dt} = j_\omega k_o - j_\omega k_o \frac{N_*}{N_0}\frac{g_0}{g_*} - \frac{N_*}{\tau} - k_q N_* N_e \left(1 + \frac{\hbar\omega}{j_*}\right) .$$

This equation is transformed into (10.2), if the following criterion holds true

$$N_e \ll N_{tra} \equiv \left[k_q \tau \left(1 + \frac{\hbar\omega}{j_*}\right)\right]^{-1} . \tag{10.22}$$

Table 10.8 contains the values of N_{tra} in the case of alkali metal vapors. Note that if this criterion is not valid, the character of the above electron processes and the electron temperature do not change, but the excitation temperature decreases in accordance with the above balance equation.

▶ **Problem 10.24** Determine a time of ionization of excited atoms in a photoresonant plasma when a decrease of the number density of atoms in the ground state makes a given plasma region to be transparent with respect to resonant radiation. The initial number density of atoms N_o exceeds significantly the number density N_{tr} of formula (5.32) above which the absorptioncoefficient is independent of the atom number density.

Ionization of atoms of an absorbing gas leads to blooming of a photoresonant plasma. Formula (10.19) gives a typical time of ionization of atoms, so that through this time the number density of electrons and ions becomes comparable with the

Table 10.8 Parameters of contribution of electron processes in establishment of electron processes. N_{tra} is a typical number density of electrons according to formula (10.22) and the efficiency of ionization η due to formula (10.26).

	N_{tra} (10^{14} cm^{-3})	η
$Li(2^2P)$	1.3	0.50
$Na(3^2P)$	1.8	0.41
$K(4^2P)$	0.78	0.62
$Rb(5^2P)$	1.1	0.62
$Cs(6^2P)$	0.69	0.70

initial number density of atoms ($N_e \sim N_o$). But since $N_o \gg N_{tr}$, such an atom ionization does not lead to blooming of an absorbing gas that takes place when the number density of atoms will be comparable with N_{tr}.

Let us denote by τ_{bl} a time after which the number density of atoms becomes $\sim N_{tr}$. We below determine this value. For this purpose, we consider the balance equation for the number density of ions

$$\frac{dN_e}{dt} = M + k_{ion} N_e N_* ,$$

which we used above for determination of the ionization time (10.19). Here the first term of the right-hand side of this equation is related with the process of association ionization (10.10) and is of importance at low number densities of electrons. Let us rewrite this equation in the form

$$\frac{dN_e}{dt} = \kappa N_e (N_o - N_e) ,$$

where $\kappa = k_{ion} N_* / (N_o - N_e)$. The solution of these equations is

$$N_e = \frac{N_o}{1 + \frac{N_o}{N_1} \exp(-\kappa N_o t)}; \qquad N_o - N_e = \frac{N_o}{1 + \frac{N_1}{N_o} \exp(\kappa N_o t)} .$$

Here N_1 is the number density at $t = 0$, when associative ionization and direct atom ionization give the same contribution to formation of free electrons ($N_1 \sim N_o k_{as}/k_{ion}$), and $N_1 \ll N_o$.

This solution allows us to find a time of blooming τ_{bl} at which the number density of atoms in the ground state is $N_0 \sim N_{tr}$. We obtain the blooming time

$$\tau_{bl} = \frac{1}{\kappa N_o} \frac{N_o^2}{N_1 N_{tr}}; \qquad \frac{\tau_{bl}}{\tau_{ion}} = 1 + \frac{\ln \frac{N_o}{N_{tr}}}{\ln \frac{N_o}{N_1}} . \qquad (10.23)$$

► **Problem 10.25** Determine the excitation temperature at high intensity of a resonant radiation beam as a function of the electron number density.

At low number densities of electrons, the excitation temperature is determined by the balance equation (10.2) that accounts for radiative processes. Let us denote by T_r the excitation temperature in neglecting collision processes that follows the balance equation (10.2) and is given by formula (10.4)

$$T_r = \frac{\hbar\omega}{\ln \frac{j_\omega + j_o}{j_\omega}} . \tag{10.24}$$

Including in the balance equation (10.2) processes (10.16) involving electrons, we obtain the number density of excited atoms

$$N_* = \frac{j_\omega k_o N_0}{k_o \frac{g_0}{g_*}(j_\omega + j_o) + N_e k_q \left(1 + \frac{\hbar\omega}{J^*}\right)} .$$

This gives the excitation temperature

$$T_* = \frac{\hbar\omega}{\ln[\frac{j_o}{j_\omega} + 1 + N_e k_q \tau (1 + \frac{\hbar\omega}{J^*})]} = \frac{\hbar\omega}{\ln(\frac{j_o}{j_\omega} + 1 + \frac{N_e}{N_{tra}})} .$$

Note that this formula is derived under the condition when the electron temperature is determined by processes (10.16), and the rate constant of ionization of excited atoms by electron impact is given by formula (10.17).

On the basis of formulas (10.4) and (10.24) we obtain the ratio of the excitation T_* and radiative T_r temperatures (T_r is the excitation temperature in the absence of collision processes)

$$\frac{T_*}{T_r} = \frac{\ln\left(\frac{j_o}{j_\omega} + 1\right)}{\ln\left[\frac{j_o}{j_\omega} + 1 + N_e k_q \tau \left(1 + \frac{\hbar\omega}{J^*}\right)\right]} . \tag{10.25}$$

Figure 10.3 shows the dependence (10.25) on the reduced electron number density. Since formula (10.25) is valid at high values of the excitation temperature, it holds true at not large electron number densities.

▶ **Problem 10.26** Analyze the character of variation of the number density of excited atoms in during evolution of a photoresonant plasma created by a high intensity beam of resonant radiation.

Along with resonant radiation that excites a gas and creates resonantly excited atoms, processes (10.16) involving electrons must be taken into account. The first stage of the process of plasma evolution consists in formation of excited atoms, and equilibrium for the number density of excited atoms is established during a time of the order of $\tau j_o / j_\omega$. When the number density of electrons reaches the value $N_{tra} j_o / j_\omega$, the number density of excited atoms decreases with time.

▶ **Problem 10.27** Determine the efficiency of plasma generation by a resonant radiation beam in the case of local equilibrium for excited atoms and electrons with a forming plasma.

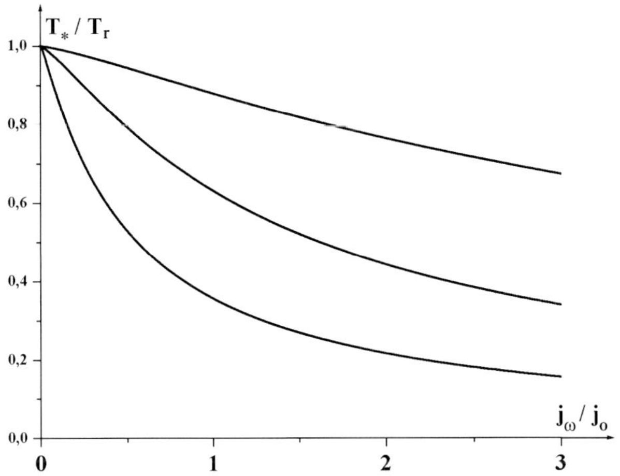

Fig. 10.3 The ratio of the excitation and radiation temperatures as a function of the reduced electron number density according to formula (10.25).

In the case of the processes under consideration excited atoms are formed as a result of absorption of incident radiation, and then the absorbed energy is consumed on ionization of excited atoms. Along with this, a part of absorbed energy releases as a result of spontaneous emission of excited atoms, and this energy is transferred to neighboring regions. Since transport processes are absent, the above processes provide all the energy release transmitted from an incident radiation beam. The ratio of the rates of atom ionization and spontaneous radiation is

$$\xi = \frac{\overline{k_{\text{ion}} N_* N_e}}{\overline{N_*/\tau}} = \frac{\tau}{\tau_{\text{ion}}} \, .$$

Here an average is made over time of plasma evolution, and τ_{ion} is the average time of ionization, $k_{\text{ion}} N_* N_e$ is the number of ionization events per unit time and unit volume, N_*/τ is the number of events of spontaneous radiation per unit time and unit volume; τ_{ion} is an ionization time that is given by formula (10.19), τ is the radiative lifetime of an isolated resonantly excited atom.

Note that because of a high excitation temperature, for a broad beam of resonant radiation, transverse photons leave an excited region due to stimulated radiation. But since an inoculating photon results from spontaneous radiation, we reduce losses due to transverse photons to spontaneous radiation. We take the energy per one forming electron and ion as $J + 3T_e/2$, where J is the atom ionization potential, $3T_e/2$ is the average kinetic energy of a forming electron that exceeds that for an ion, and the energy per spontaneous photon $\hbar\omega$. Let us introduce the portion of energy of absorbed radiation that η is consumed on ionization, and is related with the ratio of the above rates by the relation

$$\eta = \frac{(J + 3T_e/2)\xi}{(J + 3T_e/2)\xi + \hbar\omega} = \frac{(J + 3T_e/2)\tau}{(J + 3T_e/2)\tau + \hbar\omega\tau_{\text{ion}}} \, . \qquad (10.26)$$

Table 10.8 contains the values of the parameter η for a photoresonant plasma of alkali metals at the initial number density of atoms $N_0 = 1 \cdot 10^{15}\ \text{cm}^{-3}$. One can see that the efficiency of transformation of the incident radiation energy into the energy of ionization increases with an increase of N_0, the total number of atoms at the beginning, and in the end all the atoms are converted into ions and electrons.

Thus the process of absorption of incident resonant radiation in a vapor proceeds as follows. Absorption of radiation by atoms transfers atoms in resonant excited states, and these atoms can emit stimulated or spontaneous radiation or take part in processes of plasma generation. Simulated radiation restores an incident beam and does not influence on the processes of energy conversion that proceeds from spontaneous radiation and ionization of excited atoms by electron impact. Note that resonant radiation is absorbed in a small region of a vapor (or in a buffer gas with a buffer gas admixture), and after total ionization of vapor atoms this region becomes transparent for incident radiation, so that absorption takes place in a neighboring region. We assume a dimension of a vapor region L to be large in comparison with the mean free path (10.8) for resonant photons.

10.4
Propagation of Excitation and Ionization Waves

▶ **Problem 10.28** Determine the velocity of propagation for an intense incident beam of resonant radiation through the vapor when the mean free path of photons λ is small compared to a vapor dimension l.

We take the following character of absorption of incident photons. Assuming for simplicity the photon frequency to be identical to the center of a transition spectral line, we first observe radiation penetrates on a depth of $1/k_0$, where k_0 is the absorption coefficient for the spectral line center of a nonexcited gas. If we neglect emission of spontaneous radiation, which holds true for intense beams, we find according to equation (10.2) a typical time τ' of atom excitation on this path

$$\tau' \sim \frac{N_0}{k_0 j_\omega} \sim \frac{j_0 \tau}{j_\omega} .$$

Through this time a part of a beam path of a size $\sim 1/k_0$ is bloomed due to excitation of atoms, and incident radiation excites the next region. Hence the velocity of propagation of a resonant laser beam inside a gas or vapor is

$$v_p = \frac{j_\omega}{N_0} . \qquad (10.27)$$

Note that this velocity is identical for different photon frequencies, if the mean free path of photons is relatively small. Indeed, a decrease of the absorption coefficient k_ω compared to its value in the spectral line center leads to an increase of an excitation region and simultaneously to a corresponding decrease of a time τ' of atom excitation in this region. Both effects are mutually cancelled.

▶ **Problem 10.29** Find the depth of penetration of an intense resonant beam under conditions of the previous problem and a time of establishment of its equilibrium with the gas.

In considering the velocity of propagation of a resonant radiation beam through a gas, we ignore spontaneous radiation of excited atoms because of a high intensity of a beam. In the course of propagation of this beam, the excitation temperature increases according to formula (10.4), and the absorption coefficient decreases, since the number density of excited and nonexcited atoms becomes nearby and stimulated radiation occurs. The propagation process finishes when spontaneous radiation becomes of importance, and this determines the depth L of penetration for a resonant radiation beam. Comparing the number of absorbed photons per unit time and per unit area j_ω and the number of excited photons per unit time N_*/τ, we find the penetration length

$$L = \frac{j_\omega}{j_0 k_o} . \qquad (10.28)$$

Note that this value is identical for different photons frequencies if their mean free path is small compared to the penetration length L. The total time of establishment of the equilibrium, if resonant radiation penetrates in a gas such that its absorption is equalized by spontaneous radiation, is of order of τ — the radiative lifetime of an isolated atom with respect to spontaneous radiation.

▶ **Problem 10.30** Estimate the dependence of the penetration depth of an intense resonant beam inside an absorbing gas on a beam radius.

In deriving formula (10.28) for the depth of penetration of an intense resonant beam in a gas we assume implicitly a beam radius to be small in comparison the mean free path of resonant photons of the spectral line center. Then under considered conditions $j_\omega \gg j_0$, where the excitation temperature T_* exceeds significantly the photon energy $\hbar\omega$, we found the number densities of atoms in the ground N_0 and resonantly excited states N_* states almost identical $N_* \approx N_0$. Therefore, the penetration depth exceeds the mean free path of resonant photons in a nonexcited gas in $T_*/(\hbar\omega)$ times because of stimulated radiation. Thus, an incident beam creates a medium with new optical properties and propagates in this medium. Then amplification in transversal directions is absent because of relatively small beam width.

If an incident beam becomes wide, stimulated radiation will determine propagation of radiation in transversal directions. This means that if even the laser beam is enough broad, an atomic gas becomes transparent with respect to transverse photons because of stimulated radiation. Hence, the character of penetration of resonant radiation inside a gas is independent of the beam width until a surrounding gas is not excited strongly. Then each photon resulted from spontaneous radiation of excited atoms, leaves the beam region, possibly due to stimulated radiation, and the relation is conserved between incident photons and transversal photons originated as a result of spontaneous emission. Therefore, the character of propagation of an intense resonant beam does not depend on the beam width.

This consideration shows a nonstationary character of the absorption process for resonant radiation. In the first stage this radiation is absorbed in a region of order of the mean free path for photons in a nonexcited gas. Excitation of this gas leads to a decrease of the absorption coefficient, and the beam of resonant radiation propagates in a gas and clarifies it. This stage of absorption of resonant radiation lasts $\sim \tau$, the radiative lifetime of an isolated atom. Subsequently excitation of a gas proceeds in a region that surrounds a beam region.

▶ **Problem 10.31** Find the velocity of propagation on large distances for an intense incident beam of resonant radiation through the vapor when a typical time of establishment of ionization equilibrium is large compared to the radiative lifetime τ of an isolated atom.

A beam of resonant radiation propagates in a gas in the form of an excitation wave, and the velocity of propagation of this wave v_p is given by formula (10.27). In this case absorption proceeds in a narrow region on the beam path. This absorption leads to clarifying of this region and causes subsequently a displacement of the absorption region. The propagation of this beam finishes on a depth L of propagation of a resonant radiation beam that is determined by formula (10.28) and is connected with spontaneous radiation of excited atoms.

We now consider the next step of this process when a gas is clarified as a result of ionization of excited atoms by electron impact. After transformation of atoms into ions and electrons, a given region of a vapor is converted in a plasma with the total ionization of atoms and becomes transparent for incident radiation. In this way an incident radiation beam propagates in a vapor.

For determining the velocity of the photoresonant wave, we consider the energy per ionization of atoms per unit volume to be $N_0 J$, where N_0 is the initial number density of atoms and J is the ionization potential of atoms. Because in the course of plasma ionization a certain part of energy is consumed on spontaneous radiation of excited atoms, so that the energy $N_0 J/\zeta$ is inserted per unit volume from a beam of photoresonant radiation. Since the energy flux is $\hbar\omega j_\omega$, where j_ω is the photon flux, we have for the velocity w of propagation of the photoresonant radiation in the vapor—the velocity of the ionization wave

$$v_p = \frac{\hbar\omega j_\omega \zeta}{N_0 J} = \frac{w_0}{1+\xi}; \qquad w_0 = \frac{j_\omega \tau}{2\tau_{ion} N_*} . \tag{10.29}$$

On the basis of formula (10.19) we obtain this formula

$$w_0 = \frac{j_\omega \tau k_{ion}}{2 \ln \frac{k_{ion}}{k_{as}}} . \tag{10.30}$$

One can see that this value is proportional to the flux of incident photons, as well as in the regime according to formula (10.27) and is independent of the number density of atoms.

▶ **Problem 10.32** Summarize the character of propagation for a resonant radiation beam of a high intensity in a dense gas.

There are two mechanisms which lead to blooming of an absorbing gas and increase penetration of a resonant radiation beam inside the gas. The first one follows from the excitation of a gas, and then stimulated radiation increases the mean free path of resonant photons in a gas. Subsequent ionization of excited atoms decreases firstly for propagation of a region of excitation, so that the absorption coefficient due to disappearance of absorbing atoms and subsequently a gas becomes transparent with respect to resonant radiation. We now consider the series of these processes.

The first stage of these processes is excitation of a given region that proceeds through a time $\tau j_\omega / j_0$ in accordance with the balance equation (10.2). Here τ is the radiative lifetime of an isolated atom, and for an intense radiation beam the ratio j_0 / j_ω according to formula (10.3) is small. Through this time the absorption coefficient for a given region drops from k_0 to $k_0 j_0 / j_\omega$, and radiation penetrates in a more deep region.

The second stage is the formation and development of a photoresonant plasma in this region. Through a time τ_{ion} that is given by formula (10.19) a remarkable part of excited atoms is ionized. Though through a time (10.19) a region under consideration becomes not transparent, the absorption coefficient drops because of a decrease of the number density of atoms and correspondingly the parameter j_0 decreases according to formula (10.3). This region becomes transparent for resonant radiation through a time (10.23), but a time range from τ_{ion} to τ_{bl} is not essential for beam propagation because of a small part of a beam energy is absorbed on this stage. Therefore, just the parameter τ_{ion} rather than τ_{bl} determines the velocity of propagation of the ionization wave.

10.5 Heating of Atoms and Expanding of Photoresonant Plasma

▶ **Problem 10.33** Find the conditions of generation of sound from a photoresonant plasma.

This photoresonant plasma results from absorption of incident resonant radiation by a gas or vapor. As a result, an absorbed radiation energy is transformed in the excitation energy of excited atoms and in ionization of the vapor. Because these processes proceed fast, we obtain a gas with nonuniformities due to heated regions or to a heightened pressure. Such regions expand that results in generation of acoustic waves. Therefore, for sound generation a fast increase of the atom temperature is required for a photoresonant plasma or a fast increase of its pressure. Returning of these regions to an equilibrium causes generation of acoustic waves.

So, we have two mechanisms of sound generation, as a result of fast heating of regions where incident radiation is absorbed or by generation of new particles—electrons and ions that leads to a fast increase of pressure of a photoresonant plasma. Subsequent expansion of these regions leads to gas oscillations and hence to generation of acoustic waves. The condition of this process is that the rate of formation of nonuniformities $1/\tau$ would be large compared to the wave frequency $\sim R/c_s$, where R is a nonuniformity dimension, c_s is the sound speed, i. e., the criterion of sound generation has the form

$$\frac{1}{\tau} \geq \frac{R}{c_s} .$$

► **Problem 10.34** Give the criterion of sound generation resulted from heating of a photoresonant plasma.

In this case atoms of a photoresonant plasma formed by laser pumping of a gas or vapor that is located inside a buffer gas are heated in collisions, and hence some region of a photoresonant plasma has a heightened temperature. This region expands that leads to motion of a gas as a whole and causes oscillations in a gas—sound generation. Let us consider the first stage of expansion of the absorbing gas assuming adiabatic conditions to be fulfilled. This gives $T^{3/2}V$ =const for an expanding atomic gas, where $V \sim R^3$ is a heated volume occupied by the absorbing gas, and R is a typical dimension of this region. From this we obtain a temperature decrease as a result of expansion

$$\left(\frac{dT}{dt}\right)_{\exp} = \frac{T}{2R}\frac{dR}{dt} .$$

The rate of this volume is restricted by the sound speed c_s, i. e., $dR/dt \sim c_s$. Since a time of heating of this region is higher than that for expansion, we obtain the criterion for the heating rate $\left(\frac{dT}{dt}\right)_{\text{heat}}$

$$\left(\frac{dT}{dt}\right)_{\text{heat}} \gg \frac{Tc_s}{R} . \tag{10.31}$$

Thus, generation of acoustic waves is realized due to fast channels of energy transfer to atoms of a photoresonant plasma.

► **Problem 10.35** Determine the rate of gas heating in a photoresonant plasma due to elastic collisions between electrons and atoms.

One of the mechanisms of gas heating results in elastic collisions between electrons and atoms in a photoresonant plasma. Since the electron temperature is higher than the gaseous temperature, this process leads to heating of a gas. On the basis of formula (7.32)we have the balance equation for the gaseous temperature in this case

$$\frac{dT}{dt} = \frac{N_e}{N_0}\frac{m_e}{M}\nu(T_e - T) ,$$

where m_e, M are the electron and atom masses, respectively, N_e, N_0 are their number densities, T_e, T are the electron and gaseous temperatures, respectively, and

ν is the rate of elastic electron–atom collision ($\nu = N_0\langle v\sigma_*\rangle$, where v is the electron velocity, σ_* is the diffusion cross section of elastic electron–atom scattering, an average is made over electron velocities).

From this wc find thc law of variation of the gaseous temperature

$$T = T_o + (T_e - T_o)[1 - \exp(-\nu_T t)] ,$$

where T_o is the initial gaseous temperature, and we assume typical times of establishment of the electron temperature to be small compared to this for the gaseous one. A typical rate of establishment of the gaseous temperature is

$$\nu_T \sim \frac{m_e}{M} N_e\langle v\sigma_*\rangle ,$$

and a typical time τ_T of variation of the initial gaseous temperature ($T_o \ll T_e$) is estimated as

$$\tau_T \sim \frac{T}{T_e \nu_T} .$$

Since a typical rate constant of electron–atom collisions in a photoresonant plasma of alkali metals is $\sim 10^{-6}$ cm^3/s, at a typical number density of alkali metal atoms $N_o \sim 10^{16}$ cm^{-3} and at $m/M \sim 10^{-5}$ for heavy alkali metals, we have $\nu_T \sim 10^5$ s^{-1}. This value corresponds to the rate (10.31) of sound propagation.

▶ **Problem 10.36** An equilibrium between the states of fine structure of resonantly excited atoms in a plasma results from collisions with electrons whose temperature is T_e and exceeds the gas temperature T. Collisions between atoms in the ground states and resonantly excited atoms with a change of fine structure of excited atoms lead to gas heating. Under conditions $\delta_f \ll T < T_e$ (δ_f is the difference energies for states of fine structure) find the rate of gas heating due to transitions between fine structure states.

This is a mechanism of heating of a photoresonant plasma as a result of conversion of the electron energy of atom excitation into the thermal energy of colliding atoms. For definiteness, we consider a vapor consisting of alkali metal atoms with the ground state $^2S_{1/2}$ and resonantly excited states $^2P_{1/2}$ and $^2P_{3/2}$. The heat balance equation owing to the processes of transitions between fine structure states has the form

$$C\frac{dT}{dt} = \delta_f \left[k\left(\frac{3}{2} \to \frac{1}{2}\right) N_{3/2} - k\left(\frac{1}{2} \to \frac{3}{2}\right) N_{1/2} \right] N_o$$

Here $C = 3N_o/2$ is the gas heat capacity, N_o is the number density of atoms in the ground state, $N_{1/2}$, $N_{3/2}$ are the number densities of resonantly excited states with a given state of fine structure, the argument of the rate constants indicates the transition states. Because of the thermodynamic equilibrium with the electron temperature, the number densities of excited atoms are connected by the Boltzmann formula (1.12)

$$N_{3/2} = 2N_{1/2}\exp\left(-\frac{\delta_f}{T_e}\right) ,$$

and according to the principle of detailed balance, the relations between the rates of the processes with the fine structure change have the form

$$k\left(\frac{1}{2}\to\frac{3}{2}\right) = 2k\left(\frac{3}{2}\to\frac{1}{2}\right)\exp\left(-\frac{\delta_f}{T}\right) .$$

Expanding exponents of these relations, we reduce the heat balance equation to the form

$$\frac{dT}{dt} = (T_e - T) = \nu_{ef}(T_e - T), \qquad \nu_{ef} = \frac{4}{27}\frac{\delta_f^2}{T_e T}N_* k_{ex} ,$$

where $N_* = N_{1/2} + N_{3/2}$ is the total number density of resonantly excited atoms, and we use $k\left(\frac{1}{2}\to\frac{3}{2}\right) = \frac{2}{3}k_{ex}$, $k\left(\frac{3}{2}\to\frac{1}{2}\right) = \frac{1}{3}k_{ex}$, k_{ex} is the rate constant of excitation transfer, and its values for the excitation transfer cross section are given in Table 3.1.

► **Problem 10.37** Being guided by a photoresonant plasma of alkali metals, find the optimal conditions of sound generation as a result of atom ionization in a photoresonant plasma if a width of an incident radiation beam is $R \sim 1$ mm.

Fast ionization of a vapor of a photoresonant plasma leads to an increase of the pressure in a region occupied by this plasma. One can estimate this increase as $\Delta p = N_e T_e$, where N_e is the number density of electrons, and T_e is the electron temperature. The energy transformed into gas oscillations is of the order of $V\Delta p$, where V is the volume of a photoresonant plasma, and T_e is the energy per atom.

A typical time of ionization is given by formula (10.19), and we use formula (4.38) for the rate of stepwise ionization. This process starts from the ground state, and the number density of excited atoms is relatively small in the course of the ionization process. A typical time of development of acoustic oscillations is given by formula (10.31). Table 10.9 contains the electron temperature of alkali metal

Table 10.9 Parameters of a photoresonant plasma which provide the optimal conditions of sound generation at the number density of alkali metal atoms of 10^{16} cm^{-3} and at the temperature of the alkali metal vapor of 500 K.

	c_s (10^3 cm/s)	T_e (eV)	N_* (10^{12} cm^{-3})	η_0(%)
Li	4.6	0.28	12	5.2
Na	2.5	0.25	2.3	4.9
K	1.9	0.20	3.6	4.6
Rb	1.3	0.19	2.7	4.6
Cs	1.0	0.17	3.3	4.4

atoms that provides the time of total ionization of photoresonant plasma during a time (10.31) of sound propagation. This table contains also the number density of excited atoms at this electron temperature.

▶ **Problem 10.38** Estimate the coefficient of conversion of an energy of an incident radiation beam into the energy of acoustic oscillations if the total ionization of a photoresonant plasma proceeds.

The process of sound generation from a photoresonant plasma results from conversion of the energy of an incident radiation into the mechanical energy, and this causes oscillations of a gas. Assuming total ionization of atoms under the action of an incident radiation beam and subsequent ionization of the photoresonant plasma, we obtain the thermal energy per electron $3T_e/2$ that is converted subsequently into the energy of acoustic oscillations. The energy enclosed in a gas per one electron is of the order of J, the atom ionization potential. Hence, the efficiency of this process is

$$\eta \sim \eta_0 = \frac{T_e}{J} \, .$$

11
Waves in Plasma and Electron Beams

11.1
Oscillations in an Isotropic Weakly Ionized Gas

► **Problem 11.1** Analyze acoustic oscillations of an atomic gas.

In considering wave and oscillations in a gas and plasma, we assume them to be weak. This means the amplitude of oscillations is small and a perturbation due to oscillations is small. Hence any macroscopic parameter A of this system may be represented as

$$A = A_0 + \sum_{\omega} A'_{\omega} \exp[i(kx - \omega t)] .$$

Here A_0 is an unperturbed parameter (in the absence of oscillations), A'_{ω} is the amplitude of oscillations, ω is the oscillation frequency, and k is the wave number of the oscillation. We assume the oscillation to be propagated in the form of a wave in the x-direction. Next, because of a weakness of oscillations, one can neglect the interaction between oscillations of different frequencies and to separate them in this way. Therefore, in this linear approximation one can restrict by an oscillation of one frequency, i. e., instead of the above expression for a macroscopic parameter A one can use a more simple one

$$A = A_0 + A' \exp[i(kx - \omega t)] \tag{11.1}$$

In analyzing an oscillation, we shift a system from its equilibrium state weakly. This causes a returning force that tries to restore a system in an equilibrium state. As a result, oscillations occur near an equilibrium state, and our task is to find the relation between the frequency ω and the wavelength or wave vector k of the oscillation.

In the case of elastic oscillations or acoustic waves in a gas, we consider weak perturbations (11.1) for the gas density, its pressure and wave velocity. We use for this purpose a standard form of the continuity equation (6.15)

$$\frac{\partial N}{\partial t} + \operatorname{div}(N\boldsymbol{w}) = 0 ,$$

Plasma Processes and Plasma Kinetics. Boris M. Smirnov

ISBN: 978-3-527-40681-4

the Euler equation (6.19)

$$\frac{\partial \boldsymbol{w}}{\partial t} + (\boldsymbol{w} \cdot \nabla)\boldsymbol{w} + \frac{\nabla p}{\rho} - \frac{\boldsymbol{F}}{m} = 0$$

and the adiabatic conditions in the wave from the equation

$$pN^{-\gamma} = \text{const} . \qquad (11.2)$$

Here N is the number density of gas atoms, p is the gas pressure, w is the mean gas velocity, and for a motionless gas $w_o = 0$; m is the atom mass, and F is the force per atom. Next, $\gamma = c_p/c_V$ is the adiabatic exponent, that is, c_p is the specific heat capacity at constant pressure, c_V is the specific heat capacity at constant volume, and for an atomic gas $\gamma = 5/3$.

If we represent gas parameters in the form (11.1), we obtain from the continuity equation

$$\omega N' = k N_0 w' ,$$

since acoustic wave is a longitudinal oscillation, the gas velocity $\boldsymbol{w}$ is directed along the wave vector $\boldsymbol{k}$. Taking an external force F to be zero in the Euler equation, we obtain from it on the basis of formula (11.1) in the first approximation

$$\omega w' = \frac{k}{m N_o} p' .$$

Next, from the adiabatic equation in the wave we have

$$p'/p_o = \gamma N'/N_o .$$

Excluding perturbed gas parameters from the above equations, we find the relation between the wave frequency ω and its wave vector k (the dispersion relation)

$$\omega = c_s k , \qquad (11.3)$$

where the speed of sound c_s is

$$\omega = \sqrt{\frac{\gamma T}{m}} k . \qquad (11.4)$$

We below use the state equation for an ideal motionless gas $p_o = N_o T$. The sound velocity is seen to be of the order of the thermal velocity of gas particles. Figure 11.1 represents the dispersion relation (11.4) for air at $T = 300$ K in the range of centimeter wavelengths.

► **Problem 11.2** Find the criterion of validity of the dispersion relation (11.4) for acoustic oscillations in an atomic gas.

The dispersion relation (11.4) for acoustic oscillations in a motionless gas is valid under adiabatic conditions in the wave if a typical time τ of heat transport in the

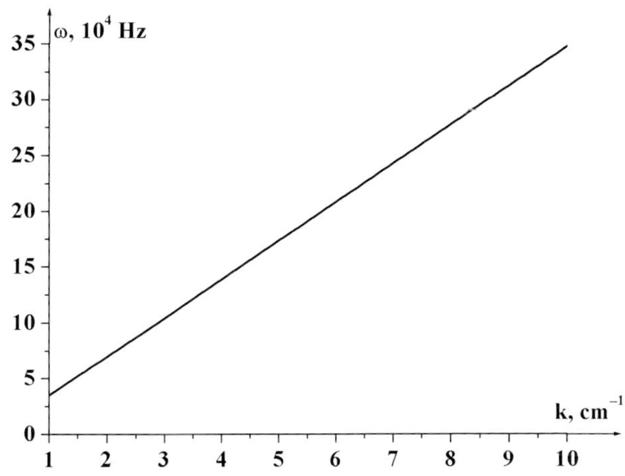

Fig. 11.1 Dispersion relation (11.4) for air at $T = 300$ K in the centimeter range of wavelengths.

wave is small compared to the period of oscillations $1/\omega$. Assuming the heat transport in the wave to be due to thermal conductivity, we obtain $\tau \sim r^2/\chi \sim (\chi k^2)^{-1}$, where a distance r is of the order of the mean free path of atoms in a gas λ, and χ is the coefficient of thermal diffusivity. This leads to the adiabatic criterion

$$\omega \ll c_s^2/\chi .$$

Let us express the coefficient of thermal diffusivity through the mean free path of atoms in a gas $\chi = \kappa/(c_p N) \sim v_T\lambda$, where v_T is a typical thermal velocity of atoms and $c_s \sim v_T$. Then the above criterion takes the form

$$\lambda k \ll 1 . \tag{11.5}$$

Thus, under adiabatic conditions, the wavelength of the acoustic oscillations is large compared to the mean free path of the gas atoms.

▶ **Problem 11.3** Determine the frequency of oscillations of plasma electrons which result from a displacement of electrons as a whole.

Let us take a uniform infinite plasma and shift all the plasma electrons. This causes a returning force due to electric fields that lead to oscillation of electrons as a whole. We will find the frequency of these oscillations.

If starting from a plane $x = 0$ all the electrons are shifted at the initial time by a distance x_0 to the right, this creates an electric field whose strength follows from the Poisson equation

$$\frac{dE}{dx} = 4\pi e(N_i - N_e) ,$$

where N_i, N_e are the ion and electron number density, respectively. Assuming the electric field strength at $x < 0$ to be zero, we obtain the electric field strength

$E = 4\pi e N_0 x_0$ for $x > x_0$ by solving the Poisson equation, where N_0 is the average number density of electrons and ions in the plasma. The motion of all the electrons under the influence of the electric field leads to a change in the boundary position. The equation of motion for each electron has the form

$$m_e \frac{d^2(x + x_0)}{dt^2} = -eE \,,$$

where m_e is the electron mass and x is the distance of an electron from the boundary. Because x is a random value and is independent of oscillations under consideration and therefore it is not depended on time. This gives the equation of electron motion,

$$\frac{d^2 x_0}{dt^2} = -\omega_p^2 x_0 \,.$$

As is seen, the electron motion proceeds now in the form of oscillations, and the plasma frequency (or the Langmuir frequency) is

$$\omega_p = \sqrt{\frac{4\pi N_0 e^2}{m_e}} \,. \tag{11.6}$$

The value $1/\omega_p$ is a typical time for a plasma response to an external signal. Note that the value $r_D \omega_p = \sqrt{2T/m_e}$ is the thermal electron velocity. Thus, considering a plasma response on an external electric signal as an electron displacement on a distance of the order of the Debye–Hückel radius r_D, that is a shielding distance for fields in a plasma, and electrons move with a typical thermal velocity, we obtain the above time of the response time.

▶ **Problem 11.4** Derive the dispersion relation for plasma oscillations.

In considering the plasma oscillations, let us take the macroscopic equations for a plasma similar to the method of derivation of the dispersion relation (11.4) for acoustic waves. Then by analogy with the case of the acoustic oscillations, we obtain from the continuity equation (6.15), Euler equation (6.19) and adiabatic equation (11.2), the relations which connect unperturbed and perturbed plasma parameters,

$$-i\omega N_e' + ikN_0 w' = 0 \,, \quad -i\omega w' + i\frac{kp'}{m_e N_0} + \frac{eE'}{m_e} = 0 \,, \quad \frac{p'}{p_0} = \gamma \frac{N_e'}{N_0} \,. \tag{11.7}$$

Here unperturbed and perturbed parameters are defined in accordance with formula (11.1), N_e' is the perturbed number density of electrons in the wave, N_0 is the average number density of electrons and ions, and in the Euler equation the force $F = -eE$ is introduced from the electric field of the wave that occurs owing to the disturbance of the plasma quasineutrality; k and ω are the wave number and the frequency of the plasma oscillations, respectively. Along with these equations, we take into account the Poisson equation that gives

$$ikE' = -4\pi e N_e' \,.$$

Using the state equation for electrons of a motionless plasma $p_0 = N_0 m_e \langle v_x^2 \rangle$, we obtain the above equations.

Eliminating the oscillation amplitudes N_e', w', p', and E' from the set of the above equations, we obtain the following dispersion relation for plasma oscillations:

$$\omega^2 = \omega_p^2 + \gamma \left\langle v_x^2 \right\rangle k^2 \,, \tag{11.8}$$

where $\omega_p = \sqrt{4\pi N_0 e^2/m_e}$ is the plasma frequency that is given by formula (11.6). Plasma oscillations are longitudinal ones similar to acoustic oscillations where a plasma motion and the wave vector have the same direction.

▶ **Problem 11.5** Find the criterion of validity of the dispersion relation (11.8) for plasma oscillations.

The dispersion relation (11.8) for plasma oscillations is valid under adiabatic conditions of wave propagation. Taking heat transport to be due to electron thermal conductivity, we obtain the adiabatic condition in the form $\omega\tau \sim \omega/(\chi k^2) \gg 1$, where ω is the frequency of oscillations, $\tau \sim (\chi k^2)^{-1}$ is a typical time for heat transport in the wave, χ is the electron thermal diffusion coefficient, and k is the wave number of the wave. Since $\chi \sim v_e \lambda$, then $\omega \sim \omega_p \sim v_e/r_D$, where v_e is the mean electron velocity, λ is the electron mean free path, and r_D is the Debye–Hückel radius. Thus, the adiabatic condition yields

$$\lambda r_D k^2 \ll 1 \,. \tag{11.9}$$

▶ **Problem 11.6** Derive the dispersion relation for plasma oscillations under isothermal conditions.

Under the isothermal condition that is reversed with respect to condition (11.9), the relation for the pressure and a perturbed pressure is the same as that for unperturbed one and has the form

$$p' = \gamma N_e' m_e \left\langle v_x^2 \right\rangle \,. \tag{11.10}$$

As a result, we obtain the dispersion relation (11.8) with another proportionality coefficient γ between the perturbed pressure p' and the perturbed number density N_e' of electrons. For neutral particles this coefficient is one since only kinetic energy of particles determines the pressure of these particles. For a plasma it is necessary to include in the plasma state equation the Coulomb interaction between charged particles, and then $\gamma = 3$. We obtain correctly this derivation of the dispersion relation for plasma oscillation in the isothermal case on the basis of the kinetic consideration below.

Note that plasma oscillations exist if their frequency exceeds remarkably by a reciprocal time between neighboring electron–atom collisions $\omega_p \gg v_e/\lambda$, where v_e is a thermal electron velocity and λ is the mean free path of electrons with respect to their collisions with atoms. This gives the condition for existence of plasma oscillations,

$$\lambda \gg r_D \ . \tag{11.11}$$

► **Problem 11.7** Derive the dispersion relation for plasma oscillations on the basis of the kinetic equation for electrons, assuming the distribution function of electrons to be the Maxwell one, and the wavelength λ exceeds significantly the Debye–Hückel radius r_D for the plasma.

We above obtain the dispersion relations for plasma oscillations (11.8) on the basis of macroscopic equations for the electron component. Now we obtain it directly from the kinetic equation (6.2) expanding the distribution function as $f(\boldsymbol{v}) = f_0(v) + f'(\boldsymbol{v})$. Here the first term relates to an equilibrium electron distribution, and the second term accounts for plasma oscillations, and from the kinetic equation (6.2) we obtain

$$\frac{\partial f'}{\partial t} + \boldsymbol{v}\nabla f' - \frac{e\boldsymbol{E}'}{m_e}\frac{\partial f_0}{\partial \boldsymbol{v}} = 0 \ .$$

The electric field strength $\boldsymbol{E}'$ due to oscillations directs along the wave and satisfies the Poisson equation

$$\operatorname{div} \boldsymbol{E}' = -4\pi e N \int f' d\boldsymbol{v} \ .$$

This distribution function is normalized as $\int f d\boldsymbol{v} = 1$, and we assume the absence of external fields in a nonperturbed plasma. Taking the time and coordinate dependence of f' and $\boldsymbol{E}'$ in the wave as $\exp(i\boldsymbol{kr} - i\omega t)$, we obtain

$$f' = i\frac{eE'}{m_e(\omega - kv_x)}\frac{\partial f_0}{\partial v_x} \ ,$$

where the x-axis is directed along the wave. From this we have the dispersion equation

$$\frac{\omega_p^2}{k^2}\int \frac{\frac{\partial f_0}{\partial v_x}}{v_x - \frac{\omega}{k}} d\boldsymbol{v} = 1 \ ,$$

where the plasma frequency ω_p is given by formula (11.6). Extracting the distribution function that depends on v_x and crossing around the pole at $v_x = \omega/k$, as it is shown in Fig. 11.2, we reduce this dispersion relation to the form

$$\frac{\omega_p^2}{k^2}\int_{-\infty}^{\infty} \frac{\frac{\partial f_0}{\partial v_x} dv_x}{v_x - \frac{\omega}{k}} + i\pi\frac{\omega_p^2}{k^2}\left.\frac{\partial f_0}{\partial v_x}\right|_{v_x=\omega/k} = 1 \ .$$

Let us expand the first integral over a small parameter $kv_x/\omega \ll 1$

$$\frac{\omega_p^2}{k^2}\oint_{-\infty}^{\infty} \frac{\frac{\partial f_0}{\partial v_x} dv_x}{v_x - \frac{\omega}{k}} = -\frac{\omega_p^2}{k}\int_{-\infty}^{\infty}\frac{1}{v_x}\frac{\partial f_0}{\partial v_x} dv_x \sum_{n=0}^{\infty}\left(\frac{kv_x}{\omega}\right)^n \approx \frac{\omega_p^2}{\omega^2}\left(1 + 3\frac{k^2}{\omega^2}\left\langle v_x^2\right\rangle\right) \ .$$

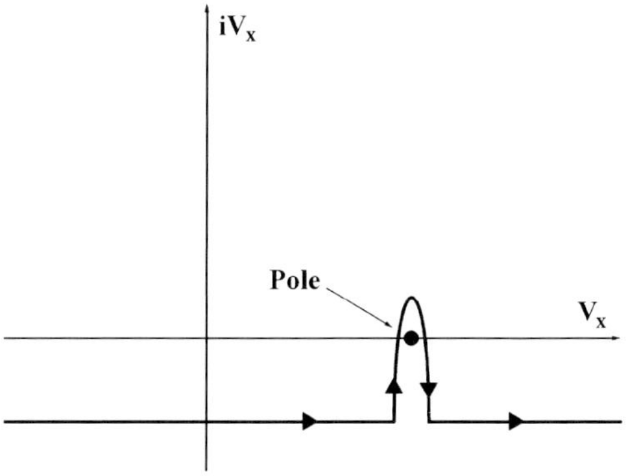

Fig. 11.2 The character of pass near the pole.

In this approximation $k^2 \langle v_x^2 \rangle \ll \omega^2$, we obtain the following dispersion equation for plasma oscillations:

$$\omega^2 = \omega_p^2 + 3k^2 \left\langle v_x^2 \right\rangle + i\pi \frac{\omega_p^4}{k^2} \left. \frac{\partial f_0}{\partial v_x} \right|_{v_x=\omega/k} . \tag{11.12}$$

This dispersion equation consider damping or amplification of waves due to electrons whose velocity v_x is equal to the wave phase velocity ω/k. In addition, this dispersion relation corresponds to the isothermal character of equilibrium, because assuming the nonperturbed distribution function of electrons to be independent of the wave, we postulate the isothermal wave conditions.

Comparing the dispersion relation (11.12) with that (11.8) from hydrodynamic considerations, we find these formulas to be identical at $\gamma = 3$. This cannot be justified. Indeed, as a longitudinal wave, the Langmuir wave includes oscillations in one direction only, and then according to formula (11.10) $\gamma = 1$. We here ignore the ion displacement as a slow component. Therefore, the contradiction between the above formulas is related with the difference between hydrodynamic and kinetic descriptions.

▶ **Problem 11.8** Derive the dispersion relation for ion sound.

Since a plasma incudes two types of charged particles, electrons and ions, two types of oscillations may be realized in a uniform plasma. The first one relates to fast oscillations due to electron motion, and ion motion is of importance for the second branch of oscillations of the plasma, that is the ion sound. Because of a small electron mass, they response fast on ion displacements that conserve the plasma quasineutrality in average, i. e., $N_e = N_i$. In addition, the electron distribution is established in accordance with plasma fields, i. e., the electron number density is determined by the Boltzmann formula

$$N_e = N_0 \exp\left(\frac{e\varphi}{T_e}\right) \approx N_0 \left(1 + \frac{e\varphi}{T_e}\right) ,$$

where φ is the electric potential due to oscillations, and T_e is the electron temperature. This allows us to express the amplitude of oscillations of the ion number density as

$$N_i' = N_0 \frac{e\varphi}{T_e} \, .$$

The relation between the ion velocity w_i in the wave and the variation of the ion number density N_i' due to the wave follows from the continuity equation (for ions $\partial N_i/\partial t + \partial(N_i w_i)/\partial x = 0$). This gives

$$\omega N_i' = k N_0 w_i \, ,$$

where ω is the frequency, k is the wave number, and w_i is the mean ion velocity due to oscillations, and we use the standard dependence (11.1) for oscillation parameters. The equation of motion for ions due to the electric field of the wave has the form $M(d\boldsymbol{w}_i/dt) = e\boldsymbol{E} = -e\nabla\varphi$, where M is the ion mass. From this it follows

$$M\omega w_i = ek\varphi \, .$$

Eliminating the oscillation amplitudes of N_i, φ, and w_i in the above equations, we obtain the dispersion relation for the ion sound

$$\omega = k\sqrt{\frac{T_e}{M}} \, . \qquad (11.13)$$

Similar to plasma oscillations and acoustic waves, ion sound is a longitudinal wave, i. e., the wave vector $\boldsymbol{k}$ of this wave and the oscillation velocity is parallel to the oscillating electric field vector $\boldsymbol{E}$. In addition, the dispersion relation for ion sound is similar to that for acoustic waves, since both types of oscillations are characterized by a short-range interaction. In the case of ion sound, the interaction is short ranged if the wavelength of the ion sound is larger than the Debye–Hückel radius for the plasma on which an electric field of the propagating wave is shielded by the plasma. This is the criterion of validity of relation (11.13).

Thus, oscillations in a two-component plasma consisting electrons and ions include two branches, plasma oscillations (11.11) and ion sound (11.13). Figure 11.3 represents these oscillations in a helium gas-discharge plasma.

▶ **Problem 11.9** Derive the dispersion relation for the ion sound taking into account a long-range interaction in a plasma.

In generalization the dispersion relation for ion sound (11.13) that we obtain below for long-wave oscillations, we use the Poisson equation for the plasma field

$$\frac{d^2\varphi}{dx^2} = 4\pi e(N_e - N_i) \, ,$$

while in the case of long-wave oscillations treated above, we took the left-hand side of this equation to be zero. We now consider this equation side to be $-k^2\varphi$ on

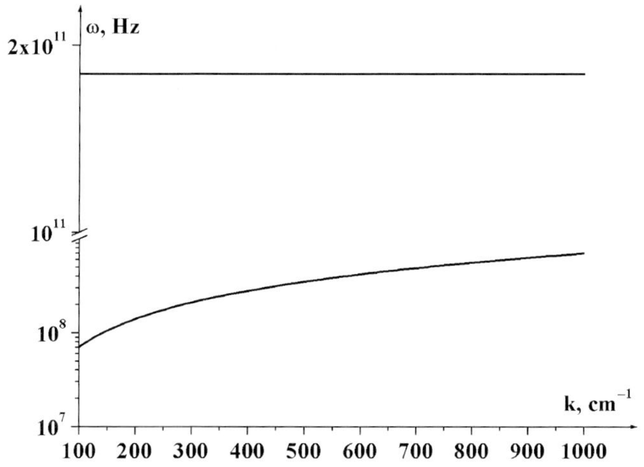

Fig. 11.3 Dispersion relations for plasma oscillations (11.12) and ion sound(11.13) in a helium discharge plasma with parameters $T = T_i = 400$ K, $T_e = 2$ eV, $N_e = 1 \cdot 10^{13} cm^{-3}$ in the centimeter range of wavelengths.

the basis of expansion (11.1), and accounting for electric fields in the wave due to violation of the plasma quasineutrality with the Boltzmann formula for the electron number density, we get from the Poisson equation

$$N_i' = N_0 \frac{e\varphi}{T_e} \left(1 + \frac{k^2 T_e}{4\pi N_0 e^2}\right) .$$

As a result, the dispersion relation (11.13) is now replaced by

$$\omega = k\sqrt{\frac{T_e}{M}} \left(1 + \frac{k^2 T_e}{4\pi N_0 e^2}\right)^{-1/2} . \qquad (11.14)$$

This dispersion relation is converted into formula (11.13) in the limit $kr_D \ll 1$ when oscillations are determined by a short-range interaction in the plasma. In the opposite limiting case $kr_D \gg 1$ we get

$$\omega_{ip} = \sqrt{\frac{4\pi N_0 e^2}{M}} . \qquad (11.15)$$

In this case a long-range interaction in the plasma is of importance, and from the dispersion relation, we see that ion oscillations are similar to plasma oscillations. Figure 11.4 represents the dispersion relation (11.14) for ion sound in a helium plasma when the wavelength range includes the transition from the case (11.14) to the case (11.15). The helium gas-discharge plasma parameters are $N_e = 1 \cdot 10^{13} \text{cm}^{-3}$, $T_i = 400$ K, $T_e = 2$ eV.

▶ **Problem 11.10** Derive the dispersion relation for the ion sound on the basis of the kinetic equation for electrons. Assume the distribution function of electrons and

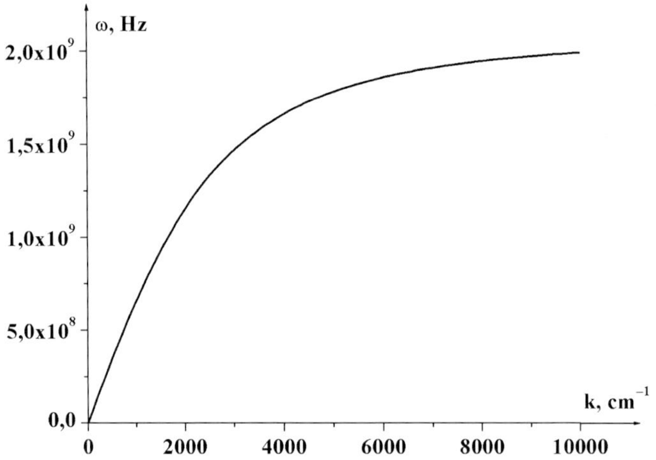

Fig. 11.4 Dispersion relations (11.14) for ion sound in a helium discharge plasma with parameters $T = T_i = 400$ K, $T_e = 2$ eV, $N_e = 1 \cdot 10^{13}$ cm^{-3} in the centimeter range of wavelengths.

ions to be the Maxwell one, and the electron temperature T_e exceeds significantly the ion T_i by one.

Since the plasma contains two charged components, and the behavior of the electron component influences the ion, it is necessary to consider both components for the analysis of oscillations of the ion component. As for the above plasma oscillations, we represent the distribution function of electrons f_e and ions f_i as

$$f_e = f_e^{(0)} + f_e', \qquad f_i = f_i^{(0)} + f_i' .$$

Here the first terms relate to nonperturbed plasma with the Maxwell distribution functions of electrons and ions, and the second terms account for oscillations in a plasma. Considering the second terms as a perturbation and taking the time and coordinate dependence for wave parameters as $\exp(ikx - i\omega t)$, we obtain on the basis of the kinetic equation (6.2) and the Poisson equation in analogy with the case of plasma oscillations

$$f_e' = i\frac{eE'}{m_e(\omega - kv_{ex})}\frac{\partial f_e^{(0)}}{\partial v_{ex}}, \quad f_i' = i\frac{eE'}{M(\omega - kv_{ix})}\frac{\partial f_i^{(0)}}{\partial v_{ix}},$$

$$ikE' = 4\pi N_e e\left(\int f_i' dv_{ix} - \int f_e' dv_{ex}\right),$$

where M is the ion mass, N_e is the number density of electrons or ions in a quasineutral plasma under consideration, and the electron temperature T_e is taken large compared to the ion temperature T_i. Taking the Maxwell distribution function of electrons and ions with the electron T_e and ion T_i temperatures

$$\frac{\partial f_e^{(0)}}{\partial v_{ex}} = -\frac{m_e v_{ex}}{T_e} f_e^{(0)}, \quad \frac{\partial f_i^{(0)}}{\partial v_{ix}} = -\frac{M v_{ix}}{T_i} f_i^{(0)},$$

we obtain the dispersion equation for the plasma

$$\frac{\omega_p^2}{k^2}\left(\frac{m_c}{T_i}\int\frac{v_{ix}f_i^{(0)}dv_{ix}}{\frac{\omega}{k}-v_{ix}}+\frac{m_e}{T_e}\int\frac{v_{ex}f_e^{(0)}dv_{ex}}{\frac{\omega}{k}-v_{ex}}\right)=1\,. \quad (11.16)$$

This equation can leads to dispersion relation of waves due to oscillations of both electron and ion plasma components. In particular, excluding the ion part from this equation, we obtain the dispersion relation (11.12) for electron oscillations. We now use this equation for the ion component taking the oscillation frequency ω in the range $kv_{ix} \ll \omega \ll kv_{ex}$ then the above equation takes the form

$$-\frac{\omega_p^2}{k^2}\frac{m_e}{T_e}+\frac{\omega_p^2}{\omega^2}\frac{m_e}{M}$$
$$-i\pi\frac{\omega_p^2}{k^2}\frac{m_e}{T_e}\left(v_{ex}f_e^{(0)}\right)\Big|_{v_{ex}=\omega/k}-i\pi\frac{\omega_p^2}{k^2}\frac{m_e}{T_e}\left(v_{ix}f_i^{(0)}\right)\Big|_{v_{ix}=\omega/k}=1\,, \quad (11.17)$$

and the imaginary part is determined by poles of the corresponding functions under the integrals. Ignoring the imaginary part, we obtain from this the dispersion relation (11.14)

$$\omega^2=k^2\frac{T_e}{M}\frac{1}{1+k^2r_D^2}\,,$$

where $r_D=\sqrt{T_e/(4\pi N_e e^2)}$ is the Debye–Hückel radius in the case if screening is determined by electrons only.

▶ **Problem 11.11** Analyze the character of shielding of an external field in a quasi-neutral plasma from the dispersion relation (11.16).

Shielding of a constant electric field penetrating in a plasma reflects a reaction of electrons on an external signal. We take the dependence of an external field directed inside the plasma as $E = E_o\exp(ikx - i\omega t)$ and use the dispersion relation (11.16) for an isothermal plasma. Taking this dispersion relation for a stationary plasma $\omega = 0$, we obtain $k = ir_D$, where r_D is the Debye–Hückel radius. From this we obtain that the electric field strength inside the plasma varies as $E = E_o\exp(-x/r_D)$ in accordance with the definition of the Debye–Hückel radius by formula (1.55).

On the basis of the above analysis, one can determine the Debye–Hückel radius from the dispersion relation (11.16) if we substitute in this relation $\omega = 0$. Such an operation gives

$$-\frac{1}{k^2}=\frac{\frac{1}{T_i}+\frac{1}{T_e}}{4\pi N_o e^2}=r_D^2\,,$$

and formula (1.57) is used for the Debye–Hückel radius in the case of different electron T_e and ion T_i temperatures.

11.2
Plasma Oscillations in Magnetic Field

▶ **Problem 11.12** Derive the dispersion relation for magnetohydrodynamic waves in a plasma, when this plasma is located in a strong magnetic field and the magnetic lines of force are frozen in the plasma.

We consider oscillations of a high-conductivity plasma that is located in a strong magnetic field, and the magnetic lines of force are frozen in the plasma. Then a change in the plasma current causes a change in the magnetic lines of force, which acts in opposition to this current. As a result, this plasma oscillates and magnetohydrodynamic waves are generated in it.

We assume the wavelength of magnetohydrodynamic waves to be small compared to the radius of curvature of magnetic lines of force, that is

$$\frac{1}{k} \ll \left|\frac{H}{\nabla H}\right| , \qquad (11.18)$$

where H is the magnetic field strength. This allows us to consider the magnetic lines of force as straightforward lines. A displacement of the magnetic lines of force causes a plasma displacement, and due to the plasma elasticity, such a motion is an oscillation. By analogy with acoustic oscillations, the velocity of their propagation follows from the continuity and Euler equations and is equal to $c_A = \sqrt{\partial p/\partial \rho}$, where p is the pressure, $\rho = MN$ is the plasma density, and M is the ion mass. Because the pressure of a cold plasma is determined by the magnetic pressure $p = H^2/(8\pi)$, we obtain

$$c_A = \sqrt{\frac{H\partial H/\partial N}{4\pi M}}$$

for the velocity of propagation of magnetohydrodynamic waves. Since the magnetic lines of force are frozen in the plasma, $\partial H/\partial N = H/N$, which gives the velocity of magnetohydrodynamic waves—the Alfvén speed.

$$c_A = \frac{H}{\sqrt{4\pi MN}} . \qquad (11.19)$$

There are two types of magnetohydrodynamic waves depending on the direction of wave propagation (see Fig. 11.5). If the wave propagates along the magnetic lines of force (Fig. 11.5a), it is called an Alfvén wave or magnetohydrodynamic wave. This wave is analogous to a wave propagating along an elastic string. In the other case (Fig. 11.5b), the vibration of one magnetic line of force causes the vibration of a neighboring line, and the wave, the magnetic sound, propagates perpendicular to the magnetic lines of force. The dispersion relations for both types of oscillations are identical and have the form

$$\omega = c_A k . \qquad (11.20)$$

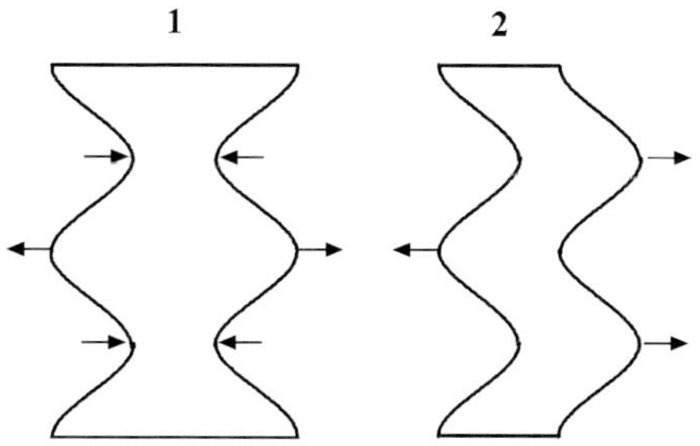

Fig. 11.5 Two types of magnetohydrodynamic waves. (1) magnetic sound; (2) Alfvén waves.

► **Problem 11.13** Analyze the wave properties of a nonuniform plasma located in an external magnetic field if the wave propagates perpendicular to the direction of variation of the plasma density. The magnetic field is directed perpendicular to the density gradient.

Under these conditions, an electron drift in the direction of the density gradient leads to rise of the electric field in the direction, which is perpendicular to the magnetic field and the density gradient. This causes a reverse motion of electrons. As a result, oscillations are originated in the form of drift waves which will be considered below.

We assume the phase velocity of drift waves to be within the limits

$$\sqrt{\frac{T_i}{M}} \ll \frac{\omega}{k} \ll \sqrt{\frac{T_e}{m_e}} .$$

The left inequality allows us to ignore the motion of ions and according to the right equality electrons follow for the wave field, so that the electron number density N_e corresponds to thermodynamic equilibrium in the wave potential φ

$$N_e = N_o \exp\left(\frac{e\varphi}{T_e}\right) ,$$

where we use the Boltzmann formula (1.12), taking into account the sign of electron charge, and N_o is the equilibrium number density of electrons. Introducing the electron number density N' due to the wave according to the relation $N_e = N_o + N'$, we obtain from the above equation

$$N' = N_o \frac{e\varphi}{T_e}$$

if we consider the wave to be weak.

In order to derive the dispersion relation for the drift wave, we take the dependence of wave parameters as $\exp(ik_y y - i\omega t)$ and use the continuity equation (6.15) for electrons

$$\frac{\partial N_e}{\partial t} + \operatorname{div}(N_e \boldsymbol{v}_e) = 0 .$$

In the absence of oscillations the plasma is motionless. Next, the gradient of the electron number density is directed along x, and the wave propagates along y. We obtain from the continuity equation

$$-i\omega N' + v_{ex}\frac{\partial N_0}{\partial x} \,,$$

where $\partial N_0/\partial x$ is the initial gradient of the electron number density. The drift of electrons under the action of the magnetic field H and the electric field E of the wave proceeds in the direction x that is perpendicular to the magnetic field direction (z) and the wave direction (y). We obtain the electron drift velocity under the action of electric and magnetic fields

$$v_{ex} = \frac{cE_y}{H} = -i\frac{ck_y\varphi}{H} \,.$$

Substitute this into the continuity equation and excluding the wave electric potential by using the Boltzmann formula, we obtain the dispersion relation for the drift wave in the form

$$\omega = -k_y\frac{cT_e}{eH}\frac{d\ln N_0}{dx} \,. \tag{11.21}$$

This dispersion relation is valid only if its phase velocity significantly exceeds a typical ion velocity and is small compared to an electron thermal velocity. This condition gives

$$1 \ll \frac{\omega_H L}{\sqrt{T_e/m_e}} \ll \sqrt{\frac{M}{m_e}\frac{T_e}{T_i}} \,,$$

where $\omega_H = eH/(m_e c)$ is the Larmor frequency for electrons and L is a characteristic distance over which the electron number density varies noticeable.

11.3 Propagation of Electromagnetic Waves in Plasma

▶ **Problem 11.14** Analyze the character of transformation of an electromagnetic wave if it propagates in a uniform weakly ionized gas.

If an electromagnetic wave propagates in a plasma, plasma motion due to fields of an electromagnetic field influences wave parameters, and therefore the plasma behavior establishes the dispersion relation for the electromagnetic wave. We shall now derive the dispersion relation for an electromagnetic wave in a plasma based on the Maxwell equations for the wave fields and include in these equations plasma currents under the action of electromagnetic fields. We have for the electromagnetic wave

$$\mathrm{curl}\,\mathbf{E} = -\frac{1}{c}\frac{\partial \mathbf{H}}{\partial t}\,, \qquad \mathrm{curl}\,\mathbf{H} = \frac{4\pi}{c}\mathbf{j} - \frac{1}{c}\frac{\partial \mathbf{E}}{\partial t}\,,$$

where $\boldsymbol{E}$ and $\boldsymbol{H}$ are the electric and magnetic fields in the electromagnetic wave, respectively, $\boldsymbol{j}$ is the density of the electron current produced by the action of the electromagnetic wave, and c is the light velocity. Applying the curl operator to the first equation of this set and the operator $-(1/c)(\partial/\partial t)$ to the second equation of the set, and then eliminating the magnetic field from the resulting equations, we obtain

$$\nabla \operatorname{div} \boldsymbol{E} - \Delta \boldsymbol{E} + \frac{4\pi}{c^2}\frac{\partial \boldsymbol{j}}{\partial t} - \frac{1}{c^2}\frac{\partial^2 \boldsymbol{E}}{\partial t^2} = 0 . \qquad (11.22)$$

We take the plasma to be quasineutral, so that according to the Poisson equation $\operatorname{div} \boldsymbol{E} = 0$. Next, the electric current due to motion of electrons is $\boldsymbol{j} = -eN_o\boldsymbol{w}$, where N_o is the average number density of plasma electrons in the absence of an electromagnetic wave, and $\boldsymbol{w}_e$ is the electron velocity due to the action of the electromagnetic field. The equation of motion for electrons is $m_e d\boldsymbol{w}_e/dt = -e\boldsymbol{E}$, which gives

$$\frac{\partial \boldsymbol{j}}{\partial t} = -eN_o\frac{d\boldsymbol{w}_e}{dt} = \frac{e^2 N_o}{m_e}\boldsymbol{E} .$$

From this we obtain the equation for the electric field of the electromagnetic wave

$$\Delta \boldsymbol{E} - \frac{\omega_p^2}{c^2}\boldsymbol{E} + \frac{1}{c^2}\frac{\partial^2 \boldsymbol{E}}{\partial t^2} = 0 ,$$

where ω_p is the plasma frequency (11.6). Taking the wave parameters in the standard form according to formula (11.1) and substituting it in the above equation, we obtain the following dispersion relation for the frequency of a propagating electromagnetic wave in a plasma:

$$\omega^2 = \omega_p^2 + c^2k^2 . \qquad (11.23)$$

When the plasma density is low ($N_e \to 0$, $\omega_p \to 0$), the dispersion relation transfers to that of an electromagnetic wave propagating in a vacuum, $\omega = kc$. According to the dispersion relation (11.23), electromagnetic waves do not propagate in a plasma if their frequencies are lower than the plasma frequency ω_p. A characteristic damping distance for such waves is of the order of $c/\sqrt{\omega_p^2 - \omega^2}$ according to formula (11.23).

▶ **Problem 11.15** Analyze the character of plasma interaction with an electromagnetic wave on the basis of the kinetic equation for electrons while ignoring ion motion.

Taking the coordinate and time dependence for the electric field strength $\boldsymbol{E}$ of an electromagnetic wave propagating in a plasma as $\exp(-i\omega t + i\boldsymbol{kr})$, and the same dependence for the electron current $\boldsymbol{j}$, we obtain equation (11.22) for these parameters in the form

$$k^2\boldsymbol{E} - \boldsymbol{k}(\boldsymbol{kE}) = \frac{\omega^2}{c^2}\boldsymbol{E} - \frac{4\pi i\omega}{c^2}\boldsymbol{j} . \qquad (11.24)$$

For determining the electric current due to plasma electrons, we consider the kinetic equation for electrons (6.3) in the tau-approximation. On taking the distribution function in this equation in the form $f^{(o)} + f'$ and the dependence of f' in the form $\exp(-i\omega t + i\boldsymbol{kr})$, we obtain this equation

$$f'\left(-i\omega + i\boldsymbol{kr} - \frac{1}{\tau}\right) - \frac{e\boldsymbol{E}}{m_e}\frac{\partial f^{(o)}(\boldsymbol{v})}{\partial \boldsymbol{v}} = 0\,.$$

Taking the electric field direction along the x-axis as well as the direction of the current density $\boldsymbol{j}$, we obtain

$$f' = \frac{ieE}{m_e}\frac{\partial f^{(o)}}{\partial v_x}\left(-i\omega + i\boldsymbol{kv} - \frac{1}{\tau}\right)^{-1}\,.$$

Taking the direction of wave propagation along the z-axis, we find from this

$$f' = \frac{ieE}{m_e}\frac{\partial f^{(o)}}{\partial v_x}\left(\omega - kv_z - \frac{1}{\tau}\right)^{-1}\,.$$

This gives the electron current density due to the wave electric field as

$$j_x = -e\int f d\boldsymbol{v} = -\frac{ie^2 E}{m_e}\int v_x \frac{\partial f^{(o)}}{\partial v_x} dv_x \left(\omega - kv_z - \frac{1}{\tau}\right)^{-1}\,.$$

Taking this integral over the parts and using the normalization condition for the distribution function ($\int f^{(o)} d\boldsymbol{v} = N_o$), we obtain

$$j_x = \frac{iN_o e^2 E}{m_e}\left\langle \frac{1}{\omega - kv_z - \frac{1}{\tau}} \right\rangle\,,$$

where brackets imply an average over the electron distribution in the absence of the electromagnetic wave. Substituting this expression for the electron current density in formula (11.22), we obtain the dispersion relation for the electromagnetic wave propagating in a plasma

$$\omega^2 = c^2k^2 + \omega\omega_p^2\left\langle \frac{1}{\omega - kv_z - \frac{1}{\tau}} \right\rangle\,. \tag{11.25}$$

In particular, for high frequencies we have the following dispersion relation for an electromagnetic wave:

$$\omega^2 = c^2k^2 + \omega_p^2\left(1 - \frac{i}{\omega\tau}\right)\,, \tag{11.26}$$

that coincides with formula (11.23) if we ignore damping due to collisions.

► **Problem 11.16** Determine the depth of penetration for electromagnetic waves of a low frequency inside a plasma, if the electron current density is determined by electron–atom collisions.

We analyze the dispersion relation (11.25) at low wave frequencies, when the left-hand side of this equation can be neglected. In addition, $\omega\tau \ll 1$, which allows us to restrict by the term $1/\tau$ in the denominator of the second term of the right-hand side of the dispersion relation (11.25). Then this dispersion relation is reduced to the form

$$\omega = -i\frac{k^2c^2}{\omega_p^2\tau} \,. \qquad (11.27)$$

This dispersion relation allows us to find the penetration depth of an electromagnetic wave in a plasma. Since the coordinate dependence is $\exp(ikx)$, we represent it in the form $\exp(-x/\Delta)$, so that $1/\Delta = \mathrm{Im}\,k$. From the above dispersion relation we obtain the depth of penetration of an electromagnetic wave in a plasma

$$\Delta = \frac{1}{\mathrm{Im}\,k} = \frac{c}{\omega_p\sqrt{2\omega\tau}} \,. \qquad (11.28)$$

One can see that the penetration depth becomes infinite in the limit of low frequencies, until this depth attains the Debye–Hückel radius. This phenomenon is the normal skin effect.

A numerical example of the skin effect can be given for the plasma of the Earth's ionosphere at an altitude of about 100 km. The plasma conductivity is $\Sigma \sim 10^9$ Hz and the plasma frequency is $\omega_p \sim 3 \cdot 10^7$ Hz. For frequencies of the order of the plasma frequency, the penetration depth is of the order of 1 m. Electromagnetic waves with frequencies smaller than the plasma frequency cannot pass through the Earth's ionosphere.

► **Problem 11.17** Determine the depth of penetration for electromagnetic waves of a low frequency inside a plasma.

This limit corresponds to the criterion

$$\omega \ll k\sqrt{\frac{T_e}{m_e}} \,,$$

where T_e is the electron temperature. We now consider the limit $\omega\tau \gg 1$ when collisions in a plasma are not essential. Taking the residue of the integrand function, we obtain the dispersion relation at low wave frequencies if

$$\omega = -i\sqrt{\frac{2T_e}{\pi m_e}\frac{c^2k^3}{\omega_p^2}} \,.$$

This dispersion relation gives the penetration depth of an electromagnetic wave in a plasma, which in this case is equal to

$$\Delta = \frac{1}{\operatorname{Im} k} = 2\left(\frac{2T_e}{\pi m_e}\right)^{1/6}\left(\frac{c^2}{|\omega|\omega_p^2}\right)^{1/3} . \qquad (11.29)$$

One can see that the penetration depth becomes infinite in the limit of low frequencies. Indeed, since $\omega\tau \gg 1$, we have $\Delta \ll \sqrt{T_e/m_e}/\tau$. But $\omega\tau \gg 1$ in this case, i. e., the frequency is restricted below. This result relates to the abnormal skin effect. Note that the normal and abnormal skin effects which are described by the dispersion relations (11.28) and (11.29) relate to different frequency ranges.

► **Problem 11.18** Derive the dispersion relation for small oscillations in an isotropic plasma on the basis of the plasma dielectric constant.

A general method of deriving the dispersion relation for small oscillations consists in obtaining equations for quantities which characterize a weak difference of each quantity from the equilibrium one. Then parameters of oscillations follow from the condition of existence of such deviations from their steady values, and this reduces to the condition that the determinant of the equation set for quantity additions is zero. We use this set of equation in the form (11.24) and represent this equation in the form

$$k^2 E_\alpha - k_\alpha k_\beta E_\beta - \frac{\omega^2}{c^2} E_\alpha - i\frac{4\pi\omega}{c^2} j_\alpha = 0 ,$$

and the summation is implied over twice repeated subscripts. Using Ohm's law (7.60) ($j_\alpha = \Sigma_{\alpha\beta} E_\beta$) and relation (7.69) between the conductivity $\Sigma_{\alpha\beta}$ and plasma dielectric constant $\varepsilon_{\alpha\beta}$, we reduce the above equation to the form

$$k^2 E_\alpha - k_\alpha k_\beta E_\beta - \frac{\omega^2}{c^2}\varepsilon_{\alpha\beta} E_\beta = 0 .$$

A nonzero solution of this set of equations exists if its determinant is zero, that is

$$\left| k^2 E_\alpha - k_\alpha k_\beta E_\beta - \frac{\omega^2}{c^2}\varepsilon_{\alpha\beta} E_\beta \right| = 0 . \qquad (11.30)$$

The dispersion equation (11.30) establishes a relation between the parameters of waves in a plasma.

► **Problem 11.19** Obtain the dispersion relation for long-wave oscillations due to electrons in a cold isotropic plasma on the basis of equation (11.30).

For long-wave oscillations in an isotropic plasma, equation (11.30) has the form $\varepsilon(\omega, k = 0) = 0$. Using the dielectric constant, expression (7.70) that takes into account motion of electrons as a whole and their collisions with atoms in the form of the tau-approximation. Then the dispersion relation for such waves has the form

$$\omega^2 = \omega_p^2 - i\frac{\omega_p}{\tau} ,$$

where we account for $\omega \approx \omega_p$. This corresponds to formula (11.8) with $k = 0$ if we ignore electron–atom collisions.

11.4 Electromagnetic Waves in Plasma in Magnetic Field

▶ **Problem 11.20** Analyze the Faraday effect—rotation of the polarization vector for an electromagnetic wave propagating in a plasma in a magnetic wave.

The Faraday effect is a rotation of the polarization vector of an electromagnetic wave propagating in a medium in an external magnetic field. This effect is due to electric currents in a medium subjected to a magnetic field, and leads to different refractive behaviors for waves with left-handed circular polarization as compared to right-handed circular polarization. Hence, electromagnetic waves with different circular polarizations propagate with different velocities, and propagation of electromagnetic waves with plane polarization is accompanied by rotation of the polarization vector of the electromagnetic wave.

We consider an electromagnetic wave in a plasma propagating along the z-axis while being subjected to an external magnetic field. The wave and the constant magnetic field $\boldsymbol{H}$ are in the same direction. We treat a frequency regime such that we can neglect ion currents compared to electron currents. Hence, one can neglect motion of the ions. The electron velocity under the action of the field is given by formulas (7.50). The electric field strengths of the electromagnetic wave corresponding to right-handed (subscript +) and left-handed (subscript −) circular polarization are given by

$$E_+ = (E_x + iE_y)e^{-i\omega t}\,, \qquad E_- = (E_x - iE_y)e^{-i\omega t}.$$

For a plasma without collisions ($\omega\tau \ll 1$), formulas (7.50) lead to

$$j_+ = -eN_e(w_x + iw_y) = \frac{iN_e e^2 E_+}{m_e(\omega + \omega_H)} = \frac{i\omega_p^2 E_+}{4\pi(\omega + \omega_H)}\,,$$

$$j_- = -eN_e(w_x - iw_y) = \frac{i\omega_p^2 E_-}{4\pi(\omega - \omega_H)}\,.$$

Let us use expansion (11.1) on time and space coordinates of wave parameters in equation (11.22), that leads to equation (11.22) in the form

$$k^2\boldsymbol{E} - \frac{4\pi i\omega \boldsymbol{j}}{c^2} - \frac{\omega^2 \boldsymbol{E}}{c^2} = 0\,. \tag{11.31}$$

We consider propagation of the electromagnetic wave along the magnetic field. Using the above expressions for the density of the electron currents, we obtain the dispersion relations for the electromagnetic waves with different circular polarizations in the form

$$k_+^2 + \omega_+\frac{\omega_p^2}{\omega_+\omega_H} - \frac{\omega_+^2}{c^2} = 0\,, \qquad k_-^2 + \omega_-\frac{\omega_p^2}{\omega_-\omega_H} - \frac{\omega_-^2}{c^2} = 0\,, \tag{11.32}$$

where subscripts + and − refer to right-handed and to left-handed circular polarizations, respectively. Based on these dispersion relations, we can analyze the propagation of an electromagnetic wave of frequency ω in a plasma in an external magnetic field. At $z = 0$ we take the wave to be polarized along the direction of the x-axis, so that $\boldsymbol{E} = \boldsymbol{i}E\exp(-i\omega t)$, and we introduce the unit vectors $\boldsymbol{i}$ and $\boldsymbol{j}$ along the x and y axes, respectively. The electric field of this electromagnetic wave in the plasma is

$$\boldsymbol{E} = \boldsymbol{i}E_x + \boldsymbol{j}E_y = \frac{\boldsymbol{i}}{2}(E_+ + E_-) + \frac{\boldsymbol{j}}{2i}(E_+ - E_-)\,.$$

We use the boundary condition

$$E_+ = E_0 e^{i(k_+ z - i\omega t)}\,, \quad E_- = E_0 e^{i(k_- z - i\omega t)}\,.$$

Introducing $k = (k_+ + k_-)/2$ and $\Delta k = k_+ - k_-$, we obtain the result

$$\boldsymbol{E} = E_0 e^{i(kz-\omega t)}\left(\boldsymbol{i}\cos\frac{\Delta k z}{2} + \boldsymbol{j}\sin\frac{\Delta k z}{2}\right)\,. \tag{11.33}$$

From the dispersion relations (11.32) it follows

$$k_+^2 = \frac{\omega^2}{c^2}\left[1 - \frac{\omega_\mathrm{p}^2}{\omega(\omega+\omega_H)}\right]\,, \quad k_-^2 = \frac{\omega^2}{c^2}\left[1 - \frac{\omega_\mathrm{p}^2}{\omega(\omega-\omega_H)}\right]\,.$$

Assuming the criterion $\Delta k \ll k$ to be valid and $\omega_H \ll \omega$, we obtain from the above equation taking into account $\omega_+ = \omega_- = \omega$

$$\Delta k = \frac{\omega_H}{c}\frac{\omega_\mathrm{p}^2}{\omega^2}\,. \tag{11.34}$$

This result establishes the rotation of the polarization vector during propagation of the electromagnetic wave in a plasma. The angle φ of the rotation in the polarization direction is proportional to the distance z of propagation. This is a general property of the Faraday effect. In the limiting case $\omega \gg \omega_\mathrm{p}$, $\omega \gg \omega_H$, we have

$$\frac{\partial\varphi}{\partial z} = \frac{\Delta k}{2} = \frac{\omega_H}{c}\frac{\omega_\mathrm{p}^2}{2\omega^2}\,.$$

For a numerical example we note that for maximum laboratory magnetic fields $H \sim 10^4$G the first factor ω_H/c is about 10 cm^{-1}, so that for these plasma conditions the Faraday effect is detectable for propagation distances of the order of 1 cm.

From the above results it follows that the Faraday effect is strong in a frequency range near the cyclotron resonance $\omega \approx \omega_H$. Then a strong interaction takes place between the plasma and the electromagnetic wave with left-handed polarization. In particular, it is possible to have the electromagnetic wave with left-handed polarization absorbed, while the wave with right-handed circular polarization passes through the plasma freely. Then the Faraday effect can be detected at small distances.

► **Problem 11.21** Analyze propagation of cyclotron waves in a plasma in an external magnetic field. These waves propagate along the magnetic field with the frequency close to the Larmor one.

We now consider equation (11.31) that follows from the Maxwell equations and has the form

$$(\omega^2 - k^2c^2)\boldsymbol{E} + 4\pi i\omega \boldsymbol{j} = 0 \ , \qquad (11.35)$$

and the current density is determined by electrons, so that $\boldsymbol{j} = -eN_0\boldsymbol{v}_e$, where N_0 is the equilibrium number density of electrons and $\boldsymbol{v}_e$ is its velocity due to the oscillation.

The motion equation for an electron in the magnetic field $\boldsymbol{H}$ and the electric field $\boldsymbol{E}$ of the wave has the form

$$m_e\frac{d\boldsymbol{v}_e}{dt} = e\boldsymbol{E} + \frac{em_e}{c}[\boldsymbol{v}_e \times \boldsymbol{H}] \ .$$

Using the time harmonic dependence for wave parameters, we obtain from this

$$-i\omega \boldsymbol{j} = \omega_H[\boldsymbol{j} \times \boldsymbol{h}] - \frac{\omega_p^2}{4\pi}\boldsymbol{E} \ . \qquad (11.36)$$

Here $\omega_H = eH/(m_e c)$ is the Larmor frequency, $\boldsymbol{h}$ is the unit vector directed along the magnetic field $\boldsymbol{H}$, and we above return to the electron current density $\boldsymbol{j}$ from the velocity of an individual electron $\boldsymbol{v}_e$. Excluding the current density from equations (11.35) and (11.36), we reduce them to the equation

$$i\omega(\omega_p^2 - \omega^2 + k^2c^2)\boldsymbol{E} + (\omega^2 - k^2c^2)\omega_H[\boldsymbol{E} \times \boldsymbol{h}] = 0 \ .$$

Writing down this equation in vector components and equalizing the determinant of the set of equations to zero, we obtain from this the dispersion relation

$$\omega^2(\omega^2 - \omega_p^2 - k^2c^2)^2 - \omega_H^2(\omega^2 - k^2c^2) = 0 \ .$$

We consider the limiting case $kc \gg \omega_p$ when this dispersion equation is divided in two branches. The first one describes an electromagnetic wave with the frequency near kc, and the second relates to the cyclotron wave whose frequency is close to the Larmor frequency and is given by the dispersion relation

$$\omega = \omega_H\left(1 + \frac{\omega_p^2}{\omega_H^2 - k^2c^2}\right) \ . \qquad (11.37)$$

► **Problem 11.22** Analyze hybrid waves resulted from mixing of an electromagnetic wave and cyclotron waves—whistlers.

Insertion of a magnetic field into a plasma leads to a large variety of new types of oscillations in it. We considered above magnetohydrodynamic waves and magnetic sound, both of which are governed by elastic magnetic properties of a cold plasma. In addition to these phenomena, a magnetic field can produce electron and ion cyclotron waves that correspond to rotation of electrons and ions in the magnetic field. Mixing of these oscillations with plasma oscillations, ion sound, and electromagnetic waves creates many types of hybrid waves in a plasma. As an example of this, we now consider waves that are a mixture of electron cyclotron and electromagnetic waves. These waves are called whistlers and are observed as atmospheric electromagnetic waves of low frequency (in the frequency interval 300–30 000 Hz). These waves are a consequence of lightning in the upper atmosphere and propagate along the magnetic lines of force. They can approach the magnetosphere boundary and then reflect from it. Therefore, whistlers are used for exploration of the Earth's magnetosphere up to distances of 5–10 Earth radii. The whistler frequency is low compared to the electron cyclotron frequency $\omega_H = eH/(m_e c) \sim 10^7$ Hz, and it is high compared to the ion cyclotron frequency $\omega_{iH} = eH/(Mc) \sim 10^2$–$10^3$ Hz (M is the ion mass). Below we consider whistlers as electromagnetic waves of frequency $\omega \ll \omega_H$ that propagate in a plasma in the presence of a constant magnetic field.

We employ expansion (11.1) for oscillatory parameters of a monochromatic electromagnetic wave. Then equation (11.22) gives

$$k^2 \boldsymbol{E} - \boldsymbol{k}(\boldsymbol{k} \cdot \boldsymbol{E}) - i\frac{4\pi\omega \boldsymbol{j}}{c^2} = 0 ,$$

when $\omega \ll kc$. We take the current density of electrons in the form $\boldsymbol{j} = -eN_e\boldsymbol{w}$, where N_e is the electron number density. The electron drift velocity follows from the electron equation in neglecting electron–atom collision and variation of the electric field strength $\nu \ll \omega \ll \omega_H$, that has the form $e\boldsymbol{E}/m_e = -\omega_H (\boldsymbol{w} \times \boldsymbol{h})$, and $\boldsymbol{h}$ is the unit vector directed along the magnetic field. Substituting this in the latter equation, we obtain the dispersion relation

$$k^2 \left[\boldsymbol{j} \times \boldsymbol{h}\right] - \boldsymbol{k}\left[\boldsymbol{k} \cdot (\boldsymbol{j} \times \boldsymbol{h})\right] - i\boldsymbol{j}\frac{\omega\omega_p^2}{\omega_H c^2} = 0 ,$$

where $\omega_p = \sqrt{4\pi N_0 e^2/m_e}$ is the plasma frequency. We introduce a coordinate system such that the z-axis is parallel to the external magnetic field (along the unit vector $\boldsymbol{h}$) and the plane xz contains the wave vector. The x and y components of this equation are

$$-\frac{i\omega\omega_p^2}{\omega_H c^2} j_x + (k^2 - k_x^2) j_y = 0 , \qquad -k^2 j_x - \frac{i\omega\omega_p^2}{\omega_H c^2} j_y = 0 . \tag{11.38}$$

The determinant of this system of equations must be zero, which leads to the dispersion relation

$$\omega = \omega_H \frac{c^2 k k_z}{\omega_p^2} = \omega_H \frac{c^2 k^2 \cos\vartheta}{\omega_p^2} . \tag{11.39}$$

Here ϑ is the angle between the direction of wave propagation and the external magnetic field.

One can see that the whistler frequency is considerably higher than the frequency of Alfvén waves and magnetic sound. In particular, if the whistler propagates along the magnetic field, according to the dispersion relation (11.39) gives $\omega = \omega_A^2/\omega_{iH}$, where ω_A is the frequency of the Alfvén wave, and ω_{iH} is the ion cyclotron frequency. Since we assume $\omega \gg \omega_{iH}$, this infers that

$$\omega \gg \omega_A \gg \omega_{iH} . \tag{11.40}$$

In addition, because $\omega \ll \omega_H$, the condition $\omega \ll kc$ leads to the inequalities

$$\omega \ll kc \ll \omega_p . \tag{11.41}$$

► **Problem 11.23** Derive the dispersion relation for helicons—a whistler propagating along the magnetic field.

Whistlers are determined entirely by the motion of electrons. To examine the nature of these waves, we first note that the electron motion and the resultant current in the magnetized plasma give rise to an electric field. This electric field, in turn, leads to an electron current according to equation (11.31). In the end, the whistler oscillations are generated. Note that because the dispersion relation has the dependence $\omega \sim k^2$, the group velocity of these oscillations, $v_g = \partial\omega/\partial k \sim \sqrt{\omega}$, grows with the wave frequency. Then the tone of a short-time signal with a wide band of frequencies is the whistler like, which explains the name of this wave.

The polarization of a whistler propagating along the magnetic field can be found from the relation between two current components, j_x and j_y, that leads to the dispersion relation (11.39). As it follows from relations (11.38), in this case ($k = k_z,\ k_x = 0$) we obtain the following relation between the current components:

$$j_y = ij_x ,\ j_x = -ij_y . \tag{11.42}$$

From this it follows that this wave, the helicon, has a circular polarization. Therefore, this wave propagating along the magnetic field has a helical structure. The direction of rotation of wave polarization is the same as the direction of electron rotation. The development of such a wave can be described as follows. Suppose electrons in a certain region possess a velocity perpendicular to the magnetic field. This electron motion gives rise to an electric field and compels electrons to circulate in the plane perpendicular to the magnetic field. This perturbation is transferred to the neighboring regions with a phase delay.

11.5 Damping of Waves in Plasma

► **Problem 11.24** Determine the attenuation coefficient for plasma oscillations due to collisions of electrons with atoms, which are taken into account in the tau-approximation (6.3).

Based on the set of equations (11.7), we introduce a friction force F_{fr} into the motion equation for electrons, which is taken as $m_e w_x/\tau$, where w_x is the drift velocity along the wave. Then the dispersion equation (11.11) takes the form

$$\omega(\omega - i/\tau) = \omega_p^2 + k^2 \langle v_x^2 \rangle ,$$

and this dispersion relation is valid if $\omega_p \tau \gg 1$. In this approximation we obtain the dispersion relation in the form

$$\omega = \sqrt{\omega_p^2 + k^2 \langle v_x^2 \rangle} - \frac{i}{\tau}. \quad (11.43)$$

As is seen, the attenuation time for this mechanism of damping is τ.

Let us analyze the condition $\omega_p \tau \gg 1$ of existence of plasma oscillations under real conditions. Taking a typical collision time for electrons as $1/\tau \sim (N_a \sigma v)^{-1}$, where N_a is the atom number density, σ is the cross section for electron–atom collisions, which is equal to a typical gas-kinetic cross section, and v is the electron energy and is ~ 1 eV. Then the above criterion is

$$\frac{N_e^{1/2}}{N_a} \gg 10^{-12} \text{cm}^{3/2} ,$$

and this condition may be valid for some plasma types with a low density of atoms and not small degree of ionization.

▶ **Problem 11.25** Analyze the decay of plasma oscillations in a collisionless plasma.

Let us represent the frequency of plasma oscillations in the form $\omega = Re\omega - i\delta$. According to formula (11.12) we obtain plasma waves with a frequency $|\omega - \omega_p| \ll \omega$

$$\delta = -\frac{\pi \omega_p^2}{2k^2} \left. \frac{\partial f^{(0)}}{\partial v_x} \right|_{v_x = \omega/k} , \quad (11.44)$$

which is the Landau damping. One can see that the attenuation of the plasma wave takes place if the electron distribution function $f^{(0)}$ decreases with a velocity increase. For the Maxwell distribution function this formula gives

$$\delta = -\frac{\pi \omega_p^2}{2k^2} \sqrt{\frac{m_e}{2\pi T_e}} \exp\left(-\frac{m_e \omega_p^2}{2k^2 T_e} - \frac{3}{2} \right) . \quad (11.45)$$

When the dispersion relation (11.11) holds true, i.e., $k r_D \ll 1$, we obtain $\delta \ll \omega_p$, i.e., decay of plasma oscillations proceeds through many oscillations.

▶ **Problem 11.26** Analyze damping of the ion sound in a collisionless plasma.

Based on the dispersion relation (11.17) and using the Maxwell distribution function for the ion velocities, we obtain in addition to the dispersion relation (11.13) for the ion sound its damping in the form

$$\delta = \sqrt{\frac{\pi m_e}{8M}} + \frac{\pi \omega^2}{2k} \sqrt{\frac{M}{2\pi T_i}} \exp\left(-\frac{M\omega^2}{2k^2 T_i}\right) ,$$

if we take the frequency of the ion sound as $\omega = Re\omega - i\delta$, assume $\delta \ll \omega$, $T_i \ll T_e$, and use the Maxwell distribution function for ions.

From this formula according to the dispersion relation (11.13), it follows that the ion sound is realized, i. e., $\delta \ll \omega$, if

$$T_i \ll T_e .$$

▶ **Problem 11.27** Analyze the Landau damping as a result of capture of electrons by a wave and give the criterion of wave damping.

In considering interaction of plasma electrons with plasma oscillations, one can divide electrons in two types, captured by a plasma wave and noncaptured ones. A strong interaction of a plasma wave proceeds with captured electrons which exchange with the wave by energy as a result of reflection from potential walls of the wave. According to the phase diagram for electrons located in the field of the plasma wave, given in Fig. 11.6, captured electrons have closed trajectories in the phase space, whereas trajectories of noncaptured electrons are infinite. Thus, electrons travel inside the potential well of the wave, and reflection of the captured electrons from its walls leads to energy exchange between a captured electrons and the wave. If a captured electron has a velocity u, its reflection from the wave potential well leads to an energy change

$$\Delta\varepsilon = \frac{1}{2} m_e (v_p + u)^2 - \frac{1}{2} m_e (v_p - u)^2 = 2 m_e v_p ,$$

where $v_p = \omega/k$ is the phase velocity of the wave. But if an electron losses and energy near one wall, it returns this energy after reflection from another wall. Thus, for effective exchange of energy between an electron and wave, collisions between electrons are necessary. These collisions take a captured electron from the potential wave, and after this event interaction of this electron with the wave finishes.

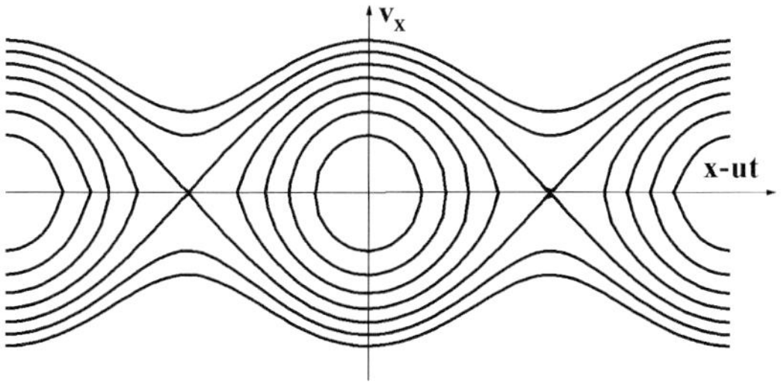

Fig. 11.6 Trajectories of electrons in the phase space when electrons interact with plasma oscillations. Captured electrons have closed trajectories.

Since the electron distribution function is restored by electron–electron collisions, one can find the direction of the energy exchange between the wave and plasma electrons. The electron distribution function is not altered by the interaction with the wave, and it is necessary to compare the number of electrons with velocity $v_p + u$ that transfer energy to the wave and the number of electrons with a velocity $v_p - u$ that take energy from the wave, where u is the wave phase velocity. The number of captured electrons is proportional to the electron distribution function $f(v)$. Hence, the wave gives its energy to electrons and is damped if $f(v_p + u)$ is larger than $f(v_p - u)$. This means that the wave is damped when

$$\left[\frac{\partial f^{(0)}}{\partial v_x}\right]\Bigg|_{v_x=\omega/k} < 0 \,. \tag{11.46}$$

Here v_x is the component of the electron velocity in the direction of the wave propagation, and the derivative is taken for the electron velocity being equal to the phase velocity v_p of the wave. When the above condition is not satisfied, the wave takes energy from the electrons, and its amplitude increases.

► **Problem 11.28** Obtain the criterion for the wave amplitude when the mechanism of wave damping due to electron capture by the wave is valid.

The above character of electron–wave interaction requires that a frequency of oscillations of a captured electron in a wave field must be small compared to the collision rate between electrons. The latter is given by formula (2.37) $\nu \sim N_e e^4 T_e^{-3/2} m_e^{-1/2} \ln \Lambda$, and the frequency of oscillations of a captured electron is of the order of $(e\varphi k/m_e)^{1/2} \sim \sqrt{eE'k/m_e} \sim \sqrt{e^2 N'/m_e}$, where k is the wave number of such an oscillation, φ, E', and N' are the corresponding parameters of this oscillation. This leads to the following criterion of the smallness of oscillations:

$$\frac{N'}{N_e} \ll \frac{N_e e^6}{T_e^3} \ln^2 \Lambda \,.$$

Note that the first factor of the right-hand side of this criterion corresponds to the ideal criterion (1.56) of a plasma. Therefore, a small part of electrons partake in plasma oscillations.

► **Problem 11.29** Determine the dependence for the attenuation factor δ of plasma oscillations on plasma parameters on the basis of the character of plasma oscillations.

We introduce the attenuation factor δ such that the energy density W for plasma oscillations drops in time as $W \sim \exp(-\delta t)$. Taking this quantity as $W \sim (E')^2 \sim \varphi^2/k^2$, where E' is the amplitude of the electric field of the wave. Let us estimate the variation of time with the energy density of the plasma oscillations as

$$\frac{dW}{dt} \sim \nu \int_{-u_0}^{u_0} f(v) \Delta\varepsilon du \,,$$

where $\nu \sim u_0 k$ is the oscillation frequency for a captured electron in the potential well of the wave, $u_0 = (2e\varphi/m_e)^{1/2}$, φ is the amplitude of electron oscillations in the wave field, and $\Delta\varepsilon = 2m_e v_p u_0$ is the maximum change in the electron energy when the direction of the electron motion is reversed. Hence, the right-hand side of the above expression is given by

$$\nu\frac{\partial f}{\partial v_x} u_0 \Delta\varepsilon u_0 \sim u_0 k \frac{\partial f}{\partial v_x} u_0 m_e v_p u_0 u_0 \sim \frac{\partial f}{\partial v_x}\frac{e^2\omega}{m_e k_2} W ,$$

where the phase velocity of the wave is $v_p = \omega/k$. Using the definition of the attenuation factor δ for the plasma wave $dW/dt = -\delta W$, we obtain the estimate for it

$$\delta \sim \frac{e^2\omega}{m_e k^2}\frac{\partial f^{(0)}}{\partial v_x} .$$

One can see that this formula is analogous to formula (11.44).

▶ **Problem 11.30** Show that electron–electron collisions in an ideal plasma cannot give the rate of damping of plasma oscillations that is comparable to the oscillation frequency.

The attenuation coefficient of plasma oscillations due to electron–electron collisions is of the order of the rate of electron–electron collisions, which is given by

$$\delta \sim N_e v\sigma \sim N_e \frac{e^4}{T_e^{3/2} m_e^{1/2}} \ln\Lambda ,$$

where we use formula (2.37) for the cross section σ of electron–electron collisions. From this we have

$$\frac{\delta}{\omega_p} \sim \left(\frac{N_e e^6}{T_e^3}\right)^{1/2} \ln\Lambda ,$$

and in this case $\Lambda \sim e\varphi/T_e$, where φ is the electric potential of the wave. Taking $\ln\Lambda \sim 1$, we obtain this ratio to be small in an ideal plasma according to the criterion (1.56).

▶ **Problem 11.31** Determine the attenuation factor δ for plasma oscillations on the basis of the motion character of an individual electron in the wave field.

Let us consider the motion equation of an individual electron that has the form

$$m_e\frac{dv_x}{dt} = -eE\cos(kx - \omega t) ,$$

where x is the wave direction, E is the electric field strength of the wave, and v_x is the electron velocity. We use the expansion for the electron velocity $v + v_0 + v_1 + v_2$, so that $v_n \sim E^n$, and use the same expansion for the electron coordinate $x = x_0 + x_1 + x_2$. This gives

$$\frac{dv_0}{dt} = 0, \qquad m_e \frac{dv_1}{dt} = -eE\cos(kx_0 - \omega t) ,$$
$$m_e \frac{dv_2}{dt} = -eE\cos[k(x_0 + x_1) - \omega t] + eE\cos(kx_0 - \omega = t) ,$$

$$eEkx_1\cos(kx_0 - \omega t) = eEk\int_0^t \sin(kx_0 - \omega t)v_1 dt .$$

In the zero-th approximation we obtain $x_0 = v_0 t + a$, where a is the electron coordinate in the beginning. The first approximation gives

$$v_1 = -\frac{eE}{m_e}\frac{\sin[(kv_0 - \omega)t + ak] - \sin ak}{kv_0 - \omega} ,$$

and

$$x_1 = \int_0^t v_1 dt \sin(kx_0 - \omega)t = -\frac{eE}{m_e}\left[\frac{\cos ka - \cos[ka + (kv_0 - \omega)t]}{(kv_0 - \omega)^2} - \frac{t\sin ka}{kv_0 - \omega}\right] .$$

From these expressions we evaluate the rate of variation of the electron energy ε with an accuracy up to $\sim E^2$ and is given by

$$\frac{d\varepsilon}{dt} = m_e v_x \frac{dv_x}{dt} = m_e v_0 \frac{dv_1}{dt} + m_e v_1 \frac{dv_1}{dt} + m_e v_0 \frac{dv_2}{dt} .$$

Based on the above expressions for electron parameters, we average over the initial electron positions, so that we have $\overline{\sin ka} = \overline{\cos ka} = 0$ and $\overline{\sin^2 ka} = \overline{\cos^2 ka} = 1/2$. As a result, we have

$$\frac{d\bar{\varepsilon}}{dt}\frac{e^2E^2}{2m_e} = \left[\frac{\sin(kv_0 - \omega)t}{kv_0 - \omega} + \frac{v_0 k\sin(kv_0 - \omega)t}{(kv_0 - \omega)^2} + \frac{v_0 kt\cos(kv_0 - \omega)t}{kv_0 - \omega}\right] .$$

Because this energy is taken from the wave, we find from this for the wave energy W per unit volume

$$\frac{dW}{dt} = -\frac{e^2E^2N_e}{2m_e}$$
$$\int f(v_0)dv_0\left[\frac{\sin(kv_0 - \omega)t}{kv_0 - \omega} + \frac{v_0 k\sin(kv_0 - \omega)t}{(kv_0 - \omega)^2} + \frac{v_0 kt\cos(kv_0 - \omega)t}{kv_0 - \omega}\right] .$$

Here the energy variation of an individual electron is averaged over electron velocities, the distribution function is normalized to one ($\int f(v_0)dv_0 = 1$), and the principal value of the integral is taken.

Let us evaluate this integral in the limit $t \to 0$. Then the first term is zero, as well as the third term, because we take the principal value of the integral, and the value $\int f(v_x)dv_x/(kv_x - \omega)$ is finite. Hence we have

$$\frac{dW}{dt} = -\frac{e^2 E^2 N_e}{2m_e} \lim_{t\to 0} \int f(v_o)dv_o \left[\frac{v_o k \sin(kv_o - \omega)t}{(kv_o - \omega)^2}\right] = \frac{e^2 E^2 N_e}{2m_e}\frac{\omega}{k}\frac{\partial f^{(o)}}{\partial v_x}\bigg|_{v_x=\omega/k}$$

$$\lim_{t\to 0} \int_{-\infty}^{\infty} dv_o \frac{\sin(kv_o - \omega)t}{kv_o - \omega} .$$

We above used that this integral is converged in the vicinity $v_o = \omega/k$, taking the integral by parts and considering the first expansion term of $f(v_o)$ that gives zero because of the symmetry of the integrand. Since

$$\int_{-\infty}^{\infty} dz \frac{\sin z}{z} = \pi ,$$

we obtain from this

$$\frac{dW}{dt} = -\frac{\pi e^2 E^2 \omega N_e}{2m_e k^2}\frac{\omega}{k}\frac{\partial f^{(o)}}{\partial v_x}\bigg|_{v_x=\omega/k} .$$

On taking the wave energy density as

$$W = \frac{E^2}{4\pi}\overline{\cos^2(kx - \omega t)} = \frac{E^2}{8\pi} ,$$

and reducing the equation for this quantity to the form

$$\frac{dW}{dt} = -2\delta W ,$$

so that the latter is the definition for the attenuation factor, we find the following expression for the attenuation factor:

$$\delta = -\frac{2\pi^2 e^2 \omega N_e}{m_e k^2} \partial f_o \partial v_x \bigg|_{v_x=\omega/k} = -\frac{\pi\omega_p^2}{2k^2}\frac{\partial f^{(o)}}{\partial v_x}\bigg|_{v_x=\omega/k} ,$$

since the wave frequency ω is close to the plasma frequency ω_p. This formula coincides with formula (11.44) that is obtained from another consideration.

▶ **Problem 11.32** Give the criterion of development of instability for plasma oscillations.

The possibility of attenuation or amplification of plasma oscillations in a plasma is determined by the sign of the distribution function derivative for electrons at the velocity corresponding to the wave phase velocity. The criterion (11.46) corresponds to damping of oscillations, and the opposite criterion

$$\left[\frac{\partial f^{(o)}}{\partial v_x}\right]\bigg|_{v_x=\omega/k} > 0 \tag{11.47}$$

relates to amplification of oscillations. In this case the oscillation amplitude increases exponentially in time, but the wave exists until this amplitude is relatively small. Amplification of an oscillation results from transfer of the energy of plasma particles to the wave, and amplification of the wave make a plasma state to be an unstable one. When an oscillation amplitude becomes large, nonlinear interactions in a plasma will lead to the formation of new distributions in plasma depending on the character of these interactions.

▶ **Problem 11.33** Consider damping of cyclotron waves due to collisions of electrons with atoms within the framework of the tau-approximation (6.3).

We modify the dispersion equation (11.37) for cyclotron waves taking into account electron–atom collisions. Then we include the term $m_e \boldsymbol{v}_e/\tau$ into the motion equation for electrons, and then equation (11.36) takes the form

$$\left(-i\omega + \frac{1}{\tau}\right)\boldsymbol{j} = \omega_H[\boldsymbol{j} \times \boldsymbol{h}] - \frac{\omega_p^2}{4\pi}\boldsymbol{E} \,. \tag{11.48}$$

Repeating the deduction of the dispersion equation (11.37), we obtain it in the form

$$\omega = \omega_H \left(1 + \frac{\omega_p^2}{\omega_H^2 - k^2c^2}\right) - \frac{i}{\tau} \,. \tag{11.49}$$

One can see that the cyclotron wave exits under the condition $\omega_h \tau \gg 1$.

▶ **Problem 11.34** Analyze damping of whistlers while taking into account electron–atom collisions within the framework of the tau-approximation (6.3).

We now repeat the deduction of the dispersion relation (11.39) by using equation (11.48) for the electron current density. The parameter τ considers electron–atom collisions, and we assume $\omega_H \tau \gg 1$. Now repeating the deduction of relation (11.39), we obtain

$$\omega = \frac{c^2k^2}{\omega_p^2}\left(\omega_H \cos\vartheta - \frac{i}{\tau}\right) . \tag{11.50}$$

As is seen, these waves can exist even if their frequency ω is less than the rate $1/\tau$ of electron–atom collisions.

▶ **Problem 11.35** Analyze damping of drift waves in a weakly ionized gas.

Drift waves have a simple nature. Usually a weakly ionized gas is supported by an external electric field that creates an electric current. If some perturbation occurs in the number density of electrons, it propagates together with the current, and hence the wave phase velocity is equal to the drift velocity w of electrons, i. e., the dispersion relation for drift waves is

$$\omega = kw \,.$$

One can derive this dispersion relation directly from the continuity equation (6.15) for electrons that has the form

$$\frac{\partial N_e}{\partial t} + \operatorname{div} \boldsymbol{j}_e = 0 .$$

Substituting the electron current density $\boldsymbol{j}_e = N_e \boldsymbol{w}$ in this equation, we obtain the above dispersion relation.

Damping of a drift wave may be resulted from expansion of an initial perturbation for electrons due to their diffusion in a gas. Including the diffusion process in the electron current, so that $j_e = \boldsymbol{w} N_e - D\nabla N_e$, where D is the diffusion coefficient for electrons in a gas, we get the dispersion relation for drift waves in the form

$$\omega = kw - iDk^2 . \tag{11.51}$$

The diffusion of electrons leads to damping of the drift wave, and the attenuation factor is equal in this case

$$\delta = Dk^2 ,$$

i. e., long-range drift waves are realized.

11.6 Dynamics of Electron Beams in Plasma

▶ **Problem 11.36** Electrons are ejected from a metallic surface as a result of thermoemission in a vacuum gap and are moving toward an electrode under the voltage U_0 between plane metallic electrodes. Assuming the distance between electrodes to be small as compared to electrode sizes, determine the dependence of the electron flux j on the voltage U_0 and distance L between electrodes.

We find the relation between parameters of this problem on the basis of general relations. The electron flux j is constant in the gap because electrons do not recombine in the gap, i. e.,

$$j = N_e(x)v_e(x) = N_e(x)\sqrt{\frac{2e\varphi}{m_e}} = \text{const} ,$$

where x is the distance from the cathode, N_e is the electron number density, $v_e = \sqrt{2e\varphi(x)/m_e}$ is the electron velocity, and the electric potential is zero at the cathode surface, i. e., $\varphi(0) = 0$. In addition, electrons in a gap creates an electric field that influences on electron parameters. The electric field strength $E = -d\varphi/dx$ follows from the Poisson equation

$$\frac{dE}{dx} = -4\pi e N_e(x) = -4\pi j e\sqrt{\frac{m_e}{2e\varphi}} .$$

It is easy to transform this equation by multiplying it by $E = -d\varphi/dx$. Solving the equation obtained, we obtain

$$E^2 = E_0^2 + 8\pi j\sqrt{2m_e e\varphi} \tag{11.52}$$

with $E_o = E(0)$.

The boundary condition on the cathode is taken from the condition that the electron current density near the cathode is lower than that can be emitted by the cathode. We consider the regime when the current density of the beam is small compared to the electron current density of thermoemission. This means that most of the emitted electrons return to the metallic surface under the action of a spatial charge. This leads to the boundary condition of solution (11.52) $E(0) = 0$, i. e., it coincides with that in the absence of an electric field near the cathode. As a result, we obtain the following equation for the voltage distribution $\varphi(x)$ inside the gap:

$$\left(\frac{d\varphi}{dx}\right)^2 = 8\pi j\sqrt{2m_e e\varphi}\,.$$

Solving this equation with the boundary condition $\varphi(0) = 0$, we obtain

$$\varphi^{3/4}(x) = \frac{3x}{4}(8\pi j\sqrt{2m_e e\varphi})^{1/2}\left(9\pi j\sqrt{\frac{m_e}{2e}}\right)^{2/3} x^{4/3}\,.$$

From this we find the relation between the electron current density $i = ej$, the voltage U_0 between electrodes, and a distance L between them

$$i = je = \frac{\sqrt{2e}}{9\pi\sqrt{m_e}}\frac{U_0^{3/2}}{L^2}\,. \tag{11.53}$$

This dependence, the three-halves power law, describes the behavior of a uniform plasma of electrons in a space between plates if the field in a space between plates is created by an electric charge of the beam plasma.

► **Problem 11.37** An electron beam of a flux j propagates between two infinite plates with a voltage U_0 and distance L between them. Determine the distribution of the electric field potential in the region between plates depending on the electric current.

We use equation (11.52) for the electric potential between plates on the basis of reduced variables $\Phi = \varphi/U_o$, $\xi = x/L$. In these variables this equation has the form

$$\left(\frac{d\Phi}{d\xi}\right)^2 = 4A(\sqrt{\Phi} + \alpha), \qquad A = \frac{2\pi j L^2}{U_0^{3/2}}\sqrt{2em_e}\,.$$

Here α is the integration constant, and this equation is added by the boundary conditions $\Phi(0) = 0$, $\Phi(1) = 1$. Using a new variable $F = \sqrt{\Phi + \alpha}$, one can solve this equation, and this solution in old variables with the boundary condition $\Phi(0) = 0$ has the form

$$\frac{3\sqrt{A}}{2}\xi = \sqrt{\sqrt{\Phi}+\alpha}\left(\sqrt{\Phi}-2\alpha\right) + 2\alpha^{3/2}, \tag{11.54}$$

and the boundary condition $\Phi(1) = 1$ gives the relation between the integration constant α and the parameter A as

$$\frac{3\sqrt{A}}{2} = \sqrt{1+\alpha}(1-2\alpha) + 2\alpha^{3/2}. \tag{11.55}$$

Let us introduce the function

$$G(\alpha) = \sqrt{1+\alpha}(1-2\alpha) + 2\alpha^{3/2}.$$

This function is given in Fig. 11.7 for positive values of the integration constant α. The function $G(\alpha)$ decreases monotonically when the argument α increases. The maximum values of this function corresponds to $\alpha = 0$ and $G(0) = 1$. The asymptotic expression at large α is $G(\alpha) = 0.75/\sqrt{\alpha}$. Correspondingly, the maximum values of the parameter A is $A = 4/9$, which is described by formula (11.54) for the maximum electron flux.

Figure 11.8 gives the dependence of the reduced electric potential $\Phi(\xi)$ on the reduced distance ξ from the first plate for some values of the parameter α in accordance with equation (11.54). In the case $\alpha = 0$ this dependence has the form

$$\Phi = \xi^{3/4},$$

in the other limiting case $\alpha \to \infty$ equation (11.54) takes the form

$$\Phi = \xi.$$

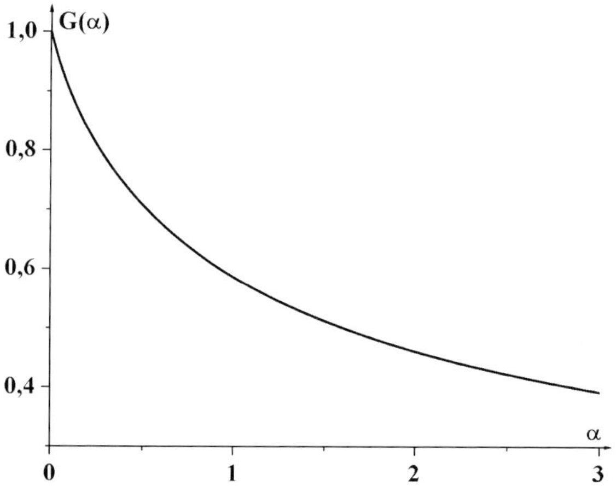

Fig. 11.7 The function (11.55) $G(\alpha) = \sqrt{1+\alpha}(1-2\alpha) + 2\alpha^{3/2}$.

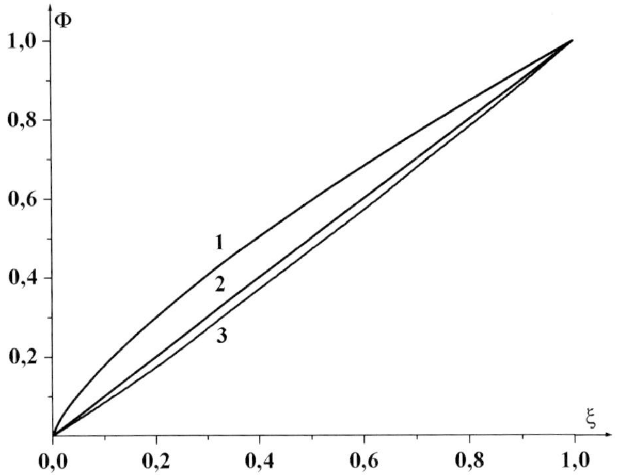

Fig. 11.8 The reduced electric potential $\Phi(\xi)$ in a gap with an electron beam as a function of a reduced distance ξ from a lower plate. (1) $\alpha = 0$, (2) $\alpha = 1$, (3) $\alpha = \infty$.

This limit corresponds to small electric current, so that the electron number density in the gap does not influence on the electric field potential between the plates. Figure 11.8 also shows the relation between Φ and ξ in the intermediate case $\alpha = 1$, when equation (11.54) has the form

$$\frac{3\sqrt{A}}{2}\xi = \sqrt{\sqrt{\Phi}+1}(\sqrt{\Phi}-2)+2\,.$$

▶ **Problem 11.38** A weak monochromatic electron beam is inserted into an atomic gas, and atoms are ionized and excited by electron impact. Decay of excited atoms result from their radiation and associative ionization with the formation of molecular ions. Express the ratio of currents of atomic and molecular ions through rates of collision processes.

The following processes proceed under these conditions:

$$e + A \longrightarrow 2e + A^+, \quad e + A \longrightarrow e + A^*, \quad A^* + A \longrightarrow A_2^+ + e, \quad A^* \longrightarrow A + \hbar\omega\,,$$

where A, A^* are atoms in the ground and excited states, A^+, A_2^+ are an atomic and molecular ion, and $\hbar\omega$ is a photon. These processes lead to the following set of balance equations:

$$\frac{dN_1}{dt} = N_e N_a k_i, \qquad \frac{dN_*}{dt} = N_e N_a k_{ex} - N^*\left(\frac{1}{\tau} + N_a k_{as}\right), \quad \frac{dN_2}{dt} = N_* N_a k_{as}\,,$$

where N_a, N_e are the atom and electron number densities, N_1, N_2 are the number densities of atomic and molecular ions, k_i, k_{ex} are the rate constants of atom ionization and excitation by electron impact, k_{as} is the rate constant of associative ionization, and τ is the radiative lifetime of an excited atom (we take such an atom excitation that gives the main contribution to formation of molecular ions). Solving the above balance equations under the assumption $N_a k_{ex} t_o \ll 1$, where t_o is the pulse duration, we obtain $t = t_o$

$$N_1(t_o) = N_a k_i \int_0^{t_o} N_e dt \ , \ N_*(t_o) = N_a k_{ex} \int_0^{t_o} N_e dt \ , \ N_2 = 0 \ .$$

Solving the balance equation at $t > t_o$ under the assumption $N_e = 0$, we obtain the ratio of the number densities of atomic and molecular ions

$$\frac{N_2}{N_1} = \frac{k_{ex}}{k_i} \frac{1 - \exp(-t/\tau - N_a k_{ex} t)}{1 + (N_a k_{ex} \tau)^{-1}} \ ,$$

where t is a time after switching of the electron beam. From this formula one can find the parameters of the above processes k_{ex}/k_i, k_{as}, τ may be found on the basis of variation of the parameters N_a and t.

▶ **Problem 11.39** A cylinder beam of electrons of a radius r propagates along an axis of the cylinder with metal walls of a radius R, $R \geq r$ and crosses in the end through a ground grid. Find the maximum electron current.

The electric field strength due to an electron charge is equal at a distance R from the cylinder axis according to the Gauss theorem

$$E = \frac{4\pi n_e}{2\pi R} = \frac{2I}{Rev_e} \ ,$$

where n_e is the number of electrons per unit beam length, I is the current of electrons, and v_e is the electron velocity in the beam. This gives on the basis of the equation $E = -d\varphi/dx$ for the voltage difference between the beam and walls

$$\varphi = \frac{2I}{ev_e} \ln \frac{R}{r} \ .$$

After passage of a ground grid (or a surface with the electric potential of walls) the electrons losses an energy $e\varphi$. If the electron energy after an entrance in the tube is eV_o (V_o is the electric potential in which electrons are accelerated), we have the electron energy $eV_o - e\varphi$ at an exit, and the velocity of electrons is

$$v_e = \sqrt{\frac{2e}{m_e}(V_o - \varphi)} \ .$$

This gives the electron current

$$I = \frac{ev_e\varphi}{2\ln(R/r)} = \frac{\sqrt{(2e/m_e)(V_o - \varphi)}\varphi}{2\ln(R/r)} \ . \tag{11.56}$$

From this we obtain the maximum current, the limiting Bursian current,

$$I_{max} = \frac{e}{3}\left(\frac{2e}{3m_e}\right)^{3/2} \frac{V_o^{3/2}}{\ln(R/r)} \tag{11.57}$$

that corresponds to the beam voltage with respect to walls

$$\varphi_{\max} = 2V_0/3, \qquad (11.58)$$

which is the voltage though which the electron beam passes.

▶ **Problem 11.40** Analyze propagation of an intense electron beam inside a cylinder metal tube.

When an electron beam propagates inside a metallic tube in a vacuum, an increase of the electron current leads to an increase of the electric potential of the beam with respect to walls. When this potential reaches the value $\varphi = 2V_0/3$, an instability occurs that increases the locking voltage up to V_0 by jump. As a result, a part of electrons is reflected and return back. In this manner the total electron current does not exceed its limiting value $I_{\max}$.

Note that in reality electrons are moving with different velocities in the beam, and the locking voltage affects in the first place on slow electrons and restricts in this way the current of passing electrons.

▶ **Problem 11.41** Show that the limiting Bursian current results from an instability, so that a random variation of the beam electric potential would be amplified.

Based on formula (11.56) that relates the beam current and its potential, we assume a fluctuation $\delta\varphi$ of the beam potential with respect to walls. We analyze the result of this fluctuation. First, it causes a decrease in the electron velocity

$$\delta v_e = -\sqrt{\frac{e}{2m_e(V_0 - \varphi)}}\delta\varphi = -\frac{v_e \delta\varphi}{2(V_0 - \varphi)} .$$

Since the current density is conserved at this fluctuation, the number density of beam electrons increases

$$\frac{\delta n_e}{n_e} = -\frac{\delta v_e}{v_e} = \frac{\delta\varphi}{2(V_0 - \varphi)} .$$

An increase of the number density of beam electrons causes an increase of the potential difference $\delta\varphi'$ between the beam and walls

$$\delta\varphi' = \varphi\frac{\delta n_e}{n_e} = \frac{\varphi}{2(V_0 - \varphi)}\delta\varphi .$$

An instability takes place, if $\delta\varphi' > \delta\varphi$, i. e., an initial potential fluctuation leads to its subsequent increase. This takes place if

$$\frac{\varphi}{2(V_0 - \varphi)} > 1 ,$$

which is the threshold of this instability determined by formula (11.58). As a result of this instability, a virtual cathode arises and due to its electric potential V_0 a part of electrons is reflected. In this manner, the electric current of passing electrons is restricted.

11.7
Beam-Plasma Instabilities

▶ **Problem 11.42** Analyze the beam-plasma instability, when an electron beam propagates in a plasma, and the beam velocity exceeds significantly a typical electron velocity in a plasma.

We below consider interaction of an electron beam with a plasma. Then the total velocity distribution function for plasma and beam electrons has the form of Fig. 11.9, and on the beam front criterion (11.47) holds true. As a result, beam electrons slow down, and the wave of this phase velocity is amplified. We derive below the dispersion relation for the wave of the total wave-plasma system, and this will give the rate of wave amplification.

Let us repeat the derivation of the dispersion relation (11.8) related to electrons of a plasma and beam. As early, on the basis of the continuity equation (6.15) and the Euler equation (6.19) for plasma electrons we derive equations for the wave amplitudes which are the first two equations of the set (11.7). Taking $p' = 0$ for a cold plasma in this set of equations and eliminating the electron velocity w' in the wave, we obtain the following relation of plasma and wave parameters:

$$N_e' = -i\frac{kE'}{m_e\omega^2}N_0 \,.$$

The same relation follows for beam parameters, if we take the electron number density in the beam as $N_b + N_b' \exp[i(kx - \omega t)]$ and the velocity of the electrons in the beam as $u + w_b \exp[i(kx - \omega t)]$, where the x-axis is parallel to the velocity of

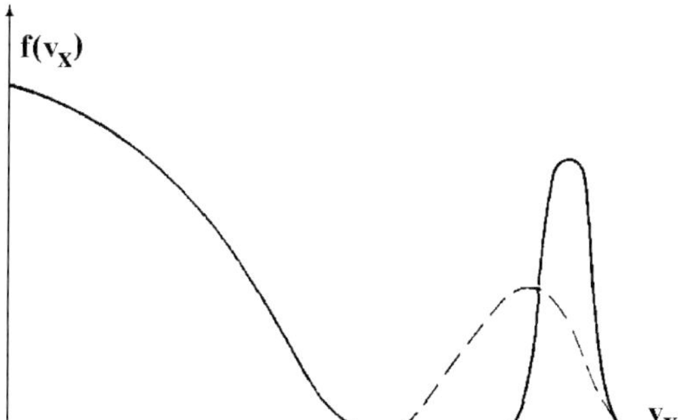

Fig. 11.9 The velocity distribution function for plasma and beam electrons when electron beam is injected in a plasma. The solid curve corresponds to the initial electron distribution in a plasma and beam. Ranges with a positive derivation are unstable, and development of the instability transforms the distribution function into that indicated by a dotted line.

the beam, and N_b and u are, respectively, the electron number density and velocity in the unperturbed beam. From this it follows

$$N'_b = \frac{-ikE'N_b}{m_e(\omega - ku)^2} .$$

Poisson's equation gives the following relation between parameters of this system:

$$ikE' = -4\pi e(N'_e + N'_b) .$$

Eliminating the wave parameters from these three equations, we finally finally the dispersion relation for the wave in this system in the form

$$\frac{\omega_p^2}{\omega^2} + \frac{\omega_p^2}{(\omega - ku)^2}\frac{N_b}{N_0} = 1 . \qquad (11.59)$$

If the number density of the beam electrons is zero ($N_b = 0$), equation (11.59) reduces to the dispersion relation (11.6) for the plasma of zero temperature.

Analyzing the beam–plasma interaction on the basis of the dispersion relation (11.59), we observe that the strongest interaction occurs when the phase velocity of the plasma waves ω/k is equal to the velocity of the electron beam u. In this case, taking the number density of beam electrons N_b to be small compared to the number density N_0 of the plasma electrons, we find the frequency of the plasma oscillations to be close to the plasma frequency ω_p. Hence, we consider below waves with the wave number $k = \omega/u$, which have the most effective interaction with the electron beam. We represent the frequency of these oscillations as $\omega = \omega_p + \epsilon$ and insert it into the dispersion relation (11.59). Expanding the result in a series in terms of the small parameter δ/ω_p, we obtain

$$\epsilon = \omega_p \left(\frac{N_b}{2N_0}\right)^{1/3} \exp\left(\frac{2\pi i n}{3}\right) ,$$

where n is an integer. One can see that $\epsilon/\omega_p \sim (N_b/N_0)^{1/3} \ll 1$, and this justifies an expansion over a small parameter ϵ/ω_p.

When the imaginary component of the frequency ω, that is equal to the imaginary component of ϵ, is negative, the wave is damped; in the opposite case it is amplified. The maximum value of the amplification factor corresponds to $n = 1$ and gives for the amplification factor δ (the wave amplitude $\sim \exp(i\omega t - \delta t)$)

$$-\delta = \frac{\sqrt{3}}{2}\left(\frac{N_b}{N_0}\right)^{1/3} \omega_p = 0.69\omega_p \left(\frac{N_b}{N_0}\right)^{1/3} . \qquad (11.60)$$

The amplitude N'_b varies with time as $\exp(\gamma t)$; this result is valid if the plasma oscillations do not affect the properties of the plasma. As a result of the beam-plasma instability, the distribution function of the beam electrons expands (see Fig. 11.9) in the velocity space, and the energy surplus is transferred to plasma oscillations.

▶ **Problem 11.43** Determine the threshold of the beam-plasma instability.

Let us analyze the dispersion relation (11.59) that is given in Fig. 11.10 in the form $f(\omega) = 1$, where

$$f(\omega) = \frac{\omega_p^2}{\omega^2} + \frac{\omega_p^2}{(\omega - ku)^2} \cdot \frac{N_b}{N_o} .$$

We assume the number density of beam electrons N_b to be small compared to the average number density N_o of plasma electrons ($N_b \ll N_o$), and the wave vector k to be given by a geometry of the plasma system. Let us consider the limiting cases with respect to ku/ω_p in the dispersion relation (11.59).

If $ku \gg \omega_p$, the dispersion relation is divided into two branches, in the vicinity $\omega = 0$ and $\omega = ku$. The corresponding expressions for the frequency have the form

$$\omega = \left(ku \pm \omega_p \sqrt{\frac{N_b}{N_o}} \right) \left[1 - \frac{\omega_p^2}{2(ku)^2} \right] , \qquad \omega = \pm\omega_p \left[1 - \frac{\omega_p^2}{2(ku)^2} \frac{N_b}{N_o} \right] .$$

All these four solutions are real and damping is absent. These solutions correspond to the curve 1 in Fig. 11.10.

In another limiting case $ku \ll \omega_p$ the oscillation frequency is close to ku that allows us to ignore unity in the dispersion relation (11.60) in comparison to ω_p^2/ω^2. This gives

$$\omega = ku \left(1 \pm \frac{N_b}{N_o} \right) .$$

These solutions are imaginary and are not represented in Fig. 11.10. Two other solutions $\omega = \pm\omega_p$ are given in Fig. 11.10 by intersections of a line 2 with a curve $f(\omega)$.

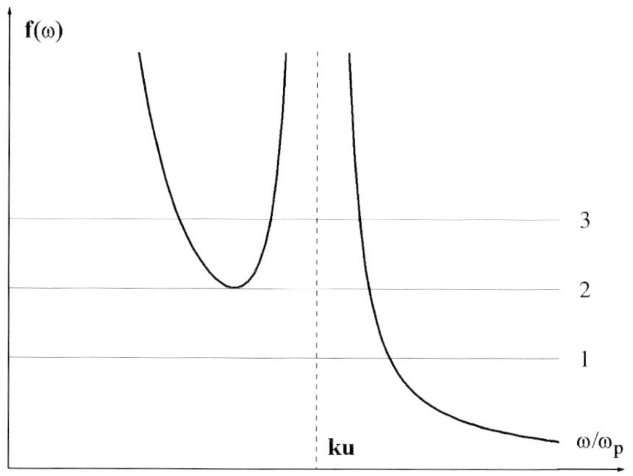

Fig. 11.10 The frequency dependence for the function $f(\omega)$ if the equation (11.59) is represented as $f(\omega) = 1$.

We now consider the complex solutions of equation (11.59), so that one of these correspond to damping of oscillations, and other one relates to their amplification. The threshold of an instability corresponds to transition from one limiting case to the other one and is described by line 3 in Fig. 11.10. We have at the threshold

$$f(\omega) = 1\,,\; f'(\omega) = 0\,,$$

that gives at the threshold

$$\omega = \omega_p \sqrt{1 + \left(\frac{N_b}{N_o}\right)^{1/3}} = \frac{ku}{1 + (N_b/N_o)^{1/3}}\,.$$

Hence, the instability can occur if the parameters of this system satisfy the relation

$$\frac{ku}{\omega_p} < \left[1 + \left(\frac{N_b}{N_o}\right)^{1/3}\right]^{3/2} \approx 1 + \frac{3}{2}\left(\frac{N_b}{N_o}\right)^{1/3}\,.$$

Under these conditions oscillations occur with the phase velocity ω/k that coincides with the beam velocity u. These oscillations can be amplified, and then the energy of the electron beam is converted in the energy of plasma oscillations.

▶ **Problem 11.44** Analyze the Langmuir paradox, when an electron beam is injected from a hot metal surface and penetrates in a plasma that borders on the metal surface. For a rare plasma a path of breaking of this beam is less than the mean free path of beam electrons colliding with plasma particles.

When electrons with a certain drift velocity propagates in a rare plasma, they slow down as a result of collision with plasma particles, and the Langmuir paradox is such that the mean free path of electrons λ with respect to collisions with plasma particles is larger than the observed value. Note that in this case the velocity of beam electrons v_b is higher than a thermal velocity of plasma electrons, and the number density N_b of beam electrons is considerably lower than the number density N_o of plasma electrons. Assume the deceleration of the electron beam to occur owing to the scattering of beam electrons on electrons and ions of the plasma. Then the mean free path of beam electrons with respect to their transformation in plasma electrons is $\lambda \sim N_o^{-1}$.

There is, however, another mechanism for deceleration of the electron beam due to the beam instability, and the amplification factor for this process is determined by formula (11.60) and gives the following dependence on the beam and plasma parameters $|\delta| \sim N_o^{1/6} N_b^{1/3}$. Since the mean free path of beam electrons with respect to this process is $\lambda \sim v_b/|\delta|$, we have a more weak dependence on the number density of plasma particles than that due to collisions processes. Therefore, for a rare plasma the mechanism of beam deceleration owing to the beam-plasma instability may be dominant, and deceleration proceeds on less distances than that due to collision processes. This explains the Langmuir paradox. Finally, transformation of the beam kinetic energy into the plasma energy proceeds through collective degrees of freedom of the beam-plasma system instead of collision processes.

▶ **Problem 11.45** Estimate the mean free path of electrons in an ideal plasma due to fluctuations of internal plasma fields.

We convince that along with collision processes, collective plasma degrees of freedom may be responsible for energy transformation in the plasma. We now consider interaction of an individual particle with a plasma through fluctuations of particle densities and fields in the plasma. Indeed, location of a charged particle in an ideal plasma causes displacement of surrounding particles. In turn this provides shielding of the particle field by plasma particles. Along with this, surrounding charged particles create random fields near a test charged particle because of large fluctuations with respect to the mean particle energy. We now determine the mean free path of a charged particle (electron or ion) in an ideal plasma as a result of scattering in a random field.

Taking into account that the electric potential of a plasma field varies by $\sim \Delta U$ on a distance of $\sim r_D$, where the average potential energy $\overline{U}$ of a test charged particle in a plasma and its fluctuation are given by formulas (1.59) and (1.60). The particle energy varies by $\sim \Delta U$ at a distance of $\sim r_D$, and since this change has an arbitrary sign, a typical energy of a charged particle $\sim T$ results from $\sim (T/\Delta U)^2$ events of scattering. From this we estimate the mean free path λ of a charged particle, that is a distance on which the particle energy varies by $\sim T$

$$\lambda \sim r_D \left(\frac{T}{\Delta U}\right)^2 = \frac{1}{2\pi N_0 (e^2/T)^2} . \qquad (11.61)$$

As is seen, the mean free path of a charged particle in an ideal plasma due to its scattering on plasma nonuniformities is inversely proportional to the mean number density of charged particles and is proportional to the temperature square.

▶ **Problem 11.46** Electrons are moving in a plasma with the drift velocity u with respect to ions. Determine the threshold of an instability that leads to growth of plasma oscillations. Assume the drift velocity of electrons to be large compared to thermal velocities of ions.

This problem is analogous to that for interaction of an electron beam with a plasma, but because of electron–ion interaction, it is necessary to change parameters of plasma electrons by ion parameters, and this dispersion relation takes the form

$$\frac{m_e}{M}\frac{\omega_p^2}{\omega^2} + \frac{\omega_p^2}{(\omega - ku)^2} = 1 .$$

Here M is the ion mass, and the plasma is quasineutral, i. e., the number densities of the electrons and ions are equal. Below we determine the maximum amplification factor of the plasma oscillations.

Taking the ratio m_e/M to be zero, we obtain the dispersion relation $\omega = \omega_p + ku$. Therefore, we represent the oscillation frequency as

$$\omega = \omega_p + ku + \delta .$$

Substituting this expression into the above dispersion relation and expanding the result in a power series in terms of the small parameter δ/ω_p, we obtain

$$\frac{2\delta}{\omega_\mathrm{p}} = \frac{m_\mathrm{e}}{M} \frac{\omega_\mathrm{p}^2}{(\omega_\mathrm{p} + ku + \delta)^2} \,.$$

From this it follows that the strongest interaction between the electron beam and the wave takes place, if the oscillation wave number is $k = -\omega_\mathrm{p}/u$. This gives

$$\delta = \left(\frac{m_\mathrm{e}}{M}\right)^{1/3} \omega_\mathrm{p} \exp\left(\frac{2\pi i n}{3}\right) \,,$$

where n is an integer. The highest amplification factor corresponds to $n = 1$ and is given by

$$-\delta = \mathrm{Im}\,\omega = \frac{\sqrt{3}}{2} \left(\frac{m_\mathrm{e}}{2M}\right)^{1/3} \omega_\mathrm{p} = 0.69\omega_\mathrm{p} \left(\frac{m_\mathrm{e}}{M}\right)^{1/3} \,. \tag{11.62}$$

As is seen, the frequency of oscillations and the amplification factor have the same order of magnitude. This type of instability of the electron beam due to interaction with plasma ions, the Buneman instability, can be realized in gas discharge plasma where electrons are drifting under the action of external electric field.

12
Relaxation Processes and Processes with Strong Interaction in Plasma

12.1
Relaxation Processes in Plasma

▶ **Problem 12.1** A plasma beam propagates in a gas, and the electron temperature T_e decreases in this process such that dT_e/dt is constant. Analyze the character of evolution of the electron number density $N_e(t)$.

In the first stage of the relaxation process, equilibrium in this system is supported by the processes

$$e + A \longleftrightarrow 2e + A^+ , \tag{12.1}$$

and the number density of electrons is $N_S(T_e)$, the equilibrium number density of electrons under a current temperature T_e that is given by the Saha formula (1.69). Taking into account the relation between the rate constants of processes (12.1), one can represent the balance equation for the number density of electrons in the form

$$\frac{dN_e}{dt} = K_{ei}\left(N_S^2(T_e) - N_e^2\right) N_e ,$$

where K_{ei} is the coefficient of three-body electron–ion recombination.

As it follows from this equation, at slow cooling or high electron temperatures T_e we have $N_e = N_S(T_e)$, while at low electron temperatures the equilibrium (12.1) is violated. Then, we get $N_e \gg N_S$ and $N_e = (2K_{ei}t)^{-1/2}$ at large t if we ignore the temperature dependence for K_{ei}. In a general case we obtain this limit

$$N_e = \left(\int_{t_o}^{t} 2K_{ei}dt\right)^{-1/2} . \tag{12.2}$$

One can give the asymptotic solution of the above equation if we take $N_S(T) = N_s \exp(-\alpha t)$, where $N_s = N_S(T_o)$, $T(t=0) = T_o$, and

$$\alpha = \frac{J}{T_e^2}\left|\frac{dT_e}{dt}\right| ,$$

Plasma Processes and Plasma Kinetics. Boris M. Smirnov

ISBN: 978-3-527-40681-4

and dT_e/dt is negative. This gives the balance equation in the form

$$\frac{dN_e}{dt} = K_{ei}\left(N_S^2(T_e) - N_e^2\right) N_e ,$$

and its solution at large t is

$$\frac{1}{N_e^2} = \frac{2K_{ei}}{\alpha}\left[\alpha t - \ln\frac{2K_{ei}N_s^2}{\alpha} - C\right] ,$$

where $C = 0.577$ is the Euler constant. Comparing this solution with formula (12.2), we find $N_s = N_S(t_o)$

$$K_{ei}N_s^2 = \frac{\alpha}{2}e^{-C} = 0.28\alpha = \frac{0.28J}{T_e^2}\left|\frac{dT_e}{dt}\right| . \qquad (12.3)$$

This relation gives the parameter t_o in formula (12.2).

▶ **Problem 12.2** A flux of an equilibrium plasma passes through an orifice and moves in a rare buffer gas such that the flux forms an angle β with respect to the perpendicular to the orifice of a radius r_o (Fig. 12.1). This adiabatic expansion of the flux leads to decrease of the plasma temperature. Taking into account the temperature dependence for the three-body rate constant of electrons and ions (4.41), find the electron number density at large times, when the equilibrium (12.1) is violated.

Assuming the flux drift velocity u to be independent of the distance from the orifice, we find a current flux radius $R = r_o + ut\tan\beta$, where t is a time of drifting to this cross section, and the number density of atoms in this cross section is

$$N = \frac{N_o r_o^2}{(r_o + ut\tan\beta)^2} .$$

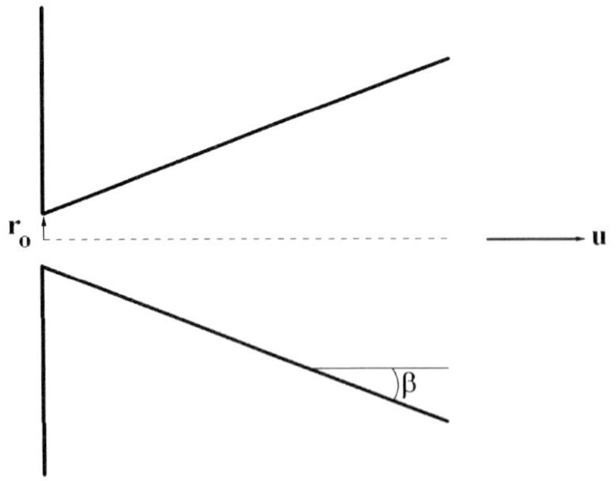

Fig. 12.1 Geometry of a plasma flow through an orifice.

Here N_0 is the number density near the orifice, and we take into account that the total flux of atoms is conserved in each cross section. Because of the adiabatic character of expanding $N \sim T^{3/2}$, the gas temperature varies as

$$T = \frac{T(0)r_0^3}{(r_0 + ut\tan\beta)^3} ,$$

and $T(0)$ is the temperature near the orifice. From this it follows

$$\frac{dT}{dt} = -\frac{3T(0)r_0^3 u\tan\beta}{(r_0 + ut\tan\beta)^4} = -\frac{3u\tan\beta}{r_0}\frac{T^{4/3}}{T^{1/3}(0)} .$$

Let us account for the temperature dependence (4.41) of the three-body recombination rate constant $K_{ei}(T_e) \sim T_e^{-9/2}$, and below we take the electron T_e and gas T temperatures to be identical. Formula (12.3) gives the temperature T_0 at time t_0 when the transition from the equilibrium to nonequilibrium regime takes place for evolution of the electron number density

$$K_{ei}(T_0)N_S^2 = \frac{0.84J}{T_0^{2/3}T^{1/3}(0)}\frac{u\tan\beta}{r_0} .$$

From this, on the basis of formula (12.3) for the electron number density at a given temperature T when the ionization equilibrium is violated, we obtain

$$N_e = \left(\int_{T_0}^{T}\frac{2K_{ei}(T')dT'}{\left|\frac{dT'}{dt}\right|}\right)^{-1/2} = \left[\frac{4}{29}\frac{r_0 K_{ei}(T_0)}{u\tan\beta}\left(\frac{T_0^{9/2}T^{1/3}(0)}{T^{29/6}} - \frac{T^{1/3}(0)}{T_0^{1/3}}\right)\right]^{-1/2} . \qquad (12.4)$$

This relation establishes a relation between the electron number density and a current temperature.

► **Problem 12.3** Relaxation of a plasma results from three-body recombination of electrons and ions (4.40) and a released energy is consumed on electron heating. Prove that under these conditions, a typical time τ_T of variation of the electron temperature is less than that τ_N for variation of the electron number density.

Let us analyze the balance equations for the number density of electrons N_e and temperature. Assuming recombination proceeds according to the three-body process (4.40), we obtain the number density of electrons in a quasineutral plasma

$$\frac{dN_e}{dt} = -K_{ei}N_e^3 ,$$

where the rate constant of three-body process according to formula (4.41) is given by

$$K_{ei} \sim \frac{e^{10}}{m_e^{1/2}T_e^{9/2}} .$$

Since $T_e \ll J$, where J is the ionization atom potential, we obtain from the balance equation for the average energy of the electron component

$$\frac{d}{dt}\left(\frac{3}{2}N_e T_e\right) = \frac{3}{2}N_e\frac{dT_e}{dt} = -J\frac{dN_e}{dt} .$$

Thus, in the course of relaxation of this plasma, recombination of electrons and ions leads to the formation of atoms and heating of electrons. From the above balance equation we have $d\ln T_e/dt \sim (J/T_e)d\ln N_e \gg d\ln N_e/dt$, i. e.,

$$\tau_T \ll \tau_N . \tag{12.5}$$

A typical time of the electron temperature variation follows from the balance equation

$$\left|\frac{dT_e}{dt}\right| \sim \frac{T_e}{\tau_T} ,$$

and we obtain the following estimation for this typical time:

$$\tau_T \sim \frac{m_e^{1/2} T_e^{5.5}}{J e^{10} N_e^2} . \tag{12.6}$$

▶ **Problem 12.4** Show that the lifetime of a two-component plasma with strong coupling is less than a typical atomic time.

The ideality parameter (1.56) of this plasma is $\beta = N_e e^6/T_e^3$, so that for an ideal plasma $\beta \ll 1$, while for a plasma with strong coupling $\beta \gg 1$. Let us compare a typical time of variation of the electron temperature τ_T that is given by formula (12.6) and a typical atomic time τ_0 that is a time for electron displacement on a distance of the order of an atom size

$$\tau_0 \sim \frac{r}{v} \sim \frac{e^2}{J}\sqrt{\frac{m_e}{T_e}} .$$

Here a typical atom size is $r \sim e^2/J$, and a typical electron velocity is $v \sim (T_e/m_e)^{1/2}$. Formula (12.6) gives

$$\tau_T \sim \frac{\tau_0}{\beta^2} ,$$

where β is the ideality parameter. From this formula, it follows that a typical time of relaxation of a plasma with a strong coupling is less than a typical atomic time. Of course, this deduction is based on formulas which are valid for an ideal plasma. Nevertheless, this deduction gives a general tendency in relaxation of a two-component plasma with strong coupling, namely such a plasma decays fast in comparison with atomic times. Of course, additional interactions in a plasma can reject this conclusion, for example, if along with electrons and ions, a neutral component affects the plasma behavior, this conclusion may be violated.

► **Problem 12.5** Consider hardening of an air plasma in which ozone is formed as a result of fast cooling of a dissociated oxygen in air.

Ozone is a metastable oxygen compound, and its formation in molecular oxygen may result from nonequilibrium conditions. We now consider a rapid cooling of partially dissociated air when ozone formation and its decay proceeds in the following processes:

$$O + 2O_2 \leftrightarrow O_3 + O_2\,, \quad O + O_2 + N_2 \leftrightarrow O_3 + N_2\,, \quad O + O_3 \leftrightarrow 2O_2\,. \tag{12.7}$$

Assume the degree of oxygen dissociation in the initial state to be small. This allows us to neglect those recombination processes that require the participation of two oxygen atoms. The cooling rate dT/dt is an important parameter in these reactions.

The hardening phenomenon is associated with equilibrium violation at rapid cooling. At high temperatures the equilibrium among O, O_2, and O_3 is supported by the processes shown in formula (12.7), but starting from a typical temperature T_0, the equilibrium between atomic and molecular oxygen is violated. We introduce the equilibrium constants from the Saha relations (1.69)

$$\frac{[O]^2}{[O_2]} = K_2(T) = C_2(T)\exp\left(-\frac{D_2}{T}\right), \quad \frac{[O][O_2]}{[O_3]} = K_3(T) = C_3(T)\exp\left(-\frac{D_3}{T}\right),$$

where $[X]$ means the number density of particles X, $K_2(T)$, and $K_3(T)$ are the equilibrium constants, so that $C_2(T)$ and $C_3(T)$ are weak functions of T, $D_2 =$ 5.12 eV is the dissociation energy of oxygen molecules O_2, and $D_3 = 1.05$ eV is the dissociation energy of ozone molecules O_3. Taking into account the principle of detailed balance relating the rate constants of these processes, we obtain the balance equation for ozone molecules

$$\frac{d[O_3]}{dt} = -k[O_3][O] + K_a[O_2]^2[O]\,, \tag{12.8}$$

where K_a, k are the rate constants for the first and third processes (12.7), respectively.

We assume that in the course of cooling the equilibrium between atomic oxygen and ozone is maintained up to temperatures of the order of T_0. Then this equation is

$$\frac{d[O_3]}{dt} = -\frac{k[O_3]^2 K_3}{[O_2]} + K_a K_3[O_2][O_3]\,.$$

At temperatures below T_0, one can neglect the second term on the right-hand side of this equation. Then the ozone number density $[O_3]$ at the end of the process is given by the relation

$$\frac{1}{[O_3]} - \frac{1}{N_0} = \frac{k(T_0)K_2(T_0)}{[O_2]}\frac{T_0^2}{D_3 dT/dt},$$

where N_o is the number density of ozone molecules at the temperature T_o, and we assume a weak temperature dependence for $k(T)$.

From this we estimate a typical ozone number density at the end of the process as

$$[O_3] \sim \frac{[O]_{eq}[O_2]_{eq}}{K_2(T_o)} \sim \frac{[O_2]\, D_3}{k(T_o)K_3(T_o)T_o^2}\frac{dT}{dt} ,$$

where $[X]_{eq}$ means the equilibrium number density at this temperature. Note that the parameters are taken at a temperature T_o at which an equilibrium violates. This temperature follows from the equation

$$\frac{dT}{dt} \sim C\exp\left[-\frac{D_2}{2T_o}\right] ,$$

where C depends weakly on the temperature. This gives

$$k[O]_{eq} \sim \frac{D_3}{T_o^2}\frac{dT}{dt} ,$$

where $[O]_{eq} \sim \sqrt{K_2[O_2]}$ is the equilibrium number density of atomic oxygen. From this we obtain the final ozone number density

$$[O_3] \sim \left(\frac{dT}{dt}\right)^{1-2D_3/D_2} \sim \left(\frac{dT}{dt}\right)^{0.6} . \tag{12.9}$$

The rate constants allow us to find parameters of this process at a given rate of cooling. In particular, T_o ranges 1700–2000 K for a cooling rate 10^4–10^5 K/s, and the ozone number density at the end of the process is $[O_3] \sim 10^{10}\ cm^{-3}$.

12.2
Thermal Phenomena and Thermal Waves in Plasma

▶ **Problem 12.6** A weakly ionized gas is formed under the action of an external electric field, and the electron number density satisfies to criterion (7.18) that leads to the Maxwell distribution function of electrons. Find the attenuation factor for damping of a heightened density of electrons originated as a perturbation at a certain point.

Under considered conditions, local ionization equilibrium is supported within the plasma, so that the Saha relation for the electron number density N_e is valid and heat transport processes are not essential. Because the electron number density and temperature T_e are connected, a perturbation of one value causes variation of the other one, which in turn leads to a change of the first quantity. In this way one can find the rate of damping or growth of an initial perturbation.

First, we find the relation between perturbations of the electron number density N_e' and temperature T_e'. Because of the local ionization equilibrium, we have according to the Saha formula (1.69) $N_e \sim \exp(-J/2T_e)$, where J is the atom ionization potential. This gives

$$\frac{N_e'}{N_e} = \frac{T_e'}{T_e}\frac{J}{2T_e}\,. \tag{12.10}$$

From this it follows

$$\frac{T_e'}{T_e} \ll \frac{N_e'}{N_e}\,.$$

We now use the balance equation for the electron energy per unit volume $W = 3N_eT_e/2$, which according to equations (7.31), (7.32), and (7.33) have the form

$$\frac{dW_e}{dt} = [-eEw_eN_e]' - 3\frac{m_e}{M}[(T_e - T)\nu N_e]'. \tag{12.11}$$

We take here for simplicity the rate ν of electron–atom collisions to be independent of the collision velocity, and below we assume $T_e \gg T$.

When the plasma is under equilibrium, the left-hand side of this equation (12.11) as well as the right-hand are zero. Note that $j = -eEw_e = \text{const}$. Inserting in equation (12.11) perturbations of the electron number density N_e' and electron temperature T_e', which vary in time, we reduce this equation to the form

$$\frac{1}{N_e}\frac{dN_e'}{dt} + \frac{1}{T_e}\frac{dT_e'}{dt} = -\frac{T_e'}{T_e}\,.$$

Expressing a perturbation of the electron temperature through a perturbation of the electron number density and using formula (7.33), we reduce this balance equation to the form

$$\frac{dN_e'}{dt} = -\delta N_e', \quad \delta = 2\frac{m_e}{M}\frac{T_e}{J}\nu. \tag{12.12}$$

Note that a typical time of establishment of local thermodynamic equilibrium assumes to be small compared to $1/\delta$.

► **Problem 12.7** A gas is located in a gap between two infinite plates with a distance L between them. The wall temperature is supported to be T_w, and the rate of heat release depends on the temperature exponentially. Find the temperature difference between the gap middle and its walls and analyze the possibility of development of a thermal instability under these conditions.

Let us take z-axis to be perpendicular to the walls and the middle of the gap to be $z = 0$, so that the coordinates of walls are $z = \pm L/2$. Denote by $f(T)$ the specific power of heat release per unit volume and take its temperature dependence in accordance with the Arrhenius law

$$f(T) = A\exp(-E_a/T)\,, \tag{12.13}$$

where E_a is the activation energy of the heat release process. This dependence of the rate of heat release is identical to that of the chemical process and represents a strong temperature dependence because $E_a \gg T$. Assume that heat transport results through thermal conductivity of a gas in the gap, so that the heat balance equation (6.29) has the form

$$\kappa \frac{d^2 T}{dz^2} + f(T) = 0 ,$$

where κ is the thermal conductivity coefficient. Let us introduce a new variable $X = E_a(T - T_0)/T_0^2$, where T_0 is the gas temperature in the center of the gap, and then the heat balance equation takes the form

$$\frac{d^2 X}{dz^2} - Be^{-X} = 0 ,$$

where $B = E_a A \exp(-E_a/T_0)/(T_0^2 \kappa)$. Solving this equation with the boundary conditions $X(0) = 0$, $dX(0)/dz = 0$ (the second condition follows from the symmetry consideration $X(z) = X(-z)$), we have

$$X = 2 \ln \cosh z .$$

This gives the temperature difference between the gap center and the walls as

$$\Delta T \equiv T_0 - T_w = \frac{2T_0^2}{E_a} \ln \cosh \left[\frac{L}{2} \sqrt{\frac{AE_a}{2T_o^2 \kappa}} \exp\left(-\frac{E_a}{2T_0} \right) \right] . \tag{12.14}$$

Figure 12.2 shows the dependence on T_0 for the left-hand and right-hand parts of this equation (curves 1 and 2, respectively) at a given wall temperature T_w. The intersection of these curves yields the center gap temperature T_0. The right-hand part of the equation does not depend on the wall temperature and depends strongly on T_0. Therefore, it is possible that curves 1 and 2 do not intersect, i. e., a stationary solution of the problem is absent. The physical implication of this result is that thermal conduction cannot ensure removal of heat release from the gas, and this leads to a thermal instability.

▶ **Problem 12.8** Find the threshold of the thermal instability in a gas located in a gap between two plane plates which are supported at a certain temperature. The specific power of heat release is determined by formula (12.13).

The threshold of the thermal instability corresponding to the curve 1′ of Fig. 12.2, along with the equality of the left-hand side and right-hand side of equation (12.14) and their derivatives are also equal. The latter gives

$$\Delta T = \frac{2T_0^2}{E_a} \ln \cosh y, \qquad 1 = y \tanh y ,$$

where

$$y = \frac{L}{2} \sqrt{\frac{AE_a}{2T_o^2 \kappa}} \exp\left(-\frac{E_a}{2T_0} \right) .$$

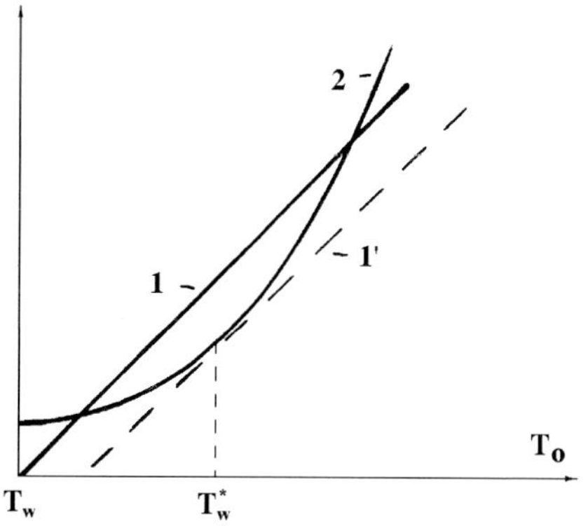

Fig. 12.2 The dependencies on the temperature at the gap middle for the left-hand side and right-hand side of equation (12.14). The right intersection of these curves gives the solution of this equation. The threshold of the thermal instability takes place when the line and curve are tangential with each other.

The solution of the above equation for y gives $y = 1.2$, which means the following relation at the threshold the thermal instability is:

$$\frac{AE_a L^2}{T_o^2 \kappa} \exp\left(-\frac{E_a}{T_o}\right) = 11.5, \qquad \Delta T = 1.19 \frac{T_o^2}{E_a}. \tag{12.15}$$

These relations show that the thermal instability starts at the following ratio of the specific power of heat release in the middle gap and at walls:

$$\frac{f(T_w)}{f(T_o)} = \exp(-1.19) = 0.30 .$$

Correspondingly, the threshold of the thermal instability in terms of the wall parameters for heat release has the form

$$\frac{AE_a L^2}{T_w^2 \kappa} \exp\left(-\frac{E_a}{T_w}\right) = 3.5 .$$

Though we considered the Arrhenius temperature dependence for the rate of heat release, in reality it is valid for arbitrary strong temperature dependence. Hence, one can rewrite the condition for the threshold of thermal instability in the form

$$\frac{L^2}{\kappa} \left| \frac{df(T_w)}{dT_w} \right| = 3.5, \tag{12.16}$$

and this criterion is based on the condition

$$T_w \left| \frac{df(T_w)}{dT_w} \right| \gg 1 .$$

Relation (12.16) gives a certain connection between rates of heat release and heat transport.

► **Problem 12.9** When the condition of thermal instability is fulfilled for a large gas system, this instability propagates inside this system in the form of a wave. Determine the velocity of thermal wave.

If a size of a gas system exceeds significantly by the parameter L in formula (12.16), thermal instability develops in the form of a thermal wave. Thus, the thermal wave propagates in a medium, which is a nonequilibrium at the beginning, and a remarkable energy part is located in internal degrees of freedom. For example, in a chemically active gas the rate of the chemical process that leads to heat release depends strongly on the temperature. This chemical process proceeds on the wave front with a heightened temperature, and the region of the chemical process propagates in the form of a wave.

Figure 12.3 shows the space temperature distribution in a gas in the course of propagation of a thermal wave. Region 1 relates to the initial gas state, the temperature rise proceeds in region 2 due to heat transport from hotter regions. Processes of heat release occur in region 3, where the gas temperature is close to its maximum. Region 4 of Fig. 12.3 is located after the passage of the thermal wave, and its temperature T_m is determined by the specific internal energy at the beginning.

On the basis of formula (12.13), in a range of a thermal process, we take the temperature dependence for the specific power of heat release in the form

$$f(T) = f(T_m) \exp\left[-\alpha(T_m - T)\right], \qquad \alpha = \frac{E_a}{T_m^2}, \tag{12.17}$$

where T_m is the final gas temperature that is determined by the internal gas energy. Taking the time and coordinate dependence of the temperature in thermal wave as $T = T(x - ut)$, where x is the coordinate, and u is the velocity of the thermal wave, we transform the heat balance equation (6.31) to the form

$$u\frac{dT}{dx} + \chi\frac{d^2T}{dx^2} + \frac{f(T)}{c_p N} = 0 .$$

We find below the wave velocity u as the eigenvalue of this equation by sewing its solution in different regions of Fig. 12.6. For this goal it is convenient to use a new

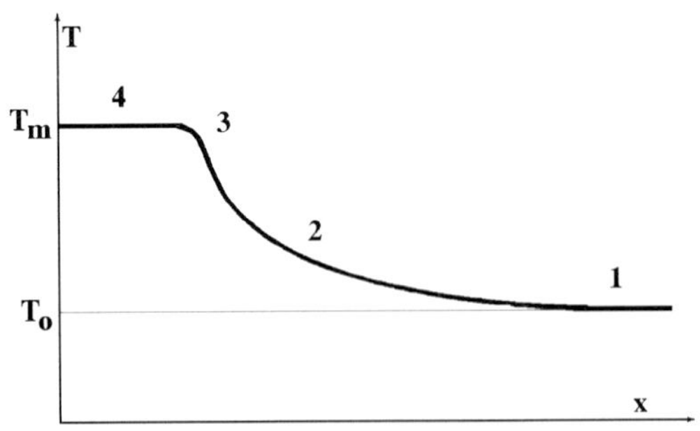

Fig. 12.3 Space distribution of the temperature in the course of propagation of a thermal wave.

variable $Z(T) = -dT/dx$, and in the temperature range $T_o < T < T_m$ we have $Z(T) \geq 0$. This gives

$$\frac{d^2T}{dx^2} = \frac{d}{dx}\left(\frac{dT}{dx}\right) = \frac{ZdZ}{dT}\,.$$

In new variables, the heat balance equation reduces to the form

$$-uZ + \chi\frac{ZdZ}{dT} + \frac{f(T)}{c_p N} = 0\,. \qquad (12.18)$$

We use a simple and rough method of solution of this equation that is illustrated in Fig. 12.4. In regions 1, 4 according to definition of Fig. 12.3 we have $Z(T_o) = Z(T_m) = 0$. In region 2 heat release is absent practically that allows us to ignore the last term of equation (12.18) and yields region 2

$$Z = \frac{u(T - T_o)}{\chi}\,, \qquad (12.19)$$

where T_o is the initial temperature. Neglecting the first term of the above equation in region 3, we obtain

$$Z = \sqrt{\frac{2}{c_p N\chi}\int_T^{T_m} f(T)dT}. \qquad (12.20)$$

One can sew the above solutions in regions 2 and 3 by equalizing them at a temperature T_* on the boundary region where these solutions are identical. As a result, we obtain the Zeldovich formula for the thermal wave velocity

$$u = \frac{1}{T_* - T_o}\sqrt{\frac{2\chi}{c_p N}\int_{T_*}^{T_m} f(T)dT}. \qquad (12.21)$$

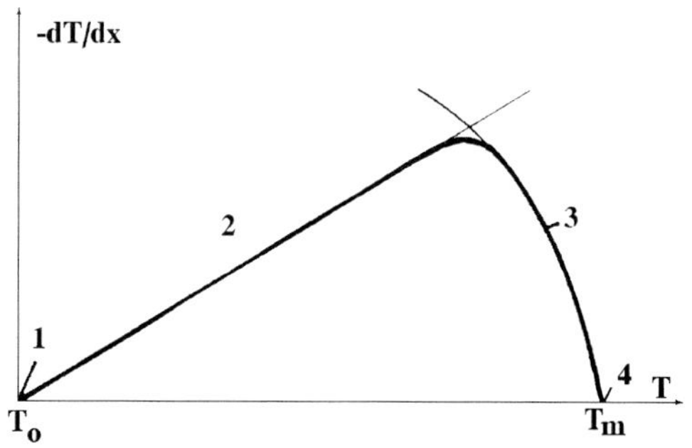

Fig. 12.4 The temperature dependence for the temperature gradient $Z(T)$ according to equation (12.18).

One can expect that though the temperature T_* is close to T_m ($T_* - T_o \gg T_m - T_*$), but $f(T_*) \ll f(T_m)$. Then the Zeldovich formula can be rewritten as

$$u = \frac{1}{T_m - T_o}\sqrt{\frac{2\chi}{c_p N}\int_{T_o}^{T_m} f(T)dT} . \quad (12.22)$$

▶ **Problem 12.10** Analyze the correctness of the Zeldovich formula (12.22) by deriving the solution of equation (12.18).

Approximating the rate of heat release by formula (12.17), it is convenient to obtain the solution of equation (12.18) as

$$Z(T) = Z = \frac{u(T - T_o)}{\chi}\sqrt{1 - \exp[\alpha(T - T_m)]} .$$

This expression gives the correct solutions in regions 1 and 4 of Fig. 12.3 $Z(T_o) = Z(T_m) = 0$, and this formula gives solution (12.19) in the region 2. If we substitute this expression into equation (12.18), one can find the thermal wave velocity. Taking the criterion $\alpha(T_m - T_o) \gg 1$ to be valid, one can find the velocity of a thermal wave, comparing the above expression and formula (12.20) in the region 3 of Fig. 12.3. As a result, we obtain the Zeldovich formula (12.22).

▶ **Problem 12.11** Analyze propagation of the wave of vibration relaxation in a nonequilibrium molecular gas in the case when the diffusion coefficient for excited molecules in a gas and the thermal diffusivity coefficient of the gas are identical.

We have at the beginning a molecular gas whose vibrational temperature exceeds remarkably the translation one. Then the relaxation process leads to quenching of excited molecules and gas heating. Since the rate constant of relaxation increases significantly with growth of the temperature, i. e., the relaxation process accelerates in the course of gas heating. Our goal is to determine the velocity of the wave of vibration relaxation. Along with thermal process as considered earlier we take in consideration the process of diffusion of excited molecules.

We consider the balance equation for the number density N_* of excited molecules

$$\frac{\partial N_*}{\partial t} = D\Delta N_* - NN_*k(T) ,$$

where N is the total number density of molecules, and $N \gg N_*$, D is the diffusion coefficient of excited molecules in a gas, and $k(T)$ is the rate constant of vibrational relaxation. Taking into account the usual dependence of wave parameters $N_*(x,t) = N_*(x - ut)$, where u is the wave velocity, we transform the above equation to the form

$$u\frac{dN_*}{dx} + D\frac{d^2N_*}{dx^2} - N_*Nk(T) = 0 . \quad (12.23)$$

In the wave front we have $N_* = N_{max}$, and after the wave we have $N_* = 0$. Introducing the mean molecule energy $\Delta\varepsilon$ released in a single vibrational relaxation event, Then the difference of the gas temperatures after (T_m) and before (T_0) the relaxation wave is

$$T_m - T_0 = \frac{N_o \Delta\varepsilon}{N c_p}, \tag{12.24}$$

where N_o is the initial number density of excited molecules.

The heat balance equation (12.18) may be represented in the form

$$u\frac{dT}{dx} + \chi\frac{d^2T}{dx^2} - \frac{\Delta\varepsilon N_* k(T)}{c_p} = 0\,. \tag{12.25}$$

The wave velocity can be obtained from the simultaneous analysis of equations (12.24) and (12.25). The simplest case occurs when $D = \chi$. Then both balance equations are identical, and the relation between the gas temperature and the number density of excited molecules is

$$T_m - T = N_* \Delta\varepsilon/(N c_p). \tag{12.26}$$

In this case we have the analogy between the heat release specific power and the rate constant of relaxation $f(T)/(c_p N) = (T_m - T)Nk(T)$. Then on the basis of formula (12.22) we obtain the relaxation wave velocity

$$u = \frac{T_m^2}{E_a(T_m - T_0)}\sqrt{\frac{2\chi}{\tau(T_m)}}, \tag{12.27}$$

where $\tau(T_m) = 1/[Nk(T_m)]$ is a typical time for vibrational relaxation at the temperature T_m. Because of the exponential temperature dependence ($\alpha(T_m - T_0) \gg 1$) for the vibrational relaxation rate constant assumption, we have

$$u \ll \sqrt{\frac{\chi}{\tau(T_m)}}\,.$$

▶ **Problem 12.12** Analyze the propagation of the wave of vibration relaxation in a nonequilibrium molecular gas in the limiting cases of the relation between the diffusion coefficient of excited molecules in a gas and the thermal diffusivity coefficient.

Let us first consider the case $D \gg \chi$ that is represented in Fig 12.5a and which gives the spatial distribution of the number density of excited molecules N_* and gas temperature T along the wave. Note that the centers of these two distributions coincide because quenching-excited molecules input heat into a gas, and the distribution of the number density of excited molecules is wider than that for the temperature. According to the balance equation (12.23) for the number density of excited molecules $N_* = N_{max} - (N_{max} - N_0)\exp(-ux/D)$, for $x > 0$, where N_0 is the number density of excited molecules at $x = 0$. Since the temperatures varies

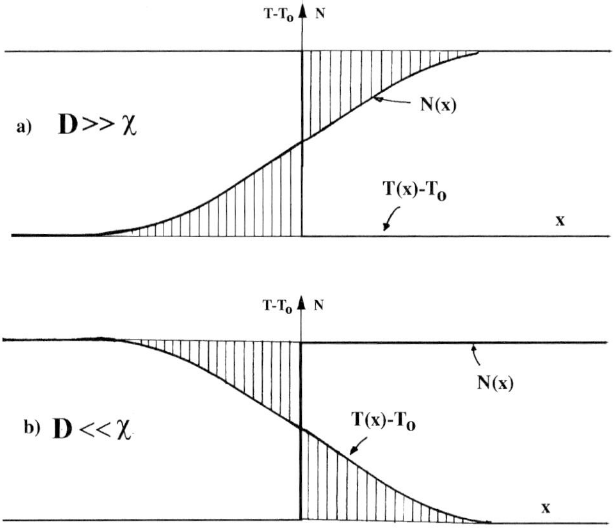

Fig. 12.5 Space distribution of the temperature and the number density of excited molecules in the course of propagation of a vibrational relaxation wave in cases of different limiting relations between the diffusion coefficient D of excited molecules in a gas and the thermal diffusivity coefficient χ of the gas.

by step, vibrational relaxation is absent at $x > 0$ and takes place at $x < 0$, and equation (12.23) gives at $x < 0$

$$N_* = N_0 \exp(\alpha x), \qquad \alpha = \sqrt{\left(\frac{u}{2D}\right)^2 + \frac{1}{D\tau}} - \frac{u}{2D},$$

where $\tau = 1/[Nk(T_m)]$. Equalizing derivatives of these expressions for N_* at $x = 0$, we get in this case when propagation of the thermal wave of vibrational relaxation is governed by diffusion of excited molecules in a hot region where vibrational relaxation takes place

$$\alpha = \frac{u}{D}, \qquad u = \sqrt{\frac{2D}{\tau(T)}} = \sqrt{2DNk(T_m)}, \quad D \gg \chi . \tag{12.28}$$

In accordance with character of relaxation process, the velocity of the vibrational relaxation wave is $u \sim \sqrt{D/\tau}$ in this case, and the width of the wave front is $\Delta x \sim \sqrt{D\tau}$.

Figure 12.5b shows the spatial distribution for the number density of excited molecules and temperature in the opposite limiting case $\chi \gg D$, when diffusion of excited molecules is not essential, and the velocity of the thermal wave of vibrational relaxation is determined by formula (12.22). Connecting the specific power of heat release in this formula with the rate of vibrational relaxation according to (12.26) and accounting the relation (12.24) between the final temperature and the initial number density of excited molecules, we obtain this approximately

$$u = \frac{T_m}{\sqrt{E_a(T_m - T_0)}}\sqrt{\frac{\chi}{\tau}}, \tag{12.29}$$

where $\tau = [Nk(T_r)]^{-1}$ and roughly $T_r \approx T(x = 0)$. Note that in this case the wave velocity is lower than that in the case $D = \chi$ since vibrational relaxation proceeds at lower temperatures.

▶ **Problem 12.13** Analyze the propagation of the wave of ozone decomposition in air.

This example is of interest because processes of diffusion transport and heat transport are comparable in wave propagation, and the rates of chemical processes are known that allows us to analyze a real phenomenon. Chemical processes of ozone decomposition in air proceed according to the following scheme:

$$\begin{aligned} &O_3 + M \xrightarrow{k_{dis}} O_2 + O + M \\ &O + O_2 + M \xrightarrow{K} O_3 + M \\ &O + O_3 \xrightarrow{k_1} 2O_2, \end{aligned} \tag{12.30}$$

and definitions of their rate constants are given above; M is an air molecule. The air temperature T_m after the thermal wave is connected with its initial temperature T_0 by the relation

$$T_m = T_0 + 48c, \tag{12.31}$$

where the temperatures are expressed in Kelvin, and c is the ozone concentration in air expressed as a percentage.

On the basis of scheme (12.30) we obtain the set of balance equations

$$\begin{aligned} \frac{d[O_3]}{dt} &= -k_{dis}[M][O_3] + K[O][O_2][M] - k_1[O][O_3] \\ \frac{d[O]}{dt} &= k_{dis}[M][O_3] - K[O][O_2][M] - k_1[O][O_3], \end{aligned} \tag{12.32}$$

where $[X]$ is the number density of particles X. Estimates show that the three-body process is weak compared to the pair one under considering conditions ($p \leq 1$ atm, $T_m > 500$ K) we have $K[O_2][M] \ll k_1[O_3]$ and the second term of the right-hand side of each equation in (12.32) is less than the third one. In addition, in reality $[O] \ll [O_3]$, which gives $d[O]/dt \ll d[O_3]/dt$, and it allows us to take below $d[O]/dt = 0$ and $[O] = k_{dis}[M]/k_1$. As a result, the first equation of (12.32) is transformed into

$$d[O_3]/dt = -2k_{dis}[M][O_3].$$

Joining this with equations (12.23) and (12.25), we obtain the number density of ozone molecules $[O_3]$ and air temperature T in the wave of ozone decomposition

$$u\frac{d[O_3]}{dx} + D\frac{d^2[O_3]}{dx^2} - 2k_{\text{dis}}[M][O_3] = 0$$

$$u\frac{dT}{dx} + \chi\frac{d^2T}{dx^2} + \frac{2}{c_p}\Delta\varepsilon k_{\text{dis}}[O_3] = 0, \tag{12.33}$$

where $\Delta\varepsilon = 1.5$ eV is the energy released from the decomposition of one ozone molecule.

We can now substitute numerical parameters for the above processes for a thermal wave in air at atmospheric pressure, namely $D = 0.16\ \text{cm}^2/\text{s}$ and $\chi = 0.22\ \text{cm}^2/\text{s}$. These quantities are almost equal and have identical temperature dependence. Therefore we take them to be equal and given by

$$D = \chi = \frac{0.19}{p}\left(\frac{T}{300}\right)^{1.78},$$

where the values D and χ are expressed in cm^2/s, the air pressure p is given in atmospheres, and the temperature is expressed in Kelvin. We also use below the dissociation rate constant $k_{\text{dis}} = 1.0 \cdot 10^{-9}\ \text{cm}^3/\text{s}\exp(-11\,600/T)$. We can observe

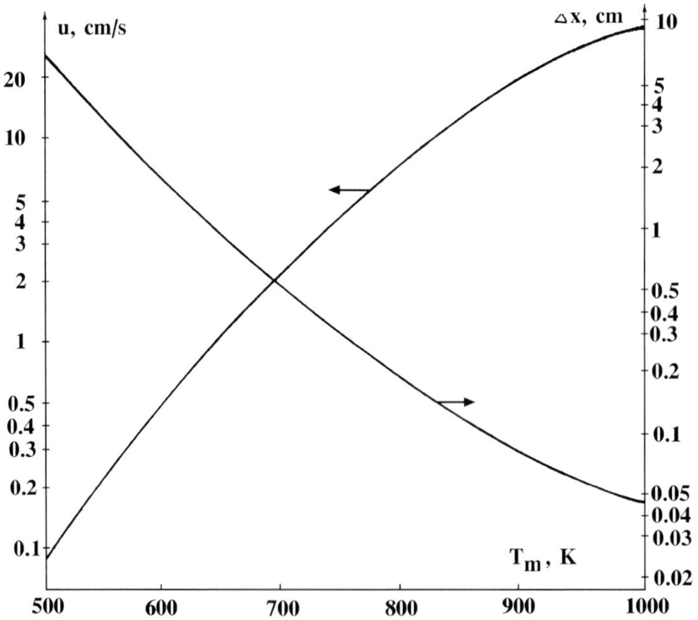

Fig. 12.6 The velocity of propagation of the wave of ozone decomposition in air depending on the initial ozone concentration (or the final air temperature), and the width Δx of this wave front.

how equating of the values D and χ simplifies the problem. Then formula (12.27) gives the thermal wave velocity expressed in cm/s

$$u = \frac{1.3T_{\mathrm{m}}^{2.39}}{T_{\mathrm{m}} - T_0} \exp\left(-\frac{5800}{T_{\mathrm{m}}}\right) .$$

Figure 12.6 shows the wave velocity obtained on the basis of this formula for $T_0 =$ 300 K, and u does not depend on the air pressure.

The width of the wave front is characterized by the value $\Delta x = (T_{\mathrm{m}} - T_0)/(dT/dx)_{\max}$, where the maximum temperature gradient is $(dT/dx)_{\max} = u(T_* - T_0)/\chi$ and the temperature T_* corresponds to the maximum of $Z(T)$ that is given in Fig. 12.4. Figure 12.6 contains Δx depending on the temperature. As it follows from the data of Fig. 12.6, the thermal wave velocity is small compared with the sound velocity, i. e., propagation of a thermal wave is a quiet process.

12.3
Plasma in Magnetic Field

▶ **Problem 12.14** Derive the set of magnetohydrodynamic equations for a plasma of a high density located in a magnetic field.

In a dense plasma located in a magnetic field the plasma motion is connected with the magnetic field and creates a self-consistent field. Hence, these parameters must be considered simultaneously and are described by the set of magnetohydrodynamic equations for a plasma. This set of equations includes the continuity equation for the number density of electrons and ions (9.5), the equation for the average momentum of electrons and ions (9.8), Poisson's equation, and the Maxwell equations. The resulting set of equations is called the set of equations of magnetohydrodynamics and has the form

$$\begin{aligned} \frac{\partial N}{\partial t} + \mathrm{div}\,(N\boldsymbol{w}) &= 0 \\ \frac{\partial \boldsymbol{w}}{\partial t} + (\boldsymbol{w}\cdot\nabla)\,\boldsymbol{w} + \frac{\nabla p}{mN} - \frac{\boldsymbol{F}}{N} &= 0 \\ \mathrm{div}\,\boldsymbol{E} &= 4\pi(N_{\mathrm{i}} - N_{\mathrm{e}}) \\ \mathrm{curl}\,\boldsymbol{H} &= \frac{4\pi}{c}\boldsymbol{j} - \frac{1}{c}\frac{\partial \boldsymbol{E}}{\partial t} \\ \mathrm{curl}\,\boldsymbol{E} &= -\frac{1}{c}\frac{\partial \boldsymbol{H}}{\partial t} \\ \mathrm{div}\,\boldsymbol{H} &= 0\,. \end{aligned} \tag{12.34}$$

The first two equations can be written both for electrons and ions. Here $\boldsymbol{w}$ and N are the drift velocity and the number density of electrons or ions in a quasineutral plasma. It is necessary to add to the equation set (12.35) the equation of plasma state and the thermodynamic equation for the processes in the plasma. The above equations relate the number density of plasma particles, their temperature and the pressure, and Ohm's law (7.60). The set of magnetohydrodynamic equations with the indicated additions and the initial conditions will give a complete description of plasma evolution.

▶ **Problem 12.15** Based on the set of magnetohydrodynamic equations for a high-density plasma (12.35), show that magnetic lines of force are frozen in the plasma.

A plasma of a high conductivity contains electrons whose velocities considerably greater than the velocities of ions. Then the electric current is determined by electrons and is given by

$$j = -eN_e w_e ,$$

where w_e is the drift velocity of the electrons, and N_e is their number density. If the motion occurs in a magnetic field, an additional electric field is produced in the laboratory frame of axes, whose strength is

$$E' = \frac{1}{c}(w_e \times H) = -\frac{1}{ecN_e}(j \times H) .$$

This field acts on the electrons, giving rise to an additional force acting on the entire plasma. The force per unit volume of the plasma is

$$eE'N_e = -\frac{1}{c}(j \times H) .$$

If the plasma conductivity is sufficiently high, its response to the electric field (according to this equation) will result in the movement of electrons. This movement will continue until separation of the electrons and ions gives rise to an internal electric field in the plasma

$$E = -\frac{1}{c}(w_e \times H) , \tag{12.35}$$

which compensates the above electric field. Inserting this into the Maxwell equation $-(\partial H/\partial t) = c \text{ curl } E$ that yields

$$\frac{\partial H}{\partial t} = \text{curl}\,[w_e \times H] . \tag{12.36}$$

Let us analyze the variation of the magnetic field in a moving plasma of a high conductivity when the electric current is created by electrons mostly. We transform equation (12.36) by using the relation

$$\text{curl}\,(w_e \times H) = w_e \text{div}\ H + (H \cdot \nabla)w_e - (w_e \cdot \nabla)H - H\text{div}\ w_e .$$

Using the Maxwell equation div $H = 0$, and taking the expression for div w_e from the continuity equation for electrons, we obtain

$$\frac{\partial H}{\partial t} - \frac{H}{N_e}\frac{\partial N_e}{dt} + (w_e \cdot \nabla)H - \frac{H}{N_e}(w_e \cdot \nabla)N_e = (H \cdot \nabla)w_e .$$

Dividing this equation by N_e, we obtain

$$\frac{\partial}{\partial t}\left(\frac{H}{N_e}\right) = \left(\frac{H}{N_e}\cdot\nabla\right) w_e\,, \tag{12.37}$$

where

$$\frac{\partial}{\partial t}\left(\frac{H}{N_e}\right) = \frac{\partial}{\partial t}\left(\frac{H}{N_e}\right) + (w_e\cdot\nabla)\frac{H}{N_e}$$

is the derivative at a point that moves with the plasma.

To analyze the motion of an element of plasma volume with the length dl and the cross section ds containing $N_e ds dl$ electrons, we assume at first that the vector dl is parallel to the magnetic field H so that the magnetic flux through this elementary plasma volume is Hds. If the plasma velocity at one end of the segment dl is w_e, then at its other end the velocity is $w_e + (dl\cdot\nabla)w_e$, so that the variation of the segment length during a small time interval δt is $\delta t(dl\cdot\nabla)w_e$. Hence, the length of the segment satisfies the equation

$$\frac{d}{dt}(dl) = (dl\cdot\nabla)w_e\,,$$

which is identical to equation (12.37). From this it follows that during plasma evolution the element dl has the same direction as the magnetic field. In addition, the length of the plasma element remains proportional to the quantity H/N_e. This means that the magnetic flux through this plasma element does not vary with time during the plasma motion. Thus, the magnetic lines of force are frozen into the plasma, that is, their direction is such that the plasma electrons travel along these lines. Note that this is valid for a high conductivity plasma.

▶ **Problem 12.16** Analyze the penetration of a slow magnetic field in a motionless plasma.

Since an electric current passes through a plasma, the electric field arises, and its strength (E) follows from Ohm's law (7.60)

$$j = \Sigma E\,,$$

where j is the electric current density and Σ is its conductivity. We insert this relation into the Maxwell equations

$$\text{curl } H = \frac{4\pi}{c} - \frac{1}{c}\frac{\partial E}{\partial t}, \quad \text{curl } E = -\frac{\partial H}{\partial t}, \quad \text{div } H = 0\,.$$

We apply the operation curl to the first equation, the operation $\partial/\partial t$ to the second equation and exclude the quantity E from these equations. We use the relation curl curl a = grad div $a - \nabla^2 a$ and assume a typical frequency ω of variation of fields to be small compared to the plasma conductivity Σ. As a result, we obtain the following equation of the magnetic field strength:

$$\frac{\partial H}{\partial t} = D_H \Delta H, \qquad D_H = \frac{c^2}{4\pi\Sigma}. \tag{12.38}$$

This is the diffusion equation (8.33), so that we have a diffusion character of penetration of a magnetic field in a plasma, and a typical time of penetration in a depth L is

$$\tau \sim \frac{L^2}{D_H} ,$$

and the penetration depth Δ for fields of a typical frequency ω is

$$\Delta \sim \sqrt{\frac{D_H}{\omega}} = \frac{c}{\sqrt{4\pi\Sigma\omega}} .$$

One can see that a decrease of a typical field frequency leads to an increase of the penetration depth for such fields in a plasma. Since the plasma conductivity (7.63) is $\Sigma \sim \omega_p^2\tau/(4\pi)4$, we find that the above formula gives the penetration depth (11.28) in the case of the normal skin effect.

▶ **Problem 12.17** Analyze the steady-state motion of a high-conductivity plasma on the basis of the relation of magnetohydrodynamics.

The property of moving plasma of a high density in a magnetic field according to which magnetic lines of force are frozen in a plasma allows us to analyze some aspects of this plasma with a strong interaction between motion of electrons and self-consistent magnetic field. In particular, according to equation (12.35), the force in this plasma per electron is equal to

$$\boldsymbol{F} = -e\boldsymbol{E} = \frac{e}{c}\left[\boldsymbol{w}_e \times \boldsymbol{H}\right] = -\frac{1}{cN_e}\left[\boldsymbol{j} \times \boldsymbol{H}\right] .$$

Taking into account the Maxwell equation that connects the current density with the magnetic field strength $\boldsymbol{j} = c/(4\pi) \cdot \operatorname{curl} \boldsymbol{H}$, we obtain

$$\boldsymbol{F} = -\frac{1}{cN_e}\left[\boldsymbol{j} \times \boldsymbol{H}\right] = \frac{1}{4\pi N_e}\left[\boldsymbol{H} \times \operatorname{curl} \boldsymbol{H}\right] = \frac{1}{4\pi N_e}\left[\frac{1}{2}\nabla H^2 - (\boldsymbol{H}\cdot\nabla)\boldsymbol{H}\right] .$$

Let us substitute this relation into the second equation (12.37). Assuming the drift velocity of the electrons to be considerably smaller than their thermal velocity that allows us to neglect the term $(\boldsymbol{w}_e \cdot \nabla)\boldsymbol{w}_e$ compared to the term $\nabla p/(mN_e)$. As a result, we obtain

$$\nabla\left(p + \frac{H^2}{8\pi}\right) - \frac{(\boldsymbol{H}\cdot\nabla)\boldsymbol{H}}{4\pi} = 0. \qquad (12.39)$$

The quantity $H^2/(8\pi)$ is the magnetic field pressure or magnetic pressure that according to its action on a plasma is analogous to the kinetic pressure.

▶ **Problem 12.18** The electric current in a plasma with frozen magnetic lines of force in the plasma occupies a cylinder region. Find the relation between parameters of this plasma.

In the case under consideration the magnetic lines of force are cylinders, and because of the axial symmetry, equation (12.39) for the direction perpendicular to the field and current has the following form:

$$\nabla_\perp \left[p + \frac{H^2}{8\pi} \right] = 0 .$$

The solution of this equation shows that the total plasma pressure, which is the sum of the gas-kinetic and magnetic pressures, is independent of the transverse coordinate,

$$p + \frac{H^2}{8\pi} = \text{const.} \tag{12.40}$$

Let the radius of the plasma column be r_0 and the current in it be I, so that the magnetic field at the surface of the column is $H_\varphi = 2I/(ca)$ and is directed perpendicular to the current direction. The total pressure outside the column near its surface is equal to the magnetic field pressure, i. e., $I^2/(2\pi c^2 r_0^2)$. If the magnetic field is absent inside the plasma, the total pressure inside the plasma column is equal to the gas-kinetic pressure p. From the equality of these pressures we obtain the radius of the plasma column,

$$r_0 = \frac{I}{c\sqrt{2\pi p}} . \tag{12.41}$$

This pinch effect establishes the relation between the current radius and the magnetic field that is created by this current.

▶ **Problem 12.19** Analyze the stability of a cylinder current in a column of a cold plasma with respect to pinching, when a linear current of the cylinder shape propagates in a gas in an external magnetic field directed along the current. Pinching involves variation of the plasma column radius, but does not change its axial symmetry.

Under these conditions, an electric current propagates in a cylinder column located in an external electric field, and if this current shrinks in some point, the radius of curvature of magnetic lines of force exceeds significantly the radius of current column. Taking z as the current direction, and let the column of a radius r_0 be changed by δr_0. Since the magnetic lines of force are frozen in the plasma, the magnetic flux across the plasma column $H_z \pi r_0^2$ is conserved. From this it follows that the variation δH_z of the longitudinal magnetic field is connected with the radius variation δr_0 by the relation

$$\frac{\delta H_z}{H_z} + 2\frac{\delta r_0}{r_0} = 0 .$$

Next, the electric current, that is $I_z = c r_0 H_\varphi / 2$, does not change as a result of a radius change. This gives the variation of the tangential component of the magnetic field

$$\frac{\delta H_\varphi}{H_\varphi} + \frac{\delta r_0}{r_0} = 0 \, .$$

Hence,

$$\frac{\delta H_z}{H_z} = 2\frac{\delta H_\varphi}{H_\varphi} \, .$$

In addition, we have that variation of the column radius leads to variation of the magnetic pressure. The variation of the magnetic field pressure inside the plasma due to variation of the magnetic field frozen inside the plasma is

$$\delta p = \delta \frac{H_z^2}{8\pi} \, .$$

In the same manner we find that the variation of a magnetic field outside the plasma column is,

$$\delta p = \delta \frac{H_\varphi^2}{8\pi} \, .$$

Thus, an increase of the magnetic pressure outside the plasma as a result of a decrease of the column radius must be compensated by the magnetic pressure of a cold plasma inside the column. Then the plasma column may be protected from pinching. From the above relations it follows that this is fulfilled if the criterion

$$H_z^2 \geq \frac{H_\varphi^2}{2} \tag{12.42}$$

holds true. Thus, the current in plasma may be stable if an external longitudinal magnetic field is used. In other case, an instability of the sausage type can develop.

▶ **Problem 12.20** A weakly ionized gas is located in crossed constant electric and magnetic fields, and an electron current supports an electron number density and temperature. Analyze the possibility of an ionization instability when a random perturbation at some plasma region is amplified in time.

In this case a plasma is restricted by electrodes, and a current flows through the plasma of the current density $\boldsymbol{j} = -eN_0\boldsymbol{w}_0$, where N_0 and w_0 are the stationary values of the electron number density and drift velocity. In order to ascertain the possibility of growth of a random perturbation in a plasma, we analyze the variation in the heat release per unit volume that is $W = -\boldsymbol{jE}$. A variation of this quantity is

$$W' = eN_e'\boldsymbol{w}_0\boldsymbol{E}_0 - eN_0\boldsymbol{w}'\boldsymbol{E}_0 - eN_0\boldsymbol{w}_0\boldsymbol{E}' \, ,$$

where unperturbed values are denoted by subscript "o," and perturbed values are supplied by superscript ′.

The motion equation for an electron is

$$m_e \frac{d\boldsymbol{w}}{dt} = -e\boldsymbol{E} - \frac{e}{c}[\boldsymbol{w} \times \boldsymbol{H}] - m_e \boldsymbol{w}\nu \; .$$

Solution of this equation in the absence of perturbations is given by formula (7.43), and from this formula it follows the electric field strength

$$\boldsymbol{E} = -\frac{m_e \nu}{e}\boldsymbol{w} - \frac{m_e}{e}[\boldsymbol{\omega}_H \times \boldsymbol{w}] \; ,$$

where $\boldsymbol{\omega}_H = e\boldsymbol{H}/(m_e c)$. Taking from this formula the electric field strength for the stationary $\boldsymbol{E}_o$ and perturbed $\boldsymbol{E}'$ distributions and substituting them in equation for variation of the specific electron energy, we obtain

$$W' = N_e' m_e w_0^2 \nu + 2N_0 m_e \boldsymbol{w}_0 \boldsymbol{w}' \nu \; . \tag{12.43}$$

The relation between perturbations w' and N_e' follows from the continuity equation (6.15) for electrons $\operatorname{div}(N_e \boldsymbol{w}) = 0$. Taking a perturbation value as $\sim \exp(i\boldsymbol{k}\cdot\boldsymbol{r})$, where $\boldsymbol{k}$ is the wave vector, we obtain

$$N_e(\boldsymbol{w}'\cdot\boldsymbol{k}) + N_e'(\boldsymbol{w}_0\cdot\boldsymbol{k}) = 0 \; .$$

Let us now find the direction of a perturbed drift velocity $\boldsymbol{w}'$. Assuming that the perturbation develops slowly, we have $\boldsymbol{E}' = -\nabla\varphi'$, which gives $\boldsymbol{E}' = -i\boldsymbol{k}\varphi'$, where φ' is the perturbation of the electric potential. Thus, the vectors $\boldsymbol{E}'$ and $\boldsymbol{k}$ are either parallel or antiparallel. Next, from formula (7.43) we obtain the direction of a perturbed drift velocity

$$\boldsymbol{w}' = \text{const}\cdot\left[\boldsymbol{k} - (\boldsymbol{k}\times\boldsymbol{\omega}_H)/\nu\right] \; .$$

Multiplying this vector by itself, we find the constant in this expression, that is

$$\boldsymbol{w}' = \pm w' \frac{\nu\boldsymbol{k} - (\boldsymbol{k}\times\boldsymbol{\omega}_H)}{k\sqrt{\omega_H^2 + \nu^2}} \; .$$

Substituting this expression into the relationship derived from the continuity equation yields

$$\frac{w'}{w} = \pm\frac{N_e'}{N_e}\left(\frac{w_0}{\nu}\right)\sqrt{\omega_H^2 + \nu^2}\cos\alpha \; ,$$

where α is the angle between vectors $\boldsymbol{w}$ and $\boldsymbol{k}$ and $\cos\alpha = k_x/k$. This gives

$$W' = W_0\left(1 - 2cos^2\alpha + \frac{\omega_H}{\nu}\sin 2\alpha\right) \; ,$$

where $W_0 = N_e m_e w_0^2 \nu$.

We now analyze equation (12.11) that has the form

$$\frac{dW'}{dt} = W' - W_0\frac{T_e'}{T_e} \; .$$

On the other hand,

$$\frac{dW'}{dt} = \frac{d}{dt}\left(\frac{3N_e T_e}{2}\right) = W_0 \frac{1}{N_e}\frac{dN_e'}{dt} + W_0 \frac{1}{T_e}\frac{dT_e'}{dt} ,$$

and according to formula (12.10), one can neglect the second term in the right-hand side compared to the first because the atom ionization potential is $J \gg T_e$.

On the basis of these relation we have

$$\frac{dW'}{dt} \equiv W_0 \frac{1}{N_e}\frac{dN_e'}{dt} = W_0 \frac{N_e'}{N_e}\left(\frac{\omega_H \sin 2\alpha}{\nu} - \cos^2\alpha - \frac{T_e}{2J}\right) .$$

The instability occurs if the right-hand side of this equation is positive. Then any random perturbation of the electron number density will grow. The right-hand side is maximum when $\tan 2\alpha = -\omega_H/\nu$. For this direction of the vector $\boldsymbol{k}$, the previous equation takes the form

$$\frac{dW'}{dt} = N_e' m_e w^2 \nu \left(\frac{\sqrt{\omega_H^2 + \nu^2}}{\nu} - 1 - \frac{T_e}{2J}\right) . \tag{12.44}$$

From this it follows that this instability has a threshold and starts if $\omega_H/\nu \geq (T_e/J)^{1/2}$. In the case of a large parameter ω_H/ν, the ionization instability develops perturbations propagating at the angle $\alpha = 45^\circ$ to the current direction. If this ratio is small, the most unstable perturbations propagate in a direction almost perpendicular to the current.

▶ **Problem 12.21** An electric current flows in a cylindrical tube, and the electric and magnetic fields are directed along the plasma column axis. A plasma is quasineutral, and the number density of charged particles varies slowly along the tube radius. Taking into account that electrons and ions are magnetized, find the conditions when an ionization instability is developed.

Under the above conditions, the plasma under consideration is quasineutral and its electrons and ions are magnetized, i. e., the Larmor radius for electrons and ions is small compared to their mean free path. Since electrons and ions recombine on the tube walls, the electric current is constant along the tube axis. Hence, if small gradient of the number density of electrons in the current direction, an additional electric field arises along the current in order to conserve its value.

Let us create an oblique perturbation of the electric field that forms a certain angle with nonperturbed electric field strength. Then an azimuthal electric field compels electrons to rotate and may enforce separation of electrons and ions. This can create an ionization instability. We below determine the threshold of this instability. The nature of this instability consists in drift of electrons and ions in different directions under the action of crossed electric and magnetic fields. This creates an electric field that can be enforced. In particular, in the case under consideration the number density of electrons and ions in the radial direction means the existence of an electric field in this direction. This causes the azimuthal drift of electrons and ions in different directions that creates an azimuthal electric field. In turn, this can strengthen the number density gradient of charged particles and lead to an ionization instability.

In order to find the conditions of this instability, we derive below the dispersion relation for such perturbations. The drift velocity of electrons and ions under the action of the magnetic field $\boldsymbol{H}$ and electric field $\boldsymbol{E}$ of the wave according to formula (7.51) is equal to

$$\boldsymbol{w} = c\frac{[\boldsymbol{E} \times \boldsymbol{H}]}{H^2} ,$$

and we neglect a magnetic field because of a small wave amplitude. Therefore, the wave electric field can be described by the potential φ, where $\boldsymbol{E} = -\boldsymbol{\nabla}\varphi$.

Taking the direction along the tube axis to be z, we have Ohm's law (7.60) for the current density $j_z = \Sigma E_z$, where Σ is the plasma conductivity. The electric field is the sum of the external (E_0) and wave fields. Since the wave parameters are proportional to $\exp(i\boldsymbol{kr} - i\omega t)$, we have $E_z = E_0 - ik_z\varphi$. The plasma conductivity is determined by electrons and is proportional to the electron number density, which gives

$$\Sigma = \Sigma_0 + \Sigma' = \Sigma_0\left(1 + \frac{N_e'}{N_0}\right) ,$$

where Σ_0 is the plasma conductivity, N_0 is the electron number density in the absence of perturbations, and Σ' and N_e' are the corresponding perturbed parameters.

The condition for conservation of the current density in the field direction has the form

$$-ik_z\varphi\Sigma_0 + \Sigma' E_0 = 0 ,$$

or

$$-ik_z\varphi + \frac{N_e'}{N_0}E_0 = 0 .$$

We add to this the continuity equation (6.15) for electrons

$$\frac{\partial N_e}{\partial t} + \operatorname{div}(N_e\boldsymbol{w}) - D_a\frac{\partial^2 N_e}{\partial z^2} = 0 ,$$

where D_a is the ambipolar diffusion coefficient for the plasma. We have ignored the diffusive flux of electrons perpendicular to the magnetic field because of its smallness. Taking the harmonic dependence of perturbed parameters on time and coordinates, we can rewrite the above equation in the first order as

$$(-i\omega + k_z^2 D_a)N_e' + w_x\frac{dN_0}{dx} = 0 ,$$

where the x-axis is in the direction of the maximum gradient of the equilibrium number density.

Let us use the above expression (7.51) for the electron drift velocity in the azimuthal direction

$$w_x = \frac{cE_y}{H} = -\frac{ick_y\varphi}{H} = \frac{k_y}{k_z}\frac{cE_o}{H}\frac{N_e'}{N_o}.$$

Substituting this in the above dispersion relation, we obtain it in the form

$$i\omega = k_z^2 D_a + \frac{k_y}{k_z}\frac{cE_o}{HL}, \tag{12.45}$$

where we introduce the characteristic length L as $1/L = -d(\ln N_o)/dx$. One can see that an instability ($\mathrm{Im}\,\omega < 0$) will develop if the ratio k_y/k_z has an appropriate sign and value. The instability has a threshold with respect to the electric field. The magnetic field must be high enough to meet the conditions described above. This ionization instability is named the current-convective instability.

Equation (12.45) gives the optimal condition for this instability when the Larmor frequency for ions is of the order of the collision frequency of ions with gas particles. Because the diffusion motion of charged particles demolishes perturbations, the threshold of this instability is connected with the diffusion of charged particles.

12.4 Nonlinear Phenomena in Plasma

▶ **Problem 12.22** Taking the dependence of the oscillation frequency ω as a function of the oscillation amplitude E as $\omega = \omega_o(k) + \alpha E^2$, show that the wave packet with a narrow range Δk of wave vectors k ($\Delta k \ll k$) can conserve a small width in time if the Lighthill criterion

$$\alpha \frac{\partial v_{gr}}{\partial k} < 0 \tag{12.46}$$

holds true, where v_{gr} is the group velocity of the wave packet.

Nonlinear phenomena in a plasma affect on the character of wave propagation, and we below consider it on the example of a one-dimensional wave packet consisting of waves concentrating a narrow range of wave numbers ($\Delta k \ll k$). The amplitude $a(x,t)$ of the signal can be composed from monochromatic waves in the form

$$a(x,t) = \sum_k a(k)\exp(ikx - i\omega t),$$

where $a(k)$ is the amplitude of the wave with a wave number k. Taking into consideration a nonlinearity wave according to formula (12.46) and the dispersion relation, we represent the wave frequency as

$$\begin{aligned}\omega(k) &= \omega(k_o) + \frac{\partial\omega(k_o)}{\partial k}(k-k_o) + \frac{\partial^2\omega(k_o)}{\partial k^2}(k-k_o)^2\\ &= \omega_o + v_g(k-k_o) + \frac{1}{2}\frac{\partial v_g(k_o)}{\partial k}(k-k_o)^2.\end{aligned}$$

Here k_o is the mean wave number of the wave packet, and v_g is the group velocity of the wave. Since wave number values are restricted to an interval of width Δk near k_o, then the wave packet is initially concentrated in a spatial region of extent $\Delta x \sim 1/\Delta k$. As the wave packet evolves, it diverges due to different group velocities of individual waves. The initial wave packet, which has a size of the order of $1/\Delta k$, diverges on a time scale given by $\tau \sim (\Delta k^2)^{-1} \left(\partial v_g/\partial k\right)^{-1}$. Thus, the wave dispersion usually leads to increasing spatial extension of the wave in time.

The interaction of waves of different k affects on the behavior of this process. Taking the wave frequency in the form $\omega = \omega_0 - \alpha E^2$, where E is a field amplitude, we below show that a nonlinearity can leads to the wave compression, if criterion (12.46) is valid. Using the above formula for the wave frequency, we obtain the amplitude of the wave packet as

$$a(x,t) = \sum_k a(k) \exp\left[i(k-k_o)(x-x_o) - ik_o x_o - i(k-k_o)^2 \frac{\partial v_g}{\partial k} t - i\alpha E^2(x) t\right] ,$$

where $x_o = v_g t$.

From this relation it follows that nonlinear wave interactions can lead to modulation of the wave packet. With certain types of modulation, the wave packet may decay into separate bunches, or it may be compressed into a solitary wave—a soliton. Because of the nature of this process, it is known as a modulation instability.

According to this formula, the compression of a wave packet or its transformation into separate bunches can take place only if the last two terms in the exponent of this relation have opposite signs. Only in this case can a nonlinear interaction compensate for the usual divergence of the wave packet. Therefore, modulation instability can occur if the Lighthill criterion (12.46) is fulfilled.

► **Problem 12.23** Derive the Korteweg–de Vries equation that includes a small nonlinearity together with a small dispersion of waves of the acoustic type, if the wave dispersion may be represented in the form

$$\omega = v_g k(1 - r_o^2 k^2), \quad r_o k \ll 1 . \tag{12.47}$$

According to the Lighthill criterion (12.46) a small wave nonlinearity in combination with a small wave dispersion can lead to compression of a wave packet. We derive the equation that describes simultaneously this combination for long waves propagated in an elastic medium, as a liquid, gas or plasma, if the dispersion relation (12.47) is applicable for these waves. This equation follows from the Euler equation (6.19) that has the form for the velocity of particles in a longitudinal wave

$$\frac{\partial v}{\partial t} + v\frac{\partial v}{\partial x} - \frac{F}{m} = 0 ,$$

where $v(x,t)$ is the particle velocity in a wave that propagates along the x-axis, F is the force per particle, and m is the particle mass. Within the framework of a linear approximation one can write the particle velocity in the form $v = v_g + v'$, where v_g

is the group velocity, and v' is the particle velocity in the frame of reference where the wave is at rest. Because $v' \ll v_g$, we have the linear approximation as

$$\frac{\partial v'}{\partial t} + v_g \frac{\partial v'}{\partial x} - \frac{F}{m} = 0 \, ,$$

and the last term is a linear operator with respect to v'. Determining this term in the harmonic approximation $v' \sim \exp(ikx - i\omega t)$, when the dispersion relation (12.47) is valid, we reduce the Euler equation to the form

$$\frac{\partial v'}{\partial t} + v_g \left(\frac{\partial v'}{\partial x} + r_0^2 \frac{\partial^3 v'}{\partial x^3} \right) = 0 \, .$$

The last term of this equation takes into consideration a weak dispersion of long-wave oscillations. In order to account for a nonlinearity of these waves, we analyze the second term of this equation. In the linear approach we replace the particle velocity v by the group velocity v_g. If we return this term to its initial form and in this manner take into account weak nonlinear effects, we obtain the Korteweg–de Vries equation

$$\frac{\partial v}{\partial t} + v \frac{\partial v}{\partial x} + v_g r_0^2 \frac{\partial^3 v}{\partial x^3} = 0 \, . \tag{12.48}$$

This equation accounts for nonlinearity and weak dispersion simultaneously, and therefore serves as a convenient model equation for the analysis of nonlinear dissipative processes. As applied to plasmas, it describes propagation of long waves in a plasma, such as sound and ion sound for which the dispersion relation (12.47) is valid.

▶ **Problem 12.24** Analyze the propagation of a solitary wave (soliton) in a plasma on the basis of the Korteweg–de Vries equation.

The above analysis shows that wave dispersion leads to divergence of the wave packet. If this divergence is weak, a weak nonlinearity is able to change its character, and we consider this for solitary waves—solitons with the dispersion relation (12.47) which are described by the Korteweg–de Vries equation (12.48). As it follows from the dispersion relation (12.47), such waves propagate more slowly than long waves, but nonlinear effects compensate for the spreading of the wave. In order to obtain this conclusion analytically, we consider a wave of a velocity u, so that the space and time dependence for the particle velocity in this wave has the form $v = f(x - ut)$. This gives

$$\partial v / \partial t = -u \partial v / \partial x \, ,$$

so that the Korteweg–de Vries equation in the frame of reference moved with the wave takes the form

$$(v - u) \frac{dv}{dx} + v_g r_0^2 \frac{d^3 v}{dx^3} = 0. \tag{12.49}$$

One can decrease the order of this equation assuming that a perturbation is absent at large distances from the wave. This gives $v = 0$, $d^2v/dx^2 = 0$ at $x \to \infty$ and transforms the above equation to the form

$$v_g r_o^2 \frac{d^2 v}{dx^2} = uv - \frac{v^2}{2} .$$

Among solutions of this equation we have the solution of the form $v = a/\cosh^2(\alpha x)$ that correspond to location of the wave in a restricted region. Substituting this solution in the above equation, we find the parameters of such a solution. As a result, we obtain

$$v = 3u \cosh^{-2}\left(\frac{x}{2r_0}\sqrt{\frac{u}{v_g}}\right) . \tag{12.50}$$

The wave described by formula (12.50) is concentrated in a limited spatial region and does not diverge in time. The wave becomes narrower with increase of its amplitude, with its extension inversely proportional to the square root of the wave amplitude.

Thus, this analysis shows the existence of stationary solutions of the Korteweg–de Vries equation in the form of nonexpansible waves or solitons. The amplitude a and extension $1/\alpha$ of solitons are such that the value a/α^2 does not depend on the wave amplitude. If the initial perturbation is relatively small, evolution of the wave packet leads to the formation of one solitary wave. Therefore, the solitons describe evolution of perturbations in a nonlinear disperse medium.

► **Problem 12.25** Derive the equation for the electric potential of a nonlinear ion sound.

For the analysis of the nonlinear ion sound we use for its description the Euler equation (6.19), the continuity equation (6.15), and the Poisson equation for ions. In the linear approach they lead to the dispersion relation (11.13) for ion sound. Taking these equations without a linear approximation, we obtain

$$\frac{\partial v_i}{\partial t} + v_i \frac{\partial v_i}{\partial x} + \frac{e}{M}\frac{\partial \varphi}{\partial x} = 0$$

$$\frac{\partial N_i}{\partial t} + \frac{\partial}{\partial x}(N_i v_i) = 0, \qquad \frac{\partial^2 \varphi}{\partial x^2} = 4\pi e(N_e - N_i) .$$

Here v_i is the velocity of ions in the wave, φ is the electric potential of the wave, N_e and N_i are the number densities of electrons and ions, respectively, and M is the ion mass. As was discussed above, electrons are in equilibrium with the field owing to their high mobility. Therefore, the Boltzmann distribution applies to the electrons, so $N_e = N_o \exp(e\varphi/T_e)$, where N_o is the mean number density of charged particles, and T_e is the electron temperature.

Let us analyze the motion of ions in the field of a steady-state wave when the time and spatial dependence for plasma parameters v_i, N_i and φ has the form

$f(x - ut)$, where u is the velocity of the wave. Then the above set of equations for plasma parameters takes the form

$$(v_i - u)\frac{dv_i}{dx} + \frac{e}{M}\frac{d\varphi}{dx} = 0 \quad \frac{d}{dx}\left[N_i(v_i - u)\right] = 0 \; ,$$

$$\frac{d^2\varphi}{dx^2} = 4\pi e\left[N_0 \exp\left(\frac{e\varphi}{T_e}\right) - N_i\right] .$$

The perturbation is assumed to be zero far from the wave, which gives $N_i = N_0$, $v_i = 0$ and $\varphi = 0$ at $x \to \infty$. From the first two equations it follows

$$\frac{v_i^2}{2} - uv_i + \frac{e\varphi}{M} = 0, \qquad N_i = N_0\frac{u}{u - v_i} \; ,$$

and the second equation gives $v_i \leq u$ because $N_i \geq 0$. This means that $\varphi \geq 0$ in the first equation, i. e., the electric potential of this ion-acoustic wave is positive. The first equation gives

$$v_i = u - \sqrt{u^2 - \frac{2e\varphi}{M}} \; ,$$

where we use the physical condition $v_i \leq u$. The second equation takes the form

$$N_i = N_0\frac{u}{\sqrt{u^2 - \frac{2e\varphi}{M}}} \; .$$

Substituting this in the last equation of the above set of equations, we obtain the electric potential of nonlinear ion sound

$$\frac{d^2\varphi}{dx^2} = 4\pi e N_0\left[\exp\left(\frac{e\varphi}{T_e}\right) - \frac{u}{\sqrt{u^2 - 2e\varphi/m}}\right] . \tag{12.51}$$

This equation describes the spatial distribution for the electric potential of the nonlinear ion-acoustic wave. This has the form of the motion equation for a particle if φ is regarded as the coordinate and x as time. A general property of this type of equations corresponds to the energy conservation law that follows from multiplication by $d\varphi/dx$ and integration of the obtaining expression. This operation gives

$$\frac{1}{2}\left(\frac{d\varphi}{dx}\right)^2 - 4\pi N_0 T_e \exp\left(\frac{e\varphi}{T_e}\right) - 4\pi N_0 M u\sqrt{u^2 - \frac{2e\varphi}{M}} = \text{const} \; .$$

Assuming that at large distances from the wave the potential φ and the electric field strength of the wave $-d\varphi/dx$ are zero, we can evaluate the constant of integration and obtain the following equation for the electric potential of a nonlinear ion sound:

$$\frac{1}{2}\left(\frac{d\varphi}{dx}\right)^2 + 4\pi N_0 T_e\left[1 - \exp\left(\frac{e\varphi}{T_e}\right)\right] + 4\pi N_0 M u\left(u - \sqrt{u^2 - \frac{2e\varphi}{M}}\right) = 0 \; . \tag{12.52}$$

This solution describes a solitary wave because, according to the boundary conditions, the perturbation goes to zero at large distances. The solution allows one to determine the soliton profile and the relation between wave parameters for various amplitudes of a nonlinear ion sound.

▶ **Problem 12.26** Derive the relation between the maximum electric potential of a nonlinear ion sound and the velocity of its propagation.

As it follows from the problem symmetry, the maximum electric potential corresponds to $x = 0$, where $d\varphi/dx = 0$ and $\varphi = \varphi_{max}$. Using the reduced variables $\zeta_{max} = e\varphi_{max}/T_e$ and $\eta_{max} = Mu^2/(2T_e)$, we represent equation (12.52) in the form

$$1 - \exp\zeta_{max} + 2\eta_{max}\left(1 - \sqrt{1 - \frac{\zeta_{max}}{\eta_{max}}}\right) = 0. \tag{12.53}$$

Figure 12.7 shows the dependence $\zeta_{max}(\eta_{max})$ in accordance with this equation.

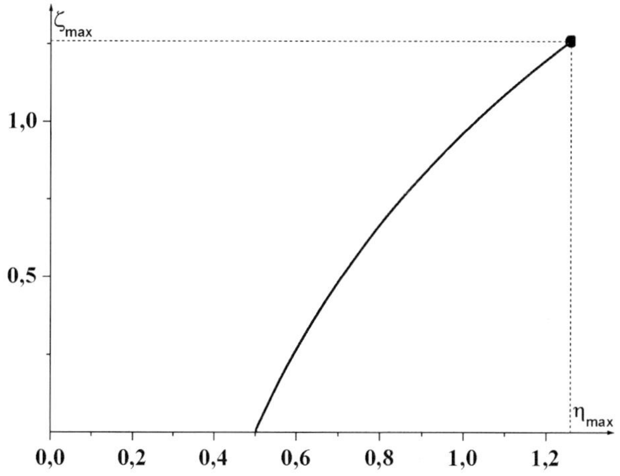

Fig. 12.7 The dependence for the reduced maximum electric potential of the ion sound soliton on the reduced maximum transversal energy.

Let us consider the limiting cases of equation (12.53). At small amplitudes of the ion-acoustic wave, $\zeta_{max} \to 0$, equation (12.52) gives $\eta_{max} = 1/2$. From this we find the phase velocity of the wave $u = (T_e/M)^{1/2}$ in accordance with the dispersion relation (11.13) for ion sound of a small amplitude. At large wave amplitudes, taking $\zeta_{max} = \eta_{max}$, we transform equation (12.53) for ζ_{max} to the form

$$1 - \exp\zeta_{max} + 2\zeta_{max} = 0$$

The solution of this equation is $\zeta_{max} = 1.26$, which yields

$$e\varphi_{max} = 1.26T_e, \qquad u = 1.58\sqrt{\frac{T_e}{M}}.$$

For larger wave amplitudes the electric potential in the wave center becomes too large, and ions are reflected from the crest of the wave. As a result, part of the wave reverses and the wave separates into parts. Thus, a solitary ion-acoustic wave exists only in a restricted range of wave amplitudes and velocities. Exceeding the limiting amplitude leads to a wave splitting into bunches—separate waves. Thus solitary waves exist in a restricted range of their parameters.

Introducing the reduced variables $\zeta = e\varphi/T_e$, $\eta = Mu^2/(2T_e)$ and the reduced length $\xi = x/r_D$, where $r_D = \sqrt{4N_0 e^2/T_e}$ is the Debye–Hückel radius for motionless ions, we rewrite equation (12.52) in the form

$$\frac{1}{2}\left(\frac{d\zeta}{d\xi}\right)^2 + \left(1 - e^{\zeta} + 2\eta\left(1 - \sqrt{1 - \frac{\zeta}{\eta}}\right)\right) = 0\,.$$

Figure 12.8 shows the spatial distribution of the soliton electric potential that is obtained according to this equation at different values of its maximal value.

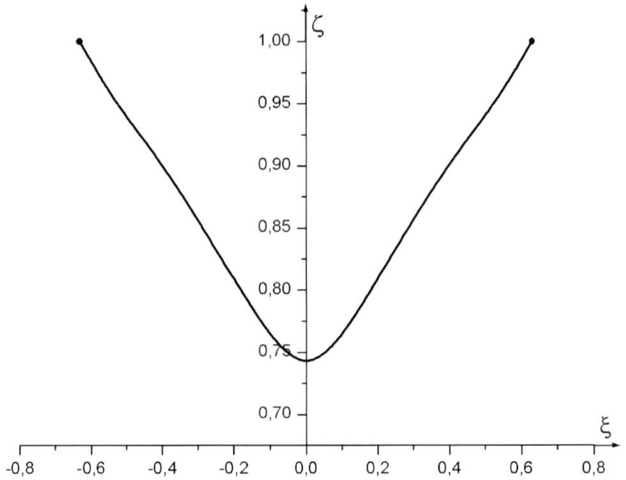

Fig. 12.8 The dependence on the reduced distance from the center of the ion sound soliton for the reduced electric potential if $e\varphi_{max}/T_e = 0.715$ that corresponds to $\eta = 1$.

▶ **Problem 12.27** Obtain the dispersion equation for nonlinear plasma oscillations under the condition that the electric field strength due to plasma oscillations varies weakly at distances of the order of the wavelength of oscillations.

The nature of solitons is such that an electric field occurs in a plasma as a result of a wave process, and this field locks a plasma perturbation in a restricted region. We considered above this phenomenon as a result of combination of the dispersion and nonlinearity of waves of an acoustic type. We now consider another mechanism of this phenomenon when a wave creates a spatial well for a plasma, and due to this well the number density of a plasma increases in the well region. This phenomenon also exists in the wave form and is typical for plasma oscillations. Therefore, we consider below plasma oscillations as a solitary wave.

Introducing the electric field strength of the plasma oscillations $E(x,t)$, we obtain the energy density W of the plasma with accounting for plasma oscillations

$$W(x) = \frac{\overline{E^2}}{8\pi} ,$$

where the bar means a time average. Assuming the equality of electron and ion temperatures $T_e = T_i = T$, we obtain the pressure of a quasineutral plasma as $p = 2N_e T$. The plasma pressure is established with a sound velocity that is larger than the velocity of propagation of long-wave oscillations. Then, because of the uniformity of plasma pressure at all points of the plasma, we have

$$2N(x)T + W(x) = 2N_0 T ,$$

where N_0 is the number density of charged particles at large distances where plasma oscillations are absent; the plasma temperature is assumed to be constant in space.

The dispersion relation for plasma oscillations (11.8) has the form

$$\omega^2 = \omega_p^2 \left(1 - \frac{\overline{E^2}}{16\pi N_0 T}\right) + \gamma \left\langle v_x^2 \right\rangle k^2 ,$$

where we use the above relation between the average and current electron number density, and the plasma frequency ω_p is taken in the absence of oscillations according to formula (11.6).

Let us transform this relation, taking the electric field strength of the wave as $E = E_0 \cos \omega t$, which gives $\overline{E^2} = E_0^2/2$. Using $\overline{v_x^2} = T/m_e$ for electrons and inserting in the above relation the Debye–Hückel radius r_D, we reduce this dispersion relation to the form in the isothermal case,

$$\omega^2 = \omega_p^2 \left(1 - \frac{E_0^2}{32\pi N_0 T} + 2\gamma r_D^2 k^2\right) . \tag{12.54}$$

The first term on the right-hand side of this dispersion relation is significantly larger than the other two.

Thus, nonlinear Langmuir oscillations can form a solitary wave—Langmuir soliton. The dispersion relation (12.52) shows that if the energy density of plasma oscillations is high enough, so that the second term of equation (12.52) is larger than the third one, then the oscillations cannot exist far from the soliton. The oscillations create a potential well in the plasma and become enclosed in this well. These oscillations propagate in the plasma together with the well and occupy a restricted spatial region. The size of the potential well, i. e., the soliton size, decreases with increase of the energy density of the plasma oscillations. Because $r_D k \ll 1$, the solitons are formed when the energy density of the oscillations is small compared to the specific thermal energy of charged particles of the plasma. Thus, this analysis demonstrates the tendency of long-wave plasma oscillations to form solitons, but the above analysis does not allow one to study the evolution of large-amplitude oscillations.

▶ **Problem 12.28** Show that the Lighthill criterion (12.46) is fulfilled for long-wave plasma oscillations which are locked in a restricted plasma region.

For isothermal plasma oscillations the dispersion relation (12.54) for plasma oscillations can be represented as

$$\omega = \omega(k) + \alpha E_o^2 ,$$

and for an isothermal plasma

$$\omega(k) = \omega_p(1 + r_D^2 k^2), \qquad \alpha = -\omega_p/(64\pi N_o T_e)$$

Since the group velocity of plasma oscillations is $v_g = \partial\omega/\partial k = 6T_e k/m_e$, we have

$$\alpha \frac{\partial v_g}{\partial k} = -\frac{3\omega_p}{32\pi m_e N_o} < 0 .$$

Thus, the Lighthill criterion (12.46) holds true for long-wave plasma oscillations, and these waves can form a solitary wave.

▶ **Problem 12.29** Show the instability of an electron drift wave for a nonmonotonic dependence of the electron drift velocity on the electric field strength.

A drift wave is a perturbation of the electron number density that propagates together with the electron current. These waves damp due to diffusion (11.21). But in the case of a nonmonotonic dependence of the electron drift velocity on the electric field strength, as it is shown in Fig. 12.9, they can grow since drift waves related to different electric fields can propagate with identical drift velocity. We prove this in consideration the stage when these waves are weak that allows us to consider this problem in the linear approximation.

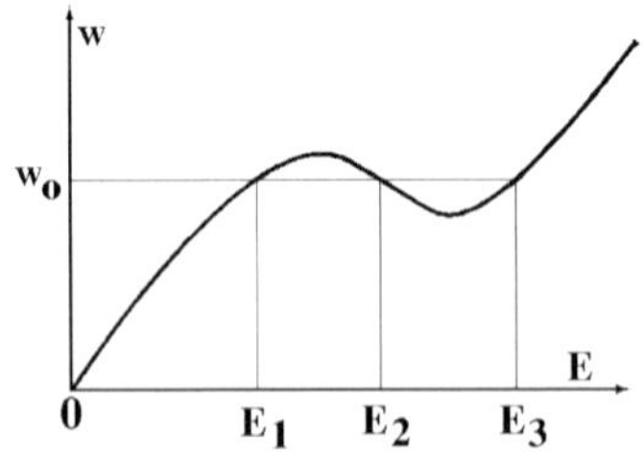

Fig. 12.9 A non-monotonic dependence of the electron drift velocity on the electric field strength that can lead to formation of a domain.

For description of the drift wave, we use the continuity equation (6.15) and the Poisson equation

$$\frac{\partial N_e}{\partial t} + \frac{\partial j_e}{\partial x} = 0, \qquad \frac{\partial E}{\partial x} = 4\pi e(N_o - N_e), \tag{12.55}$$

where N_o is the equilibrium number density of electrons. The electron flux j_e is equal to

$$j_e = -N_e w - D\frac{\partial N_e}{\partial x},$$

i. e., the electron diffusion motion is taken into account along with the drift motion (D is the diffusion coefficient of electrons). We repeat the deduction of the dispersion relation (11.21) by using the harmonic dependence on the coordinate and time for perturbation parameters, but in addition we account for the term $N_o dw_o/dx \cdot \partial E/\partial x$ in the expression for $\partial j_e/\partial x$. This leads to the dispersion relation

$$\omega = -kw - iDk^2 - i4\pi N_o e\frac{dw}{dE} \tag{12.56}$$

that coincides with equation (11.21) for monotonic dependence $w(E)$. Then perturbation of the electric field causes an additional damping of drift waves along with the diffusion one. But if $dw/dE < 0$, long drift waves with $Dk^2 < i4\pi N_0 e dw/dE$ can develop. As a result, a stable structure, the electric domain, is formed and propagates with an electric current.

▶ **Problem 12.30** Analyze the decay of a plasma oscillation in a plasma oscillation of a low frequency and ion sound.

Nonlinear phenomena are responsible for interaction between different modes of oscillation. A possible consequence of this interaction is the decay of a wave into two waves. Since the wave amplitude depends on time and spatial coordinates by the harmonic dependency $\exp(i\mathbf{k}\cdot\mathbf{r} - i\omega t)$, such a decay corresponds to fulfilling the relations

$$\omega_0 = \omega_1 + \omega_2, \qquad \mathbf{k}_0 = \mathbf{k}_1 + \mathbf{k}_2, \tag{12.57}$$

where subscript 0 relates to the parameters of the initial wave, and subscripts 1 and 2 refer to the decay waves. This results from a parametric instability for which the relations (12.57) are fulfilled. We consider below decay of a plasma oscillation that decays into a plasma oscillation of a lower frequency and an ion-acoustic wave (ion sound).

The electric field of the initial plasma oscillation is

$$E = E_0 \cos(k_0 x - \omega_0 t),$$

where x is the direction of propagation. In zero approximation we assume the electric field amplitude E_0 and other wave parameters to be real values. The equation of motion for electrons $m_e dv_0/dt = -eE$ yields the electron velocity $v_0 = u_0 \cos(k_0 x - \omega_0 t)$, where $u_0 = eE_0/(m_e\omega_0)$.

Let another plasma wave and the ion sound wave be excited in the system simultaneously with the initial plasma oscillation, and let their amplitudes be small compared to the amplitude of the initial oscillation. Consider the time development of these waves taking into account their interaction with each other and with the initial oscillation. Since ion velocities are much lower than electron velocities, one can analyze these waves separately. The equation of motion and continuity equation for ions are

$$M\frac{dv_{\mathrm{i}}}{dt} = eE, \qquad \frac{\partial N_{\mathrm{i}}'}{\partial t} + N_0\frac{\partial v_{\mathrm{i}}}{\partial x} = 0\,,$$

where M is the ion mass, v_{i} is the ion velocity, N_0 is the equilibrium number density of ions, N_{i}' is the perturbation of the ion number density due to the oscillation, and E is the electric field due to the oscillations. Elimination of the ion velocity from these equations yields

$$\frac{\partial^2 N_{\mathrm{i}}'}{\partial t^2} + \frac{eN_0}{M}\frac{\partial E}{\partial x} = 0. \tag{12.58}$$

We can find the electric field strength E that acts on ions from the equation of motion for the electrons by averaging over fast oscillations. The one-dimensional Euler equation (6.19) for electrons can be rewritten as

$$\frac{\partial v_{\mathrm{e}}}{\partial t} + v_{\mathrm{e}}\frac{\partial v_{\mathrm{e}}}{\partial x} + \frac{1}{m_{\mathrm{e}}N}\frac{\partial p_{\mathrm{e}}}{\partial x} + \frac{eE}{m_{\mathrm{e}}} = 0\,,$$

where the electron gas pressure p_{e} is expressed through electron temperature T_{e} and number density N as $p_{\mathrm{e}} = NT_{\mathrm{e}}$. Taking the electron velocity as $v_{\mathrm{e}} = v_0 + v_{\mathrm{e}}'$, where v_{e}' is the electron velocity due to the small-amplitude plasma wave. After averaging over fast oscillations, when the first term in this equation becomes zero, we transform this equation to the form

$$\overline{v_{\mathrm{e}}\frac{\partial v_{\mathrm{e}}}{\partial x}} = \frac{1}{2}\frac{\partial \overline{v_{\mathrm{e}}^2}}{\partial x} = \frac{1}{2}\frac{\partial}{\partial x}\overline{(v_0 + v_{\mathrm{e}}')^2} = \frac{\partial}{\partial x}\left(\overline{v_0 v_{\mathrm{e}}'}\right)\,,$$

where the bar denotes averaging over fast oscillations. We take the electron temperature to be constant in a space. During the ion motion, the plasma quasineutrality is supported due to fast electron motion. Hence, the deviation of the electron number density from the equilibrium one is the same as that for ions, and the third term of the above Euler equation is

$$\frac{1}{m_{\mathrm{e}}N}\frac{\partial p_{\mathrm{e}}}{\partial x} = \frac{T_{\mathrm{e}}}{m_{\mathrm{e}}N_0}\frac{\partial N_{\mathrm{i}}'}{\partial x}\,(N \approx N_0)\,.$$

The Euler equation after these operations is transformed into

$$\frac{\partial}{\partial x}\left(\overline{v_0 v_{\mathrm{e}}'}\right) + \frac{T_{\mathrm{e}}}{m_{\mathrm{e}}N_0}\frac{\partial N_{\mathrm{i}}'}{\partial x} + \frac{eE}{m_{\mathrm{e}}} = 0\,.$$

Substituting the electric field from this equation into equation (12.58), we obtain

$$\frac{\partial^2 N_i'}{\partial t^2} - \frac{T_e}{M}\frac{\partial^2 N_i'}{\partial x^2} - \frac{m_e N_0}{M}\frac{\partial^2}{\partial x^2}\left(\overline{v_0 v_e'}\right) = 0. \tag{12.59}$$

If we ignore the last term in equation (12.59) and assume harmonic dependence of the ion density on time and x, it gives the dispersion relation (11.13) for ion sound, i. e., $\omega = c_s k$, $c_s = \sqrt{T_e/M}$.

To take into account the interaction between ion sound and plasma oscillations, it is necessary to analyze the motion of electrons in the field of the small-amplitude plasma wave. For this purpose, we consider the Maxwell equation for the electric field of the small amplitude wave E', ignoring the magnetic field, which gives

$$\partial E'/\partial t + 4\pi j' = 0 \,,$$

where j' is the electric current density under the wave action. For simplicity, we shall ignore the thermal motion of the electrons, since it has only a small effect on the oscillation frequency. Hence we ignore the variation of the electron number density due to the electron pressure of the plasma wave. Assume the electron number density to be $N_e + N_i'$, where N_e includes the equilibrium electron number density and its variation under the action of the initial plasma wave, and N_i' is the variation of the ion number density owing to the motion of ions. Correspondingly, the electron velocity is $v_0 + v_e'$, where v_0 is the electron velocity due to the initial plasma wave, and v_e' is the electron velocity due to the small-amplitude plasma wave. From this we obtain the current density due to the small-amplitude plasma wave as

$$j' = -e(N_e + N_i')(v_0 + v_e') + eN_e v_0 = -eN_e v_e' - eN_i' v_0 \,,$$

and we neglect the second-order terms. Then the Maxwell equation takes the form

$$\frac{\partial E'}{\partial t} - 4\pi e N_i' v_0 - 4\pi e N_e v_e' = 0 \,.$$

The motion equation for an electron has the form

$$m_e \frac{dv_e'}{dt} = -eE' \,.$$

Eliminating E' from these equations, we obtain the equation for the electron velocity due to the small-amplitude plasma wave

$$\frac{\partial^2 v_e'}{\partial t^2} + \omega_p^2 v_e' + \frac{N_i'}{N_e}\omega_p^2 v_0 = 0 \,. \tag{12.60}$$

Here ω_p is the frequency of plasma oscillations in neglecting a thermal motion of electrons. One can see that if we ignore the interaction between the small-amplitude plasma wave with the initial plasma oscillation and the ion-acoustic wave, i. e., ignore the last term of equation (12.60), then the small-amplitude plasma wave frequency coincides with the plasma frequency in accordance with assumptions used.

For solution of the set of equations (12.59) and (12.60) we take the wave parameters in the form

$$v_0 = u_0 \cos(k_0 x - \omega_0 t), \quad v_e = a \cos(k_e x - \omega_e t), \quad N_i' = b N_0 \cos(k_i x - \omega_i t) ,$$

where a and b are slowly varying oscillation amplitudes, ω_e and k_e are the frequency and the wave number of the small-amplitude plasma wave, ω_i and k_i are the frequency and the wave number of the ion sound, and N_0 is the equilibrium number density of the charged particles. [We assume that $N_e = N_0$ in equation (12.60)]. Since the oscillation amplitudes vary slowly, the time and space dependences are identical, so we find from equations (12.59) and (12.60) that

$$\omega_0 = \omega_e + \omega_i; \qquad k_0 = k_e + k_i ,$$

in accordance with formulas (12.57). This condition is similar to that for parametric resonance of coupled oscillators, and therefore the instability that we analyze is termed a parametric instability.

Taking into account a slow variation of the oscillation amplitudes and condition (12.57), we find from equations (12.59) and (12.60) the following set of equations for the oscillation amplitudes:

$$\frac{\partial a}{\partial t} = \frac{\omega_e u_0 b}{4}, \quad \frac{\partial b}{\partial t} = -\frac{m_e k_i u_0 a}{4 M \omega_i} .$$

From solution of these equations, a growth of oscillations corresponds to the dependence $a, b \sim \exp(\delta t)$ with

$$\delta = \frac{1}{4}\sqrt{\frac{m_e \omega_e}{M \omega_i}}\, u_0 k_i = \frac{1}{4}\sqrt{\frac{m_e \omega_e}{M \omega_i}}\, \frac{e E_0}{m_e \omega_0} k_i . \tag{12.61}$$

Thus, the initial plasma wave is unstable. It can decay into a plasma wave of a lower frequency and the ion sound. This instability is also known as the decay instability. The exponential growth parameter for the new wave is proportional to the amplitude of the decaying wave.

12.5 Plasma Structures

► **Problem 12.31** A lightly ionized admixture is added to a buffer gas in which gas discharge is burnt. Ions of gas discharge belong to the admixture, and the number density of admixture ions N_i is small compared to the number density N_a of admixture atoms. Ions are drifting to the cathode under the action of the discharge electric field and transfer their momentum to atoms due to the resonant charge exchange process. Estimate the dimension of a discharge tube L where admixture atoms and ions are concentrated.

This phenomenon, electrophoresis, corresponds to separation of a lightly ionized and a buffer gas in gas discharge. As a result, a lightly ionized gas is concentrated in a restricted discharge region, and discharge glowing due to radiation of the admixture excited atoms is observed in this discharge region only. In considering this phenomenon, we assume atoms and ions of an admixture to be a unit system located in a region of a dimension L. It is valid if a typical time of transformation of a test atom into an ion as a result of the resonant charge exchange process $(N_i\sigma_{res}\overline{v_a})^{-1}$ is small compared to a typical diffusion time of atom diffusion L^2/D_a through this region, i. e.,

$$N_i\sigma_{res}\overline{v} \gg \frac{D_a}{L^2} \,.$$

Here σ_{res} is the cross section of resonant charge exchange in collisions of admixture ions and atoms, $\overline{v}$ is a typical velocity of admixture atoms or ions, D_a is the diffusion coefficient of admixture atoms in buffer gas.

Under the above conditions, we have the following balance equation:

$$-D_a\frac{dN_a}{dx} + w_i N_i = 0 \,,$$

that means that the diffusion transport of atoms in the direction of the discharge direction x is compensated by ion drift, where w_i is the ion drift velocity under the action of the discharge electric field. Assuming the discharge electric field strength E to be relatively small, we use the Einstein relation for the ion drift velocity

$$w_i = \frac{eED_i}{T} \,,$$

where T is the temperature of atoms and ions and D_i is the diffusion coefficient of ions in a buffer gas. Assuming $D_i \sim D_a$, we estimate a dimension L of a region where admixture atoms and ions are located

$$L = \left(\frac{d\ln N_a}{dx}\right)^{-1} \sim \frac{T}{eEc_i}w_i = \frac{eED_i}{T} \,, \tag{12.62}$$

where $c_i = N_i/N_a$ is the concentration of admixture ions. We assume this value L to be less than the tube length. One can see that electrophoresis influences the parameters of a gas discharge.

► **Problem 12.32** Determine the electric field distribution along the discharge tube in the discharge positive under electrophoresis condition. Estimate a typical time of the electrophoresis establishment after switching on the discharge.

In considering the electrophoresis in a long discharge tube, we assume a plasma is formed in each cross section due to ionization processes in electron–atom collisions, and electrons obtain energy from the discharge electric field. Therefore in the region, where admixture atoms are located, this field is less than that in the region occupied by buffer gas atoms, since ionization of admixture atoms proceeds at less electron energies.

For electrophoresis establishment after switching on the discharge it is necessary to transport admixture atoms to the corresponding region that proceeds under the action of ions which transfer the momentum to admixture atoms. Hence, a typical time for electrophoresis establishment is estimated as

$$\tau \sim \frac{L}{c_i w_i} ,$$

where a dimension with admixture atoms is determined by formula (12.62).

▶ **Problem 12.33** Analyze the properties of a nonlinear electric domain—a plasma structure that propagates with the drift velocity of electrons in a plasma.

A nonlinear electric domain refers to a unstable range of the electric field strengths of Fig. 12.9. It is described, as earlier, by set (12.55) of equations, and nonperturbed electric current density is

$$j_0 = -N_0 w(E_2) = -N_e w(E) - D\frac{\partial N_e}{\partial x}$$

in notations of Fig. 12.9. Combining this equation with the Poisson equation (12.55) and eliminating the electron number density from these equations, we obtain the distribution of the electric field strength in the electric domain as

$$D\frac{d^2 E}{dx^2} = w(E)\frac{dE}{dx} - 4\pi e N_0 \left[w(E) - w(E_2)\right] .$$

Considering diffusion to be of importance on the periphery of this structure and neglecting diffusion for its main part, we reduce this equation to the form

$$\frac{dE}{dx} = 4\pi e N_0 \left[\frac{w(E_2)}{w(E)} - 1\right] . \tag{12.63}$$

Figure 12.10a contains the solution of this equation with the boundary condition $E = E_2$ at $x = 0$, and Fig. 12.10b shows the corresponding distribution of the electron number density in the electric domain. From this it follows that $E(x)$ increases with x until $E_2 < E < E_3$, and $w(E_3) = w(E_2)$. At $E = E_3$ we have $dE/dx = 0$, and for subsequent values of x we obtain $E = E_3$. Thus, this solution describes the conversion of the system from the unstable state E_2 to the stable state with $E = E_3$.

This transition means that the change of the discharge regime as a result of the perturbation of the electric field strength increases up to E_3. This would require a variation of the discharge voltage that is impossible, because the discharge voltage is maintained by the external voltage. Therefore, the variation of the electric field strength from E_2 to E_3 is a perturbation that takes place in a limited region of the plasma. A return to the initial value of the field occurs as a result of diffusion, leading to decay of the perturbation. Hence, a typical size of the back boundary

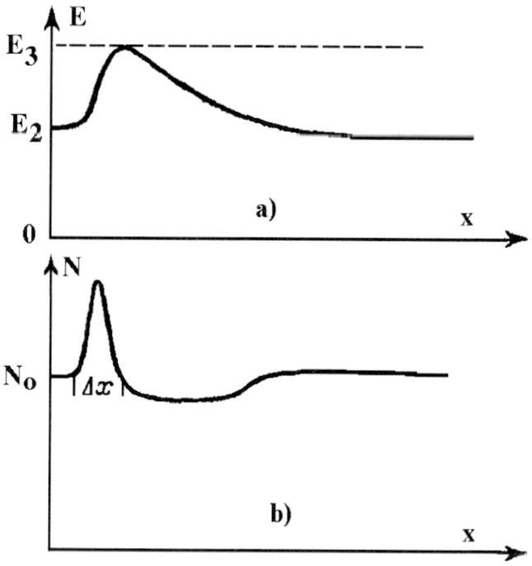

Fig. 12.10 The distribution of the electric field (a) and the electron number density (b) in the electric domain.

zone of the electric domain is of the order of D/w. The forward boundary zone size can be estimated from equation (12.63), and is $\Delta x = \Delta E/(4\pi e N_0 \Delta w)$, where $\Delta E = E_3 - E_2$, and $\Delta w = w(E_2) - w_{min}$ according to Fig. 12.9. A concomitant of the distribution of the electric field strength in the electric domain is the distribution of the electron number density shown in Fig. 12.10, arising from Poisson's equation (12.55).

► **Problem 12.34** Analyze the conditions of striation formation in a long discharge tube at small number densities of electrons and ions.

Striations or strates are ionization waves which can exist within a certain range of parameters of nonlinear processes involving electrons and transport phenomena. In the Tsendin model, under consideration, we consider the mechanism of formation of ionization waves that can be realized in a dense gas and proceeds through associative ionization involving excited atoms. Electrons with energy exceeded the atom excitation energy ε_{ex} excite atoms and then these atoms form molecular ions in collisions with atoms in the ground state. Because electrons take energy from the discharge electric field of strength E, ionization processes proceed at a distance l from a previous ionization event that satisfies the relation

$$\int_0^l eE(x)dx = \varepsilon_{ex} .$$

Here we assume that excitation takes place near the threshold, and the processes takes place in a narrow long discharge tube.

Let us introduce the ionization probability for an excited atom in collisions with ground state atoms to be ξ, i. e., the probability of its quenching is $1-\xi$, and $0 < \xi < 1$. In addition, we introduce the lifetime τ for electrons and ions with respect to their attachment to walls. In the absence of ionization, the balance equation for the electron number density N_e is

$$w_e \frac{dN_e}{dx} - \frac{N_e}{\tau} = 0 ,$$

where x is directed along the gas discharge tube, and w_e is the electron drift velocity. This balance equation has the following solution:

$$N_e(x) = N_0 \exp\left(-\int_0^x \frac{dy}{w_e \tau}\right) .$$

Within the framework of the Tsendin model, we assume that all the electrons have zero energy at the origin $x = 0$. When these electrons acquired the energy ε_{ex} it excites an atom, and this atom is then ionized in subsequent collisions with atoms. As a result, $(1 + \xi)$ slow electrons are formed at this point instead of one fast electron. Therefore, if the first ionization process takes place at $x = l$, we have the balance of attachment and ionization processes as

$$\int_0^l \frac{dx}{w_e \tau} = \ln(1 + \xi) .$$

This leads to a periodic function $N_e(x)$ with a period l, and Fig. 12.11 gives the distributions of the electron number density and the electric field strength along the x in this case. The electric field strength $E(x)$ is a periodic function also. Indeed,

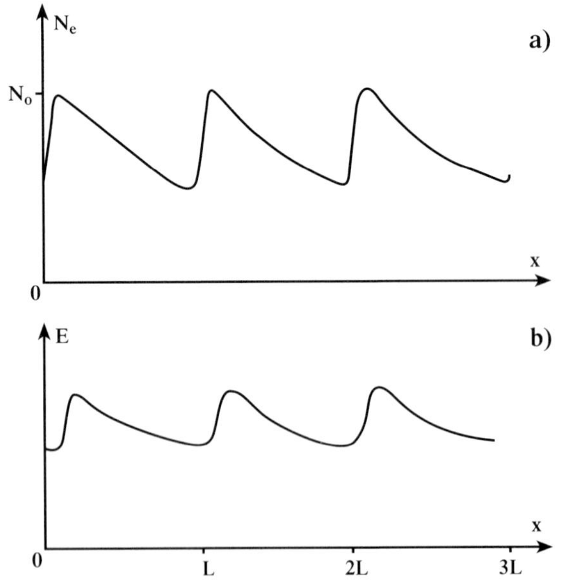

Fig. 12.11 Distributions of the number density of electrons (a) and the electric field strength (b) along a gas-discharge tube of a small radius under formation of striations.

the electric field strength $E(x)$ satisfies the Poisson equation

$$\frac{dE}{dx} = 4\pi e(N_i - N_e) ,$$

and a plasma is quasineutral on the average, i. e.,

$$\int_0^l N_i dx = \int_0^l N_e dx .$$

Hence, from the Poisson equation it follows $E(0) = E(l)$, and the period of the function $E(x)$ is l.

Under these conditions, we neglect energy exchange between electrons and atoms in elastic collisions, as well as that in electron–electron collisions, i. e., the electron number density is small. Hence, the number density of electrons is relatively small. If the number of ionization events per period of the striation is ν, the total number of electrons and ions in one striation band is $\nu\tau$. However, ions are concentrated in an ionization zone near the point where they formed, whereas electrons formed in this ionization zone are distributed over a region of several striations.

▶ **Problem 12.35** Show the stability of the Tsendin model for a periodic distribution of the electron number density and electric field under conditions of validity of this model.

In analyzing the reality of the Tsendin model, we note that it relates to atomic gases. Then the energy exchange in electron–atom elastic collisions is small because of the small parameter m_e/M, where m_e is the electron mass and M is the atom mass. As a result, electrons expand the energy obtained from the electric field on excitation of atoms mostly. Assuming that the ionization event proceeds when electrons reach the energy ε_{ex}, we obtain a jump in the electron number density and electric field strength which occurs over the distance $l = \varepsilon_{ex}/\left(e\overline{E}\right)$ from the origin ($\overline{E}$ is the average electric field strength). Correspondingly, a strong glow occurs near this point due to excitation of atoms by electron impact. In fact, the excitation cross section is zero at threshold, so that excitation proceeds in some region in space, where the electron energy leads to effective atomic excitation. But if the width of the excitation region increases with its population of electrons, the periodic structure of the plasma is destroyed. We analyze below this effect.

Let us construct the sequence of events when an electron that has zero energy at the origin excites an atom within a distance $l + \delta l_1$ from the origin, where it has the energy $\varepsilon_{ex} + e\overline{E}\delta l_1$. After excitation of k atoms this electron has traveled a distance $kl + \sum_{i=1}^{k} \delta l_i$, where δl_i is the distance required for the i-th excitation beyond l, and ε_k is the electron energy after the k-th excitation. Taking this energy to be zero, we obtain the relation

$$\sum_{i=1}^{k} \delta l_i = 0 .$$

From this it follows that the distance required for k excitations by any electron approaches kl, which infers the bunching of electrons. This effect supports a periodical distribution structure for the electron and ion number densities, as well as for the electric field strength.

Evidently, accounting for the real behavior of this plasma leads to the broadening of ionization zones. In particular, within the framework of the simplified Tsendin model being used, we assume the spatial oscillation period l to be large compared to the tube radius. Because the lifetime τ of electrons and ions is the time required for ion drift over a distance of the order of the tube radius, the width of the ionization zone exceeds this value. This means that the bunching effect refers to long discharge tubes of small radius. Bunching of electrons takes place if each electron excites many atoms during its lifetime. Note that there are different mechanisms which can lead to the formation of striations, and we consider above one of these.

13
Cluster Plasma

13.1
Equilibrium of Clusters in Vapor

► **Problem 13.1** Find the equilibrium number density of clusters of a given size in a vapor.

Growth and evaporation of clusters of a given size result from the processes of atom attachment and cluster evaporation whose rates are given by formulas (2.39) and (13.7). The kinetic equation for the size distribution function f_n of clusters according to formula (6.71), i. e., the number density of clusters consisting of n atoms, has the form

$$\frac{\partial f_n}{\partial t} = Nk_{n-1}f_{n-1} - Nk_n f_n - \nu_n f_n + \nu_{n+1} f_{n+1} \,. \tag{13.1}$$

From this it follows that if an equilibrium between clusters of a given size is supported by the processes of atom attachment and cluster evaporation, we obtain the size distribution function

$$f_{n+1}\nu_{n+1} = f_n N k_n \,.$$

Using formulas (2.39) and (13.7) for the rates of these processes, we obtain

$$\frac{f_{n+1}}{f_n} = S \exp\left(\frac{\varepsilon_{n+1} - \varepsilon_o}{T}\right) ,$$

with the supersaturation degree S defined according to formula (6.72)

$$S = \frac{N_{\text{sat}}(T)}{N} \,. \tag{13.2}$$

Within the framework of the liquid drop model for large clusters, we obtain the total binding energy of cluster atoms E_n

$$E_n = \varepsilon_o n - An^{2/3} , \quad n \gg 1 , \tag{13.3}$$

Plasma Processes and Plasma Kinetics. Boris M. Smirnov

ISBN: 978-3-527-40681-4

where ε_o is the atom binding energy for a macroscopic system, and the second term accounts for the cluster surface energy with the specific surface energy A. This gives for the atom binding energy ε_n in a large cluster $n \gg 1$

$$\varepsilon_n = \frac{dE_n}{dn} = \varepsilon_o - \frac{2A}{3n^{1/3}} \; .$$

From this we obtain the ratio of the equilibrium size distribution functions for clusters of neighboring sizes

$$\frac{f_{n+1}}{f_n} = S \exp\left(-\frac{2A}{3n^{1/3}T}\right) \; . \tag{13.4}$$

From this equation it follows that for a supersaturated vapor $S > 1$ the size distribution function of large clusters grows with an increase of a cluster size. Next, the size distribution function of clusters as a function of their sizes is minimum at the critical number of cluster atoms n_{cr}, which is according to the above formula

$$n_{cr} = \left(\frac{\Delta\varepsilon}{T \ln S}\right)^3 \tag{13.5}$$

▶ **Problem 13.2** Show that an atomic vapor can be converted in clusters in full only under nonequilibrium conditions.

According to formula (13.4), the number density of clusters as a function of their size is minimum at the critical size of clusters. From this follows an important conclusion. Clusters are an intermediate phase of matter between gaseous and condensed phases. According to formula (13.4), the majority of atoms of a supersaturated vapor $S > 1$ consisting of a gas and a condensed phase under thermodynamic equilibrium are found in the condensed phase. On the other hand, in a nonsaturated vapor $S < 1$ the majority of atoms are found in the gaseous phase. This means that clusters include a small portion of the atoms of a system under thermodynamic equilibrium. But transition from the gaseous to the condensed phase in a space proceeds through the formation and growth of clusters. From this one can conclude that a high portion of atoms in clusters corresponds to nonequilibrium conditions. Hence, methods of generation of clusters are based on fast nucleation of vapors or gases in a volume at the stage of absence of thermodynamic equilibrium between clusters and a condensed phase.

For this reason, generators of clusters or small particles formed from a supersaturated vapor are based on the expansion of a vapor in a region of low pressure. In the course of expansion, the vapor temperature decreases and the vapor comes to be supersaturated, which leads to the formation and growth of clusters or small particles. But the time for the process is not sufficient for the transformation of clusters in a condensed system, so that almost all the atoms of the incident vapor belong to clusters at the end of the process.

▶ **Problem 13.3** Determine a typical time of evaporation of metal clusters in a dense buffer gas if this process is restricted by diffusion motion of metal atoms in a buffer gas.

The concentration c of free metal atoms in a buffer gas is small. Because of processes of evaporation and attachment of atoms to a cluster, the atom concentration varies near the cluster. Let us denote their concentration far from the cluster by c_∞. The flux of attached metal atoms to the cluster surface is given by,

$$j = -DN\nabla c \, ,$$

where D is the diffusion coefficient for metal atoms in a buffer gas and N is the number density of buffer gas atoms. Since a number of atoms is conserved in a volume, we have $J = 4\pi R^2 j = \text{const}$, where R is a distance from the cluster. This is the equation for the concentration of metal atoms c in a space

$$J = -4\pi R^2 D \frac{dc}{dR} \, ,$$

and the solution of this equation is

$$c(R) = c_\infty \left(1 - \frac{r}{R}\right) \, ,$$

where c_∞ is the concentration of free metal atoms far from the cluster. Introducing the rate J_{ev} of atom evaporation and J_{at} of atom attachment to the cluster surface, we obtain the total rate of atom attachment

$$J = J_{at} - J_{ev} \, .$$

In the diffusion regime of attachment $J \ll J_{at}$

$$c(R) = c_\infty + \frac{J}{4\pi DNR} . \tag{13.6}$$

The total atom flux toward the cluster is

$$J = J_{at} - J_{ev} = J_{ev} \frac{S - S_o}{S_o} \, , \tag{13.7}$$

where J_{ev} is proportional to the attachment rate that is given by formula (13.7), $S = N/N_{sat}$ is the supersaturation degree, and S_o is the supersaturation degree when the cluster radius is equal to the critical radius. In the case of a dense buffer gas we have $J_{at} \approx J_{ev} \gg J$. Hence, near the cluster surface $S = S_o$, i. e., the number density of metal atoms of the parent vapor decreases until the cluster radius becomes equal to the critical radius. Then the concentration of free atoms of the parent gas at a distance R from the particle center is $c(R) = c_\infty + (c_* - c_\infty)r/R$, where r is a cluster radius, and the rate of cluster evaporation is given by

$$J_{ev} = 4\pi DNr(c_* - c_\infty) \, .$$

Here c_* is the concentration of metal atoms in a buffer gas if the cluster radius is equal to the critical one at which the fluxes of atom evaporation and atom attachment are equal. Assuming $c_* \gg c_\infty$, we obtain the balance equation for cluster evaporation

$$\frac{dn}{dt} = \frac{4\pi r^2 \rho}{m} \cdot \frac{dr}{dt} = 4\pi r D N_{sat}(T) \exp\left(\frac{2\gamma m}{\rho T r}\right) \, .$$

Taking $r \gg 2\gamma m/(\rho T)$, we obtain an evaporation time,

$$\tau_{ev} = \frac{r^2 \rho}{2mDN_{sat}(T)} . \tag{13.8}$$

► **Problem 13.4** Give the criterion for the diffusion and kinetic regimes of cluster equilibrium if the cluster is located in a buffer gas and the equilibrium results from processes of attachment and evaporation of cluster atoms.

In the diffusion regime of cluster equilibrium in a buffer gas, the mean free path of metal atoms is small compared to a cluster radius, while in the kinetic regime the inverse relation between these parameters is fulfilled. It is easy to characterize the competition of these regimes by the parameter

$$\alpha = \sqrt{\frac{T}{2\pi m}}\frac{\xi r}{D}, \tag{13.9}$$

where T is the temperature, m is the mass of a metal atom, D is the diffusion coefficient of metal atoms in a buffer gas, and ξ is the probability of attachment as a result of atom contact with the cluster surface.

From relation (13.6), we obtain the atom concentration c_0 near the cluster surface,

$$c_0 = c_\infty + \alpha(c_* - c_0) ,$$

and the parameter α given by formula (13.9), characterizes competition between the kinetic and diffusion regimes of cluster growth.

We have the following relation between the concentration of metal atoms at the cluster surface:

$$c_0 = \frac{c_\infty + \alpha c_*}{1 + \alpha} ,$$

where their concentration is c_∞ far from the cluster surface and the criterion of the diffusion regime of evaporation corresponds to $c_\infty \ll c_0$. Under the condition $c_\infty \ll c_*$, this corresponds to

$$\alpha = \sqrt{\frac{T}{2\pi m}}\frac{\xi r}{D} \gg 1 .$$

This criterion corresponds to the diffusion character of cluster evaporation if the cluster radius is large compared to the mean free path λ of free parent atoms in the buffer gas. Using the estimation for the diffusion coefficient of atoms $D \sim \lambda\sqrt{\frac{T}{m}}$, we get $\alpha \sim \xi r/\lambda$. Hence, the diffusion regime of cluster growth and evaporation ($\alpha \gg 1$) takes place for large clusters

$$r \gg \frac{\lambda}{\xi} .$$

▶ **Problem 13.5** Determine a typical time of cluster evaporation in a buffer gas for the diffusion regime of the process at a given number density of buffer gas atoms.

In the diffusion regime, the concentration of atoms of a parent gas near the cluster surface is c_* and it exceeds that far from the cluster surface $c_* \gg c_\infty$. Formula (13.8) is valid if $c_* \leq 1$. The above formula must be change if the number density of buffer gas atoms is less than that for metal atoms near the cluster of the critical radius. Taking near the cluster surface the concentration of metal atoms to be $c_0 = 1$, one can transform formula for an evaporation time to the form

$$\tau_{ev} = \frac{r^2 \rho}{2mDN} ,$$

where N is the number density of buffer gas atoms.

▶ **Problem 13.6** For the diffusion regime of atom attachment to the cluster surface, find a typical time of cluster growth on a nucleus of condensation in a buffer gas.

In the diffusion regime of growth of metallic clusters in a buffer gas and a parent metal vapor, the concentration of metal atoms at a distance R from the cluster surface is equal to

$$c(R) = c_\infty \left(1 - \frac{r}{R}\right) .$$

From this it follows the flux of attaching atoms toward the cluster surface is

$$J_{at} = 4\pi r^2 DN \, |\nabla c| = 4\pi r DN c_\infty .$$

Next, the balance equation for the number n of cluster atoms $dn/dt = J_{at}$ gives the law of evolution of the cluster radius

$$r = \sqrt{\frac{mDNc_\infty t}{2\rho}} .$$

13.2
Kinetics of Cluster Growth in Plasma

▶ **Problem 13.7** Growth of metallic clusters in a plasma where an atomic metal vapor is an admixture to a buffer gas, results from the formation of metallic diatomic molecules in three-body collisions and later these diatomic molecules are the nuclei of condensation. Under the assumption that the number density of metallic atoms remains constant until all the free atoms are transformed in bound atoms of clusters, find a time of this process and the cluster size in the end.

Growth of metallic clusters which are formed from atoms in a dense buffer gas proceeds according to the scheme

$$2M + A \rightarrow M_2 + A, \quad M_n + M \longleftrightarrow M_{n+1} . \qquad (13.10)$$

Here M and A are the metallic atom and atom of a buffer gas, respectively, and n is the cluster size—the number of cluster atoms. The rate constant of the pair attachment process $k_n = k_o n^{3/2}$ is given by formula (2.39), if the cluster is modelled by the liquid drop model. Because the first stage of this process has a three-body character, formation of a diatomic metallic molecule is a slow process. Subsequently forming diatomic molecules become the nuclei of condensation and are converted fast in large clusters as a result of attachment of free metallic atoms in pair collision processes. From this it follows that at the end of the nucleation process, when initially free atoms are converted into bound ones, the final clusters are large.

This character of cluster growth is governed by a large parameter

$$G = \frac{k_o}{N_a K} \gg 1 . \qquad (13.11)$$

Here N_a is the number density of buffer gas atoms, K is the rate constant of formation of a diatomic molecule in three-body collisions, and its typical value in this case is $K \sim 10^{-32}\ \mathrm{cm^6/s}$, whereas a typical rate constant of the attachment process is $k_o \sim 3 \cdot 10^{-11}\ \mathrm{cm^3/s}$ for metallic atoms in a buffer gas. Hence, the considering character of cluster growth ($G \gg 1$) takes place if $N_a \ll 3 \cdot 10^{21}\ \mathrm{cm^{-3}}$, i. e., this character of cluster growth is realized in a not dense buffer gas.

The scheme of processes (13.11) leads to the following set of balance equations for the number density N of free metallic atoms, the number density N_{cl} of clusters, the number density of bound atoms N_b in clusters, and cluster size n (a number of cluster atoms) :

$$\frac{dN_b}{dt} = -\frac{dN}{dt} = \int N k_o n^{2/3} f_n dn + K N^2 N_a , \quad \frac{dN_{cl}}{dt} = K N^2 N_a , \qquad (13.12)$$

where f_n is the size distribution function of clusters that satisfies the normalization conditions

$$N_{cl} = \int f_n dn, \quad N_b = \int n f_n dn .$$

The set (13.12) of balance equations describes the character of cluster growth for conditions under consideration. One can see that the ratio of the second and third terms of the right-hand side of the first balance equation (13.12) is of the order $Gn^{2/3}$, where n is a current cluster size n, and since $G \gg 1$, $n \gg 1$, one can neglect the third term in the first balance equation. In addition, we here assume that a typical cluster size n is large compared to the critical cluster size that allows us to neglect the processes of cluster evaporation in the set of balance equations (13.12).

We solve below this set of balance equations where we ignore cluster evaporation processes and for simplicity we assume in these equations $N = \mathrm{const}$, until N_b reaches the value N. Then relation

$$N_b(\tau) = N$$

gives the time τ of the nucleation process.

In determining the size distribution function f_n of clusters, we note that a size $n(t)$ of the cluster satisfies the equation

$$\frac{dn}{dt} = k_0 n^{2/3} N \, ,$$

if a diatomic molecule, a condensation nuclei for this cluster, is formed at $t = 0$. From this, it follows that a size of this cluster at time t is

$$n = \left(\frac{Nk_0 t}{3}\right)^3 . \qquad (13.13)$$

We have $f_n dn$, the number density of clusters with a size range between n and $n + dn$ is proportional to the time range dt, when diatomic molecules are formed which are nuclei of condensation. This corresponds to $f_n dn \sim dt$, which leads to the following size distribution function of clusters:

$$f_n = \frac{C}{n^{2/3}} \quad n < n_{\max} \, , \qquad (13.14)$$

where C is the normalization constant, $n_{\max}$ is the maximum cluster size at this time and $f_n = 0$, if $n > n_{\max}$. This formula may be obtained directly from the kinetic equation for the distribution function f_n if we neglect evaporation processes

$$\frac{\partial f_n}{\partial t} = -\frac{\partial}{\partial n}\left(Nk_0 n^{2/3} f_n\right) \, ,$$

and hence this equation confirms formula (13.14).

Using this distribution function, we determine the cluster parameters at the end of the nucleation process on the basis of the normalization condition and the cluster distribution function (13.14)

$$C = \frac{4N}{3n_{\max}^{4/3}} \, , \qquad N_{cl} = 3Cn_{\max}^{1/3} = \frac{4N}{n_{\max}} \, , \qquad \overline{n} = \frac{n_{\max}}{4} \, ,$$

where $\overline{n}$ is the average cluster size, and according to formula (13.13) we obtain the maximum cluster size at the end of the nucleation process introducing the parameter $\xi = Nk_o\tau$

$$n_{\max}(\tau) = \left(\frac{Nk_o\tau}{3}\right)^3 = \left(\frac{\xi}{3}\right)^3 .$$

Integrating the balance equations (13.12) over time of nucleation and neglecting the second term in the right-hand side of the second equation, we obtain

$$N_{cl} = KN^2 N_a \tau = N\frac{\xi}{G} \, , \qquad N_b(\tau) = \int Nk_0 C n_{\max} dt = \frac{1}{4}\xi C n_{\max} = N \, .$$

This gives,

$$n_{\max} = \frac{4G}{\xi} \, , \qquad \xi = 3\left(\frac{4}{3}G\right)^{1/4} = 3.2G^{1/4} \, ,$$

and because $G \gg 1$, the reduced nucleation time is large. This also leads to large cluster sizes at the end of the nucleation process. Finally, we obtain the cluster parameters at the end of the nucleation process

$$n_{\max} = 1.2G^{3/4}, \quad \overline{n} = 0.31G^{3/4}, \quad N_{\text{cl}} = 3.2NG^{-3/4}, \quad \tau = \frac{3.2NG^{1/4}}{Nk_{\text{o}}} . \tag{13.15}$$

► **Problem 13.8** Metal-containing molecules MX_k (M is a metal atom and X is a halogen atom) are injected in a plasma. Find the conditions when decomposition of molecules leads to the formation of metal clusters.

The chemical equilibrium for metal-containing molecules MX_k in a buffer gas is described by the scheme

$$MX_k \longleftrightarrow M + kX . \tag{13.16}$$

Denote the binding energy per halogen atom by ε_X, so that the total binding energy of atoms in the compound MX_k is $k\varepsilon_X$. In addition, the equilibrium for metal clusters has the form

$$M_n + M \longleftrightarrow M_{n+1} . \tag{13.17}$$

Introducing the binding energy per atom for a bulk metal ε_{M}, we have the following rough criterion of existence of the above chemical compound at low temperatures:

$$\varepsilon_{\text{M}} < k\varepsilon_X .$$

From the chemical equilibrium (13.16) for the component MX_k one can estimate a typical temperature T_1 when this compound is decomposed in atoms, and from the chemical equilibrium of clusters (13.17) a typical temperature T_2 follows when clusters are transformed in atoms

$$T_1 = \frac{\varepsilon_X}{\ln \frac{N_{\text{o}}}{[X]}} , \qquad T_2 = \frac{\varepsilon_{\text{M}}}{\ln \frac{N_{\text{o}}}{[M]}}$$

Here $[X]$ denotes the total number density of free and bound halogen atoms, N_{o} is a typical atomic value, and $[M]$ is the total number density of free and bound metallic atoms. Evidently, clusters are formed form metal-containing molecules and exist in the temperature range

$$T_1 < T < T_2 . \tag{13.18}$$

One can determine the temperature of destruction of clusters T_2 more precisely on the basis of the saturation number density N_{sat} from the formula

$$[M] \sim N_{\text{sat}}(T_2) .$$

Evidently, if $[X] \sim [M]$, the possibility of existence of clusters corresponds to the criterion

$$\varepsilon_X < \varepsilon_{\text{M}} .$$

► **Problem 13.9** Assuming the binding energies of bonds $MX_k - X$ for different radicals to be identical, analyze the kinetics of radical formation when metal-containing molecules MX_6 are injected in a hot dense plasma and metal clusters are produced after decomposition of these molecules.

In the course of decay, a metal-containing molecule MX_k is transformed into atoms and radicals, and then metal atoms are converted into metal clusters, whereas halogen atoms and molecule leave a region where metal-containing molecules MX_6 are located at the beginning (we assume a dimension of this region to be small in comparison with the plasma dimension). Assuming the identical binding energies ε_X for each bond $MX_k - X$ ($k = 1 \div 5$), we use the following rates of decay of molecules and radicals in collisions with buffer gas atoms:

$$\nu_d = N_a k_{gas} \exp\left(-\frac{\varepsilon_X}{T}\right). \tag{13.19}$$

Here N_a is the number density of buffer gas atoms, T is a current temperature of buffer gas atoms, k_{gas} is the gas-kinetic rate constant for collisions of molecules and radicals with atoms of a buffer gas, and we assume the rate ν_d to be independent of the number of halogen atoms k in the molecule or radicals. On the basis of this model, below we analyze kinetics of destruction of metal-containing molecules in a hot buffer gas.

Under the above conditions, the set of balance equations for the radical concentrations c_k, if a radical contains k atoms, has the form

$$\frac{dc_0}{dt} = \nu_d c_1, \quad \frac{dc_k}{dt} = -\nu_d(c_k - c_{k+1}), \quad k = 1 \div 5; \quad \frac{dc_6}{dt} = -\nu_d c_6 . \tag{13.20}$$

From the beginning, the system consists of molecules that corresponds to the following boundary conditions

$$c_6(0) = c_M; \quad c_k = 0, \quad k \neq 6 ,$$

where c_M is the total concentration of free and bound metallic atoms, and we ignore here transport processes.

This set of balance equations with given boundary conditions may be solved analytically, and the solution is

$$c_k = c_M \frac{x^{6-k}}{(6-k)!} e^{-x}, \quad c_0 = c_M [1 - f(x)]; \quad f(x) = e^{-x} \sum_{k=0}^{5} \frac{x^k}{k!}, \tag{13.21}$$

where $k = 1 \div 6$, $x = \int_0^t \nu_d dt$, $f(x)$ is a part of molecules and radicals, i.e., $1 - f(x)$ is a part of free metallic atoms. Figure 13.1 shows the dependence $f(x)$.

► **Problem 13.10** Metal-containing molecules MX_6 injected in a hot buffer gas plasma. Find cooling of a buffer gas as a result of transformation of metal-containing molecules in metal clusters.

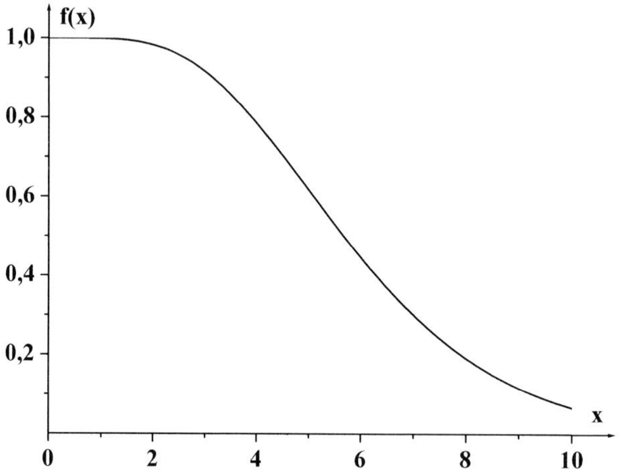

Fig. 13.1 The function $f(x)$ in accordance with formula (13.21).

Assuming a prompt cooling of a buffer gas after decomposition of molecules and radicals, we obtain from the set of balance equations (13.20) and their solution (13.21) the following heat balance equation in the course of molecule destruction

$$\frac{dT}{dt} = \delta T \sum_{k=0}^{5} \left(\frac{dc_{k+1}}{dt} - \frac{dc_k}{dt} \right) = \delta T \left(\frac{dc_6}{dt} - \frac{dc_0}{dt} \right) = -\nu_d c_M \delta T f(x) ,$$

where δT is the cooling temperature for a buffer gas per metal-containing molecule and per broken bond. We also include in this parameter the heat release resulted from conversion of metal atoms into clusters.

The solution of this set of equations gives the temperature of a buffer gas after destruction of molecules and transformation of metal atoms into metal clusters

$$T = T_0 - \delta T \sum_{k=0}^{5} (6-k) c_k = T_0 - c_M \delta T F(x) , \tag{13.22}$$

where

$$F(x) = 6 - e^{-x} \left(\frac{x^5}{120} + \frac{x^4}{12} + \frac{x^3}{2} + 2x^2 + 5x + 6 \right) . \tag{13.23}$$

Here the function $F(x)$ given in Fig. 13.2 is a number of broken bonds per molecule, $T_0 = T(0)$ is the initial temperature, and the total temperature of the buffer gas cooling is

$$T_0 - T(\infty) = 6 c_M \delta T$$

This formula is based on assumption that a buffer gas has enough high heat capacity, that is valid at low concentrations of metal-containing molecules.

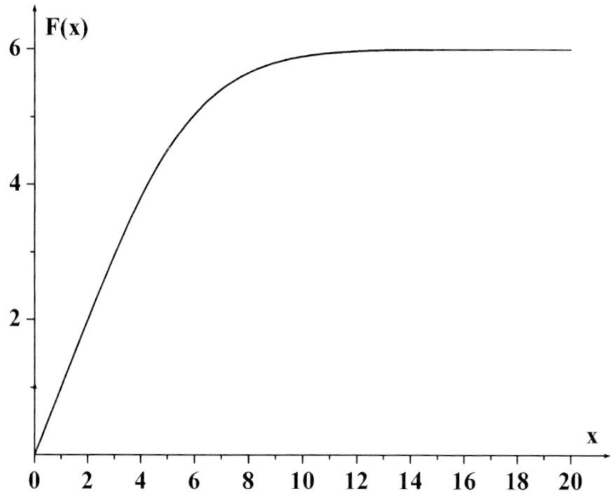

Fig. 13.2 The function $F(x)$ defined by formula (13.23).

► **Problem 13.11** Metal-containing molecules MX_6 injected in a flow of a hot dense plasma, are located in the beginning in a cylinder region of a radius ρ_0. Analyze the thermal regime of transformation of metal-containing molecules into clusters if heat release is determined by thermal conductivity of a buffer gas.

Since the binding energy of metal-containing molecules ε_X is large in comparison with a thermal energy $\sim T$, the process of molecule destruction acts strongly on the heat balance of a buffer gas with a metal admixture in spite of a small concentration of metal-containing molecules in a buffer gas. On the basis of equation (13.22), the heat balance equation considering thermal conductivity of a buffer gas is

$$\frac{\partial T}{\partial t} = \chi \Delta T - -\nu_d c_M \delta T f(x) ,$$

where χ is the thermal diffusivity coefficient of a buffer gas, and we ignore diffusion of molecules and radicals outside this region.

In considering this equation in the case, when metal-containing molecules occupy a cylindrical region in a plasma of a radius ρ_0, we reduce this equation, to a quasi-stationary regime. Let the buffer gas temperature be T_0 in the region $\rho \geq \rho_0$, and be T_* at the center. Then this equation is reduced to the form

$$\frac{T_0 - T_*}{\tau_{eq}} = \nu_d c_M \delta T f(x), \quad \tau_{eq} = \frac{0.17\rho_0^2}{\chi} , \tag{13.24}$$

where τ_{eq} is a typical time of equilibrium establishment.

Another typical time of this process is the total time of destruction of metal-containing molecules that is equal to

$$\tau_0 = \frac{6}{\nu_d(T_*)} = \frac{6}{N_a k_{gas} \exp\left(-\frac{\varepsilon_X}{T_0}\right)} .$$

Since we consider the quasi-stationary regime of destruction of metal-containing molecules and heat transport, the following criterion must be fulfilled:

$$\tau_{eq} \ll \tau_o \,.$$

Table 13.1 is a demonstration of the above results and gives both the parameters of compounds of some metal-containing molecules (ρ is the density at room temperature, T_m, T_b are the melting and boiling temperatures, respectively) and parameters of kinetics of destruction of these molecules in hot argon. We there represent an initial stage of molecule destruction $x = 0$, and take the optimal buffer gas temperature far from the destruction region such that $N_m/N_{sat}(T_*) \approx 10$ for MoF_6, WF_6, and $N_m/N_{sat}(T_*) \approx 1000$ for IrF_6, WCl_6. We assume metal-containing molecules to be located in a cylinder of radius $\rho_o = 1$ mm, the argon pressure is $p = 1$ atm, and the concentration of metal-containing molecules is $c_M = 10\,\%$ with respect to argon atoms. In addition, we take gas-kinetic cross section to be $\sigma_{gas} = 3 \cdot 10^{-15}$ cm^2 and use this value in the rate constant of destruction of metal-containing molecules and their radicals in collisions with argon atoms.

Table 13.1 Parameters of evolution of metal-containing molecules in hot argon ($p = 1$ atm, $c_M = 10\,\%$, $\rho_o = 1$ mm).

Compound	MoF_6	IrF_6	WF_6	WCl_6
ρ(g/cm^3)	2.6	6.0	3.4	3.5
T_m K	290	317	276	548
T_b K	310	326	291	620
ε_X (eV)	4.3	2.5	4.9	3.6
ε_M (eV)	6.3	6.5	8.4	8.4
T_1 (K)	2200	1200	2500	1700
T_2 (K)	4100	4000	5200	5200
$\delta T(10^3$ K)	12	3.7	13	4.8
T_o (K)	3700	6000	5000	5800
T_* (K)	3600	2900	4600	4200
τ_{eq} (10^{-5} s)	12	17	8	9
τ_o (10^{-3} s)	5	0.1	1	0.1
N_m (10^{17} cm^{-3})	2.0	2.5	1.6	1.8

Thus, the heat balance of a mixture consisting of a dense hot buffer gas and metal-containing molecules, includes heat absorption due to destruction of metal-containing molecules and heat release as a result of joining of metal atoms in clusters which are compensated by heat transport from surrounding regions.

► **Problem 13.12** Analyze nucleation in a hot buffer gas with metal-containing molecules within the framework of a model where the rate of formation of free metal atoms Q is approximated by an appropriated dependence.

If criterion (13.18) is fulfilled, thermodynamic equilibrium of a buffer gas with an admixture of metal-containing molecules corresponds to destruction of molecules and formation of a condensed metal. This leads to the following scheme of basic processes:

$$MX_k \rightarrow M + kX, \quad M + M_n \rightarrow M_{n+1} . \tag{13.25}$$

Note that in contrast to nucleation in a pure metal vapor that proceeds according to the scheme (13.10), one can ignore now slow three-body processes, and the first stage of cluster growth processes due to formation of diatomic molecules proceeds as

$$MX + M \rightarrow M_2 + X .$$

Hence, in this case a buffer gas does not account in the nucleation process, so that the nucleation rate does not depend on the number density of buffer gas atoms. But a buffer gas determines destruction of metal-containing molecules and their radicals, so that the specific rate of formation of free metal atoms as a result of molecule destruction is $Q = N_a \nu_d = N_a N_b k_d$, where N_a is the number density of buffer gas atoms, N_b is the number density of bound atoms at the end of the destruction process, and the rate ν_d of this process is given by formula (13.19).

Note that the nucleation process proceeds in a restricted space region where metal-containing molecules are located from the beginning. Free halogen atoms and molecules go out the cluster region, whereas clusters remain in this region because of a low mobility. Thus, we neglect the role of halogen atoms in cluster growth, and though the nucleation process is considered in a uniform system, for other processes it is not uniform.

Under the above conditions, we obtain the following balance equations for the number density of free metallic atoms N, the number density of clusters N_{cl} and a typical cluster size n (a number of cluster atoms)

$$\frac{dN}{dt} = Q - N_{cl} k_n N, \quad \frac{dN_{cl}}{dt} = k_{ch} N^2, \quad \frac{dn}{dt} = k_n N . \tag{13.26}$$

Here $k_n = k_0 n^{2/3}$ is the rate constant of atom attachment to a cluster, k_{ch} is the rate constant of the process of formation of a metal diatomic molecule M_2. We consider a diatomic metal molecule that is transformed later in a growing cluster as a nucleus of condensation, and attachment of atoms to it takes place in pair collisions as well for larger molecules and clusters.

The analysis of this set of balance equations shows that the cluster number density N_{cl} and cluster size n grow in time, while the number density of free atoms N grows on the first stage of the cluster growth process and drops on its second stage. Roughly, one can divide the time in three ranges, as it is shown in Fig. 13.3, so that at $t \leq \tau_{max}$ the number density of free atoms grows, and at a time $\tau \sim 1/\nu_d$ metal-containing molecules are destructed.

In the second stage of the nucleation process ($\tau > t > \tau_{max}$), we have

$$Q = N_a N_b k_d \sim N_{cl} \frac{dn}{dt} \sim k_{ch} N^2 n ,$$

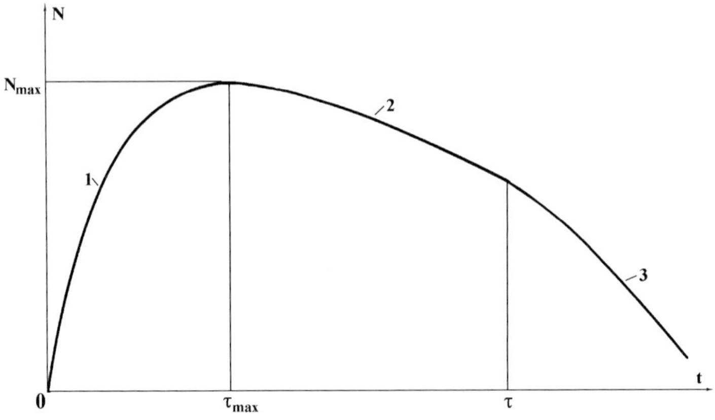

Fig. 13.3 Time dependence for the number density of free metal atoms in the course of formation of metal clusters from metal-containing molecules in a buffer gas.

and assuming $k_d \sim k_{ch}$, we obtain in the second stage of the nucleation process

$$N \sim \sqrt{\frac{N_a N_b}{n}} \,. \tag{13.27}$$

Note that from the first equation of set (13.26) the maximum N is attained at time

$$\tau_{max} \sim \frac{N}{Q} \sim \frac{N}{N_a N_b k_d} \,,$$

and since a typical time of decay of metal-containing molecules is

$$\tau \sim \frac{1}{N_b k_d} \,,$$

this gives

$$\tau \gg \tau_{max} \,.$$

This means that the time dependence for the number density of free atoms N has the form of Fig. 13.3, and the maximum of this value is achieved before total destruction of metal-containing molecules.

In addition, we have from the last equation of set (13.26) and relation (13.27) for the number density of free atoms the following estimate for the cluster size n in the end of the nucleation process

$$n \sim (k_o N \tau)^3 \sim (k_o \tau)^3 \left(\frac{N_a N_b}{n}\right)^{3/2} \,,$$

which gives

$$n \sim (k_o \tau)^{6/5} (N_a N_b)^{3/5} \sim \left(\frac{k_o}{k_d}\right)^{6/5} \left(\frac{N_a}{N_b}\right)^{3/5} \,,$$

and each factor of this product is large.

In the same manner, using the above formulas for τ_{max} and formula (13.27) for the number density of free metal atoms at this time, we obtain the cluster size at this time

$$n \sim \left(\frac{k_o}{nk_d}\right)^3 ,$$

which gives

$$n \sim \left(\frac{k_o}{k_d}\right)^{3/4} \gg 1 ,$$

and since $k_o \gg k_d$, we deal with large clusters on this stage of the nucleation process. Thus, the cluster growth process through decomposition of metal-containing molecules in a buffer gas proceeds mostly at large cluster sizes.

► **Problem 13.13** Within the framework of the liquid drop model for clusters derive the kinetic equation for evolution of the size distribution function of clusters as a result of cluster coagulation.

Coagulation is one of the mechanisms of cluster growth (see Fig. 13.4) that proceeds according to the scheme,

$$M_{n-m} + M_m \rightarrow M_n . \tag{13.28}$$

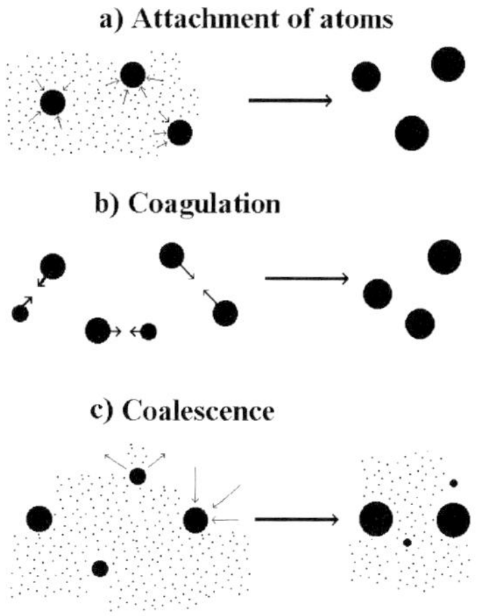

Fig. 13.4 Mechanisms of cluster growth: a) attachment of free atoms to clusters; b) the coagulation process results in joining clusters at their contact; c) the coalescence process that follows from cluster equilibrium with a parent vapor that is created by cluster evaporation and is disappeared by atom attachment to clusters; as a result, small clusters evaporate, large clusters grow, and the average cluster size increases.

Because of this character of cluster growth, evolution of the size distribution function f_n is described by the Smolukhowski equation

$$\frac{\partial f_n}{\partial t} = -f_n \int k(n,m) f_m dm + \frac{1}{2} \int k(n-m,m) f_{n-m} f_m dm \,. \tag{13.29}$$

Here $k(n-m,m)$ is the rate constant of process (13.29), the factor 1/2 considers the collisions of clusters consisting of $n-m$ and m atoms are present in the equation twice, and the distribution function is normalized as $\int f_n dn = N_{\rm cl}$, where $N_{\rm cl}$ is the number density of clusters.

▶ **Problem 13.14** Show that the total number density $N_{\rm o}$ of bound atoms is conserved in during the coagulation process.

The total number density of bound atoms is $N_{\rm o} = \int_0^\infty n f_n dn$. We obtain the equation for this value, multiplying equation (13.29) by n and integrating over dn, which gives

$$\frac{dN_{\rm o}}{dt} = -\int nk(n,m) f_n dn f_m dm + \frac{1}{2} \int nk(n-m,m) f_{n-m} f_m dn dm \,,$$

and in the second integral $n > m$. Changing $n-m$ by n in the right-hand side of this relation, we obtain the terms to be mutually cancelled, and $dN_{\rm o}/dt = 0$, i. e., with the framework of the Smolukhowski equation, the total number density $N_{\rm o}$ of bound atoms is conserved in this process.

▶ **Problem 13.15** Find the time dependence of the average cluster size when it is large compared to the mean free path of gaseous atoms and then the rate constant of joining of two clusters is independent of the cluster size.

Taking the rate constant of two clusters joining $k_{\rm as}$ to be independent of cluster size, we reduce equation (13.29) to the form

$$\frac{\partial f_n}{\partial t} = -k_{\rm as} f_n \int_0^\infty f_m dm + \frac{1}{2} k_{\rm as} \int_0^n f_{n-m} f_m dm \,.$$

Multiplying this equation by n^2 and integrating over dn, we obtain

$$\frac{d}{dt} \int_0^\infty n f_n dn = -k_{\rm as} \int_0^\infty n f_n dn \int_0^\infty f_m dm + \frac{1}{2} k_{\rm as} \int_0^\infty n dn \int_0^n f_{n-m} f_m dm = 0 \,.$$

This means that the total number density of bound atoms $N_{\rm o} = \int_0^\infty n f_n dn$ is conserved during cluster growth. If we multiply the kinetic equation by n^2 and integrate over dn, we get

$$\frac{d}{dt} \int_0^\infty n^2 f_n dn = \frac{d\overline{n}}{dt} = -k_{\rm as} \int_0^\infty n^2 f_n dn \int_0^\infty f_m dm + \frac{1}{2} k_{\rm as} \int_0^\infty n^2 dn \int_0^n f_{n-m} f_m dm = \frac{1}{2} k_{\rm as} N_{\rm o} \,,$$

where we define the average clusters size $\overline{n}$ as

$$\overline{n} = \frac{\int\limits_0^\infty n^2 f_n dn}{\int\limits_0^\infty n f_n dn} .$$

This gives for evolution of the average cluster size,

$$\overline{n} = \frac{1}{2} k_{as} N_0 t , \tag{13.30}$$

if in the beginning the average size is relatively small.

► **Problem 13.16** Find the size distribution function of clusters during their coagulation if a typical cluster size is large compared to the mean free path of gaseous atoms, so that the rate constant of joining of two clusters does not depend on the cluster size.

Let us introduce the concentration of clusters of a given size $c_n = f_n N_0$, where N_0 is the total number density of bound atoms in clusters. The normalization condition for the cluster concentration is $\sum\limits_n n c_n = 1$, and the kinetic equation (13.29) in terms of cluster concentrations has the form

$$\frac{\partial c_n}{\partial \tau} = -c_n \int\limits_0^\infty c_m dm + \frac{1}{2} \int\limits_0^n c_{n-m} c_m dm \tag{13.31}$$

where the reduced time is $\tau = N_0 k_{as} t$. The solution of this equation is

$$c_n = \frac{4}{\overline{n}^2} \exp\left(-\frac{2n}{\overline{n}}\right) . \tag{13.32}$$

This expression satisfies the normalization condition $\int_0^\infty n c_n dn = 1$ and the average cluster size $\overline{n}$ corresponds to formula (13.30). Indeed, substituting (13.32) in the kinetic equation, we confirm formula (13.30) $\overline{n} = \tau$.

► **Problem 13.17** Within the framework of the liquid drop model for association of two clusters, determine the time dependence for the average cluster size if it is small compared to the mean free path of gaseous atoms.

In this case the cross section of association of two clusters σ_{as} is similar to the association of two liquid drops and is equal to

$$\sigma_{as} = \pi r_1^2 + \pi r_2^2 ,$$

where r_1, r_2 are the radii of colliding clusters. Correspondingly, the rate constant of association of two clusters $k(n, m)$ of sizes n and m which within the framework of the liquid drop model for clusters is given by

$$k(n, m) = k_0 (n^{1/3} + m^{1/3})^2 \sqrt{\frac{n+m}{nm}} ,$$

where k_o is given by formula (2.39). Multiplying the Smolukhowski equation (13.29) by n^2 and integrating over dn by using the normalization condition $\int n f_n dn = N_b$, where (N_b is the number density of bound atoms in clusters), leads to the following relation:

$$\frac{d\overline{n}}{dt} = k_o N_b I \overline{n}^{1/6} , \qquad I = \frac{1}{2}\int_0^\infty \int_0^\infty (x^{1/3} + y^{1/3})^2 \sqrt{\frac{x+y}{xy}} \exp(-x-y) dx dy = 3.5 ,$$

where we use the notations $x = n, y = m$. This gives

$$\overline{n} = 3.6(N_b k_o t)^{1.2} . \tag{13.33}$$

Because we use the assumption $n \gg 1$, these formulas are valid under the condition

$$k_o N_b t \gg 1 .$$

▶ **Problem 13.18** A buffer gas containing liquid clusters expands in a vacuum. Taking into account nucleation of small clusters (in comparison to the mean free path) according to the scheme (13.28), determine the mean size of small clusters at the end of the process. Assume the diffusion character of the motion of clusters in a buffer gas, and take the typical size of clusters to be larger than the mean free path of gaseous atoms.

Under these conditions, the normalization condition takes the form $\sum_n n f_n = N_0 \exp(-t/\tau_{ex})$, where τ_{ex} is the typical expansion time. Correspondingly, the kinetic equation (13.31) is transformed to the form :

$$\frac{\partial c_n}{\partial \tau} = -\frac{c_n}{\tau_{ex}} - N_0 c_n \int_0^\infty k_{n,m} c_m dm + \frac{1}{2} N_0 \int_0^n k_{n-m,m} c_{n-m} c_m dm ,$$

where N_0 is the initial total number density of bound atoms. According to this equation, during expansion the rates of cluster collisions decrease, and at the end of the process the buffer gas becomes so rare, that cluster collisions cease. This gives the mean cluster size at the end of the process $\overline{n} \sim N_0 k_{dif} \tau_{ex}$.

For a more accurate determination of the mean particle size at the end of the expansion process let us multiply the above equation by n^2 and integrate the result over dn. Using the normalization condition $\sum_n n c_n = \exp(-t/\tau_{ex})$, we obtain

$$\frac{d\overline{n}}{dt} = N_0 k_o \overline{n}^{1/6} I e^{-t/\tau_{ex}} ,$$

where the integral I is

$$I = \frac{1}{2}\int_0^\infty \int_0^\infty (x^{1/3} + y^{1/3})^2 \sqrt{\frac{x+y}{xy}} \exp(-x-y) dx dy = 3.5 .$$

The solution of the above equation in the limit $t \to \infty$ is

$$\overline{n} = 3.6 \left(N_0 k_o \tau_{ex}\right)^{1.2} . \tag{13.34}$$

This formula corresponds to formula (13.33).

▶ **Problem 13.19** Determine the number of released atoms resulting from joining two large clusters because of heat release in this process.

Take the binding energy of a large cluster consisting of n atoms in accordance with formula (13.3) $E = \varepsilon_0 n - An^{2/3}$. Therefore, process (13.28) leads to the energy release

$$\Delta E = A\left[m^{2/3} + (n-m)^{2/3} - n^{2/3}\right] .$$

In particular, in the case $m = n/2$, when this function of m at a given n has a maximum, this formula gives

$$\Delta E_{\text{max}} = 0.25An^{2/3} .$$

Thus, the released energy at cluster coagulation results from the surface cluster energy, which is small compared to the total binding energy of the cluster atoms, but can exceed the binding energy of one surface atom.

Release of this energy leads to an increase of the temperature of a forming cluster compared to that of the joining clusters, which can cause the release of surface atoms. Note that an excited cluster can emit only one atom, because evaporation of molecules and fragments is characterized by a small probability. Hence, an excited cluster resulting from the joining of two clusters can subsequently emit several atoms, and we have a more complicated process compared to process (13.28), which proceeds according to the scheme

$$M_{n-m} + M_m \rightarrow M_{n-q} + qM . \tag{13.35}$$

Let us find a number of released atoms under the assumption that in the end of process (13.35) the temperature of a formed cluster is equal to the temperature of colliding clusters. This means that the released energy is spent on the liberation of atoms, so that the number of released atoms is given by

$$q = \Delta E\,(dE/dn)^{-1} ,$$

where dE/dn is the binding energy of the surface atoms. In the case of a large cluster we take $dE/dn = \varepsilon_0 - 2A/(3n^{1/3}) \approx \varepsilon_0$, so that the maximum number of released atoms as a result of the formation of a cluster containing n atoms is given by

$$q_{\text{max}} = 0.25An^{2/3}/\varepsilon_0$$

Thus, this effect of cluster heating resulting from joining of clusters can be responsible for the formation of free atoms in an expanding nucleating vapor, but the number of released atoms is small compared to the number of atoms in a formed cluster.

13.3 Charging of Clusters

▶ **Problem 13.20** Determine a current of positive ions of a plasma on the surface of a small spherical cluster located in a weakly ionized gas if its radius greatly exceeds the mean free path of atoms of a gas.

We consider that motion of ions is determined by their diffusion in a gas and that they drift under the action of an electric field of a charged cluster. Then the number density of ions equals zero on the cluster surface and tends to the equilibrium value at large distances from the cluster. Taking the cluster charge to be Z, we obtain the current I of positive ions toward the cluster at a distance R from it

$$I = 4\pi R^2 \left(-D_+ \frac{dN}{dR} + K_+ E N \right) e \,.$$

The first term corresponds to diffusion motion of ions, the second term corresponds to their drift motion in a gas, N is the number density of ions, D_+, K_+ are the diffusion coefficient and the mobility of positive ions in a gas, and e is the ion charge (for simplicity. We take the ion charge to be identical to the electron charge e) and $E = Ze/R^2$ is the electric field strength from a charged cluster. Using the Einstein relationship (8.66) $D_+ = eK_+/T$, where T is the gaseous temperature, we obtain the positive ion current on the cluster surface

$$I = -4\pi R^2 D_+ e \left(\frac{dN}{dR} - \frac{Ze^2 N}{TR^2} \right) .$$

Considering that ions do not recombine in space, we observe that the ion current is of independent R. Then one can consider the above relation for the ion current as an equation for the ion number density. Solving this equation with the boundary condition $N(r) = 0$, we obtain

$$N(R) = \frac{I}{4\pi D_+ e} \int_r^R \frac{dR'}{(R')^2} \exp\left(\frac{Ze^2}{TR'} - \frac{Ze^2}{TR} \right) = \frac{IT}{4\pi D_+ Ze^3} \left[\exp\left(\frac{Ze^2}{Tr} - \frac{Ze^2}{TR} \right) - 1 \right]$$

Let us assume that at large R the ion number density tends to the equilibrium value in a plasma N_+ far from the cluster. This leads to the following expression for the ion current:

$$I_+ = \frac{4\pi D_+ N_+ Ze^3}{T \left\{ \exp[Ze^2/(Tr)] - 1 \right\}} \,. \tag{13.36}$$

This is the Fuks formula.

▶ **Problem 13.21** Analyze the limiting case of the Fuks formula (13.36) for the ion current on the surface of a small neutral particle.

First, we consider the limit when the particle charge tends to zero. Then the Fuks formula (13.36) is transformed into the Smoluchowski formula for the diffusion flux J_0 of neutral particles on the surface of an absorbed sphere

$$J_0 = \frac{I_+}{e} = 4\pi D_+ N_+ r \,. \tag{13.37}$$

This formula is valid if the cluster radius is large compared to the mean free path of atoms that allows us to consider the diffusion character of atom motion near the cluster surface.

► **Problem 13.22** Analyze the limiting case of the Fuks formula (13.36) for the current of negative ions on the surface of a positively charged cluster.

The Fuks formula (13.36) describes the positive ion current if the particle charge has the same sign. In order to obtain the expression for negative ion current, it is necessary to substitute in this formula $Z \to -Z$, and the parameters of positive ions must be replaced by the parameters of negative ions. Then we obtain the negative ion current toward the cluster as

$$I_- = \frac{4\pi D_- N_- Ze^3}{T \cdot \left[1 - \exp\left(-\frac{Ze^2}{Tr}\right)\right]} \,. \tag{13.38}$$

In the limit $Ze^2/(rT) \gg 1$ this formula is transformed into the Langevin formula

$$I_- = \frac{4\pi Ze^3 D_-}{T} = 4\pi Ze^2 K_- \,. \tag{13.39}$$

Note that the relation between the flux of negatively charged ions on the cluster surface J_- and the current I_- of ions are connected by the relation $I_- = eJ_-$.

► **Problem 13.23** Generalize formula for the current of positive and negative ions on the surface of positively charged spherical cluster.

Introducing the reduced variable $x = |Z|e^2/(rT)$, one can represent the Fuks formula (13.36) in the form

$$I_< = \begin{cases} eJ_0 x/(e^x - 1), & Ze^2 > 0, \\ eJ_0 x/(1 - e^{-x}), & Ze^2 < 0 \end{cases} , \quad \lambda \ll r \,. \tag{13.40}$$

Here J_0 is the diffusion flux of neutral atomic particles on the surface of an absorbed sphere of radius r according to the Smoluchowski formula (13.37) $J_0 = 4\pi DNr$, where N is the number density of atoms and D is its diffusion coefficient of atoms in a gas. In the limiting case $x \gg 1$ this formula is transformed into the Langevin formula (13.39).

► **Problem 13.24** Determine the average charge of a spherical cluster located in a quasineutral plasma if its radius is large in comparison to the mean free path of gas atoms.

In a quasineutral plasma $N_+ = N_-$ the currents of positive and negative ions given by formulas (13.36) and (13.38) are identical. This yields for the cluster equilibrium charge.

$$Z = \frac{rT}{e^2} \ln \frac{D_+}{D_-} . \qquad (13.41)$$

From this it follows that the cluster has a positive charge if $D_+ > D_-$, i. e., positive ions have a greater mobility than negative ones. Note that formula (13.41) is valid under the condition

$$r \gg \frac{e^2}{T} .$$

Under fulfilment of this criterion, an individual ion captured by the cluster does not vary essentially the cluster electric potential. An additional criterion of validity of the above expressions is $r \gg \lambda$. At room temperature, the first criterion gives $r > 0.06\ \mu$, and according to the second criterion for atmospheric air we have $\lambda = 0.1\ \mu$.

▶ **Problem 13.25** Find a charge of a large spherical cluster located in a nonquasineutral plasma if its radius is large compared to the mean free path of gas atoms.

Taking the number density of positive N_+ and negative ions N_- far from a charged cluster to be different, we repeat derivation of previous problem for ion currents and the average cluster charge Z. As a result, we obtain formula (13.41)

$$Z = \frac{rT}{e^2} \ln \frac{D_+ N_+}{D_- N_-} . \qquad (13.42)$$

▶ **Problem 13.26** Determine the average charge of a small spherical particle located in a plasma of a glow gas discharge if the particle radius is large compared to the mean free path of gas atoms.

A plasma of the positive column of a glow gas discharge includes electrons and ions, so that the distribution function of electrons on velocities differs from the Maxwell distribution. The Fuks formula (13.36), without using the Einstein relation between the mobility and diffusion coefficient of attached charged particles, has the form for the current I_e of electrons on a small particle

$$I_e = \frac{4\pi K_e N_e Z e^3}{1 - \exp[-Ze^2/(T_{ef} r)]} .$$

Here we introduce the effective electron temperature T_{ef} as $T_{ef} = eD_e/K_e$ instead of the electron temperature T_e, and K_e, D_e are the mobility and diffusion coefficient of electrons in a gas. We take a cluster charge to be $-Z$, and use a typical electron energy eD_e/K_e to be large in comparison to a typical ion energy. Then according to the Langevin formula (13.39) the ion current I_+ on the cluster surface is

$$I_+ = 4\pi Z e^2 K_+ N_+ ,$$

where N_+ is the number density of positive ions far from the cluster and r is the cluster radius. Equalizing the fluxes of charged particles, taking into account the plasma quasineutrality $N_e = N_+$, we obtain the cluster charge

$$Z = -r\frac{D_e}{eK_e}\ln\frac{K_e}{K_+} . \tag{13.43}$$

According to this formula, we have $Ze^2/(rT) \gg 1$, where T can corresponds to the temperature of both electrons and ions.

▶ **Problem 13.27** Determine the current of ions on the surface of a spherical cluster if its radius is small compared to the mean free path of gas atoms.

For a neutral particle, the rate J of atom attachment to the cluster surface is

$$J_0 = \frac{1}{4}\overline{v}\cdot 4\pi r^2 N_0 = \sqrt{\frac{T}{2\pi m}}\cdot \pi r^2 N_0 ,$$

where the average thermal velocity of atoms is $\overline{v} = \sqrt{8T/(\pi m)}$, so that T is the gas temperature, m is the atom mass, N_0 is the number density of atoms. If a cluster charge is Ze, and the mean free path of atoms exceeds the parameter e^2/T ($\lambda \gg e^2/T$), then collisions with atoms are not essential for the collision of ions with the cluster. Because the distance of closest approach r_0 is connected with the impact parameter of collision ρ by relation (2.15) $1 - \rho^2/r_0^2 = Ze^2/(r_0\varepsilon)$, where ε is the collision energy in the center-of-mass system of axes, we have that the cross section of ion collision with the cluster surface under these conditions is

$$\sigma = \pi r^2\left(1 - \frac{Ze^2}{r\varepsilon}\right) . \tag{13.44}$$

If $Z > 0$, i.e., the charges of the cluster and ions have the same sign, it is necessary to consider that if $\varepsilon \leq Ze^2/r$, then the cross section equals to zero, because in this case the potential energy of repulsion of the cluster and colliding ion exceeds their kinetic energy near the cluster surface. From this it follows that the rate constant of contact between a colliding ion and cluster after average over the ion energies on the basis of the Maxwell ion distribution is

$$k = \langle v\sigma\rangle = k_r \exp\left(-\frac{|Z|\,e^2}{rT}\right) ,$$

where r is the cluster radius and $k_r = \pi r^2\sqrt{8T/(\pi m)}$. In the case of attraction of the particle and ion, the mean rate constant of their collision is given by

$$\sigma = \pi r^2\left(1 + \frac{Ze^2}{r\varepsilon}\right) , \qquad k = k_r\left(1 + \frac{|Z|\,e^2}{rT}\right) . \tag{13.45}$$

Introducing the reduced parameter $x = |Z|\,e^2/(rT)$, we combine the above formulas for the ion current to the cluster and obtain, assuming that each contact of an ion with the cluster surface leads to its attachment,

$$I_> = ek_r N_i \cdot \begin{cases} (1+x), & Ze^2 < 0 \\ \exp(-x), & Ze^2 > 0 \end{cases} \quad \lambda \gg r . \tag{13.46}$$

► **Problem 13.28** Generalize formula for the ion current to the cluster surface for arbitrary relation between a cluster radius and the mean free path of gas atoms.

In order to combine formulas (13.40) and (13.46) for the ion current on the surface of a spherical particle, let us first consider the limiting case $\lambda \gg r$ with a general boundary condition on the cluster surface $N(r) = N_1 \neq 0$. Using expression (13.46) for the current near the cluster surface, let us find the ion number density in an intermediate region where it varies from the value N_1 near the cluster surface up to the value N_i, which it has far from the cluster. Repeating the operations that we used at deduction of the Fuks formula (13.36), we have

$$N(R) - N_1 = \frac{I}{4\pi e D_i} \int_r^R \frac{dR'}{(R')^2} \exp\left(\frac{Ze^2}{TR'} - \frac{Ze^2}{TR}\right)$$
$$= \frac{IT}{4\pi D_i Ze^3}\left[\exp\left(\frac{Ze^2}{Tr} - \frac{Ze^2}{TR}\right) - 1\right].$$

Using the second boundary condition $N(\infty) = N_i$, we obtain the ion current

$$I = \frac{4\pi D_i (N_i - N_1) Ze^3}{T\left[\exp\left(\frac{Ze^2}{Tr}\right) - 1\right]}.$$

Taking the boundary value N_1 of the ion number density such that formula (13.46) is valid for the ion current leads to the following expression for the ion current

$$I = \left(\frac{1}{I_>} + \frac{1}{I_<}\right)^{-1}, \tag{13.47}$$

where the ion currents $I_<$ and $I_>$ correspond to formulas (13.40) and (13.46). Formula (13.47) is transformed into formula (13.40) in the limit $I_< \gg I_>$, and into formula (13.46) for the opposite relation $I_< \ll I_>$ between these currents. Thus, formula (13.47) includes both relations between the problem parameters. Note that the ratio of these currents is estimated as

$$\frac{I_<}{I_>} \sim \frac{\xi r}{\lambda}$$

where ξ is the probability of ion attachment to the cluster surface at their contact.

► **Problem 13.29** A spherical cluster whose radius is small compared to the mean free path of ions in a gas is charged by attachment of electrons and positive ions. Find the cluster charge if the temperature of electrons and ions is identical.

The charge of clusters in a weakly ionized buffer gas is negative because of a more high mobility of electrons. We assume the Maxwell distribution function of electrons and each contact of an electron and ion with the cluster surface leads to transferring of their charge to the cluster. We obtain the rate of electron attachment to the cluster surface of a radius $r \ll \lambda$ (λ is the mean free path of gas atoms)

$$J_e = \frac{2}{\sqrt{\pi}} \int_x^\infty z^{1/2} e^{-z} dz \sqrt{\frac{2\varepsilon}{m_e}} \pi r_o^2 = (1+x)e^{-x}\sqrt{\frac{8T}{\pi m_e}} N_e \pi r^2 .$$

Here ε is the electron energy, T is the electron temperature, N_e is the number density of electrons, m_e is the electron mass, $z = \varepsilon/T$, $x = |Z|e^2/(rT)$, Z is the negative cluster charge, and we account for the electron attachment to the cluster surface is possible if the electron energy ε exceeds the repulsion energy of charge interaction $|Z|e^2/r$. The cross section of ion contact with the surface of charged cluster is given by formula (13.45), and the ion current to the cluster surface is

$$I_i = e(1+x)\sqrt{\frac{8T}{\pi m_i}}N_i\pi r^2 ,$$

where N_i is the ion number density and m_i is the ion mass.
Equalizing the electron and ion currents on the surface of a charged cluster and assuming the plasma to be quasineutral $N_e = N_i$, we obtain the cluster charge

$$|Z| = \frac{r_0 T}{2e^2}\ln\frac{|Z|\,e^2}{r_0\varepsilon}\ln\frac{m_i}{m_e} . \tag{13.48}$$

▶ **Problem 13.30** Metal clusters are located in a glow gas discharge where the electron temperature T_e exceeds remarkably the gas temperature T. Find the temperature of internal cluster electrons T_{cl} under the assumptions that an electron and an atom after collision with the cluster has an energy $3T_{cl}/2$. A cluster radius is small compared to the mean free path of electrons and ions in a buffer gas.

Assuming that the series of temperatures is

$$T_e > T_{cl} > T ,$$

we find that the cluster obtains energy from electrons and transfers it to atoms of a buffer gas. Colliding with clusters, atoms and electrons exchange energy with them, so that the cluster temperature is determined by the temperature of the gas and of the electrons. Within the framework of a simple model we assume that the average energy of the atoms varies from $3T/2$ to $3T_{cl}/2$ after collision with the cluster, and the mean electron energy varies from $3T_e/2$ to $3T_{cl}/2$. This takes place when a strong interaction exists between colliding atoms and clusters and can proceed through the atom capture by the cluster surface. Evidently, this takes place at strong interaction of colliding particles that can result from contact of an incident atom or electron with the cluster. The cross section of such atom–cluster collisions is equal to the cluster cross section πr_0^2, where r_0 is the cluster radius. The rate constant of electron-cluster collisions is given by formula (2.16). Hence within the framework of the liquid drop cluster model, we obtain the rate constants of atom–cluster (k_a) and electron–cluster (k_e) collisions.

$$k_a = \left\langle v_a\pi r_0^2\right\rangle = \sqrt{\frac{8T}{\pi m}}\pi r_0^2,$$

$$k_e = \left\langle v_e\pi r_0^2\left(1+\frac{Ze^2}{\varepsilon_e r_0}\right)\right\rangle = \sqrt{\frac{8T_e}{\pi m_e}}\pi r_0^2\left(1+\frac{Ze^2}{r_0 T_e}\right) .$$

Here v_a, v_e are the atom and electron velocities, respectively, ε_e is the electron energy, and an average is done over the velocity distribution of atoms or electrons.

We assume the cluster radius to be sufficiently large so that collisions are governed by classical laws, but still small in comparison to the mean free path of atoms and electrons. Hence, each cluster collision is with a single particle.

Assuming an atom and electron have the average energy $3T_{cl}/2$ after collision with a cluster, we obtain the following balance equation for the cluster temperature

$$(T - T_{cl})k_a N_a + (T_e - T_{cl})k_e N_e = 0 \,,$$

where N_a and N_e are the number densities of atoms and electrons respectively. From this we find the cluster temperature

$$T_{cl} = \frac{T + \zeta T_e}{1+\zeta}\,, \qquad \zeta = \sqrt{\frac{T_e m}{T m_e}}\left(1 + \frac{Ze^2}{r_o T_e}\right)\frac{N_e}{N_a}\,. \tag{13.49}$$

This formula contains two small parameters, m_e/m and N_e/N_a, and they are govern by the cluster temperature.

▶ **Problem 13.31** Find the charge of a spherical cluster whose radius is small compared to the mean free path of electrons and ions if the cluster is located in a weakly ionized gas where the temperatures of electrons T_e and ions T_i are different.

Using the above expressions for electrons and ions, we obtain the cluster charge, equalizing the electron and ion currents to the cluster surface,

$$x \equiv \frac{|Z|\,e^2}{r_o T} = \ln\left[\frac{m_i}{m_e}(1+x)\left(1 + x\frac{T_e}{T_i}\right)^{-1}\right],$$

where $x = |Z|e^2/(rT_e)$ and in the case $x \gg T_i/T_i$ this formula gives

$$x \equiv \frac{|Z|e^2}{rT} = \ln\left[\frac{T_i}{T_e}\frac{m_i}{m_e}\left(1 + \frac{1}{x}\right)\right]. \tag{13.50}$$

▶ **Problem 13.32** Analyze the character of charging of a small spherical cluster in a unipolar plasma.

The equation of cluster charging has the form

$$\frac{dZ}{dt} = \frac{I}{e}\,,$$

where Z is the cluster charge and I is the current of ions on the cluster surface. According to this equation, the cluster charge increases monotonically in time. Using formula (13.47) for currents on the cluster surface, one can rewrite the above equation in the form

$$\frac{dx}{d\tau} = \frac{x}{e^x - 1}\,,$$

where $x = Ze^2/Tr$, $\tau = te/(TR)(1/I_> + 1/I_<)^{-1}$. The solution of this equation in the limiting cases is given by

$$x = \begin{cases} \tau, & \tau \ll 1 \\ \ln(1+\tau), & \tau \gg 1 . \end{cases}$$

Figure 13.5 gives the dependence $x(\tau)$.

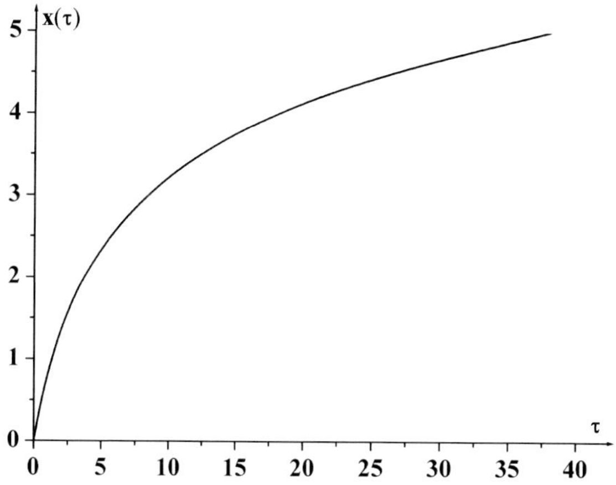

Fig. 13.5 The dependence of the reduced cluster charge x on the reduced time τ.

13.4
Cluster Transport

► **Problem 13.33** Find the diffusion coefficient of clusters in an atomic gas within the framework of the hard sphere model for atom–cluster collisions.

The diffusion coefficient of a cluster is similar to the atom diffusion coefficient (8.40) within the framework of the hard sphere model. Next, because a radius of action of atomic forces in atom-cluster collision is small compared to a cluster radius r, the diffusion cross section of atom-cluster scattering is $\sigma_* = \pi r^2$. Because the reduced mass of the atom–cluster system coincides with the atom mass, and a cluster radius $r = r_W n^{1/3}$, where r_W is the Wigner–Seitz radius for a cluster material, m is the mass for a buffer gas atom, and n is a number of cluster atoms. We obtain the diffusion coefficient of a cluster consisting of n atoms,

$$D_n = \frac{3\sqrt{T}}{8\sqrt{2\pi m} N_a r_W^2 n^{2/3}} . \tag{13.51}$$

Here N_a is the number density of buffer gas atoms and T is the gas temperature. This formula is valid if atoms collide with the cluster separately, i. e., each time a strong cluster interaction is possible with one atom only. Therefore the mean free path of buffer gas atoms λ is large compared to a cluster radius r.

From this we obtain the following dependence of the cluster diffusion coefficient on a number of cluster atoms:

$$D_n = \frac{D_0}{n^{2/3}}\,, \tag{13.52}$$

and the temperature dependence of the cluster diffusion coefficient is $D_n \sim \sqrt{T}$.

▶ **Problem 13.34** Find a force acting on the cluster that is moving in an atomic gas with a velocity $\boldsymbol{w}$. Consider atom–cluster collisions within the framework of the hard sphere model.

The force acting on a cluster consisting of many atoms is the variation of the cluster momentum per unit time, and this force is transferred to a cluster from atoms as a result of elastic atom–cluster collisions. If a cluster radius is small compared to the mean free path of atoms in a gas, two subsequent collisions of the cluster with atoms are separated, and the momentum $m\boldsymbol{v}(1 - cos\vartheta)$ is transferred to the cluster at elastic collision with an atom, where m is the atom mass that is small compared to the cluster mass, $\boldsymbol{v}$ is the relative velocity of colliding particles, and ϑ is the scattering angle. Thus, the force $\boldsymbol{F}$ acting on the cluster is given by

$$\boldsymbol{F} = \int m\boldsymbol{v}(1 - \cos\vartheta) f(\boldsymbol{v}) v d\sigma d\boldsymbol{v}\,,$$

where $f(\boldsymbol{v})$ is the velocity distribution function of atoms in the frame of reference, where the cluster is motionless, that is normalized as $\int f(\boldsymbol{v}) d\boldsymbol{v} = N$, N is the number density of atoms, and $d\sigma$ is the differential cross section of elastic scattering.

Assume the cluster velocity $\boldsymbol{w}$ to be small compared to a typical atomic velocity and taking the Maxwell distribution function of atoms on velocities ($f(\boldsymbol{v}_a) \sim \exp[-mv_a^2/(2T)]$), where $\boldsymbol{v}_a$ is the atom velocity, and transferring to the frame of reference where the cluster is motionless, we obtain $\boldsymbol{v} = \boldsymbol{v}_a - \boldsymbol{w}$. This gives for the force acting on the cluster

$$\boldsymbol{F} = \int m\boldsymbol{v}(1 - \cos\vartheta) f(\boldsymbol{v}_a - \boldsymbol{w}) v d\sigma d\boldsymbol{v} = \int m\boldsymbol{v}\sigma^* f(\boldsymbol{v}_a - \boldsymbol{w}) v d\boldsymbol{v}\,,$$

where $\sigma^* = \pi r^2$ is the diffusion cross section of atom-cluster scattering. Expanding the Maxwell distribution function over a small parameter, we have

$$f(\boldsymbol{v}_a - \boldsymbol{w}) = f(v_a)\left(1 - \frac{m\boldsymbol{v}_a\boldsymbol{w}}{T}\right).$$

This leads for the cluster resistance force with respect to a gas flow

$$\boldsymbol{F} = -\boldsymbol{w}\frac{m^2 N\sigma^*}{3T}\left\langle v_a^3\right\rangle = -\frac{8\sqrt{2\pi}}{3}\sqrt{mT}Nr^2\boldsymbol{w}\,. \tag{13.53}$$

Here brackets denote an average over atom velocities for the Maxwell velocity distribution function of atoms.

► **Problem 13.35** Determine the cluster mobility in an atomic gas.

If a cluster of a charge Z is moving in an electric field, the force acting on the cluster from the electric field $\boldsymbol{F} = Ze\boldsymbol{E}$ is equalized by the friction force (13.53). Here $\boldsymbol{E}$ is the electric field strength, and defining the cluster mobility K from the relation $\boldsymbol{w} = K\boldsymbol{E}$, we obtain the cluster mobility

$$K_n = \frac{3Ze}{8\sqrt{2\pi m T} N r^2} = \frac{3Ze}{8\sqrt{2\pi m T} N r_W^2 n^{2/3}} , \quad \lambda \gg r , \tag{13.54}$$

and this formula is valid if a cluster radius is small compared to the mean free path of atoms in a gas.

Formula (13.54) also follows from the Einstein relation (8.66) and expression (13.52) for the cluster diffusion coefficient which give

$$K_n = \frac{ZeD_n}{T} , \quad K_n = \frac{K_o}{n^{2/3}} , \tag{13.55}$$

where the cluster diffusion coefficient D_n in a buffer gas is given by formula (13.51). The reduced cluster mobility is

$$K_o = \frac{3Ze}{8\sqrt{2\pi m T} N_a r_W^2} .$$

► **Problem 13.36** Determine the force acting on a moving cluster in an atomic gas if its radius exceeds the mean free path of atoms.

The resistance force that acts on a moving cluster in a motionless gas occurs because a gas flows together with the cluster near its surface, whereas far from the cluster the average atom velocity is zero. Hence in this case the frictional force has the viscosity nature and therefore is expressed through the gas viscosity coefficient. By definition, the viscosity force F per unit surface acted from a flowing gas is given by formula (8.5)

$$\frac{F}{S} = \eta \frac{\partial v_\tau}{\partial R} ,$$

where η is the viscosity coefficient of the gas, S is the area of the frictional surface, v_τ is the tangential component of the velocity with respect to the flux, and R characterizes the normal direction to this stream. From this one can estimate the friction force, taking into account $S \sim r^2$, $R \sim r$, $v_\tau \sim v$, the cluster velocity, which gives

$$F \sim \eta r v .$$

The accurate derivation gives the Stokes formula for the friction force that has the form

$$F = 6\pi\eta r v , \qquad \lambda \ll r . \tag{13.56}$$

Taking the friction force to be equal to the electric force eE acted on the cluster from an electric field of a strength E, we obtain the following expression for the cluster mobility:

$$K = \boldsymbol{v}/\boldsymbol{E} = \frac{e}{6\pi r\eta}$$

and the Einstein relation (8.66) gives for the cluster diffusion coefficient

$$D = \frac{KT}{e} = \frac{T}{6\pi r\eta}, \qquad \lambda \ll r . \tag{13.57}$$

As is seen, the only cluster parameter of this formula is the cluster radius r.

▶ **Problem 13.37** Generalize the expression for the cluster diffusion coefficient for an arbitrary relation between the cluster radius an the mean free path of atoms.

For this goal we first express the gas viscosity coefficient through the mean free path of gas atoms λ within the framework of the hard sphere model where the mean free path does not depend on the atom velocity. After this, combining formulas (13.52) and (13.57), we get for the cluster diffusion coefficient in the gas

$$D = \frac{T}{6\pi r\eta}(1 + CKn) , \tag{13.58}$$

where the numerical coefficient is $C = 3.1$ and $Kn = \lambda/r$ is the Knudsen number. In particular, for the air at atmospheric pressure and room temperature this formula can be rewritten in the form

$$D = \frac{D_o}{r}\left(1 + \frac{0.14}{r}\right) ,$$

where $D_o = 1.3 \cdot 10^{-7}\ \text{cm}^2/\text{s}$ and the particle radius r is expressed in microns.

▶ **Problem 13.38** Compare the cluster and atom diffusion coefficients in a buffer gas.

Within the framework of the hard sphere model for atomic collisions, the mean free path of atoms does not depend on the collision velocity, and one can consider colliding atoms as rigid balls. Then the diffusion coefficient of atoms D_a is given by formula (8.40). Taking in this formula the average cross section $\overline{\sigma} = \pi\rho_o^2$, where ρ_o is the radius of action of atomic forces, we obtain the ratio of the cluster D_{cl} and atom diffusion coefficients

$$\frac{D_{cl}}{D_a} = \frac{\rho_o^2}{r^2} = \frac{\sigma_{gas}}{\pi r^2} ,$$

where r is the cluster radius, ρ_o is of the order of atomic value and $\pi\rho_o^2$ is the gas-kinetic cross section. This ratio depends weakly on the gas temperature and gas parameters. In particular, for the atmospheric air at room temperature the parameters of this formula are equal $\rho_o = 40$ nm, $D_a N = 4.8 \cdot 10^{18}\ \text{cm}^{-1}\text{s}^{-1}$, $D_{cl} N r^2 = 4.6 \cdot 10^3$ cm/s, and N is the number density of air molecules.

► **Problem 13.39** Determine the cluster mobility in a weakly ionized gas in the limit of large cluster radius.

In this limit a cluster radius exceeds greatly the mean free path of gas atoms and the cluster mobility K is expressed through its diffusion coefficient D by the Einstein relation (8.66)

$$K = \frac{ZeD}{T} ,$$

where Z is the cluster charge. On the basis of formula (13.57) for the cluster diffusion coefficient and formula (13.41) for the cluster charge in an ionized gas we get

$$K = \frac{T^2}{6\pi\eta e} \ln\left(\frac{D_-}{D_+}\right) , \quad r \gg \lambda \tag{13.59}$$

where D_-, D_+ are the diffusion coefficients of negative and positive ions of the weakly ionized gas which attach to the cluster, and we take $D_- > D_+$. Of course, we use here the criterion of a large cluster radius $r \gg \lambda$. Because the cluster charge is proportional to its radius, whereas the cluster diffusion coefficient is inversely proportional to its radius, the mobility of a large cluster does not depend on its radius.

In particular, for atmospheric air of a typical humidity for which $D_+/D_- = 0.8$, this formula gives $K = 1.8 \cdot 10^{-5}\ \mathrm{cm}^2/(\mathrm{V \cdot s})$.

► **Problem 13.40** Determine the free fall velocity for a cluster in a gas under the action of the gravitational force, if the cluster radius greatly exceeds the mean free path of gas atoms, and the Reynolds number of a moving cluster is small.

The equilibrium velocity of a falling cluster v of the particle follows from the equality of the gravitational and resistance forces, which has the form

$$\frac{4\pi}{3} \rho g r^3 = 6\pi\eta r v ,$$

where ρ is the cluster density and g is the acceleration of gravity. This gives for the cluster free fall velocity

$$v = \frac{2\rho g r^2}{9\eta} . \tag{13.60}$$

This formula is valid for small values of the Reynolds numbers $\mathrm{Re} = vr/\nu \ll 1$, where $\nu = \eta/\rho_g$ is the kinematic gas viscosity, and ρ_g is the gas density, and this criterion has the form

$$r \ll \frac{\eta^{2/3}}{(\rho\rho_g g)^{1/3}} .$$

In particular, for water clusters in atmospheric air this criterion gives $r \ll 30\ \mu$.

▶ **Problem 13.41** Determine a typical time during which a cluster injected in a gas flow obtains the flow velocity if the particle radius is larger than the mean free path of gas atoms.

This time follows from solution of the motion equation for a cluster that has the form

$$M\frac{dv}{dt} = 6\pi\eta r(v_0 - v) ,$$

where M is the cluster mass, v is its velocity and v_0 is the equilibrium cluster velocity. Solving this equation, we have

$$v = v_0[1 - \exp(-t/\tau)], \qquad \tau = \frac{M}{6\pi r\eta} .$$

Let us represent the relaxation time in the form

$$\tau = \tau_0\left(\frac{r}{a}\right)^2 , \qquad \tau_0 = \frac{9\rho a^2}{2\eta}$$

where ρ is the cluster density and a is a parameter. In particular, for a water cluster moving in atmospheric air we have $\tau_0 = 1.2 \cdot 10^{-5}$ s for $a = 1$ µm.

▶ **Problem 13.42** Determine the recombination rate constant of two oppositely charged clusters in a dense buffer gas.

Under these conditions the recombination process of two clusters of charges Z_+e and Z_-e in a dense gas is restricted by cluster approach because of a large frictional force. The velocities of positively v_+ and negatively v_- charged particles are determined by an electric field that is created by another charged particle. The electric field strength at the point where a particle is located is $E = e/R^2$, and the velocities of particles are $v_+ = EK_+$, $v_- = EK_-$, where K_+, K_- are the mobilities of clusters. Thus the number of negatively charged clusters which fall on a test positive cluster per unit time is $J_- = 4\pi R^2(v_+ + v_-)N_- = 4\pi e(K_+ + K_-)N_-$, where N_- is the number density of negative clusters. According to the definition, the recombination coefficient k_{rec} of ions is

$$\frac{dN_+}{dt} = -k_{rec}N_+N_- = -J_-N_+ ,$$

where N_+ is the number density of positively charged clusters. From this we get the Langevin formula for the recombination coefficient of positive and negatively charged clusters in a dense gas

$$k_{rec} = 4\pi e(K_+ + K_-) .$$

Using cluster parameters, one can represent the recombination coefficient for oppositely charged clusters in the form

$$k_{rec} = \frac{3\sqrt{\pi}Z_+Z_-e^2}{2\sqrt{2mT}Nr^2}\left(\frac{1}{r_+^2} + \frac{1}{r_-^2}\right) , \lambda \gg r_+, r_- , \qquad (13.61)$$

where r_+, r_- are the radii of the positive and negative clusters. As is seen, the recombination coefficient $k_{rec} \sim 1/r^2$, where $r \sim r_+, r_-$. In particular, taking $r_+ = r_- = r$ for air at room temperature and atmospheric pressure, we have $k_{rec}r^2 = 2.6 \cdot 10^{-20}\ \text{cm}^5/\text{s}$.

► **Problem 13.43** Analyze the cluster instability that develops in a buffer gas with an admixture of a metallic vapor if temperature varies in this system. As a result, clusters are formed in a cold region, and atoms are moving from a hot region and attach to clusters in a cold region.

One can expect an equilibrium in a buffer gas with an admixture of a metallic vapor if the temperature varies in a direction that we denote as z. Then metallic atoms travel to a cold region and attach to clusters there. In turn, clusters travel in a hot region and evaporate there. For the analysis of this equilibrium, we study evolution of an individual cluster that travels to a hot region for evaporation. Growth of this cluster results from attachment of free atoms and is described by the balance equation

$$\frac{dn}{dt} = k_0 n^{2/3} N ,$$

where n is a current number of cluster atoms and N is the number density of metallic atoms. Next, the cluster travels due to its diffusion in a buffer gas, and its displacement is given by the equation

$$\frac{d\overline{z^2}}{dt} = 2D_n ,$$

where the dependence of the diffusion coefficient D_n of the cluster on its size is given by formula (13.52).

On the basis of these equations, we represent evolution of the cluster size as it displaces from an initial point in the form

$$\frac{d\overline{z^2}}{dn} = \frac{2D_0}{k_0 n^{4/3} N} .$$

From this equation it follows that the average distance square $\overline{z^2}(t)$ from an initial point at the end of the process is

$$\overline{z^2}(\infty) = \frac{\Delta^2}{n^{1/3}}, \qquad \Delta^2 = \frac{6D_0}{k_0 N} . \tag{13.62}$$

One can see, the larger is a cluster, the less distance it can go. As a result, an instability occurs, and due to motion of atoms, all the metal is collected in a cold region in the form of clusters.

Table 13.2 gives values of the reduced parameter $\Delta\sqrt{N_0 N}$ that does not depend on the density of metallic atoms and atoms of a buffer gas. Under typical parameters of a dense cluster plasma $N_a \sim 10^{19}\ \text{cm}^{-3}$, and $N \sim 10^{13}$–$10^{15}\ \text{cm}^{-3}$ we have $\Delta \sim 0.01$–0.1 cm, i. e., the cluster instability is realized under typical real laboratory conditions.

Table 13.2 The reduced diffusion coefficient according to formulas (13.51) and (13.58) for metallic clusters in argon at the temperature $T = 1000$ K and the normal number density of argon atoms $N_a = 2.69 \cdot 10^{19}$ cm^{-3}. The reduced displacement of clusters $\Delta\sqrt{N_a N}$ is given by formula (13.62).

Element	D_o (cm^2/s)	$\Delta\sqrt{N_o N}$ (10^{15} cm^{-2})
Ti	0.91	1.59
V	1.05	3.51
Fe	1.17	2.13
Co	1.20	2.22
Ni	1.22	2.31
Zr	0.74	1.52
Nb	0.90	1.85
Mo	0.98	2.05
Rh	1.05	2.23
Pd	1.01	2.16
Ta	0.90	2.19
W	0.98	2.42
Re	1.01	2.49
Os	1.05	2.60
Ir	1.01	2.51
Pt	0.98	2.45
Au	0.93	2.32
U	0.81	2.11

▶ **Problem 13.44** Estimate the depth of penetration of a flux of metallic atoms in a cluster plasma consisting of a dense buffer gas and an admixture of an atomic metallic vapor and metal clusters.

There is a local thermodynamic equilibrium between clusters and the atomic vapor at each point that is governed by the processes of evaporation of clusters and attachment of atoms to their surface. In the limit of large clusters, the number density of atoms tends to the saturated vapor density at a given temperature. The gradient of the temperature of a buffer gas ∇T which is supported in the plasma creates a gradient of the number density of metallic atoms

$$\nabla N = -\frac{\varepsilon_o}{T^2} N \nabla T \, ,$$

and this number density increases toward a cold region. As a result, the flux of metallic atoms is directed to a cold region where atoms attach to clusters. Therefore, clusters in a hot region are evaporated, and forming atoms partake in the growth of clusters in a cold region. Finally, metallic atoms are gathered in a cold discharge region forming there clusters. Thus, as a result of the above processes a metal is concentrated in a cold region of a nonuniform plasma.

For estimating the depth of penetration of the atomic flux in the cluster plasma, we consider favorable conditions for cluster growth when one can neglect the cluster evaporation in a cold region. Then the atomic flux in a cold region is

$$j = -D_a \nabla N = D_a N \frac{\varepsilon_o}{T^2} \nabla T \, ,$$

where D_a is the diffusion coefficient of metal atoms in a buffer gas. Taking into account the character of cluster growth through attachment of atoms to a cluster, we obtain the depth l of penetration of free atoms into a cold region of the cluster plasma

$$l = \frac{j}{k_o n^{2/3} N_{cl} N} = \frac{D_a}{k_o N_b} \frac{\varepsilon_o}{T} \frac{\nabla T}{T} n^{1/3} . \tag{13.63}$$

Here N_{cl} is the number density of clusters, $N_b = n N_{cl}$ is the total number density of bound atoms in clusters and n is the average number of cluster atoms. In this regime of cluster evolution the following criterion holds true:

$$N_b \gg N ,$$

i. e., the most part of metal atoms is bound in clusters. This condition corresponds to intense nucleation processes in a plasma and provides in the end collection of metallic atoms in a narrow region of the plasma.

14
Aeronomy Processes

14.1
Oxygen Atoms in the Upper Atmosphere

▶ **Problem 14.1** The photon flux of the intensity of $j = 3 \cdot 10^{12}$ cm^{-2}s^{-1} in the spectral range $\lambda = 132$–176 nm propagates in the upper atmosphere from the Sun, and causes photodissociation of molecular oxygen. Determine the distribution of absorbed photons on altitudes if the altitude distribution of molecular oxygen is determined by the barometric formula.

Properties of the upper Earth atmosphere are determined by processes of photodissociation and photoionization under the action of solar radiation. Table 14.1 gives fluxes from solar radiation penetrated in the Earth atmosphere which correspond to wavelength ranges responsible for photodissociation and photoionization processes (see also Fig. 14.1). In particular, oxygen atoms are formed in the upper atmosphere as a result of electron excitation of oxygen molecules by a short-wave radiation in the range of 132 to 176 nm. As a result, molecule transfers into a repulsive electron term that leads to its dissociation. This spectral range, the Schumann–Runge continuum, corresponds to the photon energy range of 6 to 10.3 eV and is characterized by cross sections of photodissociation of molecular oxygen of 10^{-19} cm^2 to 10^{-17} cm^2, and the cross section as a function of the wavelength has an oscillation structure.

Let us consider the character of absorption for a given photon frequency, denoting the photodissociation cross section for this frequency by σ_ω and the photon

Table 14.1 The flux of solar radiation on the Earth level in corresponding spectral ranges.

Spectral range $\Delta\lambda$ (nm)	Range of photon energies (eV)	Photon flux (cm^{-2} s^{-1})
177–132	7.00–9.39	$2.7 \cdot 10^{13}$
132–103	9.39–12.0	$3.5 \cdot 10^{11}$
103–91	12.0–13.6	$1.2 \cdot 10^{10}$
91–80	13.6–15.5	$1.3 \cdot 10^{10}$
80–63	15.5–19.7	$4.6 \cdot 10^{9}$
63–46	19.7–27	$6.1 \cdot 10^{9}$

Plasma Processes and Plasma Kinetics. Boris M. Smirnov

ISBN: 978-3-527-40681-4

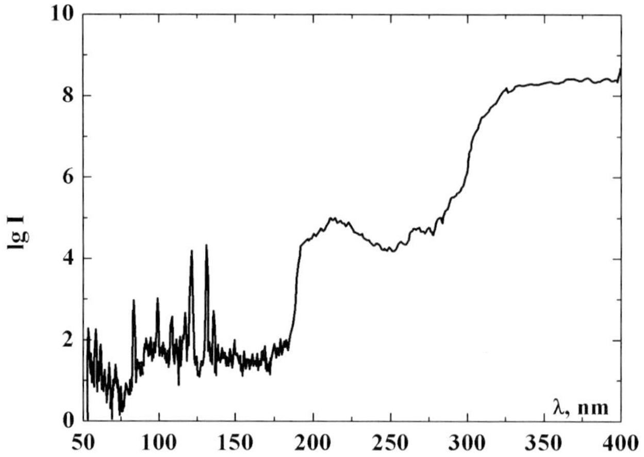

Fig. 14.1 The spectral power of Sun radiation that penetrates in the upper atmosphere of the Earth.

flux far from the atmosphere by j_ω. The balance equation for the photon flux j_ω of solar radiation of a given frequency, when it moves through the atmosphere toward the Earth surface, has the form

$$\frac{dj_\omega}{dz} = -j_\omega \sigma_\omega N, \tag{14.1}$$

where z is the altitude, $N(z)$ is the number density of molecular oxygen at a given altitude, and σ_ω is the photodissociation cross section of the oxygen molecule for these photons. We assume the solar radiation to be perpendicular to the Earth surface, and the number density of molecular oxygen varies almost according to the barometric formula $N(z) = N_0 \exp(-z/l)$, where $l = T/mg \approx 10$ km (see Fig. 14.2). Then the solution of the balance equation (14.1) is

$$j_\omega(z) = j_\omega^{(0)} \exp\left[-\exp\left(-\frac{z - z_0}{l}\right)\right], \tag{14.2}$$

where $j_\omega^{(0)}$ is the intensity of solar radiation above the atmosphere, and the altitude z_0 is determined by the relation $\sigma_\omega l N(z_0) = 1$. Figure 14.3 shows the altitude dependence for the reduced radiation flux propagates in the atmosphere.

From formula (14.2) it follows that photons of a given frequency are absorbed mostly in a layer of a thickness $\sim l$ near $z \approx z_0$. In fact, because the photodissociation cross section for the oxygen molecule is $\sigma_\omega \sim 10^{-19}$–$10^{-17}$ cm^2 in a wide frequency range, the absorption takes place in a wide altitude range $\approx 5l$. This proceeds at altitudes approximately 120–180 km where the number density of molecular oxygen ranges $N \sim 10^{11}$–10^{13} cm^{-3}. The number density of atomic oxygen is equalized to the number density of molecular oxygen at altitudes of 100 to 120 km, and to the number density of molecular nitrogen at altitudes of 150 to 200 km. In particular, the average total number densities of nitrogen and oxygen

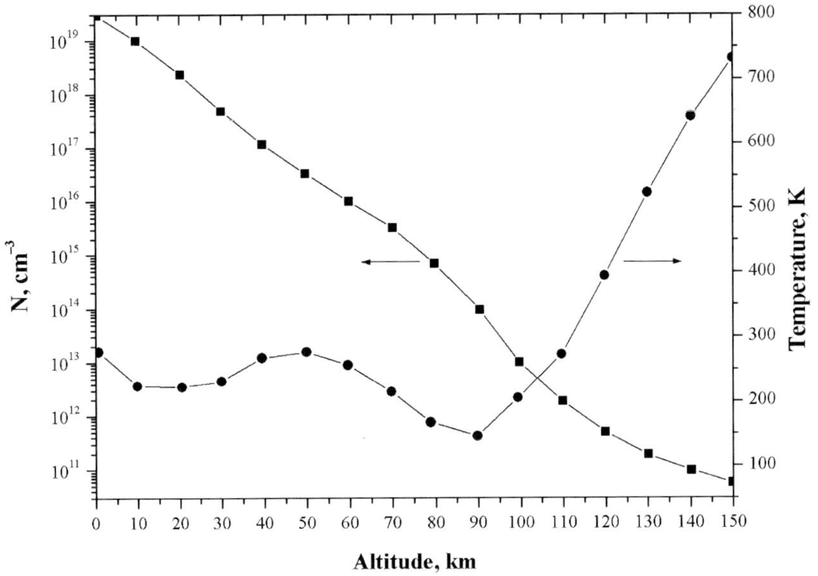

Fig. 14.2 The total number density of atmospheric molecules and atoms as a function of an altitude z; l is a typical dimension, that is expressed through the number density of atmosphere atoms and molecules N as $l = -d \ln N/dz$.

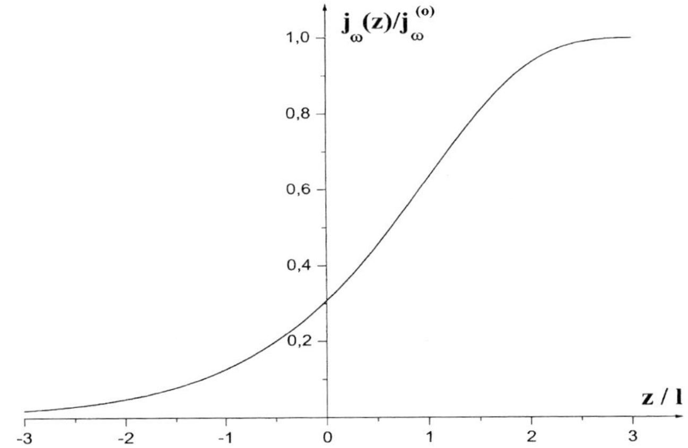

Fig. 14.3 The altitude dependence for a passing solar radiation inside the atmosphere according to formula (14.2).

molecules varies from $7 \cdot 10^{12}$ cm^{-3} to $3 \cdot 10^9$ cm^{-3}, as the altitude varies from 100 to 200 km. The average number density of atomic oxygen varies in the altitude range from $6 \cdot 10^{11}$ cm^{-3} to $4 \cdot 10^9$ cm^{-3}.

Note that along with the Schumann–Runge continuum spectrum, a weak absorption due to the weak Herzberg continuum is observed in the wavelengths range of 140 to 175 nm, where the absorption cross section is less than $2 \cdot 10^{-24}$ cm^2 and can lead to photodissociation of molecular oxygen. This determines the generation of atomic oxygen at altitudes below 80 km. The recombination rate of two oxygen atoms in three-body collisions decreases as altitude increases, and hence the concentration of atomic oxygen increases.

▶ **Problem 14.2** Estimate a typical time of the photodissociation of molecular oxygen in the upper atmosphere.

Let us introduce a photodissociation lifetime of an oxygen molecule by averaging it over the photon frequencies as

$$\tau_{\text{dis}} = \frac{1}{4}\left(\int \sigma_\omega j_\omega d\omega\right)^{-1} = \frac{1}{4 j_\omega \sigma_\omega} \approx 1 \times 10^6 \text{ s} \approx 10 \text{ days} . \tag{14.3}$$

Here j_ω is the photon flux at a given frequency ω at high atmosphere altitudes, the factor 1/4 results from averaging the radiation flux over the entire surface of the Earth, the average photodissociation cross section is $\sigma_\omega \sim 10^{-19}$ cm^2, and $\hbar\omega = 6$–10 eV is the photon energy. From this it follows that the daytime and nighttime number density of atomic oxygen is the same.

▶ **Problem 14.3** Find the drift velocity of an oxygen atom formed at a high altitude toward the Earth surface under the action of gravity force.

Motion of an individual oxygen atom is governed by the weight force mg, where m is the oxygen atom mass and g is the free fall acceleration. This gives the drift velocity

$$w_{\text{O}} = b \cdot mg = \frac{mg}{T} D_{\text{O}} ,$$

where b is the atom mobility and D is the diffusion coefficient which are connected by the Einstein relation. Taking the gas-kinetic cross section for collision of an oxygen atom and nitrogen molecule to be $\sigma_{\text{gas}} = 35$ Å^2, which gives the diffusion coefficient $D = 0.32$ cm^2/s. Note that this diffusion coefficient is reduced to the number density of nitrogen molecules $2.7 \cdot 10^{19}$ cm^{-3}. From this we obtain the reduced drift velocity of atomic oxygen in molecular nitrogen

$$w_{\text{O}} N = 5.5 \cdot 10^{12} \text{ cm}^{-2}\text{s}^{-1} , \tag{14.4}$$

where N is the number density of nitrogen molecules. One can assume this identical for molecular nitrogen and oxygen, and then transferring this formula to the upper atmosphere, one can consider N to be the total number density of molecular nitrogen and oxygen that assumes to exceed significantly the number density of atomic oxygen. Correspondingly, this gives the diffusion coefficient of oxygen atoms in air as

$$D_{\text{O}} N = \frac{T}{mg} w_{\text{O}} N = \frac{T}{300} \cdot 8.7 \cdot 10^{18} \text{ cm}^{-1}\text{s}^{-1} , \tag{14.5}$$

where the temperature is expressed in Kelvin. Under equilibrium conditions, the altitude distribution of atomic oxygen is given by the barometric formula, and then the drift flux of atomic oxygen is equalized by its diffusion flux.

▶ **Problem 14.4** Compare typical times of drift of molecular oxygen to upper layers of the atmosphere and a typical time of photodissociation of the oxygen molecule in the upper atmosphere.

If a typical transport time for molecular oxygen is small compared to a typical time of photodissociation of an oxygen molecular under the action of solar radiation, this transport restores the equilibrium number density of molecular oxygen, and it is given by the barometric formula. We first find the parameters of transport of oxygen molecules by analogy with that of transport of atomic oxygen analyzed in the previous problem, and we have in this case

$$w_{O_2}N = 6.0 \cdot 10^{12}\ \mathrm{cm^{-2}s^{-1}}, \qquad D_{O_2}N = 4.8 \cdot 10^{18}\ \mathrm{cm^{-1}s^{-1}}. \tag{14.6}$$

We ignore in these formulas the temperature dependence for the mobility and diffusion coefficient. Let us divide the upper atmosphere in layers of the thickness of the order of $l = T/mg \approx 10$ km (m is the molecule mass), so that the equilibrium number density of molecules varies remarkably according to the barometric formula when we go from one layer to the neighboring one. Let us estimate a transport time of oxygen molecule τ_{O_2} through one layer that is equal to l/w_{O_2}. At the altitude $h = 100$ km, where the total number density of molecules is $N = 7 \cdot 10^{12}\ \mathrm{cm^{-3}}$, the drift velocity of oxygen molecules is $w_{O_2} \approx 1$ cm/s, and the transport is $\tau_{O_2} \sim 10^6$ s. At the altitude $h = 200$ km, where the total number density of molecules is $N = 3 \cdot 10^9\ \mathrm{cm^{-3}}$, this time is $\tau_{O_2} \sim 3 \cdot 10^3$ s. Comparing these values with the photodissociation lifetime (14.3) of an oxygen molecule, we find that transport of oxygen molecules cannot restore their concentration at lower altitudes, i. e., the concentration of molecular oxygen there is lower than that near the Earth surface. On the contrary, at high altitudes a typical transport time becomes small and restores the equilibrium concentration of molecular oxygen between neighboring layers. Of course, this concentration is less than that near the Earth surface and it drops as the altitude increases.

▶ **Problem 14.5** At low atmosphere altitudes oxygen atoms recombine in three-body processes. Compare the rates of three-body processes of the formation of oxygen and ozone molecules.

Oxygen atoms decay in lower atmosphere layers as a result of the processes given in Table 14.2. Comparing the rates of processes 1 and 2 of this table with the rates of the processes 4 and 5, we find a portion of oxygen atoms whose decay leads to the formation of ozone molecules is approximately

$$\eta = \frac{1}{1 + 6\frac{[O]}{N}},$$

where N is the total number density of nitrogen and oxygen molecules and [O] is the number density of oxygen atoms. Because at low altitudes, where these processes are important, $[O] < 0.1N$, the main part of oxygen atoms is converted into ozone molecules (Fig. 14.4).

Table 14.2 The rate constants of the processes involving oxygen atoms at room temperature.

Number	Process	Rate constant, activation energy
1	$O + 2O_2 \to O_3 + O_2$	$7 \cdot 10^{-34}\ cm^6/s$
2	$O + O_2 + N_2 \to O_3 + N_2$	$6 \cdot 10^{-34}\ cm^6/s$
3	$O + O_3 \to 2O_2$	$1.7 \cdot 10^{-11} \exp\left(-\frac{2230}{T}\right)\ cm^3/s$ [a]
4	$2O + O_2 \to 2O_2$	$3 \cdot 10^{-33}\ cm^6/s$
5	$2O + N_2 \to O_2 + N_2$	$4 \cdot 10^{-33}\ cm^6/s$

[a] The temperature T is expressed in Kelvin.

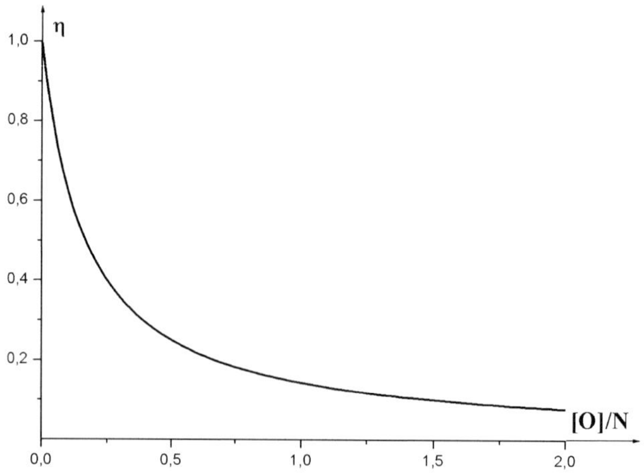

Fig. 14.4 A part of oxygen atoms that is converted in ozone molecules in the atmosphere.

▶ **Problem 14.6** Determine the concentration of oxygen atoms in the upper atmosphere, where the formation of oxygen atoms results from photodissociation of molecular oxygen in lower atmosphere layers, which is responsible for their losses.

We assume that a loss of atomic oxygen as a result of three-body processes finishes at such altitudes where three-body processes are not essential. This gives the balance equation

$$\frac{j}{4} = w_O[O] ,$$

where [O] is the number density of oxygen atoms, and the factor 1/4 accounts for the radiation flux of the Sun spent over the entire surface of the Earth. Using formula (14.4) for the drift velocity of oxygen atoms, and the value of the total radiation flux that causes dissociation of molecular oxygen, we obtain this formula

$$[O] \sim 0.14N \,. \tag{14.7}$$

This corresponds to the maximum concentration of oxygen atoms and is realized at altitudes until three-body processes are absent.

► **Problem 14.7** Determine the maximum of the number density of oxygen atoms in the upper atmosphere.

The concentration of oxygen atoms is constant in some altitude range according to formula (14.7). This means that the number density of oxygen atoms increases as we move toward the Earth surface.

The maximum number density of atomic oxygen may be found by comparison of the balance of transport of oxygen atoms $w_O[O]$ with the rate of three-body loss of atomic oxygen according to processes 1, 2, 4, 5 of Table 14.2, and the maximum number density is observed at altitudes where the rates of transport and three-body processes are comparable.

Thus, the maximum number density of atomic oxygen $[O]_{max}$ follows from the relation

$$\frac{w_O[O]_{max}}{L} \sim K[O]_{max}N\,[O_2] \,.$$

We take a dimension $L \sim 100$ km, where the concentration of oxygen atoms varies significantly, and a typical three-body rate constant we take to be $K \sim 10^{-33}$ cm^6/s. This gives that the maximum number density of oxygen atoms is attained at altitudes where

$$N^2[O_2] \sim \frac{w_{ON}}{KL} \,.$$

Taking $[O_2]/N = 1/4$, we find the number density of nitrogen and oxygen molecules in a region of the maximum number density of atomic oxygen $N^3 \sim 4 \cdot 10^{40}$ cm^{-9}, and this gives the maximum number density of oxygen atoms $[O]_{max} \sim 4 \cdot 10^{12}$ cm^{-3}.

► **Problem 14.8** Ozone molecules are formed as a result of three-body attachment of oxygen atoms to oxygen molecules and decay in collisions with oxygen atoms. Determine the maximum concentration of ozone molecules at high atmosphere altitudes.

Taking into account the processes 1, 2 of Table 14.2 for the formation of ozone molecules and the process 3 for their decay, we obtain the equilibrium number density of ozone molecules $[O_3]$

$$[O_3] = \frac{[O_2]NK}{k_3}$$

Taking $[O_2] = N/4$ and the rate constant of the pair process 3 of Table 14.2 at the temperature $T \approx 300$ K to be $k_3 = 1 \cdot 10^{-14}$ cm^3/s, we obtain

$$\frac{[O_3]}{N^2} \approx 2 \cdot 10^{-20}\ \text{cm}^3\ .$$

One can expect that the maximum ozone concentration reaches at the region of the maximum number density of oxygen atoms where $N \sim 4 \cdot 10^{13}$ cm^{-3}, and the ozone concentration is $[O_3]/N \sim 10^{-4}$. Thus, the photodissociation process in the upper atmosphere leads to a high ozone concentration (up to 10^{-4}) at altitudes from 40 to 80 km.

14.2 Ions in the Upper Earth Atmosphere

▶ **Problem 14.9** Molecular ions are formed as a result of photoionization of nitrogen and oxygen molecules and decay by dissociative recombination. Determine the maximum electron and ion number density in the upper atmosphere taking into account that the average photon flux with the wavelengths shorter than 100 nm which can ionize molecules and atoms is $j_{ion} = 2 \cdot 10^{10}$ cm^{-2}s^{-1}.

Atmospheric ions are formed as a result of photoionization of molecules N_2, O_2 and atoms O, and the photoionization cross section is $\sigma_{ion} \sim 10^{-18}$–$10^{-17}$ cm^2, so that the number density of molecules $N_m \sim (\sigma_{ion} L)^{-1} \sim 10^{11}$–$10^{12}$ cm^{-3} at altitudes where photoionization occurs, and $L \sim 30$ km is the thickness of the atmosphere layer where ionization proceeds. A typical number density of atmospheric ions is determined by the balance of ionization and recombination processes. Table 14.3 contains the values of dissociative recombination at room temperature.

Table 14.3 Rates of some recombination processes involving electrons at room temperature.

Number	Process	Rate constant (10^{-7} cm^3/s)
1	$e + N_2^+ \to N + N$	2
2	$e + O_2^+ \to O + O$	2
3	$e + NO^+ \to N + O$	4
4	$e + N_4^+ \to N_2 + N_2$	20

A typical number density of charged particles follows from the balance of ionization and recombination processes that gives the number density of molecular ions N_i at altitudes where molecular ions are formed $\alpha N_e N_i \sim j_{ion}/L$, and α is the dissociative recombination coefficient whose values are given in Table 14.2. From this we find ($\alpha \sim 2 \cdot 10^{-7}$ cm^3/s)

$$N_i \sim \sqrt{\frac{j_{ion}}{\alpha L}} \sim 2 \cdot 10^5\ \text{cm}^{-3}\ . \tag{14.8}$$

▶ **Problem 14.10** Estimate a typical number density of molecular ions for a night atmosphere.

A typical time for establishment of the equilibrium for molecular ions that leads to estimate (14.8) gives $\tau_{rec} \sim (\alpha N_e)^{-1} \sim 30$ s, whereas a characteristic ion drift time for these altitudes is $\tau_{dr} \sim 10^5$ s. Therefore, local equilibrium for molecular ions results from the competing ionization and recombination processes. In addition, from this it follows that the number density of charged particles for daytime and for nighttime atmospheres is different. The above estimate refers to the daytime atmosphere when the flux of solar radiation passes through the atmosphere. The ion number density of the nighttime atmosphere follows from the balance relation in the absence of the solar flux that takes the form $\alpha N_i t \sim 1$, where t is a time after stoppage of the solar flux. Taking this time to be equal to the night duration, we obtain the number density of molecular ions $N_i \sim 4 \cdot 10^2$–10^3 cm^{-3} for a nighttime atmosphere.

▶ **Problem 14.11** Determine the maximum number density of molecular ions in an upper atmosphere.

Atomic ions found at high altitudes are there because of the short transport time required. We can estimate the maximum number density of oxygen atomic ions by comparing a typical drift time L/w and a characteristic time for ion-molecular reactions listed in Table 14.4, measured by $(k[N_2])^{-1}$. Assuming the basic component of the atmosphere at these altitudes to be atomic oxygen, we find that the maximum number density of atomic ions O^+ occurs at altitudes where $[N_2][O] \sim 3 \times 10^{19}$ cm^{-6}. This corresponds to altitudes of approximately 200 km. The maximum number density of atomic ions N_i follows from the balance equation

$$[O] \int \sigma_{ion} dj_{ion} \sim k[N_2] N_i$$

in the case of photoionization of atomic oxygen we have $\int \sigma_{ion} dj_{ion} = 2 \cdot 10^{-7}$ s^{-1}, and the maximum number density of atomic ions is

$$N_i \sim 2 \cdot 10^5 \text{ cm}^{-3} \frac{[O]}{[N_2]} \sim 10^6 \text{ cm}^{-3}. \tag{14.9}$$

In higher layers of the atmosphere the number density of atomic ions is determined by the barometric formula, because it is proportional to the number density of primary atoms, and declines as the altitude increases.

▶ **Problem 14.12** Molecular ions transfer in a lower atmosphere as a result of drift under the action of gravitation force and decay due to dissociative recombination. Estimate the number density of molecular ions at low altitudes.

This number density follows from the balance of ions which are transported from upper atmosphere layers, where ions are formed, and dissociative recombination

Table 14.4 Rates of some ion-molecular processes involving nitrogen and oxygen ions and processes of electron attachment in three-body processes at room temperature.

Number	Process	Rate constant, activation energy
1	$O^+ + N_2 \to NO^+ + N + 1.1$ eV	$6 \cdot 10^{-13}$ cm^3/s
2	$O^+ + O_2 \to O_2^+ + O + 1.5$ eV	$2 \cdot 10^{-11}$ cm^3/s
3	$N^+ + O_2 \to NO^+ + O + 2.3$ eV	$2 \cdot 10^{-11}$ cm^3/s
4	$N_2^+ + O \to NO^+ + N + 3.2$ eV	$1 \cdot 10^{-11}$ cm^3/s
5	$e + 2O_2 \to O_2^- + O_2$	$3 \cdot 10^{-30}$ cm^6/s
6	$e + O_2 + N_2 \to N_2^- + N_2$	$1 \cdot 10^{-31}$ cm^6/s

of molecular ions and electrons. This leads to the balance equation

$$\frac{w_i N_i}{L} \sim \alpha N_i^2 ,$$

where w_i is the drift velocity of ions under the action of the gravitation force and L is a distance from a region of generation of molecular ions. This gives the number density of ions

$$N_i \sim \frac{w_i}{\alpha L} .$$

Let us find the drift velocity for ions being guided by the Dalgarno formula for the mobility of ions that results from the polarization interaction between ions and molecules of the atmosphere. Then by analogy with deduction of formula (14.6), we obtain the reduced drift velocity and diffusion coefficient of ions

$$w_i N = 1.7 \cdot 10^{12}\ \mathrm{cm}^{-2}\mathrm{s}^{-1}, \qquad D_i N = 2.1 \cdot 10^{18}\ \mathrm{cm}^{-1}\mathrm{s}^{-1}. \tag{14.10}$$

Taking a typical distance from a region of molecular ion generation $L \sim 100$ km, we obtain the ion number density

$$N_i \sim \frac{10^{12}\mathrm{cm}^{-6}}{N} , \tag{14.11}$$

and the number density of ions drops inversely proportional to the number density of atmospheric molecules, as we move toward the Earth surface.

▶ **Problem 14.13** Atomic oxygen ions are formed in the upper atmosphere by photoionization of oxygen atoms and decay in ion-molecular processes. Determine an altitude of decay of atomic oxygen ions.

Atomic oxygen ions are formed mostly as a result of photoionization of atomic oxygen, and therefore they are located mostly at more high altitudes compared to molecular ions (Fig. 14.5). In the E-layer region of the ionosphere where molecular ions are found mostly, the number density of atomic oxygen ions is less than that for molecular ions. Indeed, taking the number density of atomic oxygen to be

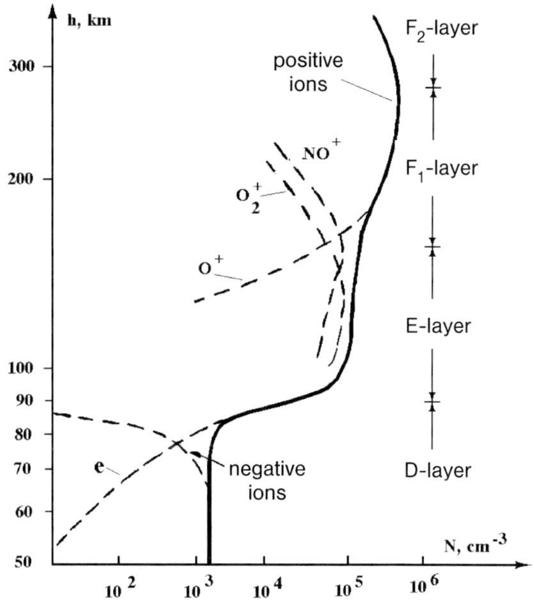

Fig. 14.5 The altitude distribution of the number density of charged atomic particles in an ionosphere.

[O] = $0.1N$ in regions, where photoionization takes place, and assuming the cross sections of photoionization to be identical for oxygen atoms and nitrogen and oxygen molecules, we obtain the rate of generation of O^+ to be less by an order of magnitude than the rate of generation of molecular ions.

At more high altitudes the number density of atomic oxygen ions becomes more than that for molecular ions for two reasons. First, the number density of atomic oxygen at these altitudes exceeds that for nitrogen and oxygen molecules. Second, molecular ions decay as a result of dissociative recombination, while the loss of atomic oxygen ions results from their chemical processes with nitrogen molecules, and the rate of this process drops with an altitude inversely proportional to the number density of nitrogen molecules. Therefore, the number density of atomic oxygen ions in the ionosphere of F-layers exceeds that of molecular ions in E-layer almost by an order of magnitude (see Fig. 14.5).

The character of formation and decay of atomic oxygen ions follows from the balance equation, which in the stationary case has the form

$$M(z) + D_{\rm i} N_{\rm i} - w_{\rm i} \frac{dN_{\rm i}}{dz} - k_{\rm ch} N_{\rm i} N = 0\,, \qquad (14.12)$$

where $N_{\rm i}$ is the number density of atomic oxygen ions, z is altitude, $M(z)$ is the rate of the photoionization process at a given altitude, N is the number density of nitrogen molecules, and $k_{ch} = 6 \cdot 10^{-13}$ cm^3/s is the rate constant of the process

$$O^+ + N_2 \rightarrow NO^+ + N ,$$

given in Table 14.4.

As it follows from the balance equation (14.12), the character of evolution of oxygen atomic ions in the upper atmosphere is determined by three times, a typical time of ion formation $\tau_i = M/N_i$ (N_i is a typical number density of atomic oxygen ions), a typical drift time τ_{dr} for oxygen ions and a typical time $\tau_{ch} \sim (k_{ch} N)^{-1}$ of chemical transformation for atomic ions. The relation between these times depends on the altitude, and we consider below the limiting cases for these times.

In particular, at lower edge of the ionosphere, F-layers where the transition takes place from atomic oxygen ions to molecular ones, we have $\tau_{dr} \sim \tau_{ch}$. Taking the drift velocity of atomic ions by analogy with formula (14.10) as $w = 2 \cdot 10^{12}$ cm^{-2}s^{-1}, we find equalizing the drifting ($\tau_{dr} \approx l/w_i$, $L \sim 10$ km) and chemical times

$$N \sim 2 \cdot 10^{10}\ \text{cm}^{-3} ,$$

which corresponds to the lower boundary of F-layer at the altitudes 150–160 km, and $\tau_{dr} \sim 10^4$ s for such altitudes.

▶ **Problem 14.14** Atomic oxygen ions are generated in F-layers of the upper atmosphere at altitudes around 250 km as a result of photoionization of oxygen atoms. Formed atomic oxygen ions are moving down and decay at the altitude of about 150 km. Determine the space distribution of the number density of atomic oxygen ions in an intermediate region.

Within the framework of a simplified model we assume that atomic oxygen ions are formed at the altitude of 250 km and then decay at the altitude of 150 km in accordance with the previous problem. At intermediate altitudes the concentration c_i of atomic ions follows from the balance equation

$$j = -D_i N \frac{dc_i}{dz} + w_i N c_i = \text{const} .$$

Here $N(z)$ is the total number density of molecules and atoms at a given atmosphere altitude z, i. e., Nc_i is the number density of atomic ions, and j is the flux of atomic oxygen ions down that characterizes the rate of photoionization events for atomic oxygen per unit square. The drift velocity of atomic oxygen ions down w_i in this equation and the diffusion coefficient D_i of ions are connected by the Einstein relation

$$\frac{w_i}{D_i} = \frac{mg}{T} \equiv \frac{1}{l} ,$$

where m is the oxygen atom mass, g is the free fall acceleration, and T is the temperature.

The drift velocity of atomic oxygen ions depends both on the atmosphere temperature and its content. In particular, if we assume the upper atmosphere consisting of atomic oxygen, we obtain the reduced drift velocity atomic oxygen ions by analogy with formula (14.10) the value $w_i N = 1.2 \cdot 10^{12}/(\text{cm}^2\,\text{s})$ at the temperature $T = 300$ K and $w_i N = 7.0 \cdot 10^{11}/(\text{cm}^2\,\text{s})$ at the temperature $T = 1000$ K. In the same manner, assuming the atmosphere to be consisted of molecular nitrogen, we obtain the reduced drift velocity of atomic ions $w_i N = 1.3 \cdot 10^{12}/(\text{cm}^2\,\text{s})$ at the temperature $T = 300$ K and $w_i N = 9.5 \cdot 10^{11}/(\text{cm}^2\,\text{s})$ at the ions depends weakly on the atmosphere content. Taking for the definiteness the temperature to be $T = 500$ K, we find in this case

$$w_i N = 1.0 \cdot 10^{12} \frac{1}{\text{cm}^2\,\text{s}}, \qquad D_i = 2.6 \cdot 10^{18} \frac{1}{\text{cm}\,\text{s}} . \tag{14.13}$$

Let us introduce a new variable

$$\xi = \int_{z_0}^{z} \frac{mg dz}{T(z)} \equiv \int_{z_0}^{z} \frac{dz}{l} ,$$

so that $\xi(z_0) = 0$. The balance equation for a new variable takes the form

$$\frac{dc_i}{d\xi} + c_i = \frac{j}{w_i N} ,$$

and its solution is

$$c_i = \frac{j}{w_i N}\left(1 - e^{-\xi}\right) . \tag{14.14}$$

Let us analyze within the framework of this model the space distribution of atomic oxygen ions. We assume that photoionization of atomic oxygen proceeds at a certain altitude also, taking it to be 250–300 km, and decay of ions proceeds at altitude of $z_0 = 150$ km. Then the variation of the ion concentration is described by formula (14.14), that more or less corresponds to the data of Fig. 14.5. According to this figure, the maximum number density of atomic ions is $N_i \sim 10^6$ cm^{-3} or the concentration of atomic ions at such altitudes to be $c_i \sim 10^{-3}$. From formula (14.14) we obtain a correct estimation for the total flux of photons whose absorption leads to photoionization of atomic oxygen in F-layer $j \sim 10^9\ (\text{cm}^2\,\text{s})^{-1}$. Thus, this simple model gives understanding of the character processes in F-layer of the ionosphere involving atomic oxygen ions.

► **Problem 14.15** In the D-layer the negative charge of the atmosphere transfers from electrons to negative ions. Estimate a typical altitude of this transition and typical ion number density.

We assume that the formation of negative ions results from processes (5) and (6) of Table 14.3. Taking the number density of oxygen molecules to be $[O_2] = N/5$, we obtain the equality of rates of ion loss by dissociative recombination and electron attachment

$$N_i \sim 6 \cdot 10^{-25}\ \text{cm}^{-3} \cdot N^2 .$$

From this according to formula (14.11) we find that the transition from electrons to negative ions takes place at altitudes where the number density of air molecules is

$$N \sim 10^{13}\ \text{cm}^{-3}\ ,$$

which corresponds according to Fig. 14.1 to the altitudes 80–100 km where the D-layer of the ionosphere is located.

▶ **Problem 14.16** Show that the ion drift to lower altitudes does not violate the ionization equilibrium in the upper Earth atmosphere.

Atomic ions formed as a result of the photoionization process participate in ion-molecular reactions given in Table 14.3. A typical time for these processes is $\tau \sim (kN)^{-1} \sim 0.01$–$10$ s, and is small compared to a typical recombination time. This explains the fact that ions of the ionosphere are molecular ions. In addition, it shows the origin of molecular ions NO^+. These ions cannot result from photoionization because of the small number density of NO molecules.

At altitudes where photoionization occurs, the negative charge of the atmospheric plasma comes from electrons. In the D-layer of the ionosphere, electrons attach to oxygen molecules in accordance with processes 5, 6 of Table 14.3, and it is negative ions that govern the negative charge of the atmosphere. At the altitudes where this transition takes place, the balance equation $w N_e/L \sim K N_e [O_2]^2$ is appropriate, where $K \sim 10^{-31}\ \text{cm}^6/\text{s}$ is the rate constant of processes 5 and 6 of Table 14.3. From this it follows that the formation of negative ions occurs at altitudes where $N[O_2]^2 \sim 3 \cdot 10^{39}\ \text{cm}^{-9}$, or $[O_2] \sim 10^{13}\ \text{cm}^{-3}$. We account for the coefficient of ambipolar diffusion of ions being of the order of the diffusion coefficient of atoms. In the D-layer of the ionosphere, recombination proceeds according to the scheme

$$A^- + B^+ \rightarrow A + B\ ,$$

and is characterized by a rate constant of about $\alpha \sim 10^{-9}\ \text{cm}^3/\text{s}$. This leads to the relation

$$N_i N \sim 10^{17}\ \text{cm}^{-6} \tag{14.15}$$

for the number density of ions N_i.

Thus the properties of the middle and upper atmosphere are established by processes in an excited and dissociated air involving ions, excited atoms and excited molecules. Due to photoionization of atomic oxygen or molecular nitrogen under the action of the hard ultraviolet Sun radiation, this atmosphere part contains charged particles and is named the ionosphere. The ionosphere is divided into a number of layers according to the character of processes involving charged particles. The lowest D-layer at altitudes in the range of 50–90 km contains a negative charge in the form of negative ions, and charged particles penetrate the D-layer from the higher E- and F-layers of the ionosphere. The E-layer of the ionosphere at altitudes of 90–140 km contains molecular positive ions and electrons, and they are formed as a result of photoionization processes involving nitrogen and oxygen molecules, and also by chemical reaction of the atomic oxygen ion and nitrogen molecule. The F-layer of the ionosphere that is usually divided into the F_1-layer (140–200 km) and the F_2-layer (200–400 km), contains atomic oxygen ions and provides the maximum number density of charged particles of the order of 10^5–10^6 cm^{-3}. Some peculiarities of the atmospheric plasma at such altitudes were considered above.

14.3
Processes in the Earth Magnetosphere

▶ **Problem 14.17** The magnetic field of the Earth is determined by the Earth magnetic moment $\boldsymbol{M}$ that is created by an internal Earth current and is directed from the north pole to the south one. Determine the position of magnetic lines of force not far from the Earth.

Within the framework of this simply model, the magnetic field strength $\boldsymbol{H}$ is expressed through the magnetic potential μ by the formula $\boldsymbol{H} = -\nabla\mu$, and the magnetic potential is equal to

$$\mu = \frac{\boldsymbol{MR}}{R^3} ,$$

where R is the distance from the Earth center. This gives the magnetic field strength

$$\boldsymbol{H} = \frac{3(\boldsymbol{MR})\boldsymbol{R} - \boldsymbol{M}R^2}{R^5} , \tag{14.16}$$

and components of this vector are

$$H_R = -\frac{2M\sin\theta}{R^3} , \qquad H_\theta = \frac{M\cos\theta}{R^3} ,$$

where θ is the latitude of a given point. The total magnetic field strength is

$$H = \frac{M}{R^3}\sqrt{\frac{5 - 3\cos 2\theta}{2}} .$$

Within the model under consideration the Earth is modeled by a ball of a radius $R_\oplus = 6370$ km with the magnetic moment of $M = 7.80 \cdot 10^{25}$ G s $\cdot$ cm^3, and the magnetic field strength near the equator is $H_o = 0.31$ G s. This model takes into account the principal properties of the Earth magnetic field.

According to the definition, the magnetic lines of force which are given in Fig. 14.6, satisfy the equation

$$\frac{dR}{H_R} = \frac{rd\theta}{H_\theta} ,$$

which gives

$$\frac{dR}{R} = -2\frac{\sin\theta d\theta}{\cos\theta} .$$

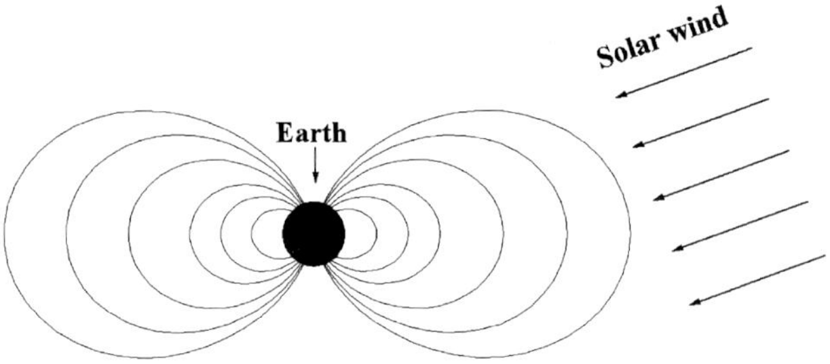

Fig. 14.6 Magnetic lines of force near the Earth.

From this it follows

$$R = R_e \cos^2\theta, \qquad (14.17)$$

where R_e is the the distance between the Earth center and magnetic field of force at the equator. Correspondingly, the latitude dependence for the magnetic field strength at a given magnetic lines of force is

$$H = \frac{M}{R_e^3 \cos^6\theta}\sqrt{\frac{5 - 3\cos 2\theta}{2}} .$$

▶ **Problem 14.18** Find the distance from the Earth for the upper boundary of the Earth magnetosphere. At this distance the pressure from the flux of the solar wind becomes equal to the magnetic pressure of the Earth.

The solar wind is a plasma stream moving from the Sun to the Earth. This results from an instability in the Sun corona in the form of individual jets and we assume the solar wind to be isotropic in average on large distances from the Sun. Accounting for a strong fluctuation of solar wind parameters, we will consider the following values on the Earth level: the number density of protons of the solar wind plasma near the Earth is $N_p \approx 10\ \mathrm{cm}^{-3}$, the average drift velocity of protons is $w = 3 \cdot 10^7$ cm/s that corresponds of the kinetic energy of protons approximately 0.5 keV, the temperature of electrons in the solar wind plasma is about 2 eV, and the proton temperature is about 0.4 eV. The subsequent estimations will be based on these parameters.

Interaction of the solar wind with the Earth magnetic field (Fig. 14.7) takes place in the region where their pressures are comparable, and the region near the Earth we call as the magnetosphere. In estimating the boundary of the magnetosphere where the above pressures are equalized, we note that electrons give a small contribution to the kinematic pressure of the solar wind because of their small momentum compared with that of protons. Hence, the pressure of the solar wind according to formula (6.18)

$$p = N_p M_p w^2 \sim 1 \cdot 10^{-6}\ \mathrm{Pa} .$$

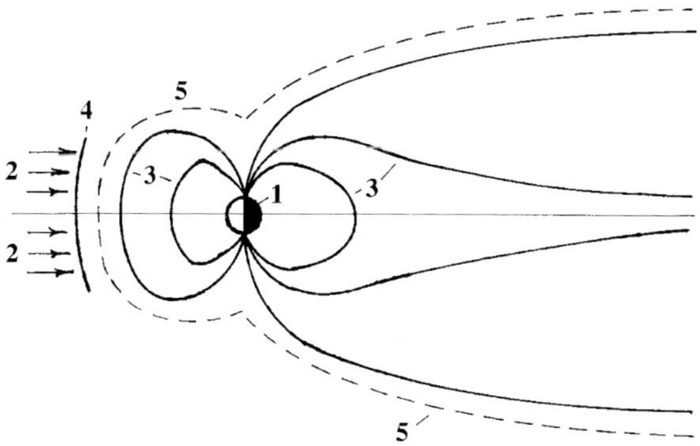

Fig. 14.7 Interaction of the solar wind with the Earth magnetic field. (1) the Earth; (2) solar wind; (3) magnetic lines of force; (4) shock waves; (5) the magnetosphere boundary where the solar wind pressure equals to the pressure of the Earth magnetic field.

This pressure is equalized by the Earth magnetic field that is estimated as

$$H = H_o \left(\frac{R_\oplus}{R} \right)^3 ,$$

where $H_o = 0.31$ Gs is the magnetic field strength, $R_\oplus = 6370$ km is the Earth radius and R is the distance from the Earth. Therefore, the upper magnetosphere boundary follows from the equality of the kinematic (6.18) and magnetic (12.40) pressure and is equal to

$$R = R_\oplus \left(\frac{H_o}{\sqrt{8\pi p}} \right)^{1/3} .$$

This formula gives $R \approx 8R_\oplus$.

At such distances, a strong interaction occurs between the solar wind and the Earth magnetic field. As a result, a plasma stream shifts magnetic lines of force, and then the plasma stream flows around the Earth along the shifted magnetic lines of force (see Fig. 14.6). Note that because the solar wind starts from a plasma instability on the Sun surface, its parameters fluctuate significantly. Correspondingly, the magnetosphere boundary oscillates in time remarkably.

► **Problem 14.19** Estimate an altitude of the lower boundary of the magnetosphere where the Larmor frequency for protons is equal to the rate of proton collision with atmosphere molecules.

The Larmor frequency of protons in the magnetic field of the Earth near its surface is equal to $\omega_H = 3 \cdot 10^3$ Hz, and the rate of proton–molecule collisions is $1/\tau = Nv_p\sigma$, and we take the gas-kinetic cross section $\sigma = 3 \cdot 10^{-15}$ cm^2 as the cross section of proton–molecule collision. This equality is fulfilled at the altitude where the number density of molecules is

$$N \sim 2 \cdot 10^{10}\ \mathrm{cm}^{-3}\ ,$$

which corresponds to the altitude of approximately 140 km. At this and lower altitudes the magnetic field is not affected on motion of protons.

► **Problem 14.20** Estimate an atmosphere altitude where aurora occurs as a result of penetration of protons of the solar wind in the Earth atmosphere.

A solar wind can penetrate deeply inside the Earth atmosphere if protons are moving to the Earth near its poles. Because the Earth during its rotation does not turn to the Sun by poles, it is possible only during magnetic storms when the plasma flow from the Sun can deviate from a straightforward way. Then protons can go close to the Earth surface and excite atoms and molecules of the atmosphere. As a result, excited atoms and molecules are formed in a rare atmospheric gas and because of a low density, they are quenched as a result of radiation. In particular, Table 14.5 contains radiative parameters of metastable nitrogen and oxygen atoms which can form as a result of proton collisions with nitrogen and oxygen molecules and oxygen atoms. The transitions indicated in this table and also radiative transitions for excited nitrogen molecule and oxygen atom are determined by glowing of the aurora that is characterized by a variety of colors.

Table 14.5 Radiation transitions for metastable nitrogen and oxygen atoms.

Transition	$\Delta\varepsilon$ (eV)	λ (μm)	τ (s)
$N(^2D_{5/2} \to ^4S_{3/2})$	2.38	0.5202	$1.4 \cdot 10^5$
$N(^2D_{3/2} \to ^4S_{3/2})$	2.38	0.5199	$6.1 \cdot 10^4$
$N(^2P \to ^2D)$	3.58	1.040, 1.041	12
$O(^1D \to ^3P)$	1.97	0.630	140
$O(^1S \to ^3P)$	4.19	0.5577	0.8

Taking a typical cross section of inelastic collision of protons with atmospheric atoms and molecules to be $\sigma \sim 10^{-17}$–10^{-18} cm^2, below we estimate the altitude where such collisions occur. We find the altitude where polar glowing proceeds from the relation

$$N\sigma l \sim 1\ ,$$

where N is the number density of atmospheric atoms and molecules, and $l \sim 10$ km is a distance where atmospheric parameters vary remarkably. This gives the number density of optimal excitation of atmospheric atoms and molecules to be $N \sim 10^{11}$–10^{12} cm^{-3}, and the altitude where this takes place is between 100 and 150 km.

► **Problem 14.21** Estimate a portion of protons of the solar wind that is scattered during motion of the solar wind near the Earth surface and can be captured by the Earth magnetic field. As a result, these protons can form an Earth radiation belt.

When the solar wind as a plasma flow encounters with the Earth magnetic field, this plasma shifts partially toward the magnetic field and in turn the plasma flows along the magnetic lines of force of the formed system. Note that this plasma is magnetized, i. e., the inverse Larmor frequency $1/\omega_H$ is small compared to a time τ_{dr} of passage of the magnetic field region by the plasma flow. In particular, for protons of the solar wind we obtain the Larmor frequency $\omega_H = eH/(m_p c) \approx 3$ kHz, and a time of plasma flow around the Earth is $\tau_{dr} \sim 2R_0/v_p \sim 300$ s, i. e., $\omega_H \tau_{dr} \sim 10^7$. Hence, electrons and protons have a circular motion around magnetic lines of force and flow along them around the Earth.

A part of electrons and protons can be captured by the Earth magnetic field, if these charged particles are moving in a nonuniform magnetic field in certain directions (Problem 7.24). These directions can result from collisions between plasma particles. Because these collisions are determined by the Coulomb interaction between colliding particles, the cross section of this process can be estimated as $\sigma \sim \pi e^4/\varepsilon^2$, where ε is the energy of particle relative motion. Taking an estimation for protons $\varepsilon \sim 1$ keV, we obtain the scattering rate $\nu \sim N v_p \sigma \sim 10^{-10}\ \mathrm{s}^{-1}$, where $N = 5\ \mathrm{cm}^{-3}$ is the number density of protons in the solar wind. Hence, a part of protons $\nu\tau_{dr} \sim 3 \times 10^{-8}$ is scattered in the course of plasma passage around the Earth. These scattered protons can be captured by the magnetic field and form a Earth radiation belt.

► **Problem 14.22** Estimate the period of oscillations for protons of the solar wind which are captured by the Earth magnetic field in the equatorial plane.

Radiation belts, regions with captured charged particles, are located at a distance of several Earth radii. Captured protons rotate around magnetic lines of force with the Larmor frequency

$$\omega_H = \frac{eH}{m_p c} = \frac{eH_o}{m_p c}\frac{R_\oplus^3}{R^3} ,$$

where R is the distance from the Earth center for proton location and $R_\oplus = 6370$ km is the Earth radius. Along with this motion, protons drift under the action of nonuniformity of the magnetic field, and the drift velocity of this drift is given by formula (7.58) and is equal to

$$w = \frac{v_\tau^2}{\omega_H}\frac{\nabla H}{H} ,$$

where the nonuniformity of the magnetic field is

$$\frac{\partial H}{\partial x} = 3H_o \frac{R_\oplus^3}{R^4} \quad \frac{\partial H}{H \partial x} = \frac{3}{R} .$$

This leads to the following drift velocity in the tangential direction to the axis:

$$w = w_o \frac{R^2}{R_\oplus^2} \quad w_o = \frac{3v_\tau^2}{\omega_{H_o} R_\oplus} \sim 1\ \mathrm{cm/s} .$$

This corresponds to a time of rotation of a captured proton of order of years.

▶ **Problem 14.23** Determine the width of a region that separates a region of the Earth magnetic field and the region of the solar wind.

The solar wind that flows around the Earth is separated from magnetic lines of force by a shock wave. Therefore, a solar wind plasma and the Earth magnetic field are not mixed. Nevertheless, one can consider the boundary between the Earth magnetic field and solar wind plasma, and on this boundary magnetic lines of force penetrates in the plasma. Let us find the width of such a boundary region.

The width of an intermediate region is narrower, the higher is the plasma conductivity. The plasma conductivity is determined by the Spitzer formula (7.65), and taking in this formula the electron temperature $T_e = 2$ eV and the Coulomb logarithm $\ln \Lambda = 10$, we obtain the plasma conductivity $\Sigma \approx 1 \cdot 10^{14}\ \mathrm{s}^{-1}$. From this, we find for the diffusion coefficient D_H of the plasma with respect to magnetic lines of force according to formula (12.38)

$$D_H = \frac{c^2}{4\pi\Sigma} \sim \cdot 10^3\ \mathrm{cm}^2/\mathrm{s}$$

Assuming the drift velocity of the solid wind to be $v = 3 \cdot 10^7$ cm/s, we find a typical time of its passage around the Earth $\tau \sim 20 R_\oplus / v \sim 400$ s, where $R_\oplus$ is the Earth radius, we find the depth of magnetic field penetration in the plasma during its motion $\Delta = \sqrt{2 D_H \tau} \sim 20$ m, that is small compared to typical plasma dimension.

14.4
Electromagnetic Waves in the Upper Atmosphere

▶ **Problem 14.24** Find the boundary wavelength of electromagnetic waves which cannot pass through the ionosphere and reflect from it.

According to formula (11.23), radiowaves of frequencies $\omega < \omega_p$ do not penetrate in the ionosphere and reflect from it. Taking the maximum number density of atomic oxygen ions and electrons in F-layer to be $N_i \sim 1 \cdot 10^6\ \mathrm{cm}^{-3}$, we obtain according to formula (11.6) for the plasma frequency $\omega_p = 6 \cdot 10^7$ Hz at the number density of electrons $N_e = N_i$. Therefore, radiowaves whose wavelengths exceed $\lambda = 2\pi/\omega_p = 30$ m do not pass through the ionosphere and reflect from it.

▶ **Problem 14.25** Determine which portion of the intensity of an electromagnetic wave is absorbed in the ionosphere as a result of reflection from it. Assume that propagation of an electromagnetic wave in the atmosphere proceeds according to the laws of geometric optics.

According to the dispersion relation (11.23), an electromagnetic wave cannot propagates in the atmosphere, where $\omega < \omega_p$. Representing the wave frequency in the form $\omega = Re\omega - i\delta$, we obtain the absorption coefficient from the dispersion relation (11.26)

$$\delta = \frac{\omega_p^2}{2\omega^2\tau} ,$$

and within the framework of geometric optics, the probability to survive for an electromagnetic wave is equal to

$$P \equiv \exp(-\zeta) = \exp\left(-\frac{\int \delta dt}{4\cos\theta}\right),$$

where the integral is taken over one half of a direct ray, and θ is the angle between the ray direction and the normal to the atmosphere layer.

Taking into account $dt = dx/v_p$, where the dispersion relation (11.23) gives the wave phase velocity $v_p = c\sqrt{\omega^2 - \omega_p^2}/\omega$, we obtain the above damping

$$\zeta = \frac{2}{c\omega\cos\theta}\int \frac{\omega_p^2 dx}{\tau\sqrt{\omega^2 - \omega_p^2}}.$$

Let us take the linear dependence of the number density of electrons on the altitude z in some range of altitudes near the turning point

$$N_e(z) = N_0 \cdot \left(1 - \frac{z - z_0}{L}\right) = N_0(1 - \xi),\ \xi = \frac{z - z_0}{L}. \tag{14.18}$$

Taking a damping of an electromagnetic wave due to electron–molecule collisions, we have the dependence of its rate $1/\tau$ on an altitude z in the form

$$\frac{1}{\tau} \sim N \sim \left(\frac{z}{l}\right),$$

where N is the total number density of atoms and molecules. This leads to the following expression for the absorption exponent:

$$\zeta = \frac{2L}{c\tau(z_0)\cos\theta}J,\quad J = \int_0^1 \frac{\xi d\xi}{\sqrt{1-\xi}}\exp\left(-\frac{L\xi}{l}\right). \tag{14.19}$$

In the limiting cases, the integral in formula (14.19) is equal to

$$J = \left(\frac{l}{L}\right)^2,\quad L \gg l;\quad J = \frac{4}{3},\quad L \ll l.$$

Thus, absorption of an electromagnetic wave proceeds in the region where electrons are located and the number density of neutral particles is enough high, and this takes place on the lower boundary of E-layer. In the above formulas ω_p is the plasma frequency on the upper boundary of E-layer. Though the absorption coefficient depends on the character of growth of the electron number density in E-layer, the considered example shows a strong absorption of an electromagnetic wave if it is reflected in E-layer. Therefore, passage of electromagnetic waves on long distances by reflection from the ionosphere is of importance in reality if this reflection proceeds in F-layer with a more high electron number density. Then the ratio $\omega^2/\omega_p^2 \sim 10$, and the electromagnetic wave is absorbed partially. The maximum electron number density in F-layer that is $\sim 10^6\ \text{cm}^{-3}$ corresponds to a reflected electromagnetic wave of the length $\lambda \approx 30$ m.

▶ **Problem 14.26** Analyze absorption of radiowaves in E-layer of the atmosphere at altitudes 90–140 km, approximating in average the number density of electrons by the linear altitude dependence in E-layer.

As it follows from the dispersion relation (11.26), the imaginary part of an electromagnetic wave is

$$\delta = \frac{\omega_p^2}{2\omega^2 \tau} ,$$

where we represent the wave frequency in the form $\omega = Re\omega - i\delta$. According to this formula, wave absorption takes place at altitudes where electrons are located. Along with this, the absorption coefficient is proportional to the number density of atmosphere molecules and atoms since it is determined by electron–molecule and electron–atom collisions.

Within the framework of the above assumptions, we obtain the absorption exponent according to formula (14.19)

$$\delta = \frac{\zeta_0}{\cos\theta} \frac{\omega_0^2}{\omega^2} ; \qquad \zeta_0 = \frac{2l^2}{cL\tau(z_0)} ,$$

where $\omega_0 = 25$ MHz is the plasma frequency for the electron number density at higher boundary of E-layer ($h = 140$ km), and absorption is determined by the lower boundary of E-layer ($h = 90$ km). At this altitude we obtain the total number of molecules $N = 5 \cdot 10^{13}$ cm^{-3} and temperature $T = 200$ K. Taking that the cross section of electron-molecule cross section is equal to the gas-kinetic cross section $\sigma = 3 \cdot 10^{-15}$ cm^2, we obtain a typical time of electron–molecule collisions $\tau(z_0) = (N v_e \sigma)^{-1} = 6 \cdot 10^{-7}$ s (the electron thermal velocity is $v_e = \sqrt{8T/(\pi m_e)} = 9 \cdot 10^6$ cm/s). Next, the depth of E-layer is $L = 50$ km, and the distance of variation of the molecule number density is $l = (d \ln N/dh)^{-1} = 7$ km. This leads to the following expression for the absorption exponent

$$\delta = \frac{10}{\cos\theta} \frac{\omega_0^2}{\omega^2} .$$

According to this result, under above assumptions any electromagnetic wave that is reflected in E-layer is absorbed in this region. In addition, this absorption is determined by parameters of the ionosphere near its edge where the electron number density is small, but the molecule number density is not small. Therefore absorption depends on the electron distribution in this region.

In reality, the electron number density distribution in E-layer varies with time. To demonstrate this, we consider two examples. In the first case the linear distribution of the electron number density (14.18) is valid, but the lower boundary of E-layer is shifted and corresponds to the altitude $h_o = 100$ km. Then the above formula for the absorption exponent takes the form

$$\delta = \frac{5}{\cos\theta} \frac{\omega_0^2}{\omega^2} .$$

In the second case we take the space distribution of the electron number density to be

$$N_e(z) - N_o \cdot \left(\frac{z - z_0}{L}\right)^2$$

instead of formula (14.18). In this case the absorption exponent is given by

$$\delta = \frac{3}{\cos\theta}\frac{\omega_o^2}{\omega^2} .$$

As is seen, the absorption coefficient for an electromagnetic wave passing through E-layer of the ionosphere can vary by an order of magnitude under varying ionosphere parameters.

► **Problem 14.27** Estimate typical altitudes at which whistlers as slow electromagnetic waves in the Earth magnetic field can be generated.

Whistlers result from electric breakdown in the atmosphere when a plasma is formed in an extent region and contains electrons of a not low density. According to the dispersion relation (11.50), whistlers exist if the criterion

$$\omega_H \tau \gg 1 ,$$

holds true. The Larmor frequency for electrons is $\omega_H \sim 10^7 \text{s}^{-1}$ near the Earth surface. Taking the cross section of electron-molecule cross section to be of the order of gas-kinetic one ($\sigma \sim 3 \cdot 10^{-15}$ cm^2), and the electron thermal velocity $v_e \sim 10^7$ cm/s, we obtain the above criterion that whistlers exist at altitudes where the number density of atmosphere molecules

$$N \ll 10^{15}\ \text{cm}^{-3} .$$

This corresponds to altitudes above 50 km. Moving in lower regions where electrons are absent, whistlers are transformed in simple electromagnetic waves.

► **Problem 14.28** Whistlers are formed in the south Earth hemisphere as a result of breakdown in the upper atmosphere, propagate along the magnetic lines of force and is detected in north hemisphere. Taking the whistler frequency to be $\omega \sim 5$ kHz and an altitude of its magnetic lines of force above the equator $h \sim$ 1000 km, estimate a delay time for low-frequency waves.

According to the dispersion relation (11.39), the group velocity of whistlers v_{gr}, which are propagated along the magnetic lines of force, is equal to

$$v_{gr} = \frac{\partial \omega}{\partial k} = 2c\frac{\sqrt{\omega_H \omega}}{\omega_p} .$$

A time of whistler propagation between two hemispheres is

$$\tau = \int \frac{dl}{v_{gr}} ,$$

where dl is an element of the magnetic lines of force that according to formula (14.17) is

$$dl = \sqrt{R^2 d\theta^2 + dR^2} = R_e \cos\theta d\theta \sqrt{1 + 3\sin^2\theta} \, .$$

In particular, taking an altitude of the magnetic lines of force near the equator $h \sim 1000$ km ($\omega_H = 5 \cdot 10^6$ Hz) and an average number density of electrons on this line of force $N_e \sim 10^4$ cm^{-3} ($\omega_p = 6 \cdot 10^6$ Hz), we have ($\omega = 5$ kHz) for the group whistler velocity $v_{gr} \approx 2 \cdot 10^9$ cm/s, and a typical time of whistler propagation $\tau \sim 6$ s. Correspondingly, a delay time for signals of different frequencies is of seconds and exceeds a typical breakdown time.

14.5
Electric Phenomena in the Earth Atmosphere

▶ **Problem 14.29** The Earth surface is charged as a result of secondary processes of water evaporation and condensation. The Earth potential is $U = 300$ kV, the average electric field strength near the Earth surface is approximately 130 V/cm. Estimate a typical thickness of the atmosphere layer near the Earth surface that is responsible for Earth charging.

The electric field strength E near the Earth surface is connected with the Earth surface charge σ by the relation $E = 4\pi\sigma$, which leads to the total Earth charge under these conditions $Q = 4\pi R_\oplus^2 \sigma = R_\oplus^2 E = 5.8 \cdot 10^5$ C, where $R_\oplus = 6370$ km is the Earth's radius, and the surface density of Earth charge is $\sigma = E/4\pi = 7 \cdot 10^7\ e/\text{cm}^2$. Let us consider the Earth as an electric system that is like a spherical capacitor with a negative charge of its lower plate. Then assuming the electric field between capacitor plates to be uniform, we find the distance l between the plates of this capacitor to be $l = U/E = 2.3$ km. This estimate gives a dimension of a region near the Earth surface where electric processes proceed with charging of the Earth. From this we find that if an unipolar plasma is located between these plates, positively charged particles are characterized by the number density of $N_i = 300$ cm^{-3}. We have from this that the processes responsible for the Earth's charge occur in the lower layers of the Earth atmosphere at altitudes of several kilometers.

▶ **Problem 14.30** Estimate a typical time of Earth charging.

We use the total average current of Earth discharging to be approximately $I = 1700$ A. Of course, a local current density is found in a wide range, and the average current density over land amounts to $2.4 \cdot 10^{-16}$ A/cm^2, and over the ocean it is $3.7 \cdot 10^{-16}$ A/cm^2. This gives an average time of Earth discharging $\tau = Q/I = 6$ min. Taking the average mobility of charged particles near the Earth surface to be $K = 2.3$ cm^2/(V · s), we find that charged particles pass a distance $s = KE\tau = 50$ m for a discharging time, i. e., the charging process proceeds close to the Earth surface. One can connect it with charged water drops and dust particles falling on the Earth surface.

▶ **Problem 14.31** Assuming that Earth charging results from falling of large negatively charged drops, find the rate of drop falling that provides the charging current ($I = 1700$ A). Compare it with the total rate of water in the Earth atmosphere.

Using formula (13.60) the velocity v of drop falling under the action of gravitational forces and formula (13.41) for the charge Z of a large drop, we obtain the current density of drops on the Earth surface

$$i = eZvN \, ,$$

where N is the number density of drops in the atmosphere.

It is essential that since the drop charge Z is proportional to a drop radius according to formula (13.41), and according to formula (13.60) the velocity of drop falling is proportional to the square of the drop radius, the current density is proportional to the total mass of drops, i. e., the current density is proportional to the total mass of water in the atmosphere. The above formula gives

$$\frac{i}{\rho(H_2O)} = 2 \cdot 10^{-11} \frac{A/m^2}{g/m^3} \, , \tag{14.20}$$

where $\rho(H_2O)$ is the water density in the form of charged drops in the atmosphere.

▶ **Problem 14.32** Find the radius of charged water drops which fall down in spite of the average electric field of the Earth surface.

Negatively charged droplets fall down under the action of gravitation forces, while the electric field of the earth acts up. Evidently, a droplet is moving down if the gravitation force $P = mg$ exceeds the electric one eEZ

$$mg > eEZ \, ,$$

where m is the drop mass, g is the free fall acceleration, $E \approx 130$ V/m is the average electric field strength near the Earth surface, and the particle charge Z is given by formula (13.41). Note that the left-hand side of this inequality is proportional to the radius cube, whereas the right-hand side of the inequality is proportional to the drop radius. Taking the average ratio of the diffusion coefficients of positive and negative ions for the drop charge to be 0.8, we find that the above criterion holds true, if the drop radius $r > 0.05$ μm. Though the assumptions used for this estimate are not valid at the low limit, this result indicates that the electric field of the Earth is not of importance for fall of micron drops.

▶ **Problem 14.33** Compare the total mass of water in the form of drops that provides an observational charging of the Earth with the total water mass located in the atmosphere.

Every year approximately $4 \cdot 10^{14}$ tonnes (metric tons) of evaporated water pass through the atmosphere, and atmospheric water is renewed 32 times per year in average. If we take an effective height $h \approx 4$ km of the atmosphere containing water the corresponding value of the water density will be 60 g/m^{-3} approximately (or about 5 % with respect to the air mass).

We now compare the total water mass in the atmosphere with that in the form of water drops whose falling can provide an observational charging of the Earth. Indeed, an observational current density over land is $2.4 \cdot 10^{-12}$ A/m^2, and over the ocean it is $3.7 \cdot 10^{-12}$ A/m^2. This value will be about $3 \cdot 10^{-12}$ A/m^2, if the total average current per Earth surface $I = 1700$ A will be divided per Earth surface area $S = 5.1 \cdot 10^{14}$ m^2. According to formula (14.20) this corresponds to the water density $\rho(H_2O)$ fin the form of drops

$$\rho(H_2O) = 1.5 \cdot 10^{-2} \text{ g/m}^3 ,$$

which is significantly less than the total water density in the atmosphere.

▶ **Problem 14.34** Taking electric phenomena in the Earth atmosphere as a secondary phenomena of water circulation between the Earth surface and atmosphere, compare the powers of these phenomena.

Thus, one can consider the Earth charging as the secondary phenomenon of water circulation in the atmosphere. Then a part of water evaporated from the Earth surface is condensed at altitudes of several kilometers, where water droplets—aerosols are formed, and, in particular, small droplets form clouds. Large droplets which are charged negatively due to different mobilities of positive and negative ions in the atmosphere fall on the Earth surface under the action of the gravity force. This leads to the Earth charging, whereas its discharging proceeds mostly as a result of lightning discharges. Note the irregular character of these processes, so that the rate of charging under optimal conditions exceeds its average value by several orders of magnitude.

The water circulation in the atmosphere causes evaporation of 13 million tonnes of water per second that requires a power of 4×10^{13} kW. The power associated with passage of electric current through the atmosphere by fall of negatively charged drops is $UI = 5 \cdot 10^5$ kW, where $U = 300$ kV is the Earth's potential and $I = 1700$ A is the total average atmospheric current. Discharging of the Earth as a result of thunderstorms associated with a more high cloud potential, roughly by a factor of 10^3. This increases the power of electric phenomena in the atmosphere by the indicated factor, but the result is less several orders of magnitudes in comparison to the power that is consumed on water circulation. Hence the contribution of electric processes to the total power of atmospheric process is very small.

▶ **Problem 14.35** Consider the sources of atmosphere ionization from the standpoint of the Earth charging processes.

For the Earth charging it is necessary in the beginning to form ions in the atmosphere and then separate these ions. There are two mechanisms of ion formation in the Earth atmosphere, due to cosmic rays and due to radioactive decays on the Earth surface.

Maximum ionization by cosmic rays, mostly generated by the Sun, is observed at altitudes of 11 to 15 km, and is characterized by ionization rates of 30–40 $\text{cm}^{-3}\,\text{s}^{-1}$. This results in an ion number density of about $6 \cdot 10^3\ \text{cm}^{-3}$. (The recombination coefficient is about $10^{-6}\ \text{cm}^3/\text{s}$ at this altitude.) To explain the observed currents of charged particles formed, it is necessary that the intensity of ionization in the lower atmospheric layers should be smaller than this maximum by a factor of 100. Another mechanism results from decay of the radioactive materials, and the rate of these processes may be changed by several orders of magnitude depending on a point of the Earth surface.

We can estimate time scales for the electric processes of the Earth. Positive and negative ions recombine under atmospheric conditions by three-body collisions, with an effective ion recombination coefficient of $\alpha = 2 \cdot 10^{-6}\ \text{cm}^3/\text{s}$ (see Table 14.3). This corresponds to a recombination time of about $\tau = (\alpha N_i)^{-1} = 0.5$ h. This contrasts with the much shorter characteristic time of $q/I = 6$ min for discharging of the Earth, where q is the Earth's charge and I is the average atmospheric current. During their life time τ, ions travel a distance $s = KE\tau = 50$ m, where K is the ion mobility and E is the electric field strength near the Earth. Because this distance is small compared to the size of the Earth's layer where electrical phenomena develop, there must be a mechanism for the generation of these ions. The intensity of this process (i. e.„ the number of ions per unit time and per unit volume) is given by $\alpha N_i^2 = 0.1\ \text{cm}^{-3}\,\text{s}^{-1}$.

14.6 Radiation of the Solar Photosphere

▶ **Problem 14.36** Find the relation between the number densities of electrons, negative ions and hydrogen atoms in the Sun photosphere.

Radiation of the Sun photosphere is governed by the process

$$e + \text{H} \longleftrightarrow \text{H}^- + \hbar\omega . \tag{14.21}$$

Because this process is relatively weak, and a photosphere plasma is dense in order to provide a high efficiency of this process, local thermodynamic equilibria are supported in the solar photosphere on the basis of the processes

$$e + \text{H} \leftrightarrow 2e + \text{H}^+ ; \qquad e + \text{H}^- \leftrightarrow 2e + \text{H} \tag{14.22}$$

From these equilibria the Saha relations (1.69) follow between the number densities of the corresponding species, which are given by

$$N_e = \left[\frac{m_e T}{2\pi\hbar^2}\right]^{3/4} N_\text{H}^{1/2} \exp\left(-\frac{J}{T}\right), \qquad N_- = \frac{1}{4}\left[\frac{m_e T}{2\pi\hbar^2}\right]^{-3/4} N_\text{H}^{3/2} \exp\left(-\frac{\varepsilon_0}{T}\right). \tag{14.23}$$

Here T is a local temperature; N_H, N_e, and N_- are the number densities of hydrogen atoms, electrons and negative ions, respectively; $J = 13.605$ eV is the ionization potential of the hydrogen atom; $\varepsilon_o = J/2 - EA = 6.048$ eV, where $EA = 0.754$ eV is the electron affinity of the hydrogen atom. We use the condition of plasma quasineutrality $N_e = N_p$, where N_p is the number density of protons. The second expression of formula (14.22) is the Saha distribution for the equilibrium between negative ions, hydrogen atoms, and electrons. In formula (14.23) for the number density of negative ions, the electron number density is taken from the Saha distribution corresponding to the equilibrium (14.22) between protons, electrons and hydrogen atoms. In the problem being examined, $N_H \gg N_e \gg N_-$.

▶ **Problem 14.37** Estimate the parameters of the layer of the solar photosphere that is responsible for its radiation.

We employ a general formula (9.13) for emission of a hot gas layer with a varied temperature writing it in the form

$$j_\omega = \frac{\omega^2}{2\pi^2 c^2} \int_0^1 d(\cos\theta) \int_0^\infty du_\omega \exp(-u_\omega/\cos\theta) F(u_\omega) ,$$

and $F(u_\omega) = [\exp(\hbar\omega/T) - 1]^{-1}$. Assuming the dependence $F(u_\omega)$ to be weak, we expand this function over a small parameter in a series

$$F(u_\omega) = F(u_o) + (u_\omega - u_o)F'(u_o) + \frac{1}{2}(u_\omega - u_o)^2 F''(u_o) ,$$

we choose the parameter u_o such that the second term is zero after integration. This yields $u_o = 2/3$, and the radiation flux is

$$j_\omega = j_\omega^{(0)} \left[1 - \frac{5F''(u_o)}{18F(u_o)}\right] , \tag{14.24}$$

where $j_\omega^{(0)} = \omega^2(4\pi^2c^2)^{-1} [\exp(\hbar\omega/T) - 1]^{-1}$ is the radiation flux of a black body at a temperature that corresponds to the point where the optical thickness is $u_\omega = 2/3$. The second term in the parentheses of formula (14.24) makes it possible to estimate the accuracy of the operation employed.

We now apply formula (14.24) to the solar atmosphere. Approximating the height dependence for the number density of negative ions by the dependence $N_-(z) = N_-(0)\exp(-z/l)$, we obtain from the relation $u_o = 2/3$ the expression

$$N_- = (2l\sigma_\omega/3)^{-1} \tag{14.25}$$

for the number density of negative ions, where σ_ω is the cross section for H^- photodetachment. The effective radiative temperature for a given frequency is taken to be the temperature of the solar atmosphere at the altitude that is defined by formula (14.24).

The photodetachment cross section of the negative hydrogen ion has a threshold at the photon energy $\hbar\omega = EA$, and has a maximum, $\sigma_{max} = 4 \cdot 10^{-17}\,\text{cm}^2$, at $\hbar\omega_{max} = 2EA = 1.51$ eV corresponding to the photon wavelength $\lambda = 0.8$ μm. Using parameters for the average quiet solar photosphere, we obtain formula (14.24) that the effective temperature for this wavelength is $T_{ef} = 6100$ K. The layer of the average solar atmosphere with this temperature contains plasma constituents with number densities $N_H = 1 \cdot 10^{17}$ cm^{-3}, $N_e = 4 \cdot 10^{13}$ cm^{-3}, and $N_- = 4 \cdot 10^9$ cm^{-3}.

▶ **Problem 14.38** Estimate a thickness of the solar photosphere that is responsible for solar emission in a visible spectrum range.

The temperature of solar plasma varies with variation of the layer altitude, and because of local thermodynamic equilibrium for plasma constituents, the number density of negative ions varies with the altitude according to their temperature dependence $N_- \sim \exp(-\varepsilon_0/T)$. This gives the thickness of the layer that is responsible for Sun radiation

$$l = \left[\frac{\varepsilon_0}{T}\frac{d\ln T}{dz}\right]^{-1} = 40 \text{ km} .$$

The effective radiation temperature T_ω for radiation at a given frequency ω depends on the photodetachment cross section σ_ω for this frequency according to formula (14.25). Taking the effective radiation temperature in the form $T_\omega = \Delta T + T_{ef}$, we obtain from this formula

$$\Delta T = l\frac{dT}{dz}\ln\frac{\sigma_{max}}{\sigma_\omega} .$$

As an example, we consider radiation of the solar photosphere at a double frequency with respect to the maximum for the absorption cross section of the hydrogen negative ion. This gives $\hbar\omega = 2\hbar\omega_{max}(\lambda = 0.4\ \mu\text{m})$. Then $\sigma_\omega = 0.65\sigma_{max}$, and we have $\Delta T = 200$ K. This corresponds to an increase in the radiation flux by 20 % compared to that of the radiation temperature T_{ef}.

To check the validity of the expansion used for the function $F(u_\omega)$, we take $F(u_\omega) = \exp(-\hbar\omega/T)$, that is valid when $\hbar\omega \gg T$. In this case the second term in the parentheses of formula (14.24) is

$$\frac{5}{18}\frac{F''(u_0)}{F(u_0)} = \frac{5}{18}\left[\frac{\hbar\omega}{T}\frac{d\ln T}{du}\right]^2 = \frac{5}{18}\left[\frac{\hbar\omega}{T}\cdot\frac{1}{u_0}\cdot\frac{d\ln T}{dz}\right]^2 = \frac{5}{18}\left[\frac{\hbar\omega}{\varepsilon_0}\right]^2 ,$$

where $u_0 = 2/3$. At the photon energy $\hbar\omega_{max}$ the second term in the brackets of formula (14.24) gives a correction of 7 % that justifies the method under consideration.

Thus, in spite of variation of the temperature of the solar photosphere depending on its altitude, one can reduce emission of the photosphere to radiation of a gas layer with a constant temperature. The method under consideration uses two parameters of the photosphere, $N_H(T_0)$ and dT/dz, where the temperature T_0 corresponds to the temperature of the layer that gives the main contribution to layer emission. In terms of this method, the effective temperature for each frequency depends on the frequency dependence for the absorption cross section of the negative hydrogen ion. The difference between the radiation flux from the solar photosphere and that from black body at the temperature T_0 can be determined within the framework of the above method and is relatively small because the value dT/dz is relatively small. Of course, this operation my be done for any current values of T_0 and dT/dz.

15
Gas Discharge Plasmas

15.1
Conditions of Self-Sustaining Gas Discharge

▶ **Problem 15.1** Compare with the Debye–Hückel radius, a distance between electrodes of a dark discharge, where a plasma charge does not influence on the field distribution in a gas gap.

Gas discharge is characterized by passage of electric currents through a gas when the gas is located between two electrodes with some voltage between them. We study the possibility of the self-sustaining regime if electrons and ions during their motion to electrodes creates new charged particles, and an electric current is maintained in this manner.

If a weakly ionized gas is located in a space between two alive electrodes, two limiting distributions are possible in a gas as it is shown in Fig. 15.1. According to the definition of the dark gas discharge, a plasma charge inside the gap is small and does not affect the field distribution, i. e., the electric field strength E is equal in this case, $E = U/L$, where U is the voltage between electrodes, and L is a distance between them. From the Poisson equation for the electric field strength inside a

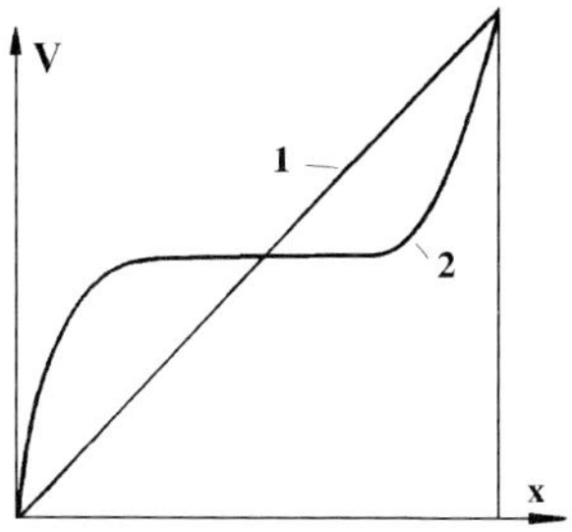

Fig. 15.1 The electric field potential distribution in a gap between two electrodes. (1) a small number density of charged particles; (2) a large number density of charged particles.

Plasma Processes and Plasma Kinetics. Boris M. Smirnov

ISBN: 978-3-527-40681-4

plane gap, we have

$$\frac{dE}{dx} = 4\pi e(N_i - N_e) ,$$

where N_i, N_e are the number density of ions and electrons, respectively, we obtain from the definition of dark gas discharge,

$$E \ll 4\pi e N_0 ,$$

where N_0 is the average number density of charged particles.

On the other hand, each electron during its motion between electrodes must create a new charged pair. Assume that a portion ξ of an energy obtained by an electron from the field is consumed on ionization. Then from the condition of current maintenance, we have

$$eEL \geq \frac{J}{\xi} ,$$

where J is the atom ionization potential. Because it exceeds an average electron energy, these two criteria give for a dark gas discharge

$$r_D \gg L ,$$

and we use the definition of the Debye–Hückel radius.

Taking estimation $L \sim 1$ cm and a typical electron energy of few eV, we obtain $N_0 \ll 10^3$ cm^{-3}, which corresponds to a typical charge number density in a real atmosphere. Hence, dark gas discharge is realized under specific conditions.

▶ **Problem 15.2** Find the condition of ignition of gas discharge in the gaseous gap where a constant electric field is supported.

Let us study evolution of an electron that is injected in a gas gap with a constant electric field. The balance equation for the electron number density in the gap has the form

$$\frac{dN_e}{dt} = \alpha N_e .$$

Here $\alpha(E) = N_a \langle v\sigma_{ion}\rangle / w_e$ is the first Townsend coefficient, N_a is the number density of atoms, σ_{ion} is the cross section for ionization of the atom by electron impact, and w_e is the electron drift velocity. The solution of this equation is $N_e = N_0 \exp(\int \alpha dx)$, where N_0 is the electron number density near the cathode, the integral is taken over the gap region, and the x-axis is perpendicular to electrodes.

The second Townsend coefficient γ is the probability for generation of an electron as a result of ion bombardment of a positively charged plate. The value of γ depends both on the gas type and on the surface material. Table 15.1 gives the values of the second Townsend coefficient for a tungsten cathode and inert gas ions at two collision energies.

Table 15.1 The second Townsend coefficient for a tungsten surface.

Ion	$E_i = 1$ eV	$E_i = 10$ eV
He^+	0.30	0.27
Ne^+	0.21	0.25
Ar^+	0.095	0.11
Kr^+	0.048	0.06
Xe^+	0.019	0.019

If one electron is formed near the positively charged plate, $e^{\alpha L}$ electrons are formed near the second plate, where L is a distance between plates, and $e^{\alpha L} - 1$ electrons are formed additionally in the gap. Then after plate bombardment by ions $\gamma(e^{\alpha L} - 1) = 1$, electrons are formed per initial electron. Hence, the condition of a self-maintained gas discharge in a gap between two plates has the form

$$\alpha L = \ln(1 + 1/\gamma) . \tag{15.1}$$

▶ **Problem 15.3** Analyze the dependence of the first Townsend coefficient of gases on the electric field strength.

We use a general expression for the first Townsend coefficient if electrons are moving in a gas in an external electric field,

$$\alpha = \frac{N_a k_{ion}}{w_e} ,$$

where k_{ion} is the rate constant of atom ionization by electron impact, and w_e is the drift velocity of an electron in a gas. As it follows from this, a general dependence of the first Townsend coefficient on the reduced electric field E/N_a has the form

$$\alpha = N_a \Phi \left(\frac{eE}{N_a} \right) ,$$

where $\Phi(x)$ is a universal function. We analyze this dependence if an average electron energy is small compared to the atom ionization potential, i. e., the ionization rate constant is determined by the tail of the electron distribution function.

We note that the function $\Phi(x)$ is sensitive to gas properties. In particular, if the number density of electrons is not small in accordance with criterion (7.18), the strongest dependence for the function $\Phi(x)$ has the form

$$\Phi \sim \exp\left(-\frac{J}{T_e}\right) , \quad T_e \ll J .$$

The electron temperature T_e is given by formula (7.26) and is proportional to $(E/N_a)^2$ at large specific electric field strengths, if the rate constant of elastic electron–atom collision is independent of the electron velocity. But even the electron number density is small, the presence of excited atoms and a small admixture of lightly ionized atoms can increase the Townsend coefficient. Below we consider the limiting case when a pure gas is not changed under the action of electrons, and the corresponding Townsend coefficient relates to the breakdown process or to other cases of a nonexcited gas.

Let us analyze the reduced electric field strength dependence for the Townsend coefficient in a pure nonexcited gas at low electric field strengths, when ionization corresponds to tail of the electron distribution function. Then it is proportional to the distribution function of electrons at the energy of the atom ionization potential, and the distribution function is a result of two factors. The first is the distribution function at energies below the atom excitation energy, if the distribution function is determined by elastic electron–atom collisions and is given by formula (7.9). The second factor is determined by the electron energies above the atom excitation energy and below its ionization potential and is given by formula (9.75). Taking into account the electric field strength dependence for each factor, we obtain the Townsend coefficient

$$\ln\alpha = \ln\alpha_0 - \frac{C_1}{(eE/N_a)^2} - \frac{C_2}{eE/N_a},$$

where α_0, C_1, C_2 are constants. Because according to formula (7.9) the constant $C_1 \sim m_e/M$, i.e., is relatively small, we have a range of small, but not very small reduced electric field strengths where the second term of this formula dominates. Then the first Townsend coefficient is approximated by the dependence

$$\alpha = AN_a \exp(-BN_a/E), \tag{15.2}$$

and Table 15.2 gives values of this formula parameters, as well as the ranges of reduced electric field strengths where this dependence holds true. These values are obtained by Townsend and are suitable for study of breakdown phenomena in a pure gas.

Table 15.2 Parameters for ionization in the cathode region [1 Td (Townsend) $= 10^{-17}$ V cm^2]. The parameters A and $(N_a L)_{min}$ are given in units of 10^{-16} cm^2.

Gas	A	B (Td)	Region of E/N_a (Td)	$(N_a L)_{min}$	U_c (V)	E_c/N_a (Td)
He	0.85	96	60–420	5.4	49	140
Ne	1.1	280	280–1100	5.0	130	390
Ar	4.0	510	280–1700	1.8	86	720
Kr	4.8	680	280–2800	2.0	130	980
Xe	7.3	990	280–2300	1.7	160	1400

At very low electric field strengths dependence (15.2) violates. Such examples are demonstrated in Figs. 15.2 and 15.3, where the first Townsend coefficient for argon and krypton is given in some range of electric fields.

▶ **Problem 15.4** Analyze the character of variation of the first Townsend coefficient of gases at low electric field strengths depending on the number density of electrons.

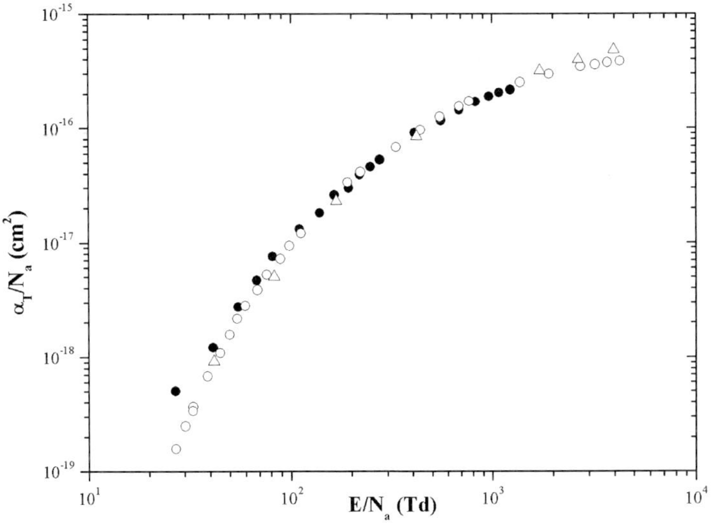

Fig. 15.2 The first Townsend coefficient for argon versus the reduced electric field strength.

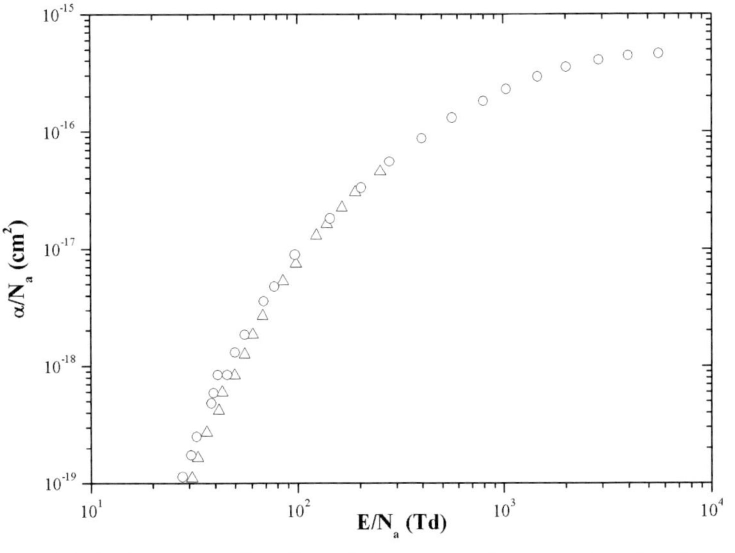

Fig. 15.3 The first Townsend coefficient for krypton as a function of the reduced electric field strength.

The first Townsend coefficient is determined by the character of ionization processes involving electrons which are moving in a gas in an external electric field. In turn, the ionization rate is connected with the tail part of the electron distribution function, and a strong drop of the wave function at energies takes place if the electron energy exceeds the excitation atom energy because of atom excitation processes. If some processes restore the electron distribution function, the first Townsend coefficient will increase at a given electric field strength. In particular, this takes at high electron number density when the tail of the distribution function is restored as a result of electron–electron collisions. In this case the electron distribution function is given by formula (7.19) and does not depend on excitation processes.

In particular, let us consider the case when electron–electron collisions dominate in establishment of the tail of the electron distribution function and along with that, the electron–atom diffusion cross section does not depend on the collision velocity. Then in accordance with dependences (7.29) for the electron drift velocity and temperature and also formula (4.45) for the ionization rate constant, we now have the following dependence for the first Townsend coefficient α on the reduced electric field strength $x = E/N_a$:

$$\ln \alpha = \frac{C_1}{x^4} \exp\left(-\frac{C_2}{x}\right) ,$$

where C_1, C_2 are constants. We show in Fig. 15.4 this dependence for neon where the diffusion cross section of electron–atom collision is $\sigma^*_{ea} = (0.57 \pm 0.03)$ Å^2 in the energy range under consideration. As is seen, a typical electric field strength responsible for ionization is shifted by more than an order of magnitude due to restoring of the tail of the electron distribution function in electron–electron collisions.

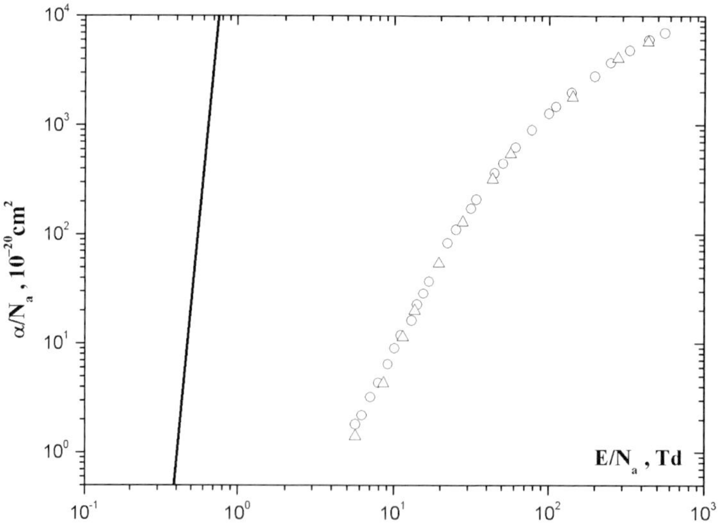

Fig. 15.4 The neon first Townsend coefficient for low (solid curve) and high (symbols) electron number densities.

We above discussed criterion (7.18) when the trunk of the electron distribution function is determined by electron–electron collisions. Note that for the tail of the distribution function this criterion is stronger.

▶ **Problem 15.5** Express the condition of ignition of gas discharge through the parameters of the Townsend coefficients.

On the basis of formula (15.1) for the first Townsend coefficient, it is convenient to represent the condition of discharge ignition (15.2) in the form

$$U_c = \frac{B(N_a L)}{\ln\left[A/\ln(1+1/\gamma)\right] + \ln(N_a L)}, \tag{15.3}$$

and the gap voltage is $U_c = EL$. This is the breakdown potential that provides subsequent development of an electric field. The gap voltage U_c is minimum in the following parameters:

$$(N_a L)_{\min} = \frac{e}{A} \ln\left(1 + \frac{1}{\gamma}\right), \tag{15.4}$$

and the optimal conditions give

$$U_{\min} = B(N_a L)_{\min} = \frac{eB}{A} \ln\left(1 + \frac{1}{\gamma}\right). \tag{15.5}$$

We found that the electric potential U in a gas-filled gap that provides gas breakdown depends on the combination $N_a L$ only. The dependence for the breakdown potential U_c

$$U_c = f(N_a L), \tag{15.6}$$

is called the Paschen law. This function (15.6) has a minimum and Fig. 15.5 represents this curve for air as an example of this curve. Note that the subsequent development of an electric current in a gas-filled gap leads to charge separation, when the charge density becomes significant. As a result, a region with a separated charge shrinks and is transformed in the cathode region of gas discharge.

▶ **Problem 15.6** Write the condition of discharge ignition if along the ionization process, attachment of electrons to gas atoms or molecules may be essential.

The process of electron attachment to atoms or molecules changes the condition of breakdown because of an additional channel of electron loss. In particular, this relates to air where the process of attachment of electrons to the oxygen molecule according to the scheme ($e + O_2 \to O^- + O$) occurs. Correspondingly, the equation of breakdown (15.1) takes the form

$$(\alpha - \eta)L = \ln(1 + 1/\gamma), \tag{15.7}$$

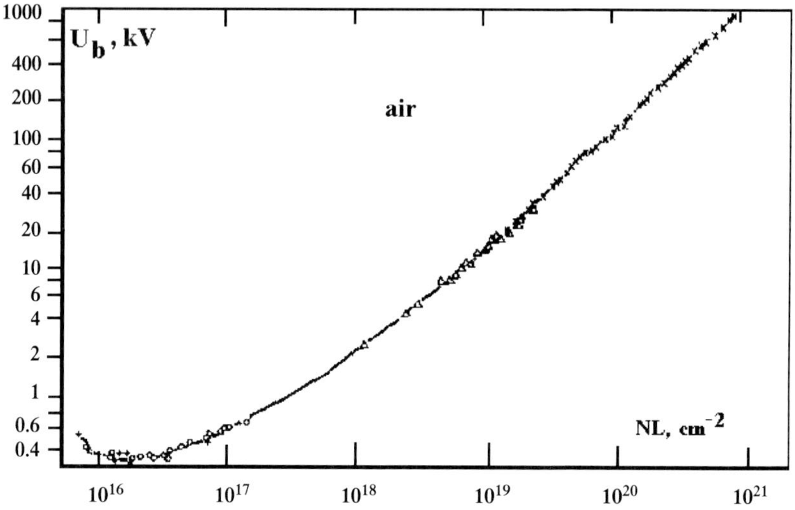

Fig. 15.5 The Paschen curve for the breakdown potential of a gas located between two parallel plates as a function of the reduced distance between plates.

where η is the rate constant for electron attachment per unit length of the electron trajectory. In the limit of large values of the parameter $N_a L$ this relation takes the form

$$\alpha(E/N_a) = \eta(E/N_a) , \tag{15.8}$$

which connects the breakdown electric field strength and the number density. For air, this relation is $E/N_a \approx 90$ Td, corresponding to an electric field of 25 kV/cm at atmospheric pressure.

► **Problem 15.7** Gas discharge is maintained in a space between two coaxial cylinders of radii ρ_1 and ρ_2, and a lightly ionized admixture through the Penning process provides ionization in this system. Obtain the condition of maintenance considering the diffusion of metastable atoms.

For this coaxial system the condition of discharge maintenance instead of condition (15.1) is given by

$$\int_{\rho_1}^{\rho_2} \alpha d\rho = \ln\left(1 + \frac{1}{\gamma}\right) , \tag{15.9}$$

where ρ_1, ρ_2 are distances from the axis to electrodes, ρ is a distance from the axis to a current point. The electric field strength at a given point is

$$\boldsymbol{E} = \frac{\Delta U}{\ln(\rho_1/\rho_2)} \frac{\boldsymbol{n}}{\rho} ,$$

where ΔU is the voltage between coaxial electrodes and $\boldsymbol{n}$ is the unit vector in the radius direction.

The ionization process starts from excitation of an atom of a basic gas. This atom displaced during its lifetime and then collides with an admixture atom that leads to the Penning process with ionization of an admixture atom. Because the excitation and ionization processes proceed in different points, this leads to a change of the ionization condition, and below we determine a correction due to this process. Indeed, if excitation proceeds at the point ρ_0 and ionization at the point ρ, this leads to variation of the potential electron energy ΔU that is given by

$$\Delta U = \frac{dE}{d\rho} \int (\rho - \rho_0) w(\rho_0, \rho) d\rho = -E(\rho_0) \frac{\overline{(\rho - \rho_0)^2}}{\rho_0} ,$$

where $w(\rho_0, \rho)$ is the probability that ionization proceeds at ρ when excitation takes place at ρ_0.

According to the definition of the first Townsend coefficient, the energy consumed per pair is eE/α, so that the variation of the Townsend coefficient $\delta\alpha$ due to this effect is given by

$$\frac{\delta\alpha}{\alpha} = \frac{\alpha \Delta U}{eE} .$$

This leads to the following condition of the discharge maintenance instead of formula (15.9)

$$\int_{\rho_1}^{\rho_2} \alpha \left(1 \pm \frac{\alpha \overline{(\rho - \rho_0)^2}}{\rho} \right) d\rho = \ln \left(1 + \frac{1}{\gamma} \right) . \qquad (15.10)$$

Here the sign "plus" relates to the case when an internal cylinder is the cathode, whereas the sign "minus" is used if the cathode is an external electrode.

According to formula (8.36) we have

$$\overline{(\rho - \rho_0)^2} = 2 D_m \tau ,$$

where D_m is the diffusion coefficient of excited atoms, and $1/\tau = N_{ad} k_P$, where N_{ad} is the number density of admixture atoms, k_P is the rate constant of the Penning process. As a result, the above condition of discharge maintenance (15.10) takes the form

$$\int_{\rho_1}^{\rho_2} \alpha \left(1 \pm \frac{\alpha D_m}{N_{ad} k_P \rho} \right) d\rho = \ln \left(1 + \frac{1}{\gamma} \right) . \qquad (15.11)$$

Thus, displacement of a point of atom ionization compared to the point of atom excitation leads to a change of the effective Townsend coefficient that is taken into account in the above formula.

15.2 Cathode Region of Gas Discharge

▶ **Problem 15.8** Find the current density in the cathode region of glow discharge, which is characterized by the ionization process by electron impact and cathode bombardment by an ion.

Three main discharge regions are the cathode region that is responsible for electron emission, the anode region, where electrons attach to the anode, and the positive column for a long distance between electrodes, and gas ionization in the positive column is independent of processes near electrodes. In fact, there are intermediate regions, but since they are not of principle, we exclude them from consideration. We now consider the cathode region of glow discharge where ions attach to the cathode and electrons are generated by the cathode.

We first analyze the spatial distribution of ions and electrons in the cathode region, and this is determined by the Poisson equation

$$\frac{dE}{dx} = 4\pi e(N_i - N_e)$$

for the electric field strength, where N_i and N_e are the number densities of ions and electrons, respectively. Assuming the mean free path of ions and electrons to be small compared to a dimension L of the cathode region, the current densities of ions i_i and electrons i_e are obtained in the following expressions:

$$i_i = eK_i N_i E, \qquad i_e = -eK_e N_e E \ ,$$

where K_i and K_e are respective mobilities of the ions and electrons, and for simplicity we assume that the electric field strength is relatively small. We assume that in the cathode region $i_i \sim i_e$ and $K_e \gg K_i$, which gives $N_e \ll N_i$ in the cathode region, where the Poisson equation takes the form

$$\frac{dE}{dx} = 4\pi e N_i = -4\pi i_i/(K_i E) \ .$$

We take the sign of the electric field strength such that the electric field hinders the approach of ions to the cathode.

Ignoring ionization processes in the cathode region, we found that the total electric current density $i = i_i + i_e$ is conserved there. The boundary condition at the cathode has the form $i_e(0) = \gamma i_i(0)$, or $i_i(0) = i/(1+\gamma)$, where γ is the second Townsend coefficient. Since ions are formed outside the cathode region, this relation $i_i = i/(1+\gamma)$ is valid in the entire cathode region. Correspondingly, the solution of Poisson's equation gives

$$E^2 = E_c^2 - \frac{8\pi i}{K_i E(1+\gamma)} x \ , \tag{15.12}$$

where x is a distance from the cathode, and $E_c = E(0)$. Taking the electric field strength to be zero on the boundary of the cathode region (L is small compared to the tube radius), we obtain

$$i = E_c^2 K_i (1+\gamma)/(8\pi L) \ . \tag{15.13}$$

▶ **Problem 15.9** Give the criterion of a normal glow discharge where a current occupies a part of the cathode.

According to formulas (15.12) and (15.13), the parameters of the cathode region follow from the condition that the cathode voltage U_c is minimal. Hence, according to formula (15.13) the current density remains constant if the total discharge current I varies. This means that variation of the discharge current leads to a change of the cathode area occupied by the current. This area is I/i and as long as this value is smaller than the cathode area πr_0^2 (r_0 is the cathode radius), this corresponds to the normal regime of glow discharge when the current occupies a part of the cathode. Then such parameters of the cathode region, as its dimension L, and the cathode voltage U_c do not vary on increasing the total discharge current.

When the discharge electric current exceeds the value $i\pi r_0^2$, the glow discharge becomes abnormal. The subsequent increase of the discharge current that leads to cathode heating will change the regime of the cathode current and give the transition to arc discharge. When heating of the cathode becomes sufficient for thermoemission of electrons, the cathode voltage drop decreases, and the transition from a glow discharge to an arc takes place. The voltage drop of the cathode for arc discharge is 20–30 V that is one order of magnitude lower than that in glow discharge.

▶ **Problem 15.10** Find the relation between the cathode voltage and parameters of the cathode region.

We now give formulas for parameters of the cathode region using the first Townsend coefficient formula (15.2). According to the Poisson equation, the electric field strength is $E = E_c\sqrt{1 - x/L}$ in the cathode region, and this value drops remarkably at transition from the cathode region to the positive column where the electric field strength is significantly low. Replacing equation (15.1) by $\int \alpha dx = \ln(1 + 1/\gamma)$ and substituting expression (15.2) for the first Townsend coefficient in equation (15.12), we obtain the condition of discharge maintenance in the form

$$Ay \int_0^1 \exp(-b/z^{1/2})dz = \ln(1 + 1/\gamma) ,$$

where $y = N_a L$, $z = \sqrt{1 - x/L}$, and $b = BN_a/E_c$. From formula for the cathode voltage $U_c = 2E_c L/3$, we have $b = 2BN_a L/(3U_c)$. The condition of the minimum of the cathode voltage $dU_c/dy = 0$ gives $db/dy = b/y$. Then expression (15.2) for the first Townsend coefficient leads to the following condition taking into account the minimum cathode voltage

$$\mathrm{J}(b) + bd\mathrm{J}(b)/db = 0 ,$$

where $J(b)=\int_0^\infty \exp(-b/z^{1/2})dz$. The solution of the above equation yields $b = 0.71$, which leads to the following expressions for parameters of the cathode region:

$$(N_a L)_{\min} = \frac{3.05}{A}\ln(1+\frac{1}{\gamma}), \qquad U_{\min} = 0.94B(N_a L)_{\min} = 2.87\frac{B}{A}\ln(1+\frac{1}{\gamma}) . \tag{15.14}$$

These equations are similar to equations (15.4) and (15.5), which use the assumption of a constant electric field strength.

Table 15.3 exhibits cathode region parameters calculated on the basis of formulas (15.14). Table 15.3 gives the values of the cathode voltage drop for a few gases and cathode materials used in glow discharges.

Table 15.3 The normal cathode drop $U_c(V)$ of glow discharges for some gases and cathode materials.

	Al	Ag	Cu	Fe	Pt	Zn
He	140	162	177	150	165	143
Ar	100	130	130	165	131	119
H_2	170	216	214	250	276	184
N_2	180	233	208	215	216	216
Air	229	280	370	269	277	277

▶ **Problem 15.11** Construct the voltage–current characteristic for the cathode region of gas discharge.

In the course of an increase of the discharge current, until this is a normal glow discharge and only a cathode part is occupied by a current, the cathode voltage is independent of the current. When the current fills all the cathode area, its subsequent increase leads to an increase of a charge in the cathode region. As a result, the voltage of abnormal glow discharge increases with a current increase, as it is shown in Fig. 15.6. But an increase of the discharge electric current leads to cathode heating, and at high cathode temperatures the character of cathode processes is changed. When the cathode temperature is enough high, thermoemission of electrons on the cathode takes place. This mechanism of generation of electrons leads to a low cathode drop that is of the order of the ionization potential of gas atoms. Including the transition to this mechanism of electron generation in the voltage–current characteristic of the voltage region, we obtain this dependence as it is shown in Fig. 15.7.

15.3
Positive Column of Glowing Discharge of High Pressure

▶ **Problem 15.12** Gas discharge is burnt in a cylindrical tube of the radius r_0, and the balance of electrons in the positive column of gas discharge is determined

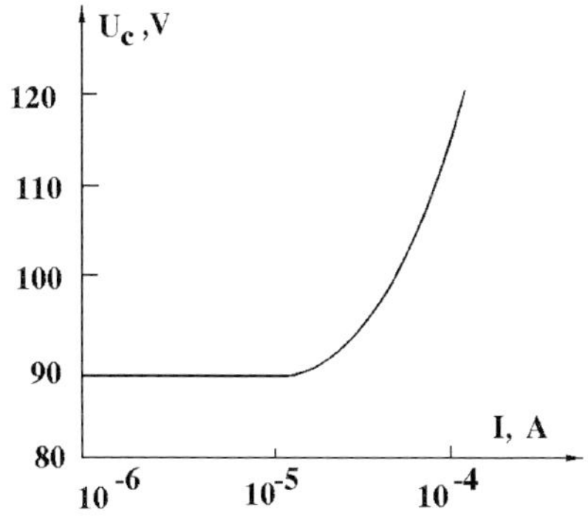

Fig. 15.6 The voltage-current characteristic for the cathode region in argon at pressure of 1 Torr, $\gamma = 0.1$ and $r_0 = 1$ cm.

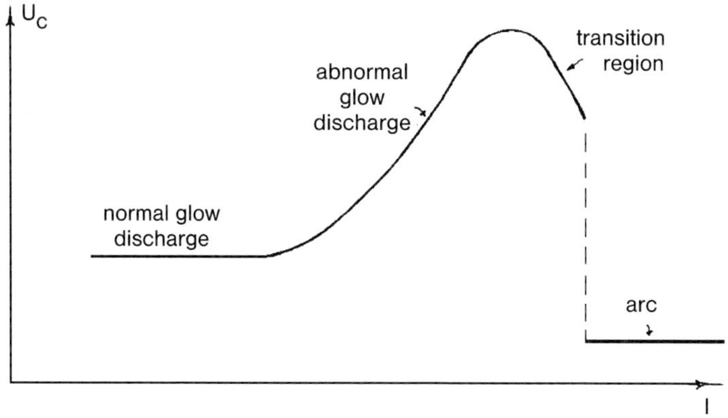

Fig. 15.7 A typical cathode voltage-current characteristic in a wide range of currents.

by direct ionization of atoms by electron impact and by attachment of electrons to walls. Taking into account electrons and ions travel to the walls by ambipolar diffusion, find the relation between the rates of processes involving electrons in the limit of a small electron number density.

A positive column contains an uniform plasma because conditions of formation of charged particles as a result of atom ionization by electron impact, as well as conditions of loss of charged particles by their attachment to the walls are identical for different cross sections of the discharge tube. From this it follows, in particular, which an increase of the distance between electrodes leads to an increase of the positive column, while dimensions of the cathode and anode regions are conserved under this operation.

While considering a self-consistent plasma of the positive column, we assume that electrons are formed by direct ionization of atoms by electron impact and decay as a result of ambipolar diffusion of electrons and ions to the walls, and the balance equation for the electron number density N_e,

$$\frac{D_a}{\rho}\frac{d}{d\rho}\left(\rho\frac{dN_e}{d\rho}\right) + k_{ion} N_e N_a = 0 \, . \tag{15.15}$$

Here ρ is a distance for the axis center, D_a is the ambipolar diffusion coefficient, N_a is the atom number density, and k_{ion} is the rate constant for atomic ionization by electron impact. This equation is added by the boundary condition $N_e(r_o) \equiv N_o = 0$, where r_o is the tube radius.

We consider the regime of weak currents when plasma parameters and rates of processes are independent of ρ, so that the solution of equation (15.15) has the form of the Schottky distribution

$$N_e(\rho) = N_o J_0\left(\rho\sqrt{\frac{k_{ion} N_a}{D_a}}\right) , \tag{15.16}$$

where $J_0(x)$ is the Bessel function. The boundary condition $N_e(r_o) = 0$ gives

$$\frac{N_a k_{ion} r_o^2}{D_a} = 5.78 \, . \tag{15.17}$$

This relation equalizes the rate of the ionization process ($\sim N_a k_{ion}$) and the rate of electron and ion transport to the walls ($\sim D_a/r_o^2$). Figure 15.8 shows dependence (15.16).

Note that these results relate to the positive column of glow discharge, where, on the one hand, the radius of the discharge tube (in reality, $r_o \sim 1$ cm) is large

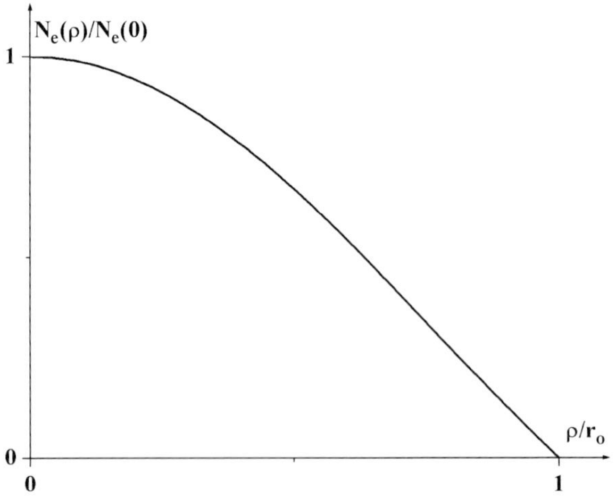

Fig. 15.8 The distribution of the electron number density over the cross section of a discharge cylinder tube for the positive column and Schottky regime of the ionization balance in accordance with formula (15.16).

compared to the Debye–Hückel radius of this plasma, and, on the other hand, the electron current does not heat remarkably the plasma and its parameters are identical over the tube cross section. In reality this relates to the number density of electrons $N_e \sim 10^8 - 10^{12}\,\text{cm}^{-3}$. This corresponds to the Debye–Hückel radius in the range $r_D \sim 1$ to $100\,\mu\text{m}$. In addition, the mean free path of electrons and ions in this plasma is large compared to the tube radius that corresponds to the gas pressures $p \gg 10^{-5}$ atm at room temperature.

▶ **Problem 15.13** Find the average lifetime for electrons located in the positive column of gas discharge of low currents.

The average lifetime τ_e of an electron in the positive column plasma is given by

$$\tau_D = \frac{1}{2\pi r_o j(r_o)} \int_0^{r_0} 2\pi\rho d\rho N_e(\rho)\,,$$

where the integral is the average number of electrons per unit length of the discharge tube, and j is the flux of electrons toward the walls. Let us represent formula (15.16) in the form

$$N_e(\rho) = N_0 J_0 \left(2.405 \frac{\rho}{r_o}\right)\,,$$

and in simplified evaluation one can use the approximation $J_0(2.405\rho/r_o) \approx 1 - (\rho/r_o)^{1.34}$ that is valid with the accuracy of 8%. The electron flux to the walls is

$$j = -D_a \frac{dN_e(r_o)}{d\rho} = \frac{1.25 D_a N_0}{r_o}\,,$$

and the total number of electrons per unit length of the discharge tube is equal to

$$\int_0^{r_0} 2\pi\rho d\rho N_e(\rho) = 1.36 N_0 r_0^2\,. \tag{15.18}$$

This gives the average lifetime of electron traveling to the walls as

$$\tau_D = \frac{1.36 N_0 r_0^2}{2\pi r_o j} = \frac{0.173 r_o^2}{D_a} = \frac{1}{N_a k_{ion}}\,, \tag{15.19}$$

which reflects the equality of the rates for electron generation in the positive column plasma and their recombination on the walls.

▶ **Problem 15.14** Determine the electric field strength in the positive column of helium in the Schottky regime of ionization processes.

The electric field strength is given by formula (15.19), and we use formula (7.16) the rate constant of ionization of a helium atom by electron impact. This approximation corresponds to assumption that the transport cross section of elastic electron–atom scattering is independent of the collision velocity, and we take it to be $\sigma^* = 6 \cdot 10^{-16}\ \text{cm}^2$. In this approximation, the electron distribution function is according to formula (7.9)

$$f_0(\varepsilon) = \frac{1.63 N_e}{\varepsilon_0^{3/2}} \exp\left(-\frac{\varepsilon^2}{\varepsilon_0^2}\right)$$

This distribution function is normalized to the electron number density, and a typical electron energy is large compared to the atom thermal energy T. The parameter ε_0 in this formula is expressed through the reduced electric field strength E/N_a as

$$\varepsilon_0 = \sqrt{\frac{M}{3m_e}}\, eE\lambda\,,$$

where M is the atom mass, and $\lambda = (N_a\sigma^*)^{-1}$ is the mean free path for electrons. Expressing the reduced electric field strength E/N_a in Townsend ($1\ \text{Td} = 10^{-17}\ \text{V} \cdot \text{cm}^2$) and the parameter ε_0 in eV, we obtain in the case of helium $\varepsilon_0 = 0.82 E/N_a$. Hence formula (15.19) gives in this case by using formula (7.16) for the ionization rate constant

$$\frac{E}{N_a} = \frac{1.22 J}{\sqrt{\ln\left(\frac{k_0 (N_a r_0)^2}{5.78 D_a N_a}\right)}} \tag{15.20}$$

This formula gives the dependence of the reduced electric field strength E/N_a on the reduced tube radius $r_0 N_a$. Figure 15.9 represents this dependence if the reduced tube radius $r_0 N_a$ ranges from 10^{17} to $10^{19}\ \text{cm}^{-2}$. Then the average electron energy $0.74\varepsilon_0$ varies from 7.1 to 4.7 eV.

▶ **Problem 15.15** Determine the voltage near the walls of a discharge tube for the positive column of gas discharge of low currents if this voltage equalizes the electron and ion currents to the walls.

Electrons and ions travel to the walls in the positive column of the discharge tube with identical drift velocities because of the mechanism of ambipolar diffusion that is realized on large distances from the walls compared to the mean free path of electrons and ions. Because of the difference of electron and ion masses, a field arises that leads to a small violation of the plasma quasineutrality and equalizes the electron and ion currents to the walls. But at the wall vicinity, where distance from the walls is comparable to the mean free path of electrons and ions, the plasma quasineutrality is violated because of equality of electron and ion currents to the walls. We below find the plasma electric potential φ_W due to this effect, the plasma sheath.

The electron j_e and ion j_i fluxes to the walls are equal to

$$j_e = N_e \sqrt{\frac{T_e}{2\pi m_e}}, \qquad j_i = N_i \sqrt{\frac{T_i}{2\pi m_i}}\,,$$

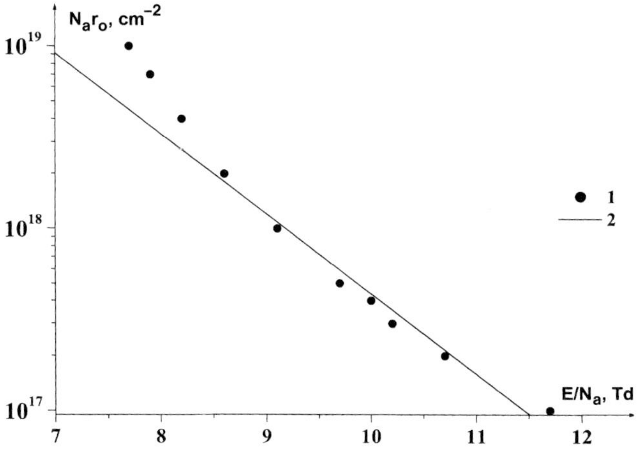

Fig. 15.9 The reduced electric field strength in the positive column for pure helium in the limit of low electron concentrations (the Schottky regime of ionization equilibrium). 1 — formula (15.20), 2 — data of Table 15.2.

where N_e and N_i are the electron and ion number densities near the walls, and m_e and m_i are the electron and ion masses, respectively, and for simplicities we consider the regime when one can introduce the electron temperature T_e that differs from the ion temperature T_i. Because the fluxes of electrons and ions on the walls are identical, an electric field arises in the double layer near the walls that slows the electrons. Assuming that the Boltzmann formula (1.12) is valid for the number density of electrons in the double layer, we find a decrease of the electron number density near the walls to be described by the factor $\sim \exp(-e\varphi_W/T_e)$. Then the equality of the electron and ion fluxes gives the difference of the electric potentials between the plasma and the walls,

$$e\varphi_W = \frac{T_e}{2} \ln\left(\frac{T_e m_i}{T_i m_e}\right). \tag{15.21}$$

A double layer also arises on any plasma boundaries.

▶ **Problem 15.16** An equilibrium in the plasma of the positive column of gas discharge results from ionization of atoms by electron impact through the atom metastable state and transport of charged particles to the walls by ambipolar diffusion. Find the relations between the number density of metastable atoms and parameters of these processes.

Taking into account the indicated processes and using the variable $x = \rho^2/r_0^2$, we obtain instead of the balance equation (15.15) the following set of the balance equations:

$$4D_a \frac{d}{dx}\left(x\frac{dN_e}{dx}\right) + k_{ion} N_e N_m r_o^2 = 0, \qquad 4D_m \frac{d}{dx}\left(x\frac{dN_m}{dx}\right) + k_{ex} N_e N_a r_o^2 = 0\,, \tag{15.22}$$

where N_m is the number density of metastable atoms, k_{ex} is the rate constant for excitation of the metastable state by electron impact, k_{ion} is the rate constant for ionization of a metastable atom by electron impact, and D_m is the metastable atom diffusion coefficient. The set of equations (15.22) is added by the boundary conditions $N_e(r_0) = N_m(r_0) = 0$.

Let us approximate the number densities of electrons and metastable atoms in the following form

$$N_e = C_1(e^{-ax} - e^{-a}), \qquad N_m = C_2(e^{-bx} - e^{-b})\,,$$

and the parameters of these dependences we obtain from the above balance equations at $x = 0$ and from these balance equations integrated over dx. In order to estimate the accuracy of this operation, we apply it to equation (15.15). Then we obtain $a = 0.842$ and

$$\frac{N_a k_{ion} r_o^2}{D_a} = \frac{4a}{1 - e^{-a}} = 5.9\,,$$

instead of (5.78) according to formula (15.17).

The above procedure gives the set of equations (15.22),

$$a = 1.410,\ b = 0.951, \quad \frac{N_a k_{ex} r_o^2}{D_m} = 1.550, \quad \frac{N_m(0) k_{ion} r_o^2}{D_a} = 1.865\,. \tag{15.23}$$

Relations (15.23) that follow from the balance equations (15.22) account for the balance of electrons and metastable atoms with respect to their formation and decay in the positive column of gas discharge. These results give the fluxes of metastable atoms j_m and electrons j_e onto the walls,

$$j_m = -D_m \frac{dN_m(r_o)}{d\rho} = \frac{1.2 D_m N_m(0)}{r_o}, \quad j_e = -D_a \frac{dN_e(r_o)}{d\rho} = \frac{0.91 D_a N_e(0)}{r_o}\,. \tag{15.24}$$

▶ **Problem 15.17** Determine the boundary number density of electrons at which metastable helium atoms $He(2^3S)$ formed in the positive column give a contribution to ionization processes in the positive column of a cylinder discharge tube in helium. The lifetime of metastable atoms is determined by its collisions with the walls.

For this goal one can use formula (15.17) in the form

$$k_{ion}^{(o)} N_o + k_{ion}^{(m)} N_m = 5.78 \frac{D_a}{r_o^2}\,, \tag{15.25}$$

where $k_{\rm ion}^{(o)}$, $k_{\rm ion}^{(m)}$ are the rate constants for ionization of an atom in the ground and metastable states, respectively, and N_o, N_m are the number densities of atoms in the ground and metastable states. In this formula we assume the number density of metastable atoms is constant over the tube cross section as well as that for atoms in the ground states. Though this assumption can violate, we ignore this because of a strong dependence of the rate constants on the field strength.

We consider the electron–atom elastic cross section independent of the collision energy in the helium case (we take it $\sigma_{ea}^* = 6$ Å^2), and then the population of metastable atoms is given by formula (7.17)

$$N_m = 1.53 N_o \left(\frac{\varepsilon_o}{\Delta E}\right)^{3/2} \exp\left(-\frac{\Delta E^2}{\varepsilon_o^2}\right),$$

where the characteristic electron energy ε_o is given by formula (7.15) and ΔE is the excitation energy for the metastable state. This formula assumes that metastable atoms are destroyed as a result of quenching by electron impact.

We consider below the case of a low electron number density when removal of metastable atoms results from their traveling to the walls, so that the following criterion holds true:

$$N_e k_q \tau \ll 1 .$$

Here k_q is the rate constant of quenching of a metastable atom by electron impact, and for metastable helium atom we take it from Table 4.1 ($k_q = 3.1 \cdot 10^{-9}$ cm^3/s); τ is the lifetime of a metastable atom due to its collision with a wall, and we take this lifetime according to formula (15.19)

$$\tau_m = \frac{0.17 r_o^2}{D_m} .$$

Here D_m is the diffusion coefficient for metastable atoms, and we give in Table 15.4 its values for metastable atoms of inert gases in a parent gase at room temperature.

Table 15.4 The reduced diffusion coefficients $D_m N_a$ for metastable atoms of inert gases in a parent gas at room temperature (D_m is the diffusion coefficient of metastable atoms, N_a is the number density of atoms).

Metastable atom	He(2^3S)	He(2^1S)	Ne(3^2P_2)	Ar(4^2P_2)	Kr(5^2P_2)	Xe(6^2P_2)
$D_m N_a$ (10^{18} (cm · s)$^{-1}$)	16	14	5.0	2.0	1.0	0.60

In this limit, the number density of metastable atoms according to formula (7.17) is

$$N_m = \frac{0.17 r_o^2 k_q N_e}{D_m} 1.53 \left(\frac{\varepsilon_o}{\Delta E}\right)^{3/2} \exp\left(-\frac{\Delta E^2}{\varepsilon_o^2}\right) .$$

This allows us to determine the boundary electron number density N_e at which both terms of the right-hand side of equation (15.25) are equal. Using formula (7.16) for the ionization rate constant, we find the boundary electron number density N_e,

$$N_e = N_0 \frac{3.8 D_m N_0}{k_q (r_0 N_0)^2} \left(\frac{\Delta E}{\varepsilon_0}\right)^{3/2} \left(\frac{J_*}{J}\right)^4 \exp\left(\frac{J_*^2 + \Delta E^2 - J^2}{\varepsilon_0^2}\right) . \quad (15.26)$$

Here $J_* = 4.77$ eV, $J = 24.59$ eV are the ionization potentials in the metastable and ground states, and $\Delta E = 19.82$ eV is the excitation energy for the helium atom.

Figure 15.10 gives the boundary values for the electron number densities starting from which ionization of the metastable state gives a contribution to helium ionization. The corresponding values of the electric field strengths and the corresponding values of the parameter ε_0 are taken from Fig. 15.8. If we express the parameter ε_0 in eV and $N_0 r_0$ in cm^{-2}, this formula for the relative boundary number density of electrons takes the form

$$\frac{N_e}{N_0} = \frac{2.4 \cdot 10^{27}}{\varepsilon_0^{3/2} (r_0 N_0)^2} \exp\left(-\frac{189}{\varepsilon_0^2}\right) .$$

As it follows from Fig. 15.10, ionization of metastable or excited states becomes essential at very low electron number densities and correspondingly at low discharge currents. If the reduced tube radius $r_0 N_0$ varies from $1 \cdot 10^{17}$ cm^2 to $1 \cdot 10^{18}$ cm^2, the reduced electron number density $r_0 N_e$ varies from $4.1 \cdot 10^8$ cm^2 to $1.1 \cdot 10^8$ cm^2.

► **Problem 15.18** Construct the current–voltage characteristic for the positive column of a cylinder discharge tube in helium if the Schottky regime is realized for the electron balance in this plasma and the ionization rate is determined by metastable states of helium atoms.

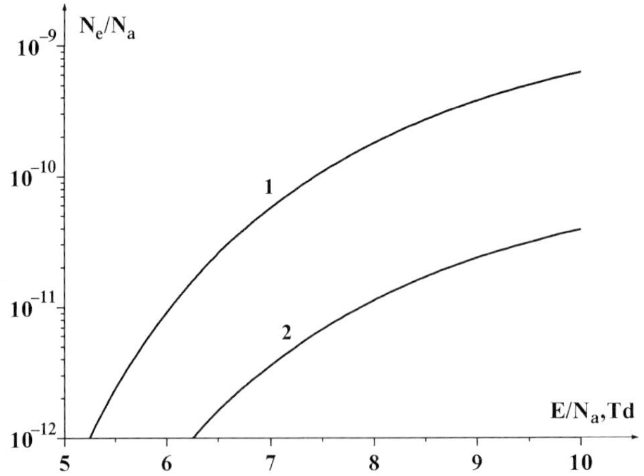

Fig. 15.10 The boundary electron concentration in a discharge helium plasma as reduced for the Schottky regime of ionization in helium when ionization of metastable $He(2^3S)$ atoms and atoms in the ground states give an identical contribution to the ionization balance. (1) $N_a r_0 = 1 \cdot 10^{17}$ cm^{-2}, (2) $N_a r_0 = 4 \cdot 10^{17}$ cm^{-2}.

The analysis shows the role of excited states in ionization processes. Along with this, it is necessary to have a large lifetime of excited atoms for their participation in ionization processes. In the above formulas this results in a high value of the quenching rate constant, as well as contribution of radiative processes in decay of these states. Therefore, only metastable states with a large lifetimes may be important in ionization processes, and we restrict by the metastable 2^3S state in the helium analysis.

We use formula (15.17) that connect rates of electron formation and loss that has the form now

$$N_m k_{ion}^{(m)} = \frac{5.78 D_a}{r_o^2} ,$$

where N_m, the number density of metastable atoms, is taken from formulas of the previous Problem, and the ionization rate constant for metastable atoms is evaluated by formula (7.16). As a result, this balance equation takes the form

$$N_e r_o \cdot (N_a r_o)^3 = f(\varepsilon_o), \tag{15.27}$$

and expressing ε_o in eV, the reduced electric field strength E/N_a in Td, and $f(\varepsilon_o)$ in 10^{60} cm^{-8}, we obtain in this approximation the following relations:

$$\varepsilon_o = 0.82\frac{E}{N_a}, \qquad f(\varepsilon_o) = \frac{1}{\varepsilon_o^3}\exp\left(\frac{416}{\varepsilon_o^2}\right) .$$

Equation (15.27) is the current–voltage characteristic of the positive column, if we change the electron number density by the electric current, and we make it in accordance with formula (15.18) $I = 1.36 w N_e r_o^2$, where w is the electron drift velocity. Expressing the latter in 10^5 cm/s and the reduced field strength E/N_a in *Td*, we take the relation between these values as $w = C\sqrt{E/N_a}$, and $C \approx 6$ in this approximation. From this we obtain the reduced discharge current in the positive column in helium, if ionization is realized through metastable atoms He(2^3S)

$$\frac{I}{r_o} = \frac{1.6 \cdot 10^{47}}{(N_a r_o)^3}\frac{1}{\varepsilon_o^{5/2}}\exp\left(\frac{416}{\varepsilon_o^2}\right) ,$$

where the current I is expressed in *A*, the tube radius r_o in cm, the number density of helium atoms N_a in cm^{-3}, and the parameter ε_o in eV. Figure 15.11 shows the current–voltage characteristic of the helium positive column according to this formula. One can see a strong current dependence on the electric field strength.

▶ **Problem 15.19** A small admixture of neon is added to helium and this gives an additional ionization channel due to the Penning process involving a metastable helium atom. Find the boundary neon concentration at which this process becomes important.

The Penning process proceeds according to the scheme

$$A^* + B \rightarrow A + B^+$$

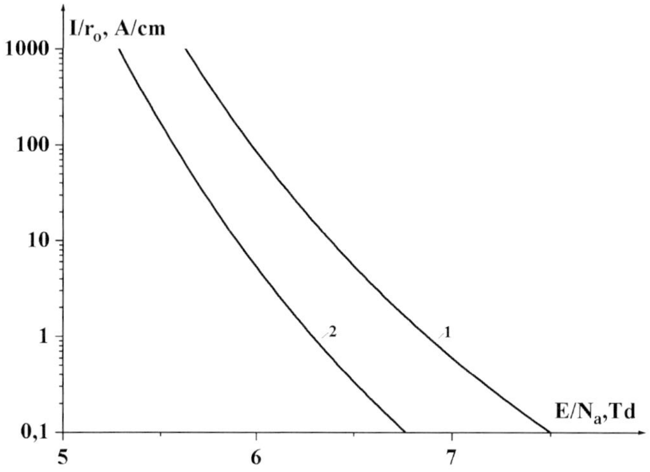

Fig. 15.11 The current-voltage characteristic of the positive column helium plasma for the Schottky regime of ionization as a function of the reduced radius of the cylinder discharge tube if ionization proceeds through metastable He(2^3S) atoms. (1) $N_a r_0 = 1 \cdot 10^{17}$ cm^{-2}, (2) $N_a r_0 = 4 \cdot 10^{17}$ cm^{-2}.

and may be responsible for generation of ions from metastable atoms if the ionization potential of an atom B is lower than the excitation energy of a metastable atom A^*. Table 15.5 gives the values of the rate constants of this process if it involves metastable atoms of inert gases.

Table 15.5 The rate constants of the Penning processes (in cm^3/s) at room temperature with participation of metastable atoms of inert gases.

	He(2^3S)	He(2^1S)	Ne($^3P_2$)	Ar($^3P_2$)	Kr($^3P_2$)	Xe($^3P_2$)
Ne	$4 \cdot 10^{-12}$	$4 \cdot 10^{-11}$	—	—	—	—
Ar	$8 \cdot 10^{-11}$	$3 \cdot 10^{-10}$	$1 \cdot 10^{-10}$	—	—	—
Kr	$1 \cdot 10^{-10}$	$5 \cdot 10^{-10}$	$7 \cdot 10^{-12}$	$5 \cdot 10^{-12}$	—	—
Xe	$1.5 \cdot 10^{-10}$	$6 \cdot 10^{-10}$	$1 \cdot 10^{-10}$	$2 \cdot 10^{-10}$	$2 \cdot 10^{-10}$	—
H_2	$3 \cdot 10^{-11}$	$4 \cdot 10^{-11}$	$5 \cdot 10^{-11}$	$9 \cdot 10^{-11}$	$3 \cdot 10^{-11}$	$2 \cdot 10^{-11}$
N_2	$9 \cdot 10^{-11}$	$2 \cdot 10^{-10}$	$8 \cdot 10^{-11}$	$2 \cdot 10^{-10}$	$4 \cdot 10^{-12}$	$2 \cdot 10^{-11}$
O_2	$3 \cdot 10^{-10}$	$5 \cdot 10^{-10}$	$2 \cdot 10^{-10}$	$4 \cdot 10^{-11}$	$6 \cdot 10^{-11}$	$2 \cdot 10^{-10}$

Evidently, the Penning process becomes important in ionization of a plasma of the positive column if a loss of metastable atoms is determined by this process. This criterion for the regime under consideration when quenching of metastable atoms by electron impact dominates until an admixture is absent, has the form

$$N_e k_q < N_{ad} k_P ,$$

where N_{ad} is the number density of neon atoms, k_q is the quenching rate constant, k_P is the rate constant of the Penning process. For the case under consideration this criterion has the form

$$\frac{N_e}{N_{ad}} < 10^{-3} .$$

Because of low electron number densities, this criterion usually may be attained.

▶ **Problem 15.20** Obtain the criterion for the number density of atoms at which molecular ions are not formed in the positive column contained in atomic gas even at low currents.

While considering this problem, we are guided by inert gases when along with atomic ions, molecular ions can be formed, and at low currents molecular ions are thermodynamically profitable. If molecular ions are located in the positive column, recombination proceeds as a result of dissociative recombination, and the Schottky regime of ionization is violated. But initial ionization of atoms by electron impact in the positive column leads to the formation of atomic ions, and subsequently they can be converted into molecular ions as a result of the three-body process (3.59). A typical time of this process is $\tau_3 \sim (K_{ia} N_a^2)^{-1}$ (K_{ia} is the three-body rate constant for the process (3.59) given in Table 3.4 for the inert gas case), and this typical time must be less than the lifetime τ of electrons and ions in the positive column (15.19) with respect to their traveling to the walls. Hence, the criterion that atomic ions are located in the positive column $\tau_3 \gg \tau$ is realized at the following atom number density:

$$N_a \ll \frac{k_{ion}}{K_{ia}}$$

▶ **Problem 15.21** Ionization equilibrium in inert gases starts from atom ionization by electron impact that leads to the formation of atomic ions. Subsequently atomic ions are converted into molecular ions if the gas temperature is not high. Analyze the ionization equilibrium in the positive column of gas discharge in inert gases if molecular ions formed during plasma evolution.

Assuming the rate of dissociative recombination of electrons and molecular ions to be large compared to the rate of plasma diffusion to the walls, we obtain the following balance equations for the number density of atomic N_i and molecular N_{2i} ions:

$$\frac{dN_i}{dt} = -\frac{N_i}{\tau_D} + k_{ion} N_e N_a - K_{ia} N_i N_a^2 = 0, \qquad \frac{dN_{2i}}{dt} = K_{ia} N_i N_a^2 - \alpha_{rec} N_{2i} N_e = 0 .$$

Here τ_D is the lifetime with respect to electron and ion traveling to the walls, K_{ia} is the rate constant of three-body process (3.59) that involves atoms and ions of inert gases, and α_{rec} is the coefficient of dissociative recombination that proceeds according to the scheme

$$e + A_2^+ \to A^* + A .$$

Assuming the portion of molecular ions to be small $N_{2i} \ll N_i$, we obtain the following equilibrium relation by using the condition plasma quasineutrality $N_e = N_i$

$$N_a k_{ion} = \frac{5.78 D_a}{r_o^2} + K_{ia} N_a^2 \,. \tag{15.28}$$

We include in the balance equation an effective time of ions and electrons τ_D for their transport to the walls in accordance with formula (15.19) such that in the limit of low number densities of atoms the balance equation (15.28) is transformed into (15.17). As the atom number density increases, the electric field strength will increase if the current occupies all the cross sections of the discharge tube. Note that in this case with respect to a combination $N_a r_o$ is not fulfilled, as it takes place in formula (15.17), so that the reduced electric field strength in the positive column depends on N_a and r_o separately.

The balance equation (15.28) holds true under the condition $N_{2i} \ll N_i$, which gives

$$N_e \gg \frac{K_{ia} N_a^2}{\alpha_{rec}} \,.$$

Since typical values $K_{ia} \sim 10^{-31}$ cm^6/s, $\alpha_{rec} \sim 10^{-7}$ cm^3/s, and N_a is found usually in a range 10^{16}–10^{18} cm^{-3}, this criterion is fulfilled if the electron concentration N_e/N_a exceeds 10^{-8}–10^{-6}. If the opposite criterion holds true, we have the following equilibrium relation:

$$N_a k_{ion} = \frac{5.78 D_a}{r_o^2} + \alpha_{rec} N_e \tag{15.29}$$

instead of (15.28). The ambipolar diffusion coefficient D_a in this formula relates to molecular ions.

▶ **Problem 15.22** Analyze the discharge current contraction for the positive column with inert gases if molecular ions are formed from the atomic ones during plasma evolution.

Under conditions of previous Problem, at large number densities of atoms the second term of the right-hand side of the balance equation (15.28) dominates and the electric field strength increases with an increase of N_a. But the electric field strength decreases if a plasma occupies a small part of the discharge tube such that a radius R_o occupied by the current region follows from the relation,

$$\frac{6 D_a}{R_o^2} \sim K N_a^2 \,.$$

Atomic ions formed in this region are transformed into molecular ions outside this region.

Analyzing the stability of this current distribution, we consider the Schottky regime of the ionization balance that relates to low currents when plasma heating may be ignored. In reality, heating of the central part of the positive column leads to an increase of its temperature, which in turn decreases the number density of atoms because the pressure is constant over the tube cross section. This causes a decrease of the electric field strength and provides the stability of this contracted discharge.

In addition we note that this contraction mechanism exists in a restricted range of discharge current. When the gas temperature reaches a limit when molecular ions become thermodynamically nonprofitable. The discharge currents expand over all the cross section of the positive column.

► **Problem 15.23** The Schottky regime of the ionization equilibrium is realized in a buffer gas with a lightly ionized admixture, and a uniform distribution of an admixture along the positive column results from gas pumping. Assuming that ionization in the positive column proceeds due to direct ionization of admixture atoms by electron impact, find the criterion when redistribution of admixture over the cross section of a discharge tube due to electrophoresis may be ignored.

The redistribution of admixture atoms over the cross section results from the electron or ion flux j_e to the walls that for the Schottky regime of ionization equilibrium is estimated as

$$j_e \sim \frac{N_e D_a}{R_o} ,$$

where D_a is the coefficient of the ambipolar diffusion, and R_o is a tube radius. The opposite flux of admixture atoms j_{ad} is

$$j_{ad} \sim \frac{N_{ad} D_{ad}}{L} ,$$

where N_{ad} is the number density of admixture atoms, D_{ad} is the diffusion coefficient of admixture atoms in a buffer gas, and L is a typical dimension on which the number density of admixture atoms vary remarkably.

Evidently, the criterion that admixture atoms are distributed uniformly over the cross section has the form $j_{ad} \ll j_e$, when $L \sim R_o$. This gives

$$\frac{N_e}{N_{ad}} \ll \frac{D_{ad}}{D_a} . \qquad (15.30)$$

Note that $D_{ad} \ll D_a$ because of a more strong interaction of ion–atom than atom–atom. Since in the regime under consideration $N_{ad} \gg N_e$, along with fulfilment of criterion the case is possible when this criterion is violated.

► **Problem 15.24** Analyze the Schottky regime of the ionization equilibrium in the positive column when electron–electron collisions dominate. Assume the cross section of electron–atom collisions to be independent of the collision velocity.

If criterion (7.8) is fulfilled for a plasma of the positive column, one can use the electron temperature T_e as the parameter of the electron distribution function that is given by formula (7.28) in this case. Correspondingly, substituting the ionization rate constant (4.45), whose data are given in Table 4.3, in the balance equation (15.17), one can find the relation between the reduced electric field strength and the reduced tube radius. In particular, one can use for this the dependence of Fig. 15.4 for the first Townsend coefficient on the specific electric field strength.

We note that the Schottky regime relates to small currents when plasma heating is weak. Hence, this relates to low number densities of atoms. But if criterion (7.18) for the electron number density holds true and the Maxwell distribution function (7.19) takes place for electrons, the electric field strength in the positive column decreases significantly in comparison to the limit of low electron number densities.

► **Problem 15.25** Find the condition when metastable atoms determine the ionization balance in the positive column of the cylinder discharge tube for a pure gas if the electron–atom rate constant is independent of the collision velocity.

As it follows from the above analysis, the ionization balance in the positive column of gas discharge includes processes of elastic and inelastic electron–atom processes as well as transport of electrons and excited atoms. In particular, the ionization rate is determined by a tail of the electron distribution function and therefore is connected in a complex manner with the character of these processes. The case under consideration is the simplest one and corresponds to the simple distribution function (7.9) of electrons

$$f_0(\varepsilon) = A \exp\left(-\frac{\varepsilon}{T_e}\right) ,$$

where the parameter T_e is equal to

$$T_e = \frac{Mw^2}{3} ,$$

and is large compared to the atom thermal energy. The electron drift velocity w according to formula (7.11) is

$$w = \frac{eE}{m_e N_a k_{ea}} ,$$

where k_{ea} is the rate constant of elastic electron–atom collision that is independent of the electron energy.

Under the above conditions we obtain the boundary electron number density N_e instead of formula (15.26),

$$N_e = N_a \frac{D_m N_a}{(N_a r_o)^2} \frac{5.78 g_*}{k_q g_o} \left(\frac{J}{J_*}\right)^2 .$$

Using the parameters of helium atoms, we obtain the boundary electron number density N_e,

$$N_e = \frac{A}{N_a r_o^2}, \qquad A = 2.4 \cdot 10^{29}\ \text{cm}^{-4} .$$

Thus, we have the same dependence of the boundary electron number density on the parameter $N_a r_0$ as it takes place in formula (15.26), but in contrast to formula (15.26), N_e is independent of the electric field strength. Next, we obtain now a more large boundary electron number density, i. e., this value is sensitive to parameters of kinetics of ionization processes in this system.

▶ **Problem 15.26** Estimate a minimal length of the positive column for a cylinder discharge tube and the Schottky regime of ionization equilibrium.

The average path $\bar{l}$ that electrons pass during their lifetime is

$$\bar{l} = w_e \tau = \frac{w_e}{N_a k_{ion}} = \frac{1}{\alpha} ,$$

where w_e is the electron drift velocity in an electric field of the positive column, τ is the electron lifetime in the positive column that is given by formula (15.19), and α is the first Townsend coefficient for the electric field strength of the positive column. From this we obtain the condition of existence of the positive column,

$$L\alpha > 1 ,$$

where L is a distance between electrodes. If this criterion holds true, ionization processes inside the positive column do not depend on processes near electrodes. Then an increase of a distance L between electrodes does not change electrode regions, but leads to corresponding increase of the positive column.

▶ **Problem 15.27** Analyze the criteria of realization of the Schottky regime of ionization equilibrium in the positive column of gas discharge.

The Schottky regime of ionization equilibrium is characterized by small atom and electron number densities in the positive column of gas discharge, i. e., with respect to small discharge currents and pressures, though the gas pressure provides the validity of the condition that the mean free path of charged particles is small compared to a discharge tube radius. Therefore, collisions between electrons are weak until the electron travels from the point of its formation to the walls. Hence in this limit the electron drift time τ_{dr} is small compared to a typical time τ_{rec} of three-body recombination of electrons

$$\tau_{dr} \ll \tau_{rec} ,$$

and this was used in the equilibrium relation (15.19) and its analogs.

A low current and pressure leads to a weak gas heating by a discharge current. Hence, a gas is uniform over the cross section of the positive column, and this fact is used for the Schottky regime of ionization equilibrium. As a result, the basic processes for ionization equilibrium in the Schottky regime are ionization of gas atoms by electron impact and recombination of electrons and ions on the walls.

15.4
Heat Processes in Positive Column of High Pressure Discharge

▶ **Problem 15.28** Determine the temperature difference between the center of a discharge tube and its walls for the positive column of a weak gas discharge.

Let us analyze the heat balance of the positive column of weak gas discharge, assuming the heat transfer will result from the thermal conductivity process. We have then the heat balance equation in the form

$$\frac{1}{\rho}\frac{d}{d\rho}\left[\rho\kappa(T)\frac{dT}{d\rho}\right]+p(\rho)=0\,, \qquad (15.31)$$

where κ is the thermal conductivity coefficient, $p(\rho) = iE$ is the specific power of heat release, so that i is the current density, E is the electric field strength. One can represent this in the form of the Elenbaas–Heller equation

$$\frac{1}{\rho}\frac{d}{d\rho}\left[\rho\kappa(T)\frac{dT}{d\rho}\right]+\Sigma E^2=0\,, \qquad (15.32)$$

where Σ is the plasma conductivity. A simple solution of this equation within the framework of the so-called parabolic model has the form

$$T = T_\mathrm{w} + \frac{\Sigma E^2(r_\mathrm{o}^2-\rho^2)}{4\kappa} = T_\mathrm{w} + (T_\mathrm{o} - T_\mathrm{w})\left(1-\frac{\rho^2}{r_\mathrm{o}^2}\right)\,, \qquad (15.33)$$

where T_w is the wall temperature, T_o is the temperature at the axis, and we assume the thermal conductivity to be constant over the cross section of the discharge tube.

We consider below the heat regime when the discharge electric current heats the positive column plasma remarkably, and the temperature T_0 is the temperature at the axis of the discharge tube exceeds remarkably the wall temperature T_w. Then introducing the variable $x = \rho^2/r_\mathrm{o}^2$ and the value $Z = \int_{\mathrm{T}}^{T_0}\kappa(T)dT$, we obtain equation (15.31) in the form

$$\frac{d}{dx}\left(x\frac{dZ}{dx}\right) - \frac{p(\rho)r_\mathrm{o}^2}{4} = 0\,. \qquad (15.34)$$

This gives the released power per unit tube length,

$$P = \int_0^{r_\mathrm{o}} 2\pi\rho d\rho p(\rho) = -4\pi\frac{dZ(x=1)}{dx}\,.$$

We first solve equation (15.31) for the Schottky regime (15.16), assuming the electric field strength to be constant and approximating the specific power by the dependence $p = p_0(1-x^{0.67})$. In this case equation (15.31) gives

$$Z = \int_{T_\mathrm{w}}^{T_0}\kappa(T)dT = 0.13P\,.$$

This relation expresses the difference between the temperatures of the axis and that of the walls of the discharge tube through the discharge power per unit length.

The above heat regime is realized at low discharge currents. In a general case of the electron distribution over the tube cross section, we will approximate the specific heat release over the cross section by formula $p = p_0[1 - (\rho/r_0)^n]$. Then the power of heat release per unit tube length is

$$P = 2\pi r_0 \left[-\kappa \frac{dT(r_0)}{d\rho} \right] = 4\pi \frac{dZ(x=1)}{dx} = \pi p_0 r_0^2 \frac{n}{n+2} .$$

Let us introduce the function $F(n) = \int_{T_w}^{T_0} \kappa(T) dT/P$. We find that $F(n)$ lies within the limits $F(0) = 0.16$ and $F = 0.13$ for the Schottky regime. This allows us to establish a simple relation between the value Z and the specific discharge power within the accuracy of 10 % has the form

$$\int_{T_w}^{T_0} \kappa(T) dT = \frac{T_0 \kappa(T_0) - T_w \kappa(T_w)}{1 + d \ln \kappa / d \ln T} = 0.14 P . \tag{15.35}$$

In particular, if the difference of the temperatures is small ($\Delta T = T_0 - T_w \ll T_0$), this relation yields

$$\Delta T = \frac{0.14 P}{\kappa(T_0)} . \tag{15.36}$$

We thus obtain that heating of a gas in a discharge tube with an electric current is not sensitive to the distribution of this current over the tube cross section, and is determined mostly by the total released power inside the tube.

▶ **Problem 15.29** Formulate the character of the electron distribution in the positive column of gas discharge of high pressure at high electric currents.

This regime is realized under the condition

$$\frac{\tau_{rec}}{\tau_{dr}} \gg 1 ,$$

and transport of electrons to the walls is not of importance in this regime. Because of a high electron number density, one can analyze this regime in terms of the electron temperature T_e, when local thermodynamic equilibrium takes place at each point of the positive column. Correspondingly, the relation between the number density of electrons N_e and atoms N_a is given by the Saha formula (1.69)

$$\frac{N_e^2}{N_a} = \frac{g_e g_i}{g_a} \left(\frac{m_e T_e}{2\pi \hbar^2} \right)^{3/2} \exp\left(-\frac{J}{T_e} \right) ,$$

where g_e, g_i, and g_a are the statistical weights of the electron, ion and atom, respectively, and J is the atom ionization potential. The electron and gas temperatures are also connected through the electric field strength by relation (7.26). In particular, in the case when the rate of electron–atom collisions does not depend on the electron velocities, this relation has the form (7.27)

$$T_e - T = \frac{1}{3} M w^2 ,$$

where M is the atom mass and w is the electron drift velocity.

The distribution of the electron temperature T_e and gas temperature T over the cross section is of importance in this regime, which is given by the Elenbaas–Heller equation

$$\frac{1}{\rho}\frac{d}{d\rho}\left[\rho\kappa(T)\frac{dT}{d\rho}\right] + \frac{1}{\rho}\frac{d}{d\rho}\left[\rho\kappa_e(T_e)\frac{dT_e}{d\rho}\right] + p(\rho) = 0 , \tag{15.37}$$

where $\kappa(T)$ is the thermal conductivity coefficient of the gas, $\kappa_e(T_e)$ is the electron thermal conductivity coefficient, and $p(\rho) = iE$ is the specific power of heat release. The role of electron and gas heat transfer, i. e., the contribution of the first and second terms of this equation, depends on the electron number density. In addition, the electron temperature under real conditions is relatively small

$$T_e \ll J, \tag{15.38}$$

where J is the atom ionization potential.

▶ **Problem 15.30** Find the distribution of the electron number density over the cross section of the positive column of high pressure gas discharge at high electron currents if heat transfer is determined by the gas thermal conductivity.

This analysis is based on relation (15.38) that leads to a sharp dependence of the electron number density on the electron temperature, which according to formula (1.69) has the form $N_e \sim \exp[-J/(2T_e)]$. One can conveniently use the variable

$$y = \frac{[T_o - T_e(\rho)]\,J}{2T_e^2(0)} , \tag{15.39}$$

where $T_o = T_e(0)$. This gives electron number density on the electron temperature $N(\rho) = N(0)e^{-y}$, which allows us to neglect the temperature dependence for other discharge parameters as, for example, the thermal conductivity coefficient of the gas. Then $p(\rho) = p(0)e^{-y}$, $\kappa_e \sim N_e \sim e^{-y}$, and, using the variable $x = \rho^2/r_o^2$, one can reduce equation (15.37) to the form

$$\frac{d}{dx}\left[x\left(e^{-y} + \zeta\right)\frac{dy}{dx}\right] - Ae^{-y} = 0 . \tag{15.40}$$

Here the parameters ζ and A are given by

$$\zeta = \frac{T\kappa(T)}{\kappa_e(T_e)}\alpha; \qquad A = \frac{p_o r_o^2 J}{8T_e^2\kappa_e(T_e)} ,$$

where $p_o = p(0)$, and the parameter $\alpha = dT(\rho)/dT_e(\rho)$. Using the dependence of the rate of electron–atom collisions ν on the electron velocity v as $\nu \sim v^\beta$, we obtain the parameter α

$$\alpha = \frac{(1+\beta-\beta T/T_e)}{2T_e/T-1} \, .$$

One can see that due to a sharp dependence of the electron number density on the electron temperature we obtain the equation for the electron number density whose parameters are defined at the discharge tube axis only.

While considering the case, when heat transfer is determined by gas thermal conductivity $\zeta \gg 1$, we ignore the first term of equation (15.40). Then the solution of this equation has the form

$$y = 2\ln\left(1+\frac{Ax}{2\zeta}\right) \, ,$$

and returning the electron number, we obtain its distribution over the tube cross section,

$$N_e(\rho) = N_e(0)e^{-y} = \frac{N_o\rho_o^2}{(\rho^2+\rho_o^2)^2} \tag{15.41}$$

where $N_o = N_e(0)$.

If the parameter ρ_o is small compared to the tube radius r_o, then contraction of the discharge current takes place, i. e., the electron current occupies a restricted region near the center of the discharge tube. This is the peculiarity of this regime of gas discharge of high pressure and at low currents. Then it is profitable for a plasma to occupy a restricted region near the tube center and to have there a high electron temperature that corresponds to the minimum of plasma resistance and hence to optimal conditions of gas discharge.

These relations give the power $P = IE$ per unit tube length (I is the total discharge power, E is the electric field strength) of arc discharge in this heat regime

$$P = IE = \int p_o e^{-y} 2\pi\rho d\rho = \frac{16T_e^2 T\kappa(T)\alpha}{J} \, , \tag{15.42}$$

where $p_o = j_o E$ is the power per unit volume at the tube center (j_o is the current density at the axis of the discharge tube). According to equation (15.42), the discharge power is expressed through plasma parameters at the center of the discharge tube.

▶ **Problem 15.31** Find the distribution of the electron number density over the cross section of the positive column of high pressure gas discharge at high electron currents when heat transfer is determined by the electron thermal conductivity.

While considering the limiting case $\zeta \ll 1$, when the electron thermal conductivity determines the heat balance of the electron current, we find that in the region

$y < \ln(1/\zeta)$, where the plasma is concentrated, equation (15.40) by using the new variable $Y = N_e(\rho)/N_e(0) = e^{-y}$ takes the form

$$\frac{d}{dx}\left(x\frac{dY}{dx}\right) + AY = 0 . \tag{15.43}$$

The solution of this equation

$$Y = J_0\left(2\sqrt{Ax}\right) ,$$

gives for the power released per unit length of the positive column

$$P = IE = \int p_0 Y 2\pi\rho d\rho = 1.36 p_0 \rho_0^2 ,$$

and the current radius ρ_0 is given by

$$\rho_0^2 = \frac{5.78 r_0^2}{A} = \frac{12 T_e^2 \kappa_e(T_e)}{p_0 J} .$$

As a result, the total discharge power per unit length of the positive column in this limiting case is

$$P = IE = \frac{16 T_e^2 \kappa_e(T_e)}{J} . \tag{15.44}$$

▶ **Problem 15.32** Find the specific power and the radius of a region occupied by the discharge current for the positive column of high pressure arc if both mechanisms of heat transfer, due to the gas and electron thermal conductivity, are realized.

Combining formulas (15.42) and (15.44), i. e., taking into account both the gas and electron thermal conductivity for heat transfer in the positive column of high pressure arc, we obtain the discharge power per unit length of the discharge tube as

$$P = IE = \frac{16 T_e^2 \kappa_e(T_e)(1 + 3.2\zeta)}{J} . \tag{15.45}$$

In the same manner one can introduce the plasma radius, defining it from expression (15.18) for the Schottky regime when the electric current occupies all the tube cross section

$$\rho_0^2 = \frac{12 T_e^2 \kappa_e(T_e)(1 + 3.2\zeta)}{p_0 J} . \tag{15.46}$$

Find the scaling law with respect to the atom number density N_a for plasma parameters of the positive column of high pressure arc in the limiting cases of different ratios between the gas and electron thermal conductivities.

In both cases of the heat transport regime we obtain the electric field strength,

$$E \sim N_a .$$

Other dependences follow from the above formulas and give in the case $\zeta \ll 1$, when heat transport is determined by the thermal conductivity of the gas

$$p_0 \sim N_a^{3/2}, \quad \rho_0 \sim N_a^{-1}, \quad P \sim N_a^{-1/2}, \quad I \sim N_a^{-3/2} .$$

In the other limiting case $\zeta \gg 1$, when heat transport is determined by the electron thermal conductivity, we obtain

$$p_0 \sim N_a^{3/2}, \; \rho_0 \sim N_a^{-3/4}, \; P \sim \text{const} , \; I \sim N_a^{-1} .$$

► **Problem 15.33** Analyze the conditions of contraction of the discharge current in the positive column of high pressure arc.

As a result of contraction, a discharge current shrinks in a narrow tube whose radius is small compared to the discharge tube radius. We analyze this phenomenon on the basis of the function

$$Z = \int^{T_e} \kappa_e(T_e')dT_e' + \int^{T} \kappa(T')dT' = \frac{2T_e^2}{J}\kappa_e(T_e) + \frac{T\kappa(T)}{1+\gamma} ,$$

that results from integration of the Elenbaas–Heller equation in the form (15.37). Here $\gamma = d\ln\kappa(T)/dT$, and we assume the electron T_e and gas T temperatures at the axis which are present in this expression, to be large compared their values at the walls. Integrating equation (15.37) twice, we obtain

$$Z = \int_0^{r_0} p(\rho)\rho d\rho \ln\frac{r_0}{\rho} \approx \frac{P}{2\pi}\ln\frac{2.3r_0}{\rho_0} .$$

Here P is the total power per unit tube length of the tube, and ρ_0 is the current radius. Below we compare this equation by expression (15.45) for the specific discharge power that allows us to find a radius of a current region.

We first analyze the case $\zeta \ll 1$, which is, when the electron thermal conductivity is dominant. Then we have

$$Z = \frac{2T_e^2}{J}\kappa_e(T_e) = \frac{P}{4\pi} . \tag{15.47}$$

In the limit of high currents, we have, on the one hand, local thermodynamic equilibrium and nearness of the electron and gas temperatures, and on the other side, electrons are concentrated in a narrow region of a high temperature where ionization and recombination processes proceed.

We derive below formula (15.47) from other consideration that allows us to understand the nature of arc contraction. The specific heat release in the heat balance equation (15.37) is proportional to the number density of electrons $p \sim N_e$, that in turn, according to the Saha relation (1.69) depends on the electron temperature as $N_e \sim \exp(-J/(2T_e))$ (we take here the electron and gas temperature to be identical, which is the parameter $\alpha = 1$). From this we have near the axis

$$p = p_0 \exp(-\beta\rho^2) ,$$

where p_o is the specific heat release at the tube axis and ρ is a distance from the tube axis, and

$$\beta = \frac{J}{2T_e}\frac{dT_e}{d\rho^2} .$$

Solving equation (15.37) with this distribution of the specific heat release, we find the temperature distribution near the axis,

$$T_e(\rho) = T_0 - \frac{p_o}{2\beta\kappa}\int_0^{\rho}\left(1 - e^{-\beta\rho^2}\right)\frac{d\rho}{\rho} ,$$

which gives

$$P = \int_0^{r_o} p \cdot 2\pi\rho d\rho = \frac{\pi p_o}{\alpha} = \frac{8\pi T_e^2\kappa}{J} ,$$

that coincides with formula (15.47). From derivation of this formula it follows that the current region decreases with an increase of the specific power p_o that in turn is proportional to the electron number density at the axis, and a current region radius is $\rho_o \sim p_o^{-1/2}$. We assume this dimension to be large compared to the electron mean free path during its lifetime

$$\rho_o \gg \sqrt{\frac{D_a}{K_{ei} N_e^2}} ,$$

where D_a is the ambipolar diffusion coefficient at the axis, $K_e i$ is the three-body recombination coefficient of electrons and ions. As is seen, compression of a current region increases with an increase of the temperature at the axis.

Let us consider the opposite limiting case $\zeta \gg 1$, if gas thermal conductivity dominates. Then according to the above formulas the current radius ρ_o that follows from the heat equation satisfies to equation

$$\left(1 + \frac{\rho_o^2}{r_o^2}\right)\ln\left(1 + \frac{r_o^2}{\rho_o^2}\right) = \frac{JT}{4T_e^2(1+\gamma)} , \qquad (15.48)$$

and contraction is possible in this case.

Thus, the arc current contraction occurs at high currents when thermal conductivity is not able to provide heat release due to electron–atom collisions. As a result, a current occupies a small part of the discharge tube, and large temperature gradients increase heat transport. Note that large temperature gradients at high pressure may cause a convective instability, so that an effective mixing of a gas results from gas convection.

15.5 Plasma of Positive Column of Low Pressure Discharge

▶ **Problem 15.34** An equilibrium in the positive column of low pressure arc results from formation of electrons by ionization of atoms by electron impact and from travelling of electrons to the walls. Assuming that the positive column plasma is almost quasineutral and electrons are locked in a self-consistent plasma field, determine the plasma current to electrodes.

Plasma parameters of low-pressure gas discharge satisfy the criterion

$$r_D \ll L \ll \lambda\,, \tag{15.49}$$

where L is a characteristic dimension of the discharge, λ is the mean free path for plasma particles, and r_D is the Debye–Hückel radius of the plasma. In spite of a large mean free path, electrons are locked in the plasma self-consistent field that increases their lifetime in the plasma and allows us to introduce the electron temperature. Below we will be guided by typical parameters of this plasma where the electron number density is $N_e \sim 10^{14}\ \text{cm}^{-3}$, the atom number density is $N_a \sim 10^{15}\ \text{cm}^{-3}$, the electron temperature $T_e \sim 1$ eV, and the positive column dimension is $L \sim 0.1$ cm. These plasma parameters lead to the mean free path of electrons $\lambda \sim 1$ cm, and $r_D \sim 5 \cdot 10^{-5}$ cm, i. e., criterion (15.49) holds true. These parameters are typical for a plasma of thermoemission converters.

The plasma under consideration is located between two infinite parallel electrodes with a distance L between them. Because of the high number density of charged particles, typical dimensions of cathode and anode regions are small, and the positive column with its quasineutral plasma occupies almost all of the space between the electrodes. This plasma is characterized by a certain self-consistent field that equalized currents of electrons and ions outside this region. Our task is to determine these electric currents. The electric potential distribution between electrodes is given in Fig. 15.12.

Denoting the electric potential of the plasma by $\varphi(x)$, where the x-axis is directed perpendicular to the electrodes, and the origin of the x coordinate is taken to be centered between the electrodes, we assume this field determines the currents of ions and electrons to electrodes.

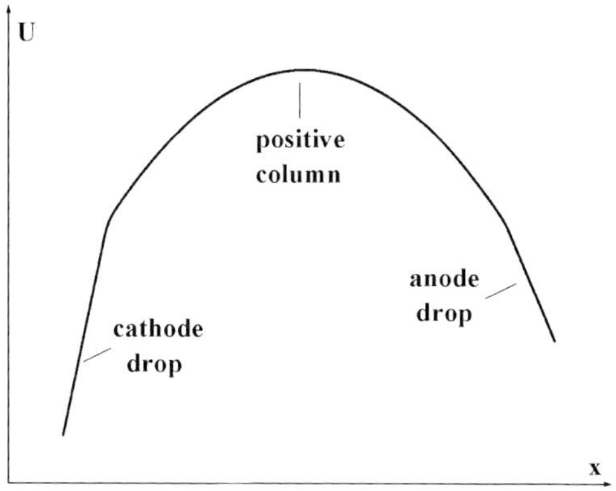

Fig. 15.12 The distribution of the electric potential in arc discharge of low pressure.

This leads to the symmetry

$$\varphi(x) = \varphi(-x)\,, \tag{15.50}$$

where the origin is taken in the middle of the positive column. Since electrons are locked inside the positive column, they are found in thermodynamic equilibrium that is characterized by the electron temperature T_e. This gives the distribution for the number density of electrons N_e inside the positive column

$$N_e = N_0 \exp\left(-\frac{e\varphi}{T_e}\right). \tag{15.51}$$

The number density of electrons almost equals to the number density of ions N_i because of the quasineutrality of the plasma. Electrons and ions are formed in the positive column in electron–atom collisions, and electrons are locked by a self-consistent plasma field, whereas ions move freely to electrodes. Hence, an ion generated at a point ξ reaches a point x with a velocity $v_x = \sqrt{2e[\varphi(\xi) - \varphi(x)]/M}$, where M is the ion mass. Denoting by $\Phi(\xi)$ a number of ions produced per unit volume per unit time at the point ξ, we obtain the number density of ions $N_i(x)$ by summation over all the ions arriving to this point

$$N_i = \int_0^x \frac{\Phi(\xi)d\xi}{\sqrt{2e[\varphi(\xi) - \varphi(x)]/M}}\,.$$

Here we consider that ions arriving to a point x are formed at points $0 < \xi < x$, because $x = 0$ corresponds to the top of the potential hump of the self-consistent plasma field for ions. Thus, the condition of plasma quasineutrality $N_e = N_i$ gives

$$N_0 \exp\left[-\frac{e\varphi(x)}{T_e}\right] = \int_0^x \frac{\Phi(\xi)d\xi}{\sqrt{2e[\varphi(x) - \varphi(\xi)]/M}}\,,$$

where N_0 is the electron number density at $x = 0$, and we define the self-consistent field potential such that $\varphi(0) = 0$. It is convenient to rewrite this equation in terms of variables $\eta(x) = -e\varphi(x)/T_e$ ($\eta > 0$) and $j_o = N_0\sqrt{2T_e/M}$

$$j_0 e^{-\eta} = \int_0^x \frac{\Phi(\xi)d\xi}{\sqrt{\eta(x) - \eta(\xi)}}\,. \tag{15.52}$$

Let us multiply this equation by $d\eta/dx[\eta(y) - \eta(x)]^{-1/2}$ and integrate over dx between ξ and y. Let us use the relation

$$\int_0^x \frac{d\eta(x)}{dx}dx \frac{1}{\sqrt{[\eta(y) - \eta(x)][\eta(x) - \eta(\xi)]}} = \int_{\eta(\xi)}^{\eta(y)} \frac{d\eta}{\sqrt{[\eta(y) - \eta][\eta - \eta(\xi)]}} = \pi\,,$$

and simplify the right-hand side of the equation

$$\int_0^{\eta(y)} \frac{d\eta(x)}{\sqrt{\eta(y)-\eta(x)}} \int_0^{\eta(x)} \frac{\Phi(\xi)d\eta(\xi)}{\sqrt{\eta(x)-\eta(\xi)}} \frac{d\xi}{d\eta(\xi)} =$$

$$\int_0^{\eta(y)} \Phi(\xi)d\eta(\xi)\frac{d\xi}{d\eta(\xi)} \int_{\eta(\xi)}^{\eta(y)} \frac{d\eta(x)}{\sqrt{[\eta(y)-\eta(x)][\eta(x)-\eta(\xi)]}} = \pi \int_0^y \Phi(\xi)d\xi \,.$$

This gives for the ion flux $j_i(y)$ at a given point y

$$j_i(y) = \int_0^y \Phi(\xi)d\xi = \frac{j_0}{\pi}\int_0^{\eta(y)} \exp(-\eta)\frac{d\eta}{\sqrt{\eta(y)-\eta}} . \tag{15.53}$$

The ion flux $j_i(y)$ toward the electrodes increases on removing from the middle point $x = 0$ of the positive column and reaches the maximum at the electrodes where the plasma quasineutrality is violated. This leads to the condition $dj_i/d\eta = 0$ at the electrode that corresponds to $d\eta/dx = \infty$, since $dj_i/d\eta = (dj_i/dx)/(dx/d\eta)$. Representing $j_i(\eta)$ in the form

$$j_i(\eta) = \frac{j_0}{\pi}\int_0^{\eta} \exp(-\eta')\frac{d\eta'}{\sqrt{\eta-\eta'}} = \frac{2j_0}{\pi}\sqrt{\eta} + \frac{2j_0}{\pi}\int_0^{\eta} \exp(-\eta')\sqrt{\eta-\eta'}d\eta' \,,$$

we find that the condition at the electrode $dj_i/d\eta = 0$ leads to the equation

$$\sqrt{\eta_o}\int_0^{\eta_o} \exp(-\eta)\frac{d\eta}{\sqrt{\eta_0-\eta}} = 1 \,,$$

where η_0 is the value of the reduced potential at the electrode. The solution of the latter equation is $\eta_0 = 0.855$. From this we find the ion flux to the electrodes.

$$j_i = \frac{j_0}{\pi\sqrt{\eta_0}} = 0.344 j_o = 0.344 N_0 \sqrt{\frac{2T_e}{M}} \,. \tag{15.54}$$

► **Problem 15.35** Find conditions for a plasma of a low pressure positive column of arc when criterion (15.49) holds true.

Let us give the criteria of validity of formula (15.54), which provide the above character of the processes in the positive column plasma. This plasma is quasineutral that according to the Poisson equation leads to the criterion

$$|N_i - N_e| = \left|\frac{1}{4\pi e}\frac{d^2\varphi}{dx^2}\right| \ll N_0, \qquad \text{or} \quad \frac{d^2\eta}{dx^2} \ll \frac{1}{r_D^2} \,.$$

Because $\eta \sim 1$ in the positive column, this condition corresponds to $r_D \ll L$, i. e., it coincides with the first criterion (15.49).

According to the second criterion (15.49) $L \ll \lambda$, ions pass the positive column without collisions. Nevertheless, because of thermodynamic equilibrium for electrons, each electron being captured by a self-consistent field of the positive column, collide with other electrons many times before it leaves this region. Hence, the total electron path during its location in the positive column is much greater than the mean free path λ_e for collisions with other electrons. This gives the criterion

$$\lambda_e \ll \sqrt{\frac{M}{m_e}} L , \tag{15.55}$$

that must be added to criteria (15.49).

Thus, the plasma of the low-pressure positive column of arc is characterized by specific properties, so that it is transparent for positive ions, while its self-consistent field creates a trap for electrons. Hence, collisions between trapped electrons establish thermodynamic equilibrium for electrons in the arc positive column, and ionization collisions of electrons with atoms determine formation of electrons and ions and influences the electron temperature of this plasma.

▶ **Problem 15.36** Derive the balance equation for formation and decay of ions in a plasma of a low-pressure positive column of arc located between two infinite parallel electrodes in order to determine the electron temperature of this plasma.

We have this balance equation in the form

$$\int_0^{L/2} \Phi(x) dx = 0.34 N_0 \sqrt{\frac{2T_e}{M}} ,$$

where the right-hand side of this equation is the ion flux in accordance with formula (15.54), the left-hand side is the number of ions per unit time and unit area which result from ionization of atoms by electron impact. Here x is a distance from the center, and the rate of atom ionization by electron impact is

$$\Phi(x) = N_e(x) N_a k_{ion}(T_e) .$$

Here we assume the number density of atoms N_a to be constant through the positive column, whereas the number density of electrons varies according to the Boltzmann formula (15.51) $N_e = N_0 \exp(-\eta)$. As a result we obtain from this balance equation

$$k_{ion}(T_e) = \frac{0.69}{N_a L I} \left(\sqrt{\frac{2T_e}{M}} \right) , \tag{15.56}$$

where the numerical coefficient I is equal

$$I = \int \frac{2dx}{L} \exp(-\eta) ,$$

and L is a distance between electrodes. Since $k_{ion} \sim \exp(-T_e/J)$, where J is the atom ionization potential, equation (15.56) allows us to determine the electron temperature.

► **Problem 15.37** Determine the minimum voltage between electrodes that can provide the balance of the positive column plasma according to the balance equation (15.56).

While considering a self-consistent field in the low-pressure positive column between two infinite electrodes, when this plasma satisfies criterion (15.49), we assume this field to be symmetric (15.50). But for operation of this system, an external source of energy is required in order to realize continuous ionization of atoms. The simplest external energy source is an additional voltage between electrodes that creates gaseous discharge. As a result, the electrons arise at the cathode, and under action of the voltage an electric current occurs. This current under the action of an additional voltage causes an energy release in the positive column that compensates energy losses due to ionization processes.

The electron flux j_e on the boundary of the positive column is

$$j_e = N_e \sqrt{\frac{m_e}{2\pi T_e}}, \qquad (15.57)$$

where N_e is the electron number density on the positive column boundary ($N_e = N_0 \exp(-\eta_0)$). Note that the electron flux does not coincide with the ion flux (15.54), and their equality for maintaining of a quasineutral plasma is attained in the anode and cathode regions by additional voltages which lead to reflection of a part of electrons.

Within the framework of this discharge scenario, we introduce an additional voltage ΔV between the left and right sides of the positive column that provides the electron current to one side. Indeed, because of thermodynamic equilibrium, the number density of electrons on the left positive column boundary is $N_0 \exp[(-e\varphi_0 + e\Delta V)/T_e]$ instead of $N_0 \exp(-e\varphi_0/T_e)$ in the absence of an additional voltage, where (φ_0 is the potential of a self-consistent field on the positive column boundary with respect to this value in the middle of the positive column). Hence, a noncompensated electric current density is

$$i = e j_e \sinh\left(\frac{e\Delta V}{2T_e}\right),$$

where the electron flux j_e in absence of an additional voltage is given by formula (15.57).

The voltage ΔV between two boundaries of the positive column follows from the energy balance in the positive column. Taking the energy consumed on formation of one pair electron–ion as cJ, where J is the atom ionization potential, $c \geq 1$, we obtain the power that is consumed per unit area of electrodes as $P = 2j_i\, cJ$, where the factor 2 accounts for propagation of ions in two directions. On the other hand, this specific power is $i\Delta V$. Assuming $\Delta V \ll T_e$, we obtain from this

$$P = i\Delta V = \frac{(e\Delta V)^2}{T_e} j_e = \frac{(e\Delta V)^2}{T_e} N_0 e^{-\eta_0} \sqrt{\frac{m_e}{2\pi T_e}}.$$

Equalizing this to $P = 2j_i\ cJ$, we find the additional voltage for the arc positive column,

$$e\Delta V = 2.38\left(\frac{m_e}{M}\right)^{1/4}\sqrt{cJT_e}. \tag{15.58}$$

According to this formula, the voltage drop in the positive column that provides the energy balance is relatively small.

▶ **Problem 15.38** Determine the anode voltage that equalizes currents of electrons and ions from the positive column.

A self-consistent field of the arc positive column allows for ions leave the positive column plasma freely, without collisions. Because the plasma remains to be quasineutral, the electron and ion currents from the positive column region must be identical, i. e., the electron current is $i_e = 2ej_i$, and the ion flux is given by formula (15.54). We assume the electron current to be directed towards to the anode, i. e., the electrons reflect from the cathode region. Next, the electron flux on the positive column boundary is given by formula (15.57), and this exceeds remarkably the ion flux. In order to equalize them, it is necessary an additional voltage drop $\Delta\varphi$ near the anode that reflects a part of electrons with a low energy. Because it leads to a decrease of the electron number density by a factor $\exp(-\Delta\varphi/T_e)$, we find from this the anode voltage

$$e\Delta\varphi = T_e \ln\left(5.67\sqrt{\frac{M}{m_e}}\right) = T_e\left(1.7 + \frac{1}{2}\ln\frac{M}{m_e}\right). \tag{15.59}$$

▶ **Problem 15.39** Determine the ion flux to the walls of a low-pressure positive column plasma located in a cylindrical tube.

This plasma satisfies to criterion (15.49) with respect to ion and electron travelling to the walls, and we use for determination of the ion flux the same method as for derivation of formula (15.53) for another geometry. Hence, we will use in this case the same assumptions, namely, ions are moving to the walls without collisions electrons are captured by the trap of a self-consistent field of the positive column plasma and their distribution over the tube cross section is given by the Boltzmann formula (15.51), and this plasma is quasineutral. Using these assumptions, we obtain the following equation that is similar to (15.52), but uses the cylindrical symmetry of this problem

$$j_0\rho(\eta)e^{-\eta(\varrho)} = \int_0^{\varrho}\frac{\Phi(r)rdr}{\sqrt{\eta(\rho)-\eta(r)}}, \tag{15.60}$$

where $\eta = -e\varphi(x)/T_e, j_0 = N_0\sqrt{(2T_e)/M}, \rho, r$ are the distances from the tube axis, and other notations are identical to those of formula (15.52). Next, the ion flux per unit tube length I_i and per unit walls area j_i are equal to

$$I_i = \int_0^{r_o} \Phi(r) 2\pi r dr; \qquad j_i = \frac{I_i}{2\pi r_o} = \int_0^{r_o} \Phi(r) r \frac{dr}{r_o} .$$

Let us multiply the both sides of equation (15.60) by $\rho[\eta(R) - \eta(\rho)]^{-1/2} d\eta/d\rho$ and integrate it over $d\rho$ from 0 to R. Since

$$\int_{\eta_r}^{\eta_R} \frac{d\eta}{\sqrt{(\eta_R - \eta)(\eta - \eta_r)}} = \pi ,$$

we obtain

$$j_o \int_0^{\eta_R} \frac{\rho(\eta) d\eta}{\sqrt{\eta_R - \eta}} \exp(-\eta) = \pi \int_0^R \Phi(r) r dr .$$

From this we find the ion flux to the walls that is equal to

$$j_i = \int_0^{r_o} \Phi(r) r \frac{dr}{r_o} = \frac{j_o}{\pi r_o} \int_0^{\eta_o} \frac{\rho(\eta) d\eta e^{-\eta}}{\sqrt{\eta_o - \eta}} , \qquad (15.61)$$

where $\eta_o = \eta(r_o)$ is the reduced potential of the positive column at its boundary.

At the positive column boundary the positive and negative charges are separated, and the plasma is not quasineutral. This means $d\eta/d\rho = \infty$, and since $dj/d\rho$ is a finite value, this means that $dj/d\eta = 0$ near the walls, where $\eta(r_o) = \eta_o$. We use the simple dependence $\eta(\rho)$, such that $\eta(0) = 0$, $\eta(r_o) = \eta_o$, and $d\eta/d\rho = \infty$ at $\rho = r_o$. In addition, $\eta(\rho) \sim \rho^2$, if $\rho \to 0$. This and inverse dependences have the form

$$\frac{\eta}{\eta_o} = 1 - \sqrt{1 - \left(\frac{\rho}{r_o}\right)^2} \quad \frac{\rho}{r_o} = \frac{\sqrt{\eta(2\eta_o - \eta)}}{\eta_o} , \qquad (15.62)$$

which is given in Fig. 15.13. This dependence together with the condition $dj/d\eta = 0$ at $\rho = r_o$ give

$$\int_0^{\eta_o} \left[\sqrt{\frac{\eta(2\eta_o - \eta)}{\eta_o - \eta}} - \sqrt{\frac{\eta_o - \eta}{\eta(2\eta_o - \eta)}} \right] e^{-\eta} d\eta = 0 .$$

From the solution of this equation we find $\eta_o = 1.145$. Correspondingly, for the ion flux to the walls we obtain

$$j_i = \frac{j_o \sqrt{\eta_o}}{\pi} \int_0^1 \sqrt{\frac{t(2-t)}{1-t}} e^{-t\eta_o} dt = 0.854 \frac{j_o}{\pi} = 0.272 N_0 \sqrt{\frac{2T_e}{M}} . \qquad (15.63)$$

This formula has the same dependence on the problem parameters as formula (15.52).

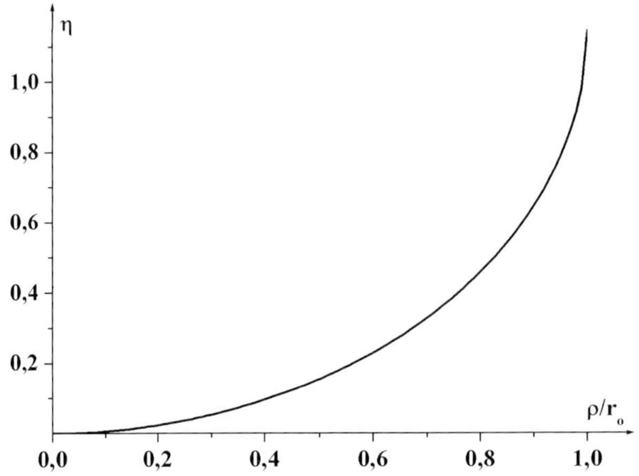

Fig. 15.13 The dependence of the reduced potential $\eta(\rho)/\eta_0$ of the positive column for arc of a low pressure on the reduced distance ρ/r_0 from the axis according to formula (15.62).

▶ **Problem 15.40** Determine the distribution function of ions on transversal energies in the positive column plasma of low pressure located in a cylindrical tube.

According to the definition, the distribution function on transversal ion energies ε has the form

$$f(\rho,\varepsilon)d\varepsilon = \text{const}\frac{\Phi(r')r'dr'}{\sqrt{\eta(\rho)-\eta(r')}}$$

We above used a single-valued relation between the ion energy at a given point and a point r' of its formation, which is $\varepsilon/T_e = \eta(\rho) - \eta(r')$. Since $\Phi(r') \sim \exp(-\eta')$, we obtain on the basis of formulas (15.62) which connect parameters η and ρ

$$f(\rho,\varepsilon)d\varepsilon = C\frac{(\varepsilon/T_e + \eta_0 - \eta)\exp(\varepsilon/T_e - \eta)}{\sqrt{\varepsilon/T_e}}d\varepsilon\ ,$$

where C is the normalization constant.

▶ **Problem 15.41** Determine the potential jump near the walls in the positive column of low pressure located in a cylindrical tube.

In a small region, near the walls, the plasma quasineutrality is violated, and as a result of arising of a charged layer, fluxes of ions and electrons to the walls become identical. According to formula the electron flux j_e on the positive column boundary is given by formula (15.57), and denoting the electric potential jump near the walls by ΔU, we obtain the electron flux to the walls as $j_e \exp(-e\Delta U/T_e)$. Equalizing this to the ion flux that is given by formula (15.63), we find for the jump of the electric potential near the walls for an arc plasma of low pressure located in a cylindrical tube

$$\frac{e\Delta U}{T_e} = \ln\left[\frac{1}{3}\ln\sqrt{\frac{M}{m_e}}\right] . \tag{15.64}$$

► **Problem 15.42** Find the relation between the atom number density and the electron temperature for the positive column plasma of low pressure located in a cylindrical tube.

We use the balance equation for ions formed and traveled to the walls of low-pressure arc that is burnt in a cylindrical tube. This balance equation has the form

$$k_{ion}(T_e)N_a\int_0^{r_o} 2\pi\rho d\rho N_e(\rho) = 2\pi r_o \cdot 0.272 N_0\sqrt{\frac{2T_e}{M}} ,$$

where N_a is the atom number density and k_{ion} is the rate constant of atom ionization by electron impact. Using dependence (15.62) for $\eta(rho)$, expression (15.51) for the electron number density, and evaluating the integral, we reduce this balance equation to the form

$$k_{ion}(T_e)N_a = \frac{0.76}{r_o}\sqrt{\frac{2T_e}{M}} . \tag{15.65}$$

This relation gives the electron temperature at a given number density of atoms. One can see the problem scaling, which is conservation of plasma parameters if the tube radius varies such that $N_a r_o$ = const. Figure 15.14 shows this dependence for argon.

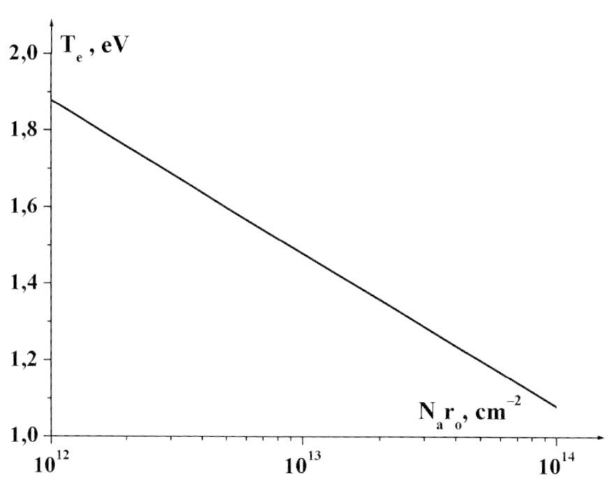

Fig. 15.14 The electron temperature T_e of the positive column plasma in arc of low pressure as a function of the reduced number density of argon atoms.

16
Appendices

Plasma Processes and Plasma Kinetics. Boris M. Smirnov

ISBN: 978-3-527-40681-4

16.1
Appendix 1: Physical Constants

Electron mass	$m_e = 9.10939 \times 10^{-28}$ g
Proton mass	$m_p = 1.67262 \times 10^{-24}$ g
Atomic unit of mass	$m_u = \frac{1}{12} m(^{12}C) = 1.66054 \times 10^{-24}$ g
Proton–electron mass ratio	$m_p/m_e = 1836.15$
Ratio of atomic and electron masses	$m_u/m_e = 1822.89$
Electron charge	$e = 1.602177 \times 10^{-19}$
	$C = 4.8032 \times 10^{-10}$ e.s.u.
	$e^2 = 2.3071 \times 10^{-19}$ erg cm
Planck constant	$h = 6.62619 \times 10^{-27}$ erg s
	$\hbar = 1.05457 \times 10^{-27}$ erg s
Velocity of light	$c = 2.99792 \times 10^{10}$ cm/s
Inverse fine-structure constant	$1/\alpha = \hbar c/e^2 = 137.03599$
Bohr radius	$a_o = \hbar^2/(m_{ee}^2) = 0.529177$ Å
Rydberg constant	$m_{ee}^4/(2\hbar^2) = 13.6057$ eV $= 2.17987 \times 10^{-18}$ J
Bohr magneton	$\mu_B = e\hbar/(2m_e) = 9.72402 \times 10^{-24}$ J/T
Avogadro number	$N_A = 6.02214 \times 10^{23} \text{mol}^{-1}$
Stephan–Boltzmann constant	$\sigma = \pi^2/(60\hbar^3 c^2) = 5.669 \times 10^{-12} \text{W}/(\text{cm}^2\text{K}^4)$
Molar volume	$R = 22.414$ l/mol
Loschmidt constant	$L = N_A/R = 2.6867 \times 10^{19} \text{cm}^{-3}$
Faraday constant	$F = N_A e = 96485.3$ C/mol

16.2
Appendix 2: Conversation Factors for Some Units

1 K = 1.38066×10^{-23} J = 8.6174×10^{-5} eV = 0.69504 cm^{-1} = 1.9859 cal/mol

1 eV = 11 604.4 K = 1.602177×10^{-19} J = 8065.5 cm^{-1} = 23 045 cal/mol = 96.4853 kJ/mol

1 cm^{-1} = 1.2398×10^{-4} eV = 1.4388 K= 1.98645×10^{-23} J= 2.8573 cal/mol

1 J = 10^7 erg

1C = 2.997924×10^9 esu

1 T = 10^4 G

1 atm = 760 Torr = 101325 Pa = 1.01325×10^6 dyn/cm^2

1 Torr = 133.322 Pa = 1.31579×10^{-3} atm

1 Pa $\equiv$ 1 N/m^2 = 10 dyn/cm^2 = 7.5007×10^{-3} Torr = 9.8693×10^{-6} atm

16.3
Appendix 3: Relations of Physics and Plasma Physics

16.3.1
Appendix 3a: Relations of General Physics

Number	Formula[a)]	Factor C	Units used
1	$v = C\sqrt{\varepsilon/m}$	5.931×10^7 cm/s	ε in eV, m in emu[a)]
		1.389×10^6 cm/s	ε in eV, m in amu[a)]
		5.506×10^5 cm/s	ε in K, m in
		1.289×10^4 cm/s	ε in K, m in amu
2	$v = C\sqrt{T/m}$	1.567×10^6 cm/s	T in eV, m in amu
		1.455×10^4 cm/s	ε in K, m in amu
3	$\varepsilon = Cv^2$	3.299×10^{-12} K	v in cm/s, m in emu
		6.014×10^{-9} K	v in cm/s, m in amu
		2.843×10^{-16} eV	v in cm/s, m in emu
		5.182×10^{-13} eV	v in cm/s, m in amu
4	$\omega = C\varepsilon$	$1.519 \times 10^{15}\text{s}^{-1}$	ε in eV
		$1.309 \times 10^{11}\text{s}^{-1}$	ε in K
5	$\omega = C/\lambda$	$1.884 \times 10^{15}\text{s}^{-1}$	λ in μm
6	$\varepsilon = C/\lambda$	1.2398eV	λ in μm
7	$\omega_H = CH/m$	$1.759 \times 10^7\text{s}^{-1}$	H in Gs, m in emu
		9655s^{-1}	H in Gs, m in amu
8	$r_H = C\sqrt{\varepsilon m}/H$	3.372 cm	ε in eV, m in emu, H in Gs
		143.9 cm	ε in eV, m in amu, H in Gs
		3.128×10^{-2} cm	ε in K, m in emu, H in Gs
		1.336 cm	ε in K, m in amu, H in Gs

a) emu is the electron mass unit ($m_e = 9.108 \times 10^{-28}$ g), amu is the atomic mass unit ($m_a = 1.6605 \times 10^{-24}$ g).

1. The particle velocity is $v = \sqrt{2\varepsilon/m}$, where ε is the energy, m is the particle mass.
2. The average particle velocity is $v = \sqrt{8T/(\pi m)}$ with the Maxwell distribution function of particles on velocities; T is the temperature expressed in energetic units, m is the particle mass.
3. The particle energy $\varepsilon = mv^2/2$, m is the particle mass, v is the particle velocity.
4. The photon frequency $\omega = \varepsilon/\hbar$, ε is the photon energy.
5. The photon frequency $\omega = 2\pi c/\lambda$, λ is the wavelength.
6. The photon energy $\varepsilon = 2\pi\hbar c/\lambda$.
7. The Larmor frequency $\omega_H = eH/(mc)$ for a charged particle of a mass m in a magnetic field of a strength H.
8. The Larmor radius of a charged particle $r_H = \sqrt{2\varepsilon/m}/\omega_H$, ε is the energy of a charged particle, m is its mass, ω_H is the Larmor frequency.

16.3.2
Appendix 3b: Relations for Physics of Gases and Plasmas

Number	Formula	Factor C	Units used
1	$N = Cp/T$	7.339×10^{21} cm^{-3}	p in atm, T in K
		9.657×10^{18} cm^{-3}	p in Torr, T in K
2	$p = CNT$	1.036×10^{-19} Torr	N in cm^{-3}, T in K
3	$\alpha = CN_e/T^3$	2.998×10^{-21}	N_e in cm^{-3}, T in eV
		4.685×10^{-9}	N_e in cm^{-3}, T in K
4	$f = Cm^{3/2}T^{3/2}$	2.415×10^{15} cm^{-3}	T in K, m in emu
		3.019×10^{21} cm^{-3}	T in eV, m in emu
		1.879×10^{20} cm^{-3}	T in K, m in amu
		2.349×10^{26} cm^{-3}	T in eV, m in amu
5	$K = C/T^{9/2}$	2.406×10^{-31} cm^6/s	T_e in 1000 K
6	$\omega_p = C\sqrt{N_e/m}$	5.642×10^4 s^{-1}	N_e in cm^{-3}, m in emu
		1322 s^{-1}	N_e in cm^{-3}, m in amu
7	$r_D = \sqrt{T/N_e}$	525.3 cm	N_e in cm^{-3}, T in eV
		4.876 cm	N_e in cm^{-3}, T in K
8	$p = CH^2$	4.000×10^{-3} Pa = 0.04 erg/cm^3	H in Gs

1. The state equation for gases $N = p/T$, p is the pressure, N is the number density of atomic particles.
2. The state equation for gases $p = NT$.
3. The parameter of an ideal plasma $\alpha = N_e e^6/T^3$, N_e is the electron number density, T is the temperature.
4. Pre-exponent of the Saha formula $f = [mT/(2\pi\hbar^2)]^{3/2}$, m is the mass of a particle, T is the temperature.
5. The coefficient of three-body electron–ion recombination $K = 1.5e^{10}/(m_e^{0.5}T_e^{9/2})$.
6. The plasma frequency $\omega_p = \sqrt{4\pi N_e e^2/m_e}$, N_e is the electron number density, m_e is the electron mass.
7. The Debye–Hückel radius $r_D = \sqrt{T/(8\pi N_e e^2)}$, N_e is the electron number density, T is the temperature.
8. The magnetic pressure $p_m = H^2/(8\pi)$.

16.3.3
Appendix 3c: Relations for Transport Coefficients

Number	Formula	Factor C	Units used
1	$D = CKT$	8.617×10^{-5} cm^2/s	K in cm$^2/(V \cdot s)$, T in K
		1 cm^2/s	K in cm$^2/(\mathrm{V} \cdot \mathrm{s})$, T in eV
2	$K = CD/T$	11 604 cm$^2/(\mathrm{V} \cdot \mathrm{s})$	D in cm^2/s, T in K
		1 cm$^2/(\mathrm{V} \cdot \mathrm{s})$	D in cm^2/s, T in eV
3	$D = C\sqrt{T/\mu}/(N\overline{\sigma_1})$	4.617×10^{21} cm^2/s	$\overline{\sigma_1}$ in Å^2, N in cm^{-3}, T in K, μ in amu
		1.595 cm^2/s	$\overline{\sigma_1}$ in Å^2, $N = 2.687 \times 10^{19}$ cm^{-3}, T in K, μ in amu
		171.8 cm^2/s	$\overline{\sigma_1}$ in Å^2, $N = 2.687 \times 10^{19}$ cm^{-3}, T in eV, μ in amu
		68.1115 cm^2/s	$\overline{\sigma_1}$ in Å^2, $N = 2.687 \times 10^{19}$ cm^{-3}, T in K, μ in emu
		7338 cm^2/s	$\overline{\sigma_1}$ in Å^2, $N = 2.687 \times 10^{19}$ cm^{-3}, T in eV, μ in emu
4	$K = C(\sqrt{T\mu}N\overline{\sigma_1})^{-1}$	1.851×10^4 cm$^2/(\mathrm{V} \cdot \mathrm{s})$	$\overline{\sigma_1}$ in Å^2, $N = 2.687 \times 10^{19}$ cm^{-3}, T in K, μ in amu
		171.8 cm$^2/(\mathrm{V} \cdot \mathrm{s})$	$\overline{\sigma_1}$ in AA^2, $N = 2.687 \times 10^{19}$ cm^{-3}, T in eV, μ in amu
		7.904×10^5 cm$^2/(\mathrm{V} \cdot \mathrm{s})$	$\overline{\sigma_1}$ in Å^2, $N = 2.687 \times 10^{19}$ cm^{-3}, T in K, μ in emu
		7338 cm$^2/(\mathrm{V} \cdot \mathrm{s})$	$\overline{\sigma_1}$ in Å^2, $N = 2.687 \times 10^{19}$ cm^{-3}, T in eV, μ in emu
5	$\kappa = C\sqrt{T/m}/\overline{\sigma_2}$	1.743×10^4 W/(cm·K)	T in K, m in amu, $\overline{\sigma_2}$ in Å^2
		7.443×10^5 W/(cm·K)	T in K, m in emu, $\overline{\sigma_2}$ in Å^2
6	$\eta = C\sqrt{Tm}/\overline{\sigma_2}$	5.591×10^{-5} g/(cm·s)	T in K, m in amu, $\overline{\sigma_2}$ in Å^2
7	$\xi = CE/(TN\sigma)$	1.160×10^{20}	E in V/cm, T in K, σ in Å^2, N in cm^{-3}
		1×10^{16}	E in V/cm, T in eV, σ in Å^2, N in cm^{-3}
8	$v = C\rho r^2/\eta$	0.2179	r in µm, ρ in g/cm^3, η in 10^{-5} g/(cm·s)
		0.01178	r in µm, ρ in g/cm^3, η for air at $p = 1$ atm, $T = 300$ K

1. The Einstein relation for the diffusion coefficient $D = KT/e$, D, K are the diffusion coefficient and mobility of a charged particle, T is the temperature.
2. The Einstein relation for the mobility of a charged particle $K = eD/T$.
3. The diffusion coefficient in the Chapman–Enskog approximation $D = 3\sqrt{2\pi T/\mu}/(16N\overline{\sigma_1})$, T is the temperature, N is the number density of gas particles, μ is the reduced mass of colliding particles, $\overline{\sigma_1}$ is the average cross section of collision.
4. The mobility of a charged particle in the Chapman–Enskog approximation $K = 3e\sqrt{2\pi/(T\mu)}/(16N\overline{\sigma_1})$, the notations are given in the previous item.
5. The thermal conductivity of a gas in the Chapman–Enskog approximation $\kappa = 25\sqrt{\pi T}/(32\sqrt{m}\overline{\sigma_2})$, m is the mass of a gas particle, $\overline{\sigma_2}$ is the average cross section of collision of gas particles, other notations are given above.
6. The viscosity of a gas in the Chapman–Enskog approximation $\eta = 5\sqrt{\pi T m}/(24\overline{\sigma_2})$, the notations are given above.
7. The parameter of ion drift in a gas in an electric field $\xi = eE/(TN\sigma)$, E is the electric field strength, T is the temperature, N is the number density of gas particles, and σ is the cross section of ion scattering on gas particles.
8. The free fall velocity for a spherical bulk particle under the action of the gravitation force $v = 2\rho g r^2/(9\eta)$, where g is the free fall acceleration, ρ is the particle density, η is the viscosity of a matter, where the particle is falling, and r is the particle radius.

16.3.4
Appendix 3d: Relations for Clusters and Nanoparticles in a Plasma

Number	Formula	Factor C	Units used
1	$r_W = C\sqrt{m/\rho}$	0.6926 Å	m in amu, ρ in g/cm^3
	$n = C(r/r_W)^3$	4.189	r and r_W in Å
2	$k_o = Cr_W^2\sqrt{T/m}$	4.5713×10^{-12} cm^3/s	r_W in Å, T in K, m in amu
3	$v = C\rho r^2/\eta$	0.2179	r in µm, ρ in g/cm^3, η in 10^{-5} g/(cm·s)
		0.01178	r in µm, ρ in g/cm^3, η for air at $p = 1$ atm, $T = 300$ K
4	$D_o = C\sqrt{T/m}/(Nr_W^2)$	1.469×10^{21} cm^2/s	r_W in Å, N in cm^{-3}, T in K, m in amu
		0.508 cm^2/s	r_W in Å, $N = 2.687 \times 10^{19}$ cm^{-3}, T in K, m in amu
		54.69 cm^2/s	r_W in Å, $N = 2.687 \times 10^{19}$ cm^{-3}, T in eV, m in amu
5	$K_o = C(\sqrt{Tm}\, Nr_W^2)^{-1}$	1.364×10^{19} cm^2/(V · s)	r_W in Å, N in cm^{-3}, T in K, m in amu
		0.508 cm^2/(V · s)	r_W in Å, $N = 2.687 \times 10^{19}$ cm^{-3}, T in K, m in amu
		54.69cm^2/(V · s)	r_W in Å, $N = 2.687 \times 10^{19}$ cm^{-3}, T in K, m in amu

1. The Wigner–Seitz radius $r_W = (3m/4\pi\rho)^{1/3}$, where m is the mass of an individual atom and ρ is the density. The radius of a spherical cluster consisting of n atoms is $r = r_W n^{1/3}$.
2. The characteristic rate constant for a cluster $k_o = \pi r_W^2 \sqrt{8T/(\pi m)}$, where r_W is the Wigner–Seitz radius, T is the temperature, and m is the mass of a free atom.
3. The falling velocity for a spherical bulk particle under the action of the gravitation force $v = 2\rho g r^2/(9\eta)$, where g is the free fall acceleration, ρ is the particle density, η is the viscosity of a matter, where the particle is falling, and r is the particle radius.
4. The diffusion coefficient of a large spherical cluster consisting of n atoms is given by $D_n = D_o/n^{2/3}$, and in the Chapman–Enskog approximation we have $D_o = 3\sqrt{2T/\pi m}/(16 N r_W^2)$, where T is the temperature, N is the number density of gas particles, m is the mass of an individual cluster atom, and r_W is the Wigner–Seitz radius.
5. The mobility of a charged cluster consisting of n atoms in the Chapman–Enskog approximation is $K_n = K_o/n^{2/3}$, where $K_o = 3e/(8 N r_W^2 \sqrt{2\pi m T})$; the notations are given in the previous item.

16.4 Appendix 4: Transport Parameters of Gases

16.4.1 Appendix 4a: Self-diffusion Coefficients of Gases

Diffusion coefficients of atoms or molecules in parent gases are reduced to the number density $N = 2.687 \times 10^{19}$ cm^{-3} corresponding to the standard conditions ($T = 273$ K, $p = 1$ atm).

Gas	D (cm^2/s)	Gas	D (cm^2/s)	Gas	D (cm^2/s)
He	1.6	H_2	1.3	H_2O	0.28
Ne	0.45	N_2	0.18	CO_2	0.096
Ar	0.16	O_2	0.18	NH_3	0.25
Kr	0.084	CO	0.18	CH_4	0.20
Xe	0.048				

16.4.2
Appendix 4b: Gas-kinetic Cross Sections

The values of the gas-kinetic cross sections are obtained from the formula $\sigma_g = T/(\mu v_T N D)$, where D is the diffusion coefficient, T is the room temperature expressed in energy units, N is the number density of atoms or molecules, $v_T = \sqrt{8T/(\pi\mu)}$ is the average velocity of particles, where μ is their reduced mass. Gas-kinetic cross sections are expressed in 10^{-15} cm^2.

Pair	He	Ne	Ar	Kr	Xe	H_2	N_2	O_2	CO	CO_2
He	1.5	2.0	2.9	3.3	3.7	2.3	3.0	2.9	3.0	3.6
Ne		2.4	3.4	4.0	4.4	2.7	3.2	3.5	3.6	4.9
Ar			5.0	5.6	6.7	3.7	5.1	5.2	5.3	5.5
Kr				6.5	7.7	4.3	5.8	5.6	5.9	6.1
Xe					9.0	5.0	6.7	6.9	6.8	7.6
H_2						2.7	3.8	3.7	3.9	4.5
N_2							5.0	4.9	5.1	6.3
O_2								4.9	4.9	5.9
CO									5.0	6.3
CO_2										7.8

16.4.3
Appendix 4c: Thermal Conductivity Coefficients of Gases

The values of thermal conductivity coefficients relate to pressure of 1 atm and are expressed in 10^{-4} W/(cm · K).

T (K)	100	200	300	400	600	800	1000
H_2	6.7	13.1	18.3	22.6	30.5	37.8	44.8
He	7.2	11.5	15.1	18.4	25.0	30.4	35.4
CH_4	—	2.17	3.41	4.88	8.22	—	—
NH_3	—	1.53	2.47	6.70	6.70	—	—
H_2O	—	—	—	2.63	4.59	7.03	9.74
Ne	2.23	3.67	4.89	6.01	7.97	9.71	11.3
CO	0.84	1.72	2.49	3.16	4.40	5.54	6.61
N_2	0.96	1.83	2.59	3.27	4.46	5.48	6.47
Air	0.95	1.83	2.62	3.28	4.69	5.73	6.67
O_2	0.92	1.83	2.66	3.30	4.73	5.89	7.10
Ar	0.66	1.26	1.77	2.22	3.07	3.74	4.36
CO_2	—	0.94	1.66	2.43	4.07	5.51	6.82
Kr	—	0.65	1.00	1.26	1.75	2.21	2.62
Xe	—	0.39	0.58	0.74	1.05	1.35	1.64

16.4.4
Appendix 4d: Viscosity Coefficients of Gases

The values of viscosity coefficients correspond to pressure of 1 atm and are expressed in 10^{-5} g/(cm · s).

T (K)	100	200	300	400	600	800	1000
H_2	4.21	6.81	8.96	10.8	14.2	17.3	20.1
He	9.77	15.4	19.6	23.8	31.4	38.2	44.5
CH_4	—	7.75	11.1	14.1	19.3	—	—
H_2O	—	—	—	13.2	21.4	29.5	37.6
Ne	14.8	24.1	31.8	38.8	50.6	60.8	70.2
CO	—	12.7	17.7	21.8	28.6	34.3	39.2
N_2	6.88	12.9	17.8	22.0	29.1	34.9	40.0
Air	7.11	13.2	18.5	23.0	30.6	37.0	42.4
O_2	7.64	14.8	20.7	25.8	34.4	41.5	47.7
Ar	8.30	16.0	22.7	28.9	38.9	47.4	55.1
CO_2	—	9.4	14.9	19.4	27.3	33.8	39.5
Kr	—	—	25.6	33.1	45.7	54.7	64.6
Xe	—	—	23.3	30.8	43.6	54.7	64.6

16.5
Appendix 5: Atomic Parameters in the Form of Periodical Tables

16.5.1 Ionization Potentials of Atoms and Their Ions
16.5.2 Electron Affinities of Atoms
16.5.3 Lowest Excited States of Atoms
16.5.4 Splitting of Lowest Atom Levels
16.5.5 Resonantly Excited Atom States
16.5.6 Polarizabilities of Atoms and Diatomics
16.5.7 Affinity to Hydrogen and Oxygen Atoms
16.5.8 Diatomic Molecules
16.5.9 Positive Ions of Diatomics
16.5.10 Negative Ions of Diatomics

16.5.1
Ionization Potentials of Atoms and Their Ions

Period \ Group	I	II	III	IV	V
1	1.008 1s 13.598 $_{1}$H $^2S_{1/2}$ Hydrogen	24.588 $1s^2$ 4.003 54.418 $^1S_0$ $_{2}$He Helium			
2	6.491 2s 5.392 Li $^2S_{1/2}$ 75.64 122.4 Lithium	9.012 $2s^2$ 9.323 $_{4}$Be $^1S_0$ 18.211 153.9 Berillium 217.7	8.298 2p 10.81 25.155 $^2P_{1/2}$ $_{5}$B 37.931 259.38 Boron	11.260 $2p^2$ 12.011 24.384 $^3P_0$ $_{6}$C 47.89 64.49 Carbon	14.534 $2p^3$ 14.007 29.602 $^4S_{3/2}$ $_{7}$N 47.45 77.47 Nitrogen
3	22.990 3s 5.139 $_{11}$Na $^2S_{1/2}$ 47.287 71.620 Sodium 98.92	24.305 $3s^2$ 7.646 $_{12}$Mg $^1S_0$ 15.035 80.144 Magnesium 109.27	5.986 3p 26.982 18.829 $^2P_{1/2}$ $_{13}$Al 28.448 119.99 Aluminium	8.152 $3p^2$ 28.086 16.346 $^3P_0$ $_{14}$Si 33.493 45.142 Silicon	10.487 $3p^3$ 30.974 19.770 $^4S_{3/2}$ $_{15}$P 30.203 51.444 Phosphorus
4	39.098 4s 4.341 $_{19}$K $^2S_{1/2}$ 31.63 45.81 Potassium 60.91	40.08 $4s^2$ 6.113 $_{20}$Ca $^1S_0$ 11.872 50.913 Calsium 67.3	44.956 $3d4s^2$ 6.562 $_{21}$Sc $^2D_{3/2}$ 12.800 24.757 Scandium 73.49	47.88 $3d^24s^2$ 6.82 $_{22}$Ti $^3F_2$ 13.58 27.49 Titanium 43.27	50.942 $3d^34s^2$ 6.74 $_{23}$V $^4F_{3/2}$ 14.66 29.31 Vanadium 46.71
4	7.726 $3d^{10}4s$ 63.546 20.293 $^2S_{1/2}$ $_{29}$Cu 36.84 57.4 Copper	9.394 $4s^2$ 65.38 17.964 $^1S_0$ $_{30}$Zn 39.72 59.57 Zinc	5.999 4p 69.72 20.51 $^2P_{1/2}$ $_{31}$Ga 30.7 64.2 Gallium	7.900 $4p^2$ 72.59 15.935 $^3P_0$ $_{32}$Ge 34.2 45.7 Germanium	9.789 $4p^3$ 74.922 18.59 $^4S_{3/2}$ $_{33}$As 28.4 50.1 Arsenic
5	85.468 5s 4.177 $_{37}$Rb $^2S_{1/2}$ 27.290 39.2 Rubidium 52.6	87.62 $5s^2$ 5.694 $_{38}$Sr $^1S_0$ 11.030 42.88 Strontium 56.28	88.906 $4d5s^2$ 6.217 $_{39}$Y $^2D_{3/2}$ 12.24 20.525 Yttrium 60.61	91.22 $4d^25s^2$ 6.837 $_{40}$Zr $^3F_2$ 13.13 23.1 Zirconium 34.419	92.906 $4d^45s$ 6.88 $_{41}$Nb $^6D_{1/2}$ 14.32 25.0 Niobium 37.7
5	7.576 $4d^{10}5s$ 107.87 21.49 $^2S_{1/2}$ $_{47}$Ag 34.8 Silver	8.994 $5s^2$ 112.41 16.908 $^1S_0$ $_{48}$Cd 37.47 Cadmium	5.786 5p 114.82 18.87 $^2P_{1/2}$ $_{49}$In 28.0 57.0 Indium	7.344 $5p^2$ 118.69 14.632 $^3P_0$ $_{50}$Sn 30.50 40.74 Tin	8.609 $5p^3$ 121.75 16.53 $^4S_{3/2}$ $_{51}$Sb 25.32 44.16 Antimony
6	132.90 6s 3.894 $_{55}$Cs $^2S_{1/2}$ 23.15 33.4 Cesium 46	137.33 $6s^2$ 5.212 $_{56}$Ba $^1S_0$ 10.004 35.8 Barium 47	138.90 $5d6s^2$ 5.577 $_{57}$La $^2D_{3/2}$ 11.1 19.18 Lanthanum 49.9	178.49 $5d^26s^2$ 6.8 $_{72}$Hf $^3F_2$ 14.9 23.3 Hafnium 33.4	180.95 $5d^36s^2$ 7.89 $_{73}$Ta $^4F_{3/2}$ Tantalum
6	9.226 $5d^{10}6s$ 196.97 20.5 $^2S_{1/2}$ $_{79}$Au 34 43 Gold	10.438 $5d^{10}6s^2$ 200.59 18.76 $^1S_0$ $_{80}$Hg 34.2 46 Mercury	6.108 6p 204.38 20.43 $^2P_{1/2}$ $_{81}$Tl 29.85 Thallium	7.417 $6p^2$ 207.2 15.033 $^3P_0$ $_{82}$Pb 31.94 42.33 Lead	7.286 $6p^3$ 208.98 16.7 $^4S_{3/2}$ $_{83}$Bi 25.56 45.3 Bismuth
7	[223] 7s 4.0 $_{87}$Fr $^2S_{1/2}$ Francium	226.02 $7s^2$ 5.278 $_{88}$Ra $^1S_0$ 10.15 Radium	227.03 $6d7s^2$ 5.2 $_{89}$Ac $^2D_{3/2}$ 11.75 20 Actinium		

Actinides

$6d^27s^2$ 6.1 232.04 11.9 $_{90}$Th $^3F_2$ 18.3 28.7 Thorium	$5f^26d7s^2$ 6.0 231.04 $_{91}$Pa $^4K_{11/2}$ Protactinium	$5f^36d7s^2$ 6.194 238.03 11.9 $_{92}$U $^5L_6$ 20 37 Uranium	$5f^46d7s^2$ 6.266 237.05 $_{93}$Np $^6L_{11/2}$ Neptunium	$5f^67s^2$ 6.06 [244] $_{94}$Pu $^7F_0$ Plutonium	$5f^77s^2$ 6.0 [243] $_{95}$Am $^8S_{7/2}$ Americium

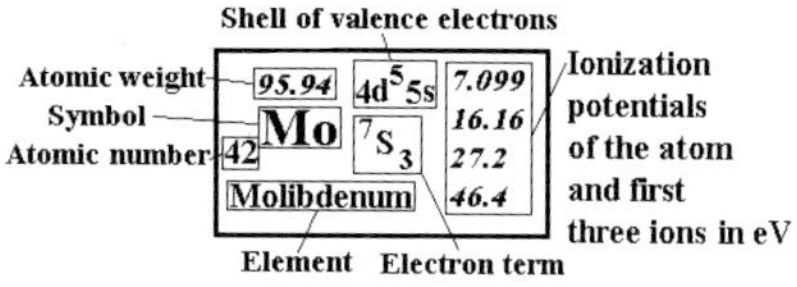

VI	VII	VIII		
15.999 $2p^4$ $^3P_2$ $_{8}$**O** Oxygen 13.618 35.118 54.936 77.414	18.998 $2p^5$ $^2P_{3/2}$ $_{9}$**F** Fluorine 17.423 34.971 62.71 87.14	20.179 $2p^6$ $^1S_0$ $_{10}$**Ne** Neon 21.565 40.963 63.46 97.12		
32.06 $3p^4$ $^3P_2$ $_{16}$**S** Sulfur 10.360 23.338 34.83 47.305	35.453 $3p^5$ $^2P_{3/2}$ $_{17}$**Cl** Chlorine 12.968 23.814 39.61 53.47	39.948 $2p^6$ $^1S_0$ $_{18}$**Ar** Argon 15.760 27.630 40.911 59.81		
51.996 $3d^54s$ $_{24}$**Cr** $^7S_3$ Chromium 6.766 16.50 31.0 49.2	54.938 $3d^54s^2$ $_{25}$**Mn** $^6S_{5/2}$ Manganese 7.434 15.640 33.67 51.2	55.847 $3d^64s^2$ $_{26}$**Fe** $^5D_4$ Iron 7.902 16.188 30.65 54.8	58.933 $3d^74s^2$ $_{27}$**Co** $^4F_{9/2}$ Cobalt 7.86 17.084 33.5 51.3	58.69 $3d^84s^2$ $_{28}$**Ni** $^3F_4$ Nickel 7.637 18.169 35.3 54.9
78.96 $4p^4$ $^3P_2$ $_{34}$**Se** Selenium 9.752 21.16 30.82 42.95	79.904 $4p^5$ $^2P_{3/2}$ $_{35}$**Br** Bromine 11.814 21.81 35.90 47.3	83.80 $4p^6$ $^1S_0$ $_{36}$**Kr** Krypton 14.000 24.360 36.95 52.5		
95.94 $4d^55s$ $_{42}$**Mo** $^7S_3$ Molibdenum 7.099 16.16 27.2 46.4	[98] $4d^55s^2$ $_{43}$**Tc** $^6S_{5/2}$ Technetium 7.28 15.26 29.5	101.07 $4d^75s$ $_{44}$**Ru** $^5F_5$ Ruthenium 7.366 16.76 28.5	102.91 $4d^85s$ $_{45}$**Rh** $^4F_{9/2}$ Rhodium 7.46 18.08 31.1	106.42 $4d^{10}$ $_{46}$**Pd** $^1S_0$ Palladium 8.336 19.43 32.9
127.60 $5p^4$ $^3P_2$ $_{52}$**Te** Tellurium 9.010 18.6 27.96 37.42	126.90 $5p^5$ $^2P_{3/2}$ $_{53}$**I** Iodine 10.451 19.131 33.0	131.29 $5p^6$ $^1S_0$ $_{54}$**Xe** Xenon 12.130 20.98 31.0 45		
183.85 $5d^46s^2$ $_{74}$**W** $^5D_0$ Tungsten 7.98	186.21 $5d^56s^2$ $_{75}$**Re** $^6S_{5/2}$ Rhenium 7.88	190.2 $5d^66s^2$ $_{76}$**Os** $^5D_4$ Osmium 8.73	192.22 $5d^76s^2$ $_{77}$**Ir** $^4F_{9/2}$ Iridium 9.05	195.08 $5d^96s$ $_{78}$**Pt** $^3D_3$ Platinum 8.96 18.56
[209] $6p^4$ $^3P_2$ $_{84}$**Po** Polonium 8.417	[210] $6p^5$ $^2P_{3/2}$ $_{85}$**At** Astatine 9.0	[222] $6p^6$ $^1S_0$ $_{86}$**Rn** Radon 10.75		

Lantanides

140.12 $4f5d6s^2$ $_{58}$**Ce** $^1G_4$ Cerium 5.539 10.8 20.20 39.76	140.91 $4f^36s^2$ $_{59}$**Pr** $^4I_{9/2}$ Praseodymium 5.47 10.6 21.62 38.98	144.24 $4f^46s^2$ $_{60}$**Nd** $^5I_4$ Neodymium 5.525 10.7 22.1 40.4	[145] $4f^56s^2$ $_{61}$**Pm** $^6H_{5/2}$ Promethium 5.58 10.9 22.3 41.0	150.36 $4f^66s^2$ $_{62}$**Sm** $^7F_0$ Samarium 5.644 11.1 23.4 41.4	151.96 $4f^76s^2$ $_{63}$**Eu** $^8S_{7/2}$ Europium 5.670 11.24 24.9 42.7	157.25 $4f^75d6s^2$ $_{64}$**Gd** $^9D_2$ Gadolinium 6.150 12.1 20.6 44.0
158.92 $4f^96s^2$ $_{65}$**Tb** $^6H_{15/2}$ Terbium 5.864 11.5 21.9 39.4	162.50 $4f^{10}6s^2$ $_{66}$**Dy** $^5I_8$ Dysprosium 5.939 11.7 22.8 41.4	164.93 $4f^{11}6s^2$ $_{67}$**Ho** $^5I_{15/2}$ Holmium 6.022 11.8 22.8 42.5	167.26 $4f^{12}6s^2$ $_{68}$**Er** $^3H_6$ Erbium 6.108 11.9 22.7 42.7	168.93 $4f^{13}6s^2$ $_{69}$**Tm** $^2F_{7/2}$ Thulium 6.184 12.1 23.7 42.7	173.04 $4f^{14}6s^2$ $_{70}$**Yb** $^1S_0$ Ytterbium 6.254 12.18 25.05 43.6	174.97 $4f^{14}5d6s^2$ $_{71}$**Lu** $^2D_{3/2}$ Lutetium 5.426 13.9 20.96 45.25

16.5.2
Electron Affinities of Atoms

Period \ Group	I	II	III	IV	V
1	$_{1}$H Hydrogen $1s^2\ {}^1S_0$ 0.75420				
2	$_{3}$Li Lithium $2s^2\ {}^1S_0$ 0.61805	$_{4}$Be Berillium $2p\ {}^2P_{1/2}$ absent	$_{5}$B Boron $2p^2\ {}^3P_0$ 0.27972	$_{6}$C Carbon $2p^3$ ${}^4S_{3/2}$ 1.2621 ${}^2D_{3/2}$ 0.033	$_{7}$N Nitrogen $2p^4\ {}^3P_0$ absent
3	$_{11}$Na Sodium $3s^2\ {}^1S_0$ 0.54793	$_{12}$Mg Magnesium $3p\ {}^2P_{1/2}$ absent	$_{13}$Al Aluminium $3p^2$ ${}^3P_0$ 0.4328 ${}^1D_2$ 0.11	$_{14}$Si Silicon $3p^3$ ${}^4S_{3/2}$ 1.3895 ${}^2D_{3/2}$ 0.5272 ${}^2D_{5/2}$ 0.5255 ${}^2P_{1/2}$ 0.029	$_{15}$P Phosphorus $3p^4\ {}^3P_0$ 0.7465
4	$_{19}$K Potassium $4s^2\ {}^1S_0$ 0.50146	$_{20}$Ca Calsium $4p\ {}^2P_{1/2}$ 0.0245	$_{21}$Sc Scandium $3d4s^24p$ ${}^3F_2$ 0.19 ${}^3D_1$ 0.04	$_{22}$Ti Titanium $3d^34s^2\ {}^4F_{3/2}$ 0.08	$_{23}$V Vanadium $4d^44s^2\ {}^5D_0$ 0.52
	$_{29}$Cu Copper $3d^{10}4s^2\ {}^1S_0$ 1.23578	$_{30}$Zn Zinc $4p\ {}^2P_{1/2}$ absent	$_{31}$Ga Gallium $4p^2\ {}^3P_0$ 0.43	$_{32}$Ge Germanium $4p^3$ ${}^4S_{3/2}$ 1.2327 ${}^2D_{3/2}$ 0.4014 ${}^2D_{5/2}$ 0.3773	$_{33}$As Arsenic $4p^4\ {}^3P_0$ 0.81
5	$_{37}$Rb Rubidium $5s^2\ {}^1S_0$ 0.48592	$_{38}$Sr Strontium $5p\ {}^2P_{1/2}$ 0.0520	$_{39}$Y Yttrium $4d5s^25p$ ${}^3F_2$ 0.31 ${}^3D_1$ 0.16	$_{40}$Zr Zirconium $4d^35s^2\ {}^4F_{3/2}$ 0.43 52	$_{41}$Nb Niobium $5d^45s^2\ {}^5D_0$ 0.89
	$_{47}$Ag Silver $4d^{10}5s^2\ {}^1S_0$ 1.30447	$_{48}$Cd Cadmium $5p\ {}^2P_{1/2}$ absent	$_{49}$In Indium $5p^2\ {}^3P_0$ 0.40	$_{50}$Sn Tin $5p^3$ ${}^4S_{3/2}$ 1.1121 ${}^2D_{3/2}$ 0.3976 ${}^2D_{5/2}$ 0.3046	$_{51}$Sb Antimony $5p^4$ ${}^3P_2$ 1.0474 ${}^3P_1$ 0.714 ${}^3P_0$ 0.700 ${}^1D_2$ 0.1308
6	$_{55}$Cs Cesium $6s^2\ {}^1S_0$ 0.47163	$_{56}$Ba Barium $6p\ {}^2P_{1/2}$ 0.1446	$_{57}$La Lanthanum $5d^26s^2\ {}^3F_2$ 0.47	$_{58}$Hf Hafnium $5d^36s^2\ {}^4F_{3/2}$ absent	$_{73}$Ta Tantalum $5d^46s^2\ {}^5D_0$ 0.32
	$_{79}$Au Gold $5d^{10}6s^2\ {}^1S_0$ 2.30863	$_{80}$Hg Mercury $6p\ {}^2P_{1/2}$ absent	$_{81}$Tl Thallium $6p^2\ {}^3P_0$ 0.38	$_{82}$Pb Lead $6p^3\ {}^4S_{3/2}$ 0.364	$_{83}$Bi Bismuth $6p^4\ {}^3P_0$ 0.94236
7	$_{87}$Fr Francium $7s^2\ {}^1S_0$ 0.46				

VI	VII	VIII		
		$2s\ ^2S_{1/2}$ absent $_2$**He** Helium		
$2p^5\ ^2P_{3/2}$ 1.46111 $_8$**O** Oxygen	$2p^6\ ^1S_0$ 3.40119 $_9$**F** Fluorine	$2p^63s\ ^2S_{1/2}$ absent $_{10}$**Ne** Neon		
$3p^5\ ^2P_{3/2}$ 2.07710 $_{16}$**S** Sulfur	$3p^6\ ^1S_0$ 3.61269 $_{17}$**Cl** Chlorine	$3p^64s\ ^2S_{1/2}$ absent $_{18}$**Ar** Argon		
$3d^54s^2\ ^6S_{5/2}$ $_{24}$**Cr** 0.6758 Chromium	$3d^64s^2\ ^5D_4$ $_{25}$**Mn** absent Manganese	$3d^74s^2\ ^4F_{9/2}$ $_{26}$**Fe** 0.151 Iron	$3d^84s^2\ ^3F_4$ $_{27}$**Co** 0.663 Cobalt	$3d^94s^2\ ^2D_{5/2}$ $_{28}$**Ni** 1.1572 Nickel
$4p^5\ ^2P_{3/2}$ 2.02067 $_{34}$**Se** Selenium	$4p^6\ ^1S_0$ 3.36359 $_{37}$**Br** Bromine	$4p^65s\ ^2S_{1/2}$ absent $_{36}$**Kr** Krypton		
$4d^55s^2\ ^6S_{5/2}$ $_{42}$**Mo** 0.7472 Molibdenum	$4d^65s^2\ ^5D_4$ $_{43}$**Tc** 0.6 Technetium	$4d^75s^2\ ^4F_{9/2}$ $_{44}$**Ru** 1.0464 Ruthenium	$4d^85s^2\ ^3F_4$ $_{45}$**Rh** 1.1429 Rhodium	$4d^{10}5s\ ^2S_{1/2}$ 0.5621 $_{46}$**Pd** $4d^95s^2\ ^2D_{5/2}$ Palladium 0.4224
$5p^5\ ^2P_{3/2}$ 1.97087 $_{52}$**Te** Tellurium	$5p^6\ ^1S_0$ 3.05904 $_{53}$**I** Iodine	$5p^66s\ ^2S_{1/2}$ absent $_{54}$**Xe** Xenon		
$5d^56s^2\ ^6S_{5/2}$ $_{74}$**W** 0.815 Tungsten	$5d^66s^2\ ^5D_4$ $_{75}$**Re** 0.2 Rhenium	$5d^76s^2\ ^4F_{9/2}$ $_{76}$**Os** 1.0778 Osmium	$5d^86s^2\ ^3F_4$ $_{77}$**Ir** 1.5643 Iridium	$5d^96s^2\ ^2D_{5/2}$ $_{78}$**Pt** 2.1251 Platinum
$6p^5\ ^2P_{3/2}$ 1.9 $_{84}$**Po** Polonium	$6p^6\ ^1S_0$ 2.8 $_{85}$**At** Astatine	$6p^67s\ ^2S_{1/2}$ absent $_{86}$**Rn** Radon		

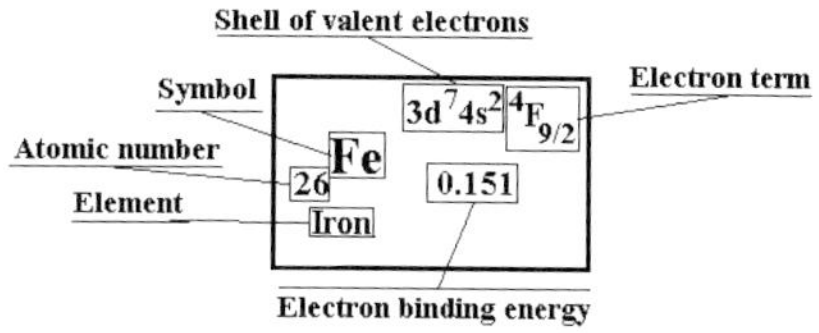

16.5.3 Lowest Excited States of Atoms

Group / Period	I	II	III	IV	V
1	$1s\,^2S_{1/2}$ $_1$H n=2 10.199 n=3 12.088 n=4 12.479 Hydrogen	$2s\,^3S_1$ 19.820 $2s\,^1S_0$ 20.616 $2p\,^3P_2$ 20.964 $2p\,^1P_1$ 21.218 $1s^2$ $^1S_0$ $_2$He Helium			
2	$2s\,^2S_{1/2}$ Li $2p\,^2P_{1/2}$ 1.848 $3s\,^2S_{1/2}$ 3.373 Lithium	$2s^2$ $_4$Be $^1S_0$ $2\,^3P_0$ 2.725 $2\,^1P_1$ 5.278 $2\,^3S_1$ 6.457 Berillium	2p $_5$B $^2P_{1/2}$ $^4P_{1/2}$ 3.579 $^2S_{1/2}$ 4.964 $^2D_{3/2}$ 5.934 Boron	$2p^2$ $^3P_0$ $_6$C $2\,^1D_2$ 1.264 $2\,^1S_0$ 2.684 Carbon	$2p^3$ $^4S_{3/2}$ $_7$N $2\,^2D_{3/2}$ 2.384 $2\,^2P_{1/2}$ 3.576 Nitrogen
3	$3s\,^2S_{1/2}$ $_{11}$Na $3p\,^2P_{1/2}$ 2.102 $3p\,^2P_{3/2}$ 2.104 $4s\,^2S_{1/2}$ 2.607 Sodium	$3s^2$ $^1S_0$ $_{12}$Mg $3\,^3P_0$ 2.709 $3\,^1P_1$ 4.316 $3\,^3S_1$ 5.108 Magnesium	3p $^2P_{1/2}$ $_{13}$Al $^2S_{1/2}$ 3.143 $^4P_{1/2}$ 3.608 $^2D_{3/2}$ 4.022 $^2P_{1/2}$ 4.085 Aluminium	$3p^2$ $^3P_0$ $_{14}$Si $^1D_2$ 0.781 $^1S_0$ 1.909 Silicon	$3p^3$ $^4S_{3/2}$ $_{15}$P $^2D_{3/2}$ 1.409 $^2P_{1/2}$ 2.321 Phosphorus
4	$4s\,^2S_{1/2}$ $_{19}$K $4p\,^2P_{1/2}$ 1.610 $4p\,^2P_{3/2}$ 1.617 $5s\,^2S_{1/2}$ 2.607 Potassium f (4s-4p) = 1.05	$4s^2$ $_{20}$Ca $^1S_0$ $^3P_0$ 1.879 $^1P_1$ 2.521 $^1D_2$ 2.709 Calsium	$3d4s^2$ $_{21}$Sc $^2D_{3/2}$ $^4F_{3/2}$ 1.439 $^2F_{5/2}$ 1.859 $^4F_{3/2}$ 1.968 $^4D_{1/2}$ 1.985 Scandium	$3d^24s^2$ $_{22}$Ti $^3F_2$ $^5F_1$ 0.813 $^1D_2$ 0.900 $^3P_0$ 1.046 $^3F_2$ 1.430 Titanium	$3d^34s^2$ $_{23}$V $^4F_{3/2}$ $^6D_{1/2}$ 0.262 $^4D_{1/2}$ 1.043 $^4P_{1/2}$ 1.195 Vanadium
	$3d^{10}4s\,^2S_{1/2}$ $_{29}$Cu $^2D_{5/2}$ 1.389 $^2D_{3/2}$ 1.642 $^2P_{1/2}$ 3.786 $^2P_{3/2}$ 3.817 Copper	$4s^2$ $^1S_0$ $_{30}$Zn $^3P_0$ 4.006 $^1P_1$ 5.796 $^3S_1$ 6.654 Zinc	4p $^2P_{1/2}$ $_{31}$Ga $^2S_{1/2}$ 3.073 $^2P_{1/2}$ 4.097 $^2D_{3/2}$ 4.312 Gallium	$4p^2$ $^3P_0$ $_{32}$Ge $^1D_2$ 0.883 $^1S_0$ 2.029 $^3P_0$ 4.643 $^1P_1$ 4.962 Germanium	$4p^3$ $^4S_{3/2}$ $_{33}$As $^2D_{3/2}$ 1.313 $^2P_{1/2}$ 2.255 6.285 Arsenic
5	5s $^2S_{1/2}$ $_{37}$Rb $5p\,^2P_{1/2}$ 1.560 $5p\,^2P_{3/2}$ 1.589 $4d\,^2D_{3/2}$ 2.400 Rubidium	$5s^2$ $^1S_0$ $_{38}$Sr $^3P_0$ 1.775 $^3D_1$ 2.251 $^1D_2$ 2.498 $^1P_1$ 2.690 Strontium	$4d5s^2$ $^2D_{3/2}$ $_{39}$Y $^2D_{5/2}$ 0.066 $^2P_{1/2}$ 1.306 $^4F_{3/2}$ 1.356 Yttrium	$4d^25s^2$ $^3F_2$ $_{40}$Zr $^3F_3$ 0.071 $^3F_4$ 0.154 $^3P_2$ 0.519 Zirconium	$4d^45s$ $^6D_{1/2}$ $_{41}$Nb $^4F_{3/2}$ 0.142 $^4P_{1/2}$ 0.620 Niobium
	$4d^{10}5s$ $^2S_{1/2}$ $_{47}$Ag $^2P_{1/2}$ 3.664 $^2D_{5/2}$ 3.750 $^2P_{3/2}$ 3.778 $^2D_{3/2}$ 4.304 Silver	$5s^2$ $^1S_0$ $_{48}$Cd $^3P_0$ 3.734 $^1P_1$ 5.417 $^3S_1$ 6.381 $^1S_0$ 6.610 Cadmium	5p $^2P_{1/2}$ $_{49}$In $^2S_{1/2}$ 3.022 $^2P_{1/2}$ 3.945 $^2D_{3/2}$ 4.078 Indium	$5p^2$ $^3P_0$ $_{50}$Sn $^1D_2$ 1.068 $^1S_0$ 2.128 $^3P_0$ 4.295 $^1P_1$ 4.867 Tin	$5p^3$ $^4S_{3/2}$ $_{51}$Sb $^2D_{3/2}$ 1.055 $^2P_{1/2}$ 2.033 Antimony
6	6s $^2S_{1/2}$ $_{55}$Cs $6p\,^2P_{1/2}$ 1.386 $6p\,^2P_{3/2}$ 1.455 $5d\,^2D_{3/2}$ 1.798 Cesium	$6s^2$ $^1S_0$ $_{56}$Ba $^3D_1$ 1.120 $^3P_0$ 1.521 $^1P_1$ 2.239 Barium	$5d6s^2$ $^2D_{3/2}$ $_{57}$La $^2D_{5/2}$ 0.131 $^4F_{3/2}$ 0.331 $^2F_{5/2}$ 0.869 $^4P_{1/2}$ 0.897 Lanthanum	$5d^26s^2$ $^3F_2$ $_{72}$Hf $^3P_0$ 0.685 $^1D_2$ 0.699 $^1G_4$ 1.306 Hafnium	$5d^36s^2$ $^4F_{3/2}$ $_{73}$Ta $^4F_{5/2}$ 0.249 $^4P_{1/2}$ 0.620 Tantalum
	$5d^{10}6s$ $^2S_{1/2}$ $_{79}$Au $^2D_{5/2}$ 1.136 $^2D_{3/2}$ 2.658 $^2P_{1/2}$ 4.632 $^2P_{3/2}$ 5.105 Gold	$5d^{10}6s^2$ $^1S_0$ $_{80}$Hg $6\,^3P_0$ 4.667 $6\,^3P_1$ 4.887 $6\,^3P_2$ 5.461 $6\,^1P_1$ 6.704 Mercury	6p $^2P_{1/2}$ $_{81}$Tl $6\,^2P_{3/2}$ 0.966 $7\,^2S_{1/2}$ 3.283 $7\,^2P_{1/2}$ 4.235 $7\,^2P_{3/2}$ 4.359 Thallium	$6p^2$ $^3P_0$ $_{82}$Pb $^3P_1$ 0.970 $^3P_2$ 1.320 $^1D_2$ 2.660 $^1S_0$ 3.653 Lead	$6p^3$ $^4S_{3/2}$ $_{83}$Bi $^2D_{3/2}$ 1.416 $^2P_{1/2}$ 2.686 $^4P_{1/2}$ 4.040 Bismuth

Legend:

- Shell of valence electrons — $4d^2 5s^2$
- Excitation energies in eV
- Symbol — Zr
- Atomic number — 40
- Electron term of the ground state — $^3F_2$
- Element — Zirconium
- Electron terms of excited states — $^3F_3$ 0.071, $^3F_4$ 0.154, $^3P_2$ 0.519

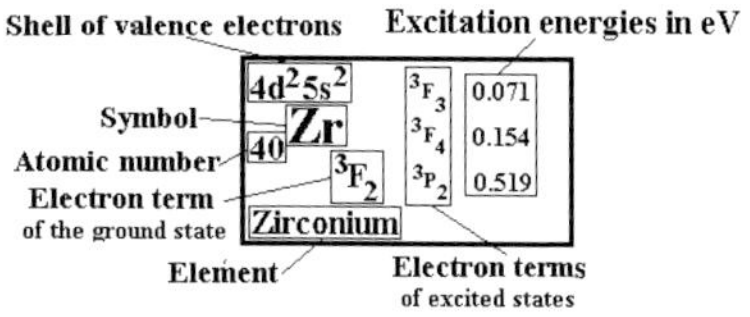

Notations of LS-coupling scheme are used for electron terms.

VI	VII	VIII		
$2\,^1D_2$ 1.967 $2p^4$ $2\,^1S_0$ 4.190 $^3P_2$ $_8$**O** Oxygen	$2\,^2P_{1/2}$ 0.050 $2p^5$ $3\,^4P_{5/2}$ 12.717 $^2P_{3/2}$ $_9$**F** $3\,^2P_{3/2}$ 12.999 Fluorine	$^3P_2$ 16.619 $^1S_0$ $2p^6$ $^3P_1$ 16.671 $_{10}$**Ne** $^3P_0$ 16.716 $^1P_1$ 16.848 Neon		
$^1D_2$ 1.145 $3p^4$ $^3P_2$ $_{16}$**S** $^1S_0$ 2.750 Sulfur	$^2P_{1/2}$ 0.109 $3p^5$ $^4P_{5/2}$ 8.922 $^2P_{3/2}$ $_{17}$**Cl** $^2P_{3/2}$ 9.282 Chlorine	$^3P_2$ 11.548 $^1S_0$ $3p^6$ $^3P_1$ 11.624 $_{18}$**Ar** $^3P_0$ 11.723 $^1P_1$ 11.828 Argon		
$3d^54s$ $^5S_2$ 0.941 $_{24}$**Cr** $^5D_0$ 0.961 $^7S_3$ Chromium	$3d^54s^2$ $^6D_{9/2}$ 2.114 $_{25}$**Mn** $^8P_{5/2}$ 2.282 $^6S_{5/2}$ Manganese	$3d^64s^2$ $^5F_5$ 0.859 $_{26}$**Fe** $^3F_4$ 1.485 $^5D_4$ $^5P_3$ 2.176 Iron	$3d^74s^2$ $^4F_{9/2}$ 0.432 $_{27}$**Co** $^2F_{7/2}$ 0.923 $^4F_{9/2}$ Cobalt	$3d^84s^2$ $^3D_3$ 0.025 $_{28}$**Ni** $^3F_4$ $^1D_2$ 1.676 $^1S_0$ 1.826 $^3P_2$ 1.935 Nickel
$^1D_2$ 1.187 $4p^4$ $^1S_0$ 2.783 $^3P_2$ **Se** $^5S_2$ 5.974 $_{34}$ $^5P_1$ 7.345 Selenium	$^2P_{1/2}$ 0.457 $4p^5$ $^4P_{5/2}$ 7.864 $^2P_{3/2}$ $_{35}$**Br** $^2P_{3/2}$ 8.329 Bromine	$^3P_2$ 9.915 $^1S_0$ $4p^6$ $^3P_1$ 10.032 $_{36}$**Kr** $^3P_0$ 10.562 $^1P_1$ 10.644 Krypton		
$4d^55s$ $^5S_2$ 1.135 $_{42}$**Mo** $^5D_0$ 1.360 $^7S_3$ $^5G_2$ 2.063 $^5P_3$ 2.260 Molybdenum	$4d^55s^2$ $^6D_{9/2}$ 0.319 $_{43}$**Tc** $^4D_{7/2}$ 1.304 $^6S_{5/2}$ Technetium	$4d^75s$ $^5F_5$ $^5F_4$ 0.148 $_{44}$**Ru** $^5F_3$ 0.259 $^5F_2$ 0.336 $^5F_1$ 0.385 Ruthenium	$4d^85s$ $^4F_{9/2}$ $^2D_{5/2}$ 0.410 $_{45}$**Rh** $^2F_{7/2}$ 0.693 Rhodium	$4d^{10}$ $^1S_0$ $^3D_3$ 0.814 $_{46}$**Pd** $^3D_2$ 0.962 $^3D_1$ 1.251 $^1D_2$ 1.453 Palladium
$^3P_0$ 0.584 $5p^4$ $^3P_1$ 0.589 $^3P_2$ $^1D_2$ 1.300 $_{52}$**Te** $^1S_0$ 2.876 Tellurium	$^2P_{1/2}$ 0.943 $5p^5$ $^4P_{5/2}$ 6.774 $^2P_{3/2}$ $_{53}$**I** $^2P_{3/2}$ 7.665 Iodine	$^3P_2$ 8.315 $^1S_0$ $5p^6$ $^3P_1$ 8.437 $_{54}$**Xe** $^3P_0$ 9.447 $^1P_1$ 9.570 Xenon		
$5d^46s^2$ $^5D_1$ 0.207 $^5D_0$ $^7S_3$ 0.366 $_{74}$**W** $^3P_0$ 1.181 Tungsten	$5d^56s^2$ $^4P_{3/2}$ 1.436 $_{75}$**Re** $^6S_{5/2}$ $^6D_{9/2}$ 1.457 $^4G_{5/2}$ 1.813 Rhenium	$5d^66s^2$ $^5D_3$ 0.340 $_{76}$**Os** $^5D_4$ $^5D_2$ 0.516 $^5F_5$ 0.638 Osmium	$5d^76s^2$ $_{77}$**Ir** $^4F_{9/2}$ $5d^86s\,^4F_{9/2}$ 0.352 $5d^76s^2\,^4F_{3/2}$ 0.508 Iridium	$5d^96s$ $^3F_4$ 0.102 $_{78}$**Pt** $^1S_0$ 0.761 $^3D_3$ $^3P_2$ 0.814 Platinum

16.5.4
Splitting of Lowest Atom Levels

Group / Period	I	II	III	IV	V
1	$_1$H Hydrogen $1s\,^2S_{1/2}$ $2p(^2P_{1/2}-{}^2P_{3/2})$ 0.365 $2p\,^2P_{1/2}-2s\,^2S_{1/2}$ 0.035	$_2$He Helium $1s^2$ $^1S_0$ $^3P_0-{}^3P_1$ 0.08 $^3P_0-{}^3P_2$ 1.06			
2	Li Lithium $2s\,^2S_{1/2}$ $^2P_{1/2}-{}^2P_{3/2}$ 0.3	$_4$Be Berillium $2s^2$ $^1S_0$ $^3P_0-{}^3P_1$ 0.6 $^3P_0-{}^3P_2$ 3.0	$_5$B Boron $2p$ $^2P_{1/2}$ $^2P_{1/2}-{}^2P_{3/2}$ 15.25	$_6$C Carbon $2p^2$ $^3P_0$ $^3P_0-{}^3P_1$ 16.4 $^3P_0-{}^3P_2$ 43.4	$_7$N Nitrogen $2p^3$ $^4S_{3/2}$ $^2D_{3/2}-{}^2D_{5/2}$ 8.72
3	$_{11}$Na Sodium $3s\,^2S_{1/2}$ $^2P_{1/2}-{}^2P_{3/2}$ 17.2	$_{12}$Mg Magnesium $3s^2\,^1S_0$ $^3P_0-{}^3P_1$ 20.05 $^3P_0-{}^3P_2$ 61.75	$_{13}$Al Aluminium $3p$ $^2P_{1/2}$ $^2P_{1/2}-{}^2P_{3/2}$ 112.1 $^4P_{1/2}-{}^4P_{3/2}$ 46.6 $^4P_{1/2}-{}^4P_{5/2}$ 122.4	$_{14}$Si Silicon $3p^2$ $^3P_0$ $^3P_0-{}^3P_1$ 77.1 $^3P_0-{}^3P_2$ 223.2	$_{15}$P Phosphorus $3p^3$ $^4S_{3/2}$ $^2D_{3/2}-{}^2D_{5/2}$ 15.6
4	$_{19}$K Potassium $4s\,^2S_{1/2}$ $^2P_{1/2}-{}^2P_{3/2}$ 57.7 $^2D_{5/2}-{}^2D_{3/2}$ 2.31	$_{20}$Ca Calsium $4s^2$ $^1S_0$ $^3P_0-{}^3P_1$ 52.16 $^3P_0-{}^3P_2$ 158.04	$_{21}$Sc Scandium $3d4s^2$ $^2D_{3/2}$ $^2D_{3/2}-{}^2D_{5/2}$ 168.3 $^4F_{3/2}-{}^4F_{5/2}$ 37.7 $^4F_{3/2}-{}^4F_{7/2}$ 90.3 $^4F_{3/2}-{}^4F_{9/2}$ 157.4	$_{22}$Ti Titanium $3d^24s^2$ $^3F_2$ $^3F_2-{}^3F_3$ 170.1 $^3F_2-{}^3F_4$ 386.9	$_{23}$V Vanadium $3d^34s^2$ $^4F_{3/2}$ $^4F_{3/2}-{}^4F_{5/2}$ 137.4 $^4F_{3/2}-{}^4F_{7/2}$ 323.5 $^4F_{3/2}-{}^4F_{9/2}$ 553.0
	$_{29}$Cu Copper $3d^{10}4s\,^2S_{1/2}$ $^2D_{5/2}-{}^2D_{3/2}$ 2042.9 $^2P_{1/2}-{}^2P_{3/2}$ 148.4	$_{30}$Zn Zinc $4s^2$ $^1S_0$ $^3P_0-{}^3P_1$ 190.1 $^3P_0-{}^3P_2$ 579.0	$_{31}$Ga Gallium $4p$ $^2P_{1/2}$ $4p(^2P_{1/2}-{}^2P_{3/2})$ 826.2 $5p(^2P_{1/2}-{}^2P_{3/2})$ 110.9	$_{32}$Ge Germanium $4p^2$ $^3P_0$ $^3P_0-{}^3P_1$ 557.1 $^3P_0-{}^3P_2$ 1410.0	$_{33}$As Arsenic $4p^3$ $^4S_{3/2}$ $^2D_{3/2}-{}^2D_{5/2}$ 323
5	$_{37}$Rb Rubidium $5s\,^2S_{1/2}$ $^2P_{1/2}-{}^2P_{3/2}$ 237.6 $^2D_{5/2}-{}^2D_{3/2}$ 0.45	$_{38}$Sr Strontium $5s^2$ $^1S_0$ $^3P_0-{}^3P_1$ 186.8 $^3P_0-{}^3P_2$ 581.0	$_{39}$Y Yttrium $4d5s^2$ $^2D_{3/2}$ $^2D_{5/2}-{}^2D_{3/2}$ 530.4	$_{40}$Zr Zirconium $4d^25s^2$ $^3F_2$ $^3F_2-{}^3F_3$ 570.4 $^3F_2-{}^3F_4$ 1240.8	$_{41}$Nb Niobium $4d^45s$ $^6D_{1/2}$ $^6D_{1/2}-{}^6D_{3/2}$ 154.2 $^6D_{1/2}-{}^6D_{5/2}$ 392.0 $^6D_{1/2}-{}^6D_{7/2}$ 695.2 $^6D_{1/2}-{}^6D_{9/2}$ 1050.3
	$_{47}$Ag Silver $4d^{10}5s\,^2S_{1/2}$ $^2D_{5/2}-{}^2D_{3/2}$ 4418 $^2P_{1/2}-{}^2P_{3/2}$ 920.7	$_{48}$Cd Cadmium $5s^2$ $^1S_0$ $^3P_0-{}^3P_1$ 542.1 $^3P_0-{}^3P_2$ 1713.0	$_{49}$In Indium $5p$ $^2P_{1/2}$ $^2P_{1/2}-{}^2P_{3/2}$ 2212.6	$_{50}$Sn Tin $5p^2$ $^3P_0$ $^3P_0-{}^3P_1$ 1692 $^3P_0-{}^3P_2$ 3428	$_{51}$Sb Antimony $5p^3$ $^4S_{3/2}$ $^2D_{3/2}-{}^2D_{5/2}$ 1342
6	$_{55}$Cs Cesium $6s\,^2S_{1/2}$ $^2P_{1/2}-{}^2P_{3/2}$ 554 $^2D_{3/2}-{}^2D_{5/2}$ 96.6	$_{56}$Ba Barium $6s^2$ $^1S_0$ $^3D_1-{}^3D_2$ 181.5 $^3D_1-{}^3D_3$ 562.6 $^3P_0-{}^3P_1$ 370.0 $^3P_0-{}^3P_2$ 1248.7	$_{57}$La Lanthanum $5d6s^2$ $^2D_{3/2}$ $^2D_{5/2}-{}^2D_{3/2}$ 1053.2	$_{72}$Hf Hafnium $5d^26s^2$ $^3F_2$ $^3F_2-{}^3F_3$ 2356.7 $^3F_2-{}^3F_4$ 4567.6	$_{73}$Ta Tantalum $5d^36s^2$ $^4F_{3/2}$ $^4F_{3/2}-{}^4F_{5/2}$ 2010.1 $^4F_{3/2}-{}^4F_{7/2}$ 3963.9 $^4F_{3/2}-{}^4F_{9/2}$ 5621.0
	$_{79}$Au Gold $5d^{10}6s\,^2S_{1/2}$ $^2D_{5/2}-{}^2D_{3/2}$ 12274 $^2P_{1/2}-{}^2P_{3/2}$ 3815	$_{80}$Hg Mercury $5d^{10}6s^2$ $^1S_0$ $^3P_0-{}^3P_1$ 1767.2 $^3P_0-{}^3P_2$ 6397.9	$_{81}$Tl Thallium $6p$ $^2P_{1/2}$ $^2P_{1/2}-{}^2P_{3/2}$ 7793	$_{82}$Pb Lead $6p^2$ $^3P_0$ $^3P_0-{}^3P_1$ 7819.3 $^3P_0-{}^3P_2$ 10650.3	$_{83}$Bi Bismuth $6p^3$ $^4S_{3/2}$ $^2D_{3/2}-{}^2D_{5/2}$ 4019

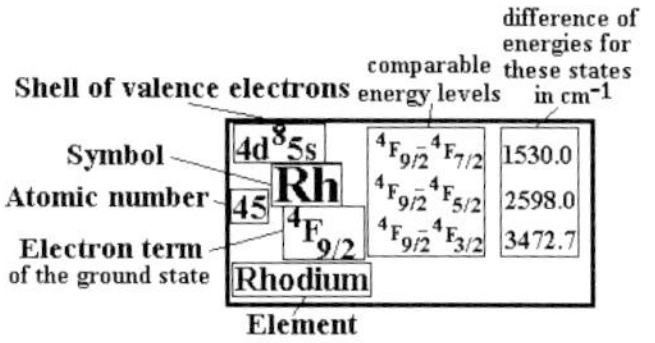

VI	VII	VIII		
$2p^4$ 8 **O** $^3P_2$ Oxygen $^3P_2$-$^3P_1$ 158.3 $^3P_2$-$^3P_0$ 227.0	$2p^5$ 9 **F** $^2P_{3/2}$ Fluorine $^2P_{3/2}$-$^2P_{1/2}$ 404.1	$2p^6$ 10 **Ne** $^1S_0$ Neon $^3P_2$-$^3P_1$ 417.4 $^3P_2$-$^3P_0$ 776.8 $^3P_2$-$^1P_1$ 1846.9		
$3p^4$ 16 **S** $^3P_2$ Sulfur $^3P_2$-$^3P_1$ 396.0 $^3P_2$-$^3P_0$ 573.6	$3p^5$ 17 **Cl** $^2P_{3/2}$ Chlorine $^2P_{3/2}$-$^2P_{1/2}$ 882.4	$2p^6$ 18 **Ar** $^1S_0$ Argon $^3P_2$-$^3P_1$ 606.8 $^3P_2$-$^3P_0$ 1409.9 $^3P_2$-$^1P_1$ 2256.1		
$3d^54s$ 24 **Cr** $^7S_3$ Chromium $^5D_0$-$^5D_1$ 60.0 $^5D_0$-$^5D_2$ 176.7 $^5D_0$-$^5D_3$ 344.4 $^5D_0$-$^5D_4$ 556.8	$3d^54s^2$ 25 **Mn** $^6S_{5/2}$ Manganese $^6D_{9/2}$-$^6D_{7/2}$ 229.7 $^6D_{9/2}$-$^6D_{5/2}$ 399.2 $^6D_{9/2}$-$^6D_{3/2}$ 516.2 $^6D_{9/2}$-$^6D_{1/2}$ 584.8	$3d^64s^2$ 26 **Fe** $^5D_4$ Iron $^5D_4$-$^5D_3$ 415.9 $^5D_4$-$^5D_2$ 704.0 $^5D_4$-$^5D_1$ 888.1 $^5D_4$-$^5D_0$ 978.1	$3d^74s^2$ 27 **Co** $^4F_{9/2}$ Cobalt $^4F_{9/2}$-$^4F_{7/2}$ 816.0 $^4F_{9/2}$-$^4F_{5/2}$ 1406.8 $^4F_{9/2}$-$^4F_{3/2}$ 1809.3	$3d^84s^2$ 28 **Ni** $^3F_4$ Nickel $^3F_4$-$^3F_3$ 1332.2 $^3F_4$-$^3F_2$ 2216.5 $^3D_3$-$^3D_2$ 675.0 $^3D_3$-$^3D_1$ 1508.3
$4p^4$ 34 **Se** $^3P_2$ Selenium $^3P_2$-$^3P_1$ 1989.5 $^3P_2$-$^3P_0$ 2534.4	$4p^5$ 35 **Br** $^2P_{3/2}$ Bromine $^2P_{3/2}$-$^2P_{1/2}$ 3685	$4p^6$ 36 **Kr** $^1S_0$ Krypton $^3P_2$-$^3P_1$ 945.0 $^3P_2$-$^3P_0$ 5219.9 $^3P_2$-$^1P_1$ 5875.0		
$4d^55s$ 42 **Mo** $^7S_3$ Molibdenum $^5D_0$-$^5D_1$ 176.8 $^5D_0$-$^5D_2$ 488.4 $^5D_0$-$^5D_3$ 892.5 $^5D_0$-$^5D_4$ 1480.3	$4d^55s^2$ 43 **Tc** $^6S_{5/2}$ Technetium $^6D_{9/2}$-$^6D_{7/2}$ 678.0 $^6D_{9/2}$-$^6D_{5/2}$ 1129.6 $^6D_{9/2}$-$^6D_{3/2}$ 1429.7 $^6D_{9/2}$-$^6D_{1/2}$ 1605.8	$4d^75s$ 44 **Ru** $^5F_5$ Ruthenium $^5F_5$-$^5F_4$ 1190.6 $^5F_5$-$^5F_3$ 2091.5 $^5F_5$-$^5F_2$ 2713.2 $^5F_5$-$^5F_1$ 3105.5	$4d^85s$ 45 **Rh** $^4F_{9/2}$ Rhodium $^4F_{9/2}$-$^4F_{7/2}$ 1530.0 $^4F_{9/2}$-$^4F_{5/2}$ 2598.0 $^4F_{9/2}$-$^4F_{3/2}$ 3472.7	$4d^{10}$ 46 **Pd** $^1S_0$ Palladium $^3D_3$-$^3D_2$ 1191 $^3D_3$-$^3D_1$ 3530 $^3D_3$-$^1D_2$ 5158
$5p^4$ 52 **Te** $^3P_2$ Tellurium $^3P_2$-$^3P_1$ 4706.5 $^3P_2$-$^3P_0$ 4750.7	$5p^5$ 53 **I** $^2P_{3/2}$ Iodine $^2P_{3/2}$-$^2P_{1/2}$ 7603.1	$5p^6$ 54 **Xe** $^1S_0$ Xenon $^3P_2$-$^3P_1$ 977.6 $^3P_2$-$^3P_0$ 9129.2 $^3P_2$-$^1P_1$ 10117.5		
$5d^46s^2$ 74 **W** $^5D_0$ Tungsten $^5D_0$-$^5D_1$ 1670.3 $^5D_0$-$^5D_2$ 3325.5 $^5D_0$-$^5D_3$ 4830.0 $^5D_0$-$^5D_4$ 6219.3	$5d^56s^2$ 75 **Re** $^6S_{5/2}$ Rhenium $^4P_{5/2}$-$^4P_{3/2}$ 2242.1 $^6D_{9/2}$-$^6D_{7/2}$ 2462.4	$5d^66s^2$ 76 **Os** $^5D_4$ Osmium $^5D_4$-$^5D_3$ 2740.5 $^5D_4$-$^5D_2$ 4159.3 $^5D_4$-$^5D_1$ 5766.1 $^5D_4$-$^5D_0$ 6092.8	$5d^76s^2$ 77 **Ir** $^4F_{9/2}$ Iridium $^4F_{9/2}$-$^4F_{9/2}$ 2835.0	$5d^96s$ 78 **Pt** $^3D_3$ Platinum $^3D_3$-$^3D_2$ 776

16.5.5
Resonantly Excited Atom States

Period \ Group	I	II	III	IV	V
1	1s $_1$H $^2S_{1/2}$ ^2P (121.56) [0.416] {1.6} Hydrogen	$1s^2$ $_2$He $^1S_0$ $^1P_1$ (58.433) [0.276] {0.56} Helium			
2	2s $_3$Li $^2S_{1/2}$ $^2P_{1/2}$ (670.79) [0.247] {27} $^2P_{3/2}$ (670.78) [0.494] {28} Lithium	$2s^2$ $_4$Be $^1S_0$ $^1P_1$ (234.86) [1.34] {1.9} Berillium	2p $_5$B $^2P_{1/2}$ $^2S_{1/2}$ (249.68) [0.12] {3.6} $^2D_{3/2}$ (208.89) [0.046] {20} Boron	$2p^2$ $_6$C $^3P_0$ $^3D_1$ (156.03) [0.1] {8.0} Carbon	$2p^3$ $_7$N $^4S_{3/2}$ 3s ^4P (120) [0.27] {2.5} $2s2p^4$ ^4P (113.4) [0.08] {7.2} Nitrogen
3	3s $_{11}$Na $^2S_{1/2}$ $P_{1/2}$ (589.59) [16] {0.318} $P_{3/2}$ (589.00) [16] {0.637} Sodium	$3s^2$ $_{12}$Mg $^1S_0$ $^1P_1$ (285.21) [1.9] {2.1} Magnesium	3p $_{13}$Al $^2P_{1/2}$ $^2S_{1/2}$ (394.40) [0.12] {6.8} $^2D_{3/2}$ (308.22) [0.18] {13} Aluminium	$3p^2$ $_{14}$Si $^3P_0$ $^3P_1$ (251.43) [0.17] {5.9} Silicon	$3p^3$ $_{15}$P $^4S_{3/2}$ $^4P_{1/2}$ (178.77) [0.05] {4.0} $^4P_{3/2}$ (178.28) [0.10] {4.0} $^4P_{5/2}$ (177.50) [0.15] {4.0} Phosphorus
4	4s $_{19}$K $^2S_{1/2}$ $^2P_{1/2}$ (769.90) [0.35] {27} $^2P_{3/2}$ (766.49) [0.70] {27} Potassium	$4s^2$ $_{20}$Ca $^1S_0$ $^1P_1$ (422.67) [1.7] {4.6} Calsium	$3d4s^2$ $_{21}$Sc $^2D_{3/2}$ Scandium	$3d^24s^2$ $_{22}$Ti $^3F_2$ Titanium	$3d^34s^2$ $_{23}$V $^4F_{3/2}$ Vanadium
	$3d^{10}4s$ $_{29}$Cu $^2S_{1/2}$ $^2P_{1/2}$ (327.40) [0.22] {7} $^2P_{3/2}$(324.75) [0.44] {7.2} Copper	$4s^2$ $_{30}$Zn $^1S_0$ $^1P_1$ (213.86) [1.5] {1.4} Zinc	4p $_{31}$Ga $^2P_{1/2}$ $^2S_{1/2}$ (403.30) [0.12] {6.2} $^2D_{3/2}$ (287.42) [0.30] {4.7} Gallium	$4p^2$ $_{32}$Ge $^3P_0$ Germanium	$4p^3$ $_{33}$As $^4S_{3/2}$ Arsenic
5	5s $_{37}$Rb $^2S_{1/2}$ $^2P_{1/2}$ (794.76) [0.32] {28} $^2P_{3/2}$ (780.03) [0.67] {26} Rubidium	$5s^2$ $_{38}$Sr $^1S_0$ $^1P_1$ (460.73) [2.0] {6.2} Strontium	$4d5s^2$ $_{39}$Y $^2D_{3/2}$ Yttrium	$4d^25s^2$ $_{40}$Zr $^3F_2$ Zirconium	$4d^45s$ $_{41}$Nb $^6D_{1/2}$ Niobium
	$4d^{10}5s$ $_{47}$Ag $^2S_{1/2}$ $^2P_{1/2}$ (338.29) [0.22] {7.9} $^2P_{3/2}$ (328.07) [0.45] {6.7} Silver	$5s^2$ $_{48}$Cd $^1S_0$ $^1P_1$ (228.80) [1.4] {1.7} Cadmium	5p $_{49}$In $^2P_{1/2}$ $^2S_{1/2}$ (410.18) [0.14] {7.4} $^2D_{3/2}$(303.94) [0.36] {7.0} Indium	$5p^2$ $_{50}$Sn $^3P_0$ Tin	$5p^3$ $_{51}$Sb $^4S_{3/2}$ Antimony
6	6s $_{55}$Cs $^2S_{1/2}$ $^2P_{1/2}$ (894.35) [0.39] {31} $^2P_{3/2}$ (852.11) [0.81] {28} Cesium	$6s^2$ $_{56}$Ba $^1S_0$ $^1P_1$ (553.55) [1.6] {8.5} Barium	$5d6s^2$ $_{57}$La $^2D_{3/2}$ Lanthanum	$5d^26s^2$ $_{72}$Hf $^3F_2$ Hafnium	$5d^36s^2$ $_{73}$Ta $^4F_{3/2}$ Tantalum
	$5d^{10}6s$ $_{79}$Au $^2S_{1/2}$ $^2P_{1/2}$ (267.60) [0.12] {6.0} $^2P_{3/2}$ (242.80) [0.26] {4.6} Gold	$5d^{10}6s^2$ $_{80}$Hg $^1S_0$ $^1P_1$ (184.95) [1.2] {1.3} $^3P_1$ (253.65) [0.024] {120} Mercury	6p $_{81}$Tl $^2P_{1/2}$ $^2S_{1/2}$ (377.57) [0.13] {7.6} $^2D_{3/2}$ (276.79) [0.29] {7.0} Thallium	$6p^2$ $_{82}$Pb $^3P_0$ $^3P_1$ (283.31) [0.21] {5.8} Lead	$6p^3$ $_{83}$Bi $^4S_{3/2}$ $^4P_{1/2}$ (306.77) [0.15] {4.6} Bismuth

VI	VII	VIII		
^3S (130.4) [0.05] {1.8} ^3D(102.7) [0.01] {25} $2p^4$ $^3P_2$ $_8$**O** Oxygen	$^2P_{1/2}$ (95.48) [0.07] {3.5} $^2P_{3/2}$ (95.19) [0.035] {3.5} $2p^5$ $^2P_{3/2}$ $_9$**F** Fluorine	$^3P_1$ (74.372) [0.01] {25} $^1P_1$ (73.590) [0.15] {1.6} $2p^6$ $^1S_0$ $_{10}$**Ne** Neon		
$^3S_1$ (180.73) [0.11] {2.8} $3p^4$ $^3P_2$ $_{16}$**S** Sulfur	$^2P_{1/2}$ (134.72) [0.11] {1.0} $^2P_{3/2}$ (133.57) [0.023] {1.0} $3p^5$ $^2P_{3/2}$ $_{17}$**Cl** Chlorine	$^3P_1$ (106.66) [0.051] {20} $^1P_1$ (104.82) [0.25] {10} $2p^6$ $^1S_0$ $_{18}$**Ar** Argon		
$3d^5 4s$ $_{24}$**Cr** $^7S_3$ Chromium	$3d^5 4s^2$ $_{25}$**Mn** $^6S_{5/2}$ Manganese	$3d^6 4s^2$ $_{26}$**Fe** $^5D_4$ Iron	$3d^7 4s^2$ $_{27}$**Co** $^4F_{9/2}$ Cobalt	$3d^8 4s^2$ $_{28}$**Ni** $^3F_4$ Nickel
$4p^4$ $^3P_2$ $_{34}$**Se** Selenium	$4p^5$ $^2P_{3/2}$ $_{35}$**Br** Bromine	$^3P_1$ (123.58) [0.15] {4.4} $^1P_1$ (116.46) [0.14] {4.5} $4p^6$ $^1S_0$ $_{36}$**Kr** Krypton		
$4d^5 5s$ $_{42}$**Mo** $^7S_3$ Molibdenum	$4d^5 5s^2$ $_{43}$**Tc** $^6S_{5/2}$ Technetium	$4d^7 5s$ $_{44}$**Ru** $^5F_5$ Ruthenium	$4d^8 5s$ $_{45}$**Rh** $^4F_{9/2}$ Rhodium	$4d^{10}$ $_{46}$**Pd** $^1S_0$ Palladium
$5p^4$ $^3P_2$ $_{52}$**Te** Tellurium	$5p^5$ $^2P_{3/2}$ $_{53}$**I** Iodine	$^3P_1$ (146.96) [0.27] {3.6} $^1P_1$ (129.56) [0.22] {3.5} $5p^6$ $^1S_0$ $_{54}$**Xe** Xenon		
$5d^4 6s^2$ $_{74}$**W** $^5D_0$ Tungsten	$5d^5 6s^2$ $_{75}$**Re** $^6S_{5/2}$ Rhenium	$5d^6 6s^2$ $_{76}$**Os** $^5D_4$ Osmium	$5d^7 6s^2$ $_{77}$**Ir** $^4F_{9/2}$ Iridium	$5d^9 6s$ $_{78}$**Pt** $^3D_3$ Platinum
$6p^4$ $^3P_2$ $_{84}$**Po** Polonium	$6p^5$ $^2P_{3/2}$ $_{85}$**At** Astatine	$6p^6$ $^1S_0$ $_{86}$**Rn** Radon		

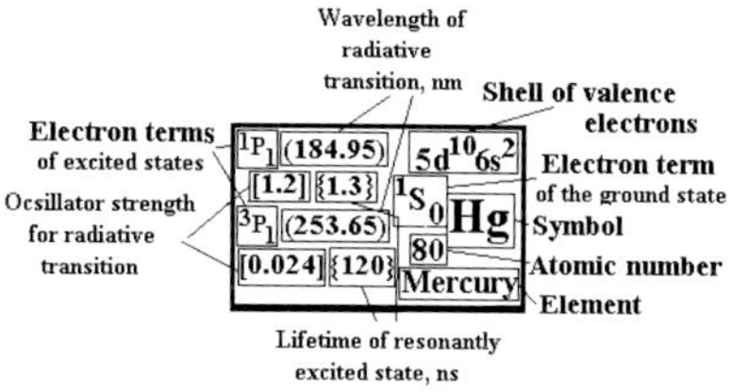

16.5.6
Polarizabilities of Atoms and Diatomics

Group / Period	I	II	III	IV	V
1	*1.008* 1s 4.5 $_1$**H** $^2S_{1/2}$ *5.417* Hydrogen	1.383 $1s^2$ *4.003* $^1S_0$ $_2$**He** Helium			
2	*6.491* 2s 162 **Li** $^2S_{1/2}$ *230* Lithium	*9.012* $2s^2$ 38 $_4$**Be** $^1S_0$ Berillium	20.5 2p *10.81* $^2P_{1/2}$ $_5$**B** Boron	11.8 $2p^2$ *12.011* $^3P_0$ $_6$**C** Carbon	7.5 $2p^3$ *14.007* *11.8* $^4S_{3/2}$ $_7$**N** Nitrogen
3	*22.990* 3s 162 $_{11}$**Na** $^2S_{1/2}$ *200* Sodium	*24.305* $3s^2$ 72 $_{12}$**Mg** $^1S_0$ Magnesium	59 3p *26.982* $^2P_{1/2}$ $_{13}$**Al** Aluminium	37 $3p^2$ *28.086* $^3P_0$ $_{14}$**Si** Silicon	24 $3p^3$ *30.974* $^4S_{3/2}$ $_{15}$**P** Phosphorus
4	*39.098* 4s 290 $_{19}$**K** $^2S_{1/2}$ *410* Potassium	*40.08* $4s^2$ 170 $_{20}$**Ca** $^1S_0$ Calsium	*44.956* $3d4s^2$ 160 $_{21}$**Sc** $^2D_{3/2}$ Scandium	*47.88* $3d^24s^2$ 150 $_{22}$**Ti** $^3F_2$ Titanium	*50.942* $3d^34s^2$ 130 $_{23}$**V** $^4F_{3/2}$ Vanadium
	40 $3d^{10}4s$ *63.546* $^2S_{1/2}$ $_{29}$**Cu** Copper	50 $4s^2$ *65.38* $^1S_0$ $_{30}$**Zn** Zinc	60 4p *69.72* $^2P_{1/2}$ $_{31}$**Ga** Gallium	41 $4p^2$ *72.59* $^3P_0$ $_{32}$**Ge** Germanium	31 $4p^3$ *74.922* $^4S_{3/2}$ $_{33}$**As** Arsenic
5	*85.468* 5s 320 $_{37}$**Rb** $^2S_{1/2}$ *460* Rubidium	*87.62* $5s^2$ 190 $_{38}$**Sr** $^1S_0$ Strontium	*88.906* $4d5s^2$ 170 $_{39}$**Y** $^2D_{3/2}$ Yttrium	*91.22* $4d^25s^2$ 122 $_{40}$**Zr** $^3F_2$ Zirconium	*92.906* $4d^45s$ 94 $_{41}$**Nb** $^6D_{1/2}$ Niobium
	67 $4d^{10}5s$ *107.87* $^2S_{1/2}$ $_{47}$**Ag** Silver	62 $5s^2$ *112.41* $^1S_0$ $_{48}$**Cd** Cadmium	69 5p *114.82* $^2P_{1/2}$ $_{49}$**In** Indium	$5p^2$ *118.69* $^3P_0$ $_{50}$**Sn** Tin	35 $5p^3$ *121.75* $^4S_{3/2}$ $_{51}$**Sb** Antimony
6	*132.90* 6s 360 $_{55}$**Cs** $^2S_{1/2}$ Cesium	*137.33* $6s^2$ 270 $_{56}$**Ba** $^1S_0$ Barium	*138.90* $5d6s^2$ $_{57}$**La** $^2D_{3/2}$ Lanthanum	*178.49* $5d^26s^2$ $_{72}$**Hf** $^3F_2$ Hafnium	*180.95* $5d^36s^2$ $_{73}$**Ta** $^4F_{3/2}$ Tantalum
	$5d^{10}6s$ *196.97* $^2S_{1/2}$ $_{79}$**Au** Gold	34 $5d^{10}6s^2$ *200.59* $^1S_0$ $_{80}$**Hg** Mercury	46 6p *204.38* $^2P_{1/2}$ $_{81}$**Tl** Thallium	49 $6p^2$ *207.2* $^3P_0$ $_{82}$**Pb** Lead	38 $6p^3$ *208.98* $^4S_{3/2}$ $_{83}$**Bi** Bismuth

VI	VII	VIII		
5.41 $2p^4$ *15.999* *10.7* $^3P_2$ $_8$**O** Oxygen	3.76 $2p^5$ *18.998* $^2P_{3/2}$ $_9$**F** Fluorine	2.68 $2p^6$ *20.179* $^1S_0$ $_{10}$**Ne** Neon		
18 $3p^4$ *32.06* $^3P_2$ $_{16}$**S** Sulfur	14 $3p^5$ *35.453* *31* $^2P_{3/2}$ $_{17}$**Cl** Chlorine	11.1 $2p^6$ *39.948* $^1S_0$ $_{18}$**Ar** Argon		
51.996 $3d^54s$ 74 $_{24}$**Cr** $^7S_3$ Chromium	*54.938* $3d^54s^2$ 100 $_{25}$**Mn** $^6S_{5/2}$ Manganese	*55.847* $3d^64s^2$ 90 $_{26}$**Fe** $^5D_4$ Iron	*58.933* $3d^74s^2$ 74 $_{27}$**Co** $^4F_{9/2}$ Cobalt	*58.69* $3d^84s^2$ 70 $_{28}$**Ni** $^3F_4$ Nickel
28 $4p^4$ *78.96* $^3P_2$ $_{34}$**Se** Selenium	30 $4p^5$ *79.904* *44* $^2P_{3/2}$ $_{35}$**Br** Bromine	16.7 $4p^6$ *83.80* $^1S_0$ $_{36}$**Kr** Krypton		
95.94 $4d^55s$ 88 $_{42}$**Mo** $^7S_3$ Molibdenum	*[98]* $4d^55s^2$ 158 $_{43}$**Tc** $^6S_{5/2}$ Technetium	*101.07* $4d^75s$ 68 $_{44}$**Ru** $^5F_5$ Ruthenium	*102.91* $4d^85s$ 51 $_{45}$**Rh** $^4F_{9/2}$ Rhodium	*106.42* $4d^{10}$ 47 $_{46}$**Pd** $^1S_0$ Palladium
27 $5p^4$ *127.60* $^3P_2$ $_{52}$**Te** Tellurium	27 $5p^5$ *126.90* $^2P_{3/2}$ $_{53}$**I** Iodine	27.4 $5p^6$ *131.29* $^1S_0$ $_{54}$**Xe** Xenon		
183.85 $5d^46s^2$ 115 $_{74}$**W** $^5D_0$ Tungsten	*186.21* $5d^56s^2$ $_{75}$**Re** $^6S_{5/2}$ Rhenium	*190.2* $5d^66s^2$ $_{76}$**Os** $^5D_4$ Osmium	*192.22* $5d^76s^2$ $_{77}$**Ir** $^4F_{9/2}$ Iridium	*195.08* $5d^96s$ $_{78}$**Pt** $^3D_3$ Platinum
$6p^4$ *[209]* $^3P_2$ $_{84}$**Po** Polonium	*9.0* $6p^5$ *[210]* $^2P_{3/2}$ $_{85}$**At** Astatine	$6p^6$ *[222]* $^1S_0$ $_{86}$**Rn** Radon		

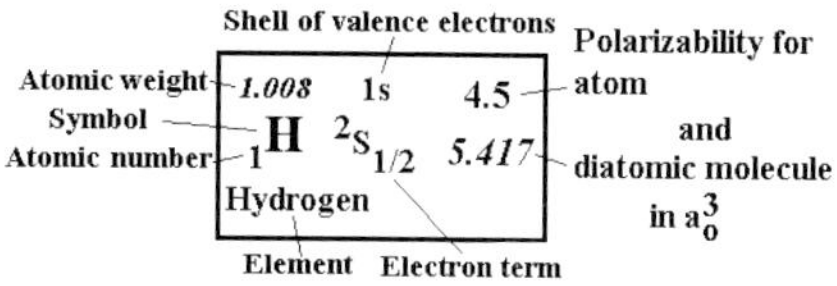

16.5.7
Affinity to Hydrogen and Oxygen Atoms

Period \ Group	I	II	III	IV	V
1	$1s$ $^2S_{1/2}$ $_1$**H** Hydrogen 4.48 *4.39*	$1s^2$ $^1S_0$ $_2$**He** Helium			
2	$2s$ $^2S_{1/2}$ $_3$**Li** Lithium 2.47 *3.5*	$2s^2$ $_4$**Be** $^1S_0$ Berillium 2.05 *4.5*	$2p$ $_5$**B** $^2P_{1/2}$ Boron 3.5 *8.4*	$2p^2$ $^3P_0$ $_6$**C** Carbon 3.46 *11.09*	$2p^3$ $^4S_{3/2}$ $_7$**N** Nitrogen 3.5 *6.50*
3	$3s$ $^2S_{1/2}$ $_{11}$**Na** Sodium 1.92 *2.6*	$3s^2$ $^1S_0$ $_{12}$**Mg** Magnesium 1.31 *3.8*	$3p$ $^2P_{1/2}$ $_{13}$**Al** Aluminium 2.95 *5.25*	$3p^2$ $^3P_0$ $_{14}$**Si** Silicon 3.1 *8.3*	$3p^3$ $^4S_{3/2}$ $_{15}$**P** Phosphorus 3.1 *6.1*
4	$4s$ $^2S_{1/2}$ $_{19}$**K** Potassium 1.81 *2.9*	$4s^2$ $_{20}$**Ca** $^1S_0$ Calsium 1.74 *4.2*	$3d4s^2$ $_{21}$**Sc** $^2D_{3/2}$ Scandium *7.8*	$3d^24s^2$ $_{22}$**Ti** $^3F_2$ Titanium 2.1 *7.0*	$3d^34s^2$ $_{23}$**V** $^4F_{3/2}$ Vanadium 2.2 *6.5*
	$3d^{10}4s$ $^2S_{1/2}$ $_{29}$**Cu** Copper 2.9 *2.8*	$4s^2$ $^1S_0$ $_{30}$**Zn** Zinc 0.89 *1.6*	$4p$ $^2P_{1/2}$ $_{31}$**Ga** Gallium 2.8 *3.8*	$4p^2$ $^3P_0$ $_{32}$**Ge** Germanium 3.3 *6.8*	$4p^3$ $^4S_{3/2}$ $_{33}$**As** Arsenic 2.84 *4.95*
5	$5s$ $_{37}$**Rb** $^2S_{1/2}$ Rubidium 1.7 *2.6*	$5s^2$ $_{38}$**Sr** $^1S_0$ Strontium 1.7 *4.4*	$4d5s^2$ $_{39}$**Y** $^2D_{3/2}$ Yttrium *7.5*	$4d^25s^2$ $_{40}$**Zr** $^3F_2$ Zirconium *8.0*	$4d^45s$ $^6D_{1/2}$ $_{41}$**Nb** Niobium *8.0*
	$4d^{10}5s$ $^2S_{1/2}$ $_{47}$**Ag** Silver 2.2 *2.3*	$5s^2$ $^1S_0$ $_{48}$**Cd** Cadmium 0.7 *2.4*	$5p$ $^2P_{1/2}$ $_{49}$**In** Indium 2.52 *3.3*	$5p^2$ $^3P_0$ $_{50}$**Sn** Tin 2.7 *5.5*	$5p^3$ $^4S_{3/2}$ $_{51}$**Sb** Antimony *4.5*
6	$6s$ $^2S_{1/2}$ $_{55}$**Cs** Cesium 1.78 *3.0*	$6s^2$ $_{56}$**Ba** $^1S_0$ Barium 1.8 *5.8*	$5d6s^2$ $^2D_{3/2}$ $_{57}$**La** Lanthanum *8.3*	$5d^26s^2$ $^3F_2$ $_{72}$**Hf** Hafnium *8.3*	$5d^36s^2$ $^4F_{3/2}$ $_{73}$**Ta** Tantalum *8.3*
	$5d^{10}6s$ $^2S_{1/2}$ $_{79}$**Au** Gold 3.1 *2.3*	$5d^{10}6s^2$ $^1S_0$ $_{80}$**Hg** Mercury 0.41 *2.3*	$6p$ $^2P_{1/2}$ $_{81}$**Tl** Thallium 1.9 *4.0*	$6p^2$ $^3P_0$ $_{82}$**Pb** Lead 1.6 *3.5*	$6p^3$ $^4S_{3/2}$ $_{83}$**Bi** Bismuth 2.7

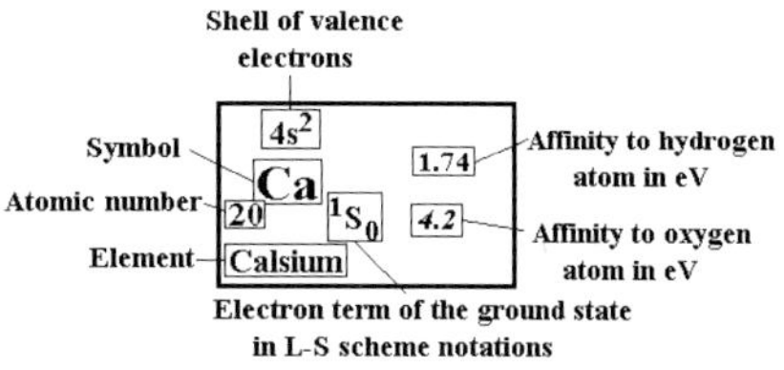

The affinity of atom M to the hydrogen or oxygen atom is the dissociation energy of radical MH or MO in the ground vibration state.

VI	VII	VIII		
$2p^4$; $^3P_2$; 8 **O**; Oxygen; 4.39; *5.12*	$2p^5$; $^2P_{3/2}$; 9 **F**; Fluorine; 5.91; *2.4*	$2p^6$; $^1S_0$; 10 **Ne**; Neon		
$3p^4$; $^3P_2$; 16 **S**; Sulfur; 3.6; *5.4*	$3p^5$; $^2P_{3/2}$; 17 **Cl**; Chlorine; 4.47; *2.75*	$3p^6$; $^1S_0$; 18 **Ar**; Argon		
$3d^5 4s$; 24 **Cr**; $^7S_3$; Chromium; 2.0; *4.2*	$3d^5 4s^2$; 25 **Mn**; $^6S_{5/2}$; Manganese; 2.4; *4.0*	$3d^6 4s^2$; 26 **Fe**; $^5D_4$; Iron; 1.9; *4.0*	$3d^7 4s^2$; 27 **Co**; $^4F_{9/2}$; Cobalt; 2.3; *4.0*	$3d^8 4s^2$; **Ni** 28; $^3F_4$; Nickel; 2.6; *4.0*
$4p^4$; $^3P_2$; 34 **Se**; Selenium; 3.26; *4.8*	$4p^5$; $^2P_{3/2}$; 35 **Br**; Bromine; 3.79; *2.44*	$4p^6$; $^1S_0$; 36 **Kr**; Krypton		
$4d^5 5s$; 42 **Mo**; $^7S_3$; Molibdenum; *6.0*	$4d^5 5s^2$; 43 **Tc**; $^6S_{5/2}$; Technetium	$4d^7 5s$; $^5F_5$; 44 **Ru**; Ruthenium; 2.4; *5.5*	$4d^8 5s$; $^4F_{9/2}$; 45 **Rh**; Rhodium; 2.6; *4.2*	$4d^{10}$; $^1S_0$; 46 **Pd**; Palladium; 2.4; *4.0*
$5p^4$; $^3P_2$; 52 **Te**; Tellurium; 2.78; *3.9*	$5p^5$; $^2P_{3/2}$; 53 **I**; Iodine; 3.07; *2.58*	$5p^6$; $^1S_0$; 54 **Xe**; Xenon; *0.4*		
$5d^4 6s^2$; $^5D_0$; 74 **W**; Tungsten; *7.0*	$5d^5 6s^2$; $^6S_{5/2}$; 75 **Re**; Rhenium; *6.5*	$5d^6 6s^2$; $^5D_4$; 76 **Os**; Osmium; *6.0*	$5d^7 6s^2$; $^4F_{9/2}$; 77 **Ir**; Iridium; *4.3*	$5d^9 6s$; 78 **Pt**; $^3D_3$; Platinum; 3.4; *4.0*

16.5.8
Diatomic Molecules

Period \ Group	I	II	III	IV	V
1	15.43; 4.478 $^{1}\Sigma_g^+$ $_{1}$H *0.741* Hydrogen *4401* *121.3* *60.85*	$^{1}\Sigma_g^+$ 22.22 ; 0.001 *2.97* $_{2}$He *0.96* Helium			
2	5.145; 1.05 $^{1}\Sigma_g^+$ $_{3}$Li *2.67* Lithium *351.4* *2.59* *0.672*	7.45 ; 0.098 $^{1}\Sigma_g^+$ $_{4}$Be *2.45* Berillium *275.8* *12.5* *0.615*	$^{1}\Sigma_g^+$ 8.8 ; 2.8 *1.60* $_{5}$B Boron *1059* *15.66* *1.216*	$^{1}\Sigma_g^+$ 12.15 ; 5.36 *1.24* $_{6}$C Carbon *1855* *13.27* *1.899*	$^{1}\Sigma_g^+$ 15.581 ; 9.579 *1.098* $_{7}$N Nitrogen *2359* *14.95* *1.998*
3	4.90 ; 0.731 $^{1}\Sigma_g^+$ $_{11}$Na *3.08* Sodium *159.1* *0.725* *0.155*	6.7 ; 0.053 $^{1}\Sigma_g^+$ $_{12}$Mg *3.89* Magnesium *51.08* *1.62* *0.093*	$^{3}\Pi_u$ 4.84 ; 0.46 *2.47* $_{13}$Al Aluminium *284.2* *2.02* *0.205*	$^{3}\Sigma_g^-$ 7.4 ; 3.24 *2.24* $_{14}$Si Silicon *510.9* *2.02* *0.239*	$^{1}\Sigma_g^+$ 10.56 *1.89* $_{15}$P Phosphorus *780.8* *2.83* *0.304*
4	4.064 ; 0.551 $^{1}\Sigma_g^+$ $_{19}$K *3.92* Potassium *92.09* *0.283* *0.165*	5.2 ; 0.13 $^{1}\Sigma_g^+$ $_{20}$Ca *4.28* Calsium *64.9* *1.09* *0.047*	1.69 $^{5}\Sigma_u^-$ $_{21}$Sc *2.21* Scandium *238.9* *0.93* *0.153*	6. 2; 1.4 $^{3}\Delta_g$ $_{22}$Ti *1.94* Titanium *407.9* *1.08* *0.187*	6.4 ; 2.62 $^{3}\Sigma_g^-$ $_{23}$V *1.78* Vanadium *537.5* *3.34* *0.209*
	$^{1}\Sigma_g^+$ 7.89 ; 1.99 *2.21* $_{29}$Cu Copper	$^{1}\Sigma_g^+$ 9.0 ; 0.034 *4.8* $_{30}$Zn Zinc *25.7* *0.60* *0.022*	$^{3}\Pi_u$ 6.5 ; 1.18 *2.76* $_{31}$Ga Gallium *158* *1.0* *0.063*	$^{3}\Sigma_g^-$ 7.2 ; 2.5 *2.44* $_{32}$Ge Germanium *259* *0.8* *0.078*	$^{1}\Sigma_g^+$? 12.0 ; 3.96 *2.103* $_{33}$As Arsenic *429.6* *1.12* *0.102*
5	3.45 ; 0.495 $^{1}\Sigma_g^+$ $_{37}$Rb *4.17* Rubidium *266.4* *1.03* *0.109*	4.74 ; 0.13 $^{1}\Sigma_g^+$ $_{38}$Sr *4.45* Strontium *39.6* *0.45* *0.019*	1.6 $^{1}\Sigma_g^+$ $_{39}$Y *2.8* Yttrium *206.5* – *0.048*	1.5 $^{1}\Sigma_g^+$ $_{40}$Zr *2.3* Zirconium *373* - *0.070*	6.37 ; 5.48 $^{3}\Sigma_g^-$ $_{41}$Nb *2.1* Niobium *424.9* *0.94* *0.084*
	$^{1}\Sigma_g^+$ 7.66 ; 1.65 *2.53* $_{47}$Ag Silver *135.8* *0.50* *0.049*	$^{1}\Sigma_g^+$ 0.013 *5.1* $_{48}$Cd Cadmium *22.5* *0.4* *0.011*	$^{3}\Pi_u$ 0.83 *3.14* $_{49}$In Indium *111* *0.8* *0.030*	$^{3}\Sigma_g^-$ 7.32 ; 2.0 *2.75* $_{50}$Sn Tin *186.2* *0.261* *0.038*	$^{1}\Sigma_g^+$ 8.8 ; 3.09 *2.34* $_{51}$Sb Antimony *269.9* *0.58* *0.050*
6	3.64 ; 0.452 $^{1}\Sigma_g^+$ $_{55}$Cs *4.65* Cesium *42.02* *0.082* *0.013*	$^{1}\Sigma_g^+$ $_{56}$Ba *4.6* Barium *84.1* *0.16* *0.009*	$_{57}$La Lanthanum	$_{58}$Hf Hafnium	$_{73}$Ta Tantalum
	$^{1}\Sigma_g^+$ 9.2 ; 2.31 *2.47* $_{79}$Au Gold *190.9* *0.42* *0.028*	0_g^+ 9.4 ; 0.055 *3.65* $_{80}$Hg Mercury *18.5* *0.27* *0.013*	$^{3}\Sigma_g^-$ 6.3 ; 0.001 *3.0* $_{81}$Tl Thallium *80* *0.5* *0.018*	$^{3}\Sigma_g^-$ 6.4 ; 0.83 *2.93* $_{82}$Pb Lead *110.2* *0.327* *0.019*	$^{1}\Sigma_g^+$ 7.4 ; 2.08 *2.66* $_{83}$Bi Bismuth *173.1* *0.376* *0.023*

VI	VII	VIII		
$^3\Sigma_g^-$ 12.071 ; 5.12 1580 *1.207* $_8$O 11.98 Oxygen 1.445	$^1\Sigma_g^+$ 15.686 ; 1.66 916.6 *1.41* $_9$F 11.24 Fluorine 0.89	$^1\Sigma_g^+$ 20.4 ; 0.037 31.3 *3.09* $_{10}$Ne 6.48 Neon 0.175		
$^3\Sigma_g^-$ 9.4 ; 4.37 725.6 *1.89* $_{16}$S 2.28 Sulfur 0.295	$^1\Sigma_g^+$ 11.50 ; 2.576 559.7 *1.99* $_{17}$Cl 2.68 Chlorine 0.244	$^1\Sigma_g^+$ 14.54 ; 0.012 30.68 *3.76* $_{18}$Ar 2.42 Argon 0.060		
6.8 ; 1.66 $^1\Sigma_g^+$ $_{24}$Cr *1.68* 470 Chromium 14.1 0.23	6.5 ; 0.79 $^1\Sigma_g^+$ $_{25}$Mn *2.52* 68.1 Manganese 1.05 0.097	6.3 ; 0.9 $^1\Sigma_g^+$ $_{26}$Fe *2.02* 412.0 Iron 1.4 0.148	6.0 ; 0.9 $^1\Sigma_g^+$ $_{27}$Co *2.0* 280 Cobalt - 0.14	7.43 ; 1.7 $^1\Sigma_g^+$ $_{28}$Ni *2.3* 250 Nickel 1.1 0.104
$^3\Sigma_g^-$ 8.88 ; 2.9 385.3 *2.16* $_{34}$Se 0.963 Selenium 0.89	$^1\Sigma_g^+$ 10.52 ; 2.05 325 *2.28* $_{37}$Br 1.08 Bromine 0.082	$^1\Sigma_g^+$ 12.97 ; 0.017 24.1 *4.01* $_{36}$Kr 1.34 Krypton 0.025		
6.2 ; 4.1 $^1\Sigma_g^+$ $_{42}$Mo *2.2* 477 Molibdenum 1.51 0.072	$_{43}$Tc Technetium	$_{44}$Ru Ruthenium	1.5 $^5\Delta_g$ $_{45}$Rh *2.67* 238 Rhodium - 0.046	7.7 ; 0.76 $^3\Sigma_g^+$ $_{46}$Pd *2.48* 159 Palladium - 0.051
$^3\Sigma_g^-$ 8.29 ; 2.7 249.1 *2.56* $_{52}$Te 0.537 Tellurium 0.040	$^1\Sigma_g^+$ 9.3 : 1.542 214.5 *2.67* $_{53}$I 0.615 Iodine 0.037	$^1\Sigma_g^+$ 11.13 ; 0.024 21.12 *4.36* $_{54}$Xe 0.65 Xenon 0.0135		
6.9 $_{74}$W 336.8 Tungsten 1.0 -	$_{75}$Re Rhenium	$_{76}$Os Osmium	$_{77}$Ir Iridium	8.7 ; 0.93 $^1\Sigma_g^+$ $_{78}$Pt *2.34* 259.4 Platinum 0.9 0.032
$_{84}$Po Polonium	$_{85}$At Astatine	$^1\Sigma_g^+$ 0.030 --- *4.68* $_{86}$Rn 0.0069 Radon		

Key	Example
Ionization potential (eV) ; Dissociation energy, eV	6.4 ; 2.62
Electron term	$^3\Sigma_g^-$
Symbol	V
Equilibrium distance, Å	*1.78*
Atomic number	23
Element	Vanadium
Vibrational energy, cm^{-1}	*537.5*
Anharmonic constant, cm^{-1}	*3.34*
Rotational constant, cm^{-1}	*0.209*

16.5.9
Positive Ions of Diatomics

Group / Period	I	II	III	IV	V
1	15.43; 2.65 $^2\Sigma_g^+$ $_1$H *1.06* *2323* *67.5* Hydrogen *30.21*	$^2\Sigma_g^+$ 22.22 ; 2.47 *1698.5* *1.08* $_2$He *35.3* *7.21* Helium			
2	5.145; 1.28 $^1\Sigma_g^+$ $_3$Li *3.12* *263.1* *1.61* Lithium *0.49*	7.45 ; 1.9 $^2\Sigma_g^+$ $_4$Be *502* *4.2* Berillium *0.752*	$^2\Sigma_g^+$ 8.8 ; 1.9 *357* *4.15* $_5$B --- *0.181* Boron	$^4\Sigma_g^-$ 12.15 ; 5.3 *1351* *1.41* $_6$C *12.1* *1.41* Carbon	$^2\Sigma_g^+$ 15.581 ; 8.713 *2207* *1.12* $_7$N *16.2* *1.932* Nitrogen
3	4.90 ; 0.98 $^2\Sigma_g^+$ $_{11}$Na *3.54* *120.8* *0.46* Sodium *0.113*	6.7 ; 1.3 $^2\Sigma_g^+$ $_{12}$Mg Magnesium	$^2\Sigma_g^+$ 4.84 ; 1.4 *178* *3.2* $_{13}$Al *2.0* *0.122* Aluminium	$^2\Pi_u$ 7.4 ; 3.24 *528* *2.19* $_{14}$Si --- *0.25* Silicon	$^2\Pi_u$ 10.56 ; 5.0 *672* *1.98* $_{15}$P *2.74* *0.276* Phosphorus
4	4.064 ; 0.81 $^2\Sigma_g^+$ $_{19}$K *4.6* *73.4* *0.2* Potassium *0.042*	5.2 ; 1.04 $^2\Sigma_g^+$ $_{20}$Ca *3.7* *119* --- Calsium *0.053*	$_{21}$Sc Scandium	6.0 ; 2.44 $_{22}$Ti Titanium	6.36 ; 3.14 $_{23}$V Vanadium
	$^2\Sigma_g^+$ 7.89 ; 1.8 *188* 2.35 $_{29}$Cu *0.75* *0.096* Copper	$^2\Sigma_g^+$ 9.0 ; 0.42 $_{30}$Zn Zinc	$^2\Sigma_g^+$ 6.5 ; 1.27 *108* *3.24* $_{31}$Ga --- *0.046* Gallium	$^4\Sigma_g^-$ 7.2 ; 2.91 *256* *2.32* $_{32}$Ge --- *0.086* Germanium	$^2\Sigma_g^+$ 9.64 ; 4.11 *430* *1.9* $_{33}$As --- *0.125* Arsenic
5	3.45 ; 0.75 $^2\Sigma_g^+$ $_{37}$Rb *4.8* *44.5* --- Rubidium *0.017*	4.74 ; 1.1 $^2\Sigma_g^+$ $_{38}$Sr *3.9* *86* *0.54* Strontium *0.025*	$_{39}$Y Yttrium	$_{40}$Zr Zirconium	6.37 ; *5.87* $_{41}$Nb Niobium
	$^2\Sigma_g^+$ 7.66 ; 1.69 *118* *2.8* $_{47}$Ag *0.05* *0.040* Silver	$^2\Sigma_g^+$ $_{48}$Cd Cadmium	$_{49}$In Indium	7.38 ; 1.96 $_{50}$Sn Tin	8.7 ; 3.2 $_{51}$Sb Antimony
6	3.76; 0.61 $^2\Sigma_g^+$ $_{55}$Cs *4.44* *32.4* *0.051* Cesium *0.013*	$^2\Sigma_g^+$ $_{56}$Ba Barium	$_{57}$La Lanthanum	$_{58}$Hf Hafnium	$_{73}$Ta Tantalum
	$^2\Sigma_g^+$ 9.2 ; $_{79}$Au Gold	$^2\Sigma_g^+$ 9.4 ; 0.96 *91.6* *2.8* $_{80}$Hg *0.301* *0.021* Mercury	$_{81}$Tl Thallium	6.1 ; 1.7 $_{82}$Pb Lead	$^2\Pi_g$ 7.4 ; 1.89 *143.4* $_{83}$Bi *0.18* --- Bismuth

VI	VII	VIII		
$^3\Sigma_g^-$ 12.071 ; 6.66 1905 1.12 $_8$O 16.3 1.689 Oxygen	$^2\Pi_g$ 15.47 ; 3.34 1073 1.32 $_9$F 9.13 1.015 Fluorine	$^2\Sigma_g^+$ 20.4 ; 1.2 586 1.75 $_{10}$Ne 5.4 0.544 Neon		
$^2\Pi_u$ 9.4 ; 5.4 806 1.82 $_{16}$S 3.33 0.318 Sulfur	$^2\Pi_g$ 11.50 ; 3.95 645.6 1.88 $_{17}$Cl 3.02 0.265 Chlorine	$^2\Sigma_g^+$ 14.54 ; 1.23 308.9 2.43 $_{18}$Ar 1.66 0.143 Argon		
6.8 ; 1.8 $_{24}$Cr Chromium	6.47 ; 1.3 $_{25}$Mn Manganese	6.3 ; 2.7 $_{26}$Fe Iron	6.0 ; 2.75 $_{27}$Co Cobalt	7.43 ; 2.35 $_{28}$Ni Nickel
$^2\Pi_u$ 8.88 ; 4.4 450 2.07 $_{34}$Se --- 0.10 Selenium	$^2\Pi_g$ 10.52 ; 2.96 376 2.3 $_{37}$Br 1.13 0.088 Bromine	$^2\Sigma_g^+$ 12.97 ; 1.15 178 2.8 $_{36}$Kr 0.82 0.051 Krypton		
6.2 ; 5.0 $_{42}$Mo Molibdenum	$_{43}$Tc Technetium	$_{44}$Ru Ruthenium	$_{45}$Rh Rhodium	7.7 $_{46}$Pd Palladium
8.2 ; 3.5 $_{52}$Te Tellurium	$^2\Pi_g$ 9.3 : 1.92 243 2.58 $_{53}$I --- 0.040 Iodine	$^2\Sigma_g^+$ 11.13 ; 1.1 123 3.25 $_{54}$Xe 0.63 0.026 Xenon		
$_{74}$W Tungsten	$_{75}$Re Rhenium	$_{76}$Os Osmium	$_{77}$Ir Iridium	8.7 ; 3.26 $_{78}$Pt Platinum
$_{84}$Po Polonium	$_{85}$At Astatine	$^1\Sigma_g^+$ $_{86}$Rn Radon		

Key:

Diatomic ionization potential (eV)	4.90
Dissociation energy, eV	0.98
Electron term	$^2\Sigma_g^+$
Symbol	Na
Atomic number	11
Element	Sodium
Equilibrium distance, Å	3.54
Vibrational energy, cm^{-1}	120.8
Anharmonic constant, cm^{-1}	0.46
Rotational constant, cm^{-1}	0.113

16.5.10
Negative Ions of Diatomics

Group / Period	I	II	III	IV
2	0.7 ; 0.88 $^2\Sigma_u^+$ $_3$**Li** *2.8* *233.1* Lithium *1.92* *0.516*	0.4 ; 2.8 $^2\Sigma_g^+$ $_4$**Be** *2.3* *620* Berillium --- ---	1.7 ; 4.2 $_5$**B** Boron	$^2\Sigma_g^+$ 3.27 ; 8.5 *1781* *1.27* $_6$**C** *11.7* Carbon *1.746*
3	0.43 ; 0.44 $^2\Sigma_u^+$ $_{11}$**Na** Sodium	$_{12}$**Mg** Magnesium	$^4\Sigma_g^-$ 1.1 ; 2.4 *335* *2.65* $_{13}$**Al** --- *0.178* Aluminium	2.19 ; 3.3 $_{14}$**Si** Silicon
4	0.49 ; $^2\Sigma_u^+$ $_{19}$**K** Potassium	$_{20}$**Ca** Calsium	$_{21}$**Sc** Scandium	$_{22}$**Ti** Titanium
	$^2\Sigma_u^+$ 0.84 ; 1.57 *196* *2.34* $_{29}$**Cu** *0.7* Copper *0.097*	$_{30}$**Zn** Zinc	$^4\Sigma_g^-$ 6.5 ; 1.18 *158* *2.76* $_{31}$**Ga** *1.0* Gallium *0.063*	$_{32}$**Ge** Germanium
5	0.5 ; 0.5 $^2\Sigma_u^+$ $_{37}$**Rb** *4.8* *28.3* Rubidium --- *0.017*	$_{38}$**Sr** Strontium	$_{39}$**Y** Yttrium	$_{40}$**Zr** Zirconium
	$^2\Sigma_u^+$ 1.03 ; 1.37 *145* *2.6* $_{47}$**Ag** *0.9* Silver *0.046*	$_{48}$**Cd** Cadmium	$_{49}$**In** Indium	$_{50}$**Sn** Tin
6	0.47 ; 0.45 $^2\Sigma_u^+$ $_{55}$**Cs** *4.8* *28.4* Cesium *0.042* *0.011*	$_{56}$**Ba** Barium	$_{57}$**La** Lanthanum	$_{58}$**Hf** Hafnium
	$^2\Sigma_u^+$ 1.94 ; 1.9 $_{79}$**Au** Gold	*3.65* $_{80}$**Hg** Mercury	$_{81}$**Tl** Thallium	$^2\Pi_g$ 1.66 ; 1.37 *129* *2.81* $_{82}$**Pb** *0.2* Lead *0.021*

V	VI	VII
$_{7}$N Nitrogen	$^2\Pi_g$ 0.45 ; 4.16 *1.35* $_{8}$O *1090* *10* *1.12* Oxygen	$^2\Sigma_u^+$ 3.08 ; 1.3 *1.92* $_{9}$F *475* *5.1* *0.47* Fluorine
$^2\Pi_g$ 0.59 ; 4.8 *1.98* $_{15}$P *640* --- *0.277* Phosphorus	$^2\Pi_u$ 1.67 ; 3.95 *1.8* $_{16}$S *601* *2.16* *0.32* Sulfur	$^2\Sigma_u^+$ 2.38 ; 1.26 $_{17}$Cl *277* *1.8* --- Chlorine
$_{23}$V Vanadium	; 0.17 $_{24}$Cr Chromium	$_{25}$Mn Manganese
0.1 ; 2.7 $_{33}$As Arsenic	$^2\Pi_g$ 1.94 ; $_{34}$Se *330* *0.86* --- Selenium	$^2\Sigma_u^+$ 2.55 ; 1.2 *2.81* $_{37}$Br *178* *0.88* *0.054* Bromine
$_{41}$Nb Niobium	$_{42}$Mo Molibdenum	$_{43}$Tc Technetium
$_{51}$Sb Antimony	1.92 ; $_{52}$Te Tellurium	$^2\Sigma_u^+$ 2.55 ; 1.05 $_{53}$I Iodine
$_{73}$Ta Tantalum	6.9 $_{74}$W *336.8* *1.0* - Tungsten	$_{75}$Re Rhenium
$^2\Pi_g$ 1.27 ; 2.8 *2.83* $_{83}$Bi *152* *0.53* *0.020* Bismuth	$_{84}$Po Polonium	$_{85}$At Astatine

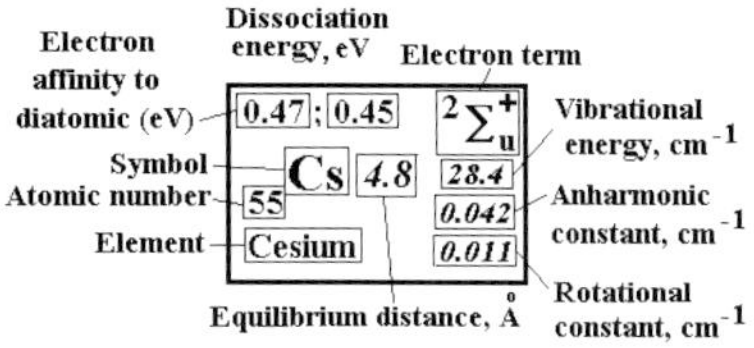

References

1 A. F. Alexandrov, L. S. Bogdankevich, and A. A. Rukhadze. *Principles of Plasma Electrodynamics*. Berlin, Springer, 1984.

2 J. A. Bittencourt. *Fundamentals of Plasma Physics*. Oxford, Pergamon Press, 1982.

3 H. V. Boening. *Plasma Science and Technology*. Itaca, Cornell University Press, 1982.

4 M. I. Boulos, P. Fauchais, and E. Pfender. *Thermal Plasmas*. New York, Plenum Press, 1994.

5 S. C. Brown. *Basic Data of Plasma Physics: The Fundamental Data on Electrical Discharges in Gases*. New York, Wiley, 1966.

6 R. A. Cairns. *Plasma Physics*. Glasgow, Blackie, 1985.

7 B. Chapman. *Glow Discharge Processes*. New York, Wiley, 1980.

8 F. F. Chen. *Introduction to Plasma Physics and Controlled Fusion*. New York, Plenum Press, 1984.

9 J. D. Cobine. *Gaseous Conductors*. Dover, New York, 1958.

10 W. Ebeling, A. Förster, V. E. Fortov, V. K. Gryasnov, and A. Yu. Polishchuk. *Thermophysical Properties of Hot Dense Plasmas*. Stuttgart, Teubner, 1991.

11 G. Ecker. *Theory of Fully Ionized Plasmas*. New York, Academic Press, 1972.

12 W. Elenbaas. *High Pressure Mercury Vapor Lamps and Their Applications*. Eindhoven, Philips Tech. Lib., 1965.

13 V. E. Fortov and I. T. Iakubov. *Physics of Nonideal Plasma*. New York, Hemisphere, 1990.

14 D. A. Frank-Kamenetskii. *Plasma—The Fourth State of Matter*. New York, Plenum, 1972.

15 V. E. Golant, A. P. Zhilinsky, and I. E. Sakharov. *Fundamentals of Plasma Physics*. New York, Wiley, 1980.

16 J. de Groot and J. van Vliet. *The High Pressure Sodium Lamp*. Deventer, Philips Tech. Lib., 1986.

17 M. F. Hoyaux. *Arc Physics*. New York, Springer, 1968.

18 J. R. Hollahan and A. T. Bell. *Techniques and Applications of Plasma Chemistry*. New York, Wiley, 1974.

19 L. G. H. Huxley and R. W. Crompton. *The Diffusion and Drift of Electrons in Gases*. Wiley, New York, 1974.

Plasma Processes and Plasma Kinetics. Boris M. Smirnov

ISBN: 978-3-527-40681-4

20 S. Ishimary. *Plasma Physics: Introduction to Statistical Physics of Charged Particles.* Benjamin, Menlo Park, 1985.

21 S. Ishimary. *Statistical Plasma Physics.* Redwood City, CA, Addison-Wesley. vol. 1. *Basic Principles,* 1992. vol. 2. *Condensed Plasmas,* 1994.

22 M. A. Kettani and M. F. Hoyaux. *Plasma Engineering.* Wiley, New York, 1973.

23 Yu. D. Korolev and G. A. Mesyats. *Physics of Pulsed Breakdown in Gases.* Ekaterinburg, URO Press, 1998.

24 A. N. Lagarkov and I. M. Rutkevich. *Ionization Waves in Electrical Breakdown in Gases.* New York, Springer, 1994.

25 M. A. Lieberman and A. J. Lichtenberger. *Principles of Plasma Discharge and Materials Processing.* New York, Wiley, 1994.

26 E. M. Lifshits and L. P. Pitaevskii. *Physical Kinetics.* Oxford, Pergamon Press, 1981.

27 V. S. Lisitsa. *Atoms in Plasmas.* Berlin, Springer, 1994.

28 F. Llewelyn-Jones. *The Glow Discharge.* New York, Methuen, 1966.

29 W. Lochte-Holtgreven. *Plasma Diagnostics.* New York, Am. Inst. Phys., 1995.

30 L. B. Loeb. *Basic Processes of Gaseous Electronics.* Berkeley, University of California Press, 1955.

31 E. A. Mason and E. W. McDaniel. *Transport Properties of Ions in Gases.* New York, Wiley, 1988.

32 H. S. W. Massey and E. H. S. Burhop. *Electronic and Ionic Impact Phenomena.* Oxford, Oxford University Press, 1969.

33 H. S. W. Massey. *Negative Ions.* Cambridge, Cambridge University Press, 1976.

34 H. S. W. Massey. *Atomic and Molecular Collisions.* New York, Taylor and Francis, 1979.

35 E. W. McDaniel and E. A. Mason. *The Mobility and Diffusion of Ions in Gases.* New York, Wiley, 1973.

36 E. W. McDaniel, J. B. A. Mitchell, and M. E. Rudd. *Atomic Collisions.* New York, Wiley, 1993.

37 G. A. Mesyats and D. I. Proskurovsky. *Pulsed Electrical Discharges in Vacuum.* Berlin, Springer, 1989.

38 E. Nasser. *Fundamentals of Gaseous Ionization and Plasma Electronics.* New York, Wiley, 1971.

39 W. Neuman. *The Mechanism of the Thermoemitting Arc Cathode.* Berlin, Academic-Verlag, 1987.

40 Y. P. Raizer. *Gas Discharge Physics.* Springer, Berlin, 1991.

41 Y. P. Raizer, M. N. Schneider, and N. A. Yatzenko. *Radio-Frequency Capacitive Discharges.* Boca Raton, CRC Press, 1995.

42 V. P. Schevel'ko and L. A. Vainstein. *Atomic Physics for Hot Plasmas.* Bristol, Institute of Physics, 1993.

43 P. P. J. M. Schram. *Kinetic Theory of Gases and Plasmas.* Dordrecht, Kluwer, 1991.

44 S. R. Seshardi. *Fundamentals of Plasma Physics.* New York, American Elsevier, 1973.

45 A. G. Sitenko. *Fluctuations and Nonlinear Wave Interactions in Plasmas.* Oxford, Pergamon Press, 1982.
46 A. Sitenko and V. Malnev. *Plasma Physics Theory.* London, Chapman and Hall, 1995.
47 B. M. Smirnov. *Physics of Weakly Ionized Gases.* Moscow, Mir, 1981.
48 B. M. Smirnov. *Negative Ions.* New York, McGraw-Hill, 1982.
49 B. M. Smirnov. *Physics of Ionized Gases.* New York, Wiley, 2001.
50 L. Spitzer. *Physics of Fully Ionized Gases.* New York, Wiley, 1962.
51 T. H. Stix. *Waves in Plasmas.* New York, American Institute of Physics, 1992.
52 P. A. Sturrock. *Plasma Physics.* Cambridge, Cambridge University Press, 1994.
53 O. Toshoro. *Radiation Phenomena in Plasmas.* Singapore, World Scientific, 1994.
54 A. von Engel. *Electric Plasmas: Their Nature and Uses.* London, Taylor and Francis, 1983.
55 J. F. Waymouth. *Electric Discharge Lamps.* Cambridge, MIT Press, 1971.
56 *Applied Atomic Collision Physics. vol. 1. Atmospheric Physics and Chemistry,* eds. H. S. W. Massey and D. R. Bates. Academic Press, New York, 1982.
57 *Basis Plasma Physics.* eds. A. A. Galeev and R. N. Sudan. Amsterdam, North Holland, 1983.
58 *Electrical Breakdown in Gases,* ed. J. A. Rees, New York and Toronto, 1973.
59 *Electrical Breakdown in Gases,* eds. J. M. Meek and J. D. Craggs. Chichester, Wiley, 1978.
60 *Plasma Science: From Fundamental Research to Technological Applications.* Washington, National Academy Press, 1995.
61 *Transport and Optical Properties of Nonideal Plasma.* eds. G. A. Kobzev, I. T. Iakubov, and M. M. Popovich. New York, Plenum Press, 1995.
62 *Vacuum Arcs.* ed. J. M. Lafferty. New York, Wiley, 1980.

Subject Index

Plasma Processes and Plasma Kinetics. Boris M. Smirnov

ISBN: 978-3-527-40681-4